Production and Neutralization of Negative Ions and Beams
(3rd Int'l Symposium, Brookhaven, 1983)

AIP Conference Proceedings
Series Editor: Hugh C. Wolfe
Number 111

Production and Neutralization of Negative Ions and Beams
(3rd Int'l Symposium, Brookhaven, 1983)

Edited by
Krsto Prelec
Brookhaven National Laboratory

American Institute of Physics
New York   1984

Copying fees: The code at the bottom of the first page of each article in this volume gives the fee for each copy of the article made beyond the free copying permitted under the 1978 US Copyright Law. (See also the statement following "Copyright" below.) This fee can be paid to the American Institute of Physics through the Copyright Clearance Center, Inc., Box 765, Schenectady, N.Y. 12301.

Copyright © 1984 American Institute of Physics

Individual readers of this volume and non-profit libraries, acting for them, are permitted to make fair use of the material in it, such as copying an article for use in teaching or research. Permission is granted to quote from this volume in scientific work with the customary acknowledgment of the source. To reprint a figure, table or other excerpt requires the consent of one of the original authors and notification to AIP. Republication or systematic or multiple reproduction of any material in this volume is permitted only under license from AIP. Address inquiries to Series Editor, AIP Conference Proceedings, AIP, 335 E. 45th St., New York, N. Y. 10017.

L.C. Catalog Card No. 84-70379
ISBN 0-88318-310-2
DOE CONF- 831180

FOREWORD

This meeting was the third in the series of symposia on the production and neutralization of negative ions and beams, organized by Brookhaven National Laboratory. The objective of the symposium was to review the progress in the knowledge and understanding of fundamental processes related to the production of negative and neutral particles and beams, to review the state of the development of devices and systems, and to try to establish guidelines for future needs and work. This time there were some changes introduced compared to the two previous symposia: heavier negative ions as well as polarized negative hydrogen ions were included in the program, both because of a recently expressed interest for fusion and other applications.

The Symposium was attended by 77 participants from nine countries, and 72 papers were presented in one poster and seven oral sessions. In addition to those, there were two panel sessions, one to consider fundamental processes for production and loss of negative ions and the other devoted to prospects and guidelines for future work. Transcripts of discussions that followed the presentation of papers, of both panel sessions, and of the concluding remarks are part of the proceedings. We have also tried to rearrange oral and poster papers according to the subject, which in some cases differs from the order of presentation.

The meeting was sponsored by the U.S. Department of Energy and Brookhaven National Laboratory; we enjoyed the full support of the Laboratory and of the BNL Accelerator Department. There were many people whose help and advice were indispensable in the organization and smooth running of the symposium. Members of the International Program Committee contributed many suggestions during the preparations of the program; members of the Local Organizing Committee shared in many responsibilities before, during, and even after the Symposium. Special appreciation is due to Th. Sluyters who served as Co-Chairman and whose experience was of great value in all stages. Mrs. Marion Heimerle served as the Symposium Secretary and was responsible for the performance of many duties required for the organization and operation of the meeting. Ms. Cheryl Conrad served as the Proceedings Secretary and in this capacity she was responsible for the transcription of the tape recordings and final preparations of the manuscript for the publication. They both did an outstanding job, with great patience under difficult circumstances.

Finally, I would like to thank participants for the presentation of papers of such high quality, session chairmen who managed to keep the program on schedule, panel members for their contributions, panel moderators for organizing two interesting sessions and for helping in the final editing of the transcripts, and K.W. Ehlers for his traditional summary of the meeting. We all hope that this meeting will further stimulate vigorous work and cooperation in this field.

Krsto Prelec, Editor
Brookhaven National Laboratory
January 1984

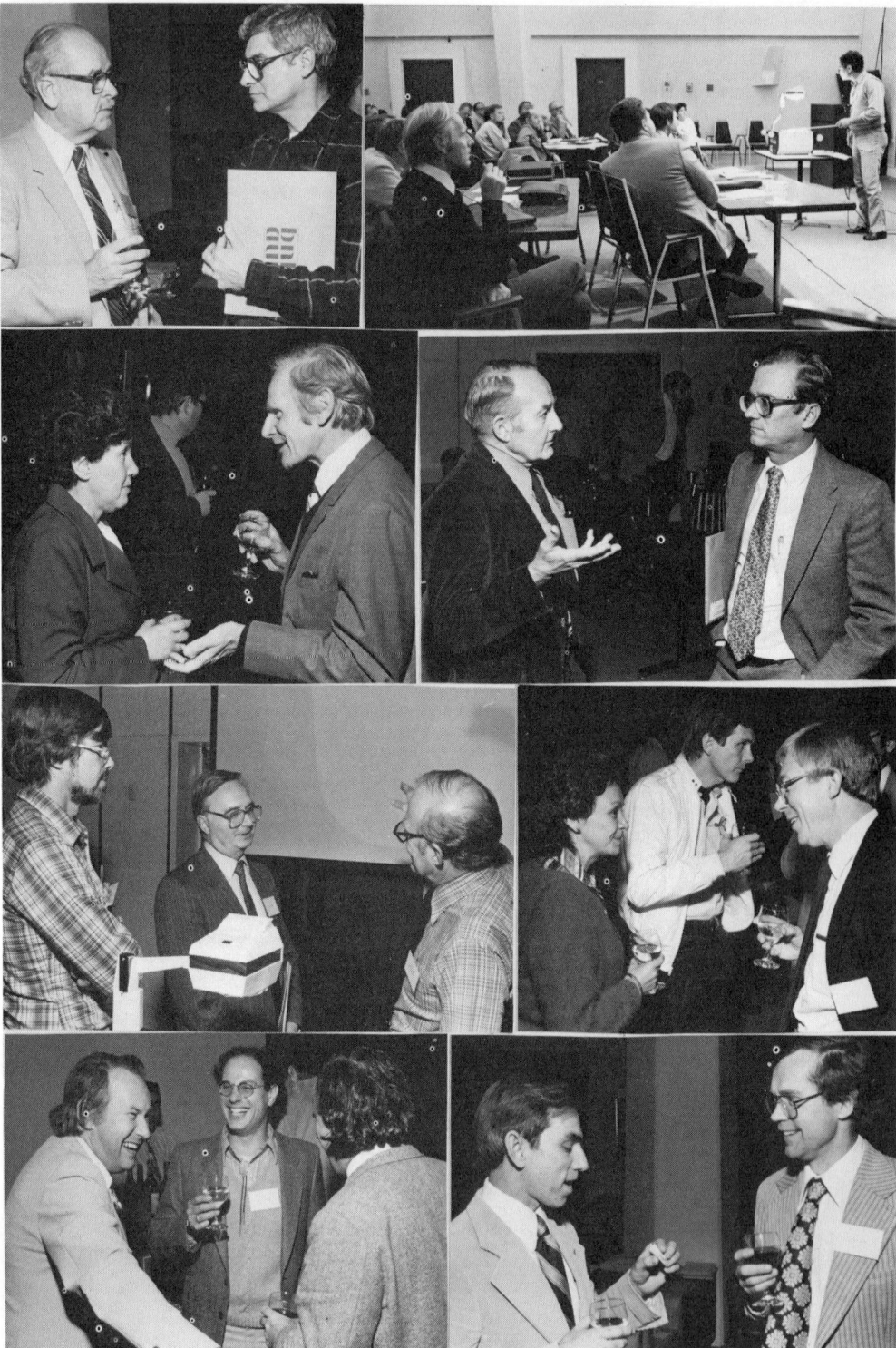

LOCAL ORGANIZING COMMITTEE

J.A. Alessi
A.I. Hershcovitch
R.A. Larson
V. Kovarik
R. McKenzie-Wilson
K. Prelec (Chairman)
Th. Sluyters (Co-Chairman)

INTERNATIONAL PROGRAM COMMITTEE

M. Bacal	Ecole Polytechnique, Palaiseau, France
G.I. Dimov	Institute for Nuclear Physics, Novosibirsk, U.S.S.R.
K.W. Ehlers	Lawrence Berkeley Laboratory, Berkeley, CA, U.S.A.
W. Haeberli	University of Wisconsin, Madison, WI, U.S.A.
J.R. Hiskes	Lawrence Livermore Laboratory, Livermore, CA, U.S.A.
K. Prelec (Chairman)	Brookhaven National Laboratory, Upton, NY, U.S.A.
Th. Sluyters (Co-Chairman)	Brookhaven National Laboratory, Upton, NY, U.S.A.
S. Staten	U.S. Department of Energy, Washington, D.C., U.S.A.

SYMPOSIUM SECRETARY

Mrs. Marion Heimerle

PROCEEDINGS SECRETARY

Ms. Cheryl Conrad

TABLE OF CONTENTS

FUNDAMENTAL PROCESSES: VOLUME 1

Volume Generation of Negative Ions in High Density Hydrogen
Discharges 3
   J.R. Hiskes and A.M. Karo

H$^-$ Production and Destruction Mechanisms in Hydrogen Low
Pressure Discharges... 31
   M. Bacal and A.M. Bruneteau

Negative Ion Production Via Dissociative Attachment to H_2 46
   J.M. Wadehra

H$^-$ and Electron Production in a Magnetic Multipole Source 55
   A.J.T. Holmes, T.S. Green, and G. Dammertz

Polar Dissociation as a Source of Negative Ions 56
   S.K. Srivastava and O.J. Orient

Volume H$^-$ Ion Production Experiments at LBL 67
   K.N. Leung and K.W. Ehlers

Negative Ion Neutralization Accompanied by Excitation 82
   T.J. Morgan

A Search for H_2^-, H_3^-, and Other Metastable Negative Ions 90
   Y.K. Bae, M.J. Coggiola, and J.R. Peterson

Research at the IKVT/KfK Related to the Development of an H$^-$
Beam Line... 96
   O.F. Hagena, P.R.W. Henkes, R. Klingelhöfer, B. Krevet, and
   H.O. Moser

H$^-$ Density in a Tandem Multicusp Discharge 105
   M. Bacal and K.N. Leung

Vacuum Ultraviolet Emission and H$^-$ Production in a Low Pressure
Hydrogen Plasma.. 113
   W.G. Graham

Dissociative-Recombination of e + H_3^+, An Analysis of Reaction
Product Channels 118
   H.H. Michels and R.H. Hobbs

Generation of Vibrationally Excited H_2 Molecules by H_2^+ Wall
Collisions 125
   J.R. Hiskes and A.M. Karo

Generation of Vibrationally Excited Hydrogen for Use in a
Negative Ion Source.. 132
   R.J. Turnbull, S.R. Walther, and J.L. Guttman

Interference Effects in Negative Ion Formation 140
   I. Alvarez, A. Morales, J. de Urquijo, and C. Cisneros

Formation of H^- by Charge Transfer in Alkaline-Earth Vapors... 149
   A.S. Schlachter and T.J. Morgan

Charge Exchange of Protons and Hydrogen Atoms in Na, K, Rb-
Vapor Targets.. 162
   F. Ebel and E. Salzborn

FUNDAMENTAL PROCESSES: SURFACE   169

Ion Backscattering from Layered Targets...................... 171
   O.S. Oen and M.T. Robinson

Cs/Transition Metal Composite Surfaces: First Principles
Calculations of High Z, Low Work Function Systems............ 184
   A.J. Freeman, E. Wimmer, S.R. Chubb, J.R. Hiskes, and A.M. Karo

De-Excitation and Equipartition in H_2-Surface Collisions 197
   A.M. Karo, J.R. Hiskes, K.D. Olwell, T.M. DeBoni, and R.J. Hardy

H^- and Li^- Formation by Scattering H^+, H_2^+ and Li^+ from Cesiated
Tungsten Surfaces... 206
   E.H.A. Granneman, J.J.C. Geerlings, J.N.M. van Wunnik,
   P.J.M. van Bommel, H.J. Hopman, and J. Los

Sputtering Yields of Negative Hydrogen Ions................... 220
   J.A. Greer and M. Seidl

Effects of Cesium in the Plasma of the Surface Conversion H^-
Source 227
   K.W. Ehlers and K.N. Leung

Systematic Investigation of Negative Ion Production from
Low-Work Function Surfaces 239
   M.J. Coggiola and J.R. Peterson

Work Function Dependence of Surface Produced H^- in the Presence
of a Plasma... 247
   M. Wada, R.V. Pyle, and J.W. Stearns

Plasma-Surface Interaction Involved in H^- Generation 254
   H.-M. Katsch and K. Wiesemann

Production of H^- Ions from Polycrystalline and Single Crystal
(110) Tungsten and Molybdenum Surfaces 258
   P.J.M. van Bommel, K.N. Leung, and K.W. Ehlers

H⁻ Production from Different Metallic Converter Surfaces 265
  K.N. Leung and K.W. Ehlers

The Negative Fraction of Deuterium and Helium Scattered from a
Sodium Surface 273
  H. Verbeek, W. Eckstein, and P.J. Schneider

Observance of H⁻ by Surface Chemi-Ionization on W(110) 281
  R.L. Palmer

PRODUCTION OF HEAVY NEGATIVE IONS   289

Experimental Investigation of Volume Li⁻ Production. 291
  M.W. McGeoch and R.E. Schlier

Formation of Negative Ions by Charge Transfer: He⁻ to Cl⁻ . . . 300
  A.S. Schlachter

PANEL: FUNDAMENTAL PROCESSES. 313

NEGATIVE HYDROGEN ION SOURCES   331

Report on the BNL H⁻ Ion Source Development. 333
  K. Prelec

The Status of ≳ 1 Ampere H⁻ Ion Source Development at the
Lawrence Berkeley Laboratory 344
  A.F. Lietzke, K.W. Ehlers, and K.N. Leung

Short-Pulse Operation With the SITEX Negative Ion Source 353
  W.K. Dagenhart, W.L. Stirling, G.M. Banic, G.C. Barber,
  N.S. Ponte, and J.H. Whealton

Pulsed Multiampere Source of Negative Hydrogen Ions. 363
  Y.I. Bel'chenko and G.I. Dimov

Production of High Brightness H⁻ Beams in Surface Plasma
Sources. 376
  G.E. Derevyankin and V.G. Dudnikov

Operational Experience with the BNL Magnetron H⁻ Source. 398
  R.L. Witkover

Development of a Multicusp H⁻ Ion Source for Accelerator
Applications 410
  R.L. York and R.R. Stevens, Jr.

Progress in Developing a 'Volume' Hydrogen Negative Ion Source . 418
  M. Bacal, F. Hillion, M. Nachman, and W. Steckelmacher

Extraction and Acceleration of H⁻ Ions from a Magnetic
Multipole Source 429
   A.J.T. Holmes and T.S. Green

Large Negative Ion Source for Energetic Neutral Beams 438
   M. Delaunay, R. Geller, C. Jacquot, P. Ludwig, P. Sermet,
   J.C. Rocco, F. Zadworny, J.B. Bergstrom, G. Hellblom,
   R. Pauli, and H. Wilhelmsson

Normalized Emittance of SITEX Negative Ion Source 450
   W.L. Stirling, W.K. Dagenhart, J.H. Whealton, and J.J. Donaghy

A Scaled, Circular-Emitter Penning SPS for Intense H⁻ Beams ... 458
   H.V. Smith, Jr., P. Allison, and J.D. Sherman

The Plasma Focus as a Source of Collimated Beams of Negative
Ion Clusters and of Neutral Deuterium Atoms 463
   V. Nardi and C. Powell

ACCELERATORS AND ACCELERATED BEAMS   471

Transverse-field Focusing Accelerator 473
   O.A. Anderson

Design Desiderata for a Laminar Flow Quadrupole-Focused
Acceleration Column 489
   A.W. Maschke

The Amsterdam "MEQALAC" RF Acceleration System 492
   E.H.A. Granneman, R.W. Thomae, F. Siebenlist, P.W. van
   Amersfoort, H.J. Hopman, J. Kistemaker, H. Klein,
   A. Schempp, and T. Weis

The BNL RFQ 505
   S.T. Giordano

Operating Experience with a 100-keV, 100-mA H⁻ Injector 511
   P. Allison and J.D. Sherman

Accelerated Beam from Cusp H⁻ Ion Source 520
   A. Takagi, Y. Mori, Z. Igarashi, K. Ikegami, C. Kubota, and
   S. Fukumoto

2D Accelerator Design for SITEX Negative Ion Source 524
   J.H. Whealton, R.J. Raridon, R.W. McGaffey, D.H. McCollough,
   W.L. Stirling, and W.K. Dagenhart

End Effects in Slot Extraction for a SITEX Negative Ion Source . 533
   R.W. McGaffey, P.S. Meszaros, J.H. Whealton, R.J. Raridon,
   and D.H. McCollough

NEUTRALIZERS   545

  Photodetachment Technology 547
    J.H. Fink

  Neutralization Efficiency of Plasma Targets for High Energy
  Negative Ions. 561
    A.I. Hershcovitch, B.M. Johnson, V.J. Kovarik, M. Meron,
    K.W. Jones, K. Prelec, and L.R. Grisham

  Technology of a Laser Resonator for Photodetachment Neutralizer. 568
    V. Vanek, T. Hursman, D. Copeland, and D.M. Goebel

BEAM SYSTEMS AND APPLICATIONS   585

  Magnetic Fusion Energy Heating Development Plan. 587
    H.S. Staten

  Requirements for Negative Ion Based Systems from Users' Point
  of View. 594
    L.D. Stewart

  Considerations Involved in the Design of Negative-Ion-Based
  Neutral Beam Systems 605
    W.S. Cooper

  Neutral Beam Injector for 475 keV MARS Sloshing Ions 617
    D.M. Goebel and G.W. Hamilton

  The Role of Multistep Collision Processes in Increasing the
  Beam Stopping Cross Section for High Energy Neutral Beams. . . . 641
    D.E. Post, R.K. Janev, and C.D. Boley

  Negative Ion Beam Requirements for Compact Tori. 647
    G.H. Miley

  Applications and Development Requirements for Multi-MeV Light
  Atom Beams 659
    L.R. Grisham, D.E. Post, and D.R. Mikkelsen

PANEL: WHERE DO WE GO FROM HERE?. 669

POLARIZED H$^-$ IONS AND SOURCES   683

  Principles of Production of H$^-$ Polarized Ions. 685
    W. Haeberli

  Production of Polarized H$^-$ Ions Using Laser Optical Pumping. . . 696
    L.W. Anderson

Production of Intense Beams of Polarized Negative Hydrogen Ions
by Double Charge Exchange in Alkali Vapours. 706
    W. Grüebler and P.A. Schmelzbach

Optics for a Spin-Polarized Hydrogen Atomic Beam 720
    D. Kleppner

Status and Future Plans for the BNL Polarized H⁻ Source. 736
    Th. Sluyters, J. Alessi, and A. Kponou

Ionization of Polarized Hydrogen Atoms 741
    J. Alessi

The Production of Polarized Negative Ion Beams by "Collisional
Pumping". 754
    L.W. Anderson, S.N. Kaplan, R.V. Pyle, L. Ruby,
    A.S. Schlachter, and J.W. Stearns

Generation of Intense Polarized Beams by Selective Neutralization of Negative Ions. 763
    A.I. Hershcovitch and E.A. Hinds

Optically Pumped Polarized H⁻ Ion Source 769
    Y. Mori, K. Ikegami, A. Takagi, Z. Igarashi, S. Fukumoto,
    W. Cornelius, and R. York

CONCLUDING REMARKS 773
    K. Ehlers

APPENDICES   779

    Appendix 1: List of Participants. 781

    Appendix 2: List of Authors 787

    Appendix 3: Symposium Program 789

FUNDAMENTAL PROCESSES: VOLUME

VOLUME GENERATION OF NEGATIVE IONS IN HIGH DENSITY HYDROGEN DISCHARGES

J. R. Hiskes and A. M. Karo
Lawrence Livermore National Laboratory
Livermore, California 94550

ABSTRACT

An optimized tandem two-chamber negative-ion source system is discussed. In the first chamber high energy (E > 20 eV) electron collisions provide for H_2 vibrational excitation, while in the second chamber negative ions are formed by dissociative attachment. The gas density, electron density, and system scale length are varied as independent parameters. The extracted negative ion current density passes through a maximum as electron and gas densities are varied. This maximum scales inversely with system scale length, R. The optimum extracted current densities occur for electron densities near $nR = 10^{13}$ electrons cm^{-2} and for gas densities, N_2R, in the range 10^{14} to 10^{15} molecules cm^{-2}. The extracted current densities are sensitive to the atomic concentration in the discharge. The atomic concentration is parametrized by the wall recombination coefficient, γ, and scale length, R. As γ ranges from 0.1 to 1.0 and for system scale lengths of one centimeter, extracted current densities range from 8.0 to 80. mA cm^{-2}.

INTRODUCTION

At the first symposium in this series Bacal and co-workers at the Ecole Polytechnique[1] reported the observation of an anomously large concentration of negative ions present in a hydrogen discharge. The intensive investigation of this anomaly both experimentally[2-5] and theoretically[6] at several laboratories has confirmed the existence of these ions. The analysis of these discharges has led to a working hypothesis for the formation process in which the hydrogen molecules are first vibrationally excited in energetic electron collisions, and, subsequently, capture thermal electrons leading to dissociative attachment.[7,8]

The theoretical description of these processes has been only partially successful in interpreting the observed large concentrations of ten percent or more that are characteristic of low and medium density discharges. The calculations instead leading to concentrations of about one or two percent. The source of this remaining discrepancy however, need not be a lack in the full accounting of the underlying atomic processes, but may instead be due to an incomplete knowledge of the high energy electron component, specifically, a failure to include non-classical processes in the calculation of the electron distribution function.

The two-step nature of the formation process, consisting of excitation followed by dissociative attachment, finds its optimum geometric expression in a tandem two-chamber confinement system maintained with different electron energy distributions in the respective chambers. In the first chamber the electron temperature may typically be kT = 5eV together with a relatively large electron density component with energies above 20 eV, while the second chamber is maintained with a thermal distribution of electrons with kT near 1 eV. A demonstration that an electron temperature differential can be supported across a magnetic filter has been shown in a two-chamber positive-ion system optimized for maximum proton extraction.[9,10] This latter system, when fitted with an extraction grid maintained with a small positive bias voltage, alters the potential distribution in the second chamber sufficiently to allow negative ions to overcome the normally positive plasma potential and be extracted.[11] This observation leads in a natural way to a three-chamber tandem-system consisting of the above mentioned two-chamber generator configuration but fitted with a gridded third chamber to optimize extraction.[12] For purposes of this paper we are concerned principally with the production processes occuring in the first two chambers and the third chamber is not considered explicitly.

To develop a source capable of providing current densities adequate for the needs of fusion research, 10 to 100 mA cm^{-2}, will require a very high density, high power hydrogen or deuterium discharge with electron densities of the order of 10^{12} to 10^{13} electrons cm^{-3}. The prototype discharges are the high-current continuously-operating sources of positive ions that have already been developed for use in positive-ion-based neutral-beam-line systems. The extensive application of experimental and computational diagnostics to these source discharges provides for a sufficient data base from which to develop a parametric study for a high-current negative ion system.

In this paper the system parameters are explored for purposes of identifying the largest possible current densities that can be extracted from an optimized high-density two-chamber system. Specifically, the hydrogen gas density, electron densities and energy distributions, and wall recombination coefficients are varied as independent variables while certain other parameters, for example gas and atom temperatures, are taken equal to the observed values. An important parameter is the system scale length, R, usually chosen to be the same for each chamber. Specifically, each discharge chamber can be assumed to be a cylinder of radius R and length 2R. This scale length is treated as an additional independent variable with values chosen to range from 0.1 to 10 centimeters. With the problem stated in these terms, we find that there exists an optimum extracted negative ion current density for each different value of the scale length.

ELECTRON ENERGY DISTRIBUTION

In the primary discharge chamber of the two-chamber system the injected electrons are slowed through the action of electron-electron Coulomb collisions and by inelastic collisions with the background gas molecules and atoms. A steady-state electron energy distribution develops that can be characterized by a Maxwellian distribution, with temperature kT, upon which is superimposed a high energy tail. The magnitude of the high energy component is proportional to the current of injected electrons. These features are displayed explicitly in the electron energy distributions measured in the positive-ion source discharges.[13,14,15]

A discussion of the electron energy distribution function for several limiting cases is presented in our earlier paper on the two-chamber system.[16] We reproduce here only that portion which is essential to the continuity of this discussion.

We derive an expression for f_s, the high energy term describing the slowing electrons. The total distribution function, f, is understood to be

$$f = f(\text{Maxwellian}) + f_s \quad . \tag{1}$$

The rate of electron energy loss is taken to be

$$\frac{dE}{dt} = -\left[\frac{an}{\sqrt{E}} + N_1 G(E) + N_2 H(E)\right] \text{ eV sec}^{-1} \tag{2}$$

the sum of electron-electron, atomic, and molecular collisional energy loss terms. Here n, N_1, N_2 are the density of thermal electrons, density of atoms, and the molecular gas density, respectively. With E is expressed in electron volts,

$$a = 4.9 \times 10^{17} \frac{4\pi e^4 \ln\Lambda}{2\sqrt{m}} \text{ (eV)}^{3/2} \text{ sec}^{-1} \quad . \tag{3}$$

Here $\ln\Lambda$ is the Coulomb logarithm defined by Spitzer.[17] The atomic and molecular energy loss rates, G(E) and H(E), are illustrated in Fig. 1, where account has been taken of ionization, electronic and vibrational excitation processes.

The function f_s is the number of slowing electrons cm^{-3} in a unit energy interval, ΔE. Electrons at an energy E are moving through the energy interval ΔE at a rate given by $f_s(dE/dt)$. More generally, particles are slowing through an energy range where injection is also occurring. The difference in the rate of particles entering and leaving an interval ΔE equals the local particle source:

$$\Delta\left(f_s \frac{dE}{dt}\right) = \left(f_s \frac{dE}{dt}\right)_{out} - \left(f_s \frac{dE}{dt}\right)_{in} = \frac{J\Delta E}{Ve} \quad . \tag{4}$$

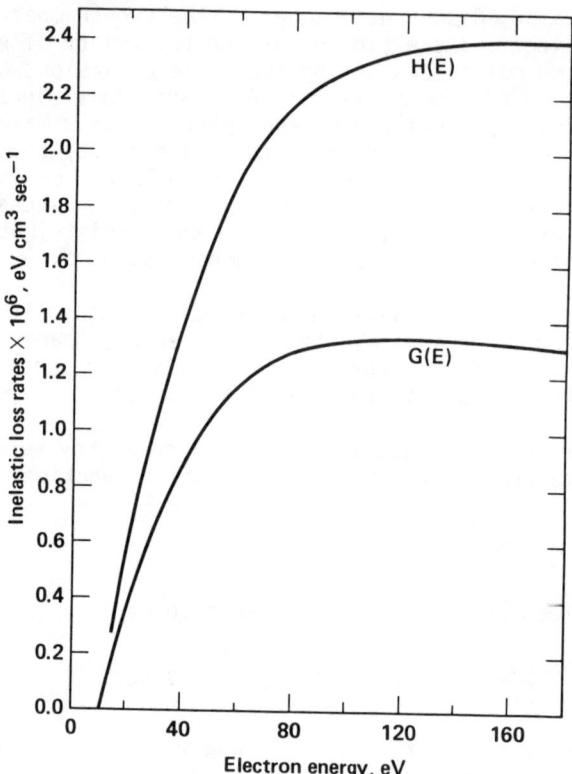

Fig. 1 The rate of energy loss per unit density versus electron energy for electrons incident on hydrogen molecules or atoms. $G(E)$ is the atomic loss rate, $H(E)$ the molecular loss rate.

Generalizing,

$$\frac{\partial}{\partial E}\left(f_s \frac{dE}{dt}\right) = \frac{J}{Ve} \qquad (5)$$

Here J is the injected electron current per unit energy interval and V the discharge volume. For injection over a range of energies E_0 to E_F, Eq. (5) integrates to

$$f_s = \frac{J}{Ve}(E_F - E_0)\left|\frac{dE}{dt}\right|^{-1}, \quad E < E_0 ;$$

$$f_s = \frac{J}{Ve}(E_F - E)\left|\frac{dE}{dt}\right|^{-1}, \quad E_0 < E < E_F ;$$

$$f_s = 0 \quad , \quad E > E_F. \tag{6}$$

We shall apply these equations to the energy distribution observed in the high-power discharge. In Fig. 2 is shown the measured electron distribution function for electrons nominally injected over the range 22.5 to 34 eV. The voltage gradient along the filaments causes most of the emission to occur from 30 to 34 eV. The solid curve is the author's fit of their data to a kT = 3.5 eV Maxwellian. Also shown in the figure are fits of function (1) appropriate to the discharge parameters of Ref. 14 but for two different electron temperatures, kT = 3.0 and 3.8 eV; the fits at these two temperatures span the range of the quoted experimental uncertainties.

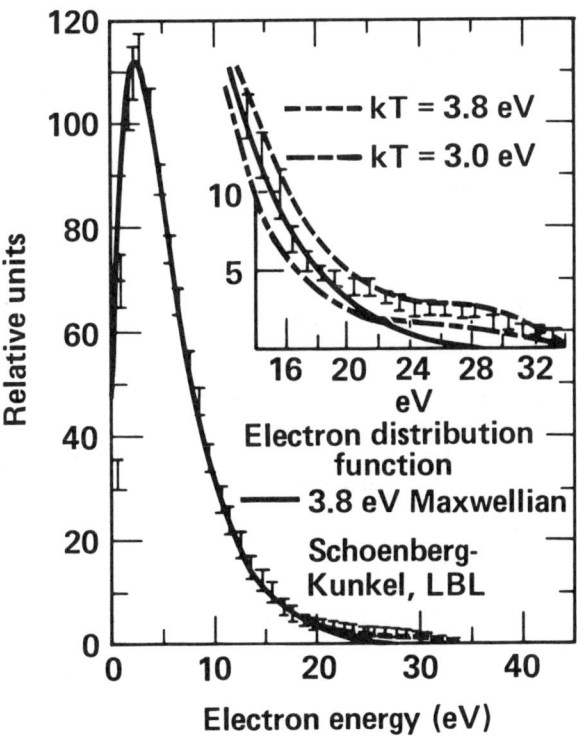

Fig. 2. The electron energy distribution versus electron energy taken from Ref. 14. The solid curve is a 3.8 eV Maxwellian drawn through the experimental data. The dashed curves are fits of Eq. 1 for kT = 3.0 and 3.8 eV.

Inspection of the figure will show that approximately 5% of the total electron population is in the high energy portion of the energy distribution above 20 eV. Eqs. (6) show that the high energy component may be increased both absolutely and relative to the Maxwellian component either by increasing the injected current, J, or by increasing the upper limit of the injection energy, E_F. To summarize, the high energy component of the observed electron energy distribution can be interpreted as due to classical energy loss processes, and may be enhanced by altering the injection current and energy.

THE FIRST CHAMBER RATE EQUATIONS

The first chamber is a high density, $10^{12} < n(e) < 10^{14}$ electrons cm^{-3}, high power discharge whose principal function is to generate vibrationally excited molecules by energetic electrons, $E > 20$ eV, according to the reaction[18]

$$e(f) + H_2(v'' = 0) \rightarrow e(f) + H_2(v'' \geq 6) \quad . \tag{7}$$

Associated with such a high power discharge is a relatively high electron thermal component with $3.0 < kT < 6.0$ eV. Negative ions generated in this first chamber will be rapidly attenuated by collisional detachment in this hot thermal bath,

$$e + H^- \rightarrow 2e + H \quad , \tag{8}$$

and their concentration will not exceed a fraction of order 10^{-3} of the thermal electron density.[16]

In an earlier paper, Bailey and Garscadden[19] had proposed enhancing the negative ion concentrations by using a two-stage tandem discharge in which the first chamber is operated at a high pressure, 10-100 Torr., thereby enchancing the vibrational population via V-V excitations. In the low-pressure system discussed here the vibrational excitation is accomplished instead by the high energy electrons.

The total electron density, $n(e)$, in the first chamber is taken to be the sum of the thermal component n, and the high energy component, $n(f)$,

$$n(e) = n + n(f) \quad . \tag{9}$$

The reaction rate for process (7) is approximately constant ($\pm$ 20%) for electron enegies ranging from 30 eV up to 500 eV. As a consequence the shape of the high energy component of the electron energy distribution is not so important as is the total number of high energy electrons. In this paper we shall assume the injection parameters are chosen to give $n(f)/n = 10\%$. To a very good approximation we find that the equilibrium concentration of $H_2(v")$ is due entirely to the $n(f)$ electron component, so that we might have chosen $n(f)$ to be the independent variable. Out of convenience however, we shall generally quote the total electron density.

The concentration of $H_2(v")$ in the discharge is sensitive to the concentration of free atoms. In Fig. 3 is plotted the ratio of atomic to molecular density as a function total electron density and for $kT = 5.0$ eV. The atomic concentration is generated by dissociation in equilibrium with ionization and wall recombination processes. Here we have taken into account dissociation rates S_2, Q_2, S_4, S_5, and S_8 quoted in Ref. 15. Atoms are lost by ionization, S_1, and recombination at the wall to form H_2 with a wall recombination coefficient, γ. The rate of wall recombination is proportional to the parameter γ/R. For each value of γ/R the ratios are shown for $N_2 = 10^{14}$ (upper curve) and $N_2 = 10^{15}$ molecules cm^{-3}. For values of γ/R of order 0.1 cm^{-1} or less the atom concentration becomes comparable to the molecular concentration.

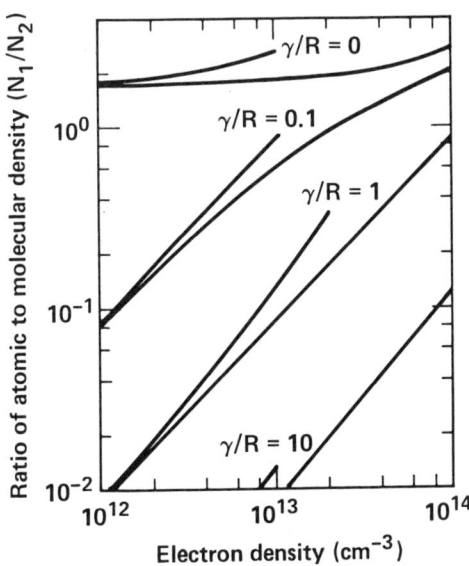

Fig. 3. The ratio of atomic to molecular concentrations versus electron density for different parametric values of γ (atom wall recombination coefficient), and R (system scale length).

The equilibrium concentration of the fourteen vibrationally excited levels, $N_2(v")$, is given by fourteen coupled equations describing the ratio of formation and loss processes for each respective level:

$$N_2(v") = \mathcal{N}(v")/\mathcal{D}(v") \text{ cm}^{-3} \quad , \quad 1 \leq v" \leq 14 \quad ; \tag{10}$$

where

$$\mathcal{N}(v") = n(f) \, N_2 \, \sigma(E-V;v")v + \tag{11}$$

$$N_2(v"+1)[N_2 \, P(v"+1, v") + W(v"+1, v") +$$

$$N_1 F(v"+1, v") \, \sigma(ADX; v"+1) \, v(A)] +$$

$$N_2(v"+2) \, [W(v"+2,v") + N_1 F(v"+2,v") \, \sigma(ADX,v"+2) \, v(A)]$$

$$\vdots$$

$$N_2(14) \, [W(14,v") + N_1 F(14,v") \, \sigma(ADX,14) \, v(A)] +$$

$$N_2(v"-1) \, [N_2 \, P(v"-1,v") + W(v"-1,v")] +$$

$$N_2(v"-2) \, W(v"-2,v") \quad +$$

$$\vdots$$

$$N_2(1) \, W(1,v");$$

and,

$$\mathcal{D}(v") = N_2[P(v",v"-1) + P(v",v"+1)] + \tag{12}$$

$$N_1[\sigma(ADX;v")v(A) + \sigma(ADS;v") \, v(A)] +$$

$$n(f) \, DR(f) + nDR(S) + \frac{3}{4}[b(v")]^{-1} \, v(v")/R \quad .$$

The successive terms in $\mathcal{N}(v'')$ are the fast-electron excitation, the E-V process[18]; the V-T rates, $P(v''+1, v'')$; the wall relaxation rates, $W(v''+1,v'')$; and the atomic de-excitation rates, proportional to the partitioning fraction, $F(v''+1,v'')$. Here v and v(A) are the fast electron and atomic velocities, respectively.

In $\mathcal{D}(v'')$ the additional terms are the atomic dissociation represented by $\sigma(ADS,v'')$, and referring to the process

$$H + H_2(v'') \rightarrow 3H \quad . \tag{13}$$

The fast-electron collisional term, $DR(f)$, refers to the sum of processes given by molecular dissociation, the S_2 rate; dissociation via the electronic excitation processes, Q_2; and ionization, S_3.[15] The thermal electron collision term, $DR(S)$, takes into account H_3^+ formation from H_2^+, S_6, charge exchange de-excitation with H_2^+, and H_3^+ collisional de-excitation collisions; these molecular species in the discharge are quoted as fractions of the thermal electron component. The thermal term, $DR(S)$, is typically an order of magnitude smaller than the fast-electron term, $DR(f)$.

The final term Eq. (12), the wall relaxation term, requires some elaboration. In a companion paper at this Symposium, Karo, et al.[20] discuss a molecular trajectory calculation for the relaxation of $H_2(v'', J = 1)$ in wall collisions on an iron lattice maintained with a temperature of 500°K. An example of the results of this calculation for a sequence of wall collisions for an initial $v'' = 8$ level is shown in Fig. 4. In this figure is plotted the vibrational energy spectrum versus energy for successive wall collisions. The initial vibrational energy is shown by an arrow at the top of the figure. The histograms are based upon a total of 300 initial trajectories.

After the first wall collision the vibrational energies are broadly distributed over the vibrationally spectrum but with a weak maximum toward the upper end of the spectrum. After successive collisions the upper portion of the spectrum is attenuated and the vibrational excitation is relaxed. If we examine the uppermost portion of the spectrum we note that it can be approximated by an exponential decay with collision number, C, and decay constant, $b(v'')$, according to

$$e^{-C/b(v'')} \quad . \tag{14}$$

For the $v'' = 8$ case shown here we find $b(v'')$ to lie in the range one to two. Similar histograms have been constructed for $v'' = 12$ and $v'' = 4$. Near the upper portion of the vibrational spectrum $b(v'')$ is unity, but for $v'' = 4$ $b(v'')$ is in the range two to four. If we select a value $b(v'')$ equal to unity for the entire active

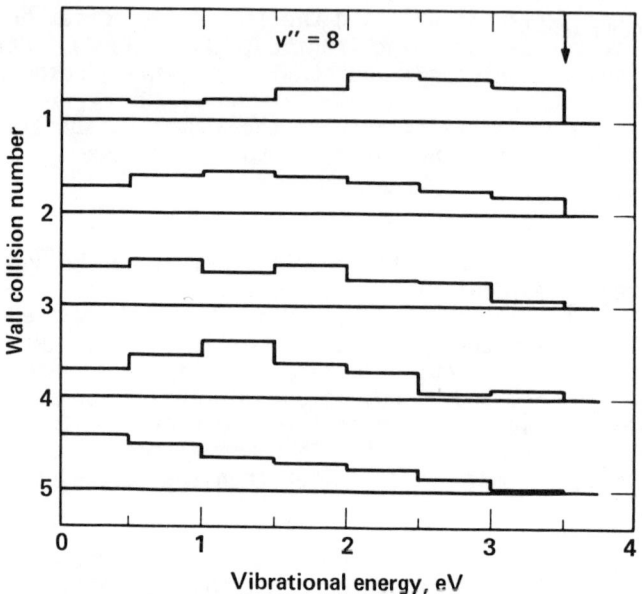

Fig. 4. Histograms of the vibrational energy distributions for successive wall collisions. These histograms are based upon 300 initial trajectories of $H_2(v'' = 8, J = 1)$ incident upon the wall. Arrow indicates initial vibrational energy.

portion of the spectrum, $v'' \geq 6$, the wall relaxation rate is overestimated.

The rate at which molecules arrive at the wall is equal to

$$\frac{3}{4} v(v'')/R , \qquad (15)$$

where $v(v'')$ is the mean molecular velocity. The wall relaxation rate given by the last term in Eq. 12 follows immediately. The relaxation of a particular level, v'', populates lower levels. The wall rate given in Eq. 11 is then expanded to read:

$$W(v'' + 1, v'') = \frac{3}{4} \omega(v'' + 1, v'') [b(v'' + 1)]^{-1} v(v'')/R , \qquad (16)$$

where the fractions $\omega(v'' + 1, v'')$ are derived from the histograms shown in Fig. 4. For the lower lying levels there is some wall excitation given by the terms $W(v'' - 1, v'')$.

Another important source of vibrational de-excitation is the atom-exchange reaction

$$H + H_2(v") \rightarrow H_2(v' < v") + H \quad . \tag{17a}$$

De-excitation can also occur through non-exchange collisions of the type

$$H + H_2(v") \rightarrow H + H_2(v' < v") \quad . \tag{17b}$$

These reactions have been studied using molecular dynamics methods for $v" = 0, 1, 2$ by Smith and Wood[21], and for all vibrational levels, $v" = 0 \rightarrow 14$, by Blais and Truhlar[22]. These latter authors have evaluated the exchange reaction rate

$$H + H_2(v",J) \rightarrow H_2(v',J') + H \tag{18a}$$

for a 300°K distribution of relative energy and for selected values of $v",J$ that span the entire vibration-rotation spectrum. Distributions of final v' are presented for $v" = 0,1,3,6,9,12$ for $J = 2$ and $J = 10$. From these distributions we have evaluated the partial rates for reaction (17a). These reduced rates are not significantly different for $J = 2$ or $J = 10$.

Cross sections for the exchange reaction, (18a), and the non-exchange reaction

$$H + H_2(v",J) \rightarrow H + H_2(v',J') \quad , \tag{18b}$$

for relative translational energies of 0.5 eV and 0.3 eV[23] have been evaluated for $v" = 2$ and 6. The channel (17a) of the total cross section for $v" = 6$ is approximately the same as for the 300°K rates. At 0.5 eV and for $v" = 4,6$ the rate for channel (17b) is about one-fourth that of (17a). Dividing the 300°K rates by the relative velocity appropriate to a hydrogen atom, 2×10^5 cm sec^{-1}, the relative cross sections at this temperature are comparable to the high energy, 0.3 to 0.5 eV, cross sections for the upper part of the vibrational spectrum. These upper level values can be jointed smoothly to the 0.5 eV cross sections of Ref. 23 to give our working values shown in Fig. 5.

The fraction of these de-excitation cross sections leading to specific final states v', $F(v",v')$, are shown in Fig. 6. These are obtained by smoothing the histograms of the $J = 2$, 300°K data. The $J = 10$ fractions lie sufficiently close so as not to be distinguished from the $J = 2$ data. Inspection of the exchanged distributions for $v" = 6$ for 0.5 eV collisions also shows similarity to the $v" = 6$ distribution shown here. The fractions $F(v",v')$ shown in Fig. 6 are linked to the parent state, $v"$, by the horizontal dashed line. The $v" = 14$ distribution is an extrapolation, the $v" = 2$ distribution is taken from Ref. 21; values for intermediate $v"$ are obtained by interpolation.

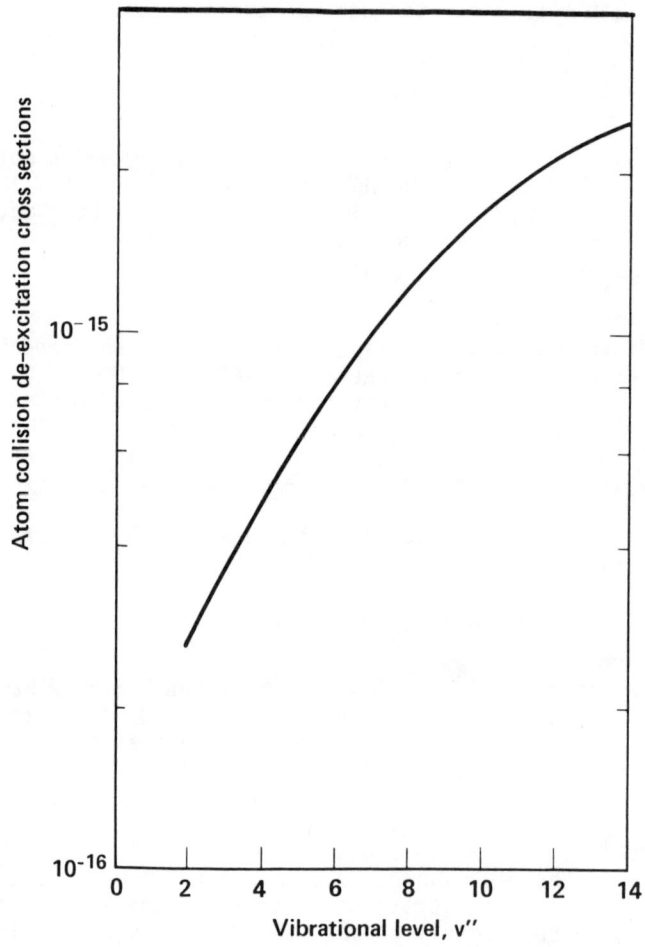

Fig. 5. The atom-molecule de-excitation cross sections versus initial vibrational level, v", for atoms with energy equal to 0.35 eV.

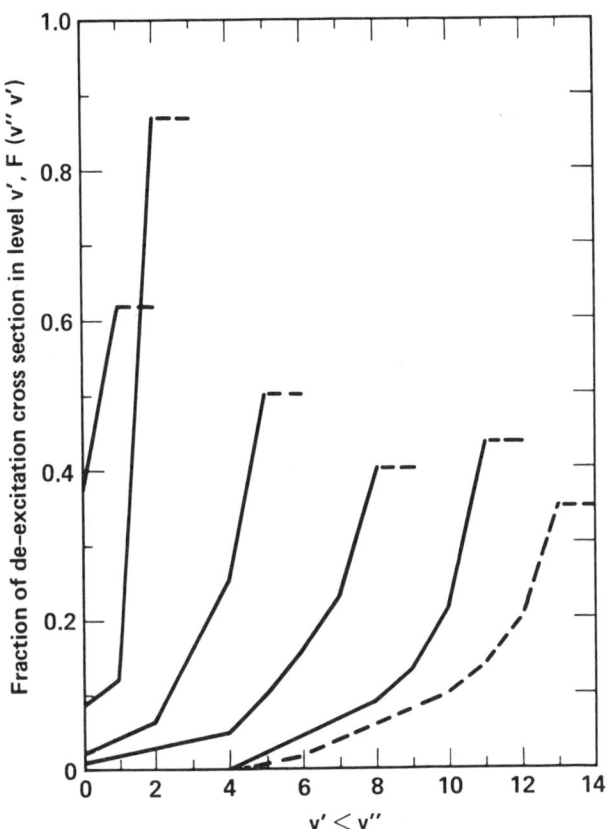

Fig. 6. Fractional transition probability to a lower level v' for a molecule in an initial level v" following an atom-H_2(v") de-excitation collision. The short horizontal dashed line joins the initial v" level to the final distribution.

A further source of vibrational de-excitation are the V-T rates. The principal de-excitations are the Δv" = -1 transitions according to

$$H_2 + H_2(v") \rightarrow H_2 + H_2(v" - 1) \quad , \tag{19}$$

with energy conservation made up in translational energy. Measured values for the de-excitation rate, P(1,0), at 500°K and at 1500°K are quoted in Refs. 24, 25. A value[26] for 3000°K is quoted in Ref. 27. An expression for the P(v",v" - 1) rates for higher v" has been derived by Keck and Carrier.[28] Using these functional expressions the rates scaled upward in v" from the empirical P(1,0) rate are shown in Fig. 2 of Capitelli, et al.[27]

We have commented earlier[6] that this scaling yields implausibly large rates near the upper end of the vibrational spectrum, and had proposed a maximum rate of 10^{-9} cm^3 sec^{-1} based upon the low-energy atom-atom elastic scattering cross sections. This conjecture is further substantiated by two different dependences exhibited in the atom-exchange rates of Ref. 22. In Fig. 5 of Ref. 22 the rates for 300°K approach 8×10^{10} cm^3 sec^{-1} near the upper part of the vibrational spectrum. In Fig. 15 and Table 14 of that paper the dissociation cross section for v" = 14 is given as 2.6×10^{-15} cm^2. For an atom exchange collision with a relative velocity corresponding to 500°K this also yields a rate of approximately 8×10^{10} cm^3 sec^{-1}. Further, Capitelli, et al.[27] have obtained better agreement with the empirical dissociation rates by reducing all V-T rates a factor of three from the scaled rates. We would expect that the V-T rates near the upper portion of the vibrational spectrum would most sensitively influence the dissociation rates. Accordingly, we have scaled the empirical v" = 0 rates upwards using the scaling as in Capitelli, et al. but limited the v" = 14 rate for 500°K to 10^{-9} cm^3 sec^{-1}. The rates at this temperature for v" = 10 and v" = 11 are reduced by a factor two and three, respectively, from the scaled rates. The maximum rate for different temperatures is assumed to increase in proportion to $[T(gas)]^{1/2}$. The rates obtained in this way are shown in Fig. 7. In the central and upper portion of the spectrum we must assume these rates remain uncertain by a factor of two or three.

For reference in later paragraphs we also show the atomic de-excitation rates for a 0.35 eV incident atom and using the cross sections of Fig. 5. The atomic de-excitation rates are large compared to the molecular rates excepting near the top of the vibrational spectrum.

The set of equations given by Eqs. 10-12 have an important scaling with R. If electron, atom, and molecular densities are scaled in proportion to R^{-1}, $\mathcal{N}(v")$ scales like R^{-2} and $\mathcal{D}(v")$ like R^{-1}, since the wall collision terms also scale as R^{-1}. It follows that the concentration $N_2(v")$ will scale as R^{-1}.

The equilibrium Eqs. 10-12 have been solved for a thermal-electron scale density of nR = 10^{13} electrons cm^{-2} (n(f)R = 10^{12} electrons cm^{-2}), for two values of the gas scale density, 10^{14} and 10^{15} molecules cm^{-2}, and for b(v") = 1, and $\gamma/R = 1$. The gas temperature and mean atom energy are taken to be 1500°K and 0.35 eV, respectively, values appropriate to high power discharges.[15] The relative distributions for the active portion of the vibrational spectrum, v" $\geq$ 6, are shown in Fig. 8. For the portion of the spectrum $6 \leq \bar{v}" \leq 10$, the populations $N_2(v")$ are seen to be proportional to the gas density N_2. Note that the total population

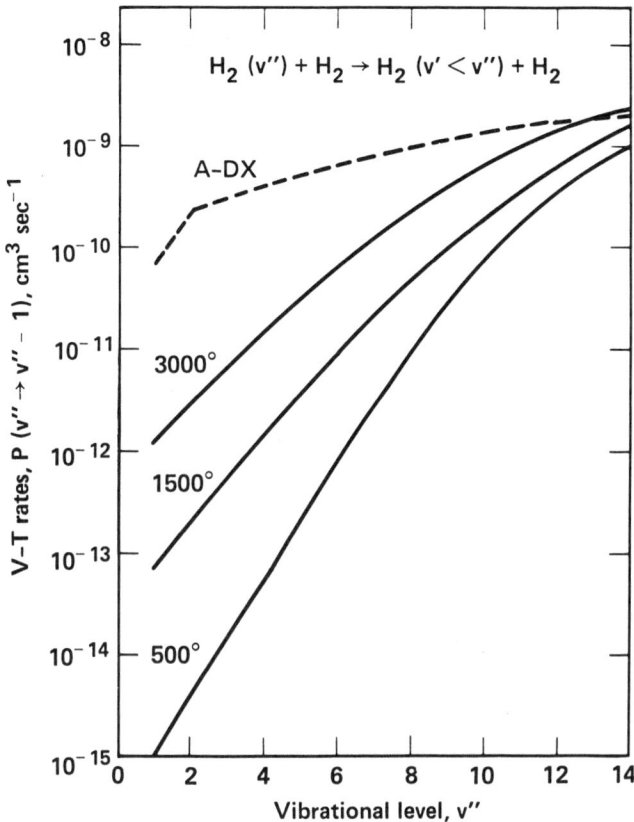

Fig. 7. The V-T rates versus v" for three different gas temperatures. The atom de-excitation rates for 0.35 eV atoms are shown for comparison by the dashed lines.

of the active portion of the spectrum is two to three percent of the overall gas density.

For a fixed value of γ and fixed ratios of the species mix H^+: H_2^+, H_3^+, the atomic ratios shown in Fig. 3 remain constant as n, N_2 are scaled with R. Small deviations apparent in Fig. 2 result from choosing different values for the species mix as these principal parameters are varied. The atomic concentrations and the corresponding atom de-excitation processes for $\gamma = 1$ are small in comparison to the other de-excitation processes for electron densities up to a few times 10^{13} electrons cm^{-3}. In their analysis

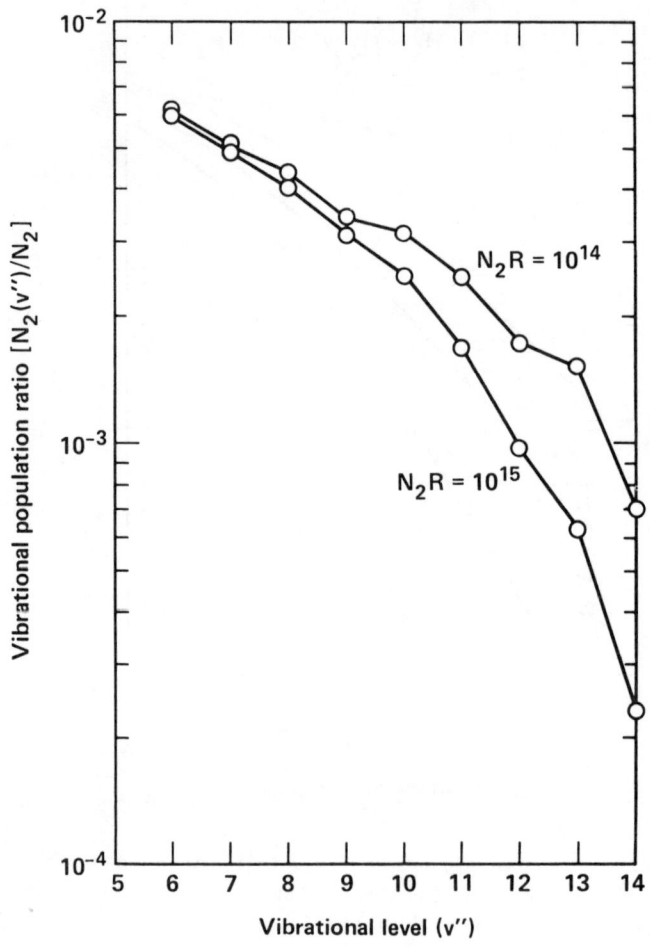

Fig. 8. Vibrational distributions, $H_2(v''\geq 6)$, versus vibrational level for molecules in the first chamber. The first chamber thermal electron density is $nR = 10^{13}$ electrons cm^{-2}, fast electron density $n(f)R = 10^{12}$ electrons cm^{-2}, and for gas densities N_2R equal to 10^{14} and 10^{15} molecules cm^{-2}.

of the high-density discharge, however, Chan et al.[15] found γ to be approximately 0.1. This will cause the atomic density to approach the molecular density. As a consequence the $v'' = 14$ population ratio will be reduced to about two-thirds the value shown in Fig. 8 and the $v'' = 6, 7$ levels a similar amount.

Taking $b(v")$ equal to unity across the spectrum causes the wall relaxation to be over-emphasized in the lower portion of the active spectrum. If $b(v")$ were increased to two, values more appropriate to the lower portion of the active spectrum, the $v" = 14$ ratios shown in the figure are little changed and $v" = 6, 7$ ratios are increased about 30 to 50%. Choosing the smaller $\gamma = 0.1$ therefore, is largely compensated for by a less restrictive and larger $b(v") = 2$ value. The possibility of selecting wall materials to enhance γ to values near 0.5 is indicated in the recombination data of Wood and Wise.[29]

THE SECOND CHAMBER

The second chamber is presumed to be free of high-energy electrons but does contain a low-temperature bath of thermal electrons. The electron temperature is chosen to enhance the dissociative attachment process,

$$e + H_2(v") \rightarrow H_2^- \rightarrow H^- + H \quad , \tag{20}$$

but also chosen low enough to suppress electron collisional detachment:

$$e + H^- \rightarrow 2e + H \quad . \tag{21}$$

These conditions are satisfied for an electron temperature $kT \simeq 1$ eV.

Both the vibrationally excited molecules and atoms generated in the first chamber pass unimpeded through the filter into the second chamber. Between the filter and the extraction end of the second chamber the $H_2(v")$ are attenuated by V-T, ADX, ADS, and dissociative attachment (DA) processes; the lack of high-energy electrons does not allow for regeneration of $H_2(v")$.

The attenuation of $N_2(v")$ at a distance z into the second chamber for a population $N_2^0(v")$ at the entrance of the chamber is then

$$N_2(v") = N_2^0(v") \exp - [an + dN_2 + (e + f)N_1]z \quad , \tag{22}$$

where

$$a = \frac{\overline{\sigma v}(DA, v")}{v(v")} \quad ; \quad d = P(v", v" - 1)/v(v");$$

$$e + f = [\sigma(ADX, v") + \sigma(ADS, v")]/v(v") \quad .$$

The rate at which negative ions are formed at a point z in the second chamber by dissociative attachment to $N_2(v")$ is given by the equation:

$$\frac{dN(-)}{dz} = n\, N_2(v")c - N(-)[n(b + h) + N_1 g] \quad , \qquad (23)$$

with

$$c = \overline{\sigma v}(DA, v")/v(-) \quad ; \quad b = \overline{\sigma v}(i-i)/v(-) \quad ;$$

$$g + \overline{\sigma v}(AD)/v(-) \quad ; \quad h = \overline{\sigma v}(CD)/v(-) \quad .$$

The b term refers to the ion-ion neutralization of H^- ions with the principal positive ion species in the discharge:

$$H^- + \begin{matrix} H^+ \\ H_2^+ \end{matrix} \to H + \text{neutrals} \quad . \qquad (24)$$

The denominator $v(-)$ is the negative ion velocity. The g term refers to the associative detachment[30,31] of H^- ions with H atoms:

$$H^- + H \to H_2^- \to H_2 + e \quad , \qquad (25)$$

For these calculations we shall take $\sigma v(AD) = 10^{-9}$ cm^2 sec^{-1}. The term h accounts for collisional detachment of H^- by the thermal electrons, reaction 8. Using Eq. (20), Eq. (21) integrates to give the negative ion concentration at the extraction end of the second chamber, $z = 2R$, due to $N_2^0(v")$ molecules entering the chamber:

$$N(-) = nCN_2^0(v") \left[n(b - a + h) - [N_2 d + N_1(e + f - g)]\right]^{-1} \times$$

$$\exp\text{-}2naR \left[\exp - 2[N_2 d + N_1(e + f)]R - \exp - 2[n(b + h) + N_1 g]R\right] \quad .$$

$$(26)$$

Inspection of the formula shows that if the densities n, N_2, and N_1 are scaled as R^{-1}, the negative ion densities also scale as R^{-1}.

Our procedure here is to vary the n, N_2 as independent variables to find those values of the densities that lead to a maximum for Eq. (26). If for a particular apparatus a functional relation connects these independent variables n, N_2, then the maximum found here may only be an upper limit for such apparatus.

It is clear that to suppress the atom concentration and the compensating loss to associative detachment we shall need to choose γ near unity. As mentioned, the collisional detachment can be made small compared to ion-ion neutralization for kT = 1 eV. Gas temperature is presumed to be 1500°K and atom energy taken equal to 0.35 eV. The σv (i-i) is approximately constant over the ion energy range 0.1 to 1 eV. In the high-power discharge H^+ is the dominant positive species and the H_2^+ concentration remains 20-30% over a broad parameter range[15]; the H_3^+ concentration is small. For a H^+: H_2^+ ratio of 70%: 30%, ions in the above energy range will exhibit an effective σv(i-i) = 1.1 x 10^{-9} cm^2 sec^{-1}.

In first approximation we set $N_1 = N_2 = 0$ and calculate N(-) as a function of nR. In Fig. 9 is plotted the ratio of N(-) per incident $N_2^0(v")$ observed at the extraction end of the second chamber as a function of nR for two typical levels v" = 6, 9. At very low electron densities this ratio rises at first because of the increased probability of dissociative attachment. As the electron density increases further, the ion-ion collisional attenuation causes the ratio to pass through a maximum near nR = 10^{13} electrons cm^{-2}. With a further increase of the electron density the ratio falls sharply both because of the attenuation of the $N_2(v")$ and the enhanced ion-ion neutralization.

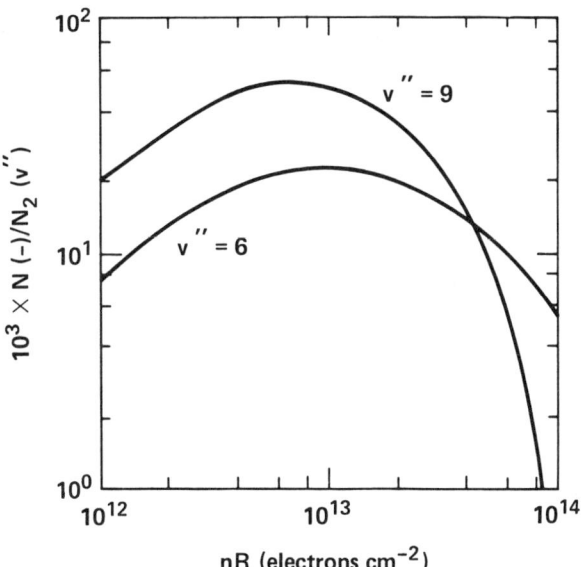

Fig. 9. Ratio of negative ion concentrations at the extraction end of the second chamber per initial $N_2^0(v"=6,9)$ at the entrance of the second chamber versus second-chamber thermal electron density calculated for no atomic or molecular attenuation of the $N_2(v")$.

Holding constant at the optimum value of $n(2)R = 10^{13}$ electrons cm^{-2} in the second chamber, but now increasing the gas density, the ratio $N(-)/N_2^0(v'')$ evaluated at the far end of the chamber falls due to the attenuation of the $N_2(v'')$ by gas and atom collisions. This decay is illustrated in Fig. 10 for two values of γ: 0.1 and 1.0. For the smaller γ the much larger concentration of atoms in the discharge for a particular value of N_2R (cf. Fig. 3) causes a much more rapid attenuation of the $N_2^0(v'')$ and hence a smaller ratio for $N(-)/N_2^0(v'')$.

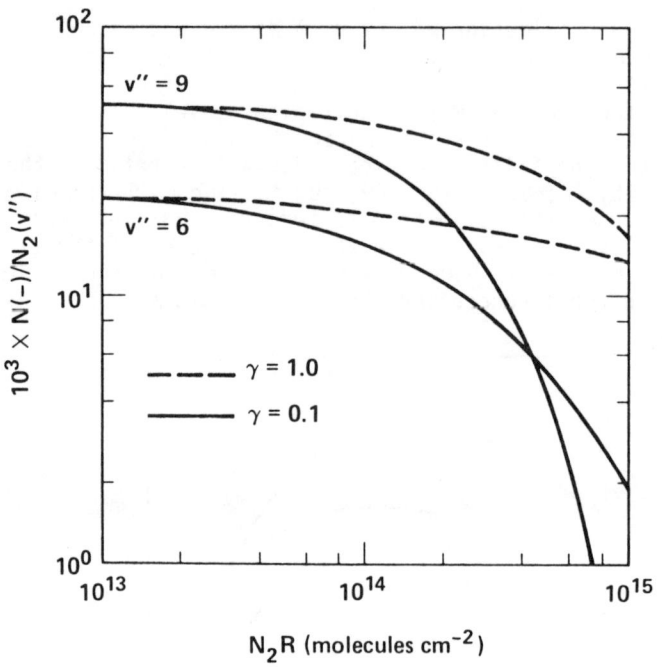

Fig. 10. Ratio of negative ion concentrations at the extraction end of the second chamber per initial $N_2^0(v''=6,9)$ at the entrance end of the second chamber versus gas density. Atomic and molecular attenuation of $N_2(v'')$ is operative. Solid curves: $\gamma = 0.1$; dashed curves $\gamma = 1.0$.

These ratios, when multiplied by the concentrations $N_2(v'')R$ calculated using Eqs. 10-12 and $n(1)R = 10^{13}$ electrons cm^{-2}, yield $N(-)R$ at the extraction end of the second chamber. The $N(-)R$ is plotted in Fig. 11 as a function of N_2R. The most distinct feature shown in Fig. 11 is the maximum as a function of N_2R. We have performed similar calculations for $n(-)R$ by varying the electron density, n(1)R, in the first chamber to locate the maxima. For a smaller n(1)R = 0.5 × 10^{13} electrons cm^{-3} the initial generation of N$_2^0$(v") is reduced, while for a larger n(1)R = 2 × 10^{13} electrons cm^{-3} a larger generation of free atoms occurs. In consequence, a broad maximum in N(-)R occurs about n(1)R = 10^{13} electrons cm^{-2}.

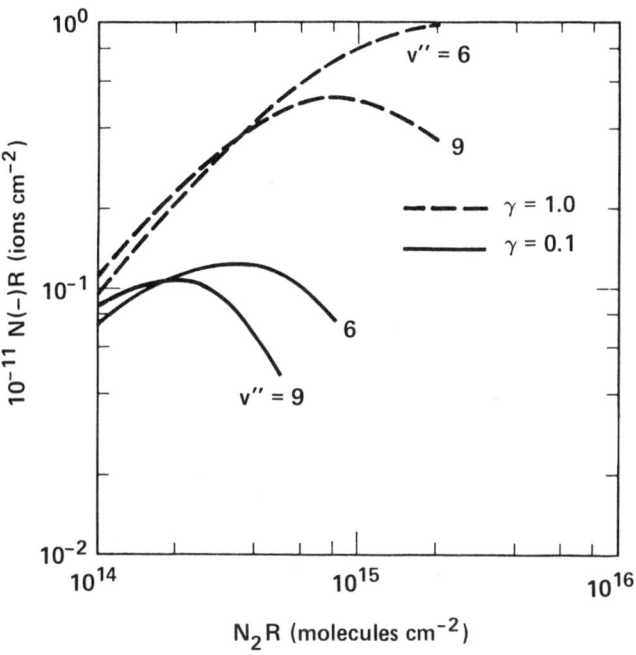

Fig. 11. Negative ion concentrations at the extraction end of the second chamber derived from initial v" = 6 or 9 levels as functions of the gas density. Solid curves: γ = 0.1; dashed curves γ = 1.0.

In summary, the optimum parameters for a two chamber system are given by

$$n(1)R = n(2)R = 10^{13} \text{ electrons cm}^{-2} \; ; \qquad (27)$$

$$N_2R = 2 - 4 \times 10^{14} \text{ molecules cm}^{-2}, \; \gamma = 0.1 \; ;$$

$$N_2R = 8 - 20 \times 10^{14} \text{ molecules cm}^{-2}, \; \gamma = 1.0 \; ;$$

with rather broad maxima about these values.

The extracted current densities are proportional to the product of the negative ion velocity and density at the extraction end of the second chamber. We shall assume that the electric potentials in the second chamber are arranged so that all negative ions upon being formed are gathered and carried in the axial direction toward the extractor electrodes. This is accomplished explicitly in the three-chamber tandem-system alluded to earlier. The extracted negative ion current density is then taken to be

$$j(-) = N(-) \, v(-) \quad , \tag{28}$$

with

$$N(-) = \frac{1}{R} \sum_{v''=1}^{14} N(-)/N_2^0(v'') \; N_2^0(v'') R \quad . \tag{29}$$

The principal contribution to the sum occurring for $v'' \geq 6$.

The velocity of the negative ions remains a somewhat uncertain quantity. Inspection of the electron energy dependence of the dissociative attachment cross sections[7] for KT = 1 eV suggests negative ions are formed with an energy near 0.2 eV. An upper limit to this energy is provided by the emittance properties of an extracted beam. Observations of transverse angular spreads show a mean perpendicular energy of 0.3 eV, indicating a total energy of 0.45 eV. For this discussion we shall choose lower and upper limits to be 0.2 and 0.45 eV, respectively.

The dissociative attachment rates used here and taken from Ref. 7 are calculated for non-rotating, J = 0 molecules. Wall relaxation collisions however, show rotational excitations enhanced through the J = 12 to 15 levels after one or two wall collisions. With these excitations included in the rotational dissociative-attachment rates reported recently by Wadehra,[8] one expects some enhancement of the dissociative attachment for levels $v'' = 6, 7$. Specifically, the $v'' = 6$ rate is doubled and the $v'' = 7$ level rate is enhanced about fifty percent, but higher v'' rates are essentially unchanged. Taking into account these rotational enhancements, the sum in Eq. (29) is increased thirty-five to fifty percent over the J = 0 value.

The optimum extracted negative ion current densities are plotted versus scale length, R, in Fig. 12. The corresponding electron density is shown on the lower abscissa. The solid curves are calculated for $\gamma = 0.1$, the value deduced for the high-power positive-ion-source discharges. With a scale length equal to one centimeter and with rotational effects included the higher energy negative ions will yield an extracted current density of 16 mA cm^{-2}.

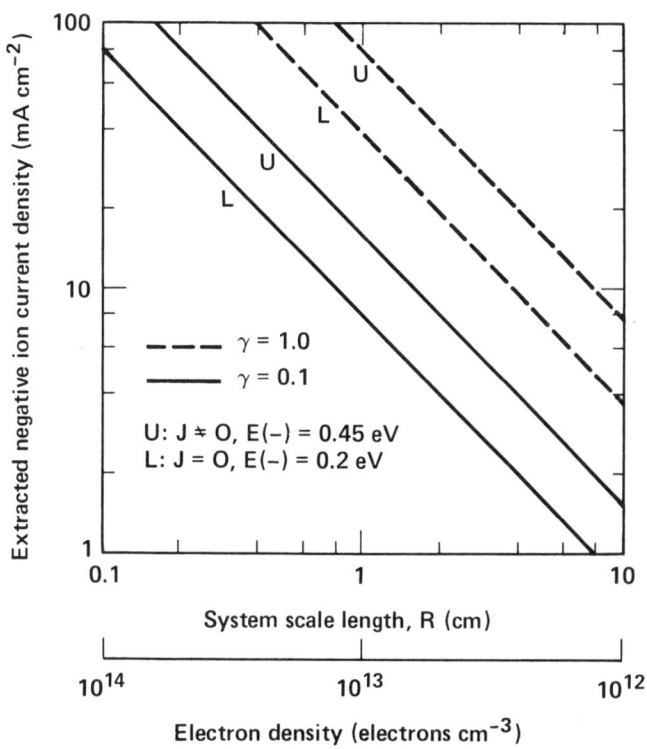

Fig. 12. Optimum extracted negative ion current densities versus system scale length or versus second-chamber thermal electron density. Solid curves: $\gamma = 0.1$; dashed curves: $\gamma = 1.0$. L: $J = 0$, $E(-) = 0.2$ eV; U: $J \neq 0$, $E(-) = 0.45$ eV.

The true optimum extracted current density will occur for a γ equal to unity and the dashed curves in Fig. 12 represent this optimum. Note that for a factor of ten increase in γ, from 0.1 to 1.0, the extracted currents increase by a factor of five.

CONCLUSIONS

The negative ion current density extracted from a two-chamber (three-chamber) tandem configuration is calculated for a system relying upon electron vibrational excitation followed by dissociative attachment. The extracted current density passes through an optimum as the electron and molecular densities are varied, and scales inversely with the scale length. Substantial increases in the current density can be achieved if the atom wall-recombination coefficient can be increased to values near unity. Since free atoms in the discharge are so effective at destroying both excited molecules by atom-de-excitation collisons and negative ions by associative detachment, a measure of their concentration is an essential component of the system diagnostic.

ACKNOWLEDGEMENTS

We are indebted to C.F. Burrell, C.F. Chan, W.C. Cooper, K.W. Ehlers, and K. Leung for valuable discussions relating to this work.

REFERENCES

1. M. Bacal, E. Nicolopoulou, and H. J. Doucet, Proc. Symp. on the Production and Neutralization of Negative Hydrogen Ions and Beams, p. 26, BNL-50727, Brookhaven, September (1977).

2. M. Bacal, Physica Scoipta. T2/2, 467 (1982). Contains an excellent bibliography through June 1982.

3. W. G. Graham, VUV Studies of a Low Pressure Hydrogen Plasma, this proceedings.

4. K. W. Ehlers, Proc. Int. Ion Engineering Congress, p. 59, Kyoto, Japan, 12-16 September (1983).

5. A.J.T. Holmes, G. Dammertz, T. S. Green, and A. R. Walker, Proc. Int. Ion Engineering Congress, p. 71, Kyoto, Japan, 12-16 September (1983).

6. J. R. Hiskes, A. M. Karo, M. Bacal, A. M. Bruneteau, and W. G. Graham, J. Appl. Phys. $\underline{53}$ (5) 3469 (1982).

7. J. M. Wadehra, Appl. Phys. Lett. $\underline{35}$, 917 (1979).

8. J. M. Wadehra, Phys. Rev. A $\underline{29}$ (1984).

9. K. W. Ehlers and K. N. Leung, Rev. Sci. Instrum. $\underline{52}$, 1452 (1981).

10. K. W. Ehlers and K. N. Leung, Rev. Sci. Instrum. $\underline{53}$, 1423 (1982).

11. K. N. Leung, K. W. Ehlers, and M. Bacal, Rev. Sci. Instrum. $\underline{54}$ (1), 56 (1983).

12. K. N. Leung, K. W. Ehlers, and J. R. Hiskes, "Concept For a Volume-Type Tandem Three-Chamber Negative Hydrogen Ion Source," LLNL Rpt. UCID-19910, October (1983).

13. K. W. Ehlers, W. R. Baker, K. H. Berkner, W. S. Cooper, W. S. Kunkel, R. V. Pyle, and J. W. Stearns, J. Vac. Sci. Technol. 10, 923 (1973).

14. K. F. Schoenberg and W. F. Kunkel, J. Appl. Phys. 50 (7), 4685 (1979).

15. C. F. Chan, C. F. Burrell, and W. S. Cooper, J. Appl. Phys. October (1983).

16. J. R. Hiskes and A. M. Karo, "Electron Energy Distributions, Vibrational Population Distributions, and Negative Ion Concentrations in Hydrogen Discharges," presented at the NATO Adv. Study Inst. on Atomic and Mol. Proc. in Contr. Thermo. Research, Palermo, Italy, July 19-30, 1982. UCRL-87779, June (1982).

17. L. Spitzer, Jr., "Physics of Fully Ionized Gases," Interscience Publishers, Inc. New York (1956).

18. J. R. Hiskes, J. Appl. Phys. 51, 4592 (1980).

19. W. F. Bailey and A. Garscadden, Proc. Second Int. Symp. on the Prod. and Neutralization of Negative Hydrogen Ions and Beams, p. 33, BNL-51304, Th. Sluyters, Ed., October (1980).

20. A. M. Karo, J. R. Hiskes, K. D. Olwell, T. M. DeBoni, and R. J. Hardy, Proc. Third Int. Symp. on the Prod. and Neutralization of Negative Ions and Beams, Brookhaven, November 14-18 (1983).

21. I.W.M. Smith and P. M. Wood, Mol. Phys. 25, 441 (1973).

22. N. C. Blais and D. G. Truhler, Potential Energy Surfaces and Dynamics Calculations, p. 431, Plenum Press (1981), D. G. Truhler, Ed.

23. N. C. Blais, private communication.

24. M. M. Audibert, C. Joffrin, and J. Ducing, Chem. Phys. Lett. 25, 158 (1974).

25. J. Dove and H. Teitelbaum, Chem. Phys. Lett. 6, 431 (1974).

26. J. H. Keifer and R. W. Lutz, J. Chem. Phys. 44, 668 (1966).

27. M. Capitelli, M. Dilonardo, and E. Molinari, Chem. Phys. 20, 417 (1977).

28. J. Keck and G. Carrier, J. Chem. Phys. 43, 2284 (1965).

29. B. J. Wood and H. Wise, J. Phys. Chem. 65, 1976 (1961).

30. R. J. Bieniek and A. Dalgarno, Astroph. Jour. 228, 635 (1979).

31. R. J. Bieniek, J. Phys. B 13, 4405 (1980).

This work was performed under the auspices of the U.S. Department of Energy by Lawrence Livermore National Laboratory under contract No. W-7405-Eng-48.

DISCUSSION

WADEHRA: In the first chamber where vibrational excitation takes place, you are requiring 40 eV electrons, but vibrational excitation cross sections usually peak at much lower energies.

HISKES: The process we are considering here is the following: a 20 or 30 eV electron excites the molecule to high singlet states, a form of electronic excitation. The electronic excitation radiates back down to the ground electronic state but leaves the molecule with vibrational excitation; these cross sections peak about 40 eV.

BARNETT: How sensitive is your model to having a boundary layer between chamber I and chamber II?

HISKES: There probably is some sensitivity and that is why I emphasized the scale length without emphasizing a specific dimension, because this is not intended to be an engineering design. Rather, it is intended to indicate that the physical dimensions do affect all these processes and do affect the scaling in a gross way. I wouldn't say that a one centimeter system gives precisely 50 mA cm^{-2} but something like that. What one would like to do eventually is a complete spacial calculation but that's an order-of-magnitude larger calculation. The physical dimensions are a very important parameter in this system.

JACQUOT: How is the density of H^- ions changing when you change the temperature of your electron distribution? For example, you have an electron temperature of 3.8 eV and you change it to 5 eV.

HISKES: The rate of vibrational excitation in the first chamber is proportional almost entirely to the density of fast electrons above 20 or 30 eV. You can, to a very good approximation, ignore the thermal distribution and what I have assumed throughout here is that the ratio of the density of fast electrons in the first chamber to the total electron density is 0.1. That seems to be a plausible upper limit to what could be achieved. You could interpret the first chamber results entirely in terms of the fast electron density.

BACAL: Did you consider the detachment of the electron from the negative ion due to collisions with high energy electrons?

HISKES: Yes, in the first chamber where we have high energy electrons the negative ion concentration is very low, like 0.1% or less. However, we aren't concerned with the first chamber density because we're not trying to make negative ions there. In the second chamber we postulate that there are no high energy electrons, just thermal electrons.

BACAL: Now, when you consider the destruction of H^- due to the atoms, do you consider the production of vibrationally excited H_2 as the result of the associative detachment?

HISKES: Yes, we consider it but it's not a large contribution because the number of vibrationally excited H_2 that resulted from the first chamber is larger than the number that resulted from the destruction of negative ions in the second chamber.

BACAL: I have read some papers by a German group in Julich on the recombination coefficient γ for the atomic hydrogen and they are the only ones who claim that this coefficient is equal to 1.

HISKES: I haven't seen the value 1 in the literature, I've seen 0.48.

BACAL: Well, I'm telling you that there are papers by Waelbroek and others from Julich who are concerned with fusion research and interactions of hydrogen atoms with the surfaces in tokamaks and they claim that a large number of surfaces they are using have the recombination coefficient equal to 1.

HISKES: Do you remember what the material was?

BACAL: Well, let's say palladium and some others. I would say that it is not really clear what is the truth about the sticking coefficient for hydrogen atoms and this looks like an important subject.

HISKES: In positive ion sources one would always emphasize a low recombination coefficient to enhance the atom concentration, but here the emphasis is just reversed.

BARNETT: Is this hard to measure with any accuracy?

HISKES: Well, I would think one could drill a hole in the side of the discharge and let the atoms come out and then measure their flux.

PETERSON: It's not unlikely that if you find a surface to optimize the wall recombination coefficient you would also optimize or tend to increase the wall relaxation coefficient.

HISKES: That could be true. I know Bill Graham is going to give a talk on varying the surfaces and varying the wall relaxation. I'm not sure what the correlation is there.

PETERSON: Do you have any feeling for the relative importance of these two?

HISKES: No, I don't. That problem can be studied with the code which Arnold Karo will discuss tomorrow morning.

BACAL: Did you consider the elastic collisions of negative and positive ions and the effects they have upon the ion temperatures?

HISKES: Only briefly. The question of the negative ion temperature is still something of an open question, but we haven't pursued that problem in any detail. Certainly the extracted current densities are proportional to the velocity of negative ions so that is really a critical parameter in evaluating the system.

H⁻ PRODUCTION AND DESTRUCTION MECHANISMS IN HYDROGEN LOW PRESSURE DISCHARGES

M. Bacal and A.M. Bruneteau
Laboratoire de Physique des Milieux Ionisés,
Groupe de Recherche N° 29 du C.N.R.S.,
Ecole Polytechnique, 91128 Palaiseau Cedex, France

ABSTRACT

The density of negative hydrogen ions and the plasma characteristics are investigated in several configurations of the multicusp plasma generator. H⁻ destruction in plasma is analyzed and two different destruction regimes are discussed. The decrease in H⁻ density with increasing gas pressure is explained in terms of an increase of the rate of mutual neutralization, due to the reduction of the ion energies when the pressure is increased.

INTRODUCTION

The study at Ecole Polytechnique of the low density hydrogen plasma produced in a glass vessel has established that negative hydrogen ions are present in sizeable fractions ($n_-/n_e \leq 0.35$) in this plasma[1]. In order to attain the higher negative ion densities required for a useful neutral beam injection source, we have investigated the H⁻ density and the extracted current from a multifilament stainless steel plasma generator, which was operated with and without a multicusp magnetic field. The purpose of this paper is to
1) review the progress made in understanding the processes leading to H⁻ production and destruction in high density plasma ;
2) describe work in progress to evaluate the performances of various multicusp configurations.

EXPERIMENTAL SET-UP

Several multicusp configurations can be produced in the stainless steel plasma vessel, described in detail elsewhere[1-3]. We denote as 'hybrid' the configuration in which there are no magnets on the end plates and the filaments are located in the multicusp magnetic field, which is produced by mounting externally on the cylindrical chamber wall 10 or 20 columns of ceramic or samarium-cobalt magnets. Measurements were make for four different configurations of the magnetic field[3] : in three of them (B, C, D) ten columns of magnets are used, while in configuration A the number of columns is increased to twenty. In C the magnetic field of the ceramic magnets is enhanced by superposing two rows of magnets. In D, an asymmetry in the magnetic field was created by doubling six of the ten columns. Each filament could be located either in the cusp plane (subscript 1) or in between two cusps (subscript 2).

A conventional multicusp ion source configuration could be produced by applying four extra rows of magnets on the top flange[4]. Two groups of filaments are placed close to the chamber axis. A samarium-cobalt magnet filter can be added to produce a tandem multicusp plasma generator[4] ; its 'driver' section is characterized by a relatively long lifetime for wall loss[2] of the primary electrons, τ_w.

The H^- ion density, n_-, is measured by the photodetachment technique[5,6] : the electron is detached from H^- by means of a pulsed laser beam (Nd Yag), crossing the plasma at half distance between the end plates. A cylindrical tungsten probe coaxial with the laser beam and biased at + 20 V relative to the anode collects the detached electrons. This results in a probe current pulse, whose height is proportional to the H^- ion density[5].

Radial and axial profiles of various plasma parameters were obtained with two movable disc probes (0.4 cm diam). Recently an H^- electron separator has been installed at the bottom end of the chamber[3]. Thus the present apparatus provides simultaneously the density of H^- ions and electrons in the center of the plasma and the respective extracted currents.

The ro-vibrational populations of H_2 molecules are measured by coherent anti-Stokes Raman scattering (CARS) in a conventional multicusp plasma generator, placed under a bell jar. In order to cool down the gas, the copper wall temperature could be lowered to 230 K.

H^- DENSITY IN VARIOUS MULTICUSP CONFIGURATIONS

Hybrid multicusp plasma. The investigation of the influence of the relative position of the filaments and magnets has shown that n_- and n_-/n_e are 50 % higher when the filaments are located in between the cusps (configuration B_2). In order to have the filaments in the same position relative to the cusps in all the configurations studied, we will present data obtained with filaments located in the cusp plane (configuration type "1"). All the data refer to ceramic magnets.

Figure 1 shows a typical dependence of n_- and n_-/n_e on I_d, for several gas pressures P. It appears that the increase in P leads to an increase in n_- and n_-/n_e up to an optimum pressure above which a decrease is observed. This behaviour is also illustrated by Fig. 2, where n_-/n_e is plotted versus P at constant I_d. The results obtained in a plasma without magnetic containment field (diffusion plasma, DP) is also shown. Figure 3 shows the variation of n_- versus I_d at optimum P in various plasma configurations.

Thus, the maximum relative negative ion density in the hybrid multicusp configuration B (0.15) is comparable to that observed in the DP at somewhat higher P. It appears that n_- and n_-/n_e are affected by the number, strength and symmetry of the columns of magnets. Both n_- and n_-/n_e depend on P and I_d.

Conventional multicusp plasma. In this multicusp configuration, n_-/n_e is lower (0.01 - 0.02) than in the hybrid ones B, C and D (0.15-0.5) for the same plasma conditions. The lowest value for n_-/n_e is observed in the configuration with the longuest τ_w, i.e. in the driver of the tandem multicusp generator[4].

Effect of wall material and wall temperature. The influence of wall material upon n_- and n_-/n_e has been tested in the plasma generator

33

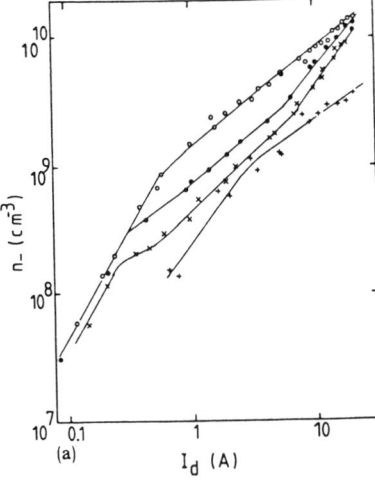

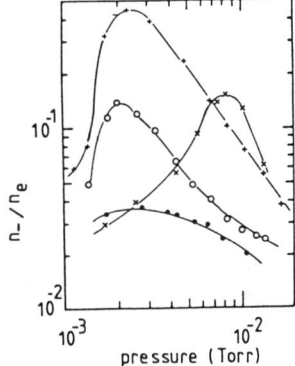

Fig. 1. Dependence upon I_d of (a) the H⁻ density and (b) the relative negative ion density in the hybrid multicusp generator (config. B) at various pressures : + - 2mTorr, O - 3.3 ; ● - 8.3 ; x 13.5 mTorr.

Fig. 2. Pressure dependence of the relative negative ion density in diffusion plasma (xx) and various hybrid multicusp configurations : ● A ; o B and C ; + D.

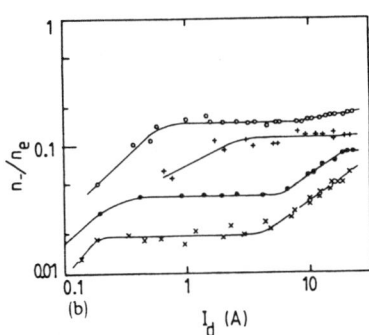

Fig. 3. Variation of the H⁻ density upon I_d in various plasma configurations, at optimum pressure.

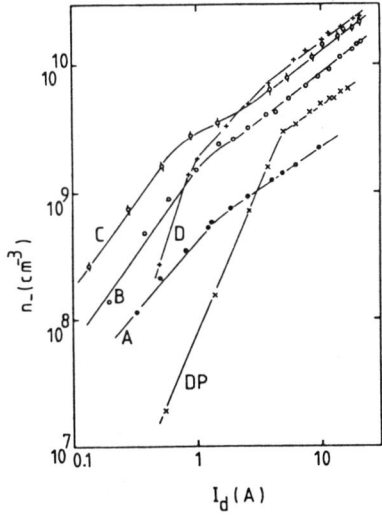

used in CARS measurements. Here the original copper wall could be covered by a stainless steel foil. The generator was operated both as a DP and in a coventional multicusp configuration. n_-/n_e goes up in the latter, at optimum pressure from 0.013 to 0.032 when the copper wall is replaced by a stainless steel wall (Figs. V.4 and V.5 in Ref. 9). This observation cannot be used in discussing the probability of deactivation of vibrationally excited H_2 in collisions with different surfaces, since secondary effects may have occurred : rise of the gas temperature when the stainless steel wall is used, possible change of kT_e (which was not measured), and of impurity content. No change in n_-/n_e is observed when the copper wall temperature is varied in the range 225-293 K.

PLASMA PARAMETERS

A detailed presentation of the plasma parameters found in the various configurations studied is given in Ref. 2 and 9. There is no significant difference in the density of the 50 eV-primary electrons in the center of the hybrid multicusp configurations A-D : n_p/I_d = 5 x 10^7 cm^{-3}/A and the corresponding τ_w (50 eV) = 10^{-7} s. The corresponding values in the driver are 7.5 times larger[9], but they are 2.5 times lower in the DP. 'Islands' in which n_p is ten times larger than in the background plasma have been observed at the periphery of the hybrid multicusp plasma, indicating that a significant fraction of the primary electrons is trapped by the surface magnetic field.

In all studied plasmas, the plasma potential is positive with respect to the anode. The presence of the multicusp magnetic field reduces V_p in the center of the plasma from 4-5 V in the DP to 1.5-3 V in configuration C ; it also produces self consistent positive potential 'islands' at the plasma periphery, in between the cusps.

Figure 4 shows the axial plasma potential profile in the hybrid multicusp plasma configuration B_2 at three values of the plasma electrode bias[4]. The axial electric field in the plasma is ~ 0.1 V/cm.

VIBRATIONAL AND ROTATIONAL POPULATION DISTRIBUTION IN HYDROGEN PLASMA

For reasons of instrumental sensitivity the CARS measurements[8] were conducted with hydrogen and deuterium gas pressure of ~40 mTorr. The main results are as follows :
1. In a plasma with $n_e \simeq 10^{12}$ cm^{-3} and kT_e = 0.85 eV, the populations of the first three vibrationally excited states follow approximately the Boltzmann law with a vibrational temperature of 2390 K.
2. The populations of the rotational substates with $J \geq 5$ of the H_2 vibrational states which we were able to probe with sufficient sensitivity (v = 0, 1 and 2) is significantly higher than expected from Boltzmann law.
3. Rotational temperature measurements taken as a function of I_d

demonstrate the rise of rotational temperature with discharge power.
4. Summing all the vibrational populations in the presence of the discharge (90 V - 10 A), we find only 61 % of the density of H_2 without the discharge, at the same temperature and pressure. The deficiency in H_2 density could be attributed to the presence of large fractions of H atoms.

At a degree of ionization of $\sim 5 \times 10^{-3}$ one would expect a Boltzmann vibrational population distribution at the low v end of the spectrum with a temperature close to kT_e, if all the excitation and de-excitation processes were due to low energy electrons. Additional deactivating processes have to be considered, in particular quenching by the hydrogen atoms and in wall collisions. The quantitative characterization of both processes suffers to some extent by the lack of information on the state-to-state quenching rates. Preliminary results show that both processes are important in deactivating the vibrationally excited molecules. This emphasizes the importance of measuring the density of hydrogen atoms in the conditions of the CARS experiment.

H^- PRODUCTION AND DESTRUCTION IN PLASMAS

It has been suggested[10,11] that the main mechanism responsible for H^- production in plasmas generated by impact ionization due to primary electrons is the dissociative attachment (DA) of low energy electrons to highly vibrationally excited molecules[12], $H_2(v^*)$. The high vibrational levels of the H_2 molecule (v > 5) are populated mainly by radiative decays from higher singlet electronic states, excited in collisions with the high energy (20-120 eV) primary electrons[11,13] (referred to as E-V singlet excitation). Thus the density of highly vibrationally excited molecules, $n(v^*)$, as well as the density of the secondary, plasma electrons, n_e, are proportional to the density of the primary electrons[2,11], n_p:

$$n(v^*) = (bL/v^*) \, n_p \, n_o \, <\sigma(EV) \, v_p> \tag{1}$$

$$n_e \approx n_+ = (L/v^+) \, n_p \, n_o \, <\sigma_i \, v_p> \tag{2}$$

where b is the number of wall collisions a $H_2(v^*)$ molecule is able to survive, L - the mean distance to the wall, $<\sigma(EV) \, v_p>$ the reaction rate for E-V singlet excitation, v^* - the average speed of the $H_2(v^*)$ molecules, $<\sigma_i v_p>$ - the reaction rate for ionization of H_2 by electron collisions, L/v^+ - the positive ion lifetime for wall losses, n_o - the neutral gas density and v_p - the speed of the primary electrons. The substitution in Eq. (2) of n_e by n_+ is justified for low relative negative ion densities (n_-/n_e). When $n_-/n_e << 2$, the positive ion flux to the wall is controlled mainly by the electron temperature, kT_e; then v^+ is equal to the ion acoustic speed:

$$v^+ = \left[\frac{kT_e + 3 kT_+}{m_+} \right]^{\frac{1}{2}} \tag{3}$$

Here m_+ is the mass of the positive ions, kT_+ is the positive ion temperature.

Taking into account collisional electron detachment (ED) due to the primary electrons as well as mutual neutralization (MN) with positive ions, the following expression is obtained for the equilibrium density of H⁻ ions in the plasma, n_- :

$$n_- = \frac{n_e \, n(v^*) \, <\sigma v \, (DA)>}{n_+ <\sigma v \, (MN)> + n_p <\sigma(ED) \, v_p> + 1/\tau_- + 1/\tau_d} \quad (4)$$

Here $<\sigma v \, (DA)>$ and $<\sigma v \, (MN)>$ are, respectively, the reaction rates for DA and MN, $\sigma(ED)$ - the cross-section for ED, τ_d is the lifetime for wall loss of the H⁻ ions and τ_- is the lifetime for H⁻ destruction in collisions with impurity atoms or molecules, such as, e.g., water vapor.

Wall neutralization has been considered as the main H⁻ destruction mechanism at low plasma density[10,11,13]. However τ_d is extremely long when V_p is positive and has a value of several kT_- ($\tau_d \simeq 10^4$s when V_p = 2 V and kT_- = 0.1 eV), while the experimentally determined lifetime of H⁻ in the low density regime is only ~ 0.5 ms.

In the case of multicusp plasma a special form of diffusion loss seems possible but its importance has to be further evaluated : H⁻ can be lost by diffusion into the positive potential 'islands' present at the plasma periphery. We assume that the negative ions are trapped in these 'islands' and are unable to return to the main plasma. The potential barrier separating the central region of the plasma from these 'islands' may be very low ; therefore small changes in the potential profile can lead to important changes in the H⁻ flux to the positive potential 'islands'.

Destruction due to H⁻-molecule reactions has not been previously considered though some of these reactions have rate constants larger than 10^{-9} cm³/s at 300 K (see Ref. 14). The fastest and most plausible reaction is

$$H^- + H_2O \rightarrow OH^- + H \quad (4)$$

With a rate constant of 3.7 x 10^{-9} cm³/s at 300 K, the lifetime for loss in this collision, τ_-, is 1 ms when the water vapor partial pressure is 10^{-5} Torr.

A special case of H⁻-molecule collision is H⁻ stripping in collisions with H_2 molecules. Recently Huq et al[15] found that the threshold for this reaction (1.45 eV) was considerably higher than the electron afinity of hydrogen. However, this type of collision can be important in the destruction of negative ions entering a positive potential island in a multicusp plasma. An important problem to be considered is whether the threshold for detachment is lowered when the hydrogen molecule is vibrationally excited.

Another special case of H⁻-atom collision is that of associative detachment collisions[16]

$$H^- + H \rightarrow H_2(v) + e \quad (5)$$

since H atoms are produced in the hydrogen plasma by several processes, and their density will vary with plasma density and other plasma conditions. However, most of the H_2 molecules that are formed by associative detachment will be in high ro-vibrational states ; reattachment to these molecules will then be likely.

In multicusp plasma devices, due to lower escape probabilities for electrons, higher densities of primary electrons may be achieved. This will lead, on the one hand, to higher $n(v^*)$ and n_e (Eq. 1 and 2), that is to increased H^- production, and, on the other hand, to higher destruction rates through ED (Eq. 4). At the same time the adverse effect on H^- production of the degradation in the primary electron energy due to collisions may become more pronounced. This effect is a consequence of the fact that, while electrons with energies as low as 10 eV may effectively destroy H^- by ED, only those with energies exceeding 20 eV are contributing to plasma production and vibrational excitation (via E-V singlet excitation). Since energy degradation is increasing with rising gas pressure, it is important for the discussion of Eq. (4) to establish the pressure at which it might become a non-negligible factor in the processes leading to H^- formation and destruction ; this pressure corresponds to a neutral gas density n_o^* :

$$n_o^* = [(\sigma_i + \sigma_e) v_p \tau_w]^{-1} \tag{6}$$

σ_e is the cross section for electronic excitation. The values of n_o^* for 50 eV-electrons correspond to a pressure of $\sim$ 10 mTorr in the DP, 3.8 mTorr in the hybrid multicusp plasma, and 0.5 mTorr in the driver. Thus, when $n_o < n_o^*$, the primary electrons are quasi-monoenergetic. It is shown[2] that in this case at a fixed gas pressure two different regimes are possible :
1. At low n_p :

$$n_- \alpha \tau_-' n_p^2 n_o^2 \tag{7}$$

whereas

$$n_-/n_e \alpha \tau_-' n_p n_o \tag{8}$$

$1/\tau_-' = 1/\tau_- + 1/\tau_d$ is the collision frequency for H^- destruction in the low density regime.

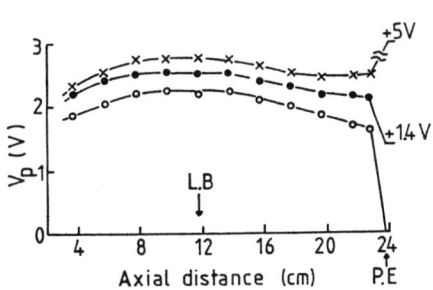

2. At high values of n_p, the H^- destruction in collision with charged particles (MN or ED) dominates over wall loss and over destruction in collisions with impurities. On the other hand MN dominates over ED when

Fig. 4. The axial plasma potential profile in the hybrid multicusp generator (config. B_2), for three different values of the plasma electrode bias. Ceramic magnets, 3 mTorr, 50 V, 2 A. The arrow indicates the position of the laser beam.

$$n_o > \frac{\sigma(ED)}{\sigma_i} \cdot \frac{(v^+/L)}{<\sigma v\,(MN)>} \approx 5 \times 10^{12} \text{ cm}^{-3} \qquad (9)$$

or $P > 0.15$ mTorr, regardless of the values of n_p. In this pressure range, for high values of n_p, we have :

$$n_- \alpha\, n_p\, n_o \qquad (10)$$

and

$$\frac{n_-}{n_e} = b\,\frac{v^+}{v^*} \cdot \frac{\sigma(EV)}{\sigma_i} \cdot \frac{<\sigma v\,(DA)>}{<\sigma v\,(MN)>} \qquad (11)$$

According to Eq. (11) n_-/n_e does not depend directly on n_o and n_p, but may be affected indirectly by n_o via its effect on the electron and ion temperatures, which in turn determine v^+, $<\sigma v(DA)>$ and $<\sigma v(MN)>$.

The change from the low density regime to the high density regime depends on the gas pressure : the higher the neutral gas pressure, the lower the n_p value for which the transition occurs.

In the case of high gas pressure and high n_p, ED can dominate the H$^-$ loss in multicusp devices with large enough τ_w (see Ref. 2). The additional destruction by ED is the reason for the lower n_-/n_e observed in the driver and the conventional multicusp plasma generator. This is not the case of the hybrid multicusp devices investigated in this paper. Therefore in these plasmas MN will dominate over ED, even under high pressure conditions. Eq. (4) will reduce to the following :

$$\frac{n_-}{n_e} = b\,\frac{v^+}{v^*}\, \frac{<\sigma(EV)\, v_p> <\sigma v\,(DA)>}{<\sigma_i\, v_p> <\sigma v\,(MN)>} \qquad (12)$$

In this case the primary electron energy degradation will affect n_-/n_e through the change in the ratio $<\sigma(EV)\, v_p>/<\sigma_i\, v_p>$.

COMPARISON OF THE THEORETICAL MODEL WITH THE EXPERIMENTAL DATA

The analysis of the various possible H$^-$ loss mechanisms indicates the existence of two different regimes : at low n_p, n_- is expected to increase in proportion to n_p^2, while $n_-/n_e \alpha\, n_p$ (Eqs. 7 and 8). At high n_p, n_- is expected to increase only linearly with n_p, while n_-/n_e = const. (Eqs. 10 and 11). The transition between these two different functional forms of the n_- dependence upon n_p will occur when :

$$(\tau_-')^{-1} = <\sigma v\,(MN)>\, n_+^{cr} \qquad (13)$$

These predictions are confirmed by the experiments with the hybrid multicusp configuration B : in the low density regime, $n_- \alpha\, I_d^2$ at all the pressures studied, except the lowest, for which this variation is somewhat slower (Figure 1a). Note that I_d is proportional to n_p. The transition from the low density regime to the high density regime occurs for a plasma density of approximately 10^{10} cm^{-3}, although it appears that both the neutral gas density and the plasma configuration affect n_e^{cr}. With $n_e^{cr} \approx 10^{10}$ cm^{-3} and assuming $<\sigma v \ (MN)> = 2 \times 10^{-7}$ cm^3/s, we find from Eq. (13) that the collision frequency for H$^-$ loss in the low density regime, $1/\tau_-'$, is approximately 2×10^3 s^{-1}.

In the high density regime n_- increases linearly with I_d, and n_-/n_e is independent of I_d, at least in a wide range of I_d values, in agreement with the theoretical analysis. Note the unpredicted linear rise of n_-/n_e with I_d observed for $I_d > 5$ A at the highest pressures (Figure 1b). This rise may be related to a new mechanism leading to vibrational excitation of hydrogen molecules, which involves two charged particles (e.g. dissociative recombination of H$_3^+$ ions) and becomes important at $n_e \geq 10^{11}$ cm^{-3}.

However, it appears from the experiments that the "high density" value of n_-/n_e is affected by n_o. In the hybrid multicusp configuration B (Fig. 1b) the high density value for n_-/n_e goes first slightly up with gas pressure, but decreases abruptly when the gas pressure exceeds 3.3 mTorr. The reduction of the 'high density' value of n_-/n_e with increasing pressure cannot be explained by changes of the rate of DA and v^+ (Eqs. 11 and 12), due to changes with pressure of the electron temperature, only. We believe that this effect is also due to an increase of the rate of MN, occurring as a consequence of the drop in the positive ion temperature when the pressure is increased[9].

Cross sections are available for MN of H$^-$, H$^+$ and for D$^-$, H$_2^+$, but we are not aware of MN data for H$^-$, H$_3^+$ collisions[17]. The MN cross section depends strongly on the positive and negative ion energies : σ (MN) goes up when the ion relative velocity goes down. In order to calculate the change in σ (MN) related to the increase in pressure, we have to find how the neutral gas density affects the positive and negative ion energies. Three factors have to be considered in discussing the variation of the ion energy :
1. The ions are moving in the electric field of the Bohm pre-sheath. The measurements (Fig. 4) indicate that this field is ~ 0.1 V/cm and extends on a significant part of the plasma.
2. The mean free path for the elastic ion-neutral scattering is an important characteristic : for a given electric field the average ion energy goes down when the mean free path is reduced as a result of an increase in neutral gas density.
3. The mean free path for MN is another important characteristic : the increase in n_- and n_+ leads to a reduction of the respective values of this mean free path for positive and negative ions, which, finally, can become shorter than the mean free path for ion-neutral collisions.

The quantitative characterization of these processes suffers due to the lack of cross section data. The cross section for the elastic scattering of H$^-$ on H$_2$ was obtained by extrapolating to lower energy (< 1 eV) the cross section for the production of slow ions, measured by Huq et al[15] : at energies below 1 eV, the major part of this cross section is due to the elastic scattering. For a barycentric energy of 0.4 eV, we obtained $\sigma(D^- - D_2) = 4.5 \times 10^{-16}$ cm^2. The cross section for the elastic scattering of H_3^+ on H_2 was obtained from experimental mobility data by Simons et al[18] and Miller et al[19]. For a barycentric energy of 0.4 eV, we found $\sigma(H_3^+ - H_2) = 3.4 \times 10^{-15}$ cm^2. The reaction rate for the MN of $H^- - H_3^+$ was calculated[9] using the scaling formula proposed by Hickman.

The positive ions are accelerated towards the wall and acquire in the electric field some energy which is in part lost in collisions. Therefore the positive ion temperature is significantly larger than the gas temperature at low pressure, but approaches the latter at higher pressure.

Recently Wadehra[20] calculated the average energy carried by the H^- ions at their formation in plasma by DA, and showed that this energy was dependent on the electron temperature and the internal energy of the molecule. Since the H^- ions are formed by DA essentially to molecules with v > 5, the average energy carried by the H^- ions at their formation is in the range 0.2 - 0.5 eV.

Since the plasma represents a potential well for the negative ions, those formed out of the plasma center will be accelerated to this center and continue to oscillate until they are destroyed. In the low density regime, this oscillation will last the time $\tau' \simeq$ 0.5 ms, while in the high density regime this time is $[n_+ \langle \sigma v(MN) \rangle]^{-1}$. As long as the collision frequency for H^-, H_2 elastic scattering is larger than the collision frequency for MN, the negative ions will loose their energy in collisions with the neutrals and will be thermalized at the gas temperature. When these collision frequencies become comparable, the negative ions will have an average energy larger, or at least equal to their average energy at formation. This condition has been calculated for the case when the average energies of H_3^+, H^- and H_2 are respectively 0.6, 0.3 and 0.067 eV :

$$\frac{n_o}{n_+} = \frac{\langle \sigma v (MN) \rangle}{\langle \sigma v (H^-, H_2) \rangle} = \frac{1.7 \times 10^{-7}}{2.7 \times 10^{-10}} = 630 \quad (14)$$

Thus the H^- temperature is equal to the gas temperature (in this case, 500 K) when $n_+ < 1.7 \times 10^{11}$ cm^{-3} and $n_o = 10^{14}$ cm^{-3}.

The change in positive ion energy in the pressure range 3.3 - 13.5 mTorr leads to an increase of $\langle \sigma v(MN) \rangle$ by a factor 2.4 and to a reduction in n_-/n_e by a factor of five.

CONCLUSIONS

1. It has been shown that certain hybrid magnetic multipole configurations, characterized by a relatively low lifetime for wall losses of primary electrons ($\sim 10^{-7}$ s) contain high densities and high proportions (10 to 50 %) of hydrogen negative ions.
2. The negative-ion density has been shown to depend strongly on the density of primary electrons and the existence of two different H^- loss regimes has been demonstrated. The transition bet-

ween these two regimes has been observed for a plasma density of approximately 10^{10} cm^{-3}.
3. It is shown that the existence of plasma potential gradients is important in defining the ion temperatures : the positive ion temperature is larger than the gas temperature at lowest pressures in the studied range, but approaches the gas temperature at the highest pressures. The negative ions are thermalized at the gas temperature, when the plasma density is lower than $\sim 10^{11}$ cm^{-3}.
4. The reduction of kT_+ when the gas pressure is increased leads to an enhancement in H$^-$ destruction by mutual neutralization, which explains in part the reduction of the negative ion relative density when the pressure increases in the range 3.3 - 13.5 mTorr.

ACKNOWLEDGMENTS

The authors are grateful to J.R. Hiskes, M. Capitelli, A. Garscadden, J.M. Wadehra and J. Taillet for many enlightening discussions. This work was supported in part by Ecole Polytechnique and NATO Research Grant N° 060.81.

REFERENCES

1. M. Bacal, A.M. Bruneteau, H.J. Doucet, W.G. Graham and G.W. Hamilton, Proc. Second Int. Symp. on the Production and Neutralization of Negative Hydrogen Ions and Beams, October 6-10, 1980, B.N.L., Report B.N.L. 51304, p. 95 (1981).
2. M. Bacal, A.M. Bruneteau and M. Nachman, J. Appl. Phys., 54 (1983) to be published.
3. M. Bacal, F. Hillion, M. Nachman and W. Steckelmacher, Proceedings of this Symposium.
4. M. Bacal and K.N. Leung, Proceedings of this Symposium.
5. G.W. Hamilton, M. Bacal, A.M. Bruneteau, H.J. Doucet and M. Nachman, in Ref. 1, pg. 90.
6. M. Bacal, Physica Scripta, T2/2, 467 (1982).
7. M. Péalat, J.P.E. Taran, J. Taillet, M. Bacal and A.M. Bruneteau, J. Appl. Phys., 52, 2687 (1981).
8. M. Péalat, J.P.E. Taran and M. Bacal, 16th Int. Conf. on Phenomena in Ionized Gases, Dusseldorf, August 29-Sept. 4, 1983, Invited Papers (to be published).
9. A.M. Bruneteau, 'Etude sur l'ion négatif d'hydrogène dans des décharges à basse pression' ; thèse de doctorat d'Etat ès-sciences, soutenue le 5 juillet 1983 à l'Université de Paris-Sud, PMI Rep. 1310 (1983).
10. M. Bacal and G.W. Hamilton, Phys. Rev. Lett., 42, 1538 (1979).
11. M. Bacal, A.M. Bruneteau, W.G. Graham, G.W. Hamilton and M. Nachman, J. Appl. Phys., 52, 1247 (1981).
12. J.N. Bardsley and J.M. Wadehra, Phys. Rev. A20, 1398 (1979).
13. J.R. Hiskes, A.M. Karo, M. Bacal, A.M. Bruneteau and W.G. Graham J. Appl. Phys., 53, 3469 (1982).
14. D.L. Albritton, Atomic Data and Nuclear Tables, 22, 1 (1978).
15. M.S. Huq, L.D. Doverspike and R.L. Champion, Phys. Rev. A27, 2831 (1983).

16. R.J. Bieniek and A. Dalgarno, Astrophys. J., 228, 635 (1979).
17. J.T. Moseley, R.E. Olson and J.R. Peterson, Case Stud. At. Phys. 5, 1 (1975).
18. J.H. Simons, C.M. Fontana, E.E. Muschlitz and S.R. Jackson, J. Chem. Phys., 11, 307 (1943).
19. T.M. Miller, J.T. Moseley, D.W. Martin and E.W. McDaniel, Phys. Rev., 173, 115 (1968).
20. J.M. Wadehra, Proceedings of this Symposium.

DISCUSSION

COOPER: Another heating mechanism you might consider are ion-ion collisions which can be very important at low ion energies. In that case the ions are thermalized and their temperature is related to the electron temperature in the plasma. Here I mean ion-ion collisions, negative ion-positive ion collisions.

BACAL: You mean collisions not leading to mutual neutralization, ion-ion scattering? Well, here is another thing we have to take into account. Looking at the motion of negative and positive ions in the potential trap for negative ions indicates that negative ions cool down in mutual collisions and positive ions are becoming warmer. The positive ions acquire energy in the field and they leave so their average energy in the plasma is much higher than the one of the negative ions. If negative ion-positive ion scattering has a reasonable mean free path compared to mutual neutralization, which I do not know, perhaps negative ions could be heated by positive ions because positive ions are hot. I was astonished somehow by the figure quoted this morning for the temperature of negative ions to be about 0.4 eV which indicates that we are in the regime where mutual neutralization dominates the mean free path. However, we do not know the plasma density but it's probably very high. But in the examples I will show in the next paper we are not in that situation and nevertheless we observe some very high fluxes of negative ions which should be explained.

WADEHRA: In the viewgraphs that you showed about mutual neutralization, one was H^- with H^+ and the other was D^- with H_2^+. Is there a scaling with the mass of the isotope, H^- to D^-, and secondly what would be the dependence on the internal rotation-vibration energy of H_2^+?

BACAL: Well, about the scaling, indeed I had a transparency which was taken from the work of Hickman from SRI who in 1979 published a paper on the subject of a scaling formula for mutual neutralization. He didn't treat H but he included many other ions and he gave the scaling formula which indicated how to calculate the rate if the reduced mass, the electron affinity and the temperature are known. Well, it was not very clear which temperature should be used, which temperature of the ions in their interactions. We used this Hickman formula to obtain the reaction rate for $H^--H_3^+$ and you will see in the text that we used this formula to get the reaction rates for suitable relative energies of ions.

MICHELS: The ion-molecule process, H^- on H_2, that you discussed as a destruction process, is very interesting because the H_3^- surface lies entirely below the H_3 surface so there's no direct mechanism for that process although there are lots of indirect mechanisms for H^- on H_2. It follows that the process which leads to electrons would have to be an indirect mechanism. It is a very interesting problem.

BACAL: Well, I did not discuss this problem. I just raised it because it seems a very important question to be answered. When you excite molecules, and you think you do a good job because you produce negative ions, the question is whether these excited molecules don't increase the destruction of the negative ions you already have. I don't know the answer, but we have distinguished atomic physicists here to study it in the future because, I think, it hasn't been studied. It is possible one could give an answer from some general considerations.

GRANNEMAN: You mentioned the formation of negative impurity ions, like OH^-. If I remember well, the electron affinity of OH^- is something like 1.8 eV. Shouldn't it be possible to measure that quite easily with a laser?

BACAL: Yes, it is possible to measure, but we developed our diagnostics for measuring H^- and we chose a laser for that purpose. We know how to measure OH^-, we just take a mass spectrometer. We have seen some impurities of OH^- but at that time we had a more complicated problem to solve and we didn't pay much attention to the small OH^- impurity. However, it's always there and when you look at the current, it is a small current but when you take into account the masses, you come to nonnegligible densities. To reduce this impurity one should take away the water vapor. Our photodetachment method is not suitable to measure the density of OH^-, because the Nd YAG laser I am using has 1.2 eV photons. You could measure it with other lasers but you may measure other things too. One of the devices which probably measures the density of OH^- is the electrostatic probe which we used in 1977. It measured all the negative ions and my thought is that one of the inconsistencies we had in comparing the electrostatic probe results with the laser photodetachment may be solved by taking into account these facts. Indeed, we do have a certain proportion of negative ions which are detected by the probe but are not detected by the photodetachment. It was our observation in 1980 that in the low density regime the response of these two diagnostics was different. In the high density regime the agreement looked OK. Anyway, we have to treat this problem by all the possible means, although it is not an important problem from the point of view of processes because it appears in the low density regime. Still, it's good to know that there is no diffusion loss because negative ions are cold and because the potential barrier is so high--the plasma potential is almost never lower than 2 volts and for a population of 0.05 eV ions to escape through a 2 volt barrier a very long time is required. It's a very small part which can escape so, in fact, what I want to say is that you can write it on the paper but the real loss is not there.

HISKES: Did I understand you to say that in the multipole discharge there are regions where negative ions can preferentially concentrate?

BACAL: Yes. Well, that's a general statement and it has been known since the work of Mr. Taillet ten years ago in oxygen. Negative ions will concentrate in the regions with the most positive potential, similar to the water in a cup that will go to the bottom of the cup. Negative ions will stay in the center of the plasma if this is the most positive point, but if they find another more positive point, they will go there. They will go to the front of the extraction slit because when the extraction voltage is applied the potential may change locally. You can get a point which is more positive than the plasma center and negative ions from all the sides can now accumulate there. What I will try to show in the other paper we have, is that you can't explain the currents we extract with a negative ion density in the center. We extract much larger currents. My personal opinion is that we have to assume that we build up a higher density in a different point, for example, in front of the slit.

NEGATIVE ION PRODUCTION VIA DISSOCIATIVE ATTACHMENT TO H_2

J. M. Wadehra
Department of Physics and Astronomy, Wayne State University,
Detroit, Michigan 48202

ABSTRACT

In this paper, the enhancement of cross sections for dissociative electron attachment to H_2 resulting from the rotational and/or vibrational excitation of the ground electronic state of the neutral molecule is reviewed. A simple physical picture explains this enhancement. General expressions for approximately calculating the cross sections and rates for attachment to various (v,J) levels of H_2 are provided.

INTRODUCTION

High energy neutral beams are currently being used for the heating of magnetically confined plasmas[1]. Such beams are formed by accelerating ion beams (positive and negative) to high energies and then neutralizing them by either charge exchange or by photodetachment. Negative ion beams are usually favored over positive ion beams because at high energies the negative ion beams can be neutralized more efficiently.

One of the major sources for the generation of hydrogen negative ions is the process of dissociative electron attachment to molecular hydrogen in its ground electronic state that has been rotationally and/or vibrationally excited[2]. The cross sections for dissociative attachment are strongly dependent upon the internal rovibrational energy of the molecule and are dramatically increased if the molecule is rovibrationally excited. In this paper, we will review this enhancement for attachment cross sections to H_2 and will investigate the effect of such enhancement on the rates of negative ion production via the dissociative electron attachment process.

METHOD OF CALCULATIONS

In our calculations the dissociative attachment of electrons to molecular H_2 occurs by formation of an intermediate resonant state of H_2^-, namely, the $X\ ^2\Sigma_u^+\ (1\sigma_g)^2(1\sigma_u)$ state:

$$e + H_2(X\ ^1\Sigma_g^+;\ v,J) \rightarrow H_2^-(^2\Sigma_u^+) \rightarrow H(1s) + H^-(1s^2).$$

It is also possible to form the $B\ ^2\Sigma_g^+\ (1\sigma_g)(1\sigma_u)^2$ electronic configuration of the resonant state to obtain H^-:

$$e + H_2(v,J) \rightarrow H_2^-(^2\Sigma_g^+) \rightarrow H + H^-.$$

Our previous calculations[3-5] have demonstrated that at low electron energies ($\lesssim 5$ eV) the contribution of the $^2\Sigma_g^+$ resonance to the dissociative attachment cross sections and attachment rates is much smaller than the contribution of the $^2\Sigma_u^+$ resonance. Consequently, in the present paper, we will confine our discussion to the negative ion production via the formation of only the $^2\Sigma_u^+$ resonant state of H_2^-.

The potential energy and the width of the resonant state as a function of internuclear separation R were obtained in a semi-empirical fashion.[3] For example, some of the typical factors used in obtaining the potential curve $V^-(R)$ and the width $\Gamma(R)$ of the resonant state were (i) the electron affinity of H, (ii) the energy separation between the potential curves of H_2 and H_2^- at the equilibrium separation of H_2 (note that this energy value effectively determines the incident electron energy at which the contribution of the resonant state to the attachment process is maximum), (iii) the internuclear separation R_s (called stabilization radius) at which the potential curves of H_2 and H_2^- cross, (iv) the change in angular momentum of the electron captured in the resonant state, and (v) the isotope effect; that is, the change in attachment cross sections on replacing the nuclei by their isotopes. More recent calculations for the potential curve and the width of the $^2\Sigma_u^+$ state of H_2^- using the self consistent field method[6] agree fairly well with our model results for these quantities and therefore provide further confidence in our cross section values. Another calculation[7] at one internuclear separation using the complex scaling method has given a far smaller width.

RESULTS AND DISCUSSION

The cross sections for dissociative electron attachment to the ground electronic state of H_2 depend strongly on the internal rovibrational energy of the neutral molecule. The cross sections for attachment to various (v,J) rovibrational levels of H_2 are shown in figure 1. The excited levels shown in the figure have roughly the same internal rovibrational energy.

From this figure, the following trends are obvious: (a) The dissociative electron attachment cross sections exhibit a peak at the threshold electron energy and decrease monotonically as the electron energy is increased. The rate of attachment at low electron temperatures is determined primarily by this peak cross section σ_{peak}; (b) As seen on the semilogarithmic plot of figure 1 the rate at which the cross sections decrease on increasing the incident electron energy does not significantly depend on the initial (v, J) level of the molecule. It thus might be possible to obtain a single universal form for analytic expressions for cross sections for attachment to all (v,J) levels of H_2; (c) For a given incident electron energy, the attachment cross section is increased if the molecule has internal energy in the form of rotational or vibrational

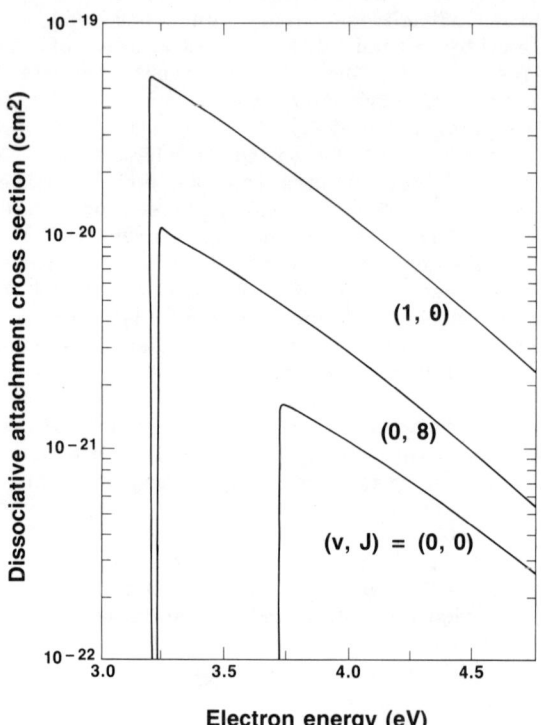

Figure 1 Cross sections for dissociative electron attachment to various rovibrational levels of H_2.

excitation. A further increment in cross sections results because of the lowering of the threshold energy of electrons that can dissociatively attach themselves to the internally excited molecule. The net result is that an enhancement of cross sections by several orders of magnitude can occur if the molecule is initially rovibrationally excited; (d) For a given internal energy the vibrational excitation of the molecule results in a higher peak attachment cross section than the rotational excitation. This feature is particularly important for low (v,J) excitations. For high (v,J) excitations, the total enhancement in cross sections is determined almost by the total internal energy and not by the exact partitioning of the internal energy in the rotational or vibrational modes.

The enhancement of cross section for dissociative electron attachment to the molecule that has been internally excited can be understood by a rather simplified physical picture involving the formation of an intermediate resonant state of H_2^-. The range of internuclear separations R over which the electron capture for resonance formation occurs is increased when the molecule is internally rovibrationally excited. The most probable internuclear separation at which an incident electron is captured is the one at which the energy difference between the potential curves of H_2 and H_2^- is equal to the incident electron energy. Low energy electrons are then captured at internuclear separations close to the crossing point R_s of the potential curves of H_2 and H_2^- while the higher energy electrons are dominantly captured at smaller values of R. Immediately after its formation, the resonant H_2^- state begins to dissociate due to the repulsive nature of its potential curve. As long as the internuclear separation is less than R_s, the possibility of H_2^- undergoing autodetachment and leaving behind a vibrationally excited H_2 exists. On the other hand, if R exceeds R_s, the resonant state of H_2^- turns into a true bound state, autodetachment becomes impossible and the dissociative attachment becomes inevitable. If the molecule is rovibrationally excited, the range of applicable internuclear separations R is increased due to an increased amplitude of vibration for vibrationally excited molecules and due to the centrifugal stretching for rotationally excited molecules. If the range of applicable R includes the crossing point R_s, then even the very low energy electrons can be captured close to R_s. For such captures, the probability of autodetachment is quite small since very little further stretching is necessary to exceed R_s to assure dissociative attachment. For higher energy electrons, since the capture occurs at smaller values of internuclear separation R, the competition from autodetachment is relatively significant and the cross section for dissociative attachment is reduced. This simple physical picture explains both the rapid decrease of the cross section as the electron energy is increased for attachment to any particular (v, J) level of the neutral molecule and the enhancement of attachment cross sections if the molecule is internally excited.

This strong enhancement of attachment cross sections has been verified experimentally[8] - the observation of a four orders of magnitude increase in peak cross sections from v = 0 to v = 4 and a fivefold increase from J = 0 to J = 7 is consistent with our calculations. The experiment has been confined to the electron energy range of 1-5 eV and only the low rovibrational levels (v = 0 - 5, J = 0 - 7) have been studied.

The fact that the rate at which the cross sections for dissociative electron attachment decrease with increasing electron energy is almost the same for any (v,J) levels of H_2 suggests that all cross sections, σ_{DA}, can be fitted to a simple analytic expression of the form

$$\sigma_{DA}(E) = \sigma_{peak} \exp[-(E - E_{th})/E_0], \quad E_0 = 0.42 \text{ eV}.$$

Here E_{th} is the threshold or minimum energy of the electrons that can be attached to the molecule in a given (v,J) level.

Assuming that the low rovibrational energy spectrum of H_2 can be fairly well represented by that of a Morse potential curve, E_{th} can be obtained from

$$E_{th} = 3.994 \text{ eV} - E_{vJ}, \text{ where}$$

$$E_{vJ} = (7.542 \times 10^{-3} \text{ eV}) \left[72.26 \left(v + \frac{1}{2}\right) - 2.074 \left(v + \frac{1}{2}\right)^2 \right.$$

$$+ J(J + 1)$$

$$\left. -3.654 \times 10^{-2} \left(v + \frac{1}{2}\right) J(J+1) - 1.609 \times 10^{-4} J^2(J+1)^2 \right].$$

As long as the internal energy E_{vJ} is less than 3.994 eV (that is $E_{th} > 0$), the attachment process is endoergic.

The attachment rate is obtained by folding the attachment cross section with the incident electron energy distribution.

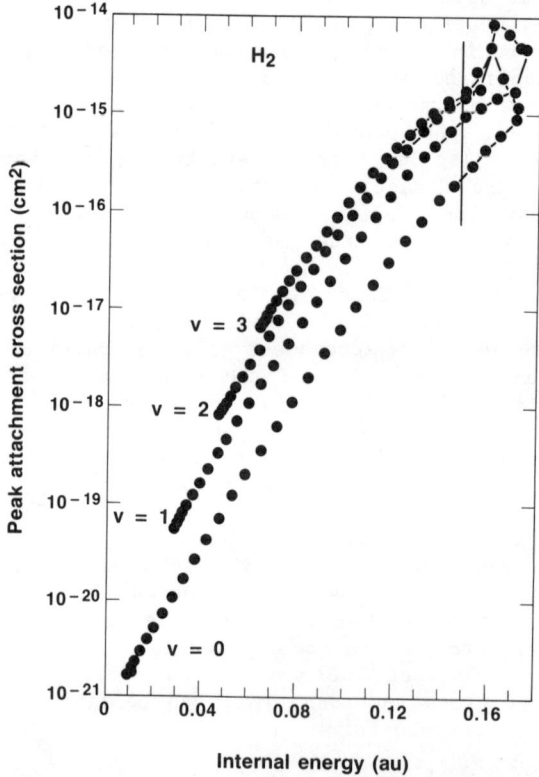

Figure 2 The peak cross sections for electron attachment to all possible rotational levels of H_2 for v = 0, 1, 2 and 3.

Assuming a Maxwellian distribution for electron energies, the rate of electron attachment to any (v, J) level is obtained from

$$k(\bar{E}) = (\frac{27}{\pi})^{1/2} \sigma_{peak} (\frac{\bar{E}}{m})^{1/2} \exp(-1.5\, E_{th}/\bar{E})$$
$$\cdot \left[(\frac{3}{2} + \frac{\bar{E}}{E_0})^{-2} + \frac{E_{th}}{\bar{E}} (\frac{3}{2} + \frac{\bar{E}}{E_0})^{-1} \right].$$

The average electron energy $\bar{E}$ is related to the electron temperature T by $\bar{E} = 3\,kT/2$. Two quantities needed for attachment rates are the peak attachment cross section and the internal energy E_{vJ}. Figures 2 and 3 show these peak cross sections for attachment to all possible J levels of several vibrational levels as a function of the internal energy of the molecule.

The velocity and energy of the negative ion formed depend, by conservation of energy, on the energy of the incident electron and on the internal energy of the molecule. For a given distribution of electron energies (for example,

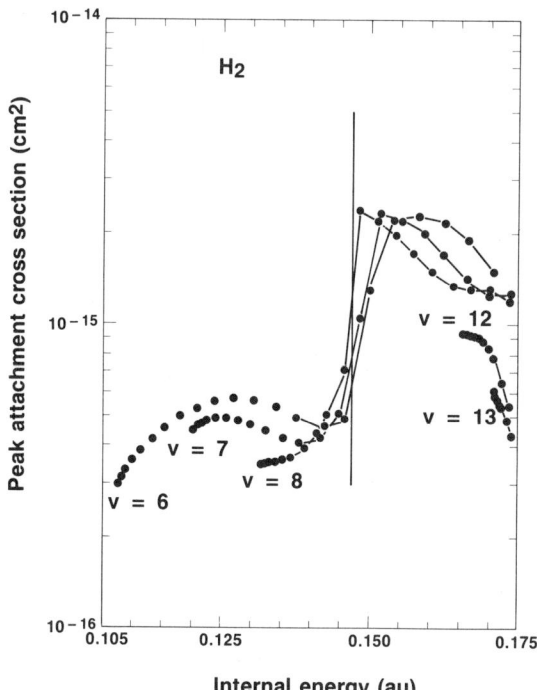

Figure 3 The peak cross sections for electron attachment to all possible rotational levels of H_2 for v = 6, 7, 8, 12 and 13.

Maxwellian), the average ion velocity depends upon the internal energy E_{vJ} of H_2. In figure 4 are shown the average velocities of H^- ions formed by attachment to various rotationally excited levels of H_2 in its lowest vibrational level. As long as the attachment process is endoergic, the average ion velocity remains close to 5.2 km/sec. This corresponds to the average ion energy of 0.14 eV. High velocity ions are formed by attachment to higher (v,J) levels of H_2 when the process becomes exoergic. In figures 2, 3 and 4, the vertical line at 3.994 eV corresponds to the internal energy above which the attachment process becomes exoergic.

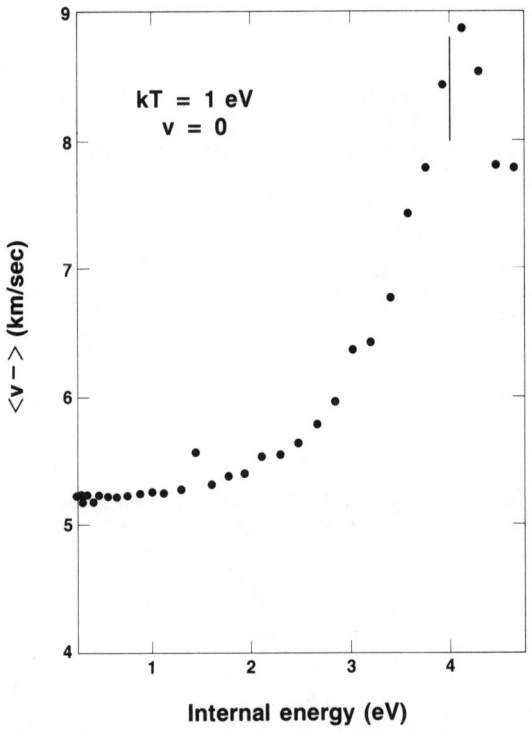

Figure 4 The average velocity of H^- ions formed by electron attachment to all possible rotational levels of H_2 for v = 0 at electron temperature of 1 eV.

FUTURE POSSIBILITIES

The cross sections and rates of dissociative electron attachment to H_2 depend strongly on the internal rovibrational energy possessed by the molecule. This obviously raises the question[9] whether the process of dissociation of H_2 by electron impact

$$e + H_2 (v,J) \rightarrow H + H + e$$

is also strongly affected by the internal rovibrational energy of H_2. Since this process can be viewed as "dissociative attachment to the continuum", a sensitive dependence of cross sections for dissociation by electron impact on the internal energy of the molecule might exist, at least for the resonant contributions. However, a detailed calculation is necessary before any final conclusion can be reached.

At low electron temperatures ($E_{e^-} \lesssim 1$ eV), one needs to consider the formation of only the $^2\Sigma_u$ resonant state of H_2^-. However, at higher electron energies, the contributions of higher resonances will become important. Large peaks and a strong isotope effect in dissociative attachment cross sections are observed[10] around 10.5 eV and 14 eV. However, no information about the rovibrational dependence of cross sections in this energy range is available.

Finally, the negative ions can also be formed by polar dissociation of H_2

$$e + H_2(v,J) \rightarrow H^+ + H^- + e.$$

In order to obtain the resonant contributions to this process, one will have to model the resonant states of H_2^- that dissociate into the ion-pair. Some preliminary information about such states of H_2^- has recently become available.[11,12]

This research has been supported by WSU Research Award, AFWAL and LBL.

REFERENCES

1. For recent reviews see, M. Bacal, Physica Scripta T2/2, 467 (1982); K.W. Ehlers. J. Vac. Sci. Technol. A1, 974 (1983); J.R. Hiskes, J. Phys. (Paris) 40, C7-179 (1979).
2. M. Bacal, A.M. Bruneteau, W.G. Graham, G.W. Hamilton and M. Nachman, J. Appl. Phys. 52, 1247 (1981).
3. J.M. Wadehra and J.N. Bardsley, Phys. Rev. Lett. 41, 1795 (1978); J.N. Bardsley and J.M. Wadehra, Phys. Rev. A20, 1398 (1979).
4. J.M. Wadehra, Appl. Phys. Lett. 35, 917 (1979).
5. J.M. Wadehra, Phys. Rev. A29, XXX (1984).
6. C.W. McCurdy and R.C. Mowrey, Phys. Rev. A25, 2529 (1982).
7. N. Moiseyev and C.T. Corcoran, Phys. Rev. A20, 814 (1979).
8. M. Allan and S.F. Wong, Phys. Rev. Lett. 41, 1791 (1978).
9. A. Garscadden, private communication.
10. D. Rapp, T.E. Sharp and D.D. Briglia, Phys. Rev. Lett. 14, 533 (1965).
11. J.N. Bardsley and J.S. Cohen, J. Phys. B 11, 3645 (1978).
12. A. Huetz and J. Mazeau, J. Phys. B 16, 2577 (1983).

DISCUSSION

BARNETT: You mentioned the work of Rapp several years ago. As I recall, that was done to explain the vibrational excitation of H_2. How does that compare with the dissociative attachment cross section?

WADEHRA: They also did experiments on dissociative attachment to H_2 around 14-15 eV but there is no theoretical information available on dissociative attachment in that energy range because in this energy range higher resonances of H_2^- can form and there is no information available on those. Recently there was an experiment done in France on dissociative attachment of H_2 by Huetz and Mazeau and the paper was published just this summer in J. Phys. B in which they looked at some of these high resonances and I think perhaps that paper might be the starting point. I'm not aware of any rovibrational state dependence of cross sections for dissociative attachment in that energy range.

PETERSON: I have a question about the details of some of the physics. When you have a higher rotational state for a vibrational level you stretch the bond distance around it which decreases the coupling or decreases the probability of the attachment. But it also at the same time increases the lifetime of this association and these two things are competing against each other. One tends to increase the probability of dissociative attachment and the other decreases it.

WADEHRA: Overall our calculations indicate that if you take into account these factors, the rotational excitation of the molecule leads to an enhancement of cross section and we attribute that enhancement to the stretching of the molecule and therefore the range of internuclear separation over which the electron could be picked up by the neutral molecule. It is the enhancement that we get in cross sections that leads to enhancement of attachment rates.

BARNETT: Do you have any idea about what the induced dipole moment of HD might do to these calculations?

WADEHRA: No, I don't. We have not considered explicitly the effect of the dipole moments.

H⁻ AND ELECTRON PRODUCTION IN A MAGNETIC MULTIPOLE SOURCE*

A.J.T. Holmes and T.S. Green
Euratom/UKAEA Fusion Association, Culham Laboratory
Abingdon, Oxon, OX14 3DB, England

G. Dammertz
KFK, Karlsruhe, PB 3640, D-7500 Karlsruhe 1, West Germany

ABSTRACT

This paper describes an investigation of the yield of H⁻ ions and of electrons from a modified magnetic multipole source. The data show a dependence of H⁻ yield on magnetic field structure in the source as well as on normal parameters (gas pressure, arc voltage, etc.), and also relationship to the positive ion density. Under optimum conditions the H⁻ current density can attain a value equal to one-half of the positive ion current density measured at the extraction plasma.

* Only abstract submitted.

POLAR DISSOCIATION AS A SOURCE OF NEGATIVE IONS

S. K. Srivastava and O. J. Orient[*]
Jet Propulsion Laboratory, California Institute of Technology
Pasadena, CA 91109

ABSTRACT

This paper discusses the process of polar dissociation of molecules. It is suggested that it can be utilized in the fabrication of a source of negative ions. Laboratory data on the electron impact induced polar dissociation for several molecules are presented.

INTRODUCTION

When an electron collides with a molecule it can produce a negative ion either by the process of dissociative attachment or by polar dissociation. The former can be represented by the following relation:

$$XY + e^-(E_o) \rightarrow XY^- \rightarrow X + Y^- \qquad (1)$$

where XY is a molecule consisting of two components X and Y, $e^-(E_o)$ is an electron with kinetic energy E_o, XY^- is a short lived molecular negative ion which subsequently dissociates into a neutral X and a negative ion Y^- components. The dissociative attachment (Eq. (1) takes place only for certain fixed energies E_o of the colliding electron and therefore it is a resonance process. Most molecules are characterised by such energies. A detailed discussion on this subject can be found in a book by Massey.[1] Cross sections for the dissociative attachment are usually in the range of 10^{-20} to 10^{-18} cm^2. However, it has been shown in the past that these cross sections increase by many orders of magnitude if the attachment takes place with vibrationally excited states of O_2,[2] H_2,[3] D_2,[3] CO_2,[4,5] $HC\ell$,[6] HF,[6] and N_2O.[7]

In general, the attachment energies, E_o, are less than 15 eV. Figure 1 shows a typical case where the production of O^- by dissociative electron attachment to O_2 has been studied. Experimental details on the measurements can be found in Ref. 8. As is clear from this figure, O^- ions are produced when the electron beam energy E_o is within a narrow range around 6.2 eV. However, as the electron beam energy is increased one more onset is found at which O^- from O_2 begins to appear. The rate of O^- production increases as the electron beam energy increases, reaches to a maximum and then decreases (in other molecules it may remain constant) slowly with increasing energy of the electron beam. This appearance of negative ions at higher energies has been

---

[*]NRC-NASA Senior Research Associate.

interpretted to be due to the process of polar dissociation. It can be represented by the following relation:

$$XY + e^- \rightarrow X^+ + Y^- + e^- ,\qquad (2)$$

$$XY + h\nu \rightarrow X^+ + Y^- ,\qquad (3)$$

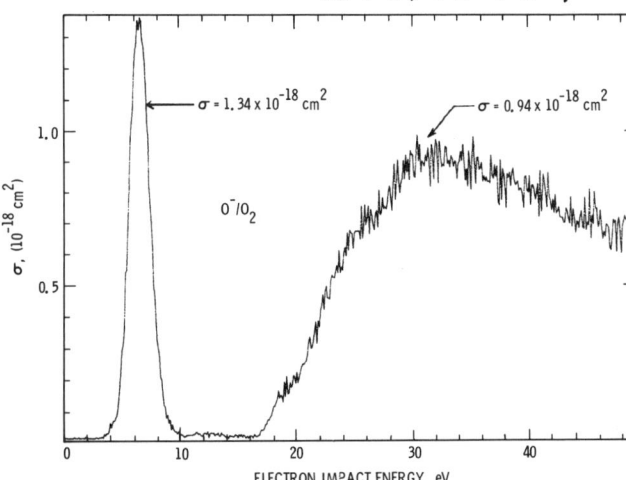

Fig. 1. O$^-$ production from O$_2$ by electron impact.

where $h\nu$ is the energy of the photon and other symbols have been defined in Eq. (1). Polar dissociation takes place through an ionic state of the molecule which is formed by a positive and a negative ion. The potential energy curve for such a state in the case of a diatomic molecule can be constructed by the following relation:[9,10]

$$V = -k_e \cdot \frac{e^2}{r} + Be^{-r/\rho} ,\qquad (4)$$

where $k_e = 8.988 \times 10^9$ (N.m^2/C^2), e is the electronic charge, B and ρ are constants. ρ is a measure of the sum of the radii of the ions under consideration. The first term of Eq. (4) represents the Coulomb attraction between two point like charges and the second term takes into account the repulsion between them due to the electron clouds at small internuclear distances. The energy of the asymptotic limit E_a of the ionic state can be obtained from the following:

$$E_a = D + I(X) - E(Y) ,\qquad (5)$$

where D is its dissociation energy at which neutral X and Y appear, I(x) is the ionization energy of the component X, and E(y) is the electron affinity of the component Y.

Ionic states of molecules are formed by the Coulomb potential (Eq. (4)) between the ion pair. They are generally crossed by (in the zero order approximation) a number of their covalent excited states. These covalent states are essentially flat at large internuclear distances. Figure 2 shows, qualitatively, the forms of the various potential energy curves for a "typical" molecule.

The ionic ($X^+ + Y^-$) and neutral ($X + Y$) states are represented by broken lines. As the two states approach each other an interaction between them takes place and the crossing is avoided. This results in a set of potential energy curves shown by solid lines. When such a molecule is excited from its ground state to any of the higher states, above their dissociation limits, then as the internuclear separation increases the crossing point (such as A in Fig. 2) between the ionic and neutral states is reached. At this point the molecule has some probability, p, to make a transition from the neutral to the ionic state and dissociate into an ion pair, X^+ and Y^-. The value of the probability p is higher if the crossing point lies at a large internuclear separation.

One can use Landau-Zener[11-13] theory to calculate the value of p. However, although these calculations for the reverse process (i.e. $X^+ + Y^- \rightarrow XY + \Delta E$) have been performed for several molecules at large crossing points A (Fig. 2) they are not readily available for the process of polar dissociation. Experimental work has been, in most cases, limited to the measurement of the threshold energy for the ion pair formation. Table I summarizes all the data that are available to date for a negative atomic ion formation. As is clear from this table, the experimental values of the threshold energies are in good agreement with the calculated values obtained by using Eq. (5). In most cases,

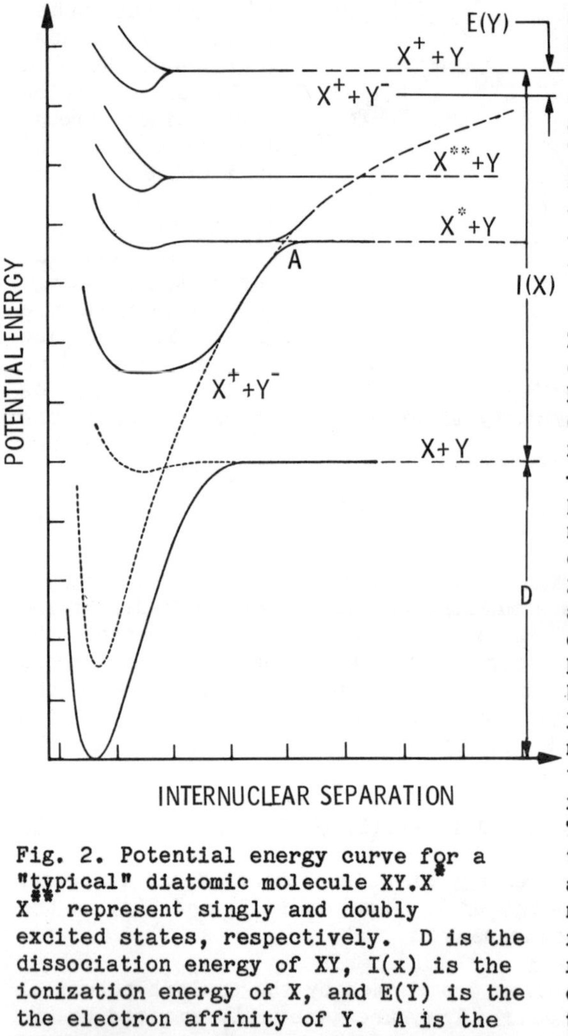

Fig. 2. Potential energy curve for a "typical" diatomic molecule XY. X^* X^{**} represent singly and doubly excited states, respectively. D is the dissociation energy of XY, I(x) is the ionization energy of X, and E(Y) is the the electron affinity of Y. A is the crossing point between an ionic and an atomic state. Broken lines represent unperturbed states and solid lines give the final potential energy curves.

the experimentally determined threshold is not very well defined. This can be readily seen in Fig. 1. It is due to the fact that the source that causes dissociation (a photon beam or an electron beam) has a certain energy spread. In order to obtain fairly accurate values of threshold energies from the experimental data one can use the various deconvolution techniques such as described by Macneil and Thynne.[14]

The molecular states which dissociate into an ion pair can be excited by both a photon and an electron beam. In the case of e-beam excitation the variation of cross section for the polar dissociation as a function of incident electron energy is similar to the excitation of any other electronic level of the molecule; i.e. the cross section rises from the threshold to a maximum and then either remains constant or slowly decreases to a lower value. Although the thresholds for polar dissociation have been measured in the past for several molecules (see Table I) the absolute values of cross sections for most of them have not been determined. Values of cross sections for polar dissociation by electron impact are available only for H_2, CO, O_2, CO_2, and N_2O. From Fig. 1 it is clear that cross sections for this process are comparable with the cross sections for dissociative attachment. In some cases former cross sections may even be larger than the latter. This is at least true in the case of CH_4. The cross sections for dissociative attachment as well as polar dissociation for this molecule are shown in Fig. 3.

At the present time it is not possible to predict, on any empirical basis, which molecule may posses a large cross section for polar dissociation. We have started an experimental program in this direction and have studied polar dissociation from H_2, O_2, CO, HCl, CO_2 H_2O, CH_4, CH_3Cl and N_2H_4. The data on O_2 and CH_4 is presented in Figs. 1 and 3. Data on other molecules are shown in Figs. 4 through 8. In general, the peak cross sections for polar

Table I
A survey of previously available data on the polar dissociation where a negative atomic ion is formed.

Ion	Source	Other Products	Method	Threshold (eV) Exp.	Threshold (eV) Calc.	Reference
H^-	H_2	H^+	RPD,EI	17.30	17.27	16
			EI	17.0±0.5		Present
	CH_4	CH_3^+	PI	13.50	13.52	17
			EI	17.0±0.5		Present
	C_3H_8	$C_3H_7^+$	PI	11.0	10.78	18
	$n-C_4H_{10}$	$C_4H_9^+$	PI	10.9	10.85	18
	C_2H_6	$C_2H_5^+$	PI	12.0	11.85	18
C^-	CO	O^+	RPD,EI	23.41	23.52	19,20
			EI	23.2±0.5		Present
O^-	O_2	O^+	PI,EI	17.25	17.33	21,22
			EI	17.0±0.1		Present
	CO	C^+	RPD,EI	20.89	20.98	19,20
			EI	20.2±0.5		Present
	NO	N^+	RPD	19.94	19.86	23
	CO_2	CO^+	EI	20.0±0.5	18.00	Present
	H_2O	$H+H^+$	EI	23.0±0.5	22.5	Present
F^-	HF	H^+	PI	16.06	15.99	24
	DF	D^+	PI	16.13	16.06	24
	F_2	F^+	PI	15.48	15.56	24
	CH_3F	CH_3^+	PI	12.56	12.80	25
	NF_2	NF^+	EI	11.80	11.55	26
	N_2F_4	$N_2F_3^+$	EI	12.00	12.15	26
	ClF	Cl^+	PI	12.04	12.05	27
	TlF	Tl	PI	7.59	7.72	28
Cl^-	Cl_2	Cl^+	RPD	11.86	11.83	29
	HCl	H^+	RPD	14.50	14.36	30
	CH_3Cl	CH_3^+	PI	10.07	10.257	25
			EI	10.1±0.5		Present
	TlCl	Tl^+	PI	7.01	6.31	28
	$PbCl_2$	$PbCl^+$	EI	7.1	8.08	31
Br^-	Br_2	Br^+	PI	10.31	10.42	32
	BrCN	CN^+	EI	14.60	14.94	33
	TlBr	Tl^+	PI	6.30	6.17	28
	CH_3Br	CH_3^+	PI	9.60	9.41	25
I^-	TlI	Tl^+	PI	5.80	5.81	34

dissociation from these molecules are found to be of the order of 10^{-19} cm^2. HCl and N_2H_4 did not give rise to negative ions through polar dissociation.

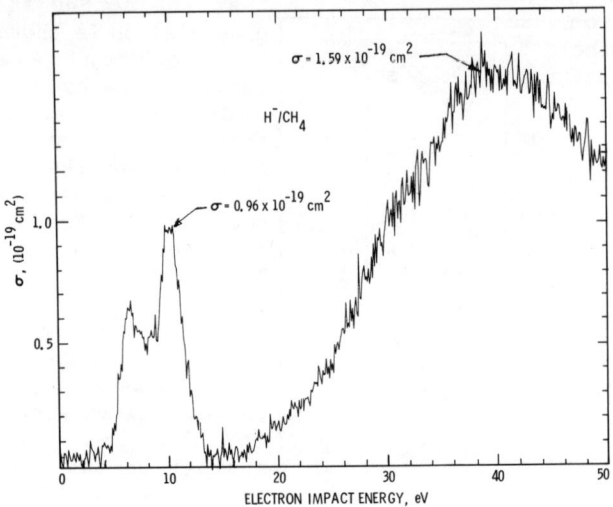

Figure 3. H$^-$ production from CH$_4$ by electron impact.

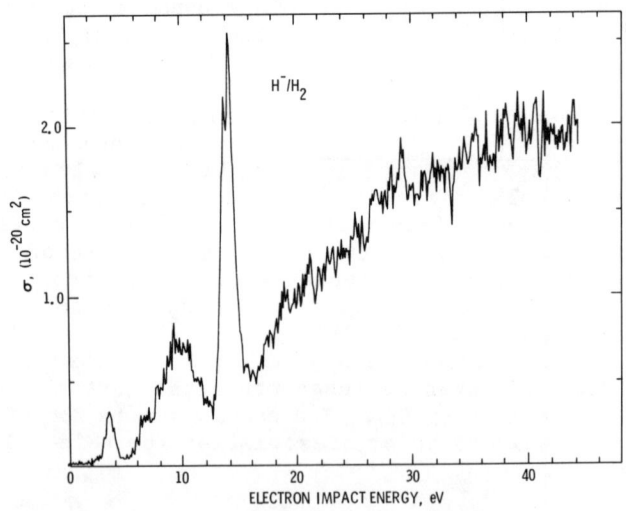

Fig. 4. H$^-$ production from H$_2$ by electron impact.

The process of dissociative attachment has been considered in the past for the purpose of generating negative ion beams. Much research has been done in this direction. We would like to suggest that the production of negative ion beams by utilizing the property of polar dissociartion of molecules should also be considered. This is due to two reasons. First, there may be molecules such as CH_4 (Fig. 3) which have large cross sections for polar dissociation. Second, integrated cross section for this process (from threshold to higher electron impact energies) is much larger than the integrated cross section for dissociative attachment which

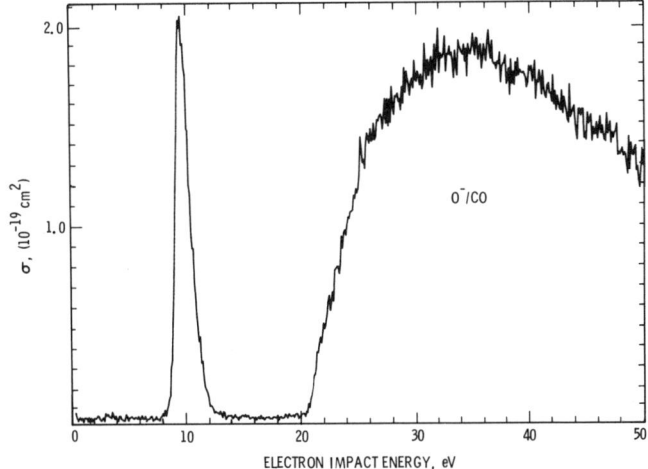

Fig. 5. O^- production from CO by electron impact.

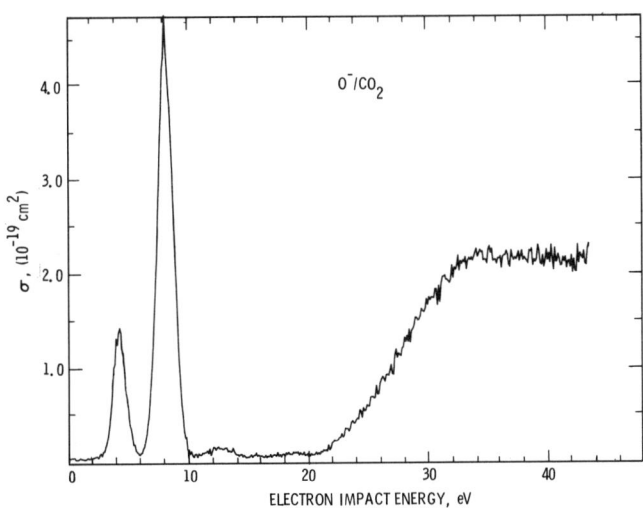

Fig. 6. O^- production from CO_2 by electron impact.

takes place over a very narrow range of electron impact energies. One additional advantage of using polar dissociation as a source of negative ions is that it can be produced by photodissociation. Thus, intense laser beams can be used for generating negative ions. However, most molecules reported here have high thresholds for polar

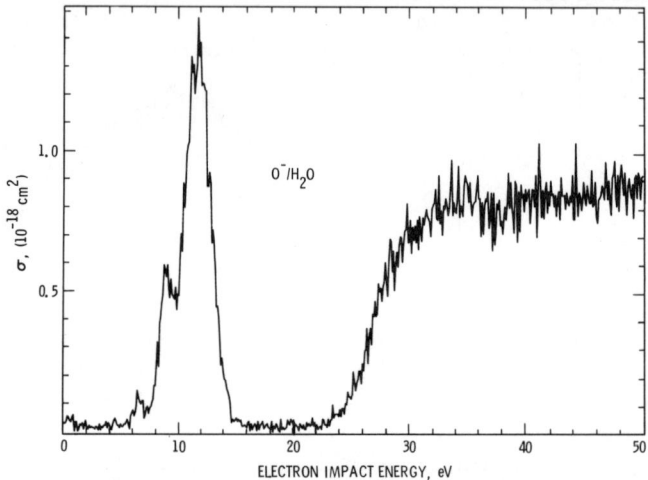

Fig. 7. O^- production from H_2O by electron impact.

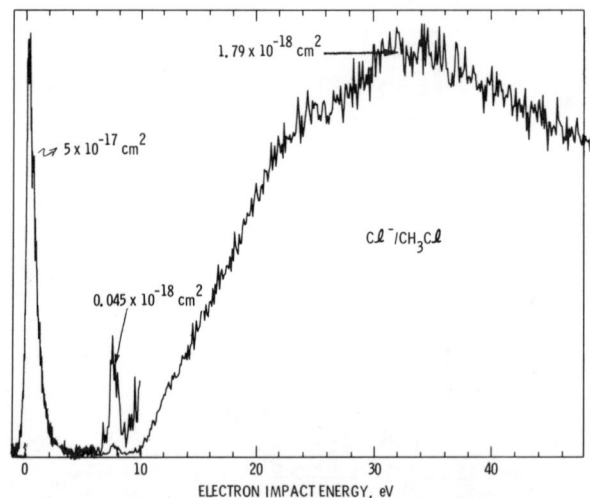

Fig. 8. Cl^- production from CH_3Cl by electron impact.

dissociation. Thus, they are beyond the reach of present day laser wavelengths. This problem can be solved by finding molecules such as alkali hydrides or halides. For example, the threshold for H^- production from LiH is 7.04 eV. This threshold will be much lower for lithium halides.

In the past, Sluyters and Prelec[15] have reported a "Duoplasmatron negative ion source" in which they suggested that the source of H^- was polar dissociation i.e. $e + H_2 \rightarrow H^+ + H^- + e$. It seems that the process of polar dissociation is a good candidate for generating negative ion beams and further research should be carried out to realize its full potential.

Acknowledgement:

The research described in this paper was carried out by the Jet Propulsion Laboratory, California Institute of Technology, and was sponsored by AFOSR.

References

1. H. Massey, "Negative Ions", Cambridge University Press, Cambridge 1976.
2. W. R. Henderson, W. L. Fite, and R. T. Brackmann, Phys. Rev. 183, 157 (1969).
3. M. Allan and S. F. Wong, Phys. Rev. Lett. 41, 1791 (1978).
4. D. Spence and G. J. Schulz, Phys. Rev. 188, 280 (1969).
5. S. K. Srivastava and O. J. Orient, Phys. Rev. A27, 1209 (1983).
6. M. Allan and S. F. Wong, J. Chem. Phys. 74, 1687 (1981).
7. P. J. Chantry, J. Chem. Phys. 57, 3180 (1972).
8. O. J. Orient and S. K. Srivastava, J. Chem. Phys. 78, 2949 (1983).
9. M. Born and J. E. Mayer, Z. Physik 75, 1 (1932).
10. G. Herzberg, "Molecular Spectra and Molecular Structure: Spectra of diatomic molecules", D. Van Nostrand Company, Inc., New York.
11. L. Landau, J. Phys. (USSR) 2, 46 (1932).
12. C. Zener, Proc. Roy. Soc. (London) A137, 696 (1933).
13. E. C. G. Stuckelberg, Helv. Phys. Acta 5, 369 (1932).
14. K. A. C. Macneil and J. C. J. Thynne, Int. J. Mass Spectrom. Ion Phys. 3, 35 (1969).
15. Th. Sluyters and K. Prelec, Nucl. Instr. Meth. 113, 299 (1973).
16. R. K. Curran, Scientific Paper 62-908-113-P7, Westinghouse Research Laboratories, Pittsburgh (1962).
17. W. C. Chupka, J. Chem. Phys. 48, 2337 (1968).
18. W. A. Chupka and J. Berkowitz, J. Chem. Phys. 47, 2921 (1967).
19. M. A. Fineman and A. W. Petrocelli, J. Chem. Phys. 36, 25 (1962).

20. M. A. Fineman and A. W. Petrocelli, Planetary Spoace Sci. 3, 187 (1961).
21. V. H. Dibeler and J. A. Walker, Advan. Mass. Spectrom. 4, 767 (1967).
22. V. H. Dibeler and J. A. Walker, J. Opt. Soc. Am. 57, 1007 (1967).
23. P. M. Hierl and J. A. Franklin, J. Chem. Phys. 47, 3154 (1967).
24. J. Berkowitz, W. A. Chupka, P. M. Guyon, J. H. Holloway, and R. Spohr, J. Chem. Phys. 54, 5165 (1971).
25. M. Krauss, J. A. Walker, and V. H. Dibeler, J. Res. NBS 72A, 281 (1968).
26. J. T. Herron and V. H. Dibeler, J. Res. NBS 65A, 405 (1961).
27. V. H. Dibeler, J. A. Walker, and K. E. McCulloh, J. Chem. Phys. 53, 4414 (1970).
28. J. Berkowitz and T. A. Walker, J. Chem. Phys. 49, 1184 (1968).
29. D. C. Frost and C. A. McDowell, Can. J. Chem. 38, 407 (1960).
30. R. E. Fox, J. Chem. Phys. 26, 1281 (1957).
31. J. E. Hastie, H. Bloom, and J. D. Morrison, J. Chem. Phys. 47, 1580 (1967).
32. K. Watanabe, J. Chem. Phys. 26, 542 (1957).
33. J. T. Herron and V. H. Dibeler, J. Am. Chem. Soc. 82, 1555 (1960).
34. J. Berkowitz and W. A. Chupka, J. Chem. Phys. 45, 1287 (1966).

DISCUSSION

PETERSON: Are you certain that in each case when you have the polar dissociation peak, you actually are creating simultaneously negative and positive ions from one electron and that there are no effects due to secondary electrons from just pure ionization? All of these crossings are arising but at the same time you exceeded the ionization threshold which means that for every electron that goes in you may be producing slow electrons with a cross section about two orders of magnitude larger than these cross sections, electrons with an energy that could cause a resonance.

SRIVASTAVA: We are now making an arrangement where we will study the positive and negative ions in coincidence. They are produced simultaneously and if we get the signal we will know that this is polar dissociation.

PETERSON: This effect might explain why you don't see negative ions formed by photoexcitation whereas you do from the electron impact.

SRIVASTAVA: No, in the works by Chupka and others, it was found that a large number of negative ions were produced by photoexcitation.

PETERSON: I'm not saying that all molecules, but some.

SRIVASTAVA: There are quite a few photon impact processes where they have found polar dissociation in molecules. We are aware of the fact that we are not producing negative ions because of secondary processes. Let's discuss this point. First of all, what we did as an initial step is to draw the excitation function for O^-. Since CO is falling apart into O^- due to the polar dissociation, there should be C^+ simultanously produced here. We have studied the excitation function of C^+ and the onset for C^+ was exactly at the same place where it was for O^-. That gave us the confidence. Moreover, suppose we have about 10 eV of secondary electron energy to form the dissociative attachment. Then it should be around here on the shoulder of the excitation function curve for the polar dissociation where there should be some increase in the negative ion intensity. Thus, there should actually be a distortion in the shape of this excitation function for the polar dissociation. But there is no real indication of this distortion in the present excitation function curve.

PETERSON: I think the answer to this question is very simple. If the relative size is really independent of pressure then it probably is a first order process which means that the aspect of secondary electrons is not important. Have you looked at that?

SRIVASTAVA: In our experiments the pressures are generally 10^{-4} Torr or less. The mean free path of secondary electrons is much larger. It was a cross beam experiment and not a static gas collision geometry. We did not find any pressure effect.

EHLERS: I've been a strong believer in polar dissociation production of H^- ions for years. By the same token the cross sections are so ridiculously low it's hard to incorporate it. However, all these cross sections were taken with ground state gas, I would assume. Do you have any guess as to what happens to this cross section in terms of its exact value if you had a few states of vibrational excitation?

SRIVASTAVA: Well that's a very good question as a matter of fact, and as J. Wadehra pointed out this morning that he would be very much interested in looking at what happens to vibrationally excited H_2. It will be quite interesting to see how the cross section for polar dissociation compares with the cross section for dissociative attachment for the vibrationally excited H_2. I think the behavior of the two should be the same. Actually it is the increase in the Frank-Condon overlap between the two electronic states which accounts for the increase in the cross section. So you should have the same effect. One thing I forgot to mention that the generation of negative ions through polar dissociation will be enhanced in a plasma because very low energy electrons in the plasma will generate vibrationally excited states of the molecules. Vibrational excitation is higher at low electron impact energies. So you could produce vibrationally excited molecules in a discharge and use a laser to pump it up to cause the polar dissociation. You can have a two step process, therefore you can enhance the polar dissociation and it would be a much cleaner process for producing negative ions. It is only a suggestion.

<u>JACQUOT</u>: I have a comment about the energy of the electrons. In an ECR source we have high energy electrons and probably the high ratio of protons in the ECR source, that we measure, is due to vibrational states of molecules. I don't observe H^- production in this source. That means we can produce vibrational states of molecules due to high energy tail of electrons but we don't observe H^- because the electron has a high temperature,

<u>EHLERS</u>: But your V_p is so high the H^- couldn't get out even if you had 100% of them. That doesn't say it's not the answer.

VOLUME H⁻ ION PRODUCTION EXPERIMENTS AT LBL*

K. N. Leung and K. W. Ehlers
Lawrence Berkeley Laboratory, Berkeley, Ca. 94720

ABSTRACT

H⁻ ions formed by volume processes have been extracted from a multicusp ion source. It is shown that a permanent magnet filter together with a small positive bias voltage on the plasma grid can produce a very significant reduction in electron drain as well as a sizable increase in H⁻ ions available for extraction. A further reduction in electron current is achieved by installing a pair of small magnets at the extraction aperture. An H⁻ ion current density of 38 mA/cm^2 was obtained with a discharge current of approximately 350 A. Different techniques to increase the H⁻ ion yield have also been investigated.

INTRODUCTION

In recent years, neutral beam injection has proven to be an effective way to heat plasmas in fusion devices to thermonuclear temperatures.[1,2] Higher energy neutral beams will be required for plasma heating and for current drive in some future fusion reactors.[3] The high neutralization efficiency of H⁻ or D⁻ ions makes these negative ions the favorite to form atomic beams with energies in excess of 150 keV.[4] There are several approaches for producing H⁻ or D⁻ ions. A self-extraction negative ion source based on surface conversion of positive ions has already been operated successfully to generate a steady-state H⁻ ion beam current greater than 1 A.[8] Several experiments have been conducted to extract H⁻ ions directly from a hydrogen discharge plasma.[9-11] This technique of generating H⁻ ions has the advantage over the surface and double charge exchange processes in that it requires no cesium and it uses presently developed large area positive ion source geometries which have demonstrated good ion optics for large positive ion current. However, one must find methods to handle the electron problem in order to make the negative ion source practical.

This paper presents a novel method of extracting volume produced H⁻ ions directly from a multicusp ion source. This type of plasma generator has demonstrated its ability to produce large volumes of uniform and quiescent plasmas with a high H⁺ or D⁺ ion fraction and with good gas and electrical efficiency.[12,13] We show that the addition of a magnetic filter[14] to the source will not only enhance the H⁻ ion yield but sizably reduce the extracted electron component. It is also shown that the electron current can be further reduced substantially by installing a pair of thin permanent magnets at the extraction aperture. The H⁻ ion beam current was found to increase almost linearly with the discharge current. An H⁻ current density of 38 mA/cm^2 was obtained with a discharge current of approximately 350 A. Attempts have also been

0094-243X/84/1110067-15 $3.00 Copyright 1984 American Institute of Physics

made to increase the H⁻ ion yield by optimizing the discharge voltage; by injecting very low energy primary electrons into different chambers; by increasing the electrostatic containment potential; and by using some special source geometries.

EXPERIMENTAL SETUP

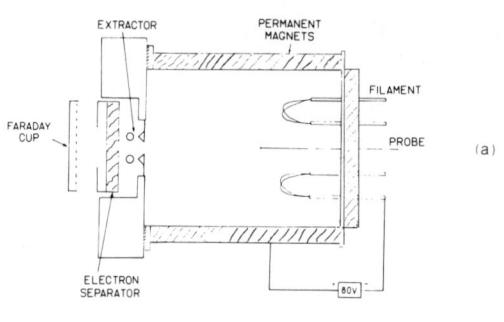
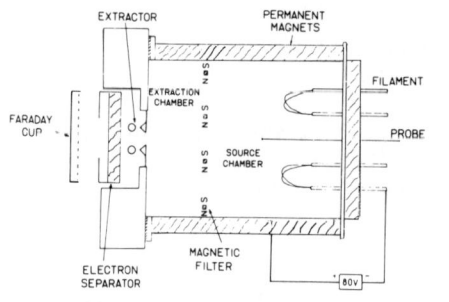
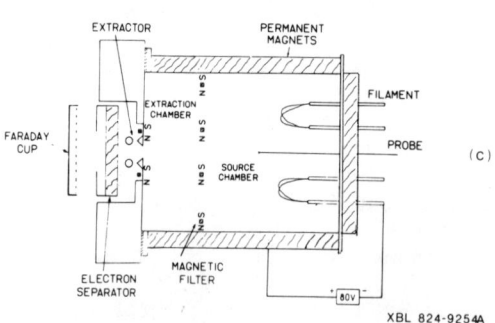

Fig. 1 Schematic diagram of (a) the multicusp ion source, (b) the multicusp ion source with a filter, and (c) a magnetically filtered multicusp source equipped with an E X B electron suppressor.

Figure 1(a) is a schematic diagram of the test multicusp ion source geometry. The device is a cylindrical stainless-steel chamber (20-cm diameter by 24-cm long) surrounded externally by ten columns of samarium-cobalt magnets ($B_{max} \approx 3.6$ kG). These magnets form continuous line-cusps parallel to the source axis and they are connected at the end plate by four extra rows of magnets. The open end of the chamber is enclosed by a three-grid extraction system. When operated as a H⁻ ion source, the second and third grids were connected to ground. The first or plasma grid was masked down to a small (0.15 x 1.3 cm²) extraction slot and was biased at a potential equal to or more positive than the anode. A steady-state hydrogen plasma was produced by primary electrons emitted from four 0.05-cm diam. tungsten filaments and the entire chamber wall served as the anode for the discharge. The data presented in the next section were obtained with a modest discharge current of 1 A and a discharge voltage of 80 V.

Figure 1(b) shows a

schematic of the source when a permanent magnet filter is included. This filter provides a limited region of transverse magnetic field which is made strong enough to prevent all energetic primary electrons in the source chamber from crossing into the extraction chamber. Ions, because of their much larger mass, can penetrate the filter, and the interesting feature of the magnetic filter is that the electrons which accompany these ions to form the plasma in the extraction region are very cold ($T_e \approx 0.4$ eV).

Recent experiments have shown that the addition of a magnetic filter to a multicusp ion source can improve the H⁻ fraction, the source operability, and the plasma density profile at the extraction plane.[15,16] In order to increase the geometric transparency and to provide adequate cooling, the filter was constructed by inserting square permanent magnet rods into copper tubes through which a square broach had been passed. Since the magnets rested on the four broached grooves, their orientation was fixed and water was then run through the remaining space to provide cooling.

During normal operation, the pressure outside the source was maintained at 1×10^{-4} Torr as measured by an ionization guage. The actual pressure inside the source chamber was approximately 1.5×10^{-3} Torr. Plasma parameters were obtained by Langmuir probes located at the center of the source chamber and in front of the plasma grid in the extraction chamber.

To extract H⁻ ions, a negative potential (~1 kV) was applied to the source chamber with respect to ground. The small extracted beam was then analyzed by two diagnostic techniques. (1) A compact magnetic deflecton mass spectrometer,[17] located just outside the extractor was used for relative measurement of the extracted H⁻ ions and for the analysis of the extracted ion species. However, this measurement could not determine the total current of H⁻ ions or electrons in the extracted beam. (2) A permanent magnet mass separator, located just behind the last grid of the extractor was used with a Faraday cup to measure the extracted H⁻ ion and electron currents. After passing through the slit, the electrons were deflected onto a grooved graphite collector by the weak magnetic field produced by a pair of thin ceramic magnets. The H⁻ ions which were only slightly affected by the magnetic field of the electron separator proceeded into the Faraday cup. A small positive bias potential on the cup was used to suppress secondary electrons. With this arrangement, it was possible to measure the ratio of extracted H⁻ ion current to electron current as well as to determine the extracted H⁻ ion current density for various operational conditions.

EXPERIMENTAL RESULTS

(a) <u>Source without filter</u>

The source was first operated with a hydrogen plasma in the geometry shown in Fig. 1(a). It was found that the extractable H⁻ ion current was extremely small when the plasma grid was left floating electrically. The Langmuir probe characteristics in Fig. 2 illustrate that the potential V_p at the source center is

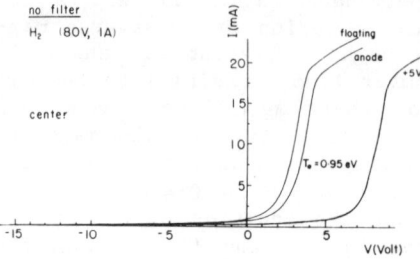

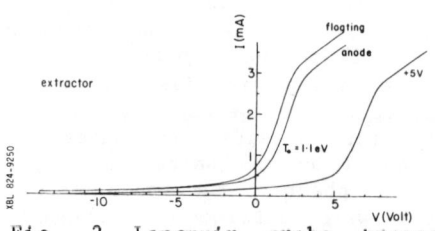

Fig. 2 Langmuir probe traces obtained at the center and near the extractor of a conventional multicusp ion source for three different plasma grid potentials.

positive (~4 V) with respect to the chamber wall or anode, and that V_p at the source center is about 1.5 V more postive than that near the extractor. This potential gradient together with the fact that the plasma grid was floating approximately -50 V relative to the anode, made it impossible for H ions formed in the center of the source to reach the extractor. These negative ions are electrostatically trapped within the source plasma by the positive potential well.

When the grid was connected to the anode, the potential of the plasma near the extractor was +3 V relative to the grid and V_p at the source center was still 1.5 V more positive than that near the extractor. As a result, only 2 μA of H$^-$ ion current was extracted from the source with a discharge current of 1 A (Fig. 3(a)). However, this negative ion current was accompanied by about 17 mA of electron current as shown in Fig. 3(b). Thus, the ratio of extracted negative ion current to electron current is 1/9000. By reversing the polarity of the extractional power supply, 245 μA of positive hydrogen ion current was extracted for the same arc conditions (Fig. 3(c)). Therefore, the ratio of I$^-$/I$^+$ was approximately 1/120 or 0.8%.

When the plasma grid was biased at a potential more positive than the anode ($V_b > 0$), the plasma potentials also increased by about the same amount as shown by the probe traces in Fig. 2. Thus, the potential difference between the plasma and the grid and the gradient of the source plasma potential remained unchanged. There was no significant change in other plasma parameters such as T_e and density at the center or near the extractor of the source chamber, or in the extracted H$^-$ ion or electron current.

(b) Source with magnetic filter

Figure 1(b) is a schematic of the source with the magnetic filter in place. Two kinds of permanent magnets have been used for the same filter geometry. Figure 4 shows the measured magnetic field across the plane of the filter between two adjacent magnet rods. It can be seen that both the magnitude and the integrated flux of the B field increase by a factor of two when the filter magnets are changed from ceramic to samarium-cobalt of the same size and shape.

When the source was operated with the "weak" ceramic magnet

71

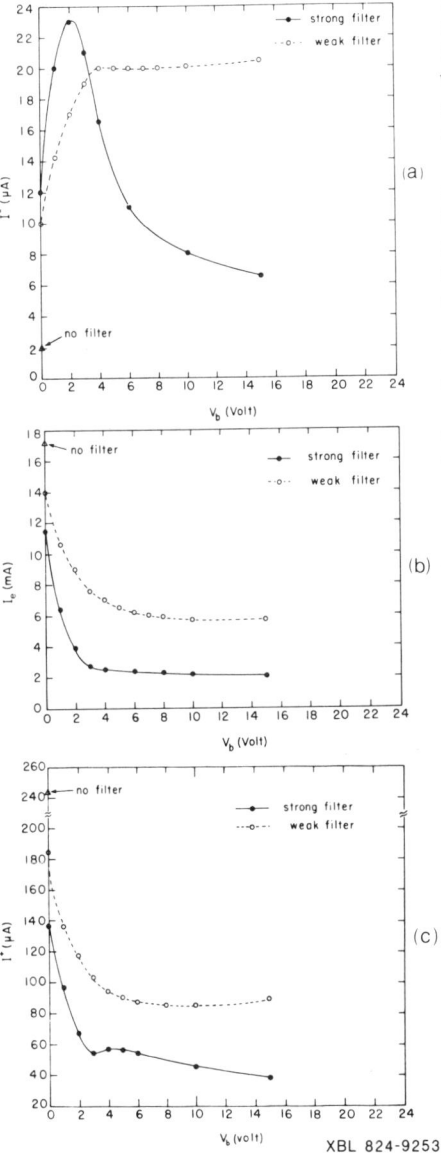

Fig. 3 (a) The extracted H⁻ ion current, (b) the extracted electron current, and (c) the extracted positive hydrogen ion current as a function of the plasma grid bias voltage for the strong and weak filters.

filter, the data in Fig. 3(a) shows that the extracted negative ion current I^- was about 10 μA when the plasma grid was biased at anode potential, but I^- increased with grid bias voltage V_b, saturating at about 20 μA at V_b = +5 V. On the other hand, the extracted electron current I_e was 14 mA with the grid at anode potential, but it decreased to 6.5 mA at V_b = +5 V (Fig. 3(b)). As a result, the ratio of I^-/I_e became 1/300 at V_b = +5 V.

When the "strong" samarium-cobalt magnet filter was used, it was found that the extracted negative ion current I^- increased from 12 to 23 μA as the grid bias V_b was changed from 0 to +2.5 V. However, unlike the "weak" filter, I^- decreased rapidly as the plasma grid was biased higher than +2.5 V as illustrated in Fig. 3(a). The extracted electron current I_e was about 11.5 mA with the grid biased at anode potential and it then decreased to 2.8 mA at V_b = +2.5 V (Fig. 3(b)). Thus,

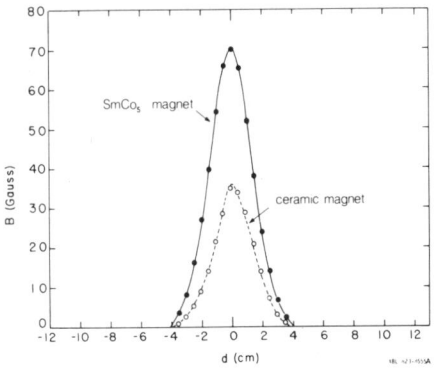

Fig. 4 A plot of the magnetic fields across the plane of the samarium-cobalt and the ceramic magnet filter in between two magnet rods.

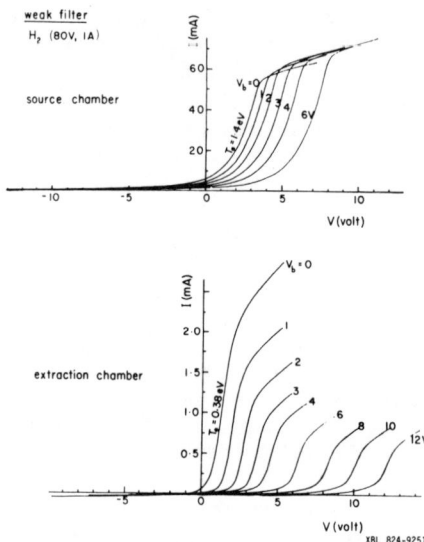

Fig. 5 Langmuir probe traces obtained in the source and extraction chambers for different plasma grid bias voltages for the case of a weak ceramic magnet filter.

by employing the strong samarium-cobalt magnet filter and by biasing the plasma grid at +2.5 V, the ratio of I^-/I_e was improved from 1/9000 to 1/200. The extracted negative ion current density J^- for discharge current of 1 A was estimated to be 0.12 mA/cm^2.

We have previously shown that the density of electrons in the extraction chamber and, therefore, the extractable electron current I_e is closely related to the number of positive ions that can penetrate the magnetic filter.[14,15] The exact process as to how these low energy electrons penetrate the filter is not yet fully understood, but we are convinced that the relative plasma potentials of the two chambers is most important. For both the weak and strong filters, the Langmuir probe measurements in Fig. 5 and 6 show that the plasma potential V_p of the source chamber is approximately 1.5 V more positive than that of the extraction chamber when $V_b = 0$. This potential gradient tends to draw positive ions into the extraction chamber and they in turn are able to bring the cold electrons with them. On the other hand, H^- ions formed in the source chamber will find it difficult to cross this potential barrier. It is also possible that some H^- ions are produced in the extraction chamber via processes such as dissociative attachment of vibrationally excited H_2 [18,19]

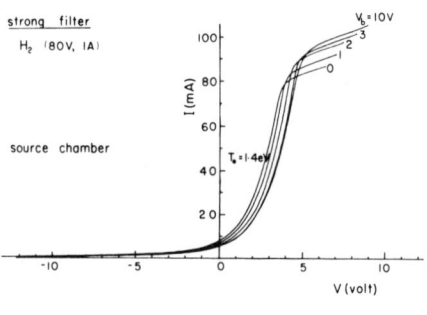

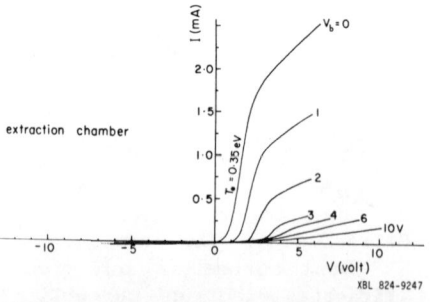

Fig. 6 Langmuir probe traces obtained in the source and extraction chambers for different plasma grid bias voltages for the case of a strong samarium-cobalt magnet filter.

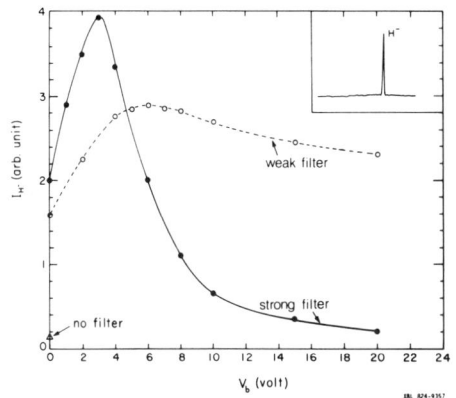

Fig. 7 The spectrometer signal for the H⁻ ion as a function of the plasma grid bias voltage for the strong and weak filters.

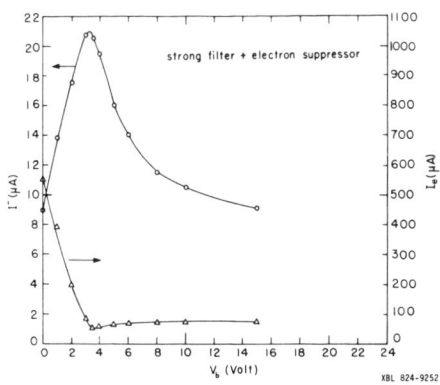

Fig. 8 The extracted H⁻ ion current as a function of the plasma grid bias voltage in the presence of a strong filter and an E X B electron suppressor.

or the dissociative recombination of H_2^+ and H_3^+.[20,21] These H⁻ ions, instead of moving towards the extractor, would be accelerated back into the source chamber by the plasma potential gradient.

As the grid bias voltage V_b increases, the difference in plasma potential between the two chambers decreases and it then becomes more difficult for the positive ions and their associated cold electrons to cross the filter into the extraction region. As a result, the extracted positive ion current I^+ (Fig. 3(c)) and the extracted electron current (Fig. 3(b)) are reduced simultaneously for both filters. No further reduction in I^+ or I_e is observed once the potentials of the two chamber plasmas become almost identical and this happens when V_b is increased to approximately +8 and +4 V for the weak and strong fliter, respectively. At this point, the H⁻ ions formed in the source chamber can cross the filter with ease while those produced in the extraction chamber are not attracted to the source chamber side. In addition, there is almost no difference in potential between the grid and the plasma. Consequently, volume-produced H⁻ ions are no longer electrostatically trapped and can now be extracted.

In the case of the strong filter, a further increase in the grid bias voltage V_b results in a drop of I^- as shown in Fig. 3(a). It is difficult to determine the plasma potential in the extraction chamber for V_b +4 V from the probe traces in Fig. 6. This potential may become positive relative to the potential of the source chamber plasma. In that case, the number of positive ions and electrons that leak through the filter is much reduced. The extracted H⁻ ion current could then decrease due to the lack of positive ions needed to satisfy the requirement for charge neutrality. It is also possible that the production of H⁻ ions in the extraction chamber, which requires the presence of thermal electrons, is reduced.

The extracted beam has also been analyzed by means of a compact magnetic-deflection mass spectrometer when the source was operated with and without a magnetic filter. A typical H⁻ ion signal is displayed on the top corner of Fig. 7. The amplitude of this sharp H⁻ ion peak is plotted as a function of the plasma grid bias voltage in Fig. 7 for the weak and strong filters and also for operation without the filter. It can be seen that the results show almost the same characteristics as the Faraday cup current presented in Fig. 3(a).

When the spectrometer was tuned for higher masses, we found that the impurities were mainly OH⁻ and O_2^- ions. The total impurity content in the extracted beam was about 1%.

(c) Strong filter with E X B electron suppression

With the use of the strong samarium-cobalt magnet filter, the electron current extracted from the multicusp source is still about 100 times the H⁻ ion currrent. Thus, an attempt was made to further reduce the electron current by means of an E X B geometry. Figure 1(c) shows two tiny ceramic magnets (0.2 cm x 0.25 cm x 3 cm) placed on each side of the plasma grid electrodes that approximate a Pierce configuration. The maximum B field produced by this pair of magnets is about 350 G with thickness of approximately 0.5 cm. The extraction electric fields cause the electrons to E x B drift vertically in a cycloidal motion, and they are thus separated from the beam. The much heavier H⁻ ions pass through the extraction gap with little effect.

Figure 8 shows the extracted H⁻ ion current and the electron current when this E X B electron suppression scheme was used with the strong filter. There is no significant change in the H⁻ ion current when the plasma grid is biased at +3 V, but the electron current is reduced by nearly a factor of 50. A further reduction in the electron current was obtained by placing a small wire (biased at +12 V relative to the plasma grid) at one end of the extraction slot to intercept the drifting electrons so that they would not leak out of the extractor. With this arrangement, the ratio of I^-/I_e has been improved to almost unity.

SCHEMES TO INCREASE THE H⁻ ION YIELD

(a) By optimizing the discharge voltage

The dependence of the extracted H⁻ ion current on the discharge voltage has been investigated for the case of the strong filter. With the plasma grid biased at anode potential, and for a constant discharge current of 5 A, Fig. 9 shows that the extracted H⁻ ion current (as indicated by the mass spectrometer signal) increases almost linearly as the discharge voltage was varied from 40 V to 100 V. Above 110 V, the H⁻ ion current levels off and

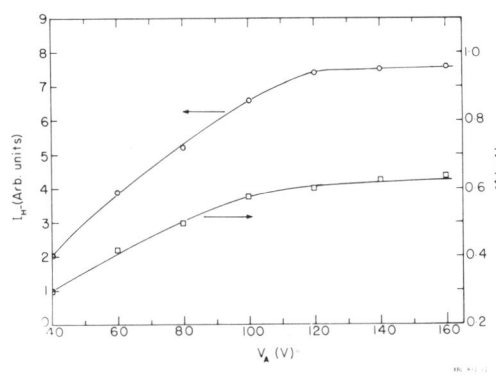

Fig. 9 H⁻ ion current and the extracted postive ion current as a function of the discharge voltage with a constant discharge current of 5 A.

remains essentially constant for discharge voltages as high as 160 V. Thus the H⁻ ion yield increases by ~40% as V_d is changed from 80 V to 120 V. When the power supply polarities were reversed for positive ion extraction, the positive hydrogen ion current also increased in the same manner as the H⁻ ion current when the discharge voltage was varied (Fig. 9). This result shows that the increase in H⁻ ion yield for high discharge voltage is mostly due to the increase in plasma production in the source chamber. Without the magnetic filter, the optimum discharge voltage for this source normally occurs at ~80 V. This change in the optimum discharge voltage in the case of the strong filter may be due to a better containment of high energy primary electrons in the source chamber.

(b) By using a magneto-electrostatic containment scheme
By including a magnetic filter, the multicusp ion source becomes essentially a "complete" multicusp generator. As a result, The plasma potential distribution in the source chamber is uniform both in the radial and axial directions.[22] The axial uniform V_p profile extends to the extraction chamber if a positive bias voltage is applied to the plasma grid. Thus, the positive ions will no longer "free-fall" down to this grid as in the case of positive ion source operation. However, there will be a substantial plasma leakage to the side walls due to a higher plasma potential. In a previous experiment, we demonstrated that the ion loss to regions between the line-cusps can be reduced by

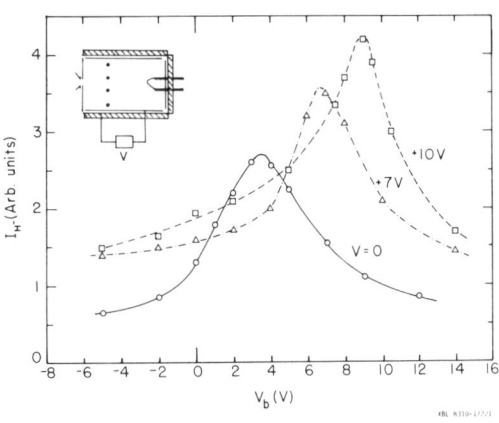

Fig. 10 H⁻ ion yield as a function of plasma grid bias for three different bias potentials on the strip electrodes.

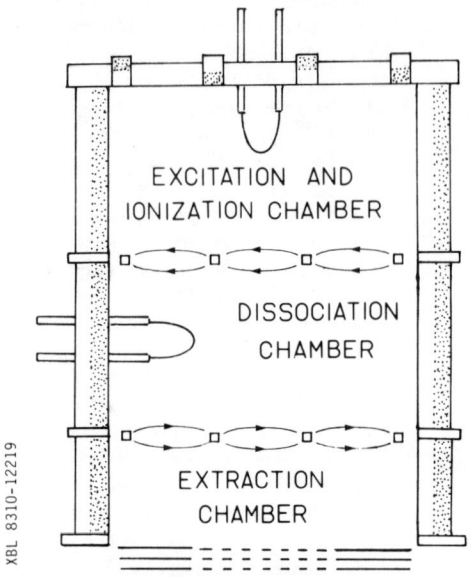

Fig. 11 A triple chamber H⁻ ion generator.

installing electrode-strips in between magnet columns.[23] When these electrodes are biased positively with respect to the anode, they reflect the positive ions back to the plasma volume, resulting in an increase of plasma density. This magneto-electrostatic plasma containment scheme was investigated in the test source operated with a strong filter. Figure 10 shows the H⁻ ion yield (as indicated by the spectrometer signal) as a function of the plasma grid bias voltage for three different strip electrode bias voltages. When the strips are biased at +10 V relative to the anode, the optimum H⁻ yield is about 53% higher than the case when they are biased at the anode potential. From the probe traces obtained in the source chamber, we find the source plasma density also increases by approximately the same amount. Thus the increase in the extracted H⁻ ion current is mainly due to better confinement of the plasma.

(C) By using a triple plasma system

Hydrogen molecules can be excited into a high vibrational state by collision with energetic electrons. It is generally believed that these vibrationally excited H_2 molecules play an important role in the formation of H⁻ ions. In a two chamber system, the plasma electron density in the extraction chamber is low when a positive bias is applied on the plasma grid. Thus the rate of H⁻ production in the extraction chamber due to the reaction $e + H_2(v^*) \rightarrow H^- + H$ is reduced. For this reason, most of the H⁻ ions extracted from the filtered source are formed in the source chamber.

A new triple chamber H⁻ ion source has been proposed both by LBL and LLNL.[24] In this geometry (Fig. 11), the neutral molecules are vibrationally excited by energetic electrons in the first (excitation and ionization) chamber. These molecules together with some plasma ions and electrons then migrate into the second (dissociation) chamber. It has been determined theoretically that cold thermal electrons with $T_e \approx 1$ eV are required to optimize the H⁻ ion production. This condition can be easily achieved by injecting a large quantity of very low energy electrons from

PLASMA PHYSICS LABORATORY

TO:

DATE

FROM:

MEMORANDUM

cathodes installed in this chamber. H⁻ ions formed are then drawn into the third (extraction) chamber by applying the proper bias voltage on the plasma grid of the extractor in the same manner as the two chamber system. The idea of this triple plasma device will be tested experimentally in the near future.

HIGH ARC POWER SOURCE OPERATION

Recently, H⁻ ions were extracted from this same magnetically-filtered cusp source but it was operated at much higher current and in a pulse mode on a high voltage test-stand at Los Alamos National Lab.[25] The accelerated beam was mass-analyzied for accurate H⁻ current and emittance measurements. A plot of the H⁻ beam current and current density as a function of the discharge current is shown in Fig. 12. The H⁻ ion beam current increases almost linearly with I_d. At an arc current of 350 A, an H⁻ current density of 38 mA/cm^2 was obtained. The source pressure was adjusted between 2.5 to 4.5 x 10^{-3} Torr. The total impurity content was found to be less than 2% and the I^-/I_e ratio varied from 1/3 at low extracted current to 1/12 at the highest current observed. The brightness of the source is 0.58 A/cm^2-mrad^2 at a current density of 38 mA/cm^2.

These results demonstrate that volume-produced H⁻ ions extracted from a magnetically-filtered, multicusp source can provide high quality H⁻ beams with sufficient current density to be useful for both neutral beam injection and accelerator applications. Work will continue to scale the source operation up to larger extraction area and perhaps with multiple beamlets. In addition, the magnet geometry at the exit aperture must be optimized to achieve electron suppression without degeneration of ion optics. Should this succeed, one could operate a large area multicusp source to provide H⁻ ions in much the same manner as is now used to provide positive hydrogen ions for neutral beam systems.

*This work is supported by U.S. DOE under contract number DE-AC03-76SF00098.

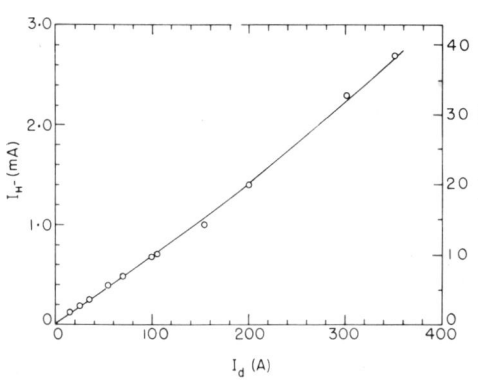

Fig. 12 The extracted H⁻ beam current and current density from a 3-mm diam. aperture as a function of discharge current.

REFERENCES

1. F.H. Coensgen, W.F. Cummins, C. Gormezano, B.G. Logan, A.W. Molvik, W.E. Nexsen, T.C. Simonen, B.W. Stallard, and W.C. Turner, Phys. Rev. Lett. $\underline{37}$, 143 (1976).
2. H.P. Eubank et al., Phys. Rev. Lett. $\underline{43}$, 270 (1979).
3. L.D. Stewart, A.H. Boozer, H.P. Eubank, R.J. Goldston, D.L. Jassby, D.R. Mikkelsen, D.E. Post. B. Prichard, J.A. Schmidt, C.E. Singer, Proc. 2nd Int. Symp. on the Production and Neutralization of Negative Hydrogen Ions and Beams, Brookhaven, p. 321 (1980).
4. K.H. Berkner, R.V. Pyle, and J.W. Stearns, Nucl. Fusion $\underline{15}$, 249 (1975).
5. K. Prelec and Th. Sluyters, Rev. Sci. Instrum. $\underline{44}$, 1451 (1973).
6. E.B. Hooper, Jr., Lawrence Livermore Laboratory Report UCID-18067 Feb. 1979.
7. J.R. Hiskes, J. Phys. (Paris) $\underline{40}$, C7-179 (1979).
8. K.N. Leung and K.W. Ehlers, Rev. Sci. Instrum. $\underline{53}$, 803 (1982).
9. M. Bacal, A.M. Brunéteau, H.J. Doucet, and A.M. Marechal, Bull. Am. Phys. Soc. $\underline{23}$, 846 (1978).
10. K.N. Leung and K.W. Ehlers, J. Appl. Phys. $\underline{52}$, 3905 (1981).
11. A.J.T. Holmes, T.S. Green, M. Inman, A.R. Walker, and N. Hampton, 3rd Varenna-Grenoble Symp. on Heating in Toroidal Plasmas, (Grenoble, 1982) p. 95.
12. M.M. Menon, Proc. IEEE $\underline{69}$, 1012 (1981).
13. K.W. Ehlers and K.N. Leung, Rev. Sci. Instrum. $\underline{50}$, 1353 (1979).
14. K.N. Leung, K.W. Ehlers and M. Bacal, Rev. Sci. Instrum. $\underline{54}$, 56 (1983).
15. K.W. Ehlers and K.N. Leung, Rev. Scie. Instrum. $\underline{52}$, 1452 (1981).
16. K.W. Ehlers and K.N. Leung, Rev. Sci. Instrum. $\underline{53}$, 1423 (1982).
17. K.W. Ehlers, K.N. Leung and M.D. Williams, Rev. Sci. Instrum. $\underline{50}$, 1031 (1979).
18. M. Bacal, A.M. Bruneteau, H.J. Doucet, W.G. Graham, and G.W. Hamilton, Proc. 2nd Int. Symp. on the Production and Neutralization of Negative Hydrogen Ions and Beams, Brookhaven, p. 95 (1980).
19. J.R. Hiskes, A.M. Karo, M. Bacal, A.M. Breneteau, and W.G. Graham, J. Appl. Phys. 53, 3469 (1982).
20. B. Peart and K.T. Dolder, J. Phys. B$\underline{8}$, 1570 (1975).
21. B. Peart, R.A. Forrest, and K. Dolder, J. Phys. B $\underline{12}$, 3441 (1979).
22. K.W. Ehlers, K.N. Leung, P.A. Pincosy, and M.C. Vella, Appl. Phys. Lett. 41, 517 (1982).
23. K.W. Ehlers and K.N. Leung, Rev. Sci. Instrum. $\underline{53}$, 1429 (1982).
24. K.N. Leung, K.W. Ehlers and J.R. Hiskes, Lawrence Livermore National Lab. Report UCID-19910 (Oct. 1983).
25. R.L. York, R.R. Stevens Jr., K.N. Leung, and K.W. Ehlers, Los Alamos National Lab. Report LA-9931 (1983).

DISCUSSION

QUESTION: What are the chances of carrying it on until you find an optimum?

LEUNG: So far we haven't seen any saturation. We're running out of filaments because we use only two, 0.06" tungsten filaments and 350 A is a lot for them. You can operate a source very easily in steady state up to 120 A but it gets awfully hot. For the emittance measurements, we operated the source in a pulsed mode and the current density was up to 40 mA/cm^2 and still going up. We haven't seen any saturation yet.

BARNETT: On your mass spectrometer scans you had a low energy component of H$^-$ which, if integrated, is rather large compared to the total H$^-$. What is that?

LEUNG: That has to do with the bias on the Faraday cup of the spectrometer because electrons produced by stripping of negative ions can easily get into the cup. If the Faraday cup inside the mass spectrometer is biased properly, you can get rid of that bump and make the base line very smooth. We also looked at the negative impurity ions at the higher arc current. The results showed about 1% impurities, mainly OH$^-$, O$_2^-$, O$^-$.

PETERSON: In the Los Alamos experiment you had a circular aperture, but the magnets were parallel. What about the emittances in the two different directions?

YORK: There was a very small variation. We used two different magnet configurations, two different strength magnets, and in the stronger magnet configuration there was a very small effect, just barely measureable. We are talking about 2% or something like that, just enough so that you could tell the difference between the vertical and horizontal plane. As you go to stronger magnets to decrease the electron loading, there is a possibility of effecting the emittance but we did not reach that point.

COOPER: What was the pulse length in these emittance measurements at high current densities?

YORK: We started at 800 μs, but towards higher arc currents we turned the pulse length down. I can't remember the actual pulse length, but I think we were down to about 1% duty factor for higher arc currents.

COOPER: That was the arc pulse length?

YORK: That's what we were pulsing. We started with very low arc currents and a pulse length of 800 μs. Above about 50 A of arc current we cut the duty factor down. I think that most of the data was taken with a pulse length of about 200 μs but we cut the repetition rate down to reduce the power load. I'd like also to comment about the last two points on the graph, at high arc currents. As K. Leung mentioned, we found a dependence of the H$^-$ yield on arc voltage at higher arc currents which was not the case at lower arc currents. When we were running at low arc currents, an increase in the arc voltage was not followed by an increase in the H$^-$ beam current, but as we got to the very high arc currents there was a significant dependence on arc voltage. As we raised the arc voltage, we got more H$^-$ current. I'd like some theorist to look into this.

COOPER: You may not be in steady state at such short pulse lengths.
YORK: We looked at the arc current pulse and it seemed to be constant over the pulse length.
HISKES: Did you get a measure of electron density in the discharge when you were near the maximum?
LEUNG: No, that is only the extracted electron current, we did not have probes to look at the density.
YORK: As we raised the arc voltage, the extracted electron current also went up.
HISKES: What is the electron density inside the discharge?
YORK: That we didn't measure, but as we raised the arc voltage the H^- current went up as well, and also the electron current went up significantly.
DAGENHART: What was the arc power efficiency and, secondly, could you explain again how many electrons and H^- you had?
LEUNG: The arc parameters are, let's say, 150 V, 350 A and that would determine the arc power. The ratio of the electron to the negative ion current varies from about 3:1 down here to about 10:1 up there, or something like that.
BACAL: Is that true with the E x B suppression?
LEUNG: Yes, this is taken with a pair of magnets at the exit but we did not cool those magnets properly. They have to be cooled because if ceramic magnets are heated they can lose the field very easily. After the experiment we discovered that one of the magnets was burned and the field dropped from 250 G to 100 G.
EHLERS: I want you to realize that the main purpose in doing this experiment was that we wanted to know how much H^- current density might be available. We weren't really interested in how many electrons we had to put up with. We were going to live with that just to get the number. Now I'm amazed that it turned out that it was as low as it was but you must remember we've done absolutely nothing in this program really to cut their numbers down. That's a program of its own. The problem I'm sure is solvable. It's just one we have to put some time and effort in. That will be one of the problems we have to solve in this game. I'm sure that number can be brought down to considerably less than 1:1 in the long run.
ALESSI: Did you look at the emittance at different arc currents?
LEUNG: Yes, we did.
ALESSI: Is it constant with arc current or does it get worse as the current goes up?
LEUNG: It changes, but the brightness stays almost the same.
JACQUOT: How do you measure, at high currents in the arc, the proportion of H^-? How do you measure the H^- and electron signals?
LEUNG: The negative ions were measured after the mass analysis. The electrons were measured from the drain current of the power supply.
JACQUOT: What is the accuracy of your measurement, because the ratio between the electrons and ions is very high at large currents, it is 40:1. What is the resolution of your measurements?
LEUNG: We have been particularly careful with the negative ion measurements because we want to see exactly what the current density is. That is the main point. Electron current we would infer from the drain current of the power supply.
YORK: We did do an experiment where we turned the high voltage down

and actually tried to look at the electron current straight out of the source. We were within, say, 50% between the actual extracted electron current and the power supply drain, but the figures K. Leung is quoting are actually the power supply drain.

BACAL: Did you make any test to prove that the current measured on the collector is H^-, a test like running the discharge on argon? You had some measurements without a mass spectrometer and you can prove it in that way. Why don't you make a discharge in argon?

LEUNG: We didn't have too much time for this experiment but the H^- measurement is a very solid one, measured with a mass spectrometer.

BACAL: Sure, but you had some other measurements where you didn't have a mass spectrometer so you could make a test in argon just to prove that when you have no H^-, there is no current on the collector. That's what you suggested for the photodetachment method sometime ago.

LEUNG: We didn't do this, but we are convinced that this is H^-.

NEGATIVE ION NEUTRALIZATION ACCOMPANIED BY EXCITATION

T. J. Morgan
Physics Department, Wesleyan University, Middletown, CT 06457

ABSTRACT

Progress on the study of negative ion neutralization accompanied by excitation (NINE) is reviewed. A table of all known theoretical and experimental investigations of the cross section for NINE is presented. Experimental and theoretical techniques are discussed briefly and the main conclusions from previous NINE studies are summarized. Recent experimental results for $H^- \rightarrow H(12 \leq n \leq 30)$ in Ar and H_2 over the energy range 10-60 keV are presented.

INTRODUCTION

Collisional destruction of negative ions proceeds via single or multiple electron detachment

$$A^- + B \begin{matrix} \rightarrow A + B + e \\ \rightarrow A^{q+} + B + (q+1)\,e. \end{matrix} \qquad (1)$$

A large number of subtle effects associated with the target B and/or the negative ion projectile core A, A^{q+} are hidden in the above reactions. For instance, if the target has a positive electron affinity, a detached electron may bind to the target. Also, either or both the target and the remaining projectile core may be left in an excited state, including the continuum. What happens in a given collision depends on the collision partners A^- and B and on their relative velocity.

An important collision channel in electron detachment is negative ion neutralization accompanied by excitation (NINE)

$$A^- + B \begin{matrix} \rightarrow A^*(n\ell) + B + e \\ \rightarrow A^*(n\ell) + B^- \\ \rightarrow A^*(n\ell) + B^*(n'\ell') + e. \end{matrix} \qquad (2)$$

In the present paper, experimental and theoretical studies of the cross section for NINE are reviewed and some new experimental results presented. Since the majority of atoms have a positive electron affinity, there are a large number of possible choices for the collision system. A substantial body of information is available on total stripping (neutralization of a negative ion without consideration of the final internal state of the atom). Data on total stripping, both experimental and theoretical, span several orders of magnitude in collision energy using a variety of negative ions. However, there have been only nine experimental studies of the cross section for NINE, all except one using the H^- negative ion. Calculations of the cross section for NINE are

0094-243X/84/1110082-08 $3.00 Copyright 1984 American Institute of Physics

limited to five, and are not reliable over the energy range where experimental data exist. Table I lists the experimental and theoretical studies of the cross section for NINE, in chronological order.

EXPERIMENT

Published data on NINE have been limited to low-lying states ($n \leq 5$), except for one measurement at one energy in a water vapor target.[11] New results for high-lying states ($12 \leq n \leq 30$) in Ar and H_2 targets over the energy range 10-60 keV will be presented.[14] Neither detailed comparative nor systematic information is yet available.

Experimental techniques are conventional. Production of negative ion beams utilizes either direct extraction from an ion source, e.g. H^- from a duoplasmatron, or charge transfer of positive ions in a gas or metal vapor cell. The question of excited state content of the incident negative ion beam must be considered. For example, unexplicable discrepancies in experimental results using H^- beams may be connected to the predicted existence of the $(2p^2)^3P$ state of H^-.[15] An experimental study to confirm the existence of this state, with predicted lifetime of $\sim 10^{-7}$ sec, is needed.

Detection of the excitation produced in a NINE collision depends on the excited level under study. For low-lying states radiative decay is used to measure the cross section. Since radiative lifetimes for atoms scale as $Z^{-4}n^3\ell^2$, this technique is not feasible for highly-excited states with low Z. In this case electric field ionization provides a well-tested valuable detection mechanism.

Figure 1 shows a schematic representation of an apparatus used to study NINE. P1 detects photons from the target region. This observation yields B^* and short lived states of A^*. P2 detects radiative decay of A^* and, if movable, P2 observes the spatial distribution of radiation. This allows for the separation of the contributions from different angular momentum states since substates decay with different lifetimes. Under certain conditions, this type of separation is also possible within the target.[6] P3 detects electric field induced decay of A^* metastable states. The field ionizer is used for the detection of long-lived high-lying excited states of A^*. Either or both A^+ and e^- may be measured after field ionization.[14] By appropriate electronic coincidence of target signals with post-target signals all three NINE channels of equation (2) may be determined separately. To date, coincidence techniques have not been used to study NINE and consequently the cross sections cited in Table I represent the sum of the partial cross sections for the collisions shown in equation (2).

Table I. Experimental (E) and theoretical (T) investigations of the cross section for NINE.

Authors	Ref.	Type	Energy (kev/amu)	Collision System	Excitation
McDowell and Peach (1959)	1	T	$2.5-10^4$	$H^- + H$	2p
Drukarev (1970)	2	T	unspec.	$H^- + H$	2s
Orbeli et al (1970)	3	E	5-40	$H^- +$ He, Ne, Ar, Kr, Xe	2s, 2p
Dose and Schmocker (1975)	4	E	3-60	$H^- + N_2, O_2$	2s, 2p
Harnois et al (1975, 1977)	5	E	1-5	$H^- +$ He, Ne, Xe, N_2	3s, 4s, 5s, 3d
Risley et al (1975, 1978)	6	E	1-6	$H^- +$ He, Ar	2p, 3p, 4p, 5p
Greenland et al (1977)	7	E	1-8.5	$H^- +$ He, Ar, H_2, N_2	2p
Dewangan and Walters (1978)	8	T	$15-10^5$	H^-, $He^- +$ He, Ne Ar, Kr, Xe	$\overset{\infty}{\underset{2}{\Sigma}}n, 2^3S$
Genoni and Wright (1980)	9	T	2000	$H^- + H$	2s
Geddes et al (1981)	10	E	3-25	$H^- +$ He, Ne, Ar, N_2, H, H_2	2s, 2p, 3s, 3p, 3d
Kim and Meyer (1982)	11	E	25	$H^- + H_2O$	9≤n≤24
Andersen et al (1983)	12	E	.35-11.4	Li^-, $\bar{Na} +$ He, Ne, Ar	2p
Gillespie et al (1983)	13	T	10^3-10^6	$H^- +$ H, H_2, He, Li, C, N, Ne	2s, 2p
Stone and Morgan (1983)	14	E	5-60	$H^- +$ Ar, H_2	12≤n≤30

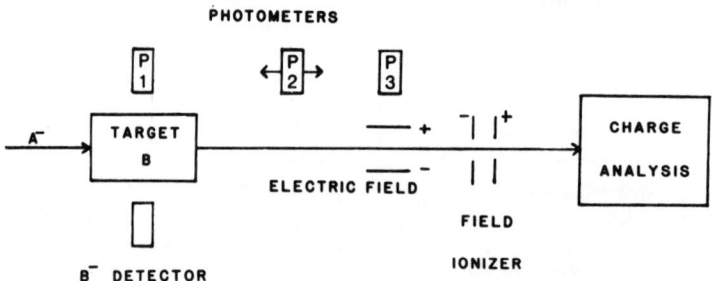

Fig. 1. Principal features of an apparatus used to study NINE collisions.

THEORY

Calculations of the cross section for NINE are limited in number and in theoretical approach. Of the five published results, four are Born calculations and the fifth is a free-electron collision model. These approximations, which have little validity below 100 keV/amu should be accurate in the high energy asymptotic limit. Low-energy theoretical calculations of NINE are needed. For simple detachment without excitation, many calculations exist at both high and low energies.[16] At energies below about 10 keV/amu the collision may be characterized as molecular. A calculation of the cross section requires knowledge of the appropriate potential curves of the collision system. In many cases good agreement exists between theory and experiment for simple electron detachment at both high and low energies.[17]

The cross section for NINE has been shown to be quite sensitive to the choice of wave function for H^-.[1,8] The diffuse nature of H^- compared to $H(1s)$ introduces a new degree of complexity when modeling collision processes. All theoretical studies except one listed in Table I (Ref. 13) used a two-parameter H^- wave function of the form

$$\Psi(r_1, r_2) = N_i (e^{-\alpha r_1} e^{-\beta r_2} + e^{-\alpha r_2} e^{-\beta r_1})$$

where N_i is the normalization constant and $\alpha = 1.04$, $\beta = 0.28$. With $\beta << \alpha \simeq 1$ the wave function is intuitively appealing since, in a sense, it represents H^- as containing one strongly bound electron and one weakly bound electron. Recent studies[8] support this description of H^- in terms of a split-shell $1s1s'$ model.[18] However, the H^- ion is much more complicated and configuration interaction is known to be important. It is not clear what role correlation effects associated with electron-electron repulsion play in NINE collisions and theoretical guidance is needed in this area. Ref. 13 used a three-state close-coupling calculation to compute wave functions. In this case, the unbound state of H^- after the collision is, for large distance, a plane wave and a $H(2s$ or $2p)$ state. For small distances the wave function is obtained numerically and is quite complicated. Results using this approach are expected to be more accurate. In any case, caution should be employed when comparing theoretical results, and when choosing a value for the cross section in the absence of experimental data. A quantitative general theory of NINE collisions is needed.

DISCUSSION AND RESULTS

The sensitivity of NINE to the choice of H^- wave functions was demonstrated in early calculations.[1] Choosing a one-parameter wave function with $\alpha = \beta = .69$, that is, treating the two electrons as equivalent, instead of using $\alpha = 1.04$ and

$\beta = 0.28$, has a dramatic effect on the cross sections for NINE. Recently it has been demonstrated[8] that for semiclassical calculations better agreement with experiment is obtained when theory views H⁻ as having a strongly bound electron with an ionization potential of a 1s electron of H together with a weakly bound electron with ionization potential of the electron affinity (0.75 eV). In 1970 it was predicted[2] that for high energy collisions

$$\sigma_{-n}/\sigma_i = \text{constant}$$

where σ_i is the cross section for ionization of hydrogen

$$H + X \rightarrow H^+ + e + X$$

and σ_{-n} is the cross section for the NINE collision

$$H^- + X \rightarrow H(n) + e + X.$$

More recently, using the free collision model,[8] it has been predicted that for high energies the cross section for neutralization resulting in an excited hydrogen atom is equal to the H(1s) ionization cross section. That is,

$$\sum_{n>1} \sigma_{-n} = \sigma_i.$$

These results suggest that at high energies the NINE collision is dominated by inner-electron ejection, which leaves the resultant H atom in an excited state after adjustment of the outer electron to the new field. High-energy calculations also predict a substantial fraction of the detachment collisions will result in excitation. For example, at 2 MeV in H⁻ + H collisions about 20% of the resultant H atoms are in the 2s state[9] while in Ne at 10 MeV theory predicts about 30%.[13] Since experimental information is limited to energies below 60 keV/amu, there is presently no verification of the predicted 1/E asymptotic behavior of the cross section for NINE.

The n dependence of NINE has been studied experimentally for $n \leq 5$ for H⁻ + He, Ar collisions for energies between 1 and 6 keV. For np states, results indicate n^{-3} scaling for $3 \leq n \leq 5$ and n^{-5} for $2 \leq n \leq 3$.[6] The effect of an applied electric field of a few hundred volts/cm at the collision region changes the n^{-3} scaling to $n^{-2.6}$ for He with no apparent change for Ar. In the case of ns state excitation, experimental data for $3 \leq n \leq 5$ indicate scaling closer to n^{-4} over comparable collision energies.[5]

A comparison of excitation in H⁻ NINE collisions with direct excitation in H° collisions is difficult at low energies due to the lack of experimental data. However, in the fundamentally important case of an atomic hydrogen target (H⁻,H°+H°) over the energy range 3-30 keV a comparison can be made. In this case, over the entire energy range H⁻ is more effective at producing H°(2s) excitation.[10] At higher energies (5≤E≤60 keV) experimental data indicate that H atom excitation via H⁻ collisions is 2 to 9 times greater (depending on state, target, and energy) than direct H atom excitation. This implies that at higher energies the weakly bound electron plays an active role in the excitations by readjusting to an excited state of H after the inner-electron is ejected. Recently, experimental results for Li⁻ + He,Ne have been compared with theoretical results for Li°(2s) + He,Ne.[12] The comparison shows that Li⁻ is more effective at producing Li(2p) above about 1 keV/amu and less effective below this energy. In Ref. 12 an interpretation based on the Massey criterion is presented. It is argued that the attachment of a weakly bound electron increases the energy defect for Li(2p) excitation and consequently results in an increase in the velocity at which the cross section maximizes. For Na(3p) excitation, a similar comparison exists. In order to permit a more detailed and careful study, data for H(Li) excitation using both H⁻(Li⁻) and H°(Li°) projectiles and using the same apparatus employing the same detection techniques are needed.

We have begun an experimental investigation of H and Li atom excitation in H⁻(Li⁻) and H°(Li°) collisions with H, H_2 and several other targets over the energy range 1-100 keV. We are studying excitation to H(2s) and to highly-excited states with principal quantum numbers in the range $12 \leq n \leq 30$. For the high-lying states, we are aware of no previous published papers on excitation using either H⁻ or H° projectiles. Our results for H⁻ + H_2, Ar collisions over the energy range 10-60 keV indicate a n^{-3} dependence. In Figure 2 we present the reduced cross section, $n^3 \sigma_{-n}$, as a function of energy.

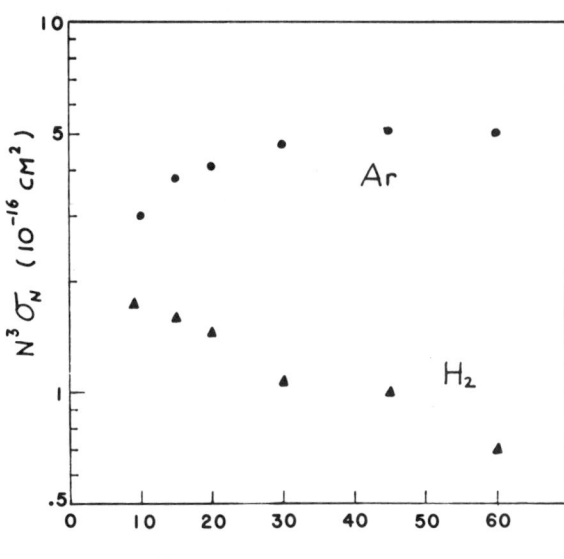

Fig. 2. Reduced cross section for highly excited state formation in H⁻ → H(N) collisions in Ar and H_2 targets.

CONCLUSION

Both experimental and theoretical information on negative ion neutralization accompanied by excitation (NINE) have been reviewed. NINE is a nascent field with limited data available. Experimental measurements and theoretical results do not share a common energy range, and several discrepancies exist between different published experimental data. These facts coupled with limited final excited state information, suggest that several comprehensive experimental and theoretical studies are needed before complete and comparative information about NINE collisions will be available.

This work was supported by the Office of Fusion Energy of the Department of Energy under contract DE-AC02-76ET53048.

REFERENCES

1. M.R.C. McDowell and G. Peach, Proc. Phys. Soc. $\underline{74}$, 463 (1959).
2. G.F. Drukarev, Sov. Phys. JETP $\underline{31}$, 1193 (1970).
3. A.L. Orbeli, E.P. Andreev, V.A. Ankudinov, and V.M. Dukelskii, Sov. Phys. JETP $\underline{31}$, 1044 (1970).
4. V. Dose and U. Schmocker, Z. Physik $\underline{A275}$, 325 (1975).
5. M. Harnois, R.A. Falk, R. Geballe, and J. Risley, Phys. Rev. $\underline{A16}$, 2256(1977), and Abst. IX ICPEAC, Seattle (1975).
6. J.S. Risley, F.J. deHeer, and C.B. Kerkdijk, J. Phys. $\underline{B11}$, 1783 (1978), and Abst. IX ICPEAC, Seattle (1975).
7. G.B. Greenland, R.A. Falk, M. Harnois, and R. Geballe, Abstracts of X ICPEAC, Paris (1977).
8. D.P. Dewangan and H.R.J. Walters, J. Phys. $\underline{B11}$, 3983 (1978).
9. T.C. Genoni and L.A. Wright, J. Phys. $\underline{B13}$, L61 (1980).
10. J. Geddes, J. Hill, and H.B. Gilbody, J. Phys. $\underline{B14}$, 4837 (1981).
11. H.J. Kim and F.W. Meyer, Phys. Rev. $\underline{A26}$, 1310 (1982).
12. N. Andersen, T. Andersen, and L. Jepsen, Bull. Am. Phys. Soc. $\underline{28}$, 805 (1983) and submitted to J. Phys. B.
13. G.H. Gillespie and R.S. Janda, Abstracts of XIII ICPEAC, Berlin (1983).
14. J.A. Stone and T.J. Morgan, present paper.
15. D.R. Beck and C.A. Nicolaides, Chem. Phys. Letts. $\underline{59}$, 525 (1978), and R. Jáuregni and C.F. Bunge, J. Chem. Phys. $\underline{71}$, 4611 (1979).
16. Two recent reviews on electron detachment collisions are available. See J.S. Risley, Invited Papers of XI ICPEAC, Kyoto (1979) and J.S. Risley, Comments At. Mol. Phys. $\underline{12}$, 215 (1983).
17. See for example Ref. 16. Also, H.B. Gilbody, Proc. US-Mexico Joint Seminar on the Atomic Physics of Negative Ions, Notas de Fisica $\underline{5}$, 146 (1982) eds. C. Cisneros and T.J. Morgan.
18. J.N. Silverman et al, J. Chem. Phys. $\underline{32}$, 1402 (1960).

DISCUSSION

SCHLACHTER: Have you compared excitation by neutrals and excitation with electron removal by negative ions? There are no measurements, I presume, of direct excitation by the negative ion where it remains negative after the collision and just excites the target.

MORGAN: There is no information available on the excitation of the target in negative ion collisions.

WADEHRA: Since there is a lot of results available on H^-, is it possible to somehow scale them and get the results for Li^-, and Na^- which are something like one-electron systems?

MORGAN: I think it probably is. Actually, there has been scaling for total detachment not considering excitation, and a scaling law there seems to work quite well. When you take the full body of information on total one electron stripping you can scale that pretty well so I think the answer is yes. I haven't tried it yet but I think one might be able to find scaling.

HISKES: If I want to enhance the excited state population, the $1/n^3$ proportion, it is still better to go charge exchange than to detach the electron from negative ions.

MORGAN: The data indicate that's true. Over the limited range we've compared it, in the case of argon and hydrogen, it appears that that's true.

A SEARCH FOR H_2^-, H_3^-, AND OTHER METASTABLE NEGATIVE IONS

Y. K. Bae, M. J. Coggiola, and J. R. Peterson
Molecular Physics Laboratory
SRI International, Menlo Park, CA 94025

INTRODUCTION

The existence of metastable excited negative ions (lifetimes $\gtrsim 10^{-6}$s) whose autodetaching decays are forbidden due to their electron spin configurations has been a subject of both theoretical and experimental interest for more than a decade. Among such ions, only He^- is well established, however H_2^-, H_3^-, N^-, and N_2^- have been either apparently observed or predicted by theory but have had no strong supporting confirmation or study.

One reason for the lack of experimental study is that even if these ions exist, their production in ordinary ion sources is likely to be very meager because they are all based on metastable excited neutral parents which are easily quenched by two-body electron collisions, while (except perhaps for N^-) their formation requires three-body electron attachment or other complicated processes. This difficulty can be overcome by a two-step capture technique, which has been used successfully for He^- production[1,2] and should be relatively effective in producing the other ions. In this method, which starts with a positive ion beam, the first capture proceeds through a near-resonant channel to the appropriate metastable neutral state. The second electron is then picked up from a neutral alkali atom with low ionization energy (typically Cs). Thus the formation is accomplished in a straightforward manner via fully-allowed transitions.

We discuss here the results of a series of careful experiments that were undertaken using this technique, to verify the existence of these ions and to determine some of their basic properties. Disappointingly, we have failed to find a convincing trace of any of them. Nevertheless, for reasons explained below, we arrive at fairly firm conclusions only about the nonexistence of metastable H_2^- and H_3^-.

EXPERIMENTAL APPARATUS

A schematic diagram of the experimental arrangement is shown in Fig. 1 A momentum-analyzed positive parent beam (marked as He^+) was directed through an alkali metal vapor oven to form the negative ions as described above. For most of these ions it appeared to us that both electrons could be effectively captured in a single alkali target vapor, and Na, K, and Cs were all used to compare different charge transfer targets. Xe gas was also added in the case of N. A wide range of vapor densities was also used. After traversing the alkali oven the +, 0 and − charged beam components were separ-

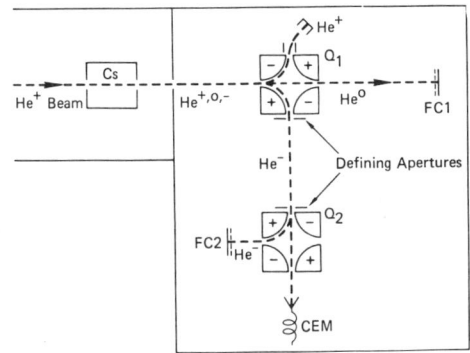

Figure 1. Experimental Apparatus

ated by an electrostatic quadrupole deflector[3] Q_1, which serves as an energy analyzer as well as charge separater. After passing two 1.4 (horizontal) x 2.2 mm (vertical) definingapertures, the negative component with the selected energy entered a second deflector Q_2, where it could again be directed either to a Faraday cup FC2, or pass undeflected to a channel electron multiplier (CEM) for counting particle currents too small ($< 1 \times 10^{-14}$ A) to measure on an electrometer.

The ability to either count individual ions or measure negative ion currents made it possible to detect ions over a very wide range of currents. With the ions deflected by Q_2, the CEM was used to detect neutral particles formed by either autodetachment or collisional detachment (by background gas) as the negative ions travelled between Q_1 and Q_2, and in this way the ions could be examined for their metastability.

About 10 nA of the parent positive beams were obtained for the production of the various negative ions. The properties of the mass analyzer, quadrupole deflector and ion detectors were well characterized using beams of well established negative ions (H^-, He^-, C^-, O^-, etc). Extreme care was needed to avoid confusing small currents of the desired negative ions with the secondary particles and other minor currents of contaminant ions. At first, the CEM detected considerable background peaks and continua as the voltage on Q_1 was varied, with Q_2 either on or off. This voltage-dependent background was apparently due to electrons or fast neutrals generated by the deflected positive ions as they struck different parts of the apparatus. A spurious peak near the proper voltage for deflecting the negative ions into Q_2 (and the CEM) was eliminated by installing a second positive ion deflector where the ions exited Q_1, and a carefully placed off-axis positive ion Faraday cup. We thus reduced the spurious signal to the continuum level of 200 counts/sec from a parent positive beam (oven cold) of 10 nA.

The search for the various negative ions was carried out by first optimizing the positive ion current to FC1 with Q_1 off and the vapor oven cold. The oven was then heated and the beam refocused to optimize the neutral current to FC1 with Q_1 on (measured by secondary electron emission). We found that these conditions optimized the He$^-$ currents from He$^+$. The (+,-) voltages on Q_1 were set to the proper value for that ion beam energy, and the voltage on Q_2 was slowly swept across the proper setting. If no negative ion peak was observed at FC2, and no corresponding detached neutral product peak was observed by the CEM, the voltages on Q_2 were turned off and a direct negative ion peak was searched for, using the high sensitivity of the CEM as the voltages on Q_1 were swept through the proper value.

Because the yields of negative ions from this two-step process cannot be expressed by a single cross section, we can only compare the efficiency of producing the ions of this study to the relative yield of He$^-$ from a current of He$^+$ (or the appropriate parent ions in the other cases) measured at FC1 before the oven was heated from room temperature. The yields of He$^-$ ions in our apparatus at the ion energies used here, was about 10^{-3} of the transmitted He$^-$ beam.

HISTORICAL BACKGROUND AND PRESENT RESULTS FOR THESE IONS

a) H_2^- and H_3^- (D_2^- and D_3^-)

The mass spectroscopic detection of H_2^- was first reported by Khvostenko and Dukel'skii[4] in 1958, and in 1974 Hurley[5] reported the observation of H_3^-. However, those results were never confirmed in experiments with isotope substitution until Aberth, Schnitzer, and Anbar[6] reported observations of HD$^-$, D_2^-, H_3^-, H_2D^-, and D_3^- with lifetimes greater than 10μs. These ions were extracted directly from a duo-plasmatron ion source. Although the observed currents of these ions were exceedingly small ($10^{-8} - 10^{-7}$ x the D$^-$ current), convincing isotopic tests were made, as well as mass-resolution of HD$^-$ from ^{3}He$^-$. As suggested by Aberth et al.[6], we expect that a metastable H_2^- would most likely exist in a quartet state such as He$^-$, possibly $^4\Sigma_g^-$, with at least the ground vibrational level lying below H_2 ($c^3\Pi_u$). However, to our knowledge, no calculations have yet been undertaken on quartet H_2^- states so there is no theoretical evidence of their stability. Some speculation has been made about the existence of a metastable doublet state of H_2^- ions ($^2\Sigma_g^-$). However, Barnett[7] has recently made an extensive search for doublet H_2^- ions (formed by simultaneous two-electron capture collisions of H_2^+ in H_2 and Xe gases), with negative results. He estimates an upper limit to the production cross sections of about 10^{-20}cm^2.

From D_2^+, we observed a large D$^-$ current peak at FC2 with half the "proper" voltages on Q_1, and Q_2, but no measurable current at the full D_2^+ energy. Using the CEM with Q_2 turned off, for higher sensitivity, we were able to find a minute peak of 500 counts/sec above the background with Q_1 tuned to the full D_2^+ energy. But because we would expect such a number of He^- from a small ($\sim 4 \times 10^{-5}$) contaminant of He in the D_2 gas (Mathesson 99.5% purity) we must attribute this peak primarily to He^-.

We could observe no D_3^- ions from D_3^+ above the ~ 200 counts/sec background. There was a very small peak observed by the CEM at 2/3 the D_3^+ energy, which can be attributed to He^- from HeD^+, also arising from the small He contaminant.

b) N^-

The most direct experimental evidence of N^- has been reported by Hiraoka, Nesbet and Welsh[8], who also summarized the previous evidence. They used a quadruple mass analyzer to identify both mass 14 and 15 negative ions extracted from N_2 and NO mixtures in a discharge sustained by electrons from a hot filament. They observed the minimum appearance potentials for N^- in these gases. These ions were presumed to be formed directly by dissociative attachment of the N_2 and NO molecules. Theoretical calculations have shown that the $2p^3$ ^4S ground state of N has a negative electron affinity (unstable). However, Thomas and Nesbet[9] found, from an isoelectronic extrapolation, that the $2p^4$ ^1D and ^1S states of N^- should lie below the ^2D and ^2P metastable states of N by 0.9 and 0.7 eV, respectively. Thus, they would be 1.44 and 2.88 eV above the N (^4S) ground state, and metastable with respect to radiative decay and autodetachment.

We were unable to find any N^- produced by two-step electron capture in Na, K, or Cs. However, forming N^- in pure alkali vapor is not straightforward because the first electron captured by N will most likely go to some $2p^2$ ^3S, ^3P, ^3D state, which must then radiatively decay to $2p^3$ ^2P or ^2D before the second electron can be captured. To solve the problem of forming the ^2P and ^2D states directly, we added Xe gas to the Cs oven. Xe with an ionization potential of 12.1 eV can facilitate near-resonant charge transfer (usually large cross sections) with N (I.P. = 14.56 eV) to form the N(^2D) state at 2.4 eV, and the reaction forming N(^2P) at 3.6 eV would be only 1.2 eV endothermic. Thus, these two metastable parents of N^- (^1D and ^1S) should dominate the products of N^+ + Xe charge transfer, providing the N(^2D and ^2P) states needed for subsequent eletron capture from Na, K, or Cs to form N^-. However, even with the addition of Xe, still no N^- could be detected.

It is possible that the capture of the second electron was improbable because the maximum fields (6 kG) of our small mass-analyzing magnet limited the speed of N^+ to $\sim 6 \times 10^6$ cm/s where the

σ_{0-} cross section may be quite small. (σ_{0-} for H + Cs reaches a maximum at about 3×10^7 cm/s.) Nevertheless, the fact that the upper limit for N^- production was at least 10^5 times smaller than for He^- from He^+, limits its amenability for experimental study.

We looked briefly for N_2^- (from N_2^+), which had also been reportedly seen in small quantities by Hiraoka et al.[8] We found no evidence of this ion produced in our apparatus.

SUMMARY

TABLE I summarizes the estimated upper limits on the production ratio of the various negative ions relative to the initial positive ions transiting our apparatus (beam energies ~2 keV).

Species	He^-	D_2^-	D_3^-	N^-	N_2^-
$\dfrac{X^-}{X^+}_{max.}$	$>10^{-3}$	4×10^{-9}	1×10^{-9}	4×10^{-10}	2×10^{-10}

Inasmuch as this technique should be very efficient (compared to direct production in an ion source) for producing D_2^- in a metastable quartet state bound below $H_2(c^3\Pi_u)$, our results indicate that this state does not exist with a lifetime greatly exceeding 2×10^{-11} s. It is similarly unlikely that D_3^- exists in a metastable state based upon a metastable Rydberg state of D_3.

Because of the limited beam energies and scope of this work, the evidence against N^- and N_2^- is somewhat less conclusive, but the results are not encouraging regarding the production of useful beams.

It is difficult to quantify the reliability of negative results, but certainly we have produced abundant quantities of all known negative ions from their positive parents by this method. In addition, very recently during the course of this work we have discovered the metastable He_2^- ion, which was hitherto unpredicted, and produced it in usable quantities.[10] Thus, the technique has an ample experimental as well as conceptual justification as a method of producing metastable negative ions efficiently. The fairly efficient production of He_2^- ($>10^{-4} \times He_2^+$), of which probably only one or two vibrational levels are bound against autodetachment to He_2^* ($^3\Sigma_u^-$), is additional support for our capability of producing a metastable H_2^-, to which similar constraints would apply.

This work was supported by the Air Force Office of Scientific Research under Contract F49620-82-K-0030 and by the National Science Foundation under grant PHY 81-11912.

REFERENCES

1. B. L. Donnally and G. Thoeming, Phys Rev. **159**, 87 (1967)

2. R. V. Hodges, M. J. Coggiola and J. R. Peterson, Phys. Rev. A 23, 59 (1981)
3. H. D. Zeman, Rev. Sci. Instr. 48, 1313 (1977).
4. V. I. Khvostenko and V. M. Dukel'skii, Zh. Eksp. Theo. Fiz. 34, 1026 (1958) [Sov. Phys. JETP 7, 709 (1958)].
5. R. E. Hurley, Nucl. Inst. Methods 118, 307 (1974)
6. W. Aberth, R. Schnitzer, and M. Anbar, Phys. Rev. Lett. 34, 1600 (1975).
7. C. F. Barnett, ORNL/TM 8693 (1983).
8. H. Hiraoka, R. K. Nesbet, and L. W. Welsh, Jr., Phys. Rev. Lett. 39, 130 (1977)
9. L. D. Thomas and R. K. Nesbet, Phys. Rev. A 12, 2369 (1975).
10. Y. K. Bae, M. J. Coggiola, and J. R. Peterson, to be published.

DISCUSSION

LEUNG: You couldn't detect a N^- by double charge exchange. Does that mean that you cannot see N^- in volume process and surface process devices?

PETERSON: I think not. I should mention that Ka-Ngo Leung and Ken Ehlers have seen the production of He^- on surfaces and this is similar, it is a fully allowed process, I don't know how efficient it is but they have seen it coming off in their cesium ion sources.

RESEARCH AT THE IKVT/KFK RELATED TO THE DEVELOPMENT OF AN H⁻ BEAM LINE

O.F. Hagena, P.R.W. Henkes, R. Klingelhöfer,
B. Krevet, H.O. Moser

Institut für Kernverfahrenstechnik, Kernforschungszentrum
Karlsruhe, D-7500 Karlsruhe 1, Federal Republic of Germany

ABSTRACT

Cluster ions of hydrogen are accelerated by 500 kV on a Cs jet-target. H⁻ are formed from H_2 molecules having an energy of the order of 100 eV/atom. The intensity of the H⁻ is rather low, most probably due to strong scattering at this low energy. In the interaction of cluster ions with cesiated metal surfaces H⁻ are formed from molecules in the energy range of 3-30 eV/atom.

1. INTRODUCTION

At the IKVT of the Nuclear Research Center at Karlsruhe we are testing different methods for the production of hydrogen negative ions with the ultimate goal of constructing a demonstration megawatt beamline with an energy exceeding 300 keV. Up to now our activities were concentrated on evaluating the properties of cluster ion beams for the production of negative ions. Cluster ion beams are particularly useful for the production of high particle current densities at low energies, where conventional methods fail because of space charge limitations.

There are two different schemes under investigation: the interaction of accelerated cluster ions with a Cs vapor jet-target and with cesiated surfaces.

2. INTERACTION OF CLUSTER IONS WITH A CS JET

At the 2nd Brookhaven Symposium on Negative Ions we reported about the design and construction of a supersonic Cs vapor jet-target, which is now mounted at the accelerator. Its performance has been described elsewhere[1] in detail. Fig. 1 shows a schematic of the system. The jet could be operated pulsed or continuously. The pulsed mode is best suited for our experiments because we normally pulse the cluster beam in order to keep the base pressure in our accelerator low. The rise time for the Cs jet to reach steady state is 100 ms, the decay time after shutoff 300 ms. Fig. 2 shows the density profile for different source conditions. The FWHM at the axis of the cluster ion beam is only 4.7 cm, the width in the direction of the nozzle slit is 7 cm.

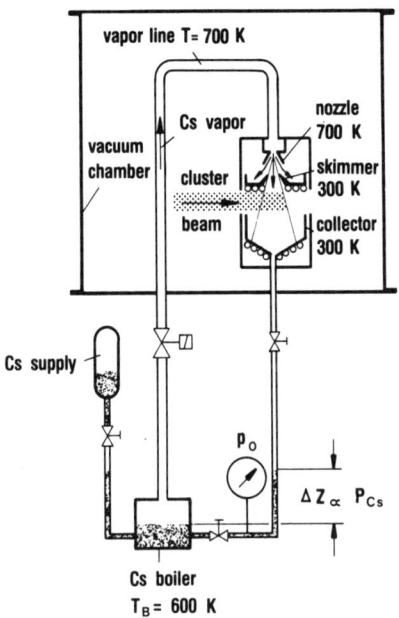

Fig. 1. Schematic of the Cs supersonic jet.

The cluster ion system was still the one that was used in investigations related to refuelling of plasmas and is not optimized for the present purpose. Crossed beam experiments were carried out using the following diagnostics:
The momentum of the cluster beam pulse was measured by a pendulum. A TOF measurement using a secondary electron detector (Cu-Be) was employed to determine the size distribution of the cluster ions. Negative ions were detected by a miniaturized momentum analyzer, using permanent magnets (Fig. 3), which was mounted in a distance of 30 cm from the center of the Cs jet. Its entrance slit is only 0.1 mm wide. This is less than the Debye length of a beam plasma one may expect at this position. Thus plasma is prevented from flowing into the analyzer. Five electrodes of different size accept H^- in the energy range from 6-40000 eV.

The distribution among the electrodes of the ion currents was not much dependent on the experimental parameters. Taking into account the various sizes of the electrodes, the highest current density was always on electrode 2, that is in the energy range from 60-190 eV. These H^- must originate from beam particles in this energy range. Since the acceleration voltage for the cluster ions is of the order of 500 kV, these particles cannot be atomic

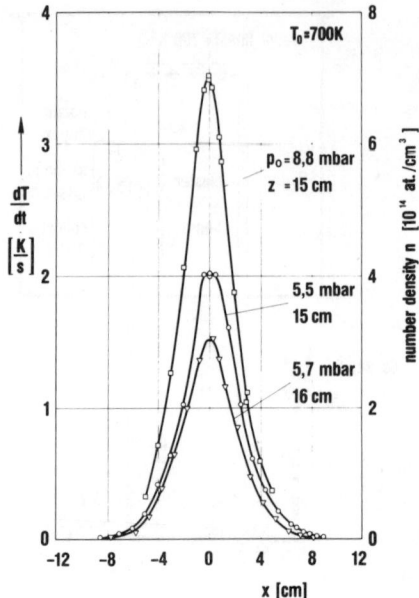

Fig. 2. Number density profiles of the Cs jet for different stagnation pressures at the nozzle.

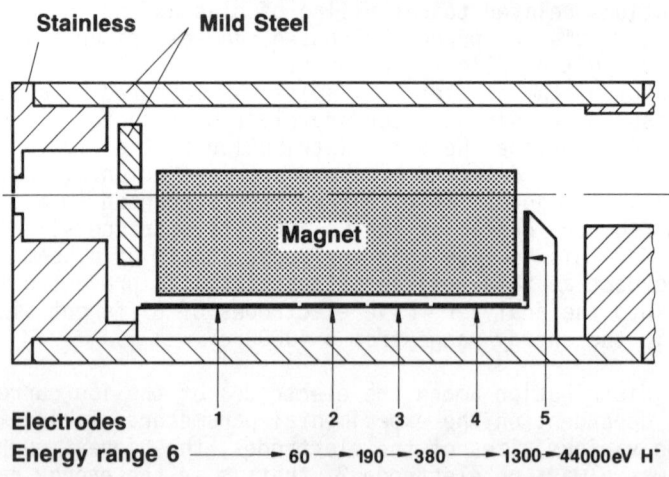

Fig. 3. Schematic of the momentum analyzer. The size of the magnet is 1x4 cm^2.

or molecular ions but must be hydrogen molecules contained in the cluster ions. This result supplements our old measurements [2,3], where we found a fairly high cross section for the production of H^- from molecular hydrogen, but could not determine the energy of those molecules producting them. Thus we have shown that H^- is not only produced by collisions of Cs with neutral hydrogen atoms, but also with hydrogen molecules. In the mean time Salzborn et al[4] found a cross section of 10^{-16} cm^2 for the production of H^- from H_2 interacting with Rb at 1 keV. However, the H_2 were produced by neutralizing accelerated H_2^+ by charge exchange, resulting in a high population of vibrationally excited molecules. In contrast the molecules contained in cluster ions are almost all in their ground state. Unfortunately our primary particle current is not known well enough in the energy range in question to give a figure for the conversion efficiency. From a practical point of view the intensity was disappointingly low - less than 10μA/cm^2 at the entrance slit of the analyzer. Although the space charge limited current density for H^- of ≈100 eV is of the same order, measurements with Langmuir probes show a plasma expanding out of the Cs jet which should neutralize the H^- space charge. Thus the main causes for the low intensity seem to be:
1. Only a small fraction of the beam particles have the proper energy, the bulk having energies below 10 eV/atom.
2. At these low energies there is a strong divergence due to transverse momentum transfer by scattering and dissociation of H_2.

The momentum of the cluster beam pulse, as measured by the pendulum, decreases by a factor of > 100 when the jet is turned on, which shows that scattering is an important process for reducing the intensity.

3. INTERACTION OF CLUSTER IONS WITH CESIATED SURFACES

At a test stand we conducted preliminary experiments with alcali-coated metal targets. The cluster ions were accelerated by 60 kV. The mean size of the cluster ions could be determined by a combination calorimeter-microbalance. First we used a commercial Ba dispenser cathode as target. The arrangement is shown in Fig. 4. The cluster ion beam, which was limited by an aperture of 0.5 cm in diameter, impinged on the target surface, which was inclined by 45° with respect to the beam. Negative ions and electrons were accelerated towards the momentum analyzer by biasing the target negative with respect to ground potential. After cleaning the cathode by ion bombardment it was activated by heating. After cooling down sufficiently, in order to avoid thermal emission of electrons, the cathode was bombarded with cluster ions. The impurity content of the negative ion current was < 2 %. The H^- were predominantly formed from a group of beam particles having a mean energy of about 17 eV/atom, whereas the main fraction has a mean energy of < 1 eV/atom. To determine total currents the momentum

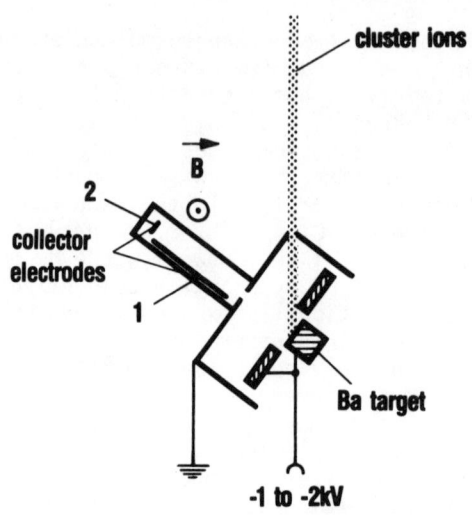

Fig. 4. Schematic of the Ba target. For the Cs experiment the Ba cathode was replaced by Mo which was coated with Cs from a dispenser.

analyzer was replaced by a detector which collected the ion and electron current seperately over a cross section of 3x3 cm². The maximum ion and electron current obtained was 370μA and 160μA resp. The Ba cathode was then replaced by a molybdenum plate which could be coated with Cs by vapor deposition from a Cs dispenser. Using essentially the same experimental conditions the measured currents increased by an order of magnitude. The maximum ion current was 4.9 mA, the electron current going with it was 3.2 mA. The current density of H⁻ at the target surface was 10 mA/cm². The total particle current hitting the target is estimated to have been 1.5 A. Thus the conversion efficiency is only 0.3 %. The reason is obviously that most H_2 of the beam have to low an energy. The Cs deposited on the Mo surface is heavily sputtered by the cluster ion beam and must be replenished continuously. Without doing so the H⁻ yield drops to half of its original value after only 0.7 s of integrated beam time.

The target was then installed at the accelerator where a higher acceleration voltage is available. The molybdenum was replaced by a tungsten single crystal (110 surface). To investigate the energy dependence of H⁻ formation the target was used as detector for TOF measurements. The results may be compared with those of the SE detector. Fig. 5 shows two typical pairs of TOF spectra that differ in the mean energy per atom of the cluster ions. The curves are normalized at the saturation current. They show between 20μs and 80μs after switching on the beam the rise of the

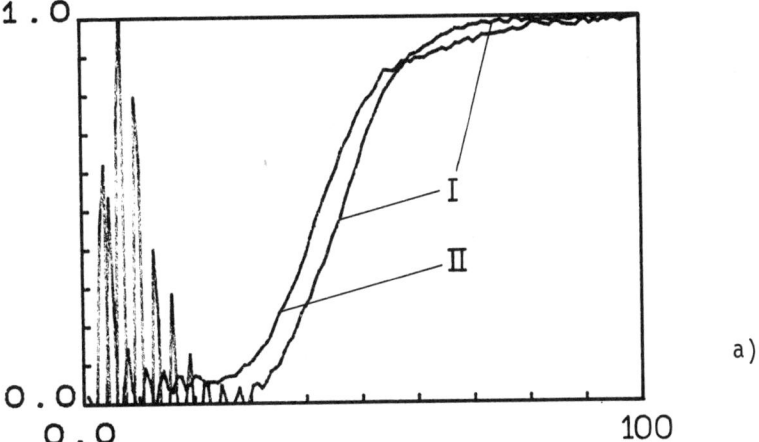

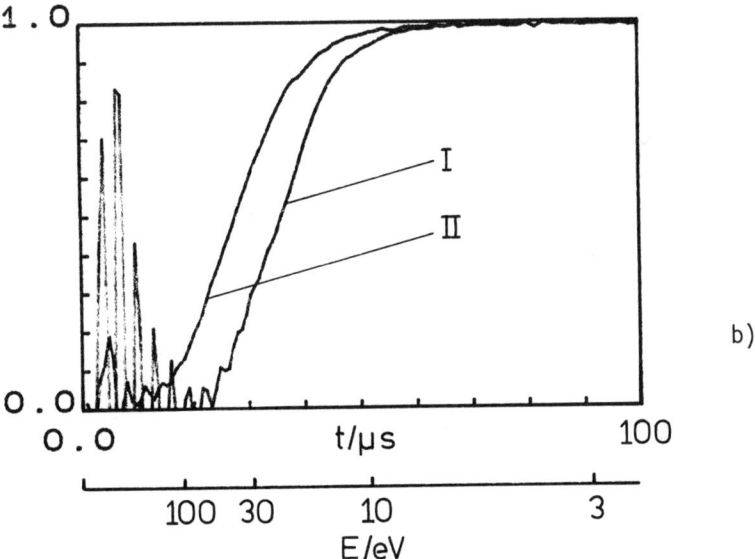

Fig. 5. Integral time of flight spectra of cluster ions measured by
I the H⁻ current produced at the cesiated tungsten surface
II the current (cluster ion current + SE current) produced on a polycrystalline Cu-Be surface.

timeintegrated current. The oscillations seen in the first 20µs are due to an interference from the switching circuitry. By comparing the curves for H⁻ with those for SE one gathers that, on a relative basis, H⁻ are formed at lower energy than SE. In Fig. 5a the current of both the H⁻ and the SE saturates at 75µs, indicating that H⁻ are formed at energies as low as 4.2 eV/atom. From Fig. 5b one concludes that cluster ions of > 60 eV/atom do not conribute significantly to the H⁻ production, although there are particles in this energy range as witnessed by the SE curve. Thus it can be concluded that the formation of H⁻ by cluster ions, hitting the cesiated tungsten 110 surface at an angle of incidence of 45°, happens predominantly in the energy region of 4-60 eV/atom. The energy corresponding to the velocity component normal to the surface is 2-30 eV/atom. Experiments to determine the conversion efficiency are in progress.

4. FUTURE PLANS

Since some time difficulties of applying cluster ion acceleration to H⁻ production became apparent or have been anticipated. Therefore, in order to have at hand a source of H⁻ for further work, we initiated the construction of an H⁻ source following the principle of the 'Selfextraction Source' developed by Ehlers and Leung[5] at Berkeley. We do not intend to do research to any extent at the source itself, since at Karlsruhe Piosczyk and Dammertz are already doing research on this type of surce as well as that of sources with volume production of H⁻.

Fig. 6 shows the design of our source. Although it follows in general that of the Berkely bucket source, it will be different in some respects: In order to better control the loss of Cs from the source its walls will run hot to prevent any deposition. The whole source will be placed into vacuum. Its anode is fabricated from Mo-sheet and is cooled by radiation only. The converter is made from copper, explosion plated with Mo, and all its joints are welded. The extraction slit can be closed by a shutter before switching off the discharge. The cooling of the converter will be turned off to let it run hot also to take off the Cs deposited on its surface. Ths Cs in the discharge volume will be removed by cryopumping. The magnetic cusp field is produced by Co-Sm magnets canned in stainless steel tubes and cooled by water. The tubes also serve as supports for thermal shields - not shown in Fig. 6 - that protect the environment from heat radiation. The source is expected to be ready for testing early next year.

After testing, the source will be installed in the accelerator. First we are going to use existing apparatus. We will convert our power supply from +1 MV/120 mA to -400 kV/> 240 mA, which will give us ca. 100 kW of power for testing acceleration and neutralization at high energy.

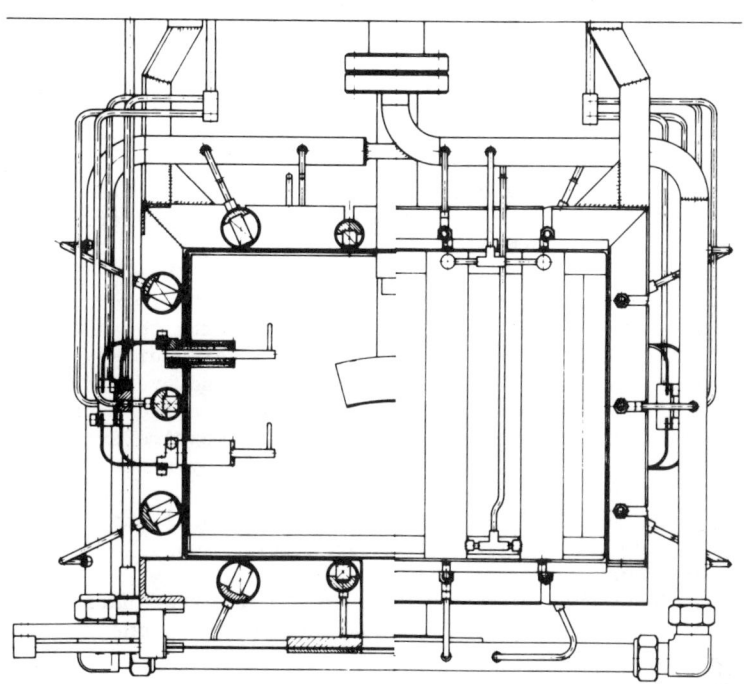

Fig. 6. Partial cross section of the Self Extraction Source.

REFERENCES

1. A. Athanasiou, O.F. Hagena, Proc. 13th Int. Symp. on Rarefied Gas Dynamics, Novosibirsk, 1982, to be published.
2. O.F. Hagena, U. Pfeiffer, Fusion Technology 1978, Pergamom, 1979, p. 295.
3. P.R.W. Henkes, Int. J. of Mass spectr. and Ion Phys. 41 (1981), 55.
4. E. Salzborn, private communication.
5. K.N. Leung, K.W. Ehlers, Rev. Sci. Instrum., 53 (1982), 803.

DISCUSSION

GRANNEMAN: What is the conversion efficiency in your work on surfaces?

HENKES: If you're asking for the efficiency with regard to the total beam, it is of the order of 0.3%. The efficiency would be higher if you are looking at only those particles that are in the right energy range. The energy range in the ion beam is very wide and in our case most of the particles are not in the right energy range. If you pick out the right energy range you get a rather high efficiency, but we have not done very much in the area of conversion efficiency studies. Measurements are still in progress.

HISKES: You seem to imply that negative ions you see are from hydrogen atoms that came in with the cluster and then were reflected back and not from desorbed particles.

HENKES: We believe it. The processes are fairly long. They are of the order of 100 milliseconds, and we know from the impurities that because of a very high current density, even if the total current is not high, that these particles are coming off very quickly, in the first few milliseconds. We would expect if we had hydrogen adsorbed at the surface and this adsorbed hydrogen would be the source of the negative ions, that it would come off at the same time scale and this is not the case. Negative ion flux is constant over the whole length.

HISKES: I think you said that about 1% of the total number of atoms coming in the cluster are converted to negative ions.

HENKES: Less than 1%.

HISKES: But those that are converted are in a rather narrow energy range, around 20-30 eV. If you just select out that energy range what is the efficiency?

HENKES: It would be higher.

HISKES: Is it like a factor of 10 higher, or 100 higher?

HENKES: The difficulty with these time of flight measurements is that we can only give relative distribution function from this curve. The uncertainty is the response function of the copper-beryllium secondary electron detector for low energy beams. Nobody has done any measurement at energies below 100 eV and we are well below this and there are effects for electron production that may be entirely different. If you have a cluster hitting a surface, that may produce a microplasma and electrons come out of this. There is evidence for this kind of effect from other studies.

H⁻ DENSITY IN A TANDEM MULTICUSP DISCHARGE

M. Bacal and K.N. Leung*
Laboratoire de Physique des Milieux Ionises
Groupe de Recherche No 29 du C.N.R.S.
Ecole Polytechnique, 91128 Palaiseau Cedex, France

ABSTRACT

The H⁻ ion density is measured by photodetachment in the source and target sections of a tandem multicusp discharge. The extracted negative ion and electron currents are measured separately and compared with the respective densities in the plasma.

INTRODUCTION

Recently Leung et al.[1] presented the results of extracting volume produced H⁻ ions from a tandem multicusp plasma generator. It was shown that a small positive bias, V_b, on the plasma electrode enhanced the H⁻ ion yield and reduced the extracted electron component.

By employing the photodetachment technique, we have investigated the effect of biasing the plasma electrode upon the H⁻ density in the source (driver) and extraction (target) chambers of a magnetically filtered multicusp source. The variation with V_b of the extracted H⁻ ion and electron current as well as that of other plasma parameters is also reported. The injection of cold electrons into the driver section leads to a significant increase of the relative density of H⁻ ions in the target section.

EXPERIMENTAL SET UP

The experiment was performed in a stainless steel vessel (diameter, 25.4 cm and height, 23.6 cm) as shown schematically in Fig. 1. A more detailed description of the multicusp discharge device is given in Refs. 2 and 3. The ten columns of ceramic or samarium-cobalt magnets on the side wall together with four rows of the same type of magnets on the top flange generate a multi-linecusp configuration for plasma and primary electron confinement. The other end of the vessel is enclosed by a stainless steel plate which simulates the plasma electrode of an extraction system. This plate is electrically isolated from the chamber wall. Thus a bias voltage, V_b, can be applied between the plate and the vessel. A hydrogen plasma is generated by primary electrons emitted from three thoriated tungsten filaments which extend about 6 cm into the vacuum chamber. In normal operation these filaments are biased at -50 V with respect to the vessel wall which forms the anode for the discharge. An additional set of filaments is also available for injecting low-energy (E < 16 eV) primary electrons into the plasma.

---

*Permanent address: Lawrence Berkeley Laboratory, Berkeley, California 94720.

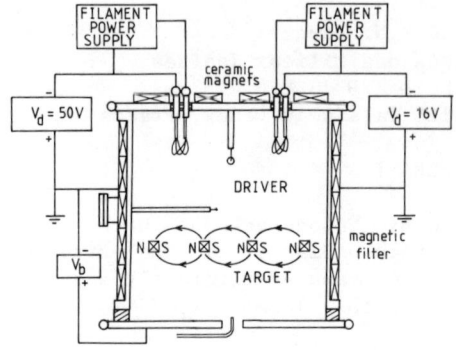

Fig. 1. Schematic diagram of the experimental setup.

Fig. 2. H⁻ relative density as a function of the plasma electrode bias voltage (a) in the target chamber (b) in the driver chamber.

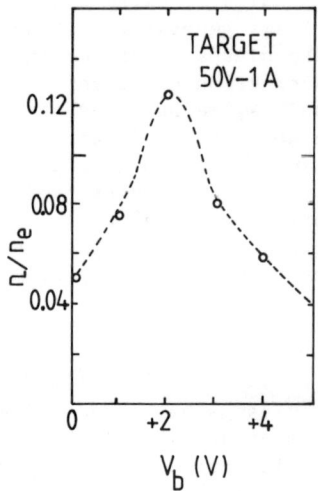

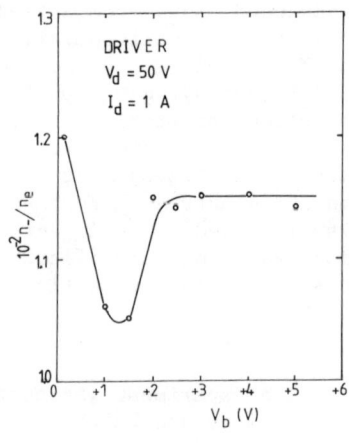

Fig. 2a   Fig. 2b

In this experiment, a magnetic filter can be installed at the center of the vessel. This filter was constructed by inserting square samarium-cobalt magnet rods (0.35 x 0.35 cm) into copper tubes through which a square broach had been passed. Since the magnets rested on the four broached grooves, their orientation was fixed and water was run through the remaining space to provide cooling. This filter is similar to that described as a strong filter in Ref. 1. With the filter in position, the vessel is divided into a 'driver' (source) and 'target' (extraction) chamber. The location of the magnetic filter can be changed over a range of 7 cm. The plasma parameters in the driver and target regions were obtained using a moveable axial Langmuir probe. The H⁻ density in the plasma was measured by the photodetachment technique, described in Ref. 4. The density in the driver section was studied with the magnetic filter placed below the photodetachment probe, while the measurement in the target section was made with the magnetic filter located above this probe. Therefore these two measurements correspond to two slightly different configurations, denoted as configuration with short and long target, respectively.

In order to study the extraction of negative ions, we replaced the stainless steel plate which encloses the bottom end of the vessel, with the system shown on Fig. 1 of Ref. 2: the bottom end of the vessel is enclosed in part by the plasma electrode, PE, of the extractor, and in part by a grid G. This configuration provides the possibility to test the effect of the area of the PE upon the extracted negative ion current, since this area can be increased by connecting G to PE. The result of this test has been reported in Fig. 5 of Ref. 2: in the presence of the magnetic filter, the extracted negative ion current exhibits a maximum for a positive value of V_b with any of the two tested values of the PE area.

EFFECT OF PLASMA ELECTRODE BIAS UPON H⁻ AND ELECTRON DENSITIES IN A TANDEM MULTICUSP GENERATOR WITH CERAMIC MAGNETS

The H⁻ ion densities in the two chambers of the tandem multicusp plasma generator were studied as a function of V_b by means of the photodetachment technique. A steady state hydrogen plasma was produced in the driver chamber by 1 A of discharge current. Figure 2 shows n_-/n_e in the target and in the driver chambers as a function of V_b. It can be seen that in the target chamber n_-/n_e increases with V_b, reaches a maximum of 13% at V_b = +2 V and then decreases. The H⁻ fraction in the driver chamber decreases with V_b and reaches a minimum of 1.05% approximately at the same V_b at which the concentration of H⁻ in the target side achieved a maximum. This result indicates that the increase of H⁻ ion fraction in the target chamber as V_b is varied from 0 to +2 V may arise from the flow of H⁻ ions from the driver side. The injection of 4 A of 16 eV electrons in the driver chamber leads to an increase of the maximum relative density of H⁻ ions in the target chamber from 13% to 22%.

The plasma electron density in the two chambers has also been measured as a function of the plate bias. Fig. 3 shows that the electron density n_e in the target side decreases by a factor of five as V_b is changed from 0 to +5 V, while n_e in the driver increases only by 10%. Comparing this result with that of Fig. 3b in Ref. 1, it can be seen that both the extracted electron current and the electron density in the target chamber decrease as the PE is biased positively with respect to the anode. The axial plasma potential profile has been obtained by using a moveable disc probe. When the PE is at anode potential (V_b = 0), Fig. 4 shows that the plasma potential V_p in the driver chamber is 0.7 V more positive than that in the target chamber. Since the average energy of the H⁻ ions is lower than 0.7 eV[6,7], the H⁻ ions formed in the driver region will be trapped electrostatically and they cannot escape either to the wall or to the target chamber. However, when the PE is biased positive with respect to the wall, only V_p on the target side increases. Figure 4 shows that axial potential gradient across the two chambers is much reduced with $V_b \approx$ +2.5 V. Some of the H⁻ ions in the driver chamber can now cross the filter into the

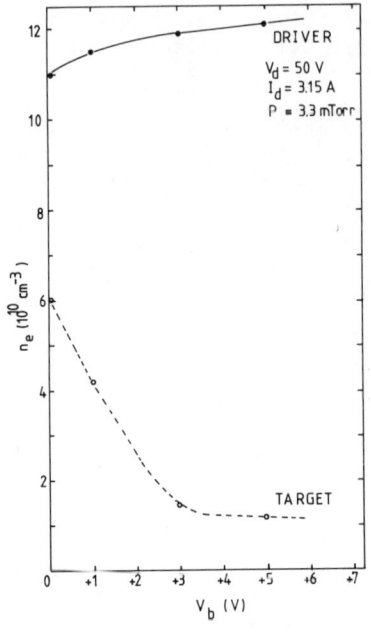

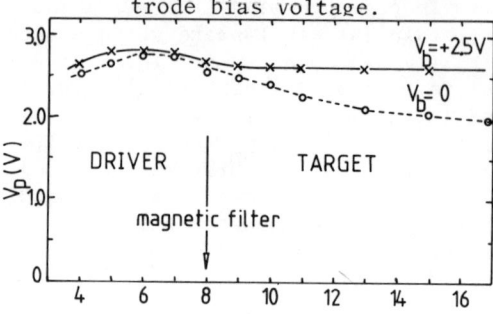

Fig. 3. Plasma electron density in the driver and target chambers as a function of the plasma electrode bias voltage.

Fig. 4. The axial plasma potential profile with two different plasma electrode bias voltages. Ceramic magnets. 50 V-1 A discharge, 3 mTorr.

target chamber. The addition of these H^- ions to the target region will cause an increase in the local H^- ion concentration.

A further increase in V_b results in the drop of n_-/n_e in the target region (so is the extracted negative ion current I^- in Fig. 3a of Ref. 1). When $V_b > 2.5$ V, the amount of positive ions and electrons that leak through the filter is much reduced. It is possible that the H^- density in the target region could decrease due to the reduced production of H^- ions in the target chamber, which requires the presence of thermal electrons.

EFFECT OF PE BIAS AND MAGNETIC FILTER POSITION UPON THE EXTRACTED NEGATIVE ION AND ELECTRON CURRENTS

In these experiments the stainless steel end plate, shown on Fig. 1, was replaced by the extraction system described in Ref. 2. The grid was connected to the PE; thus the effective diameter of the PE is 25 cm and is equal to the diameter of the plasma generator (see Figs. 1 and 2 in Ref. 2).

Fig. 5 shows the dependence upon V_b of the extracted H^- ion and electron currents in the tandem multicusp generator with samarium-cobalt magnets. In order to evaluate the effect of secondary electron emission from the collector surface of the extractor, which was discussed in detail in Ref. 2, we investigated the effect of biasing positive the collector with respect to the selector. Note that due to the secondary electron emission from the collector the negative ion current is reduced by 20% when the measurement is made with the collector and the selector at the same potential; this is the case of the data presented in all the other figures of this paper and in Ref. 2.

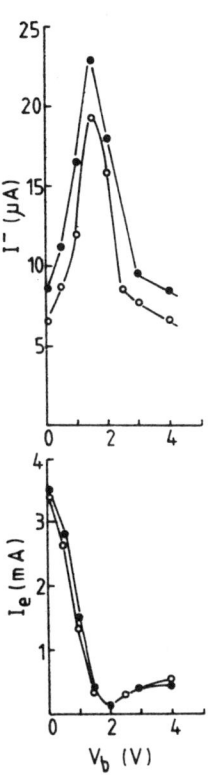

Fig. 5. Tandem multicusp plasma generator, with short target configuration. (a) The extracted H ion current, (b) the extracted electron current as a function of the plasma electrode bias voltage. 50 V-2 A, 3.5 mTorr. $V_g = V_b$. This figure also illustrates the effect of secondary electron emission from the collector. oo - collector and selector at the same potential ; ●● - collector 19V positive with respect to the collector/

In this experiment the magnetic filter is in the lowest position, as illustrated by Fig. 1, at a distance of 8 cm from the PE ('short target'). The other conditions of this experiment (50 V, 2 A, 3.5 mTorr) are close to those of the extraction experiment reported for a 'long target' configuration in Ref. 2; there the magnetic filter was at about 15 cm from the PE. Thus, we can evaluate on this example the effect of the position of the magnetic filter. It appears that with a 'short target' configuration at optimum pressure, the negative ion current at optimum V_b is only slightly larger (10%) compared to the 'long target' configuration, with no change in the electron current. The ratio I^-/I_e at optimum V_b is 1/18.

EFFECT OF THE STRENGTH OF THE MULTICUSP MAGNETIC FIELD

Fig. 6 presents the variation versus V_b of the extracted H⁻ ion and electron currents, and that of the corresponding densities in the target chamber (at 12 cm from the PE) for a 'long target' configuration. Fig. 6 illustrates the effect of the strength of the multicusp magnetic field (since both the results obtained with ten columns of weak, ceramic, magnets, and with ten columns of stronger, samarium-cobalt, magnets on the side wall of the vessel, are plotted). The four columns of magnets on the top end plate were strong magnets

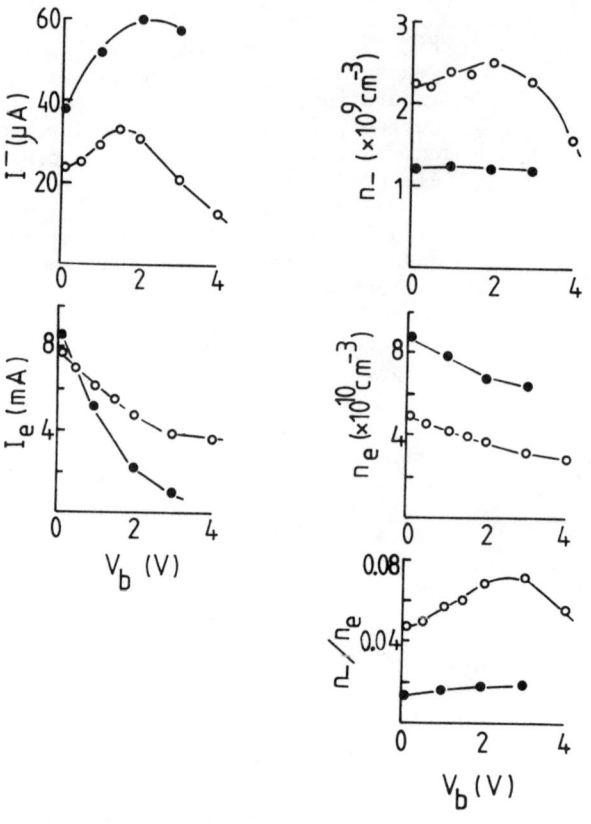

Fig 6. The extracted H⁻ ion and electron current variation versus V_b is compared to that of H⁻ ion and electron density in the target chamber at 12 cm from the plasma electrode. Long target 50V-10A discharge at 3 mTorr.
o o weak magnets
● ● strong magnets.

in both cases. The measurements are made at 3 mTorr (close to the optimum pressure) and a discharge current of 10 A. As in the case of the data presented on Fig. 5, the grid was connected to the PE. The extracted negative ion current increases when the PE is biased positive, and attains a maximum at +2 V in the case of the strong magnets, and at +1.5 V in the case of the weak magnets. With strong magnets, the extracted negative ion current is almost twice higher than that observed with weaker magnets. At optimum V_b, I⁻ is only 50% higher than its value at V = 0. This increase is much less than observed at lower discharge currents, where the corresponding increase in I⁻ was by a factor of 3.5 (see Fig. 5).

The extracted electron current is reduced when the PE is biased positive, both when the magnets are strong and weak, but this reduction is larger when the magnets are strong. The ratio I^-/I_e at optimum V_b is 1/37 for the case of strong magnets, but only 1/167 for the case of weak magnets. Thus it appears that a stronger multicusp magnetic field enhances the extracted negative ion current and the ratio I^-/I_e.

A striking feature of the data shown on Fig. 6 is that the enhancement of the multicusp magnetic field has the opposite effect upon the negative ion end electron densities in the target chamber, than upon the extracted negative ion and electron currents. Thus n_- in the target chamber goes down and n_e goes up when the multicusp magnetic field becomes stronger. This behaviour indicates that the enhancement of the multicusp magnetic field leads to a redistribution of the negatively charged particles.

Note on Fig. 6 that n_- and n_-/n_e exhibit a maximum when the magnets are weak (as also shown on Fig. 2). When the magnets are strong, n_- does not exhibit a noticeable maximum, while n_-/n_e only indicates a slight increase with V_b, due to the reduction of n_e. This shows that the observed maximum in I^- (Fig. 6) is not related to a maximum in the density of H^- in the target plasma, but to other causes.

The axial profile of V_p was measured with strong magnets for a 50 V, 2 A discharge produced at 2.7 mTorr, with the grid connected to the anode ($V_b = 0$), and is shown on Fig. 7. The extracted currents

Fig. 7. The axial plasma potential profile in the 'long target' configuration, with three different plasma electrode bias voltages. Note that $V_b = 0.96V$ corresponds to the maximum of the extracted negative ion current. The plasma electrode diameter is 25 cm ($V_g = V_b$). 50 V-2A discharge at 2.7 mTorr.

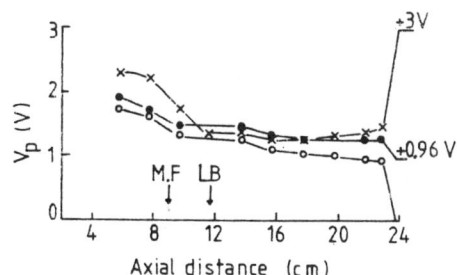

measured in these experimental conditions are shown on Fig. 5 in Ref. 2. With $V_b = 0$ the plasma potential in the driver chamber is only 1.7 V more positive than the anode (instead of +2.7 V with weak magnets, Fig. 4), but is 0.8 V more positive than at the extraction end of the target chamber. When the PE is biased at the optimum value for H^- extraction, which in this case is $V_b = +0.96$ V, the driver is still positive with respect to the target by 0.6 V, while the latter is 0.3 V positive with respect to the PE. For even higher PE bias, the plasma potential increases mainly on the driver side. Thus, in the case of strong magnets the axial plasma potential gradient across the two chambers is not reduced significantly, as it is the case when the magnets are weak. Therefore, the H^- ions from the driver chamber cannot cross the filter into the target chamber. On the other hand, at optimum PE bias for H^- extraction, the driver is 0.9 V positive with respect to the PE. It is not conceivable that the low energy H^- ions would flow under these conditions from the driver into the extractor.

CONCLUSION

The simultaneous measurement of negative ion and electron densities in the target section of the multicusp plasma generator, and of the extracted H$^-$ ion and electron currents gives a better insight into the operation of a tandem multicusp discharge. It is found that the extracted negative ion current can be optimized and the electron current reduced by a suitable choice of the positive bias of the PE. The reduction of the extracted electron current at optimum V_b is a consequence of the reduction of the plasma electron density. However the maximum of I$^-$ at optimum V_b is not always associated with a maximum of n_- in the target section of the plasma: the data in Fig. 6 indicate that n_- exhibits a maximum when the multicusp magnetic field is weak, and does not exhibit a significant maximum when this magnetic field is strong. Furthermore, it appears that the use of strong magnets leads to an increase in the extracted H$^-$ current, but to a reduction of the negative ion density in the target chamber. This leads to the conclusion that the density of H$^-$ ions in the target plasma does not control the extracted negative ion current. This conclusion is supported by the calculation (similar to that reported in Ref. 2) of the thermal flux of negative ions through the extraction slit, using measured values of n_-.

The extracted negative ion current cannot be controlled by the negative ion density in the driver either, since in the case of the strong magnets, at optimum PE bias, the plasma potential in the driver is 0.9 V positive with respect to the PE.

These considerations lead to the conclusion that when V_b and the multicusp magnetic field are varied, the negative ions are redistributed inside the plasma and possibly form a region of maximum H$^-$ density in front of the extraction slit, which effectively controls the extracted H$^-$ current.

ACKNOWLEDGMENTS

We would like to thank Mr. F. Hillion for his technical assistance. This work was supported in part by Ecole Polytechnique and by the Office of Fusion Energy of the U.S. Department of Energy.

REFERENCES

1. K.N. Leung, K.W. Ehlers, and M. Bacal, Rev. Sci. Instrum., 54, 56 (1983).
2. M. Bacal, F. Hillion, M. Nachman and W. Steckelmacher, Proceedings of this Symposium.
3. M. Bacal, A.M. Bruneteau and M. Nachman, J. Appl. Phys. 54, Dec. issue (1983).
4. M. Bacal, Physica Scripta, T 2/2, 467 (1982).
5. K.N. Leung and M. Bacal, Rev. Sci. Instrum., to be published.
6. M. Bacal and A.M. Bruneteau, Proceedings of this Symposium.
7. J.M. Wadehra, Proceedings of this Symposium.

VACUUM ULTRAVIOLET EMISSION AND H⁻ PRODUCTION
IN A LOW PRESSURE HYDROGEN PLASMA

W.G. Graham
Physics Department, The New University of Ulster,
Coleraine, Northern Ireland

ABSTRACT

The vacuum ultraviolet emission from a low pressure hydrogen plasma has been used to study the production mechanism for vibrationally excited ground state molecules, thought to be important in negative hydrogen ion production. The present results are in agreement with theoretical calculations by Hiskes and provide support for previous theoretical modelling of negative ion concentrations in such plasmas.

INTRODUCTION

It has been established[1] that there is a sizeable concentration of H⁻ ions in low pressure hydrogen plasmas, produced by electron bombardment. Recent experimental work[2] on such plasmas has suggested that dissociative attachment of thermal electrons to highly ($v \geq 5$) vibrationally excited hydrogen molecules is the most probable H⁻ production process i.e.

$$H_2(v^*) + e \rightarrow H^- + H \tag{1}$$

Theoretical modelling[3] of these plasmas seems to confirm the suggestion by Kunkel[4] that a principal source of this vibrational excitation is through singlet excitation by fast primary electrons

$$e + H_2(X^1\Sigma_g, v''=0) \rightarrow e + H_2^*(B^1\Sigma_u, C^3\Pi u) \tag{2a}$$

$$H_2^*(B^1\Sigma u, C^3\Pi u) \rightarrow H_2(X^1\Sigma_g, v'') + h\nu. \tag{2b}$$

The vibrationally excited hydrogen molecule population is then modified by thermal electron-molecule, molecule-molecule and molecule-wall interactions, as described by Hiskes et al[3].

This production mechanism for $H_2(v^*)$ can be explored by observing the photon emission associated with process 2. The transitions from the $B^1\Sigma u$ and $C^3\Pi u$ levels to the ground state, $X^1\Sigma g$ of H_2 are termed the Lyman and Werner Bands respectively and are found in the vacuum ultraviolet region.

In the present study the V.U.V. emission from a low pressure hydrogen plasma has been investigated. At the same time plasma parameters such as the positive and negative ion densities have been measured. The experimental results have been compared with theoretical calculations for process 2 by Hiskes[5].

APPARATUS

Since most of the apparatus used in the present experiment has been described in detail previously[6], only a brief description will be presented here.

The plasma was created, in a cylindrical pyrex vacuum vessel 12 cm in diameter and 22 cm high, by electron bombardment of hydrogen gas which was flowed through the system. The plasma parameters were controlled by varying the hydrogen gas pressure, the discharge current and the discharge voltage. A thin tungsten Langmuir probe was used to measure the positive and negative ion densities in the plasma, using a technique proposed by Doucet[7]. When in use the probe was positioned 6 cm from the electron accelerating grid, in the region viewed by the V.U.V. spectrometer. It was withdrawn from this region when light emission measurements were being made.

The only significantly different feature of the present apparatus from that previously described, was the addition of a Hilger Watt 1 m. normal incidence, grating vacuum spectrometer. This spectrometer was connected directly to the vacuum chamber. The spectrometer was normally operated at a resolution of 2.5 nm. The light, after passing through the spectrometer, was detected by measuring the current from a photomultiplier tube used in conjunction with a sodium salicylate scintillator.

RESULTS

A typical spectrum is shown in Figure 1. The spectrum shows the same features which have been observed in hydrogen discharge lamps[8] and in studies of electron collisions with gas phase hydrogen[9].

The extent of the Werner and Lyman band systems, obtained from the identifications of Dieke and Hopfield[10] and Herzberg and Howe[11], are shown in Fig. 1. Also indicated are the regions in which "strong" transitions leading to ground state molecules with $v \geq 5$ are expected. The relative intensity of particular transitions can be determined from the Franck-Condon factors for those transitions. These have been calculated for the Werner and Lyman bands of hydrogen by Spindler[12,13]. In the present case the transition to a particular vibrational ground state is considered to be strong if the Franck-Condon factor for the upward transition $q(o-v') \geq 0.025$ and for the downward transition, $q(v'v'') \geq 0.10$. Using this criterion all "strong" lines resulting in H ($v \geq 5$) occur between 117.5 and 165 nm. The only strong transitions leading to $v \leq 5$ ground states occuring in this region are the 1-3 (125 nm) and 2-2 (118 nm) transitions in the B-X series.

The spectral features as a whole appear unchanged by variation

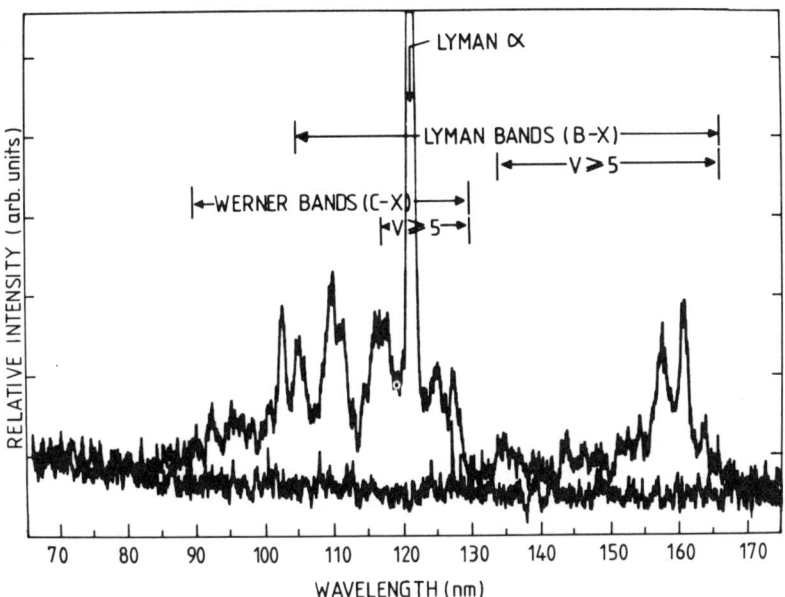

Fig. 1. A typical vacuum ultraviolet spectrum from the present hydrogen plasma. Plasma operating conditions were discharge current 1.5 A, discharge voltage 50 V and hydrogen gas pressure 1×10^{-2} Torr. The positive ion density was 1×10^{-10} cm^{-3} and the negative ion density 5×10^8 cm^{-3}. The extent of the Werner and Lyman bands and the regions in which strong transitions leading to $H_2(v \gtrsim 5)$ are expected are shown.

in the plasma parameters. The intensity of representative features of the C-X series (116 nm) and B-X series (161 nm) and also the Lyman α line of atomic hydrogen (122 nm) were found to vary linearly with the discharge current and also with the hydrogen gas density, up to 1.5×10^{-2} Torr, above which the gas density dependence became less pronounced.

The intensity of the lines at 161 nm and 122 nm were studied as a function of discharge voltage, with a constant discharge current of 0.5 A and a gas pressure of 1×10^{-2} Torr. The dependence followed closely the energy dependence of rate coefficients predicted from Lyman α[14] and $B^1\Sigma u$[15] emission cross section measurements.

DISCUSSION

From the point of view of the present work the most interesting observation is that there is a substantial contribution to the high vibrational levels of the ground state from both series. For example the structure around 157 nm can be identified with the strong B-X transitions (2-9), (7-13) and (8-14) while the structure around 124 nm can be identified with the strong C-X transitions (4-8) and (5-9). The other main features can likewise be identified with other strong transitions to high vibrational levels of the ground state. The resolution is, at present, not good enough to distinguish individual lines.

A measurement of the absolute of the C-X and B-X series emission rate would require a knowledge of the quantum yield and the optical geometry of the detection system. While a direct calibration has not been made, the significance of process 2 in the present plasma can be obtained by comparing the photon emission from the C-X and B-X series with the Lyman α emission.

In order to measure the emission leading to $H_2(v \geq 5)$ the detector current was integrated from 117.5 to 165 nm. The contribution from atomic hydrogen Lyman α emission at 121.6 nm was measured and subtracted from the total, within the resolution of the spectrometer this also included contributions from C-X (1-5), (2-6) and (3-7) transitions. The measurements were repeated many times with the same plasma parameters of discharge current 2A, discharge voltage 50V and hydrogen gas pressure 1×10^{-2} Torr.

The ratio of the $H_2(v \geq 5)$ current to Lyman α current was found to be 2.9 ± 0.3.

Hiskes[5] has calculated the cross sections for the production of each hydrogen ground state vibration level through process 2. Summing the cross section for the production of the ten highest vibrational levels yields the cross section for the production of levels with $v \geq 5$. The cross section for Lyman α emission from electron bombardment of molecular hydrogen has been measured by Mumm and Zipf[14].

The ratio of the calculated $H_2(v \geq 5)$ cross section to the Lyman α cross section at an energy of 50 eV is 2.7 ± 0.4, which is in excellent agreement with the present measurement, confirming experimentally the calculations of Hiskes[5] at 50 eV.

The actual rate of $H_2(v \geq 5)$ production can be _estimated_ by normalizing to the Lyman α rate, which can be found from the known cross section. The current, I, measured at the detector is related to the emission rate, R, by

$$R = C \times \frac{I}{e}$$

where C is a calibration factor for the detection system, and e is the charge on the electron.

For Lyman α emission

$$R = nf(e)n(v=o)\sigma v$$

where σ is the emission cross section, n(v=o) is the hydrogen gas density, v is the primary electron velocity and nf(e) is the density of fast electrons in the plasma. The fast electron density can be deduced indirectly by comparing the positive ion density in the plasma, measured using the Langmuir probe with the positive ion density calculated from rate equations as in Bacal et al[2]. The other parameters are known and so the rate of Lyman α production can be calculated, and hence the calibration factor for the detection system at 121.6 nm can be deduced. This can then be applied to the integrated detector current from 117.5 to 165 nm to obtain an estimate of the production rate of $H_2(v \geq 5)$.

For the plasma conditions described above the rate of $H_2(v \geq 5)$ production through process 2 is found to be $1(+1,-015) \times 10^{14} cm^{-3} s^{-1}$. This is similar to the value used by Hiskes et al[3] in their modelling of negative ion concentrations in medium density hydrogen discharges.

CONCLUSION

The present study confirms that production of highly vibrationally molecules through process 2 is significant in low pressure hydrogen plasmas. The present results are in agreement with the theoretical calculations by Hiskes[5] and provide support for previous theoretical modelling[3] of negative ion concentrations in such plasmas.

REFERENCES

1. M. Bacal and G.W. Hamilton Phys. Rev. Lett. **42**, 1538 (1979).
2. M. Bacal, A.M. Bruneteau, W.G. Graham, G.W. Hamilton, and N. Nachman, J. Appl. Phys. **52**, 1247 (1981).
3. J.R. Hiskes, A.M. Karo, M. Bacal, A.M. Bruneteau and W.G. Graham J. Appl. Phys. **53**, 3469 (1982).
4. W. Kunkel, Private Communication.
5. J.R. Hiskes J. Appl. Phys. **51**, 4592 (1980).
6. W.G. Graham J. Phys. D. (Appl. Phys.) **16**, 1907 (1983).
7. H.J. Doucet Phys. Lett. **33A**, 283 (1970).
8. J.A.R. Samson "Techniques of Vacuum Ultraviolet Spectroscopy" (Wiley, New York, 1967).
9. J.M. Ajello, S.K. Srivastava and Y.L. Yung, Phys. Rev. A**25**, 2485 (1982).
10. G.H. Dieke and J.J. Hopfield Phys. Rev. **30**, 400 (1927).
11. G. Herzberg and L.L. Howe Can. J. Phys. **37**, 636 (1959).
12. R.J. spindler J. Quant. Spectrosc. Radiat. Transfer **9**, 627 (1969).
13. R.J. Spindler J. Quant. Spectrosc. Radiat. Transfer **9**, 597 (1969).
14. M.J. Mumma and E.C. Zipf, J. Chem. Phys. **55** 1661 (1971).

DISSOCIATIVE RECOMBINATION OF $e + H_3^+$.
AN ANALYSIS OF REACTION PRODUCT CHANNELS*

H. H. Michels
R. H. Hobbs
United Technologies Research Center, East Hartford, CT 06108

ABSTRACT

Accurate <u>ab initio</u> calculations of the ground and excited H_3 hypersurfaces have been carried out within a configuration-interaction (CI) framework. These surfaces have been examined for geometries appropriate for an analysis of the reaction products of dissociative-recombination (DR) of $e + H_3^+$. The primary purpose of this work was to examine the various product channels for an energy range of 1-8 eV, which is of interest in the analysis of volume dependent reactions in magnetron and other similar hydrogen ion source devices. Direct dissociative recombination of H_3^+ ions in their ground vibrational state is unlikely for low collisional energies (≤ 5 eV) and indirect capture processes must be examined for this case.

INTRODUCTION

Recent successes in the tokamak program and in other controlled thermonuclear research programs[1] have focused attention on the problem of developing an efficient high-energy particle beam source. For several applications, a neutral beam with energies above 200 keV is desired. The acceleration of negative ions (primarily H^- or D^-) to such energies, followed by neutralization through a stripping reaction, appears at the present time to be the most efficient approach for producing a high energy neutral particle beam.

In another area, the feasibility of particle beams (both charged and neutral) as military weapons has been under study. The proposed endoatmospheric applications require high intensity sources and mainly focus on the problems of beam stability and propagation characteristics. Exoatmospheric applications require lower intensity sources but of very high quality. Design goals are highly collimated beams with a narrow energy spread.

A magnetron-type negative H^- source has been reported by Belchenko, Dimov and Dudnikov (BDD)[2] that has produced H^- current densities of several A cm^{-2}. This device operates as a plasma discharge in an atmosphere of hydrogen gas with cesium or other alkalis present at $\sim$.01 percent. The mechanism for the production of H^- is believed to involve an alkali catalyzed surface reaction whereby H^- ions are produced by backscattering or desorption.[3,4] The detailed kinetic mechanisms of such surface reactions are still uncertain and paramatric experimental studies are currently underway at LANL, LBL,

*Supported in part by AFOSR under Contract F49620-83-C-0094.

IRT and at other laboratories to elucidate the mechanisms and operating characteristics of BDD and similar devices and to develop information for their scale-up to higher current densities. Diagnostics of H⁻ source devices are also underway at Brookhaven[5] using beam probe and spectroscopic techniques.

Concurrent with these surface-plasma reactions are several volume-dependent processes that may lead either to the production of H⁻ or, in reverse, may act as important destructive processes of the negative ions after they are formed. Photodetachment experiments by Bacal and Hamilton[6] in hydrogen plasmas indicate H⁻ densities 100 times larger than that predicted from simple electron attachment mechanisms. Further, these experiments indicate a nonlinear dependence of the production of H⁻ on electron density, at least for densities less than 10^{10} cm⁻³. Several mechanisms[7] have been proposed to explain these volume-dependent H⁻ production processes. A general review of this subject has recently been given by Bacal.[8]

In contrast to surface catalyzed reactions, the mechanisms of most of the known volume-dependent gas phase reactions involving H are relatively simple to analyze, although much remains to be done to identify the state distributions for the products. A major problem has been to identify the most likely reaction kinetics leading to the formation and destruction of H⁻ atoms.

Hiskes[9] has proposed a mechanism for the production of vibrationally excited H_2 and the subsequent production of H⁻ via dissociative-attachment of electrons. The initial step involves electron excitation of ground state H_2 to an electronically excited state:

$$e \text{ (fast)} + H_2 \ [X \ ^1\Sigma_g^+] \rightarrow H_2^* \ [B \ ^1\Sigma_u^+, \ C \ ^1\Pi_u] + e \quad (1)$$

Both of these states have allowed radiative transitions to the ground state of H_2 but the displaced equilibrium separations of these excited states yield vibrationally excited H_2 upon radiative decay according to the most favorable Franck-Condon transitions:

$$H_2^* \ [B \ ^1\Sigma_u^+, \ C \ ^1\Pi_u] \rightarrow H_2^\ddagger \ [X \ ^1\Sigma_g^+, \ v \geq 3] + h\nu \quad (2)$$

Vibrationally excited H_2 has been shown[10,11] to have a greatly enhanced cross-section for electron dissociative-attachment:

$$e \text{ (slow)} + H_2^\ddagger \ [X \ ^1\Sigma_g^+] \rightarrow H \ [^2S] + H^- \ [^1S \ (1s^2)] \quad (3)$$

This proposed mechanism is critically dependent on the magnitude of the overall cross-sections for reactions (1)-(3) and both experimental and theoretical studies are currently underway to examine these processes.

DISSOCIATIVE RECOMBINATION OF $e + H_3^+$

As an alternative mechanism for producing vibrationally excited H_2, we have undertaken an analysis of the branching and product distribution for dissociative recombination of $e + H_3^+$. The reaction scheme is as follows:

$$e + H_3^+ [^1A_1'] \rightarrow H_3 [^2A_1 + {}^2B_2] \rightarrow H_2^{\ddagger} [X \, {}^1\Sigma_g^+] + H$$
$$\rightarrow H_2^+ [X \, {}^2\Sigma_g^+] + H^- \quad (4)$$

Detailed quantum mechanical calculations of these reaction surfaces have been carried out. The H_3 correlation diagram and possible low-lying dissociative recombination paths are indicated in Figs. 1 and 2, respectively. Dissociative recombination of $e + H_3^+$, as shown in reaction (4), is exothermic by 9.3 eV to form $H_2 + H$ and endothermic by 5.4 eV to form the ion pair, $H_2^+ + H^-$. However, the detailed branching of reaction (4) depends critically on the shape of the $^2A_1$ and $^2B_2$ hypersurfaces for H_3 for interatomic separations in the vicinity of 2-3 Å since the coulomb attraction of the ion pair gives rise to a degeneracy in the hypersurfaces of H_3 in this region. We have undertaken a series of <u>ab initio</u> calculations of the potential energy surfaces for H_3 in both C_{2v} and $D_{\infty h}$ symmetries to ascertain the character of the low-lying states. These calculations were carried out using optimized Slater-type orbital (STO) basis functions and a full CI within each symmetry group. The resultant potential energy surfaces are shown in Fig. 3. Based on these results, we find a reaction path for dissociative recombination which involves the electron attachment of $e + H_3^+$ to form a symmetric $^2A_1'$ state of H_3 which is nearly energy resonant for recombination of H_3^+ ions with three of four quanta of vibrational energy:

$$e + H_3^+ (\nu = 3,4) \rightarrow H_3 [^2A_1'] \quad \Delta E \approx 0 \quad (5)$$

This state of H_3 correlates diabatically with the ion pair, $H_2^+ + H^-$, but exhibits many curve-crossings of the lower-lying Rydberg states of H_3. In particular, the lowest-lying recombination pathway yields vibrationally excited H_2 molecules and electronically excited H with n = 2:

$$H_3 [^2A_1'] \xrightarrow{\text{curve crossing}} H_2^{\ddagger} [X \, {}^1\Sigma_g^+] + H^* (n = 2) \quad (6)$$

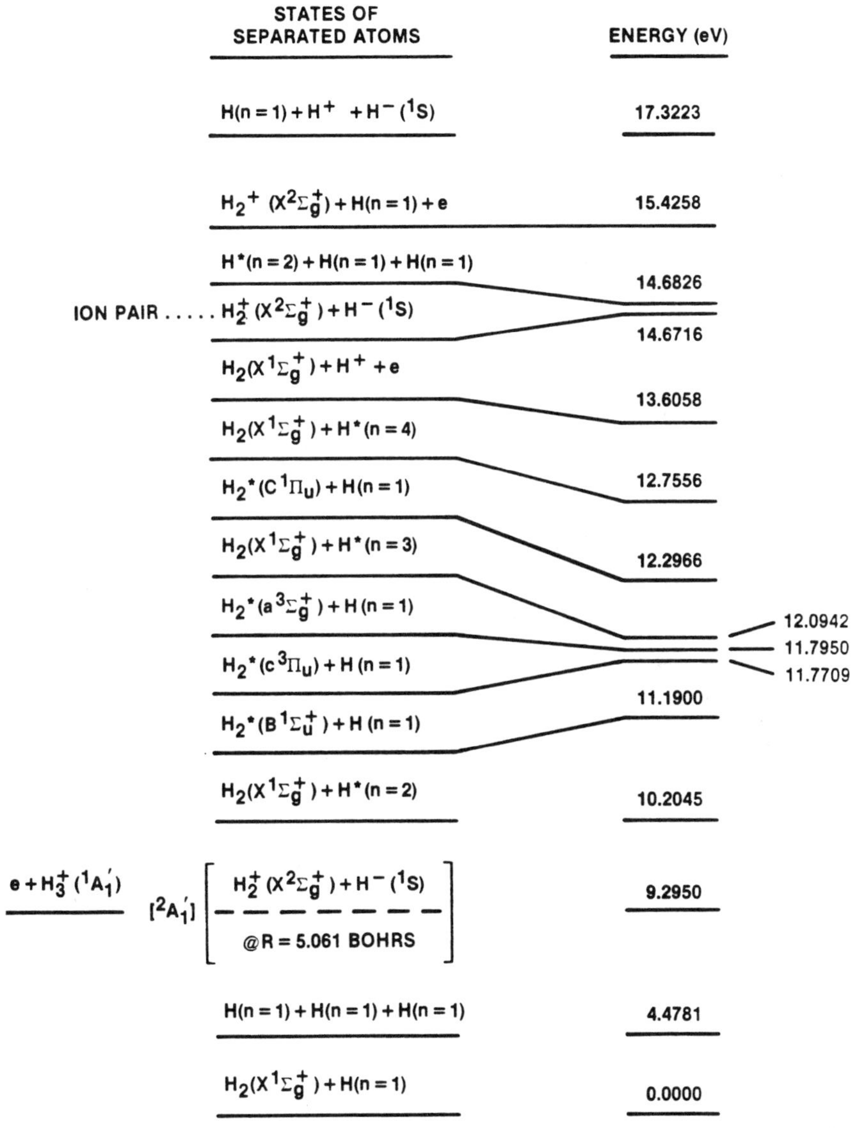

STATES OF SEPARATED ATOMS	ENERGY (eV)
$H(n=1) + H^+ + H^-(^1S)$	17.3223
$H_2^+ (X\,^2\Sigma_g^+) + H(n=1) + e$	15.4258
$H^*(n=2) + H(n=1) + H(n=1)$	14.6826
ION PAIR ..... $H_2^+ (X\,^2\Sigma_g^+) + H^-(^1S)$	14.6716
$H_2(X\,^1\Sigma_g^+) + H^+ + e$	13.6058
$H_2(X\,^1\Sigma_g^+) + H^*(n=4)$	
$H_2^*(C\,^1\Pi_u) + H(n=1)$	12.7556
$H_2(X\,^1\Sigma_g^+) + H^*(n=3)$	12.2966
$H_2^*(a\,^3\Sigma_g^+) + H(n=1)$	12.0942
$H_2^*(c\,^3\Pi_u) + H(n=1)$	11.7950
	11.7709
$H_2^*(B\,^1\Sigma_u^+) + H(n=1)$	11.1900
$H_2(X\,^1\Sigma_g^+) + H^*(n=2)$	10.2045
$e + H_3^+(^1A_1')$ $[^2A_1']$ $\left[\begin{array}{c} H_2^+(X\,^2\Sigma_g^+) + H^-(^1S) \\ ------ \\ @R = 5.061\ BOHRS \end{array}\right]$	9.2950
$H(n=1) + H(n=1) + H(n=1)$	4.4781
$H_2(X\,^1\Sigma_g^+) + H(n=1)$	0.0000

Fig. 1. H_3 Correlation Diagram.

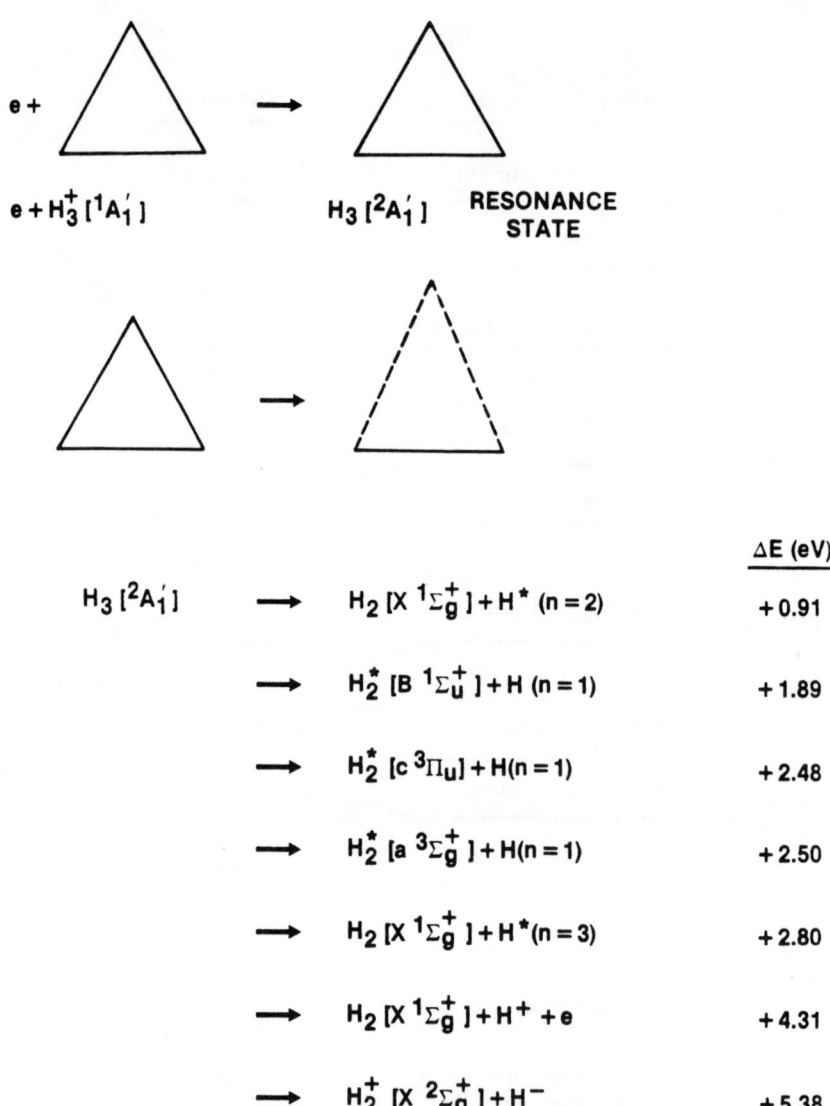

Fig. 2. Dissociative Recombination Reaction Paths.

The overall recombination pathway is illustrated in Fig. 4. We find that the overall cross-section for this route to $H_2^{\ddagger}$ (v=1) is $10^{15} - 10^{14}$ cm^2, and thus the rate of reaction (3) alone becomes rate determining.

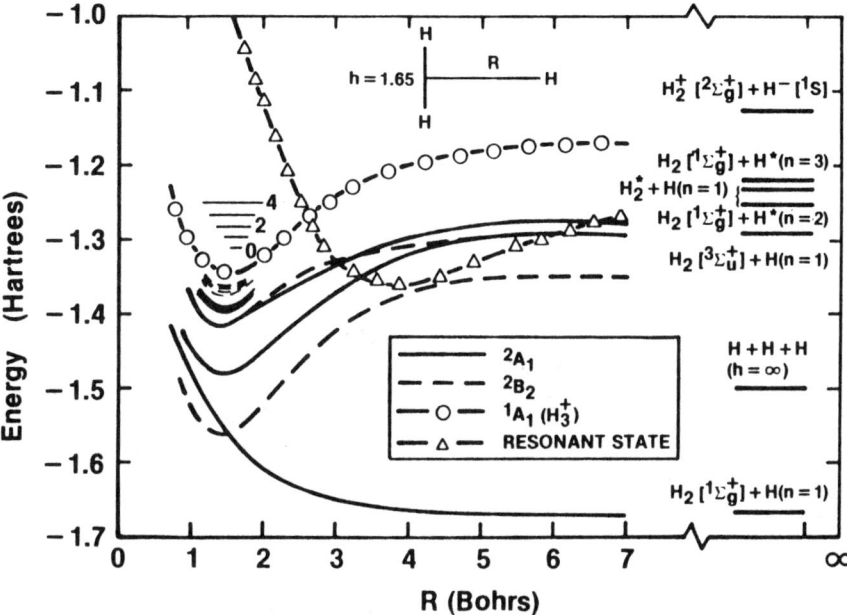

Fig. 3. Potential Energy Curves for H$_3$.

$$e + H_3^+ (\nu) \rightarrow H_3 \rightarrow H_2[X\ ^1\Sigma_g^+ (\nu)] + H^*(n=2) \qquad \Delta E\ (eV)$$

ν	ν	ΔE (eV)
0	0	+0.91
1	0	+0.52
2	0	+0.13
3	0	−0.26
4	0	−0.64
4	1	−0.12

$$H_2^+ + H_2 \rightarrow H_3^+ (\nu = 0) + H \qquad \Delta E = -1.73\ eV$$
$$ \rightarrow H_3^+ (\nu = 4) + H \qquad \Delta E = -0.17\ eV$$

Fig. 4. e + H$_3^+$ Recombination Energetics.

CONCLUSION

This study indicates that dissociative recombination of $e+H_3^+(v)$ can efficiently produce H_2 molecules with low vibrational quanta. The results in C_{2v} symmetry, as shown in Fig. 3, indicate that direct recombination occurs through the $^2A_1$ resonance state which connects diabatically to $H_2^+ + H^-$. This resonance state lies ~5.4 eV above H_3^+ in its ground vibrational state but couples effectively with H_3^+ with three quanta of vibrational energy. Under these conditions the predicted products are $H^*(n=2) + H_2$ ($X^1\Sigma_g^+$), where any excess vibrational energy in the H_3^+ ions is transferred to the product H_2 molecule.

The formation of H_3^+ via the reaction, $H_2^+ + H_2 \rightarrow H_3^+ + H$, is exothermic for H_3^+ ($v \leq 4$). Thus, a likely recombination pathway is $e + H_3^+$ ($v=3,4$) $\rightarrow H_3$ ($^2A_1$ resonance) $\rightarrow H_2^\ddagger$ ($v=0,1$) $+ H^*(n=2)$. Direct dissociative recombination of H_3^+ ions in the ground vibrational state is unlikely for low collisional energies (≤ 6 eV) and indirect capture processes must be examined for this case. A similar result has been found for this system by Kulander and Guest.[12]

REFERENCES

1. Proceedings of the Eighth Symposium of Engineering Problems of Fusion Research, San Francisco, CA, November 1979.
2. Y. I. Belchenko, G. I. Dimov and V. G. Dudnikov, Nuc. Fus. **14**, 113 (1974).
3. G. I. Dimov, Proceedings of the Second Symposium on Ion Sources and Formation of Ion Beams, Berkeley, CA, 1974.
4. J. R. Hiskes, A. Karo and M. Gardner, J. Appl. Phys. **47**, 3888 (1976).
5. M. W. Grossman, Proceedings of the Symposium on the Production and Neutralization of Negative Hydrogen Ions and Beams, Brookhaven National Laboratory, New York, 1977.
6. M. Bacal and G. W. Hamilton, Phys. Rev. Lett. **42**, 1538 (1979).
7. E. Nicolopoulou, M. Bacal and H. J. Doucet, J. Phys. (France), **38**, 1399 (1977).
8. M. Bacal, Phys. Scripta, **T2/2**, 467 (1982).
9. J. R. Hiskes, J. Appl. Phys. **51**, 4592 (1980).
10. J. M. Wadhera and J. N. Bardsley, Phys. Rev. Lett. **41**, 1795 (1978).
11. M. Allan and S. F. Wong, Phys. Rev. Lett. **41**, 1791 (1978).
12. K. C. Kulander and M. F. Guest, J. Phys. B: At. Mol. Phys. **12**, L501 (1979).

GENERATION OF VIBRATIONALLY EXCITED H_2 MOLECULES BY H_2^+ WALL COLLISIONS

J. R. Hiskes and A. M. Karo
Lawrence Livermore National Laboratory
Livermore, California 94550

ABSTRACT

The H_2^+ ions from the volume of a hydrogen discharge will strike the discharge chamber walls with a kinetic energy equivalent to the plasma potential. A three-step process is described in which the H_2^+ ions are neutralized in a two-stage Auger process followed by a third stage wall relaxation collision, with the net result that the incident ions are converted to ground state molecules having a broad vibrational excitation spectrum. For kinetic energies ranging from a few electron volts up to twenty electron volts a substantial fraction, $\simeq 2/3$, of these ions will reflect as molecules, and of this population a fraction as large as twenty percent will have vibrational excitation $v'' \geq 6$. This large vibrational population will provide a contribution to the total excited level distribution that is comparable to the E-V process. Implications for negative ion generation in an optimized tandem configuration are discussed.

INTRODUCTION

In the first of our companion papers[1] presented at this Symposium we have discussed the negative ion formation by dissociative attachment[2] to vibrationally excited molecules in a high density hydrogen discharge. The vibrational excitation was presumed to be generated by high-energy electron collisions, the E-V process, in the first chamber of a tandem system. Associated with the ionization of the discharge by the high energy electrons is the formation of H_2^+ molecular ions whose equilibrium concentration in a high-power, high-density discharge is typically twenty to thirty percent of the total electron density.[3] RF source experience has further shown, however, that under certain conditions the H_2^+ concentration can be as large as 80 percent of the electron density.[4]

The discharge normally assumes a positive potential with respect to the confining discharge walls and whose value is some multiple of the electron temperature. For the high-density discharges of interest here the plasma potential will nominally be in the range from a few volts up to about 20 volts. The molecular ions in the discharge are accelerated across this potential and strike the wall with the equivalent kinetic energy causing the formation of both vibrationally excited molecules and the dissociation of the H_2^+ ions into free atoms. The vibrationally excited molecules generated in this way will be an additive contribution to the fore-mentioned E-V excitation, and will contribute to the subsequent formation of negative ions.

Although this vibrational excitation mechanism has been recognized previously, only a qualitative description of the process was possible due to the lack of quantitative data for the wall excitation and dissociation processes.[5,6] In the second of our papers presented at this Symposium,[7] the vibrational relaxation of the diatomic system in wall collisions has been studied in sufficient detail that a quantitative description of the contribution of H_2^+ to negative ion formation is now possible.

THE NEUTRALIZATION PROCESS

When the H_2^+ ion approaches to within 10Å of the discharge wall, Auger neutralization occurs producing a vibrationally excited, $H_2(v")$, molecule in the ground electronic state. The analogue for H_2^+ Auger neutralization is the neutralization of the He^+ ion in wall collisions, a process that has been studied extensively both theoretically[8] and experimentally.[9,10] The principal neutralization mechansim for He^+ ions is a two-step Auger process wherein an electron is first captured into the 2s-level of He at a distance of 5-10 Å from the surface. The metastable 2^3s helium atom formed in this first capture then continues to drift toward the surface. At a distance of approximately 3-5 Å from the surface, a second electron is captured directly into the He ground state orbital and the 2s electron that was initially captured is ejected from the atom. The helium atom, upon reaching the surface, is now in its ground electronic configuration.

Since the electronic orbital configurations of He^+ and H_2^+ are identical, we shall postulate that the neutralization of H_2^+ proceeds by a similar two-step process. We first take note of the fact that the energies of the electronic states of H_2 are displaced upwards an amount

$$+ \frac{1}{4} \frac{e^2}{z} \, , \qquad (1)$$

as a consequence of the image forces acting on the electron. The distance is the separation of center-of-mass of the H_2^+ ion from the image plane. At close separation from the image plane the magnitude of the image shift is bounded,[11] and for purposes of discussion we shall take this upper bound to be 1.5 electron volts. The work function of most simple metals is 4.5 electron volts or more. As a consequence, only one of the excited electronic states of H_2, namely the first excited $^3\Sigma_u$ state, is accessible for Auger capture in the first step of the capture process.

The H_2^+ ions in the discharge are formed in a broad spectrum of vibrational levels with the majority population distributed over the lowest five levels.[12] As a consequence the H_2 ($^3\Sigma_u$) will in turn be formed over a rather broad range of internuclear separations, R_N, ranging from less than 0.5 Å to more than 1.2 Å. Because of the repulsive shape of the $^3\Sigma_u$ potential, the two nuclei of the excited molecule will begin to separate as the molecule continues to drift inwards toward the surface. After moving inward an amount 2-5 Å, second electron capture directly to the ground state occurs, ejecting the $^3\Sigma_u$ orbital electron. At this point however, the molecule has expanded in size (larger R_N) and the ground electronic state is formed in a high, v" > 8, level.

There is in fact an optimum "matching time," or "matching velocity" of the incident H_2^+ ion such that the initially formed H_2 ($^3\Sigma_u$) state will separate along R_N by an amount equal to ΔR, which is just the proper amount to form a high v" level when the second electron capture occurs. If Δz is the drift distance separating the two points of electron capture, v(2+) the velocity of the incoming H_2^+ ion, v the mean "rollout velocity" in the $H_2(^3\Sigma_u)$ configuration, the optimum "matching velocity" is given by

$$v(2+) = v \frac{\Delta z}{\Delta R} \quad . \tag{2}$$

Because the initial distribution in R_N will be rather broad and because the point of separation between the first and second electron capture is not sharply defined, the optimum v(2+) will have a rather broad range of values for high v" formation. Inspection of the H_2^+, $H_2(^3\Sigma_u)$ potentials indicates that the optimum formation of high, v" > 8, level formation will occur for incident H_2^+ energies ranging from a few electron volts up to twenty electron volts.

The H_2(v" > 8) formed in the two-step capture process continues to drift toward the wall eventually colliding with it. According to the results of our calculations discussed in Ref. 7, two processes predominate: vibrationally excited molecules, H_2(v"), rebound from the surface, now with a broad spectrum of vibrational levels, or, dissociation of the H_2 occurs. The fraction of incident molecules that survive the wall collisions as bound diatomic systems is denoted by f_1 and listed in Table I for several initial H_2^+ ion energies. Of interest for negative ion production is the fraction of f_1 that is formed with v" $\geq$ 6. We denote this fraction by f_2; the fraction of incident H_2^+ that survive as H_2(v" $\geq$ 6) is then $f_1 f_2$.

In order to evaluate the contribution of the three-step neutralization process we shall need to take some kind of average over levels v" since a relatively broad portion of the upper spectrum will be populated at the time of the second electron capture. In Table I is listed the fractions f_1, f_2 for incident levels v" = 2, 8, and 12, and for several incident H_2^+ ion energies.

	TABLE I			
E	v"	f_1	f_2	$f_1 f_2$
1 eV	12	0.68	0.22	0.15
4 eV	2	0.68	0.21	0.14
	8	0.45	0.59	0.27
	12	0.66	0.33	0.22
10 eV	2	0.50	0.20	0.10
	12	0.50	0.40	0.20

IMPLICATIONS FOR NEGATIVE ION FORMATION

The production of negative ions in the tandem discharge is proportional to the population of $H_2(v" \geq 6)$ generated in the first chamber. In Ref. 1 we have identified the electron and gas densities for optimum negative ion formation. To examine the consequences of H_2^+ wall neutralization, we compare this rate of $H_2(v" \geq 6)$ formation with the E-V rate used in Ref. 1. The rate for H_2^+ neutralization is given by

$$0.25 \, cnv(2+) \, f_1 f_2 \, A/V \, , \qquad (3)$$

where c is the sheath factor taken equal to 0.60, v(2+) is the velocity of H_2^+ ions moving toward the sheath, $v(2+) = (kT/M)^{1/2}$, and n the electron density in the discharge. The ratio A/V is the surface to volume ratio and is equal to 3/R, where R is the system scale length. Here we have chosen the density of H_2^+ to be equal to one-fourth the electron density.

The rate for $H_2(v'' \geq 6)$ formation by the E-V process is

$$n(f)N_2 \sum_{v''=6}^{14} \sigma(E-V, v'') v \equiv n(f)N_2 S \qquad (4)$$

Here, $n(f)$ is the fast electron ($E > 30$ eV) density, N_2 the gas density, and S the sum of E-V rates.[13] The ratio of Eqs. 3 and 4 is then

$$12 \times 10^{14} f_1 f_2 / N_2 R \quad , \qquad (5)$$

where we have taken $kT = 5$ eV, $S = 8 \times 10^{-9}$, and assumed the fast electron density contribution is one-tenth the total electron density. In Ref. 1 we found that the optimum value for N_2R ranged from 2 to 10×10^{14} molecules cm^{-2} as the wall recombination coefficient γ ranged from 0.1 to 1.0. Taking the product $f_1 f_2$ to be 0.15, the ratio (5) is near unity for the lower gas densities. It follows that the wall recombination of H_2^+ to form $H_2(v'' \geq 6)$ can make an important contribution to the total vibrational excitation and negative ion yield, comparable to that of the E-V process for the lower gas pressures. If the H_2^+ concentration can be enhanced compared to the positive ion species, the wall recombination process may dominate.

THE H_2^- STATES

The two lowest H_2^- states, $^2\Sigma_u$ and $^2\Sigma_g$, lie a few volts above the H_2 $^1\Sigma_g$ ground state for internuclear separations, R_N, near the potential minimum and intersect the $^1\Sigma_g$ state near $R_N = 2$. In the presence of the wall the image potential will lower the energy of the negative ion states an amount $- 1/4\ e^2/z$, and raise the $^1\Sigma_g^+$ state an amount $+ 1/4\ e^2/z$. If the H_2 $^1\Sigma_g$ state has been formed in the two-electron capture process discussed in the previous section, the question remains as to whether or not the H_2^- state can form by capture of an additional wall electron with subsequent direct dissociation to H^-. Again we shall assume the maximum image shifts, IM, amount to 1.5 eV.

At a nuclear separation R_N equal to one Å and for the maximum image potential the $^1\Sigma_g$ state will lie approximately one electron volt above the $^2\Sigma_u$ state but 3.5 volts below the $^2\Sigma_g$ state. An electron bound to the metal with an energy greater than or equal to a work function of $\phi_2 = 4.5$ electron volts cannot make a transition to the higher lying $^2\Sigma_u$, $^2\Sigma_g$ states.

At larger separations the energy difference between the metal-bound electron and negative ion states is narrowed. Asymptotically the two negative ion states merge and differ from the molecular ground state by twice the image potential plus the H⁻ affinity, A, and, thus, by an amount equal to 3.0 + .75 = 3.75 eV, a difference still too small to overcome the image potential. In general H_2^- formation is energetically not allowed unless,

$$2IM + A > \phi \quad .$$

This condition can be achieved with composite alkali/transition-metal minimum-work-function surfaces but not with simple metal surfaces.

Finally, simultaneous electron capture of two electrons to form H_2^- $^2\Sigma_u$ directly from H_2^+ is energetically possible. The rate of double capture would be expected to be small compared to the single capture rate. To compete with the single capture the double capture must occur when the diatomic system is still at relatively small R_N. The H_2^- ions formed at small R_N are highly susceptible to electron shakeoff rather than passing through the dissociation mode. This double capture process would not be expected to be competitive with the process described in the section entitled: Implications For Negative Ion Formation.

REFERENCES

1. J. R. Hiskes and A. M. Karo, "Volume Generation of Negative Ions In High Density Hydrogen Discharges," Proceedings of this Symposium, 1983, and UCRL-89433, Rev. 1 (1983).

2. J. M. Wadehra, Appl. Phys. Lett. 35, 917 (1979).

3. C. F. Chan, C. F. Burrell, and W. S. Cooper, J. Appl. Phys. February (1984).

4. K. W. Ehlers and K. N. Leung, Rev. Sci. Instr. 54, 677 (1983).

5. J. R. Hiskes, Journ. de Physique (Paris) 40, Colloque C7, Suppl. 7, C7-179 (1979).

6. J. R. Hiskes, M. Bacal, and G. W. Hamilton, "Atomic Reaction Rates In H⁻ and D⁻ Plasmas," Lawrence Livermore Laboratory Report UCID-18031, January 1979.

7. A. M. Karo, J. R. Hiskes, T. DeBoni, K. D. Olwell and R. J. Hardy, "De-Excitation and Equipartition in H_2^- Surface Collisions, "Proceedings of this Symposium, 1983 (UCRL-89586).

8. A. Cobas and W. E. Lamb, Phys. Rev. 65, 327 (1944).

REFERENCES (Continued)

9. R. F. Stebbings, Proc. Roy. Soc. $\underline{A241}$, 270 (1957).
10. H. D. Hagstrum, Phys. Rev. $\underline{89}$, 244 (1953).
11. N. D. Lang and W. Kohn, Phys. Rev. $\underline{B3}$, 1215 (1971).
12. D. Villarejo, J. Chem. Phys. $\underline{49}$, 2523 (1968).
13. J. R. Hiskes, J. Appl. Phys. $\underline{51}$ (9) (1980).

This work was performed under the auspices of the U.S. Department of Energy by Lawrence Livermore National Laboratory under contract No. W-7405-Eng-48.

GENERATION OF VIBRATIONALLY EXCITED HYDROGEN FOR USE IN A NEGATIVE ION SOURCE

R. J. Turnbull, S. R. Walther & J. L. Guttman*
Department of Electrical Engineering
University of Illinois
Urbana, Illinois 61801

ABSTRACT

The production of negative hydrogen ions in a discharge is greatly enhanced if the hydrogen is vibrationally excited. In the work presented here is a study of a method of producing vibrationally excited hydrogen. The technique used is to heat dense hydrogen hot enough to produce vibrational excitation and then allow it to expand thus cooling it while maintaining the vibrational excitation. Both theoretical calculations and experimental results on this technique are presented.

INTRODUCTION

The generation of negative hydrogen (or deuterium) ions in a volume process is due mostly to the dissociative attachment process. In this process an electron impacts a molecule producing a negative ion and a neutral atom. In hydrogen the cross section for this process increases by five orders of magnitude as the molecule is raised from the ground state to the fifth vibrationally excited state. In addition the peak of the cross-section occurs at a lower energy thus allowing a larger fraction of the electrons in a discharge to undergo the reaction. If the molecular hydrogen in a discharge is vibrationally excited the production of negative hydrogen can be greatly enhanced. (In deuterium the effect of vibrational excitation on the cross-section is even more pronounced.)

This research is designed to test if negative hydrogen ion production can be enhanced by producing vibrationally excited hydrogen and flowing it through an electrical discharge. In the next section the proposed source is described. Then predictions from theoretical calculations are presented. Finally, results from two experiments designed to produce vibrationally excited hydrogen are presented.

PRODUCTION TECHNIQUE

The negative ion source envisioned in this work is a three-stage device as shown in Figure 1. In the first stage dense hydrogen is heated to a temperature such that an appreciable fraction of it is vibrationally excited. In the second stage the hydrogen gas expands

*Current Address: Lockheed Palo Alto Research
3251 Hanover Street, Palo Alto, CA 94394

and cools while maintaining its vibrationally excited state and the final stage is a discharge through which the gas flows producing negative hydrogen.

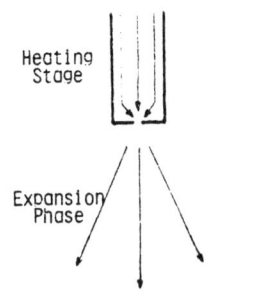

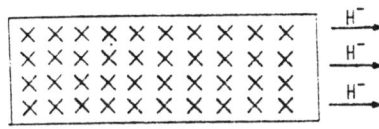

H⁻ producing discharge

Figure 1. Negative Ion Source Using Pre-Discharge Vibrational Excitation of the Hydrogen

In order to produce vibrationally excited hydrogen by heating, it is necessary to use a dense gas because at the temperatures needed to produce a high excited fraction most of the hydrogen will be dissociated if the gas is not dense (several atmospheres or better). The excited and dissociated fraction can be calculated from equilibrium equations since equilibrium is established rapidly in a dense medium. The heating can be done either by an electrical discharge in the gas as shown in Figure 1 or by a laser on a small solid hydrogen sample. Both techniques will be discussed in later sections of this paper.

The expansion will be flow out of a nozzle if the heating is done by a discharge or free expansion if the heating done by a laser on a solid. As the gas expands it cools with the degree of cooling calculated using isentropic expansion equations. If the mass flow rate is not too large the time needed for the vibrational energy to be converted to translational energy is long compared to the flow time and the molecules remain excited. Calculations in the next section will illustrate this. The negative ion forming discharge stage will be treated in a later paper.

THEORETICAL CALCULATIONS

Calculations were done on the flow expansion to determine the resulting level of vibrational excitation and the resulting negative ion formation rate. A hot dense gas in thermal equilibrium is allowed to expand. As it expands collisions between molecules result in vibrational-to-translational energy transfers as well as vibrational-vibrational energy transfers. Because of the anharmonic nature of the vibrational energy levels and because the higher levels (4-6) are extremely important in negative ion production, it was necessary to calculate the population of each level separately rather than assume a vibrational temperature. The resulting populations of the vibrational levels were calculated using a Runge-Kutta technique and the results were shown to depend on the initial temperature

(T_0) and the product of the initial density and the radius of the nozzle from which the gas emerged(nr). Results are shown in Figure 2. In Fig.(2a) the initial temperature was 3000 K and in Fig.(2b) the initial temperature was 5000 K. The equilibrium populations are shown as well as the populations after the expansion takes place. The density times radius units are in cm^{-2}. From this graph it is seen that the higher levels tend to depopulate more than the lower levels. A density radius product of about 10^{18} cm^{-2} is about the upper limit which preserves most of the vibrational excitation. Above that most of the vibrational energy is converted to translational energy.

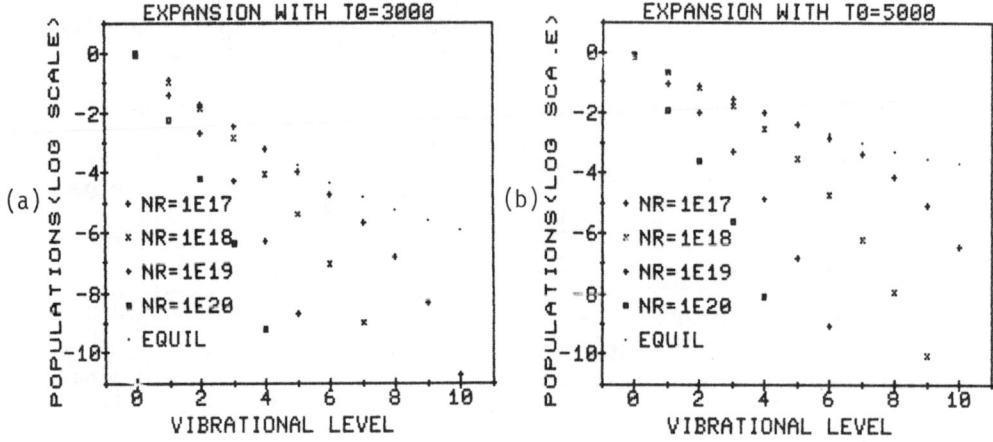

Figure 2. Vibrational State Populations Resulting from Expansion of a Heated Gas. (a) Initial Temperature - 3000 K; (b) Initial Temperature - 5000 K.

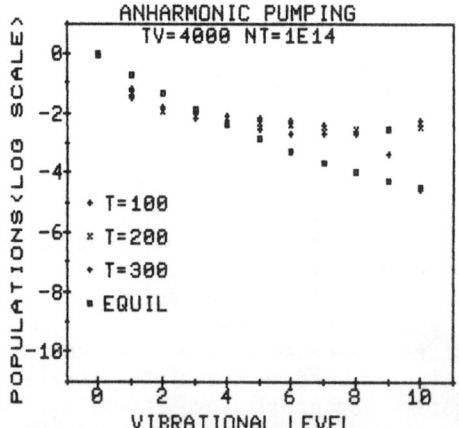

Figure 3. Enhancement of Higher Vibrational State Population by Anharmonic Pumping

An effect which does not appear to be important in the expansion is anharmonic pumping. This phenomena occurs in a cool gas with a large vibrational energy. Because of the anharmonic nature of the vibrational levels the energy tends to go into the higher states populating them higher than would be expected if the levels were in equilibrium. Calculations were also done on this effect assume a starting point a gas with a vibrational equilibrium at a much higher temperature than the translational temperature. The density was assumed constant and some results are shown in Figure

3 with an initial vibrational temperature of 4000 K and translational temperatures of a few hundred degrees. (The results are similar to those of Reference 1 which uses this effect in a proposed source.) A product of density and time (nt) equal to 10^{14} cm^{-3}-sec was needed to see an appreciable effect. As translational temperatures are raised anharmonic pumping becomes less important.

In order to determine if negative ion production can be enhanced by this technique, the vibrational populations calculated from the expansion were assumed to go into a discharge of 40 Townsend (Tn) and the resulting negative ion production rates were calculated. The electron distributed was calculated using the upflux method of Reference 2. This is shown in Figure 4 for various initial temperatures and density-radius products. An nr product of 10^{17} cm^{-2} produces results close to those of equilibrium vibration populations at the given temperature while an nr of 10^{20} cm^{-2} gives results similar to unexcited H_2. Enhancement of H$^-$ production rates of 100-1000 are predicted here. For the purposes of comparison Figure 5 is presented. It shows H$^-$ production rates which result taking into account the vibrational excitation which takes place in a discharge. It shows that a product of electron density times time of ~10^8 cm^3-sec is necessary to reach the same H$^-$ production rates as the heated expansion.

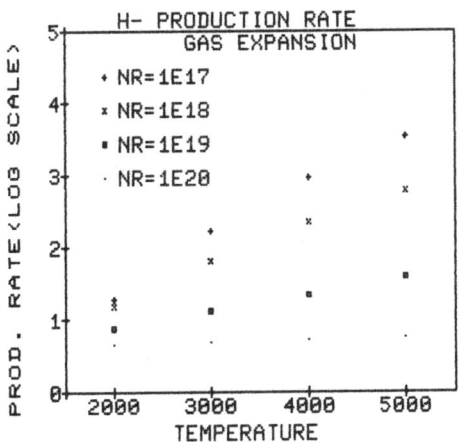
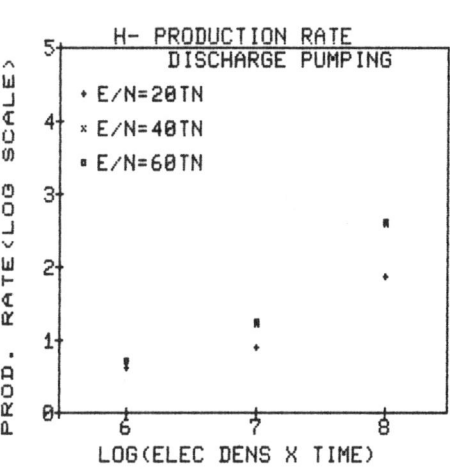

Figure 4. Negative Hydrogen Ion Production Rates for Vibrational Excited Hydrogen Produced by the Expansion Process

Figure 5. Negative Hydrogen Ion Production Rate from Discharge Pumping of Hydrogen Vibrational States

EXPERIMENT USING A DISCHARGE TO HEAT THE GAS

Experimental work on vibrationally exciting hydrogen molecules has been done using a pulsed arc discharge. The experimental apparatus is shown in Figure 6. The electronic fast gas valve pulses hydrogen gas into the small arc chamber. The presence of the hydrogen gas between the needle and nozzle electrodes initiates the arc discharge. The discharge heats the gas which then exits through the nozzle and expands into a vacuum region. The fast expansion of the gas is important to limit molecular collisions and consequently retain vibrational energy.

The important parameters to vibrational excitation are gas density and gas temperature in the discharge region. The transient pressure produced by the gas pulse was measured using a piezoelectric crystal as part of the arc chamber. With a known pressure and temperature, gas density in the arc region can be calculated. A desired gas density is obtainable by changing the back pressure on the gas valve. A linear relationship was found between valve back pressure and resulting gas density in the arc. The gas temperature in the arc was calculated using time of flight studies, since the heating temperature is related to the final directed velocity (V_m) of the gas molecule by gas dynamic equations. Since V_m is reached very quickly after the gas exits the nozzle, the velocity calculated from the time of flight is used as V_m. The time of fight for the gas was measured using a second sustained discharge downstream from the nozzle exit. The second discharge current is dependent upon gas pressure between the discharge electrodes, thus the arrival of gas from the arc discharge is detected by a sharp increase in second discharge current. Time of flight is measured as the time delay between initiation of the arc discharge and detection of the gas at the second discharge. With this measurement gas temperature in the arc can be calculated. Arc energy, and thus gas temperature in the arc, is controlled by the arc capacitor's voltage and capacitance. Figure 7 shows the resulting arc temperatures for various arc voltages and capacitances. Hence both gas temperature and gas density in the arc discharge can be controlled. This allows optimum experimental conditions for vibrational excitation to be achieved. The temperatures reached are in the range desired for this source. Details of these experiments are in Reference 3.

EXPERIMENT USING A LASER TO HEAT A SOLID

Another method studied for the first stage heating was to form a thin slab of solid hydrogen which then was heated by a laser. The heated hydrogen vaporized and then expanded into the vacuum surrounding it where it cooled and hopefully maintained its vibrational excitation.

The thin slab of solid hydrogen was mounted on a metal substrate and laser light with power density $10^7 - 10^8$ W/cm^2 was used to heat the target. The hydrogen could not be heated directly because the laser power necessary would heat the hydrogen to much too high a temperature. The laser light passed through the hydrogen and was absorbed by the metal substrate which in turn transferred its energy to the solid hydrogen. Hydrogen is ablated off the target at the hot metal surface and the resulting gas is heated by conduction. At some point the pressure in the gas becomes large enough so that it blows a hole in the solid hydrogen. The hot gas then flows out the hole and expands into the surrounding vacuum.

Figure 8 contains framing pictures of the results of laser

137

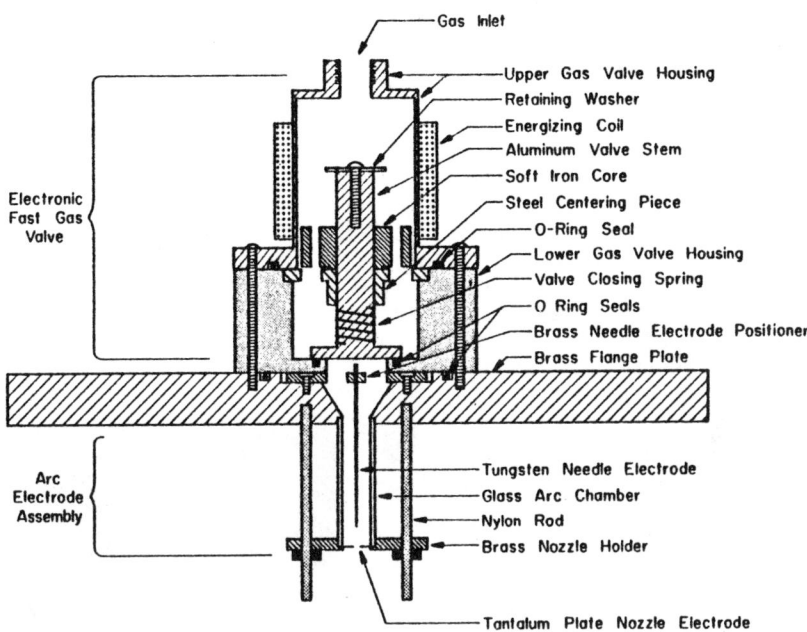

Figure 6. Experimental Apparatus for Discharge Heating of Hydrogen.

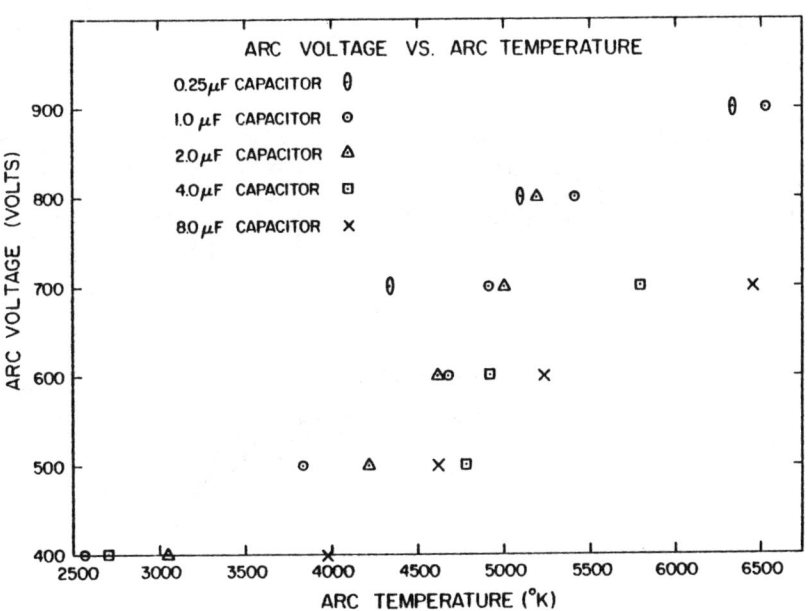

Figure 7. Gas Temperature Resulting from Discharge Heated Hydrogen.

heating of a solid hydrogen slab. In these photographs one can clearly see the hydrogen plasma expanding away from the substrate. This expanding plasma is first seen in the photographs taken at t = 0. Since the exposure time here is 200 ns, all that can be said is that the hydrogen started expanding sometime in the first 200 ns. The velocity of propagation of the luminous boundary is of the order of 2×10^3 m/s. It also appears that the luminous plasma is surrounded by the expanding cooler hydrogen. The hot gas escaping from the hole in pellet was estimated to be in the 5000-10000° K based on the velocity of expansion. In order to achieve a large degree of vibrational excitation, it will be necessary to reduce the amount of mass being heated. These experiments were done using a 250 μm thick target and calculations based on the results indicate the optimum thickness to be about 50 μm. Details of the experiments can be found in Reference 4.

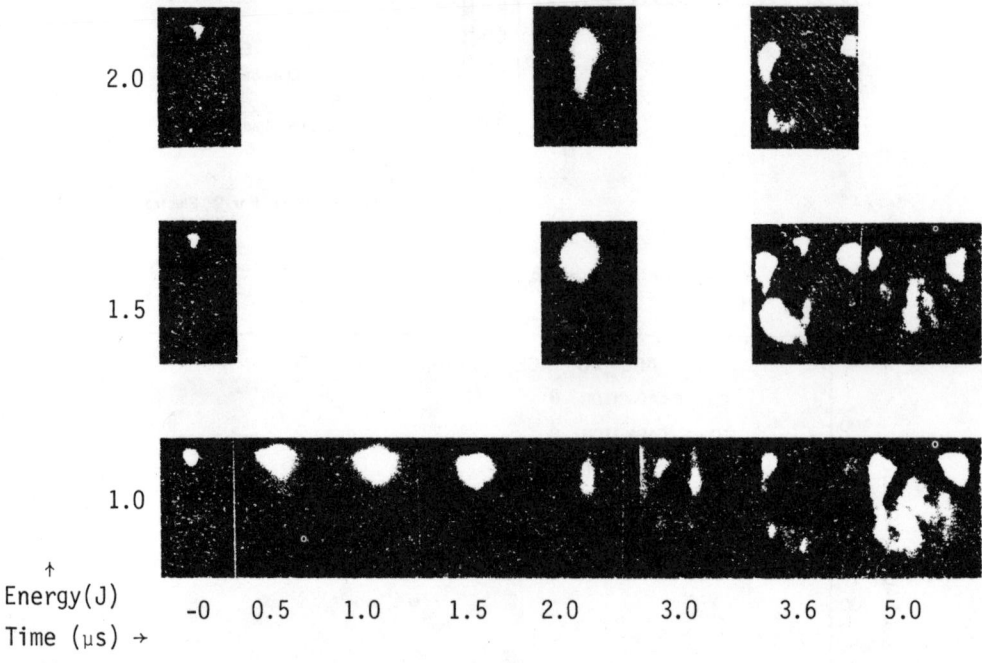

Figure 8. Framing Pictures Taken at Various Times and Energies for Laser Heating of Solid Hydrogen on a Stainless Steel Substrate

ACKNOWLEDGMENT

This work was supported by the Air Force Office of Scientific Research under Grant 81-0160.

REFERENCES

1) A. Garscadden and W. F. Bailey, Progress in Astronautics and

Aeronautics, Vol. 74, p. 1125, 1980.

2) W. P. Allis and H. A. Haus, <u>J. Appl. Phys.</u>, <u>45</u>, 781 (1974).

3) J. L. Guttman, Ph.D. Thesis, Department of Electrical Engineering, University of Illinois, Urbana, 1982.

4) S. R. Walther, M.S. Thesis, Department of Electrical Engineering, University of Illinois, Urbana, 1983.

INTERFERENCE EFFECTS IN NEGATIVE ION FORMATION

I. Alvarez, A. Morales, J. de Urquijo and C. Cisneros

Instituto de Física, UNAM
México, D.F. 01000

ABSTRACT

This paper presents recent data on differential cross sections for H^- formation from collisions of H^+ and $H°$ with Ar in the energy range 1.0 to 4 keV. Experimental data exhibit a sharp maximum at 0° scattering angle as well as an oscillatory structure. The functional form and scaling properties strongly indicate that there is a glory maximum which occurs when the classical deflection function changes over from attractive to repulsive at some finite impact parameter. The oscillations are predicted by the Bessel function and may be said to arise from interference of the contributions from the two branches of the deflection function near to a glory.

INTRODUCTION

While much work has been published on the total cross sections for negative ion formation, very few measurements of the differential cross sections have been reported. It is felt that a better understanding of the physical process leading to negative ion formation could be obtained by studying differential cross sections as well as total cross sections. With this motivation, differential cross sections were reported in previous papers for H^- formation in collisions of H^+ on C_s[1]. The measurements of Berkner et al[2] and of Morgan and Eriksen[3,4] of total cross sections for negative ion formation from H^+ on Mg and the total cross sections for H^- formation in collisions of H^+ and $H°$ on Ar reported by B. Van Zyl et al[5,6] and Morgan and Eriksen[3,4], stimulated a good deal of interest concerning the collisional dynamics of these systems. We have already reported measurement for H^- formation from H^+ on Mg[7,8] and $H°$ on Mg[9]. Differential cross sections are reported here for the processes

$$H^+ + Ar \rightarrow H^- \quad \text{and} \quad H° + Ar \rightarrow H^-$$

at energies from 1.0 - 4.0 keV.

gas pressure as measured with a MKS Baratron and the differential cross section was obtained using the relation

$$\frac{d\sigma}{d\Omega} = \frac{I^-(\theta)}{I_0 \, nl} \tag{1}$$

where l is the effective path length and n is the Ar particle density. The total cross section is found by integration over angles θ and ϕ and since scattering is symmetrical about ϕ, then

$$\sigma_- = 2\pi \int_0^\pi \frac{d\sigma}{d\Omega} \sen\theta \, d\theta \tag{2}$$

SCALING LAW

The behavior of the deflection function responsible for glory scattering in negative ion production can be understood in a considerably simplified picture. The interaction potentials can be reasonably approximated by a constant interaction in the incoming channel while the outgoing portion of the trajectory is governed by an attractive Coulomb force. The Ford and Wheeler[11] description for a quantum mechanical treatment of glory scattering has been discussed in previous papers[7,12] and a scaling law has been derived in order to interpret the data. Although their model does not apply to an excitation process such as that presented here, it is useful in interpreting the data.

According to the scaling law presented in previous papers[7,12] the Ford and Wheeler model predicts a universal curve, independent of incident energy, when the cross sections scaled as $E^{-3/2} \, d\sigma/d\Omega$ are plotted as a function of $\sqrt{E'}\,\theta$, such as

$$E^{-3/2} \frac{d\sigma}{d\Omega} = \frac{2\pi b g^2}{\alpha} J_0^2 (kbg \sen\theta)$$

where k is the relative momentum in a.u. and α is a constant independent of the energy. For small angles the right hand side is a function of $k\theta$ or $\sqrt{E'}$ only. This model also predicts a strong peak in the forward direction followed by oscillations in the wing of the peak, which may be said to arise from interference of the contributions from the two branches of the deflection function near a glory[11].

The most interesting feature in the angular distributions for these systems is a clear indication of a glo̲ry maximum at 0° scattering angle. This arises because the deflection function[10] θ(b) passes through 0° as it changes from negative (attractive scattering at large impact parameter) to positive (repulsive scattering at small impact parameter), the classical cross section becoming infinite. However, in the semiclassical descrip̲tion, the singularity in the classical cross section is replaced by a finite peak in the forward or backward directions. It has been shown[7,12] that the quantum mechan̲ical model of Ford and Wheeler[10] predicts a universal curve for glory scattering if $E^{-3/2} \, d\sigma/d\Omega$ is plotted as a function of $E^{1/2}\theta$. The measured angular distributions are well represented by these reduced variables.

EXPERIMENTAL

The experimental scattering apparatus is essentially the same as described in previous papers[7,9] and consists basically of four parts: ion source, neutralization chamber, collision chamber and detection system. The H^+ ions, extracted from a colutron type ion source are accelerated, focused and analyzed by a Wien velocity filter. This beam is then bent 10° by cylindrical electrostatic deflection plates in order to prevent photons created in the ion source from reaching the detection system.

The energy selected and collimated H^+ beam entered the neutralization cell, where electron capture collisions converted a fraction of the beam to $H^°$. The beam on emerg̲ing from the gas cell was passed to a set of transverse electrostatic deflection plates, which deflected the charged component of the beam. For H^+ measurements on Ar the neutralization cell had to be evacuated. Either H^+ or $H^°$ projectiles were then allowed to enter the Ar-inter̲action cell, which was rotated within the useful range of $\pm 7°$. Between the Ar cell and the detector assembly, a retractable Faraday cup was placed so that it could measure either the current due to charged particles or the secondary emission due to neutrals. The detector chamber houses a parabolic electrostatic analyser with two funnel channel electron multipliers.

The procedure was the same as described in Ref. 7. The H^- current $I^-(\theta)$, scattered in the solid angle $d\omega$, was determined as a function of θ as the detector assembly was rotated about the scattering cell center. The number of atoms present in the target was determined by the Ar

RESULTS AND DISCUSSION

The angular distributions for H⁻ formed from single and double electron capture of H⁺ and H° on Ar at several collision energies have been measured, and the absolute differential cross sections were obtained by using Eq.(1). The H⁻ ions were formed mainly in the forward direction, however a well defined and periodic structure at angles bigger than 0° was found. In order to gain more insight into the collision process, the angular distributions were replotted in terms of the scaled variables and are shown in Figs. 1 and 2. It can be seen that the Ford and Wheeler model is not strictly applicable for our present results, since excitation processes were not allowed for.

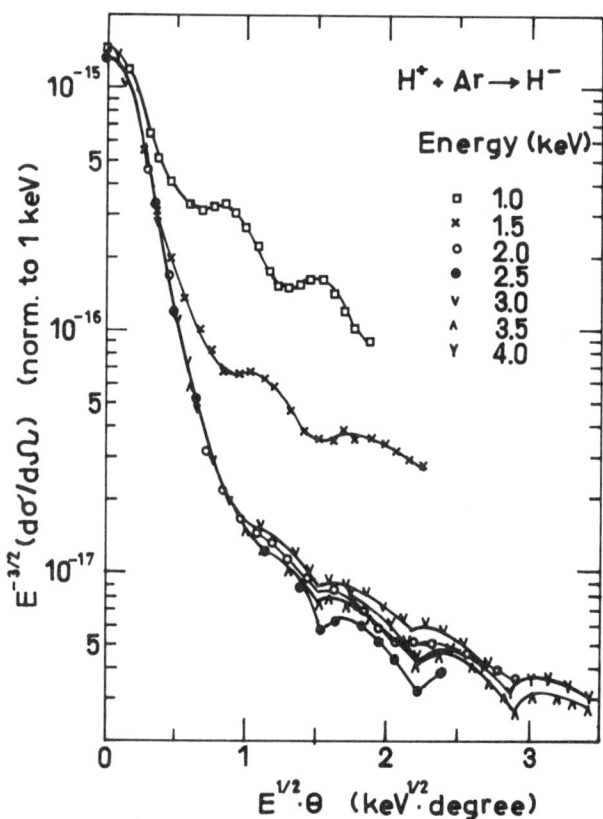

Fig.1.- Scaled differential cross section for the collision H⁺ + Ar → H⁻

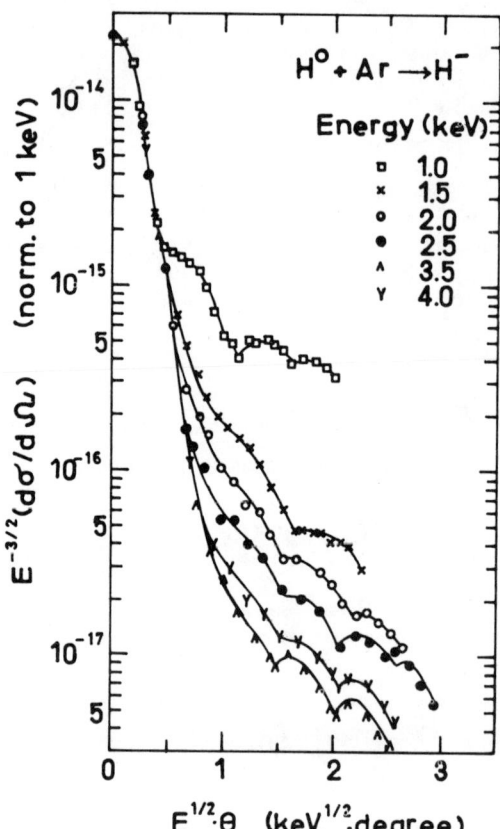

Fig. 2.- Scaled differential cross section for the collision $H^° + Ar \rightarrow H^-$.

A more appropiate model to describe the differential cross sections for negative ion production is based on a multichannel treatment of the scattering process[13]. We have also compared our results with this model. As an example, Fig. 3 shows the computed differential cross sections for $H^° + Ar$, using simplified potentials[12], and are plotted in terms of the scaled variables for three different energies. For comparison, the experimental data at 1.0 keV have been plotted, and show good agreement with the calculated maxima and minima. Furthermore, the calculated peaks are seen to decrease monotonically with increasing energy, which is consistent with the results shown in Figs. 1 and 2.

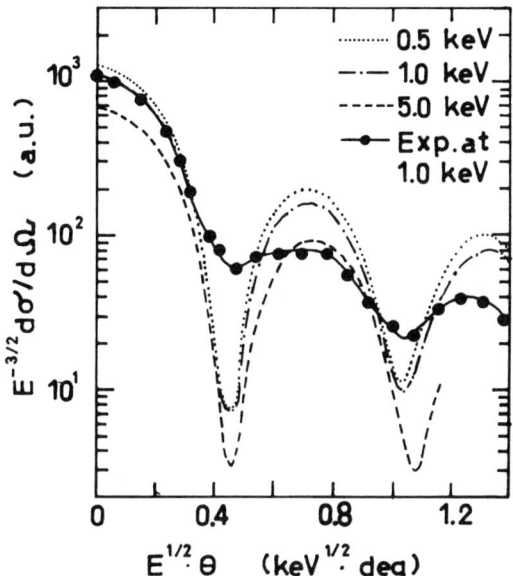

Fig. 3.- Calculated differential cross sections for a model considering diffraction effects for a multichannel process[12], and plotted in terms of scaled variables. The experimental curve at 1.0 keV is also shown.

Detailed calculations based in a more realistic model will be presented elsewhere. Finally, the total cross sections were obtained by integrating the differential cross sections (Fig. 2), and are shown in Figs. 4 and 5, together with those of other workers[4,6,14-19], and are seen to be in overall agreement over the measuring range.

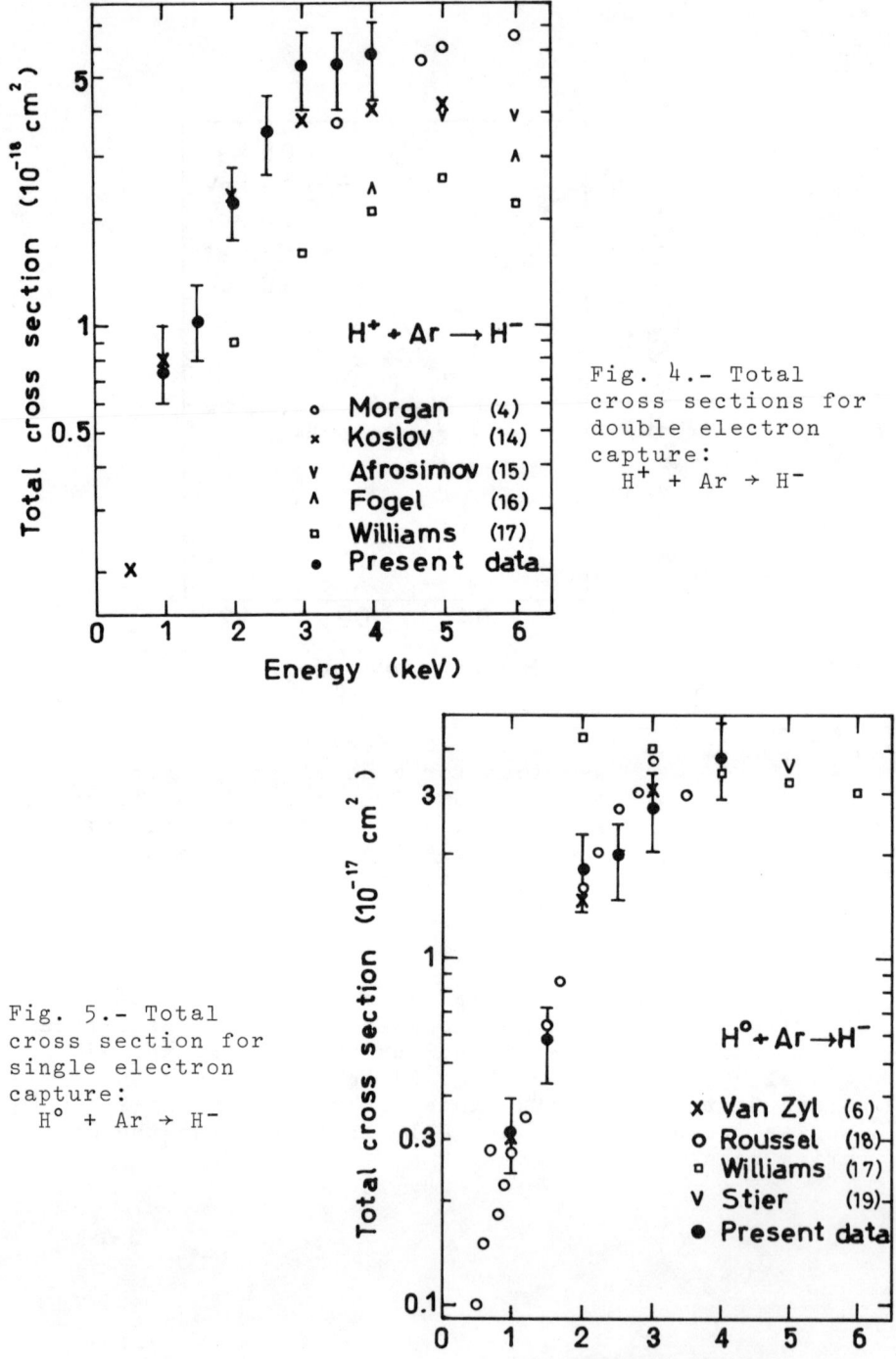

Fig. 4.- Total cross sections for double electron capture:
$H^+ + Ar \rightarrow H^-$

Fig. 5.- Total cross section for single electron capture:
$H^0 + Ar \rightarrow H^-$

ACKNOWLEDGEMENT

This work was partially supported by CONACYT, Mexico Grant PCCBBEU-0102238.

REFERENCES

1) C. Cisneros, I. Alvarez, C.F. Barnett and J.A. Ray, Phys. Rev. A**14**, 76 (1976).

2) K.H. Berkner, R.V. Pyle and J.W. Stearns, Phys. Rev. **178**, 248 (1969).

3) T.J. Morgan, and F.J. Eriksen, Phys. Lett. A**66**, 198 (1978).

4) T.J. Morgan and F.J. Eriksen, Phys. Rev. A**19**, 1448 (1979).

5) B. Van Zyl, T.Q. Lee, H. Newmann and R.C. Amme, Phys. Rev. A**15**, 1871 (1977).

6) B. Van Zyl, H. Newmann, H.L. Rothwell, Jr., and R.C. Amme, Phys. Rev. A**21**, 716 (1980).

7) I. Alvarez, C. Cisneros and A. Russek, Phys. Rev. A**26**, 77 (1982).

8) M. Mayo, J. Stone, J.T. Morgan, I. Alvarez and C. Cisneros, Proceedings of the Second International Symposium on the Production and Neutralization of Negative Ions and Beams, 1980, edited by T. Sluyters (Brookhaven National Laboratory, 1980).

9) I. Alvarez, C. Cisneros, R. Castillo, A. Morales and J.T. Morgan to be published.

10) G.A.L. Delvine and J. Los, Physica **67**, 166 (1973).

11) K.W. Ford and J.A. Wheeler, Ann. Phys. (N.Y.) **7**, 259 (1959).

12) A. Russek, Proceedings of the U.S. Mexico Joint Seminar on the Atomic Physics of Negative Ions, edited by I. Alvarez and C. Cisneros (México, 1981).

13) A. Russek, Phys. Rev. A**20**, 113 (1979).

14) V.F. Koslov, Ya. M. Fogel, and V.A. Stratienko, Sov. Phys. JETP **17**, 1226 (1963).

15) V.V. Afrosimov, R.N.Il'in, and E.S. Solov ev, Sov. Phys. Tech. Phys. **5**, 661 (1961).

16) Ya. M. Fogel, V.A. Ankudinov, D.V. Pilipenko, and N.V. Topoglia, Sov. Phys. JETP **7**, 400 (1958).

17) J.F. Williams, Phys. Rev. 153, 116 (1967).

18) F. Roussel, P. Pradel and G. Spiess,
    Phys. Rev. A$\underline{16}$, 1854 (1977).

19) P.M. Stier and C.F. Barnett,
    Phys. Rev. $\underline{103}$, 896 (1956).

FORMATION OF H⁻ BY CHARGE TRANSFER IN ALKALINE-EARTH VAPORS*

A. S. Schlachter† and T. J. Morgan‡

†Lawrence Berkeley Laboratory, University of California,
Berkeley, CA 94720

‡Wesleyan University, Middletown, CT 06457

ABSTRACT

Progress since the last symposium on the study of H⁻ formation by charge transfer in alkaline-earth vapors is reported. High yields are obtained at low energies, in agreement with theoretical predictions.

INTRODUCTION

Considerable progress has been made in the study of H⁻ formation by charge transfer in alkaline-earth vapors since the 1980 Brookhaven Symposium.[1] Olson[2] wrote at that time:

". . . we predict the heavier alkaline earths, and in particular Sr or Ba, will surpass the 35% maximum yield realized using Cs."

Measurements[3] have since confirmed that prediction, showing a maximum H⁻ equilibrium fraction of 50% for 250 eV/amu H in strontium vapor. The behavior of the cross sections[4] indicates that this large yield at low energies arises because the electron-detachment cross section σ_{-10} is small and the electron-attachment cross section σ_{0-1} is large in heavy alkaline-earth vapors.

The subject of H⁻ and D⁻ production by charge transfer in metal vapors was extensively reviewed at the 1980 Brookhaven Symposium.[5] At that time little information was available for alkaline-earth targets, and the review dealt primarily with alkali-vapor targets. During the past 3 years considerable progress has been made on H⁻ formation by charge transfer in alkaline-earth vapors. The present review discusses and summarizes the progress made during the past 3 years. We limit the discussion to cross sections and equilibrium yields, and assume that results for H and D projectiles are the same at the same velocities. Results are available for the energy range 0.15 to 100 keV/amu.

Formation of intense beams of D⁻ by charge transfer has been considered as a means of producing energetic neutral beams of D⁰ for heating fusion plasmas. It is not an active candidate in the USA at present because it is considered too complex for fusion applications and because surface and volume production of D- seem capable of furnishing beams for fusion applications.

Figure 1 shows the equilibrium H- yield, F_-^∞, for H in a variety of targets, to show the energy dependence of an alkaline-earth target (strontium vapor) by comparison with other targets.

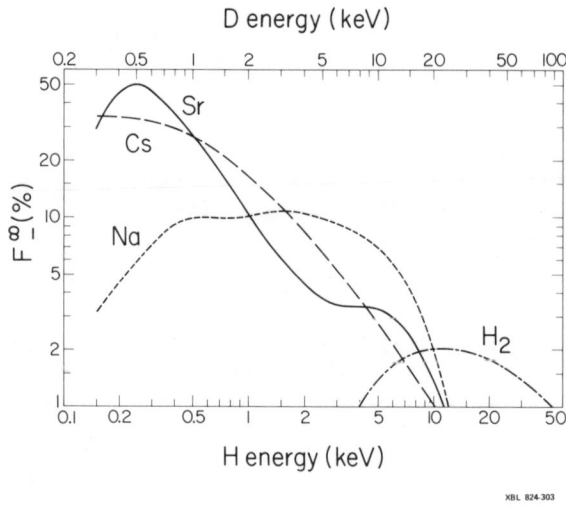

Fig. 1. Summary of equilibrium yields F_-^∞ for H- formation in typical targets.

EXPERIMENTAL CONSIDERATIONS

Measurements reported here were made primarily by experimenters at LBL (Lawrence Berkeley Laboratory) and at Wesleyan University over a period of several years.[6] A diagram of the apparatus used[3,4] is shown in Fig. 2. Similar targets were employed: a steel oven heated by electrical resistance or by quartz lamps, with temperature measured by thermocouples, and vapor pressure inferred from the temperature. A heat-pipe target, employed for alkali vapors, is not suitable for use with alkaline earths[3] at the temperatures and pressures usual for charge-transfer measurements.

Two methods have been employed for the measurement of H^0 flux: the Wesleyan group used secondary-electron emission, while experimenters at LBL used pyroelectric detection. Faraday cups were used for the measurement of H^+ and H- fluxes. Agreement is good between yields measured in the various experiments.[3,7,8]

151

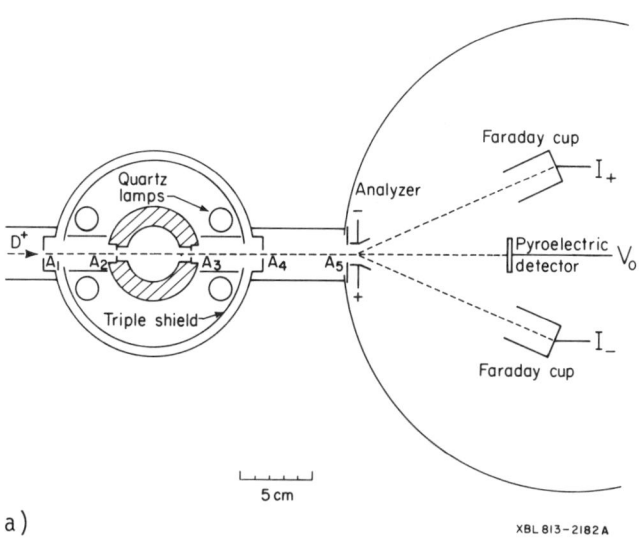

a)

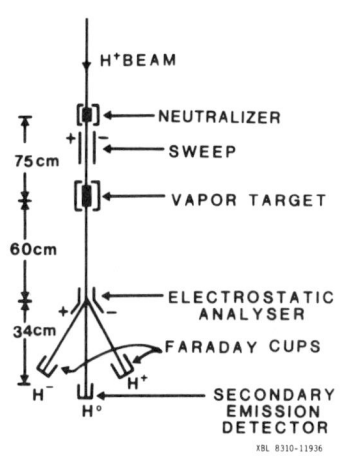

b)

Fig. 2. Schematic diagram of apparatus used by McFarland et al.[3] at LBL (2a) and by Mayo et al.[4] at Wesleyan (2b) to measure charge-state fractions in alkaline-earth vapors.

Measurements of cross sections with an H⁰ beam incident were done by partial neutralization of an H⁺ beam in a gas neutralizer, followed by deflection of residual ions and quenching of H(2s) in a transverse electric field.

Typical data for charge-state fractions as a function of target thickness are shown in Fig. 3: 1500 eV/amu D⁺ incident on barium vapor, from measurements by McFarland et al.[3]

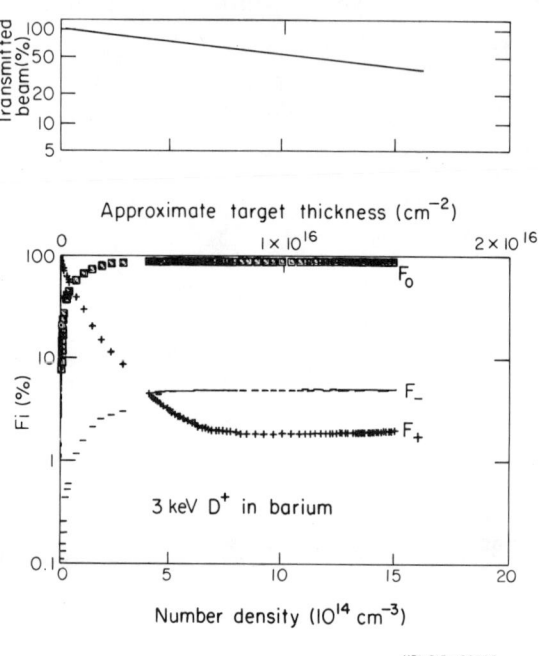

Fig. 3. Charge-state fractions F_i and total transmitted beam as a function of target number and line densities for 1500 eV/amu D⁺ incident on barium vapor, from measurements by McFarland et al.[3]

THEORETICAL CONSIDERATIONS

Olson[2] and Liu[3] have provided most of the theoretical calculations for H⁻ formation in alkaline-earth vapors, as well as much of the impetus for the experimental measurements. They pointed out that the alkaline earths do not have a bound and stable negative ion. Electron detachment must therefore be by direct ionization

$$H^- + X \rightarrow H^0 + X + e^- \qquad (1)$$

rather than charge transfer (X is an alkaline-earth atom). Olson and Liu used an ab initio molecular-interaction-energy calculation on the neutral and negative-ion CaH system to determine the lack of strong coupling between the negative-ion and neutral molecular states, and to thus predict a small cross section for electron detachment of H- in Ca at low energies.

For the case of H- formation by 2-electron capture by a proton in a single collision with a Mg atom, Olson and Liu[9] have employed a Landau-Zener calculation using ab initio potential-energy curves to obtain the cross section σ_{1-1}. The results are in good agreement with the experiment.[10]

H- EQUILIBRIUM YIELDS

The equilibrium H- yield in an alkaline-earth vapor heavier than magnesium was first reported by Berkner et al.[8] in 1977. They measured F_-^∞ for 1.65 to 19.5 keV/amu D^+ in Sr vapor, and noted a feature of the energy dependence unlike that observed at low energies for alkali-vapor or gas targets: a plateau between 2.5 and 5 keV/amu and a rise in F_-^∞ for lower energies. Morgan et al.[7] extended those measurements to several alkaline-earth vapors (Mg, Ca, Sr, Ba) in the energy range 1.25 to 100 keV/amu. They observed the same behavior in all 3 heavy alkaline-earth vapors, with F_-^∞ reaching 10%, and their results were in excellent agreement with the previous LBL result in Sr. Various hypothesis were advanced to explain the rise in F_-^∞ at low energies.

Measurements were extended to lower energies by McFarland et al. in 1982: results were published[3] for F_-^∞ in alkaline-earth vapors in the energy range 150 eV/amu to 1.5 keV/amu. The lowest energy was sufficient to observe a maximum in F_-^∞ in all alkaline-earth vapors except Ba; the highest yield observed was 50% in Sr vapor at 250 eV/amu. D^+ equilibrium yields were also measured, and were found to be negligible (<1%) at energies below 0.75 keV/amu. Results are in excellent agreement between all measurements at energies where there is overlap. Results for F_-^∞ in Mg, Ca, Sr, and Ba vapors[3,7,8,11] are shown in Figs. 4-7. A composite result is shown in Fig. 8.

Scattering in the target can limit the usefulness of charge transfer in metal vapors as a means of producing H- beams. Note in Fig. 3 the reduction in total transmitted beam as target thickness is increased; this result, of course, is specific to the geometry of the target.

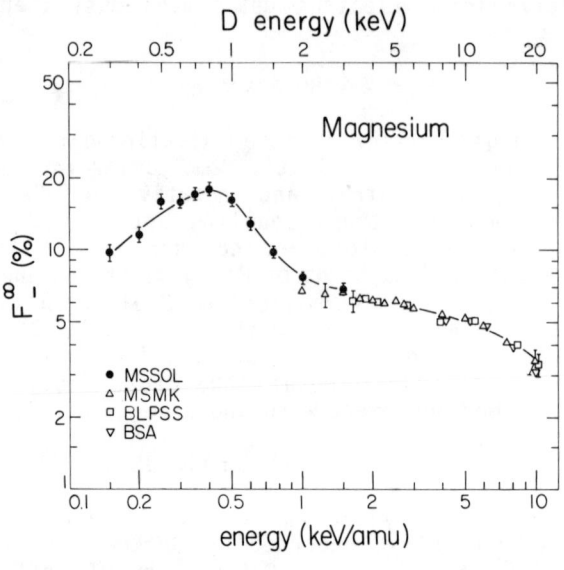

Fig. 4. Equilibrium yield F_-^∞ for H⁻ formation in magnesium vapor.

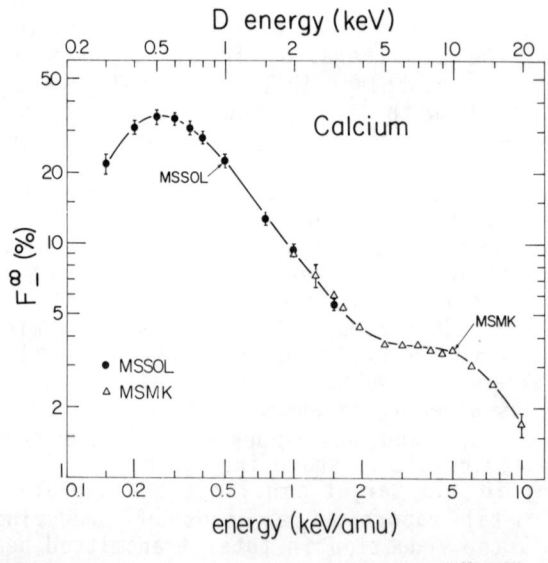

Fig. 5. Equilibrium yield F_-^∞ for H⁻ formation in calcium vapor.

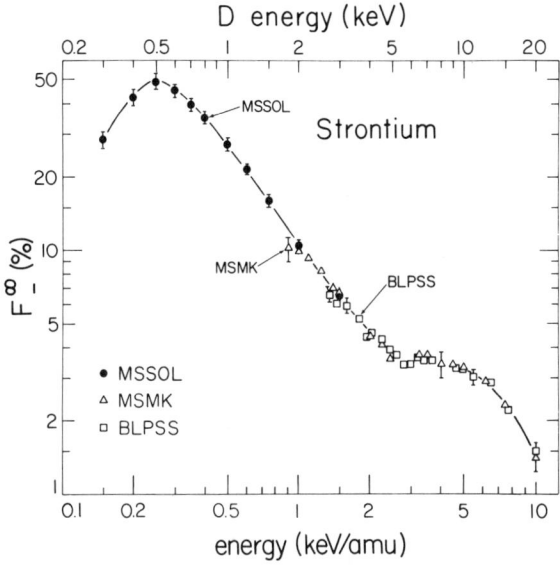

Fig. 6. Equilibrium yield F_{-}^{∞} for H⁻ formation in strontium vapor.

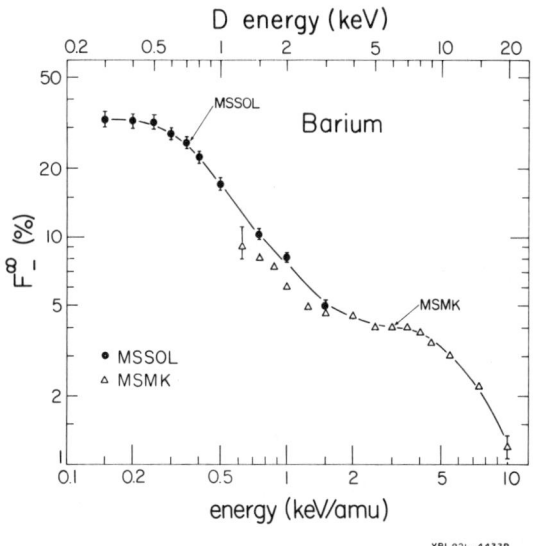

Fig. 7. Equilibrium yield F_{-}^{∞} for H⁻ formation in barium vapor.

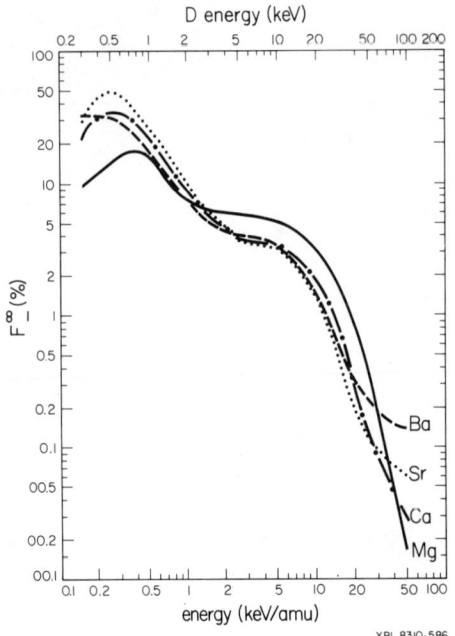

Fig. 8 Summary of equilibrium yields F_{-}^{∞} for H⁻ formation in alkaline-earth vapors.

CROSS SECTIONS

Electron-attachment cross sections, σ_{0-1} for H⁰ in Ca and Sr vapor targets have recently been published[4] by Mayo et al. for the energy range 1-70 keV/amu. These cross sections have been used with measured H⁻ equilibrium yields to infer the electron-loss cross section, σ_{-10}, by use of the formula

$$\sigma_{-10} = \sigma_{0-1} \left[\frac{1}{F^{\infty}_{-}} - 1\right]$$

which is true at low energies (H⁺ must be small in a thick target and 2-electron transfer processes must be negligible). A more complicated expression is used at higher energies, where those conditions are not met. Results for measurements[4] of the electron-attachment cross section σ_{0-1} in Ca and Sr vapors are shown in Figs. 9 and 10. Figure 11 shows the electron-detachment cross section σ_{-10} inferred from measurements of σ_{0-1} and F^{∞}_{-}; this cross section has a

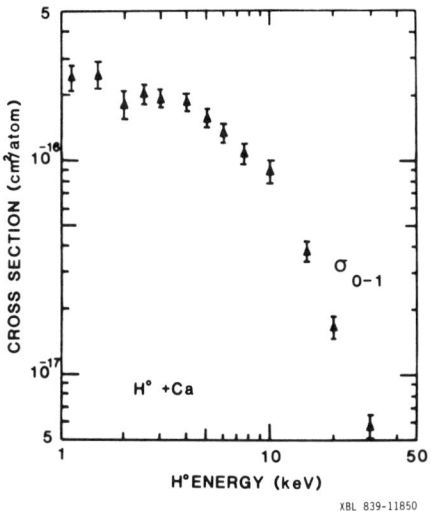

Fig. 9. Electron-attachment cross section σ_{0-1} for collisions of H^0 with calcium vapor.

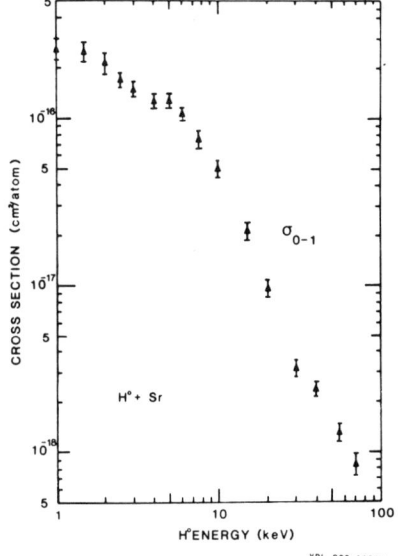

Fig. 10. Electron-attachment cross section σ_{0-1} for collisions of H^0 with strontium vapor.

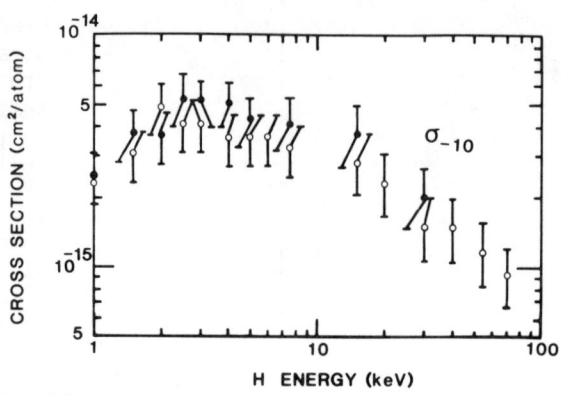

Fig. 11. Electron-detachment cross section σ_{-10} for collisions of H⁻ with calcium (●) and strontium (○) vapors, deduced from measured σ_{0-1} and F_-^∞.

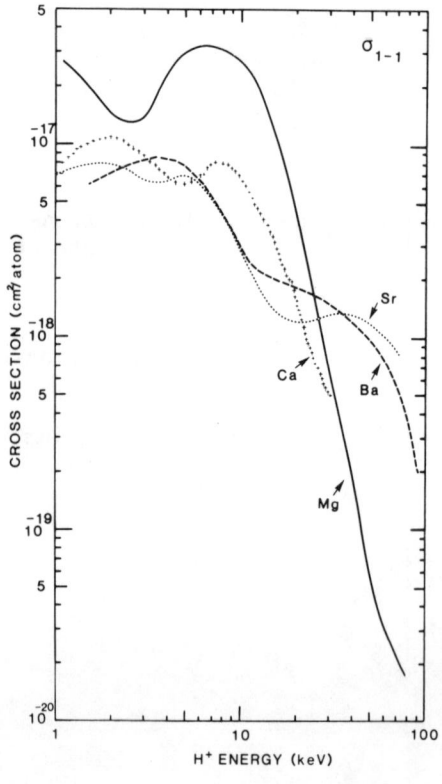

Fig. 12. Double-electron-capture cross section σ_{1-1} for collisions of H⁺ with alkaline-earth-vapor targets.

maximum around 2.5 keV/amu, and decreases with decreasing energy for lower energies.

The Wesleyan group has also studied H^- formation by 2-electron capture in a single collision ($H^+ \rightarrow H^-$). Figure 12 shows a summary[4] of their results for alkaline-earth targets. Molecular-curve-crossing effects are important in a Mg-vapor target at low energies.

One measurement of a differential cross section has been reported[12] for hydrogen ions in an alkaline-earth target: H^+ H^0, H^+ H^-, and H^0 H^- for 0.5 to 5 keV projectiles in Mg vapor. Interference effects are observed in the 2-electron-transfer case.

CONCLUSION

Recent measurements of H^- formation by charge transfer of H^+ in alkaline-earth-vapor targets are reasonably comprehensive; quantitative agreement is found between results measured by different experimenters, and there is qualitative agreement between experiment and theory. Heavy alkaline-earth-vapor targets are the most effective media for H^- production by charge transfer, although their usefulness is limited by scattering of a low-energy beam in the target. Cross sections for some charge-transfer processes have been calculated and/or measured; more work is needed, especially for electron attachment and detachment at energies below 1 keV/amu.

ACKNOWLEDGMENTS

This work was supported by the Director, Office of Energy Research, Office of Fusion Energy, Applied Plasma Physics Division of the U.S. Department of Energy under Contract No. DE-AC03-76SF00098 and DE-AC02-76ET53048.

REFERENCES

1. Proceedings of the Second International Symposium on the Production and Neutralization of Negative Hydrogen Ions and Beams, BNL 51304, edited by Th. Sluyters (Brookhaven National Laboratory, 1980).

2. R. E. Olson, ref. 1, p. 51.

3. R. H. McFarland, A. S. Schlachter, J. W. Stearns, B. Liu, and R. E. Olson, Phys. Rev. A 26, 775 (1982).

4. M. Mayo, J. A. Stone, and T. J. Morgan, Phys. Rev. A 28, 1315 (1983).

5. A. S. Schlachter, ref. 1, p. 42.

6. Lawrence Berkeley Laboratory: K. H. Berkner, D. Leung, R. H. McFarland (on assignment from University of Missouri/Rolla), R. V. Pyle, A. S. Schlachter. and J. W. Stearns. Wesleyan University: F. J. Ericksen, J. Kurose, M. Mayo, T. J. Morgan and J. A. Stone,.

7. T. J. Morgan, J. Stone, M. Mayo, and J. Kurose, Phys. Rev. A <u>20</u>, 54 (1979).

8. K. H. Berkner, D. Leung, R. V. Pyle, A. S. Schlachter, and J. W. Stearns, Phys. Lett. <u>64 A</u>, 217 (1977); Nucl. Instrum. Methods 143, 157 (1977).

9. R. E. Olson and B. Liu, Phys. Rev. A <u>20</u>, 1366 (1979).

10. T. J. Morgan and F. J. Eriksen, Phys. Rev. A <u>19</u>, 1448 (1979).

11. R. A. Baragiola, E. R. Salvatelli, and E. Alonso, Nucl. Instrum. Methods 110, 507 (1973).

12. I. Alvarez, C. Cisneros, and A. Russek, Phys. Rev. A <u>26</u>, 77 (1982); I. Alvarez and C. Cisneros, Notas de Fisica <u>5</u>, 517 (1982) [Proceedings of the U.S. Mexico Joint Seminar on the Atomic Physics of Negative Ions, April 1-4, 1981; Instituto de Fisica, UNAM, Mexico].

Symbols in Figures:

MSSOL = McFarland et al.[3]
MSMK = Morgan et al.[7]
BLPSS = Berkner et al.[8]
BSA   = Baragiola et al.[11]

* This work was supported by the Director, Office of Energy Research, Office of Fusion Energy, Applied Plasma Physics Division of the U.S. Department of Energy under Contract No. DE-AC03-76SF00098 and DE-AC02-76ET53048.

DISCUSSION

MICHELS: I'm a little surprised the cross sections from theory are so good because in this region, several tens of keV, you would think translational factors would be important, as they are in direct excitation, and yet they don't seem to be.
MORGAN: That's right. They haven't been put into the calculation. They just haven't been put in and it doesn't appear to make any difference. The theory only goes up to a few keV/amu so it's maybe borderline, but you are indeed right. There's no inclusion of translation factors but yet there seems to be very good agreement.

CHARGE EXCHANGE OF PROTONS AND HYDROGEN ATOMS IN Na, K, Rb-VAPOR TARGETS

F. Ebel and E. Salzborn
Institut für Kernphysik, Strahlenzentrum
Universität Giessen, D-6300 Giessen, W.Germany

ABSTRACT

Total cross sections σ_{if} (i=+,o; f=0,+,-) for charge exchange of protons and hydrogen atoms in Na, K, and Rb-vapor targets have been measured in the energy range 0.2-5 keV. Some of the data show considerable discrepancies to previous measurements available.

INTRODUCTION

The present measurements were motivated by the current interest in cross sections for charge exchange of hydrogenic projectiles in collisions with alkali atoms. A detailed knowledge of such processes is relevant for instance to the production of intense H⁻ beams for heating and fueling fusion plasmas[1].

We have measured cross sections for the following reactions

$$H^+ \xrightarrow[Na,K,Rb]{\sigma_{+o},\ \sigma_{+-}} H^{o,-}$$

$$H^o \xrightarrow[Na,K,Rb]{\sigma_{o+},\ \sigma_{o-}} H^{+,-}$$

in the energy range between 0.2 and 5 keV.

EXPERIMENTAL TECHNIQUE

Fig. 1 shows the experimental arrangement. The apparatus and the measurement technique have been described already in detail in a previous publication[2]. In short, protons formed in a RF discharge and accelerated to the desired energy are deflected by a 90° magnet. The proton beam can be neutralized in a gas cell. Behind the cell the projectile beam is collimated to an angular divergence of ± 0.4°. The beam intensity is measured either with a Faraday cup or by means of a secondary electron emission (SEE) detector in case of

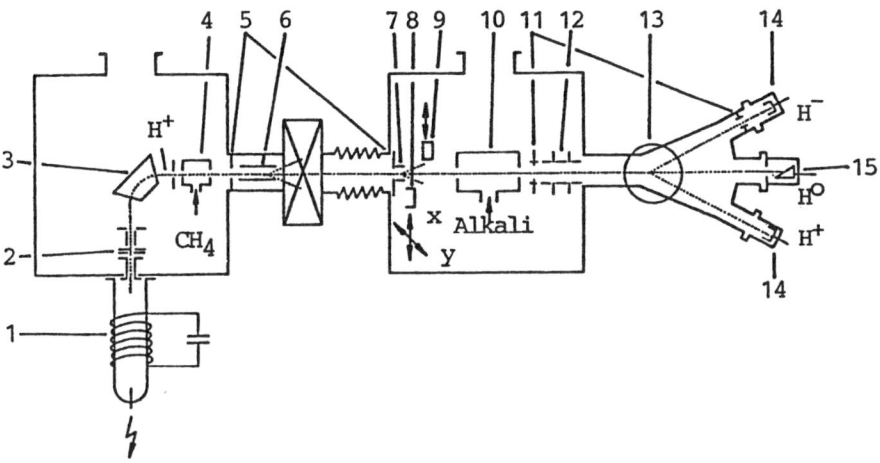

Fig. 1. Schematic diagram of the apparatus
1 RF ion source
2 ion optics
3 90° magnet
4 neutralizer cell
5 collimators
6 deflection plates
7 deflection plates
8 neutral particle detector
9 Faraday cup
10 charge exchange cell
11 iris aperture
12 ion optics
13 30° magnet
14 Faraday cup
15 Neutral particle monitor

H^0 projectiles. The SEE-detector was calibrated by measurements of the secondary-electron-coefficient γ_+^- with a H^+ beam and using the ratio $\gamma_0^-/\gamma_+^- = 1.17 \pm 0.12$ [3]. The charge exchanged particles behind the alkali vapor target cell are again measured in a second Faraday cup and SEE-detector, respectively. The charged particles produced in the target cell could be focused by an Einzellens resulting in detection of all product ions scattered up to angles of $\pm 3.9°$ in the charge changing collision.

The alkali target thickness was varied and determined by the temperature of the target cell which was regulated within less than 1.0 K. The vacuum-sealed glass ampoules of the respective target materials were broken in situ under high vacuum after baking of the cell at high temperatures. The temperature uncertainty results in an uncertainty in the particle density of the target vapor of $(\Delta n/n) = \pm 0.10$. We used the particle density formulae of Nesmeyanov[4] to calculate n.

The target thickness μ was calculated from $\mu = n(l_0+d_1+d_2)$ with $l_0 = 10.4$ cm being the geometrical length of the cell and $d_1 = 6.0$ mm and $d_2 = 10.0$ mm being the diameters of the entrance and exit apertures, respectively. An evaluation of all errors contributing to the uncertainty of the target thickness results in a root mean square value of $\Delta\mu/\mu = 0.15$. The total uncertainty of the measured cross sections is estimated to $\pm$ 20% for σ_{+0}, σ_{0-} and $\pm$ 25% for σ_{+-}, σ_{0+}, respectively.

RESULTS

All cross sections were measured by the "growth rate" method. The validity of the single collision condition was carefully checked by the linear dependence of the product-particle intensity on the target thickness (Fig. 2). The attenuation of the primary beam was less than 1% due to thin target thicknesses below $\mu \leq 2\cdot 10^{13}$ atoms/cm². For the measurements of the two electron capture cross sections σ_{+-} target thicknesses only below $\mu \leq 5\cdot 10^{12}$ atoms/cm² were used in order to avoid errors due to two step processes $H^+ \to H^0 \to H^-$ which result in increased apparent cross sections.

Fig. 3 shows the measured electron capture cross section σ_{+0} for protons in collisions with potassium atoms together with previous experimental[5,6] and theoretical[7,8] results. The present data are in good agreement with the previous results except the data of Grüebler et al.[5] at low energies. For the sodium target there is a discrepancy with the results of Grüebler et al., too.

The results for the two electron capture cross section σ_{+-} for protons in potassium are shown in Fig.4. The present results are about a factor of 4 lower than earlier data by Grüebler et al[5]. For the sodium target our data are up to a factor of about 30 lower. We think these discrepancies can be explained by a too large target thickness used in earlier experiments. Then two step processes $H^+ \to H^0 \to H^-$ increase the apparent cross section. The cross section σ_{+-} for the Rb target resembles the K-data in shape but is a factor of 2 larger. In this case there are no previous data to compare with.

Fig. 5 shows the electron capture and loss cross sections σ_{0-} and σ_{0+}, respectively, for H^0 projectiles incident on potassium atoms together with previous results for σ_{0-}. The data are in fair agreement with the measurements of Nagata[6], except at low and high energies.

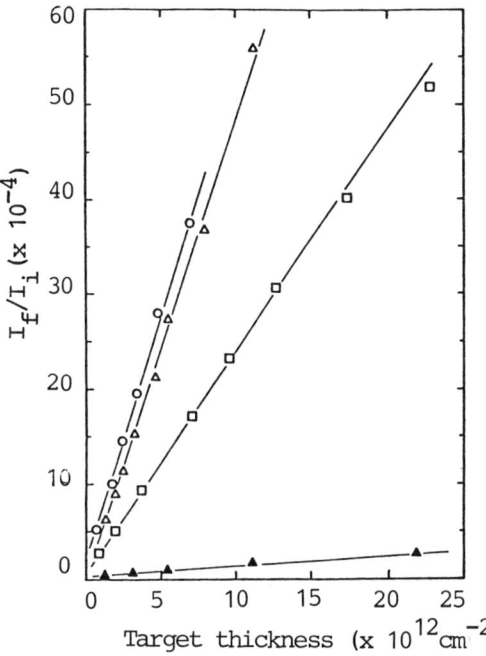

Fig. 2. Thin target values of product/primary intensity ratios I_f/I_i for 1 keV H^o particles in Na, K and Rb vapor.

$H^o \rightarrow H^-$: □Na, △K, ○Rb
$H^o \rightarrow H^+$: ▲K

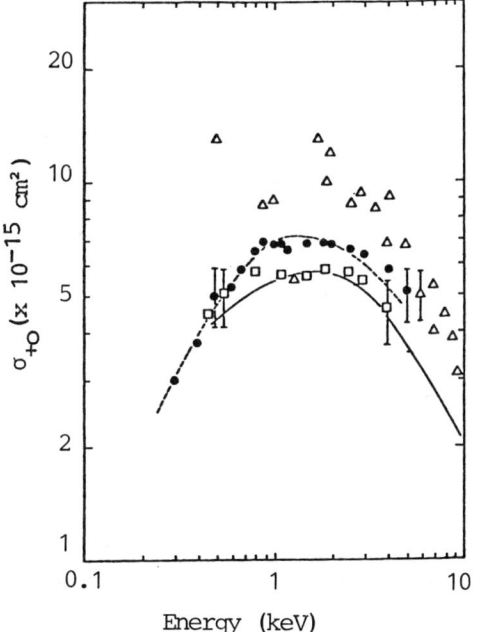

Fig. 3. Electron capture cross sections σ_{+0} for H^+ projectiles incident on potassium atoms, as a function of the projectile energy.

● this work;
△ Grüebler et al(1970)[5];
□ Nagata (1980)[6];
--- Kubach et al(1981)[7];
— Kimura et al(1982)[8].

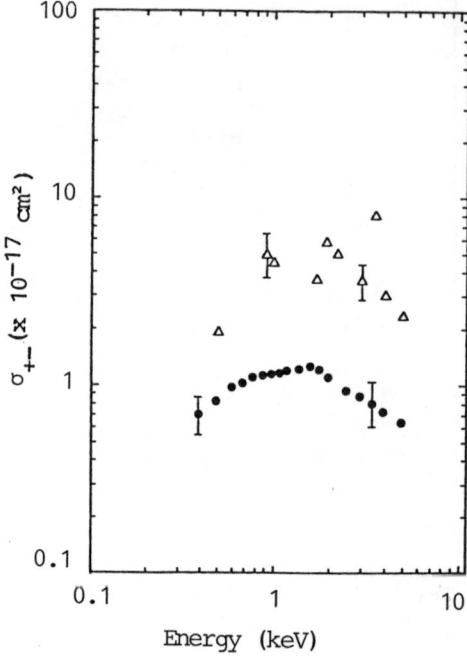

Fig. 4. Two electron capture cross sections σ_{+-} for H^+ projectiles incident on potassium atoms, as a function of the projectile energy.

● this work;
△ Grüebler et al (1980)[5]

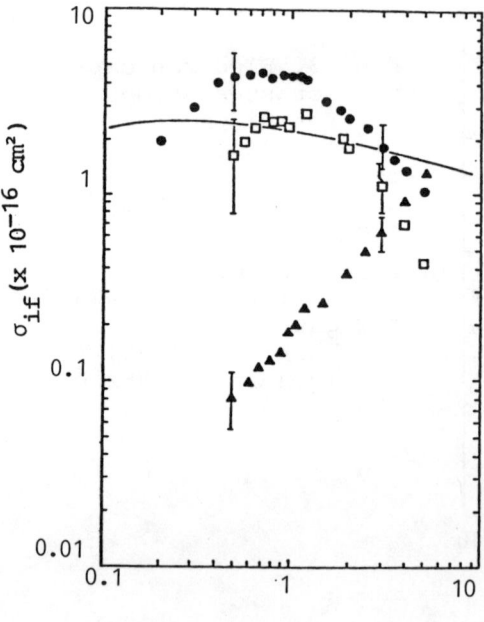

Fig. 5. Electron capture and loss cross sections σ_{o-} and σ_{o+}, respectively, for H^o projectiles incident on potassium atoms, as a function of the projectile energy.

σ_{o-}: ● this work;
□ Nagata(1980)[6];
— Janev et al (1978)[9]
σ_{o+}: ▲ this work.

The electron loss cross section σ_{o+} increases rapidly with increasing impact energy showing almost the same shape and size for the Na and Rb targets, too. We are not aware of any previous measurements with which to compare these data.

REFERENCES

1. K.H.Berkner, R.V.Pyle and J.W.Stearns,
   Nucl. Fusion 15, 249 (1975)
2. K.Miethe, T.Dreiseidler and E.Salzborn,
   J.Phys.B: Atom.Molec.Phys. 15, 3069 (1982)
3. J.A.Ray, C.F.Barnett and B.van Zyl,
   J.Appl.Phys. 50, 6516 (1979)
4. A.Nesmeyanov, Vapor Pressure of the Elements,
   1963 Infosearch Ltd, London
5. W.Grüebler, P.Schmelzbach, V.König and P.Marmier,
   Helv.Phys.Acta 43, 254 (1970)
6. T.Nagata, J.Phys.Soc.Japan 48, 2068 (1980)
7. C.Kubach and V.Sidis, Phys.Rev. A23, 110 (1981)
8. M.Kimura, R.E.Olson and J.Pascale,
   Phys.Rev. A26, 3113 (1982)
9. R.K.Janev and Z.M.Radulović, Phys.Rev. A17, 889 (1978)

FUNDAMENTAL PROCESSES: SURFACE

ION BACKSCATTERING FROM LAYERED TARGETS*

Ordean S. Oen and Mark T. Robinson
Solid State Division, Oak Ridge National Laboratory
Oak Ridge, TN 37830

ABSTRACT

The reflection of H atoms from amorphous layered targets were studied using a new version of the computer simulation code MARLOWE. Targets with low Z overlayers on high Z substrates, and vice versa, were investigated using normally incident atoms in the energy range from 0.01 to 1 keV. For very low energies, the backscattering is characteristic of the surface overlayer; at higher energies, which depend on the overlayer thickness, it becomes characteristic of the underlying substrate. In addition to the particle and energy reflection coefficients, the reflected atom energy distribution and the correlation between the emergent energy and angle have been examined.

INTRODUCTION

During the past decade there has been considerable experimental and theoretical work on the reflection of low and medium energy (0.01-10 keV) light ions from solids. This interest has stemmed mainly from the magnetically confined fusion energy program, where it is important to understand the recycling of H between the plasma and the walls of the confinement vessel.[1] The main experimental problem in reflection measurements involves the detection of the reflected particles, most of which are electrically neutral. For this reason little experimental work has been done at ion energies below ~1 keV. In this important energy region, the approach has been to refine the computational models to fit experimental data at higher energies and to rely on the calculations at the lower energies. A recent review article[2] and two compilations[3,4] of data cover much of the work done on light atom reflection from solids.

An additional motivation for reflection studies is the recent development by Hiskes and Schneider[5] of a model treating the formation of H^- and D^- by particle backscattering from alkali-metal surfaces. Important features of their model are the reflected fraction, energy and angular distribution. Of special interest are reflection from layered targets, for instance, targets with an alkali-metal covering another metal substrate. Although there have recently been some experimental studies on reflection from layered targets,[6-9] as far as the authors are aware there have been only two theoretical calculations[10,11] and they were done for oxide covered surfaces.

---

*Research sponsored by the Division of Materials Sciences, U.S. Department of Energy under contract W-7405-eng-26 with Union Carbide Corporation.

The present work investigated the reflection from layered targets of hydrogen atoms whose incident energy ranged from 0.01 to 1 keV. The calculations used the binary collision computer program MARLOWE[12-14] (Version 11.9), modified to treat layered target structures. Briefly, the projectile ion strikes the surface normally and is followed collision-by-collision until it leaves the surface again or until its energy falls below a preset value (1 eV). Each collision consists of an elastic and an inelastic part. The elastic part is treated by classical scattering mechanics using the Molière[15] approximation to the Thomas-Fermi interatomic potential with the screening lengths proposed by Firsov.[16] The inelastic part is described by the (nonlocal) electronic stopping theory of Lindhard et al.[17] The calculations were made using MARLOWE to simulate amorphous solids.[18] A typical run consisted of following the motions of 1000-2000 incident particles and recording the desired statistical information on the reflected particles.

2. RESULTS AND DISCUSSION

Figure 1 shows the effects of an overlayer of a low Z element (Li) on a high Z substrate (W) on the fraction R_N of H reflected. High Z targets such as W have much larger reflection coefficients than do low Z targets like Li at all incident energies. For the thinnest Li overlayer the particle reflection coefficient R_N is depressed at the lowest energies, but rapidly increases to that of pure W targets as the incident energy rises. The reason is that with increasing energy it is more probable for a larger fraction of the hydrogen to be reflected from the substrate. That is, it becomes energetically feasible for an incident particle to transverse the Li overlayer, be backscattered in the substrate, and still have sufficient energy to pass again through the Li overlayer and emerge as a reflected particle. As the thickness of the Li overlayer is increased, the particle reflection coefficient remains similar to that of pure Li over a wider energy region. With increasing energy, R_N exhibits more and more of the characteristics of the reflection from pure W, indicating that the substrate is playing the dominant role in the reflection. The results of Fig. 1 are qualitatively similar to those of Jackson and Eckstein[11] who calculated the reflection of H and He from a W target with an overlayer of WO_3. Figure 2 shows R_E, the fraction of energy reflected from the targets considered in Fig. 1. The results for R_E are quite similar to those for R_N in Fig. 1. One difference is that for a target with a Li overlayer, R_E never reaches values as large as those of pure W targets. This may be explained by noting the additional energy lost by the reflected particles in traversing the overlayer.

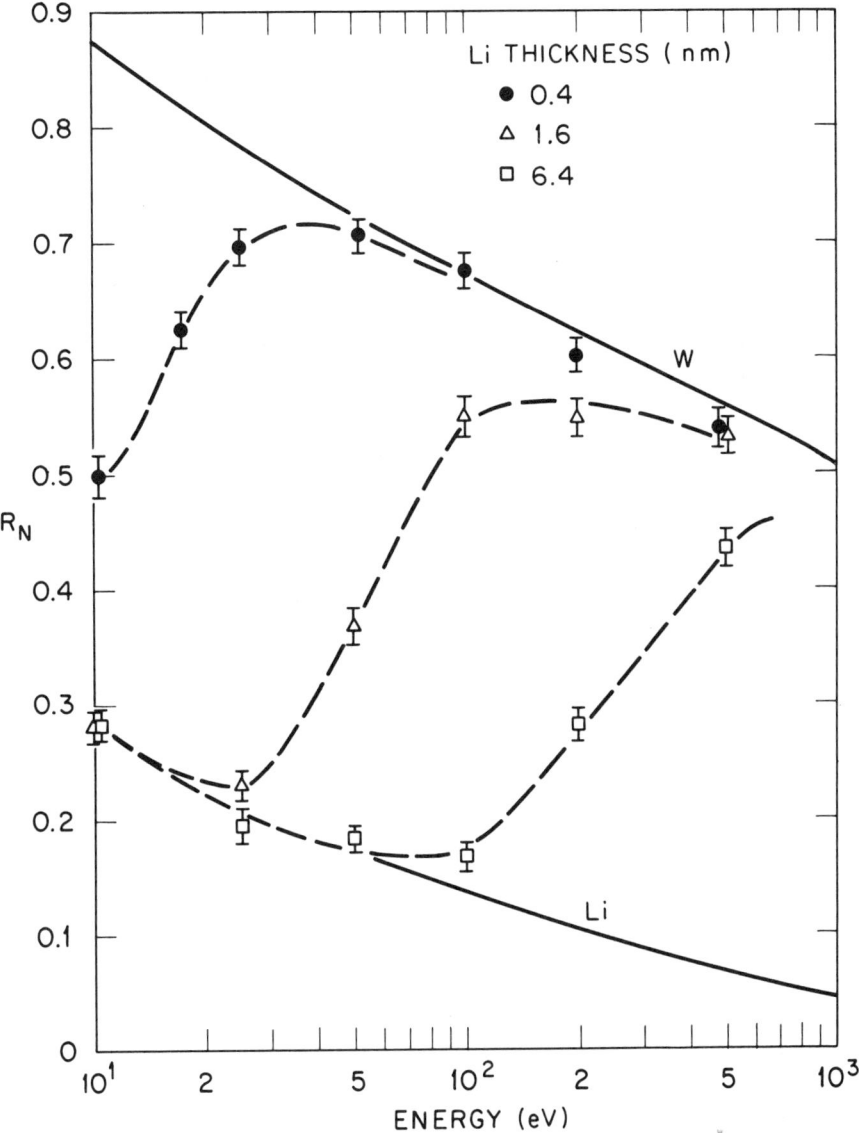

Fig. 1. Fraction of hydrogen, R_N, reflected from tungsten, lithium, and tungsten with lithium overlayers of various thicknesses as a function of the incident H ion energy. The error bars represent the standard deviation of the values plotted based on the Monte Carlo calculations.

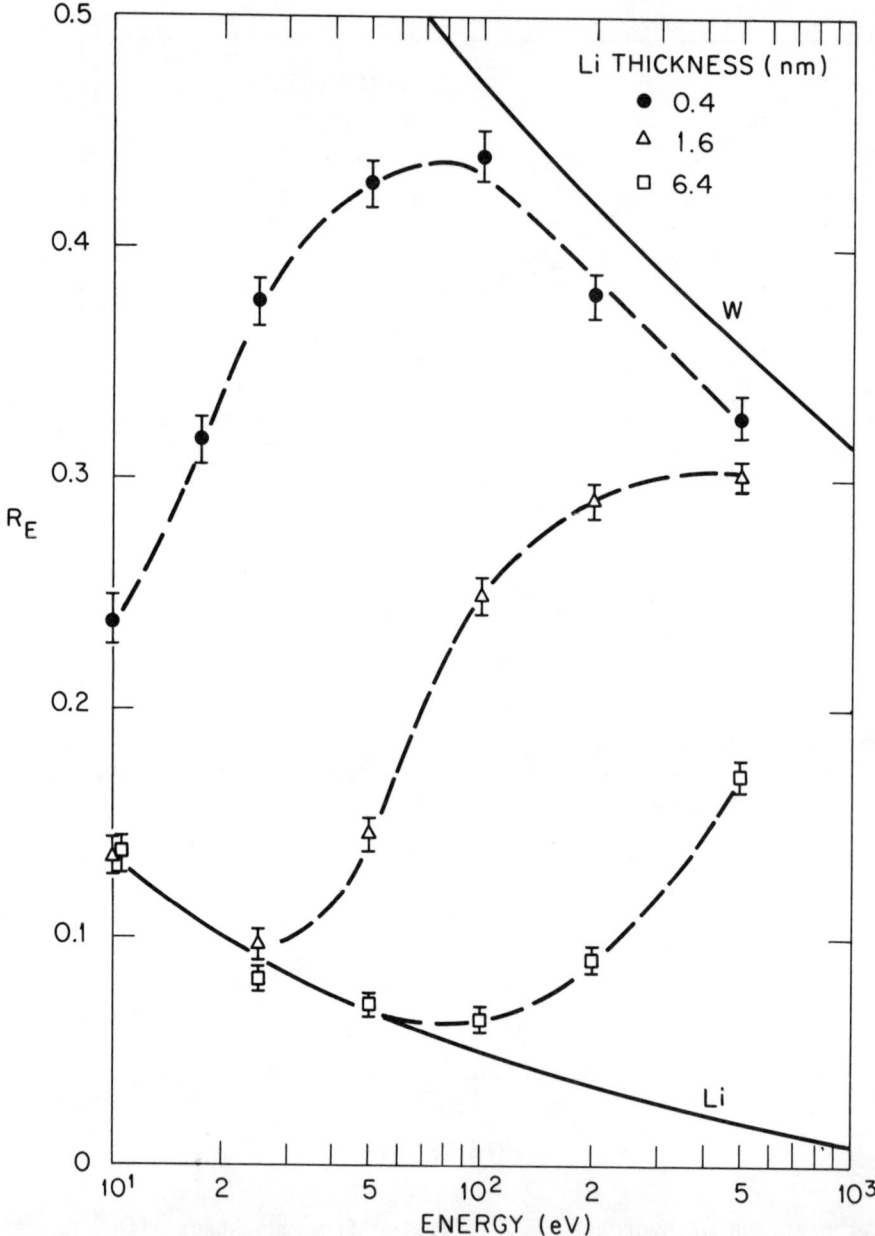

Fig. 2. Fraction of energy, R_E, reflected from tungsten, lithium, and tungsten with lithium overlayers of different thicknesses. The ratio R_E/R_N (R_N from Fig. 1) gives the average fractional energy of a reflected particle.

A comparison of the reflection of H and D from Li is made in Fig. 3. For this light target material, the H reflection coefficient R_N is some 50% larger than that for D. This is in contrast to our previous results for medium weight targets (Cu) where it was found that the reflection coefficients of the isotopes of hydrogen were approximately the same. The main reason that the reflection of H is greater than that of D in the present case is the lower total stopping power of Li (nuclear plus electronic) for H than for D. Thus, more collisions are needed to slow the H down. That is, since the particles undergo a random walk in the solid from the many large angle nuclear deflections that occur, the probability for H to be reflected is larger than for D because of the longer total path length of the H before it slows down to rest.

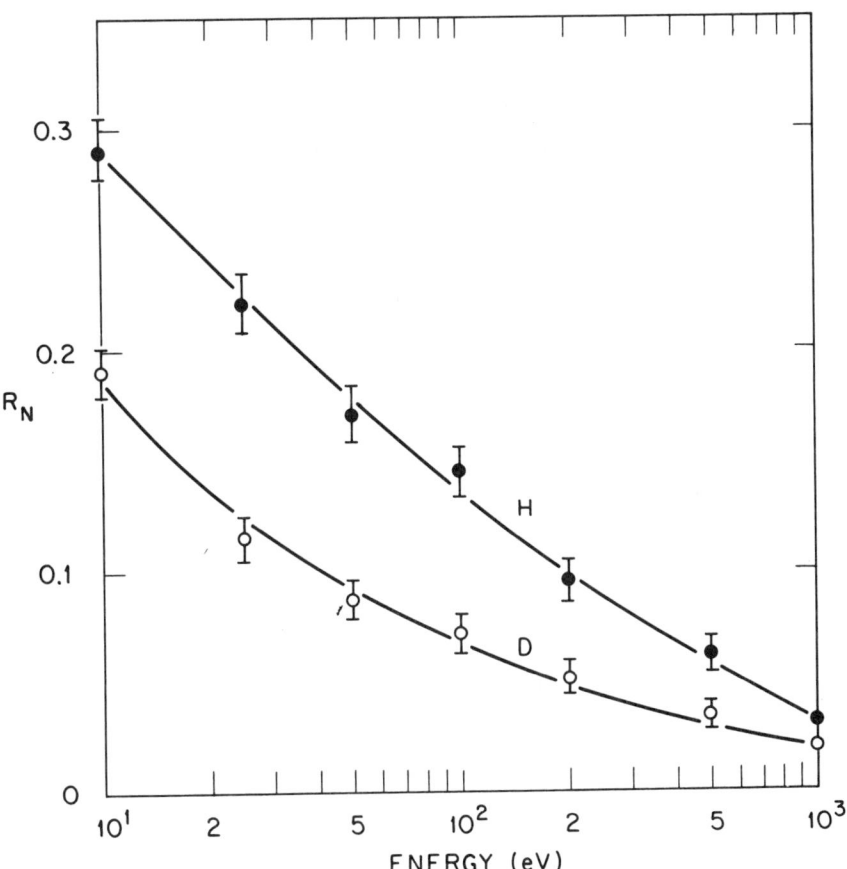

Fig. 3. Comparison of the particle reflection coefficients of hydrogen and deuterium from the low Z element lithium.

Figure 4 shows the particle R_N and energy R_E reflection coefficients for 1 keV Ar from W with various overlayer thicknesses of Li. For large overlayer thicknesses, there is very little particle or energy reflection. This indicates that Ar is backscattered only by the W atoms in the substrate, as is expected since the mass ratio of Ar to Li is almost 6. It is also seen that R_N is fairly constant for small Li thicknesses which is also expected if the Ar is reflected only by the W substrate. The very rapid decrease of R_E with small Li coverage is due to the loss of energy undergone by the reflected particles in passing through the Li overlayer.

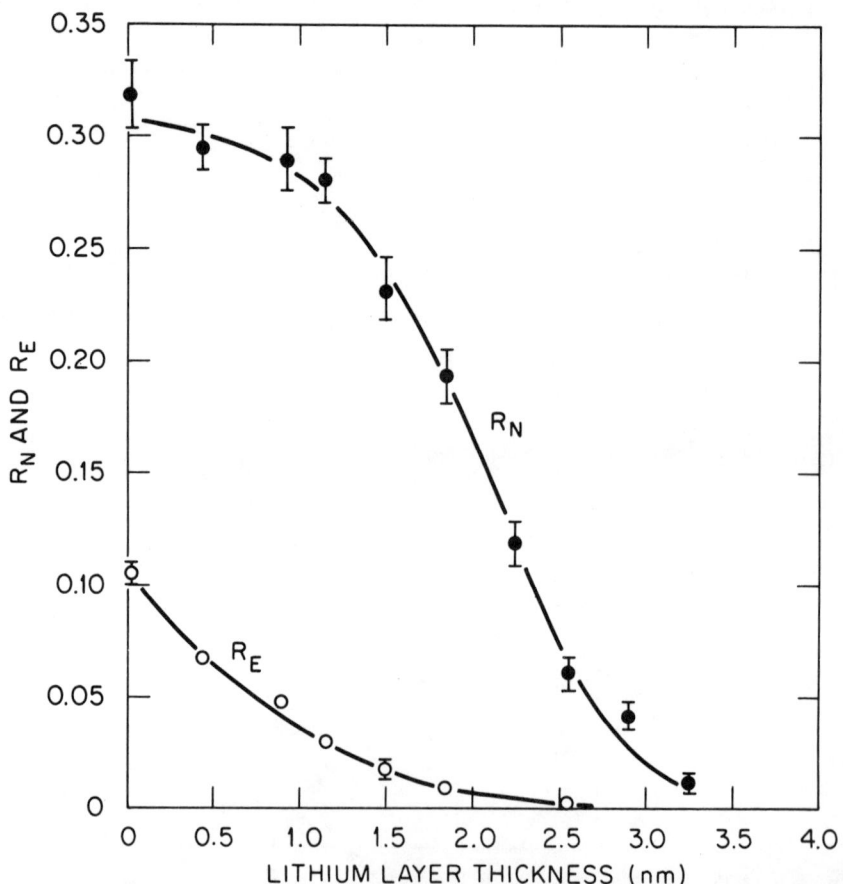

Fig. 4. Particle and energy reflection coefficients of 1 keV argon from tungsten with lithium overlayers of different thicknesses.

Figure 5 shows the particle reflection of H from Au, Al and a layered target of Au on Al. At very low incident energies the layered target has the reflection characteristics of pure Au whereas at higher energies it is similar to that of the underlying Al substrate. The dashed curve gives the average maximum penetration depth of the particles that later are reflected. When this quantity is equal to the overlayer thickness, the reflection from the composite target is about midway between the values for pure Al and pure Au.

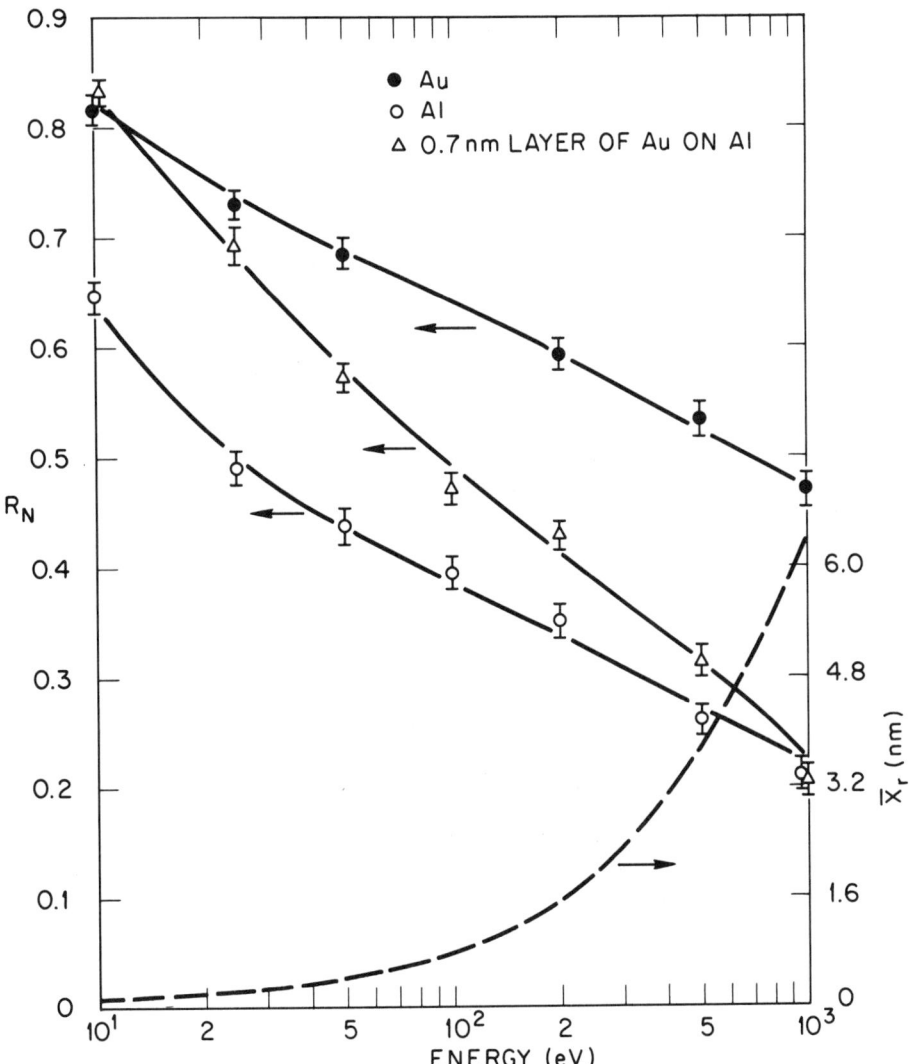

Fig. 5. Reflection of H from Al, Au and Al with Au overlayer of 0.7 nm. The dashed curve is the average maximum penetration depth reached for those particles later reflected and refers to the Au with an Al overlayer case.

Figure 6 shows the reflection of 1 keV H from amorphous Ni. The histogram in the upper left hand corner of Fig. 6 gives the angular distribution of those particles reflected with energies between 1 keV and 0.75 keV. For this energy group, which contains 37% of the reflected particles, there is a significant deviation from a cosine distribution (dashed line); in fact, it is more nearly an isotropic (uniform horizontal) distribution. The reason is that particles in this energy group are reflected from the near surface region after undergoing relatively few collisions. At lower reflected energies, the shape of the angular distribution changes and for the lowest energy group becomes over-cosine, that is more particles are reflected near the target normal direction than predicted by a cosine distribution. This over-cosine behavior becomes more pronounced as the reflected energy becomes less. The reason is that the slower moving particles are deflected toward the surface normal more effectively by the surface scattering as they leave the target. Figure 6 shows that the angular distribution integrated over all reflected energies is very close to cosine. The distribution giving the number of particles reflected per unit energy has a rather broad peak centered at about 0.6 of the incident energy. Also shown in Fig. 6 is a histogram of the maximum penetration depth of those particles that are later reflected. This distribution has a broad peak centered about 2.5 to 3.0 nm below the target surface.

Figure 7 shows the reflection of 200 eV H from amorphous Cs. As in the example of Ni discussed above, the angular distributions show marked deviations from a cosine distribution. For the high energy group (between 200-180 eV), the distribution is more nearly isotropic. For the particles reflected at lower energies, the angular distributions are again over-cosine. The angular distribution summed over all reflected energies is very close to cosine. The histogram giving the maximum penetration depth of the reflected particles is a broad distribution of rather large depths. This reflects the low number density of Cs whose lattice constant a_0 = 0.614 nm. The distribution of reflected H energies tends to peak near the incident energy which is characteristic of low energy bombardment.

Figure 8 shows the reflection of 100 eV incident H from an amorphous target consisting of an overlayer of Cs on Ni. Again the distribution of reflected H is under-cosine at the highest energies and becomes over-cosine at the intermediate and lower energies. The distribution of the maximum penetration depth reached by the reflected particles peaks around 6 a_0 (here the depth scale is in units of the bcc lattice constant of Cs, a_0 = 0.614 nm). This unusual feature is readily explained by the large difference in the atomic density of the Cs overlayer compared to that of the Ni substrate. The density of Ni is 91.4 atoms/nm^3, whereas for Cs it is only 8.64 atoms/nm^3. The energy distribution of reflected particles has a double peak which can be explained as follows: The peak near the incident energy is due to H reflected from the near surface of Cs. The broader peak at lower energy results from the H reflected from the denser Ni substrate. It occurs at a lower energy because of the energy lost by the H in traversing the Cs overlayer.

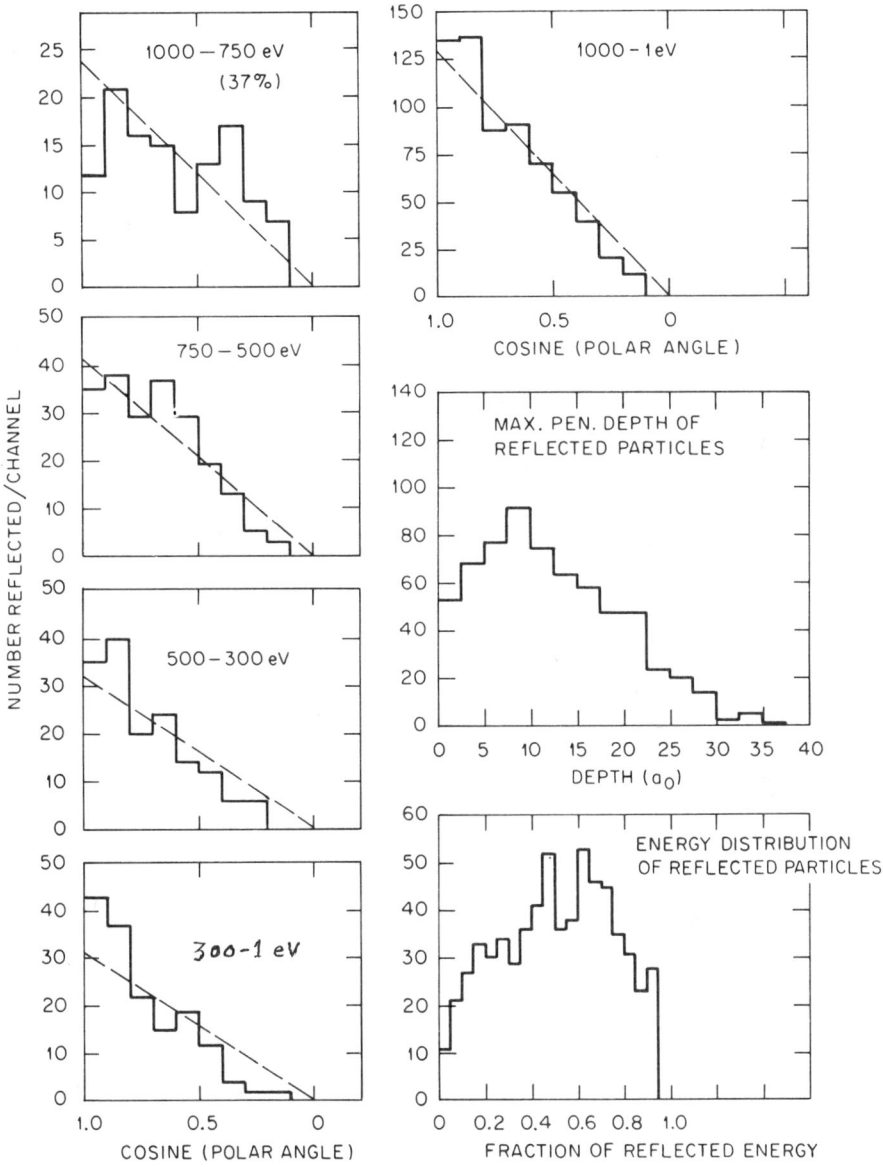

Fig. 6. Angular, energy, and maximum penetration depth of 1 keV H reflected from amorphous Ni. Here R_N = 0.320 ± 0.012, and R_E = 0.164 ± 0.007. The angular distributions are for particles reflected in different energy groups, that is, the one labeled 750-500 eV refers to the energy of the reflected particles in that interval. The polar angle is the angle of reflection and is measured from the surface normal. The lattice parameter for fcc nickel is a_0 = 0.3524 nm.

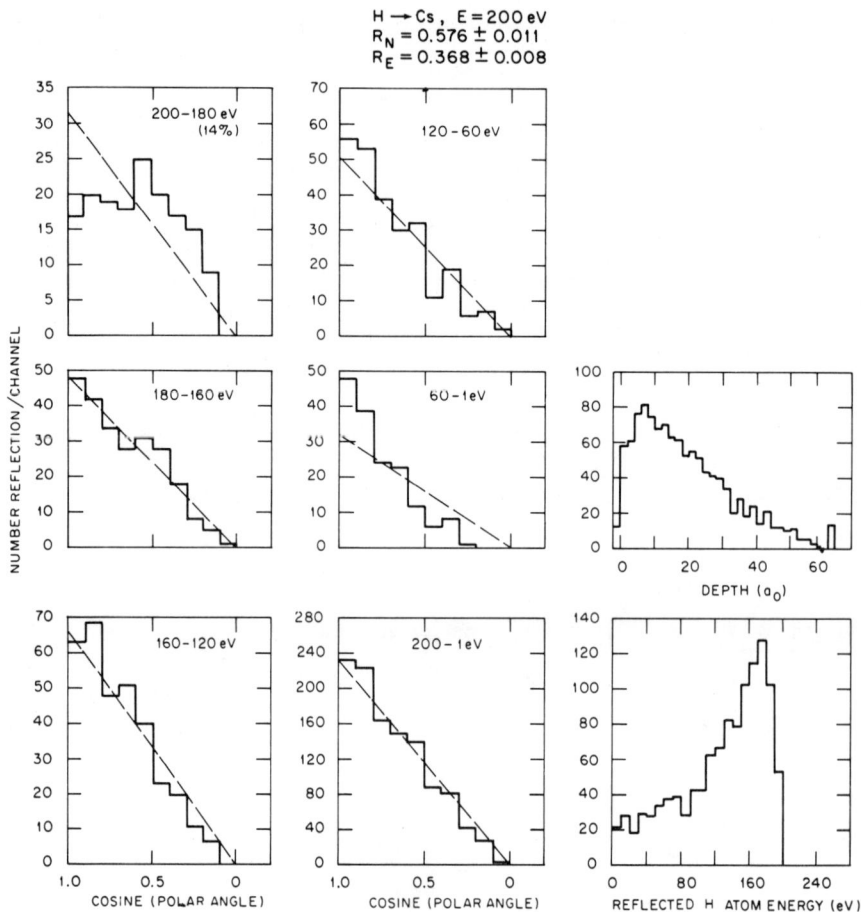

Fig. 7. Angular, energy and maximum penetration depth of 200 eV hydrogen reflected from amorphous cesium. The right top and bottom inserts are distributions giving the maximum penetration depth of particles later reflected and reflected particle energy, respectively. The lattice constant of bcc cesium is $a_0 = 0.614$ nm.

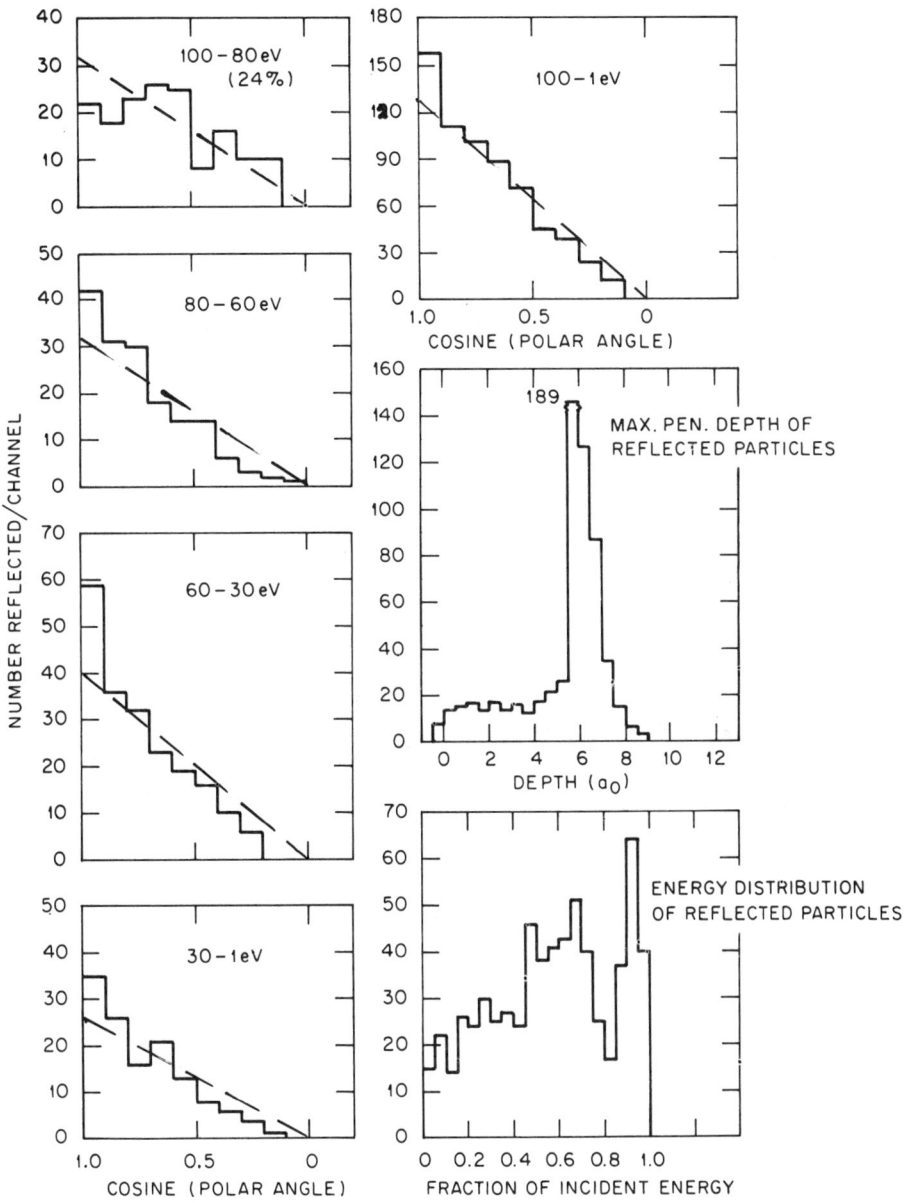

Fig. 8. Angular, energy and maximum penetration depth of 100 eV hydrogen reflected from amorphous nickel having an amorphous cesium overlayer of 5 a_0 (3.1 nm) thickness. Here $R_N = 0.649 \pm 0.015$, and $R_E = 0.370 \pm 0.011$.

CONCLUSIONS

The targets studied were chosen to have large differences between the atomic numbers of the overlayer and the substrate in order to emphasize possible reflection differences from that of monatomic targets. For very low incident energies, the back reflection is characteristic of the surface overlayer; at higher energies it becomes characteristic of the underlying substrate. The energy at which the reflection becomes characteristic of the substrate increases with increasing overlayer thickness. The reason is simply that the maximum penetration depth reached before a particle is reflected increases with increasing energy. This leads to the result, that for low Z overlayers on high Z substrates, the reflection coefficient may actually increase with an increase in energy, at least over a limited range.

The angular distributions of the reflected particles integrated over all reflected energies were found to be approximatley cosine. However, the particles reflected in the highest energy groups have an under-cosine angular distribution that is more nearly isotropic than cosine. The results are similar to our previous amorphous and polycrystalline simulations of reflection from monatomic targets.[19]

REFERENCES

1. G. M. McCracken, P. E. Stott, Nucl. Fusion 19, 889 (1979).
2. E. S. Mashkova, Radiat. Eff. 54, 1 (1981).
3. W. Eckstein, H. Verbeek, IPP Report 9/32, 1979.
4. T. Tabata, R. Ito, Y. Itikawa, N. Itoh, K. Morita, Report IPPJ-AM-18, 1981.
5. J. R. Hiskes, P. J. Schneider, J. Nucl. Mater. 93/94, 536 (1980). J. R. Hiskes, P. J. Schneider, Phys. Rev. B 23, 23 (1981).
6. R. S. Battacharya, W. Eckstein, H. Verbeek, J. Nucl. Mater. 79, 420 (1978).
7. W. Eckstein, H. Verbeek, R. S. Battacharya, Surface Sci. 99, 356 (1980).
8. P. J. Schneider, K. H. Berkner, W. G. Graham, R. V. Pyle, J. W. Stearns, Phys. Rev. B 23, 941 (1981).
9. J. N. M. Van Wunnik, J. J. C. Geerlings, E. H. A. Granneman, J. Los, Surface Sci. 131, 17 (1983), Ibid., page 1.
10. J. E. Robinson, D. P. Jackson, J. Nucl. Mater. 76/77, 353 (1978).
11. D. P. Jackson, W. Eckstein, Nucl. Instr. Methods 194, 671 (1982).
12. M. T. Robinson, in: Sputtering by Particle Bombardment I, ed. by R. Behrisch (Springer Verlag, Heidelberg, 1981) p. 73.
13. M. T. Robinson, I. M. Torrens, Phys. Rev. B 9, 5008 (1974).
14. M. T. Robinson, Phys. Rev. B 27, 5347 (1983).
15. G. Molière, Z. Naturforsch. 22, 133 (1947).
16. O. B. Firsov, Sov. Phys. JETP 6, 534 (1958).
17. J. Lindhard, M. Scharff, H. E. Schiott, Mat. Fys. Medd. Dan. Vid. Selsk. 33, No. 14 (1963).
18. O. S. Oen, M. T. Robinson, Nucl. Instr. Methods 132, 647 (1976).
19. O. S. Oen, M. T. Robinson, J. Nucl. Mater. 111/112, 789 (1982).

DISCUSSION

COOPER: You showed that at 10 eV the reflection coefficient of hydrogen from lithium was about 30% and for deuterium about 20%. Do you see a similar difference for the reflection from heavier substrates like copper at low energies of the incident particles?
OEN: No. We found the reflection of hydrogen and deuterium from heavy elements like copper to be nearly the same, within a few percent of each other. Large reflection differences between the hydrogen isotopes were found only when the mass of the target atom was comparable to that of the incident projectile.
VERBEEK: In your last example with a 5 a_0 layer of cesium on nickel you have a double peak structure in the energy distribution. Is this due to the electronic energy loss in this cesium layer or is it due to the difference in the nuclear loss of surface scattered particles?
OEN: In our model most of it is due to the electronic loss in passing through the cesium overlayer. A smaller portion is due to the increased nuclear loss in the backscattering from nickel rather than the heavier cesium.

Cs/TRANSITION METAL COMPOSITE SURFACES: FIRST PRINCIPLES CALCULATIONS OF HIGH Z, LOW WORK FUNCTION SYSTEMS

A.J. Freeman, E. Wimmer*, and S.R. Chubb
Physics Department, Northwestern University
Evanston, Illinois 60201

and

J.R. Hiskes and A.M. Karo
Lawrence Livermore National Laboratory
Livermore, California 94550

ABSTRACT

Some salient features of the experimental and theoretical data pertaining to hydrogen negative ion generation on minimum-work-function composite surfaces consisting of Cs/transition metal substrates are reviewed. Cesium or hydrogen ion bombardment of a cesium-activated negatively-biased electrode exposed to a cesium-hydrogen discharge results in the release of hydrogen negative ions. These ions originate through desorption of hydrogen particles by incident cesium ions, desorption by incident hydrogen ions, and by backscattering of incident hydrogen. Each process is characterized by a specific energy and angular distribution. The calculation of ion formation in the crystal selvage region is discussed for different approximations to the surface potential. Results of ab initio, all-electron, local density function calculations for the composite surface electronics of Cs on W(001) and Mo(001) are presented and discussed.

INTRODUCTION

As is well-known to this audience, the generation of hydrogen negative ions in particle-surface collisions on composite surfaces that are selected for their minimum work function, Φ_m has provided an effective mechanism for the development of surface-type negative ion sources. These sources take the form of a negatively-biased cesium-activated surface (the convertor) exposed to a hydrogen discharge[1,2].

---

*Permanent address: Institute für Physikalische Chemie, Universitat Wien, Wahringerstr. 42, A-1090 Vienna, Austria

The discharge provides a source of hydrogen and cesium ions that impinge on the convertor with a kinetic energy equivalent to the convertor potential. The discharge also provides a flux of energetic neutrals, of uncertain energy distribution, that contributes an additional component to the total convertor particle flux. The return hydrogen flux emitted by the convertor has several components each with its own peculiar energy and angular distribution: the cesium-desorbed component, a desorption component caused by the incident hydrogen particles, and an energetic component consisting of backscattered incident particles. If the sample of convertor substrate materials includes metals that occlude hydrogen to form metal hydride crystals, e.g. TiH_2, FeH_2, versus materials that retain hydrogen principally as an adsorbed layer, e.g. Mo, the desorption component can be subdivided into a sputtering yield of hydrogen particles emitted from the interior of the crystal, and, a surface desorption yield of particles released from the adsorbed surface layer. Experimental data is accumulating that distinguishes these return flux components and their relative magnitudes and angular distributions.

The return flux of atoms from the convertor captures electrons from the metal for conversion to negative ions within a distance of 10 Å of the surface, but the final concentration of negative ions relaxes to the equilibrium concentration over a somewhat larger distance. The electron capture and loss processes in the crystal selvage region are sensitive to the details of the surface electron density distribution. Some progress on the calculation of these processes has been achieved using approximate potentials for the surface region.

In the following sections we review briefly some aspects of negative ion formation, survival and production probabilities and discuss recent ab initio calculations that deal with the evaluation of the surface potential. An earlier review together with several recent papers dealing with backscattering and desorption are given in the reference section[3-6].

Formation, Survival, and Production Possibilities

The production probability is defined as the fraction of return flux atoms that are converted to negative ions in the selvage and survive as negative ions to infinity. In theoretical discussions for the calculation of the production probability it is sometimes convenient to factor this probability into two factors, a formation and a survival probability. The formation of negative ions occurs relatively close to the surface and is sensitive to the details of the surface electronics. The survival of negative ions as they recede to infinity is a less uncertain quantity since it relies mainly on details of the surface potential at relatively large distances from the surface where the potential is approximated by the image potential. The calculation of the survival probability for negative ions moving away from a cesium metal surface[7] is in fair agreement with the experimental observations[8].

For a composite surface consisting of a minimum work function coverage of cesium over a transition metal substrate we have adopted the schematic potential configuration for the active electron shown in[3] Fig. 1. Referring to the lower portion of the figure, at close separations the energy level of the active electron is comparable to or lower than the Fermi level of the composite surface and electron capture from the surface to form the ion can occur. At large distances from the surface the negative ion "looks back" to see an image potential and a surface dipole potential that impedes the loss of the electron back to the metal. The magnitude of the dipole layer, $\Delta\Phi$, is taken to be the difference in work function of the clean substrate and the composite surface at minimum work function, Φ_m. The production probability is clearly a function of Φ_m and $\Phi\Delta$. The effect of the dipole potential layer is to profoundly affect the magnitude of the survival probability, enhancing the survival probability by a factor of ten or more for few-volt ions[7]. Criteria have been mentioned for maximizing $\Delta\Phi$ by the selection of appropriate substrates[9]. For a particular substrate material, a polycrystalline sample that exhibits several crystal faces will give a poorer $\Delta\Phi$ than a monocrystalline faced crystal selected for optimum $\Delta\Phi$.

Using the concept of a production probability factored into a formation and survival probability, a model has been developed[10] for the interpretation of negative ion formation by backscattering of normally incident particles on alkali metals[11]. The model has also been applied to a composite surface, Cs/Ni. Using the model as a basis for extrapolation to lower energies a production probability of 58%[12] is projected for 24 eV backscattered particles[12].

An alternative formulation of the problem has been developed by the FOM group in which the production probability is calculated directly[13]. In their model, the image potential is extended downward to the bottom of the conduction band and the dipole potential term is ignored. By retaining a single adjustable parameter, the experimental data can be reproduced. In a parallel experimental program the production probability is measured directly in a low angle glancing collision of protons on Cs/W(110); production probabilities up to 45% are observed[14,15]. This experiment also demonstrates an enhancement, by a factor of two to three, of yields from a monocrystalline W(110) substrate compared with a polycrystalline W substrate[16].

With reference to the backscattering yield, an important distinction occurs for the case of particles incident normal to the convertor surface compared to the case of glancing incidence. In the former case the analysis[16] requires that the total negative ion yield be factored into a particle reflection probability and a production probability. While in the case of glancing collisions the reflection probability can be approximated by unity and the production probability is observed directly. For comparison with theoretical models the glancing collision data is often more immediately applicable, while for ion source technology needs the full normal incidence yield is essential.

Ab Initio Calculation of Surface Electronics for Composite Surfaces

All previous theoretical approaches to the alkali adsorption problem have in common the feature that they do not attempt an all-electron treatment which incorporates the full atomistic aspect of the system, since the task seemed monumental. In the past decade it became more and more evident that the local approximation to the density-function theory[17,18] gave a consistently accurate and realistic description of the electronic[19,20] structure and the energetics of condensed and also molecular[21] systems. This factor has stimulated the development of methods which allow local density calculations for free surfaces and surfaces with adsorbed atoms that include the atomistic nature of the surface[22-25]. One of the most accurate and efficient theoretical/computational approaches for condensed systems is the linearized-augmented plane-wave (LAPW) method[26-28]. In recent years this LAPW method, originally, designed for bulk systems has been adapted for a film geometry[29-31]. It has been demonstrated that this single slab geometry provides a promising approach to the theoretical/computational treatment of the electronic structure of clean surfaces[30-33] and surfaces with overlayers[34-36].

Because of the reduced symmetry and reduced coordination number of atoms at the surface, a realistic quantum mechanical treatment has to allow for a general charge density and potential, since shape (such as muffin-tin) approximations not only could influence the results in an uncontrolled way but could cause confusion concerning the strengths and limits of the local density approach to density-function theory. Recently, we have presented the full-potential linearized-augmented plane-wave (FLAPW) method for thin films[37] in which no shape approximations are made to the density and the potential, hence solving the local-density-functional (LDF) one-particle equations fully self-consistently.

As a step towards calculating a precise value for the production probability we have initiated ab initio calculations for cesium overlayers on W and Mo. The first step in this series is the calculation of a c(2x2) structure of cesium overlaying a W(001) or a Mo(001) slab. The c(2x2) structure corresponds to a full monolayer of cesium, twice the coverage of the p(2x2) configuration that yields the work function minimum. The selection of the larger coverage considerably simplifies the numerical work while providing important insights into the more general problem. These calculations are described in detail in a recent paper[38]. We shall summarize here only the principal features.

We have studied the electronic structure of a c(2x2) Cs overlayer on Mo(001) and W(001) surfaces by means of highly accurate, fully self-consistent, all-electron calculations using our full-potential linearized-augmented plane-wave (FLAPW) method[39] for thin film within local density-functional theory. The clean W(001) or Mo(001) surface was represented by a single slab of 5 layers of W or Mo. Cs was deposited in the form of c(2x2) overlayers on both sides of the 5-layer W or Mo slab resulting in a unit cell with 12 atoms. For W(001) independent self-consistent calculations were performed for 3 distances between the planes of the Cs and surface W atoms, namely d = 2.60, 2.75, and 2.90 Å. For comparison, a self-consistent calculation for the Cs monolayer with the same Cs-Cs distance as in the c(2x2) overlayer on W(001) was also carried out.

We found that upon cesiation that the occupied part of the energy band structure of W (or Mo) is not changed drastically. The characteristic high density of W surface states and surface resonance states near the Fermi energy persists for the cesiated W surface. The interaction between the Cs valence states (originating from the atomic Cs 6s levels) and the W-d surface states leads to a hybridization of these states and causes these predominantly W-d like states to be stabilized energetically. The most striking effect is seen for the d_{z^2}-like W surface state at $\bar{\Gamma}_1$ just below the Fermi energy: this state is lowered in energy by 1 eV upon cesiation[40]. The Cs-5p semi-core states are found to interact markedly with the W-s like states particularly those at the bottom of the W-s band (and similarly for Mo).

The valence charge density in the surface region is dominated by the W(Mo) states. The spill-out of W(Mo) electrons into the vacuum leads, for the cesiated surface, to an increase in valence charge near the Cs nuclei compared with the isolated Cs monolayer. Upon cesiation the Cs valence electrons are found to be polarized towards the W surface resulting in a depletion of electronic charge between and outside the Cs nuclei and an increase of electronic charge in the Cs/W interface region. The Cs-5p semi-core electrons show a polarization opposite to that of the Cs valence electrons. These multiple polarizations within the Cs overlayer amount to a net reduction of the spill-out dipole and hence bring about the lowering of the work function upon cesiation. As an example of these results we show in Fig. 2 the effective one-electron potential. The net result of these multiple surface dipoles is a lowering of the work function upon cesiation from a 4.66 Ev (clean 5-layer W slab) to 2.77, 2.55, and 2.34 eV depending on the height of the Cs atoms above the surface of 2.60, 2.75 and 2.90 Å, respectively. The Cs-induced changes in the charge density and in the surface dipole are essentially confined to the region <u>outside</u> the surface W(Mo) atoms. For the case of Cs on Mo(001), we used the single Cs-Mo distance d = 2.90 Å. Here the work function is lowered from 4.47 eV to 2.37 eV upon cesiation.

The simple classical picture of Cs donating an electron _into_ the metal, becoming a Cs^+ ion, and forming a dipole with its negative image charge is inadequate. Presumably also for lower coverages the charge redistribution will take place essentially _outside_ the W surface atoms. however, the fact that the minimum of the work function for the CsW(001) system occurs at half the Cs coverage studied in the present work shows that at lower coverages the polarization of the Cs atoms becomes more pronounced.

In conclusion, we find that Cs on a W(001) or Mo(001) surface at a coverage of 1 Cs atom for 2 W(Mo) atoms forms a polarized-metallic rather than ionic overlayer. Important for a realistic picture of the surface is the fact that the valence charge density in the surface region is strongly dominated by the d-like states of the substrate. It is the interaction of the surface W(Mo) atoms with their d-like states and the Cs valence electrons originating from the atomic 6s states, which results in the polarization of the Cs valence electrons. Surprisingly, the Cs-5p semi-core electrons participate greatly in the bond formation due to their interactions with the electrons at the bottom of the W(or Mo) s-band. The net redistribution of the Cs-5p states upon cesiation can be described as a counter-polarization to the polarization induced by the Cs valence electrons.

It is crucial that most of the surface states and surface resonance states of the W(001) and Mo(001) surface persist even after cesiation and retain the high density of surface states and surface resonance states so characteristic of the W(001) and Mo(001) surface. This means that the electronic structure of the W(Mo) substrate remains important even if the surface is covered with a Cs overlayer. It will be extremely interesting and relevant to investigate if these surfaces are a special case or if this importance of the substrate is a general feature of cesiated transition metal surfaces. Experimental and theoretical effort is needed to settle the question of the height of the Cs atoms which was shown to be important for quantitative statements about the work function.

SUMMARY

The surface generation of negative ions is now established as an essential component in the development of direct-extraction type negative hydrogen ion sources. Several processes contribute to the negative ion generation including cesium and hydrogen desorption and hydrogen backscattering. The enhancement of the specific yield of negative ions per incident particle onto the convertor is possible through the proper selection of substrate materials in the Cs/substrate complex. The optimum system has probably not yet been identified. A precise value for the production probability depends on the details of the surface electronics. Ab initio calculations for the composite Cs/W and Cs/Mo systems have indicated the origin of the work function lowering in the c(2x2) overlayer configuration as due to multiple surface dipoles generated by the polarization of the cesium overlayer.

ACKNOWLEDGEMENT

This work was performed under the auspices of the U.S. Office of Naval Research under grant number N00014-81-K-0438 and the U.S. Department of Energy by the Lawrence Livermore National Laboratory under grant number W-7-405-ENG-48.

REFERENCES

1. K.N. Leung and K.W. Ehlers, Proc. Second Int. Symp. on the Production and Neutralization of Negative Hydrogen Ions and Beams, Brookhaven, NY, Oct. 1980, p. 65, BNL 51304 (1980).
2. K. Prelec, Ibid, p. 145.
3. J.R. Hiskes, J de Phys. (Paris) 40, C7-179 (1979).
4. W. Eckstein, H. Verbeek, and R.S. Bhattacharya, Surf. Sci. 99, 356 (1980).
5. W. Eckstein and H. Verbeeck, J. Nucl. Mat. 93 and 94, 518 (1980).
6. P.J. Schneider, W. Eckstein, and H. Verbeeck, Fifth Symp. on Particle-Surface Interactions, Gatlinburg, Tn (1982).
7. J.R. Hiskes and A.M. Karo, Proc. Symp. on the Production and Neutralization of Negative Hydrogen Ions and Beams, Brookhaven, NY, BNL-50727, Sept. (1977).
8. A Pargellis and M. Seidl, Phys. Rev. B 25, 4356 (1982).
9. J.R. Hiskes and P.J. Schneider, Proc. Second. Int. Symp. on the Production and Neutralization of Negative Hydrogen Ions and Beams, Brookhaven, NY, BNL-51304, p. 15 (1980).

10. J.R. Hiskes and P.J. Schneider, Phys. Rev. B **23**, 949 (1981).
11. P.J. Schneider, K.H. Berkner, W.G. Graham, R.V. Pyle, and J.W. Stearns, Phys. Rev. B **23**, 941 (1981).
12. J.R. Hiskes and P.J. Schneider, J. Nucl. Matl. **93**, and **94**, 536 (1980).
13. B. Rasser, J.N.M. van Wunnik, and J. Los. Surf. Sci. **118**, 697 (1982).
14. J. Los, E.A. Overbosch, and J.N.M. van Wunnik, Proc. Second Brookhaven Symp. Brookhaven, NY, BNL-51304, p. 23, (1980).
15. J.N.M. van Wunnik, B. Rasser, and J. Los, Phys. Rev. Lett. A **87**, 288 (1982).
16. E.H.A. Granneman, P.J. van Bommel, P. Wassman, H.J. Hopman, F. Siebenlist, J.N.M. van Wunnik, J.J.C. Gaerlings, and J. Los, Bull. Am. Phys. Soc. **27**, No. 8, pt. II, 1135 (1982).
17. P. Hohenberg and W. Kohn, Phys. Rev. B **1346**, 864 (1964).
18. W. Kohn and L.J. Sham, Phys. Rev. A **140**, 1133 (1965).
19. V.L. Moruzzi, J.F. Janak, and A.R. Williams, Calculated Electronic Properties of Metals (Pergamon, NY 1978).
20. D.D. Koelling, Reports on Progress in Physics **44**, 139 (1981).
21. J. Harris and R.O. Jones, J. Chem. Phys. **70**, 830 (1979).
22. J.A. Appelbaum and D.R. Hamann, Phys. Rev. B **6**, 2166 (1972).
23. C.S. Wang and A.J. Freeman, Phys. Rev. B **19** 793 (1979); Phys. Rev. B **19**, 4930 (1979).
24. J.R. Smith, J.G. Gay, and F.J. Arlinghaus, Phys. Rev. B **21**, 2201 (1980).
25. D.R. Hamann, L.F. Mattheiss, and H.S. Greenside, Phys. Rev. B **24**, 6151 (1980).
26. P.M. Marcus, Int. J. Quant. Chem. S **1**, 567 (1967).
27. O.K. Andersen, Phys. Rev. B **12**, 3060 (1975).
28. D.D. Koelling and G.O. Arbman, J. Phys. F **5**, 2041 (1975).
29. O. Jepsen, J. Madsen, and O.K. Andersen, Phys. Rev. B **18**, 605 (1978).
30. H. Krakauer, M. Posternak and A.J. Freeman, Phys. Rev. B **19**, 1706 (1979).
31. M. Posternak. H. Krakauer, A.J. Freeman, and D.D. Koelling, Phys. Rev. B **21**, 5601 (1980).
32. H. Krakauer, M. Posternak, and A.J. Freeman, Phys. Rev. Lett. **43**, 1885 (1979).

33. D.-S. Wang, A.J. Freeman, H. Krakauer, and M. Posternak, Phys. Rev. B 23, 1685 (1981).
34. D.-S. Wang, A.J. Freeman, and H. Krakauer, Phys. Rev. B 24, 1126 (1981); Phys. Rev. B 24 3092 (1981); Phys. Rev. B 24, 314 (1981).
35. G.A. Benesh, H. Krakauer, D.E. Ellis, and M. Posternak, Surf. Sci. 104, 599 (1981).
36. D.-S. Wang, A.J. Freeman, and H. Krakauer, Phys. Rev. B 26, 1340 (1982).
37. E. Wimmer, H. Krakauer, M. Weinert, and A.J. Freeman, Phys. Rev. B 24, 864 (1981).
38. E. Wimmer, A.J. Freeman, J.R. Hiskes, and A.M. Karo, Phys. Rev. B (1983) (in press).
39. E. Wimmer, H. Krakauer, M. Weinert, and A.J. Freeman, Phys. Rev. B 24, 864 (1981).
40. Very recent measurements by P. Soukiassian, R. Riwan, C. Guillot, J. Lecante, and Y. Borensztein (private communication) seem to confirm our prediction of the 1 eV shift of the SS upon cesiation.

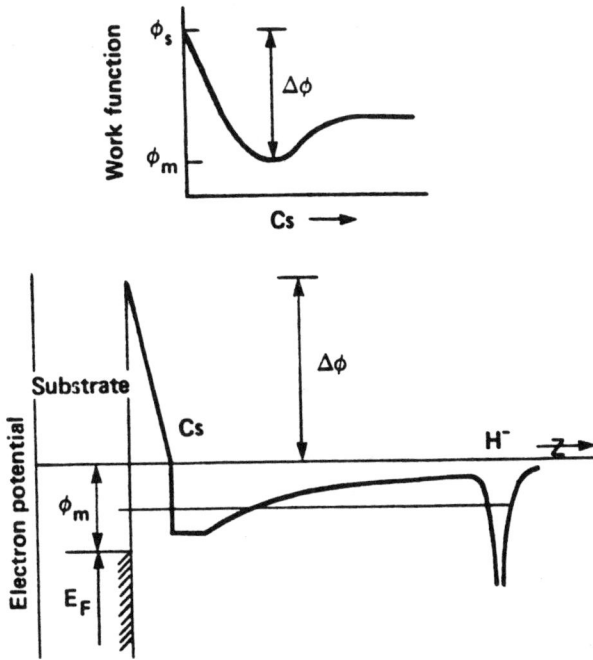

Figure 1. Upper portion: Variation of composite surface work function with cesium coverage. Lower portion: schematic of effective potential seen by active electron, Refs. 3, 10.

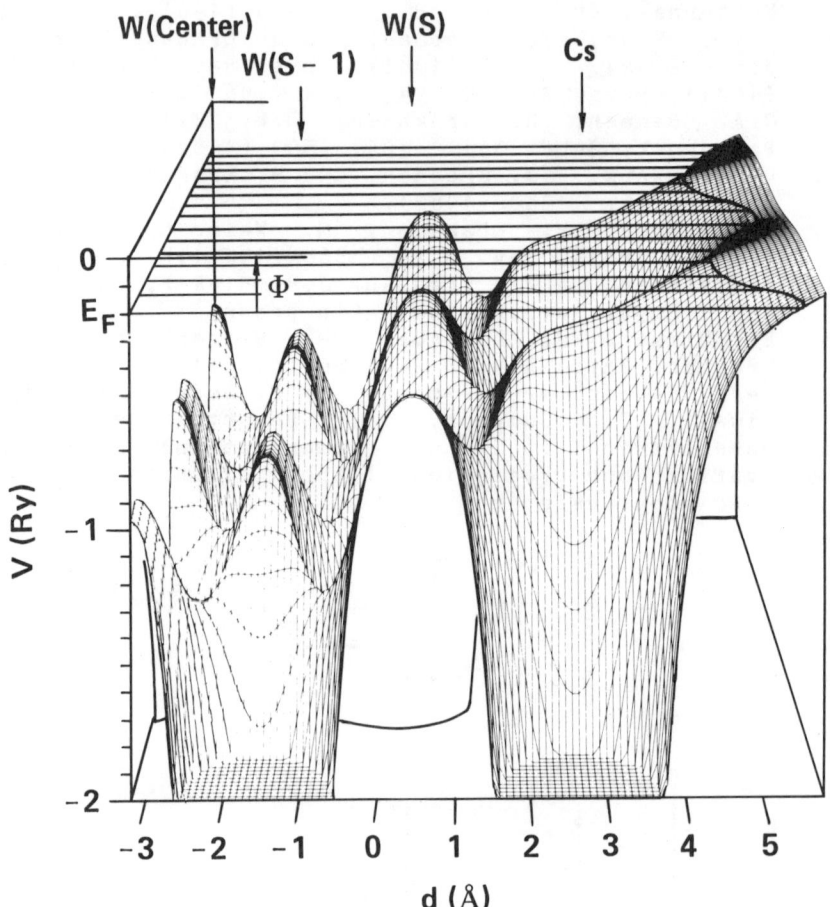

Figure 2. Effective one-electron potential for c(2x2) Cs on W(001) (d = 2.60 Å) in the (110) plane perpendicular to the surface. Φ denotes the work function of the system.

DISCUSSION

EHLERS: I am rather impressed with your work, John, but I have a question. Is it conceivable that you can extend this work to include adsorbed hydrogen onto the cesiated tungsten?
HISKES: Well yes, the code is very general. One can do most anything in principle, including fractures or multilayers, etc. The problem is computational time.
BARNETT: How much does hydrogen coverage change the work function? Do you have any idea?
HISKES: It does change the work function if the hydrogen coverage is no more than the cesium coverage. Hydrogen is electro-negative and it's known that it buries itself underneath the cesium regardless of whether you apply the hydrogen first or the cesium first, the hydrogen always gets under there. If it's no more than the cesium coverage, the p(2 x 2) coverage, it reduces the composite work function by 0.05 eV, so it's just a barely noticeable factor. Oxygen probably has a much bigger effect.
BARNETT: How big was the effect?
HISKES: About 0.05 eV is the number I've seen in the literature. It's just barely measureable.
LEUNG: Why is the cesium atom bigger than the tungsten atom?
HISKES: The spacing of the atoms in a bulk cesium crystal turns out to be precisely double the spacing of tungsten atoms in a tungsten crystal. This is rather handy because you have a very symmetrical relationship between tungsten and cesium. Cesium is bigger because it has a single valence 6s electron which is not paired with another electron and that 6s shell is just a very large shell. The 6s electron of the free cesium atom has a large RMS radius compared to that of tungsten which is closer to being a closed shell. For tungsten the 6s shell is closed and the 5d shell about half full.
PALMER: I was wondering if you would comment on the effect that your strong dipole layer has on the resonant tunnelling process.
HISKES: That's what we're trying to get at but we need an accurate potential and we haven't done that calculation yet. What we're emphasizing now is doing the p(2 x2) calculation which is the configuration for the minimum work function and the coverage where you get the maximum negative ion yield.
PALMER: But qualitatively the effect of the layer would be to inhibit that tunnelling process by a factor of 10.
HISKES: The formation process occurs when the atom is very close to the surface. It is not clear that the dipole layer is active at that time. After the negative ion is moved a good deal away from the surface, it looks back at the surface and then sees the dipole layer formation. It is then that the dipole layer inhibits the loss of the electron from the negative ion back to the metal.
PALMER: Is it the negative ion to start off or is it a neutral?
HISKES: It is almost certainly not a negative ion until it is at least 4 or 5 angstroms from the surface. It emerges through the cesium layer as a hydrogen atom and after it is about 4 or 5 angstroms off the surface and until it is almost 10 angstroms off the surface it is in the process of evolving into a negative ion.

PALMER: How far away is the dipole layer?
HISKES: The dipole layer is between the cesium layer and the surface. Our picture is that as it comes through the dipole layer it is a hydrogen atom.
PALMER: But the dipole layer then seems to me would inhibit the formation of negative ions.
HISKES: The dipole layer need not have as large an effect on the formation process as it has on the survival probability. When the atom is close to the surface the formation probability is near unity even with the dipole layer acting. Increasing the dipole layer a certain amount would not reduce the formation probability a correspondingly amount because the formation process is near saturation. The net effect is that the formation probability is near unity almost independent of the presence of the dipole layer. At large separations the survival probability is sensitive to the dipole layer because the electron must tunnel through both the image-coulomb barrier and in addition the dipole layer in order to penetrate into the interior of the crystal. It is the presence of these two barriers acting together on the outgoing electron wave, or current, that causes the sensitivity. Once we have the whole potential for all atom spacings from the p(2 x 2) calculation the next step is to watch the atom evolve into a negative ion at the various spacings; then we will have a complete potential description and from that we can generate the formation probability in a consistent way.

DE-EXCITATION AND EQUIPARTITION IN H_2-SURFACE COLLISIONS*

A. M. Karo, J. R. Hiskes, K. D. Olwell, and T. M. DeBoni
Lawrence Livermore National Laboratory
Livermore, California 94550

R. J. Hardy**
University of Nebraska
Lincoln, Nebraska 68588

ABSTRACT

We have used computer molecular dynamics to study de-excitation of vibrationally-excited H_2 molecules undergoing repeated collisions with a wall maintained at some temperature T(W). We have calculated the average loss or gain of vibrational, rotational, translational, and total molecular energy as a function of the number of collisions for a statistically significant number of molecules having some initial vibrational state v" ranging from v" = 2 to v" = 12. For initial translational and rotational temperatures around 500°K, we have obtained a consistent picture of rapid vibrational de-excitation during the first collision, with a corresponding increase in translational kinetic energy and with rotational excitation from an initial J = 1 state up through values greater than J = 14. Wall collisions are found to provide an effective and rapid kinematic mechanism for V-T, V-R, and R-T energy transfer. The rate of loss of total molecular energy to the wall (accommodation) is discussed and compared with the rate of energy redistribution (equipartition) among the vibrational, rotational, and translational degrees of freedom of the molecules.

INTRODUCTION

During the past several years we have carried out an extensive series of computer molecular dynamics experiments designed to study the transient response of vibrationally-excited H_2 molecules as they collide with an appropriate surface[1,2]. By analyzing a large number of classical trajectories, we are able to obtain a qualitative estimate of the average loss of vibrational energy as a function of the number of wall collisions. From these calculations we also obtain qualitative estimates of the rates of energy exchange, and therefore of equipartition, as well as estimates of the accommodation coefficient for energy transfer between the molecules and the

---

* Work performed under the auspices of the U. S. Department of Energy by the Lawrence Livermore National Laboratory under contract number W-7405-ENG-48.

**Consultant to Lawrence Livermore National Laboratory

wall. A number of parameters have been examined that may contribute to the rate of energy transfer from the vibrational to other degrees of freedom of the molecule or to the wall. These include the incident translational and rotational energies of the molecule, the wall temperature, and the nature of the physical or chemical interaction of the molecule with the wall.

Molecular dynamics has evolved to a flexible and useful modeling approach because of the availability of modern high-speed computers and because of the complementary development of sophisticated software, particularly computer graphics. Simply by setting up and solving the Newton's Second Law equations of motion for each of the particles in any given system, we monitor the detailed time evolution of the system. The initial positions and velocities of the particles represent the initial conditions on these equations. It remains only to define the forces acting on each of the particles. This has been done in the present calculations by means of the usual sum-of-pair-potentials approximation. Thus, we have a powerful method of examining the microscopic dynamics of a wide variety of phenomena in both homogeneous and heterogeneous systems.

DISCUSSION

Vibrotational states of H_2 have been selected corresponding to $v'' = 2$, $J = 1$, through $v'' = 14$, $J = 1$. Molecular energies found in typical hydrogen discharges range from 500 K to 1500 K; we have taken 500 K as the initial translational temperature. Several series of calculations at 1500 K showed vibrational de-excitation to be relatively insensitive to this parameter in this temperature range. Initial rotational energies were chosen to correspond to $J = 1$, the state predominating at 500 K. Again, we found in exploratory calculations little sensitivity of vibrational de-excitation to the choice of lower-lying rotational states. As in our earlier work[1,2], the H_2 potential is obtained from a Morse potential fit to spectroscopic data. The two-body potential for FeH has been adapted from the work of Scott and Richards[3], and the Fe potential is a Morse potential that approximates the two-body potential derived by Johnson using experimental elastic-constant data[4]. The molecule-wall potential is least well-characterized; therefore we have varied the parameters defining the FeH pair potential over wide ranges to verify that the qualitative aspects of our results remain essentially unchanged.

In order to present a representative sample of our results, we have chosen initial vibrotational states representing H_2 ($v'' = 8$, $J = 1$) and H_2 ($v'' = 12$, $J = 1$). These states lie 28246 cm^{-1} and 35,231 cm^{-1}, respectively, above the potential energy minimum of the Morse curve. For this potential the left and right classical turning points are 0.423 Å and 1.757 Å for the $v'' = 8$ vibrational level and 0.392 Å and 2.397 Å for the $v'' = 12$ level. The Fe atoms forming the wall have been given a local kinetic temperature corresponding to 500 K.

Our studies have been carried out for both two- and three-dimensional systems, relying on the former to obtain a broad qualitative understanding of vibrational relaxation and of the redistribution of vibrational energy into other states following repeated wall collisions. A smaller number of three-dimensional trajectory calculations have been carried out to ensure that the qualitative nature of our conclusions is not greatly affected by the lower dimensionality. Three-dimensional trajectories add new translational, vibrational, and rotational degrees of freedom to the dynamics, and it is necessary to determine to what extent these might affect the rate of vibrational de-excitation, the rate of approach of the mean energies of vibration, rotation, and translation to statistical equilibrium (i.e., equipartition among the degrees of freedom), and the rate of accommodation of the total molecular energy with the temperature of the wall.

Figure 1 shows schematically the initial configurations for the 2D and 3D situations. It can be seen that the 2D case represents a restriction to trajectories lying in a plane that includes one of the (100) planes of the face-centered-cubic Fe lattice.

Two computer codes have been used for these studies: IMPACT for both the 2D and 3D calculations, and FLUX, a 3D molecular dynamics code that continuously monitors the center-of-mass energy for an arbitrary region of the ensemble (here, the H_2 molecule), its rotational energy about the center-of-mass, its vibrational energy, and its total energy. The region being monitored can be easily extended to include the portion of the wall affected by the impact, so that we may examine in more detail not only the redistribution of energy within the molecule itself as it interacts with the wall, but also the influence of wall structure on the change of total molecular energy.

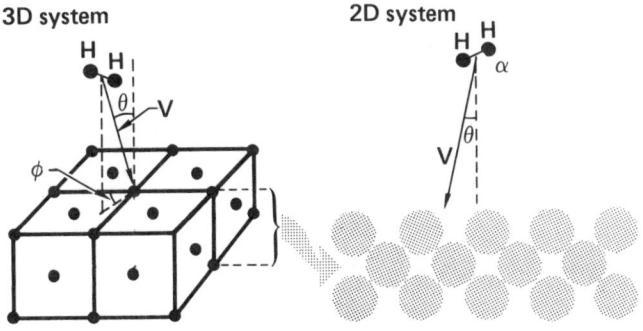

Fig. 1 Schematic representation of typical starting configurations for 2D or 3D computer simulations of an H_2 molecule moving toward a surface at velocity v and angle θ from the normal.

CALCULATIONS FOR 500 K THERMAL VELOCITIES

Initial conditions for our model system include the vibrotational and translational energies of the molecule, the wall temperature, and the geometry defining the initial part of the trajectory. A complete computer "run" consists of repeated molecular collisions with the wall for a specified number of bounces unless capture or dissociation occurs. A new H_2 molecule is then randomly selected with respect to position and phase, and a new set of trajectories obtained. This sequence is repeated until the run is completed. In our earlier work, we considered a maximum of 10 bounces and 10 sets of trajectories, so that a run in which no capture or dissociation occurred would consist of 100 individual trajectories made up of 10 "first bounce" collisions with the wall, 10 "second-bounce" collisions, and so on[1,2]. Because of the rapid approach to equipartition, a 5 bounce limit subsequently proved sufficient for initial states with $v'' > 6$.

In Figures 2, 3, and 4, we show results for the first 5 bounces for several hundred molecules initially in the $v'' = 12$, $J = 1$ vibrotational state and with a translational temperature of 500 K. In Figure 2, it can be seen that successive wall collisions result in a de-excitation and broadening of the vibrational population distribution. The collisions with the wall are accompanied by a small fractional loss of energy to the wall and a considerable increase in rotational and translational energy. As shown in Figures 2-4, the mean values for vibrational, rotational, and translational energies approach a statistical equilibrium (and, thus, equipartition among the degrees of freedom) while maintaining a broad distribution in each of these degrees of freedom. In Figure 5, we show the vibrational redistribution occurring during the first 5 wall collisions for a large ensemble of molecules in an initial $v'' = 8$, $J = 1$ state, where we again note a similar de-excitation and broadening of the distribution.

We have also analyzed these results by considering at each bounce the deviations D of the molecular energy from equipartition. We assume that, on the average, at each bounce the deviation from equipartition is reduced by a fixed fraction f of the deviation. This fractional adjustment is obtained by a least squares fitting of the following expressions for D to data from the first 5 wall collisions:

$$D_n = D_0(1-f)^n, \qquad (1)$$

where for the total energy,

$$D_n^{total} = E_n^{total} - \tfrac{5}{2}kT^{wall}, \qquad (2)$$

and for one of the individual degrees of freedom

$$D_n^{indiv} = E_n^{indiv} - p\tfrac{1}{2}kT_n^{H_2}, \qquad (3)$$

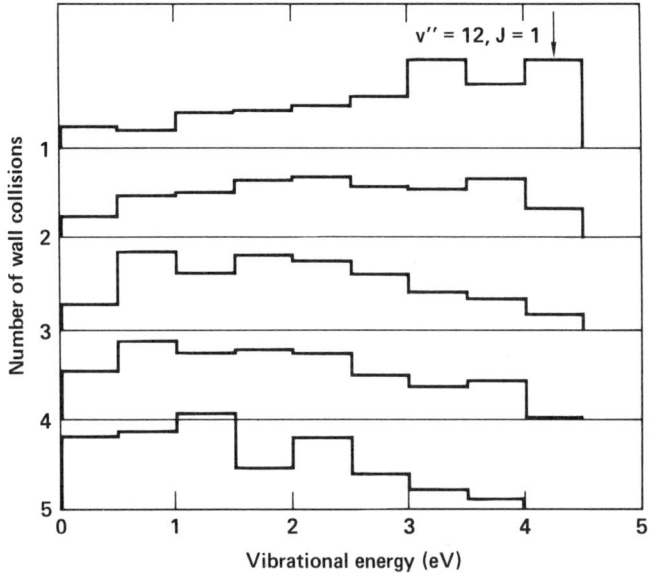

Fig. 2 Vibrational energy distributions after successive wall collisions. Initial vibrotational state of the H_2 molecules corresponds to $v'' = 12$, $J = 1$.

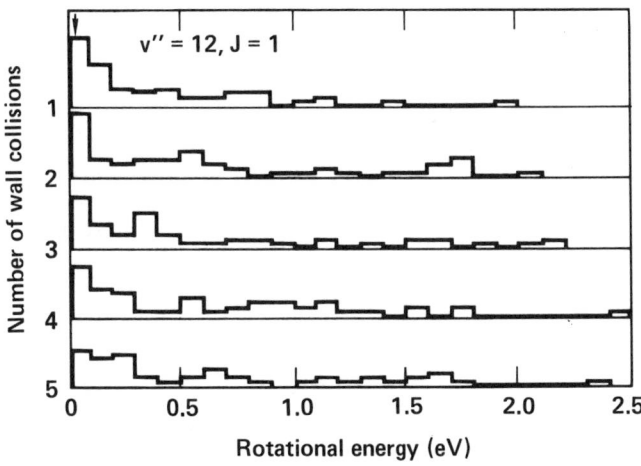

Fig. 3 Rotational energy distributions after successive wall collisions. Initial vibrotational state of the H_2 molecules corresponds to $v'' = 12$, $J = 1$.

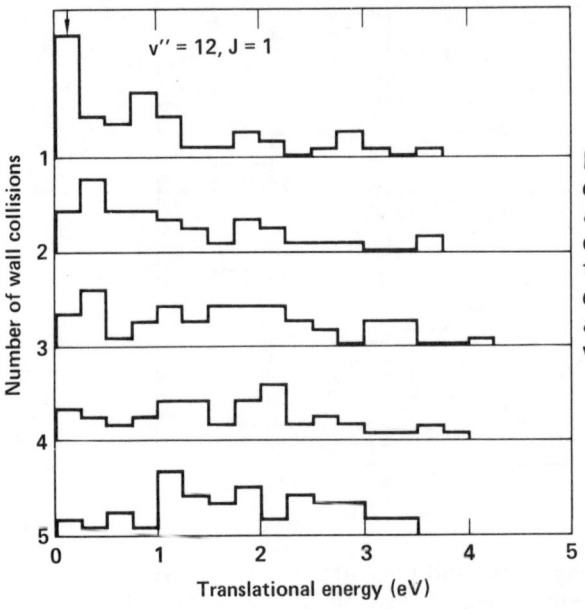

Fig. 4 Translational energy distributions after successive wall collisions. The initial translational velocity corresponds to the average molecular velocity at 500 K.

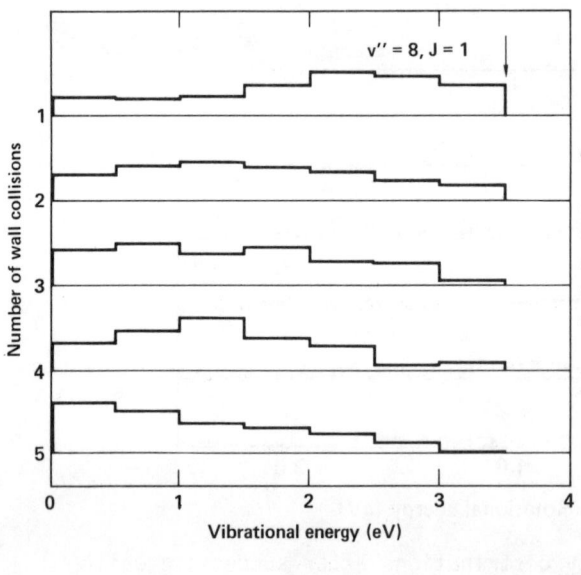

Fig. 5 Vibrational energy distributions after successive wall collisions. Initial vibrotational state of the H_2 molecules corresponds to v" = 8, J = 1.

where p is the number of degrees of freedom for vibration, rotation, or translation, and where the temperature of the molecule is given by

$$T_n^{H_2} = T^{wall} + \frac{2D_0^{total}(1-f^{total})^n}{5k} \quad (4)$$

Results are given in Table I for the total energy and for the vibrational, rotational, and translational energies. Values for f(total), the fractional adjustment of the total energy, give essentially the rates of accommodation. The relatively large values for the other fractional adjustments indicate that equipartition has occurred to a large extent after 3 to 5 bounces; the relatively small values for f(total) indicate that for molecules that survive repeated collisions, the molecular energy is being only slowly transferred to the wall.

Table I. Fractional adjustments of the deviation from equipartition for molecules undergoing five wall collisions. The mean absolute error is also given.

	v" = 8, J = 1 (1500 trajectories)		
f(total)	f(vibration)	f(rotation)	f(translation)
0.0352 ± 0.0095	0.645 ± 0.071	0.628 ± 0.146	0.715 ± 0.110
	v" = 12, J = 1 (500 trajectories)		
f(total)	f(vibration)	f(rotation)	f(translation)
0.0325 + 0.0072	0.535 ± 0.091	0.468 ± 0.084	0.624 ± 0.107

CALCULATIONS FOR 1 - 10 eV IMPACT VELOCITIES

Another source of vibrationally excited H_2 can come from the Auger neutralization of the H_2^+ molecular ion[5]. However, such molecular ions will have translational velocities normal to the wall that can range up to and exceed 10 eV. Thus, H_2 (v" > 6) molecules will be present with velocities normal to the wall in these higher energy ranges[5].

We have begun an initial survey, as shown in Table II, of one-bounce collisions in order to examine the survivability of molecules undergoing such high-energy impacts. The preliminary results given in Table II for a 2D model indicate that perhaps as many as 50-70% of such wall collisions do not lead to capture or dissociation, and that from 20-30% of these surviving molecules are in higher vibrational states corresponding to v" = 6 or greater. Results from a smaller number of 3D calculations are inconclusive, but appear to support these qualitative results.

Table II. Percent of molecules surviving a wall collision at higher impact energies. The percent of the surviving molecules with v" $\geq$ 6 is given in parentheses.

E_{normal}	θ	v"		
		2	8	12
1 eV	30°			68(22)
4 eV	30°	68(21)	45(59)	66(33)
10 eV	30°	50(20)		50(40)

CONCLUSIONS

From an extensive series of 2D classical trajectory calculations and supported by 3D calculations, we have found on the average that:
- the total energy of H_2 molecules, initially vibrationally highly excited, remains high, with wall collisions providing an effective mechanism for energy transfer to rotation and translation;
- rotational excitation from the initial J = 1 state up through very large J values that may occasionally exceed J = 14 occurs in nearly every case and is exceedingly rapid;
- vibrational de-excitation is most marked during the first collision as vibrational energy is redistributed into other degrees of freedom: during subsequent bounces some trajectories result in vibrational re-excitation to intermediate states;
- the loss of total molecular energy to the wall (accommodation) is much slower than energy redistribution (equipartition) among the vibrational, rotational, and translational degrees of freedom of the molecule. For most collisions, the H_2 molecule strongly interacts with the wall for only a few vibrational periods; and
- there is preliminary evidence that molecules with translational energies considerably in excess of 1 eV and lying in high vibrational states can undergo surface collisions with a significant survival rate and with a significant fraction retaining large vibrational energies.

Further work involving more complex surface structures and molecules with translational energies in the 1-10 eV range is now in progress.

REFERENCES

1. A. M. Karo, J. R. Hiskes, and T. M. DeBoni, <u>Proceedings of the Second International Symposium on the Production and Neutralization of Negative Hydrogen Ions and Beams</u>, edited by Th. Sluyters (Brookhaven National Laboratory Report BNL 51304, January 1981) p. 74.
2. A. M. Karo, <u>Proceedings of the U.S. - Mexico Joint Seminar on the Atomic Physics of Negative Ions</u>, edited by I. Alvarez and C. Cisneros (Notas de fisica, UNAM, Mexico, Vol. 5, No. 1, 1982) p. 305.
3. P. R. Scott and W. G. Richards, J. Chem. Phys. <u>63</u>, 1690 (1975).
4. R. A. Johnson, Phys. Rev. <u>134A</u>, 1329 (1964).
5. J. R. Hiskes and A. M. Karo (A.I.P. Conference Proceedings No. 000 (this issue)) p. 000.

H⁻ AND Li⁻ FORMATION BY SCATTERING H^+, H_2^+ AND Li^+ FROM CESIATED TUNGSTEN SURFACES

E.H.A. Granneman, J.J.C. Geerlings, J.N.M. van Wunnik,
P.J. van Bommel, H.J. Hopman and J. Los

FOM-Institute for Atomic and Molecular Physics,
Kruislaan 407, 1098 SJ Amsterdam, The Netherlands

ABSTRACT

The efficiency of converting H^+, H_2^+, H_3^+ and Li^+ ions into H^- and Li^- by scattering the ions from cesiated single crystal and polycrystalline tungsten has been determined experimentally as well as theoretically. The ions were reflected under grazing angles of incidence. Maximum efficiencies for hydrogen and lithium were measured of 40% and 23%, respectively. By determining the negative ion fraction of the reflected particles as a function of the angle of reflection, theoretical predictions on conversion are tested without the need to correct for the reflection probability. It was found that at particle velocities $\geq 3.10^5$ m/sec the conversion efficiency depends strongly on the magnitude of the velocity component parallel to the surface. In general, the experimental data show a good agreement with the theoretical predictions.

1. INTRODUCTION

Production of negative ions via collisions of positive ions with low work function surfaces is one of the most promising methods for generation of intense negative ion beams. The most advanced sources in this respect are the ones developed by the negative ion groups at the Lawrence Berkeley Lab.[1,2] and the Brookhaven Natl.Lab.[3,4]. In both types of sources a so-called converter electrode is inserted into the hydrogen plasma. The addition of small amounts of cesium to the plasma leads to a substantial decrease of the surface work function. By biasing the converter negative with respect to the plasma, positive ions are accelerated across the sheath and made to collide with the surface. At the surface these ions are either directly reflected or they induce the desorption of atoms adsorbed to the surface. In the above mentioned sources ion impact desorption is the dominant process [5,6]. However, also experiments are done in which reflection of positive ions on low work function surfaces is used as a basis for negative ion production [7]. The physics underlying negative ion production through ion impact desorption as well as through reflection of the primary ion is in principle identical. The main parameters of interest are the surface work function, the electron affinity of the atom and, at low velocities, the perpendicular velocity component of the particle as it leaves the surface. This is confirmed by a theoretical model developed by Van Wunnik et al.[8,9]. A short summary of this model is presented in section 2.

Because in negative ion sources it is hard to separate the various processes while at the same time it is extremely difficult to

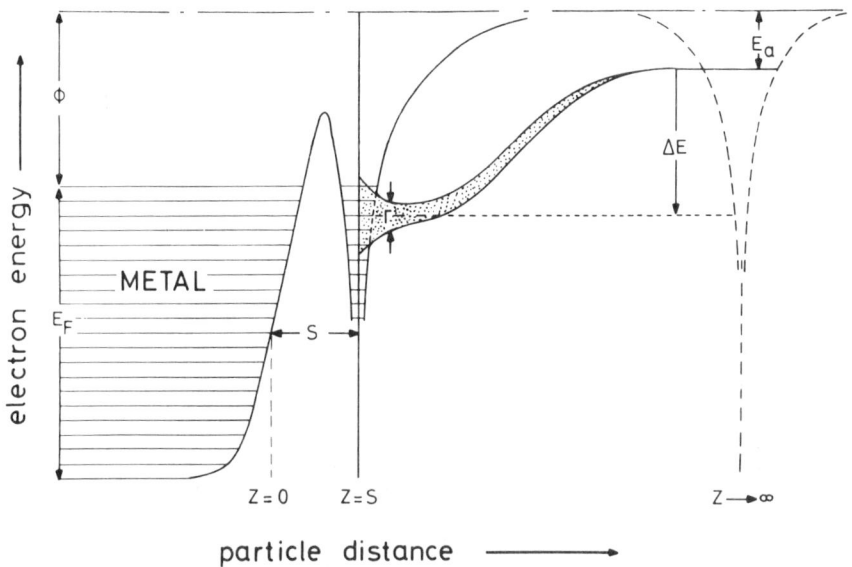

Fig. 1. A schematic energy level diagram of the atom-metal system. E_a is the electron affinity at infinity. ΔE is the shift of the affinity level. Φ and E_F are the metal work function and Fermi level energy, respectively. s is the metal-atom distance.

measure all relevant parameters, a number of experiments were set up to study the conversion process [10,11,12]. In the present experiment, described in section 3, the surface work function and the orientation of the velocity vector of the particles leaving the surface are accurately controlled. In section 4 a comparison is made between theoretical and experimental results. Section 5 deals with molecular ions as incoming particles while in section 6 mono- and polycrystalline tungsten surfaces are compared. Because also other light negative ions can serve as a basis for intense neutral beams [13], a first series of measurements on the production of Li^- is presented in section 7.

2. THEORETICAL CONSIDERATIONS

We now shortly summarize the theory on negative surface ionization developed by Rasser et al.[14] and Van Wunnik et al.[8,9]. First of all it is assumed that the incoming positive ion is neutralized at a large distance from the surface through a resonant transition followed by an Auger de-excitation. Therefore in the outgoing path only neutral and negatively charged particles need to be considered. It has been confirmed experimentally that no H^+ ions are reflected in the parameter range discussed in this paper. Close to the surface image charges are induced in the metal leading to a shift of the affinity level; see figure 1. This shift is in first approximation given by the classical image potential: $\Delta E = (e^2/4\pi\epsilon_o) \cdot (1/4Z)$. Z is the metal-particle distance. At small distances from the surface the wave

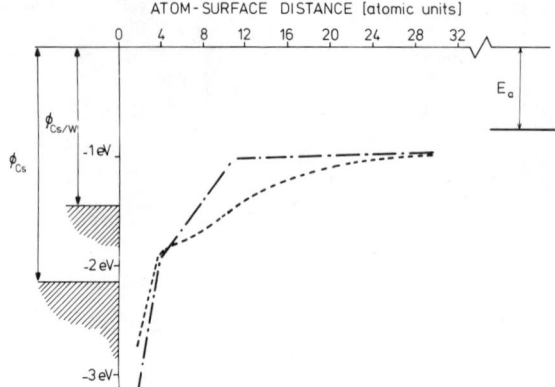

Fig. 2. The affinity level shift as a function of atom-surface distance. For large distances: $E_a = 0.75$ eV.
(-----): work function $\Phi_{Cs} = 2.15$ eV
(-·-·-): work function $\Phi_{Cs/W} = 1.45$ eV
For further details, see text.

functions of atom and metal overlap; this induces resonant transitions through the potential barrier between metal and atom. The finite lifetime of the state leads to Heisenberg broadening: $\Gamma = \hbar\omega$, where ω is the rate of transitions between metal and atom. From the position and broadening of the affinity level one can calculate the fraction N^- of negatively charged particles for every (static) distance Z from the surface. Figures 2 and 3 show the level shift and the level broadening as calculated with this model. As for the level shift the classical image potential is believed to be correct for distances $Z \gtrsim 10\ a_o$. For $Z \lesssim 3\ a_o$ Lang and Williams [15] showed that for jellium-like metals the affinity level follows the effective potential of the metal. For $3\ a_o \lesssim Z \lesssim 10\ a_o$ the precise location of the level is not known. Interpolated level shifts are used such that the calculated H^- yields fit the measured ones (see section 4). The theoretical results were found to be rather insensitive to the exact location of the levels in the interpolated range.

From this point onwards two models were developed [8,9] to calculate the charge state of the particle at infinite distance from the surface. This charge state is determined by two velocity components,

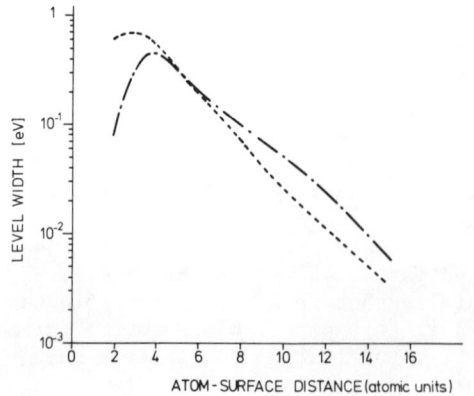

Fig. 3. The affinity level broadening as a function of atom-surface distance.
(-----): $\Phi_{Cs} = 2.15$ eV
(-·-·-): $\Phi_{Cs/W} = 1.45$ eV
These broadenings follow from the level shifts given in fig. 2.

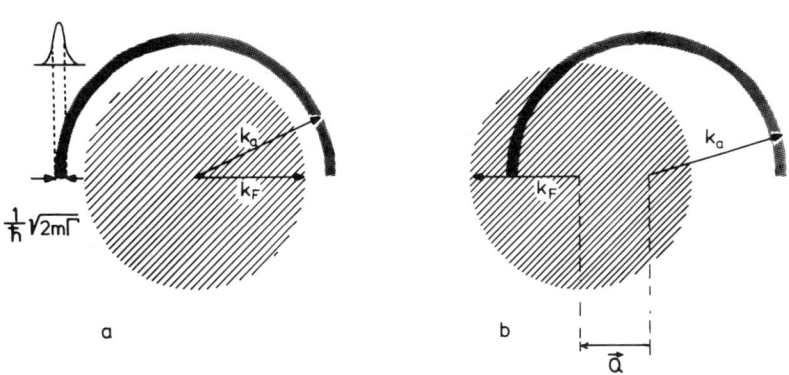

Fig. 4. Schematic diagram of the metal $\vec{k}$ space; for details, see text.

the one normal to the surface ($v_\perp$) and, at high incident velocity, the component parallel to the surface ($v_\parallel$). In the so-called probability model [8,14] the time evolution of the probability of being in the considered state is followed. In the more sophisticated amplitude model the wave function amplitude of that state is followed. In both models it is assumed that close to the surface only neutral particles are present. This is based on Rogers et al.[16] who showed that an atomic state cannot exist in the metal if the screening length of the metal is smaller than the radius of the atomic state. In a low electron density material like Cs (screening length $\sim 1.5\ a_0$) this means that a neutral hydrogen state can exist but no negative ion state (radius H^- ion $\sim 2\ a_0$; Li^- ion $\sim 3\ a_0$). Therefore the models assume that negative ion formation occurs once the atom has left the Cs layer ($Z > 2\ a_0$ for H^- and $Z > 3\ a_0$ for Li^-).

The models predict that, for low incident velocities, for a given particle and surface, the yield is a function of $v_\perp$ only. Furthermore, the H^- yield decreases sharply with an increasing value of ($\Phi-E_a$). In sections 4 and 7 some results of the amplitude model are presented in combination with measured data.

The models also include the situation in which the velocity component parallel to the surface is large, i.e. comparable to that of the metal electrons. In that case considerably less metal electrons are able to make the transition. The charged fraction N^- will then be much smaller than for $v_\parallel = 0$. The width of the conduction band of surfaces with half a monolayer Cs coverage ($\Phi = 1.45$ eV) and with thick Cs coverage ($\Phi = 2.15$ eV) is equal to 1.57 and 1.95 eV [8], respectively. This corresponds to maximum electron velocities of 7.4×10^5 and 8.3×10^5 m/sec, respectively. These velocities are comparable to that of a 3000 eV hydrogen atom: $v = 7.7 \times 10^5$ m/sec. Consequently a substantial decrease in conversion efficiency is expected at these large particle energies. The same effect can also be explained in different words; see figure 4. This figure schematically shows the metal $\vec{k}$ space.

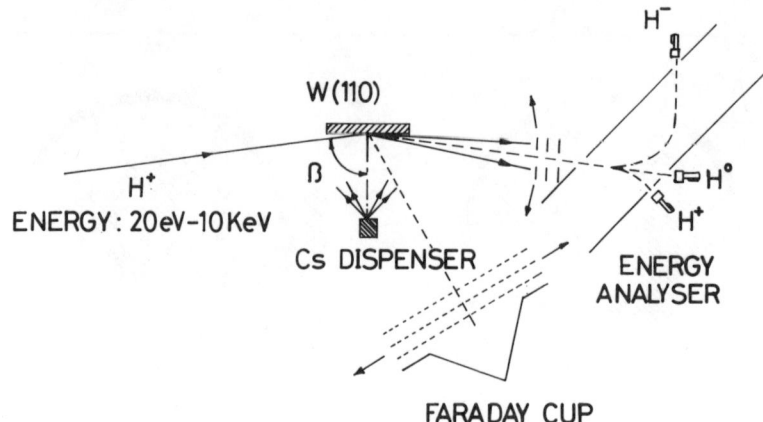

Fig. 5. The experimental arrangement.

The shaded circle is a cross section through the Fermi sphere. The dotted half ring denotes the metal states which are in resonance with the affinity level and which have a $\vec{k}$ vector pointing out of the surface. Position and band width of the affinity level correspond to radius and width of the dotted ring, respectively. If the atom-metal distance decreases, the position of the affinity level shifts to lower energies and the width of the level becomes larger. So the radius of the ring decreases while its width increases. The charged fraction N^- is determined by the overlap between the ring and the sphere. In figure 4a $\hbar^2 k_f^2/2m$ and $\hbar^2 k_a^2/2m$ give the position of the Fermi level and the affinity level with respect to the bottom of the conduction band, respectively. Further $v_{||} = 0$. In figure 4b, $v_{||} \neq 0$. The overlap between the dotted ring and the sphere is changed with respect to the case where $v_{||} = 0$. Consequently N^- will be different. $\vec{Q}$ corresponds to the parallel velocity.

3. EXPERIMENTAL SET-UP

As mentioned before, it is essential that the orientation and magnitude of the velocity vector of the particles leaving the surface are accurately known. Therefore it was decided to reflect positive ions with a relatively large energy ($\geq$ 100 eV) under grazing angles of incidence from well-polished single crystalline surfaces. This ensures that the major fraction of the particles is reflected specularly and little energy is lost [17].

The experimental arrangement is shown in figure 5. From an ion source 400 eV positive hydrogen or alkali ions are extracted. After mass selection the ions are decelerated/accelerated to energies variable between 20 eV and 10 keV. The diameter of the beam hitting the W(110) target (2×15 mm²) is approximately 1 mm, the opening angle $\sim$ 0.5° and the current $10^{-12} - 10^{-9}$ A, depending on the species. By rotating the target the angle of incidence (β) is varied between 45° and 88°. The reflected beam is either detected with a Faraday cup accepting all reflected particles or analyzed by an energy analyzer accepting only particles leaving the surface within a narrow opening angle of 0.5°. The energy analyzer (and the Faraday cup) can be rotated about the target. By accurately setting the acceptance solid angle of the analyzer with respect to the target and by measuring the particle energy, the velocity vector of the reflected particles is fully determined.

The work function of the target can be varied between 1.45 and 5.20 eV by covering the W(110) crystal with a known amount of Cs. The Cs is deposited by heating a Cs-dispenser placed ~ 1 cm in front of the surface. The work function is measured by means of a Kelvin probe as well as by means of the photoelectric method. In order to determine the total positive ion current hitting the surface, the target can be withdrawn and replaced by a thin diafragm plate which has a slit with exactly the same dimensions and alignment as the front surface of the target. Through the slit the Faraday cup measures that fraction of the current that would otherwise hit the target. The background pressure of the scattering chamber is better than 10^{-9} torr. For more details about the set-up, see Van Wunnik et al.[18].

4. REFLECTION OF H^+ ON CESIATED W(110)

Figure 6a shows the surface work function and the efficiency of converting 400 eV H^+ into H^- as a function of Cs coverage. The H^- signal (as well as the H^+ signal) is measured with the Faraday cup, i.e. the H^- yield is integrated over all angles of reflection. As expected a high H^- production coincides with a low work function. In the following we only deal with two coverages, the one which yields the lowest work function (1.45 eV; half a monolayer) and one with a thick Cs layer (2.15 eV). In figure 6b the total $H^+ \to H^-$ conversion efficiency is plotted as a function of the angle of incidence for various particle energies. It can be seen that the yield decreases for high energies of the incident particles; this is in accordance with the model presented in section 2. At higher energies (and smaller angles of incidence, $\beta \leq 75°$) the yield also decreases because the particle reflection probability becomes progressively smaller. Figure 6b also shows the survival probability of H^- as incident particle. It is clear that the reflected H^- flux is independent of the initial charge state.

Since in total conversion efficiency measurements one integrates over a certain range of exit (perpendicular) velocities while also the reflection probability is to be taken into account, the theoretical predictions are difficult to test in this way. Therefore a "differential" conversion efficiency is determined in the following way: With the energy analyzer set at a certain angle of reflection, first the H^- and H^0 particles are both measured with a channeltron detector in the rectilinear trajectory (the so-called H^0 detector, see figure 5). No deflection field is present in the analyzer. Subsequently the deflection field is switched on and only the H^0 contribution remains on this detector. After a (small) correction for the different detection efficiencies for H^- and H^0 the differential H^- yield, i.e. $\eta_D = H^-/(H^-+H^0)$ is determined. The energy of the reflected particles can be measured by analyzing the energy spectrum of the reflected H^- particles. It was found that in the range of interest of the angle of

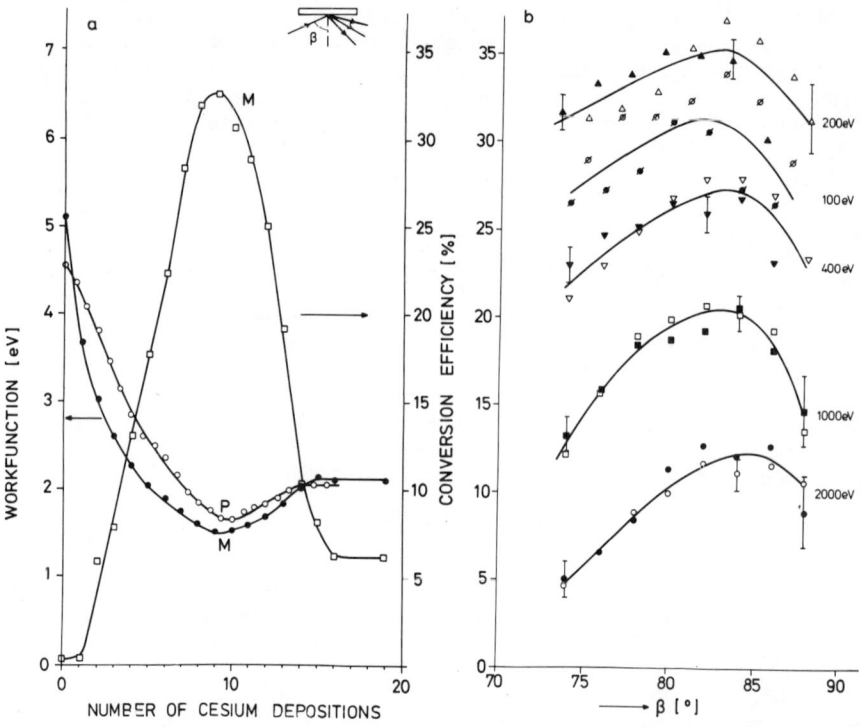

Fig. 6. a. The work function of monocrystalline (M) and polycrystalline tungsten (P) as a function of the Cs coverage on the surface. 10 depositions correspond roughly with half a monolayer. Also the total H^- yield on W(110) is shown as a function of Cs coverage for 400 eV particles. b. The total $H^+ \rightarrow H^-$ conversion efficiency (open symbols) and the total $H^- \rightarrow H^-$ survival probability (filled symbols) as a function of the angle of incidence. The surface is W(110) covered with half a monolayer of Cs. The lines are drawn to guide the eye.

incidence (and reflection) the energy loss is small ($\lesssim 10\%$). For a given energy and angle of reflection η_D was found to be practically independent of the angle of incidence β. Consequently the charge state of the reflected particle does not depend on the parameters of the incident particle. Figure 7 gives η_D as a function of $v_\perp$ of the outgoing particle for three energies and two values of the work function. The curves show the theoretical prediction based on the amplitude model. For $\Phi = 1.45$ eV it is expected that η_D is a function of $v_\perp$ only for energies up to $E \lesssim 400$ eV. For $\Phi = 2.15$ eV the decrease of η_D with energy starts at higher energies. Unfortunately too low detection efficiency did not allow to do measurements at energies < 400 eV. Therefore it was not possible to show that for such energies η_D is a function of $v_\perp$ only. However, from the data shown for Li^- in section 7 (figure 10), in which case the parallel velocity effects are predicted to play a role for $E \gtrsim 3$ keV, it is clear that at lower energies η_D is indeed a function of $v_\perp$ only.

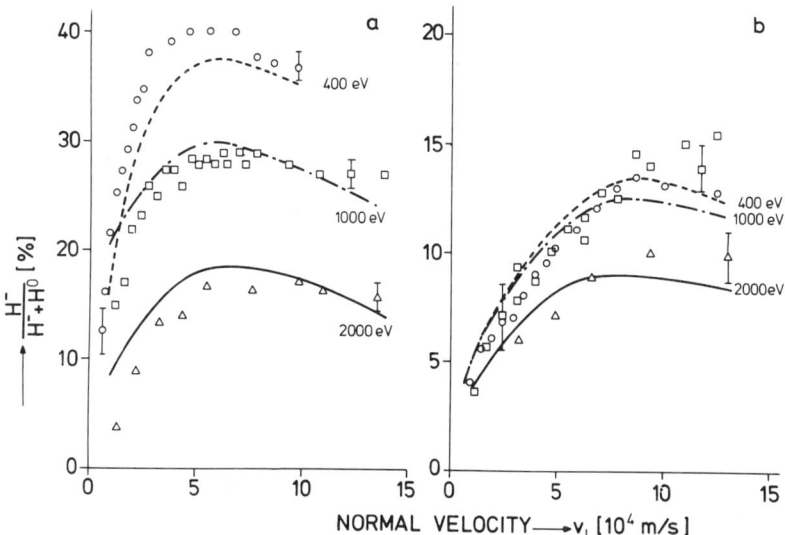

Fig. 7. The H⁻ fraction of the reflected particles as a function of normal velocity. The curves are the predictions of the amplitude model. a. Work function $\Phi = 1.45$ eV; b. $\Phi = 2.15$ eV.

5. MOLECULAR IONS AS INCIDENT PARTICLES

Figure 8 shows the total conversion efficiency per nucleon for H^+, H_2^+ and H_3^+ as a function of the angle of incidence for E = 400 eV/nucleon. In figure 9 the ratio $\eta_D = H^-/(H^-+H^0)$ is given as a function of the angle of reflection α for the same incident energy and an angle of incidence $\beta = 80°$. From these figures it can be concluded that, once reflected, the individual hydrogen particles present in H^+, H_2^+ and H_3^+ behave indentically. This leads to the conclusion that H_2^+ and H_3^+ dissociate (shortly) before they hit the surface, leading to the reflection of two and three individual neutral atoms, respectively. Each of these has the same probability of becoming negatively charged. For H_2^+ the picture is [19] that close to the surface the molecular ion resonantly neutralizes into the repulsive $b^3\Sigma_u^+$ state of H_2, after which first dissociation of H_2 and subsequently scattering takes place. At higher incident energies [19] the conversion efficiency per nucleon becomes less for H_2^+ as compared to H^+. This is explained by assuming that the two neutral particles present in the repulsive $b^3\Sigma_u^+$ state have no time to separate as individual atoms, before, close to the surface, Auger de-excitation from this state to the stable $^1\Sigma_g^+$ ground state occurs. This stable H_2 molecule can survive the reflection process. Note that no stable H_2^- ions exist. At present detailed information on the dissociation of H_3^+ is unavailable.

6. FORMATION OF H⁻ ON POLYCRYSTALLINE TUNGSTEN

Since in most present day negative ion sources based on negative

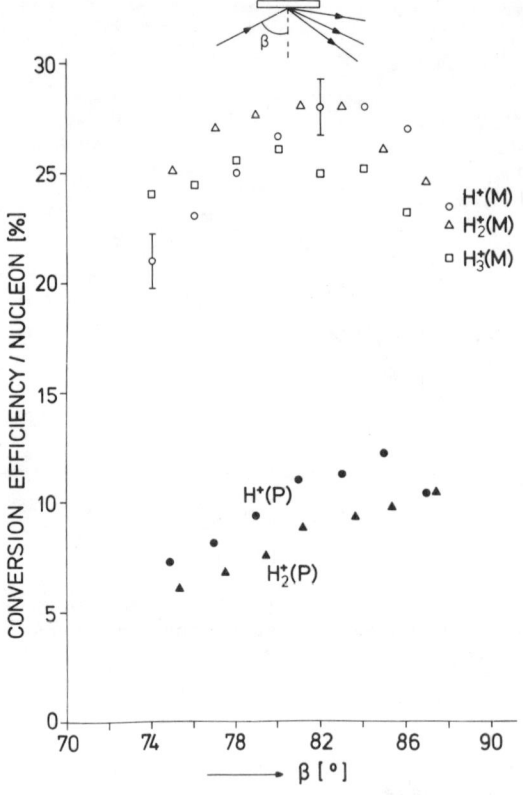

Fig. 8. The total conversion efficiency per nucleon as a function of the angle of incidence for H^+, H_2^+ and H_3^+ reflected on W(110) covered with half a monolayer of Cs (M, $\Phi = 1.45$ eV) and for H^+ and H_2^+ reflected from polycrystalline W also covered with half a monolayer (P, $\Phi = 1.68$ eV). The incident energy is 400 eV/nucleon in all cases.

Fig. 9. The negative ion fraction η_D of the reflected particles as a function of the angle of reflection for H^+, H_2^+ and H_3^+ reflected from W(110) covered with half a monolayer of Cs (M, $\Phi = 1.45$ eV) and for H^+ and H_2^+ reflected from polycrystalline W also covered with half a monolayer (P, $\Phi = 1.68$ eV). The angle of incidence is $\beta = 80°$ and $E = 400$ eV/nucleon.

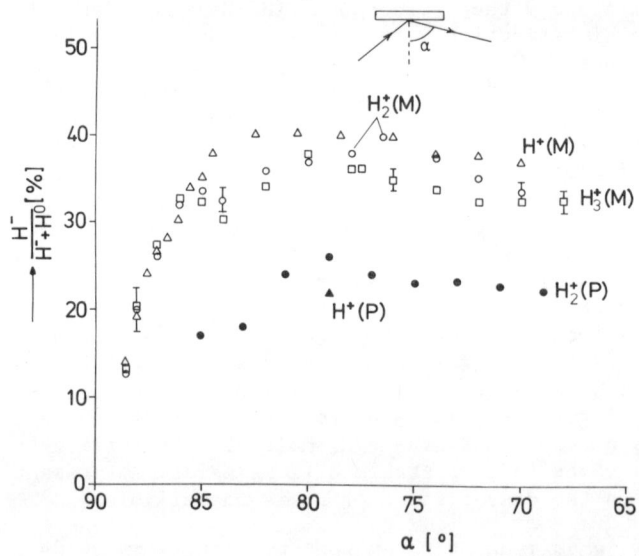

surface ionization the converter materials are polycrystalline in nature, also measurements were done on cesiated polycrystalline tungsten. Results for H^+ and H_2^+ incident particles with an energy of 400 eV/nucleon are given in figures 8 and 9. The work function of the polycrystalline materials used is shown in figure 6a. More information on these experiments can be found in Van Bommel et al.[20] For the minimum work function case ($\Phi = 1.68$ eV) the differential yield η_D, figure 9, is found to be roughly a factor two lower for the polycrystals. When η_D is calculated for $\Phi = 1.68$ eV we find theoretical values approximately 40% higher than the measured values. This discrepancy is not yet fully understood [20]. For the total conversion efficiency (figure 8) the difference between poly- and monocrystalline tungsten (~ a factor 3) is found to be larger than that for the differential measurement (figure 9). This can be accounted for by the larger reflection probability of grazingly incident particles on monocrystalline surfaces. Mono- and polycrystalline surfaces covered with thick Cs layers show much smaller differences in yields [20]. The main reason for that is that the work functions are similar for these two types of surfaces (2.15 eV and 2.05 eV, respectively; see figure 6a). From photoelectric measurements done with a HeNe laser ($h\nu = 1.96$ eV) it can be concluded that polycrystalline W surfaces covered with thick Cs have areas with work function $\Phi < 1.96$ eV. At the photon energy $h\nu = 1.96$ eV large amounts of photoelectrons were produced, this contrary to the corresponding monocrystals.

7. PRODUCTION OF NEGATIVE LITHIUM IONS

Because the electron affinity of the alkali atoms is only a few tenths of an eV smaller than that of hydrogen (0.75 and 0.62 eV for H^- and Li^-, respectively) it is expected that reasonably large Li^- currents can be produced through negative surface ionization. For that reason the hydrogen source was replaced by a Li^+ source. The results for the differential conversion efficiency $\eta_D = Li^-/(Li^- + Li^o)$ are shown in figure 10 for two incident energies, 100 and 1000 eV. A maximum efficiency of ~ 23% is obtained for a perpendicular energy $E_\perp \approx 50-100$ eV. It should be mentioned that at a work function $\Phi = 1.45$ eV no Li^+ ions are reflected. Reflection of Li^+ ions was found to occur for $\Phi \gtrsim 3.5$ eV. The Li^- production decreases rather sharply for increasing work functions. In case of a thick Cs layer ($\Phi = 2.15$ eV) the Li^- yield dropped to practically zero. As expected for these low velocities η_D is found to be only a function of $v_\perp$ and not of the incident energy. In figure 10 also the theoretical prediction for η_D is given. In the calculation all parameters are kept identical to that for hydrogen except for the electron affinity ($E_a = 0.62$ eV) and the distance from the surface beyond which negative Li ions can exist (3 a_o in stead of 2 a_o; see section 2).

8. CONCLUSIONS

From the data presented in this paper the following conclusions can be drawn at low incident particle energies (≤ 400 eV/nucleon):

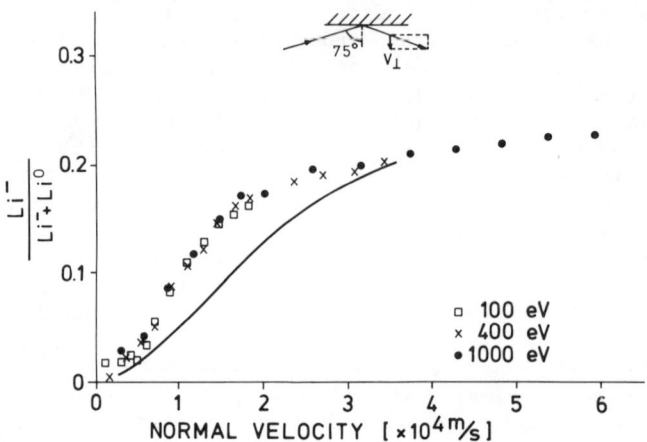

Fig. 10. The negative (Li⁻) fraction of the reflected particles (Li⁻ + Li⁰) as a function of the perpendicular velocity component for two different incident energies. The work function is $\Phi = 1.45$ eV; the angle of incidence $\beta = 75°$. The solid curve gives the theoretical prediction.

- The negative ion production efficiency η_D is in first order only a function of the difference in work function and electron affinity ($\Phi - E_a$) and the perpendicular velocity component ($v_{\perp,out}$) of the reflected particle. For a given $v_{\perp,out}$, η_D does not depend on the charge state nor on the orientation of the velocity vector of the incident particle.
- H^+, H_2^+, H_3^+ have the same probability per nucleon to produce H^-.
- For monocrystals large conversion efficiencies are found for $v_\perp = (3-10) \times 10^4$ m/sec ($E_\perp = 5-50$ eV/nucleon); $\eta_{D,max}(H^-, \Phi = 1.45$ eV$) = 40\%$; $\eta_{D,max}(H^-, \Phi = 2.15$ eV$) = 15\%$; $N_{D,max}(Li^-, \Phi = 1.45$ eV$) = 23\%$.
- For polycrystals η_D is a factor 2-3 lower than for monocrystals.

For particle energies $\gtrsim 400$ eV/nucleon η_D decreases with increasing particle energy. This is caused by the fact that fewer metal electrons have a sufficiently high velocity to be able to make a transition.

It can furthermore be concluded that the theoretical model of Van Wunnik et al.[8,9,14] quantitatively agrees with the measured data, also for high $v_{\parallel,out}$. It is shown that the above mentioned results are also of interest for surface plasma sources; in those types of sources the particles are desorbed from the converter surface with (perpendicular) energies comparable to those dealt with in this paper.

ACKNOWLEDGEMENT

This work is sponsored by FOM with financial support by ZWO/EURATOM.

REFERENCES

1. K.W. Ehlers and K.N. Leung, Rev.Sci.Instr. 51, 721 (1980).
2. K.W. Ehlers and K.N. Leung, these Proceedings.
3. K. Prelec et al., Proc. of 3rd Joint Varenna-Grenoble Int.Symp. on heating in toroidal devices, vol. III, 1039 (1982). Eds. C. Gormezano, G.G. Leotta and E. Sindoni (Commission of the European Communities, Brussels).
4. K. Prelec, these Proceedings.
5. P.J. van Bommel, K.N. Leung, K.W. Ehlers, these Proceedings.
6. K. Wieseman, K. Prelec and Th. Sluyters, J.Appl.Phys. 48, 2668 (1977).
7. H.J. Hopman, P.J. van Bommel, P. Massman and E.H.A. Granneman, Proc. 2nd Int.Symp.on Prod.and Neutr.of Neg.Hydrogen Ions and Beams, Brookhaven, USA, 1980. Ed. Th. Sluyters, Brookhaven Natl. Lab. report nr. BNL 51304, page 233.
8. J.N.M. van Wunnik, J.J.C. Geerlings and J. Los, Surf.Sci. 131, 1 (1983).
9. J.N.M. van Wunnik, R. Brako, K. Makoshi and D.N. Newns, Surf. Sci. 126, 168 (1983).
10. J. Los, E.G. Overbosch and J.N.M. van Wunnik, ref. 7, page 23.
11. J.R. Hiskes and P.J. Schneider, ref. 7, page 15.
12. P.J. Schneider, W. Eckstein and H. Verbeek, J.Nucl.Mat. 111 & 112, 795 (1982).
13. L.R. Grisham, D.E. Post, D.R. Mikkelsen, H.P. Eubank, Nucl. Techn./Fusion 2, 199 (1982).
14. B. Rasser, J.N.M. van Wunnik and J. Los, Surf.Sci. 118, 697 (1982).
15. N.D. Lang and A.R. Williams, Phys.Rev. B18, 616 (1978).
16. F.J. Rogers, H.C. Graboske and D.J. Harwood, Phys.Rev. A1, 1577 (1970).
17. O.S. Oen and M.T. Robinson, Nucl.Instr.Meth. 132, 647 (1976); J.Nucl.Mat. 63, 210 (1976).
18. J.N.M. van Wunnik, J.J.C. Geerlings, E.H.A. Granneman and J. Los, Surf.Sci. 131, 17 (1983).
19. J.N.M. van Wunnik, J.J.C. Geerlings and J. Los, to be published.
20. P.J.M. van Bommel, J.J.C. Geerlings, J.N.M. van Wunnik, P. Massmann, E.H.A. Granneman and J. Los, J.Appl.Phys. 54, 5676 (1983).

DISCUSSION

HISKES: Did your theoretical model show a limit which is less than unity? Is there an ultimate limit there? You showed a maximum conversion efficiency of 35% to 45%. Do you believe that there is any reason this couldn't be unity? Is there a maximum value that comes out of the theory?

GRANNEMAN: Yes, I think so. At any distance from the surface you have a certain charge fraction, if you look at a static situation. It is unity, of course, close to the surface. When a particle leaves the surface there is a competition between how fast it leaves the surface trying to keep its charge and the transition of the electron back to the metal. This leaves us with something like 40% at best.

HISKES: My second question is how do you determine the equilibrium charge density that you called $N(Z)$?

GRANNEMAN: We assume a Lorentz distribution of the broadened affinity level; for a given temperature you can take a Fermi distribution in the metal. The overlap between the two distributions at a certain position gives you the equilibrium charged state fraction at that position. The transition probability doesn't come in because it is a static case (it may take a year before it reaches it).

PETERSON: When you say that the outgoing relative amount of negative ions does not depend on the incoming angle, isn't this true only as long as the outgoing velocities are independent of the incoming angle?

GRANNEMAN: Well it is true that if you come in with very large velocities, so large that the transition probability close to the surface at the turning point is too small in relation to the velocity with which the particle moves away, you make less negative ions.

PETERSON: What I am thinking is to get to larger incoming angles, at some point it is going to affect the velocity distribution, at each given angle it goes out into smaller angles and therefore these velocities will change as the incoming angle changes and that must change these characteristics.

GRANNEMAN: Of course the incoming energy and the incoming velocity are important if you look at a total yield but if you set your experiment such that you look at the certain angle of reflection, and you select only those particles which have a certain energy, then it doesn't really matter how you start.

VERBEEK: It does.

GRANNEMAN: Well, OK, there is conflicting evidence apparently.

PETERSON: You said you only select particles of a given energy coming out. That is true. What I am saying is the energy will change and you change the angle of the incoming particle.

GRANNEMAN: Let me put it this way. If you come in with a 10 keV particle and you come out with a 500 eV particle and you select those particles which have 500 eV outcoming energy at a certain angle with respect to the surface then the theory should be able to predict what comes out.

VERBEEK: We did observe dependence on the angle of incidence. That means we observed difference in the negative charge fraction whether the particles are scattered from the surface or if they penetrate into it. This doesn't say that this is in contradiction to the theory but it might indicate that you have different states into which the primary particles are neutralized. If you have a level of rather high excitation, say 2s state, you cannot capture this extra electron. You didn't see it because you didn't vary your angle of incidence over a range which was sufficiently large.

GRANNEMAN: Why should that process be more efficient for large angles of incidence than for small angles of incidence? If there is any excitation process to metastable states I do not quite see how it depends on the angle of incidence.

VERBEEK: One should measure that. We see so many neutrals, we see so many excited neutrals and probably there is a difference.

SPUTTERING YIELDS OF NEGATIVE HYDROGEN IONS

J. A. Greer and M. Seidl
Department of Physics, Stevens Institute
of Technology, Hoboken, N. J. 07030

ABSTRACT

A polycrystalline molybdenum target partially covered with cesium and hydrogen is bombarded with Cs^+ ions in the energy range 150 to 1000 eV. The workfunction of the target is continuously monitored. Sputtering yields of H^- ions and electrons are measured as function of Cs^+ ion energy, target workfunction and hydrogen pressure.

INTRODUCTION

Negative hydrogen ions are produced in surface conversion sources[1] by several processes, such as backscattering hydrogen ions or atoms from the converter surface, or by sputtering adsorbed or implanted hydrogen with cesium ion or hydrogen ion bombardment. In actual sources all these mechanisms may be acting simultaneously. In order to study the basic physics of these processes it is advantageous to set up experiments in which only a single process can occur.

In a previous experiment[2] we have studied the production of negative hydrogen ions by sputtering adsorbed hydrogen from a cesiated molybdenum surface bombarded with Cs^+ ions. The lowest bombarding energy was 500 eV, well above the operating range of surface conversion sources. In this work we present sputtering yield measurements with a new high perveance diode which allowed to extend the energy range down to 150 eV.

EXPERIMENTAL APPARATUS

A cross-section of the main part of the experimental apparatus is shown in Fig. 1. The apparatus consists of a planar diode, cesium manifold, Faraday cup with lens, and an electron gun.

The cathode of the diode is a polished polycrystalline molybdenum plate. Its temperature can be varied from 10°C to 500°C. The anode consists of a fine tungsten mesh (180 wires per inch with 0.0008" wire diameter) mounted 2 mm from the cathode. During operation the mesh is heated to about 1000°C by passing current through it.

A cesium manifold provides a uniform flux of cesium vapor to the diode region. Some of the cesium is surface ionized at the hot tungsten mesh which acts as a source of Cs^+ ions.

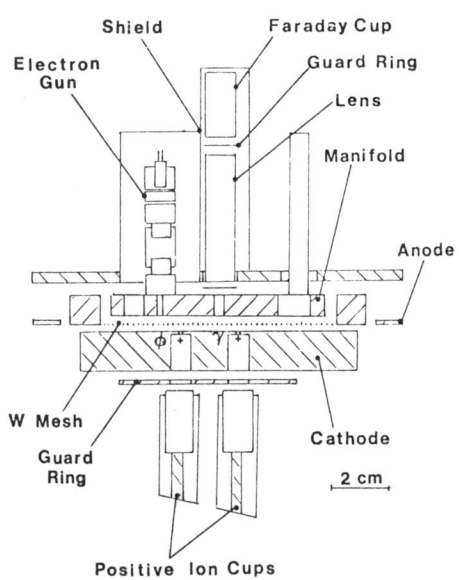

Fig. 1. Diagram of the experimental diode.

The ions are accelerated onto the negatively biased cathode. The Cs^+ ion current density is measured by two positive ion cups facing two holes drilled in the cathode block. The specific perveance of the cathode is large enough to provide a space-charge limited Cs^+ ion current density of 100 μA/cm² at 100 V. This corresponds to a flux of Cs^+ ions almost two orders of magnitude larger than the residual water vapor flux. The lens effect of the mesh adds an intrinsic angular spread of 7.5 milliradians to the negative ions accelerated by the mesh.

Cesium coverage of the cathode is varied by changing the ratio of the cesium ion to atom fluxes. The average workfunction of the cathode is continuously monitored by means of the retarding field technique[3] using a collimated electron beam produced by the electron gun shown in Fig. 1.

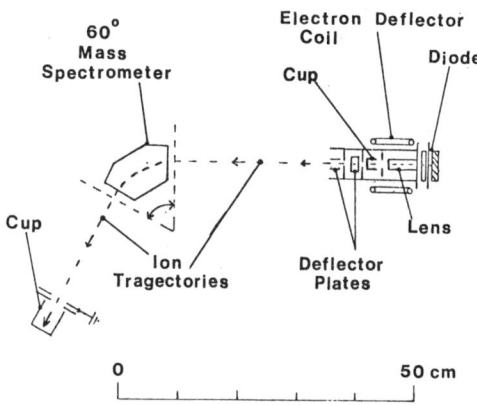

Fig. 2. Setup for analyzing the sputtered beam.

Hydrogen gas admitted into the vacuum system is chemisorbed on the cathode surface and can be sputtered by Cs^+ ions bombardment. The sputtered negative ions and electrons are accelerated back across the diode gap, partially (26.7%) attenuated by the mesh and collected by the cesium manifold. A small sample of the sputtered beam passes through two apertures, defining a cone of 14° half angle, and is then presented to the diagnostic stack. The stack consists of a lens which focuses the beam through a guard ring into a Faraday cup. When activated, a small Helmholtz coil (shown in Fig. 2) deflects the electrons out of the cup. Total ion yields are obtained by dividing the Faraday cup ion current by the transmission of the mesh and the Cs^+ cup current.

The mass spectrum of the ion beam is obtained by means of a 60° magnetic sector mass spectrometer shown in Fig. 2. A small hole in the center of the Faraday cup provides about 1% of the ions for analysis.

WORK FUNCTION MEASUREMENT

The cathode of the diode is exposed to a continuous flux of cesium vapor, hydrogen gas and to cesium ion bombardment. At optimum operation the Cs^+ ion current density is typically 100 µA/cm², the $Cs°$ flux is about 3 times larger and the H_2 flux about three orders of magnitude larger. The flux of residual water vapor is less than 5% of the Cs^+ ion flux. The average work function corresponding to the dynamical equilibrium of the cathode surface is measured by means of the retarding field technique.[3] Differences in the work function with respect to a thick cesium coating are measured. This reference coating is obtained when the Cs^+ ion bombardment is switched off by reducing the mesh temperature. The work function of the reference surface has not been measured. However, we believe that it is close to 2.3 eV measured for full coverage of coadsorbed cesium and hydrogen on tungsten (100).[4]

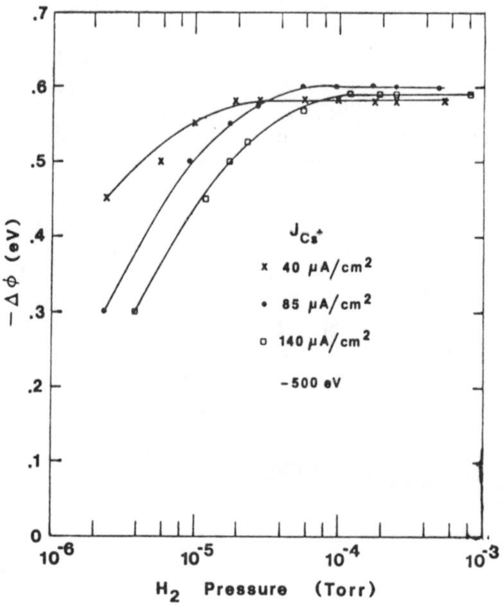

Fig. 3. Dependence of work function change on hydrogen pressure.

The work function of the reference surface is well reproducible and it depends only slightly on the hydrogen pressure, decreasing by 0.06 eV when the hydrogen pressure changes from 10^{-6} to 10^{-4} torr. When the Cs^+ ion bombardment is switched on the work function drops by about 0.6 eV for high enough hydrogen pressure, as shown in Fig. 3. The most interesting observation is the stability of the minimum work function surface which remains almost unchanged provided there is sufficient flux of

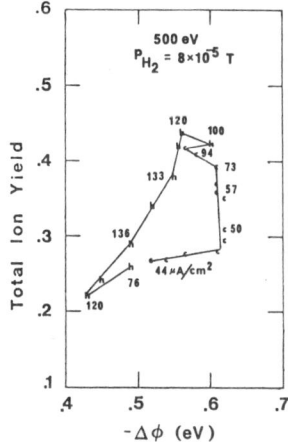

Fig. 4. Work function change and ion yield as function of Cs^+ ion current density in $\mu A/cm^2$.

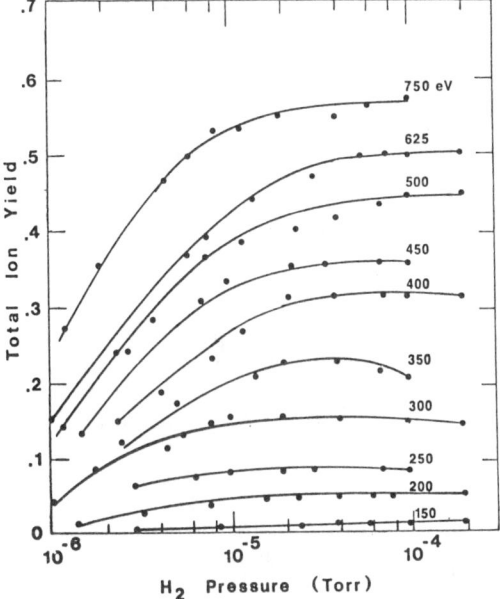

Fig. 5. Total ion yield as function of hydrogen pressure and Cs^+ ion energy for optimum cathode work function.

neutral cesium and hydrogen to fill in the vacant sites caused by sputtering. Fig. 4 shows that once the minimum work function surface has been established it remains unchanged for a wide range of Cs^+ ion bombardment current density.

YIELD MEASUREMENTS

Ion yield is defined as the number of ions produced per incident Cs^+ ion. As has been reported previously [5,2], maximum yield generally occurs when the work function is close to its minimum value. However, Fig. 4 shows that the yield is not a single valued function of the work function.

The optimum total ion yields are plotted in Fig. 5 as function of hydrogen pressure and Cs^+ ion energy. The corresponding electron yields are plotted in Fig. 6. Mass spectrograms of the ion beam indicate that about 94% of the beam consists of H^- ions and about 6% of Mo^- ions. Impurity ions, including Cs^- ions, amount to less than 1% of the ion beam. Yields were measured with an acceptance half angle of 14° which is believed to be considerably larger than the beam angular spread. However, detailed measurements of angular spread have not yet been completed.

Fig. 7 shows the dependence of the optimum total ion yield on the energy of the Cs^+ ions. The electron yield is also

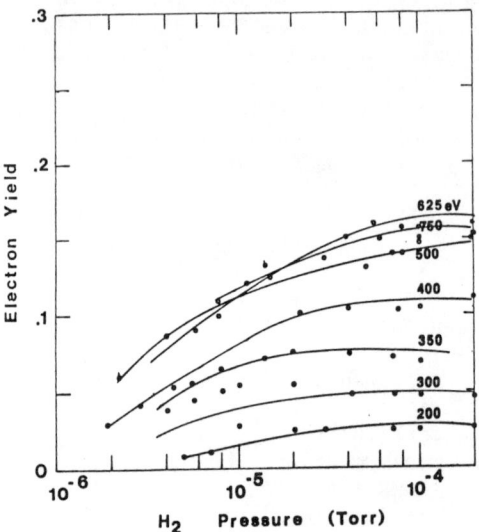

Fig. 6. Electron yield measured under the same conditions as the ion yield in Fig. 5.

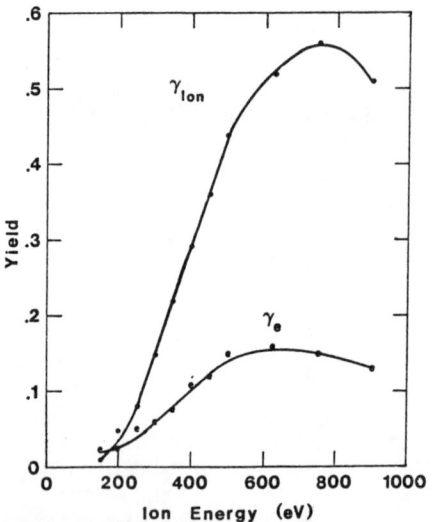

Fig. 7. Optimum total ion yield and electron yield as function of the Cs^+ ion energy.

shown. The ion yield reaches a maximum of 0.56 (corresponding to an H^- yield of 0.52) at a bombarding energy of 750 eV. The optimum bombarding energy agrees with the value obtained previously.[2] The maximum yield measured in the previous work[2] was 0.4 which has to be compared to the new value of 0.5. The ion yield decreases with bombarding energy, approaching zero at an energy of 100 eV.

DISCUSSION

The experimental data will be qualitatively discussed by means of a model described below.

In optimum steady state operation the molybdenum surface is assumed to be covered, in the average, with two monolayers of hydrogen that in turn are partially covered with cesium. Such a double layer was studied on W(100) by Papageorgopoulos and Chen.[4] The authors have found that the double layer is created spontaneously in saturated coadsorption of hydrogen with cesium and that it forms a stable configuration with minimum work function of 1.45 eV.

It is further assumed that sputtering of the adsorbed hydrogen atoms by Cs^+ ions occurs in binary elastic collisions. An incoming ion hits a hydrogen atom that in turn is reflected from the molybdenum lattice. The maximum energy that can be transferred from the cesium ion to the hydrogen atom in binary elastic collision is only 3% of the cesium ion energy. The binding energies of hydrogen to molybdenum are given as 0.7 eV, 0.87 eV, and 1.17 eV for the three different binding sites.[6] Although it is known that the cesium overlayer will increase the hydrogen binding energy[4], no published data are available for the binding energy. Nevertheless, it is plausible to assume that the hydrogen atom will have a kinetic energy less than 1 eV when sputtered with a 100 eV cesium ion.

There are no data on sputtering yields of adsorbed hydrogen atoms sputtered by any ions. However, Winters and Sigmund[7] studied sputtering of nitrogen adsorbed on tungsten. The highest observed sputtering yield was 0.8 for bombardment with 500 eV Xe^+ ions. A reasonable agreement was found with the knock-off sputtering model.[7]

The outgoing hydrogen atom can pick up an electron from the metal by resonant electron transfer and leave as H^- ion. Thus the sputtering yield of H^- ions can be written as $\gamma(H^-) = \gamma(H°)P$ where $\gamma(H°)$ is the sputtering yield of hydrogen atoms and P is the ionization probability (also called production probability or negative charge fraction).

The ionization probability has been theoretically studied by several authors.[8-10] Direct measurements of the ionization probability have been obtained in recent FOM experiments[11-13] on scattering H^+ ions on cesiated tungsten surfaces. The negative charge fraction reaches a maximum value of P = 0.4 for cesiated monocrystalline W(110) surface (work function 1.45 eV) and a maximum value of P = 0.25 for cesiated polycrystalline tungsten (work function 1.68 eV). Both maxima occur for a kinetic energy about 10 eV in the direction perpendicular to the surface. The results are reported in good agreement with theory.[10]

In our experiment the largest sputtering yield $\gamma(H^-) = 0.5$ occurs at Cs^+ ion energy of 750 eV. In a previous work we found that most H^- ions leave the surface with a perpendicular energy equal to 1.3% of the Cs^+ energy. Thus we find the optimum H^- ion energy equal to 9.75 eV in agreement with FOM experiments.[11-13] However, if we use P = 0.25 for the ionization probability we must assume a very large total sputtering yield of $\gamma(H°) = 2$.

The small yields of H^- ions for Cs^+ ion energies below 200 eV indicate that sputtering of adsorbed hydrogen by Cs^+ ion bombardment can account for only a fraction of the H^- ions produced in surface conversion sources. The hydrogen plasma in contact with the converter surface may cause at least three additional effects. Hydrogen ion implantation may substantially increase the hydrogen concentration close to the converter surface.[14] This may increase the total sputtering yield $\gamma(H°)$. Additional sputtering of hydrogen will be caused by H^+ ion bombardment. Finally, backscattering of H^+ ions will also produce H^- ions.

ACKNOWLEDGEMENT

This research was sponsored by the National Science Foundation, Grant PHY 8205886 and by the Air Force Office of Scientific Research, Grant 83-0230.

REFERENCES

1. Proceedings of the Symposium on the Production and Neutralization of Negative Hydrogen Ions and Beams, BNL Report No. 51304, edited by Th. Sluyters (Brookhaven National Laboratory, 1980) p. 137-243.
2. M. Seidl and A. Pargellis, Phys. Rev. $\underline{B26}$, 1 (1982).
3. A. G. Knapp, Surface Sci. $\underline{34}$, 289 (1973).
4. C. A. Papageorgopoulos, J. M. Chen, Surface Sci. $\underline{39}$, 283 (1973).
5. M. Yu, Phys. Rev. Letters $\underline{40}$, 574 (1978).
6. H. R. Han, L. D. Schmidt, J. Phys. Chem. $\underline{75}$, 227 (1971).
7. H. F. Winters, P. Sigmund, J. Appl. Phys. $\underline{45}$, 4760 (1974).
8. J. R. Hiskes, A. M. Karo, and M. A. Gardner, J. Appl. Phys. $\underline{47}$, 3888 (1976).
9. J. R. Hiskes, J. Phys. (Paris) $\underline{40}$, C7-179 (1979).
10. B. Rasser, J. N. M. Wunnik and J. Los, Surface Sci. $\underline{118}$, 697 (1982).
11. J. N. M. van Wunnik, J. J. C. Geerlings and J. Los, Surface Sci. $\underline{131}$, 1 (1983).
12. J. N. M. van Wunnik, J. J. C. Geerlings, E. H. A. Granneman and J. Los, Surface Sci. $\underline{131}$, 17 (1983).
13. P. J. M. van Bommel, J. J. C. Geerlings, J. N. M. van Wunnik, P. Messmann, E. H. A. Granneman, and J. Los, J. Appl. Phys. $\underline{54}$, 5676 (1983).
14. M. Seidl, A. N. Pargellis, J. Greer, Proc. of the U.S.-Mexico Joint Seminar on the Atomic Physics of Negative Ions, Edited by C. Cisneros and T. J. Morgan (Instituto de fisica, UNAM, Mexico, 1982) p. 393.

DISCUSSION

YORK: You showed a plot of yield vs. the energy of the incoming cesium ion. Did you try that for different cesium conditions to see if the distribution changed with the amount of cesium you had on the surface?

SEIDL: This plot is for optimum cesium coverage. We have measured yields for different than optimum coverages as well but I am not so sure how wide a range we have investigated. In all the cases the maximum of the yield vs. Cs^+ ion energy curve stayed close to 750 eV with a deviation not exceeding 50 eV.

EFFECTS OF CESIUM IN THE PLASMA OF THE SURFACE CONVERSION H⁻ SOURCE

K. W. Ehlers, K. N. Leung
Lawrence Berkeley Laboratory, University of California
Berkeley, California 94720

ABSTRACT

The usual method for replacing the partial monolayer of cesium which is removed by sputtering from the surface of the converter electrode of a surface production negative ion source, is to allow cesium atoms to condense or adsorb onto this surface. While this method is easily employed for short pulsed source operation, it becomes increasingly difficult in the dc case because of the high probability of ionization of the cesium atoms as they pass through the discharge. In this paper, we attempt to analyse the severity of this problem in which cesium arrives at the converter surface as an energetic ion rather than as a thermal atom.

INTRODUCTION

Perhaps the most difficult problem encountered by those working with surface production sources which must produce continuous, multi-ampere beams of H⁻ and D⁻ ions, is that of maintaining a uniform, low work-function coating of cesium on large converter surfaces. For pulsed source operation, this problem is minimized as cesium atoms can be allowed to adsorb onto the converter surface during the time the discharge is off. The optimum coverage[1] of approximately 0.67 monolayers can be obtained by supplying ample cesium vapor and then allowing the converter electrode to heat up to the temperature where only a partial monolayer will exist. When the discharge is struck, this optimum coverage will be eroded away at a rate depending on the energy and the number of ions which strike the surface. In this manner one obtains the very short pulses of very high ion current density that are characteristic of the Magnetron and Planotron.[2] In the case of dc operation, this ability to coat the converter with cesium atoms essentially disappears as the cesium must travel through the plasma and in doing so, it becomes ionized.

Fig. 1 is the cross-section for the ionization of cesium by electron impact.[3] Because of the low ionization potential of cesium atoms (3.89 eV), the cross-section remains large for even very low electron energies. Thus a very large portion of the electrons contained in a plasma are capable of ionizing cesium atoms as they enter the plasma from either a wall or from a cesium feed system located outside the plasma. Because of the effectiveness of the bulk of the plasma electrons to ionize, plus the large mass of the cesium atoms (A = 133), we can expect the ionization mean free path to be quite short.

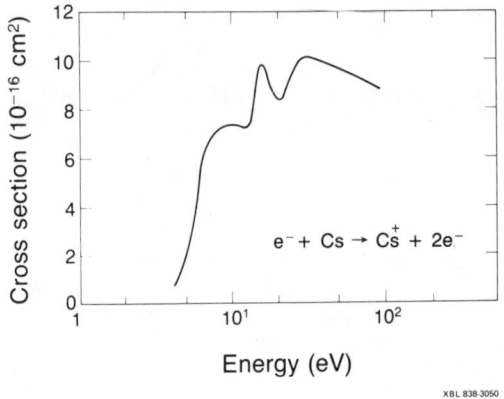

Fig. 1. Cross-Section for Ionization of Cesium by Electron Impact

If we assume that the temperature of the cesium neutrals is about 500°K ($T_0 = \sim .04$ eV), and that the average effective electron energy is 6 eV ($\sigma = 5 \times 10^{-16}$ cm^2), we then find:

$$\lambda_{mfp} = \frac{1}{\sigma \times n_e} \times \sqrt{\frac{m_e}{M_{Cs}} \times \frac{T_0}{T_e}}$$

$$= \frac{3.6 \times 10^{11}}{n_e}$$

Thus at plasma densities of 5×10^{11}/cm^3, which is about our minimum plasma density of interest, the mean free path for ionization of cesium is ~ 1.0 cm. As the density is increased, the path length becomes progressively shorter, and for a given geometry, there will be a limiting plasma density at which insufficient neutral cesium will reach the converter. When this density limit is exceeded, then more cesium will be sputtered from the converter than can be resupplied, and it is our present conviction that it is this situation that limits the output of surface sources that must operate dc.

Fig. 2 shows the output signals from a spectrometer that was tuned to monitor several Cs 1 (neutral) and Cs 11 (Cs$^+$) lines while viewing a hydrogen discharge with cesium added. While maintaining a constant arc voltage, the arc current was increased from 10 to 40 A, and as shown in Fig. 2-A, the cesium ion lines increased greatly with the increase in plasma density. The relative intensity of the 4555.28 Å cesium neutral line and the 4603.76 Å cesium ion line are plotted in Fig. 2-B. Because the intensity of the cesium ion line is not linear with arc current, we suspect that the average effective electron energy also increased as the arc current was raised.

Thus it becomes apparent that we should expect a sizable difference in H⁻ output from short pulse and dc operated systems. We know the converter can be coated by cesium atoms, but what happens to this loading when the vast majority of the cesium strikes the converter in the form of energetic ions We have found it very difficult to answer this question as much of the basic physics data has never been taken. But it does seem important to use our experience and that data we can find to make a best guess as to the magnitude of the problem and of the important parameters involved.

Specifically, we wish to know just what happens when a Cs^+ ion hits a surface, preferably molybdenum, with an energy of about 100 to 300 eV.

When a positive ion strikes a solid surface, it may give rise to a variety of phenomena:

1. Atoms can be sputtered from the surface.

2. The ion can be neutralized at the surface and then reflect as a neutral atom.

3. The ion can reflect as a positive ion with some loss in energy.

4. The ion may be neutralized and adsorbed on the surface.

5. The ion may penetrate the surface and be adsorbed.

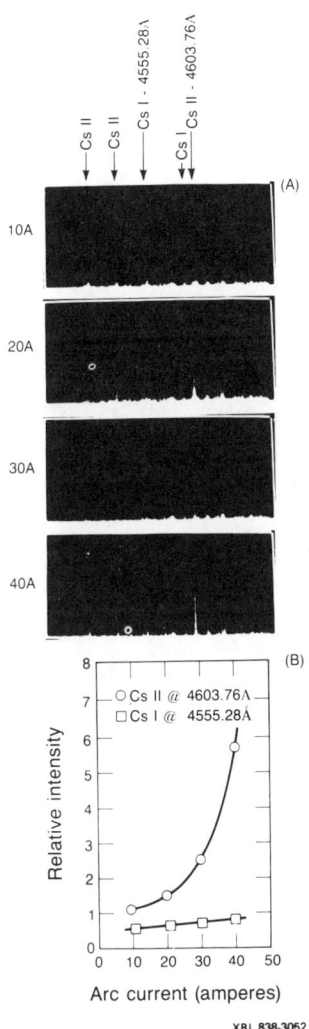

Fig. 2. Cesium Light from a Discharge vs. Arc Current

6. The positive ion can release secondary electrons from the surface.

7. The positive ion may capture two electrons and leave the surface as a negative ion.

SPUTTERING

The fractional monolayer coverage of cesium on tungsten vs. temperature was rather extensively studied by Taylor and Langmuir[4] who found that a minimum work-function coverage of cesium could be maintained if tungsten was heated to about 550°K, and that all adsorbed cesium could be removed in the form of positive ions, if the temperature was increased to ~1200°K. Although this method of cesium control is important for pulsed sources, it is not pertinent to dc sources, where the converter electrode is well cooled and the problem is one of providing enough cesium rather than too much.

For dc sources, the principal mechanism by which cesium is removed from the converter surface is the sputtering away of the cesium layer by the cesium ions present in the discharge. Experimental data for determining this sputtering ratio is not available, but as this ratio is important for this consideration, an attempt has been made to determine its approximate magnitude.

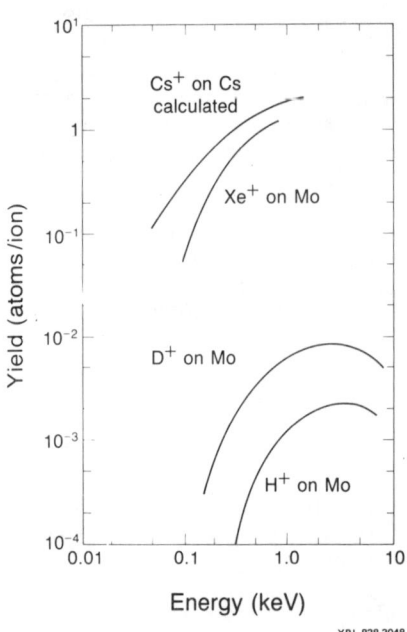

Fig. 3. Sputtering Yields for H^+, D^+, Xe^+, and Cs^+ (calculated) on Molybdenum.

Fig. 3 shows the experimentally obtained sputtering ratio of xenon, with a mass similar to that of cesium, on molybdenum.[5] This ratio should represent a lower limit to the ratio we wish to know. Needless to say, if the xenon, or cesium ions, can sputter the moly substrate, which has a high binding energy of 6.83 eV, they certainly will have removed adsorbed cesium which has a much reduced heat of sublimation. Also shown in Fig. 3 is the sputtering ratio of hydrogen and deuterium on molybdenum. The sputtering effects from these two light ions can be ignored as they are two to three orders of magnitude lower than that of a cesium mass ion. Bay[6] predicts that at 150 eV, sputtering of moly by hydrogen should be nill as this energy is about his predicted sputtering threshold.

Theoretical sputtering ratios and thresholds are related to the heat of sublimation or the heat of desorption of the target surface. Thus the sputtering of cesium from the surface should be expected to be greater than that of moly as the heat of desorption is less. In addition, the heat of desorption increases as the cesium coverage decreases, hence the

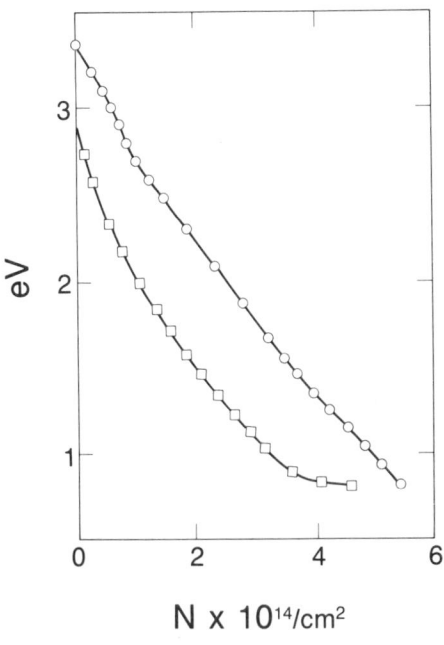

Fig. 4. Heat of Desorbtion for Cesium on the □100 and ○110 Faces of Tungsten versus Coverage.

sputtering ratio should not stay constant. Fig. 4 shows this sizable increase in the heat of desorption as the cesium coverage on a tungsten substrate decreases.[7] This indicates a thick cesium layer is more easily sputtered away than a partial monolayer. The minimum work function for the 100 face of tungsten requires about 2.5×10^{14} cesium atoms per cm^2, and the 110 face requires about $3.2 \times 10^{14}/cm^2$.

Using the Garching TRIM code for sputtering, Eckstein[8] has calculated the sputtering ratio of Cs^+ on thick cesium, and these results are also plotted in Fig. 3. Because of unknown factors, these yields can have an error as large as a factor of two. In addition the code shows a factor of two reduction in the sputtering rate when the heat of desorption is changed from .75 eV to 1.75 eV. Thus our data is limited as to the actual sputtering rate of the partial coverage of cesium by cesium ions, however the indications are that it approaches 1 in our energy range of interest, and very likely doubles as the ion energy increases from 150 to 300 eV.

We have repeatly observed that as the plasma density is increased, which results in less cesium neutrals and more cesium ions reaching the converter surface, the converter bias must be reduced in order to reduce the sputtering ratio. In our test stand with plasma densities of $\sim 10^{12}/cm^2$, we find the optimum converter potential is often 100 volts, where as we would rather operate with a higher bias as the H⁻ output would likely be increased.

REFLECTION

One normally concludes that when 150 to 300 eV ions strike a surface at normal incidence, they would neutralize, impart much of their energy to the surface, and then rebound from the surface as lower energy neutrals. However, this is not true when the ionization potential of the impacting ion is very low. Arifov[9] has investigated the processes which occur when alkalai ions strike a surface, and he finds that a remarkably high fraction of impacting Cs^+ ions scatter from the surface as low energy positive ions.

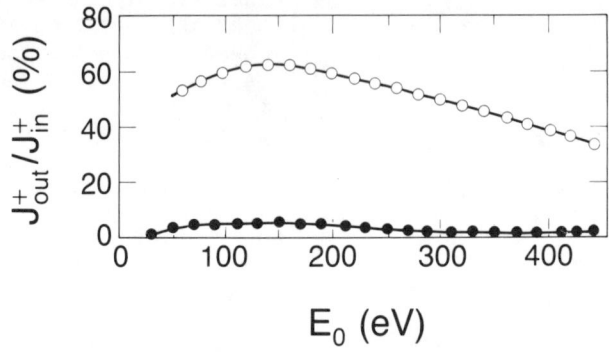

Fig. 5. Ratio of Scattered Secondary Ions vs. Energy for Similar Mass Impacting Ions. ○ = Cesium   ● = Barium

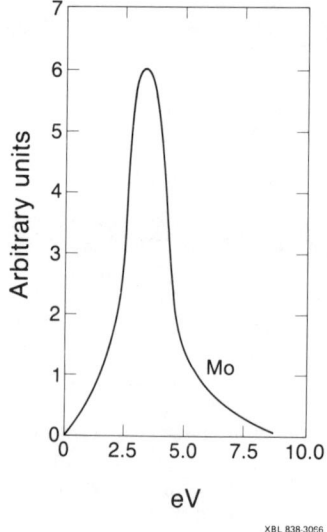

Fig. 6. Energy Spectrum of Secondary Cs^+ Ions, for 300 eV Primary Cs^+ Ions on Molybdenum

Fig. 5 is a plot of Arifov's data which shows that when 150eV Cs^+ ions (ionization potential = 3.89 eV) impact at normal incidence on molybdenum, about 63% reflect as Cs^+. For the case of barium, which has a mass similar to cesium but a slightly higher ionization potential (5.81 eV), less than 10% scatter from the surface as positive ions. Thus one can see that for gases like hydrogen and deuterium which have a considerably higher ionization potential, essentially no scattered ions would result, and only low energy neutrals would leave the surface.

Fortunately, Arifov was also able to measure the energy spectra of the reflected positive ions and these results are shown in Fig. 6. Essentially all the reflected ions are contained in the spectrum shown which ranges from near zero to less than 10 eV, and with a maximum at ~ 3.5 eV. This is a most fortunate effect for surface conversion sources. The sheath immediately adjacent to the converter surface which provides the cesium ion impact energy, will be effective in returning these low energy cesium ions back to the converter surface. Their next encounter with the surface will now be at such a reduced energy that they should now neutralize and stick to the surface as neutral atoms. Therefore, with a converter potential of 150 volts, we should expect nearly 60% of the incoming energetic cesium ions to be retained on the converter surface. In turn, it is interesting to note that one retains nearly 20% more of the incoming cesium ions by operating the converter bias at -150 rather than -300V. Both this effect as well as the reduced sputtering ratio favor reduced converter potentials.

ADSORPTION AND ABSORPTION

If Arifov's data is correct, and at 150 eV we are able to retain 60% of the impacting cesium ions, the remaining 40% must either be absorbed on the converter surface, penetrate the surface and be adsorbed, or scatter from the surface as a low energy neutral. Arifov[9] did an interesting experiment, the results of which are shown in Fig. 7. By heating a molybdenum target after it had been bombarded for some time with Cs^+ ions, he was able to detect a group of thermally produced ions which had been absorbed by the target and which required time to diffuse to the surface and become ionized. The first weak indication of cesium ions having penetrated the target was observed at ~ 200 eV. These currents were very weak, and thus we can conclude that only a few percent of the cesium ions in our energy range of interest are retained by penetration or absorption.

Although it is not conclusive, the same experiment seems to indicate that not too many of the cesium ions remained on the surface when the target was initially heated. It is much more probable that the large majority of the remaining 40% are neutralized and then leave the surface as neutral cesium atoms with energies similar to that shown in Fig. 6.

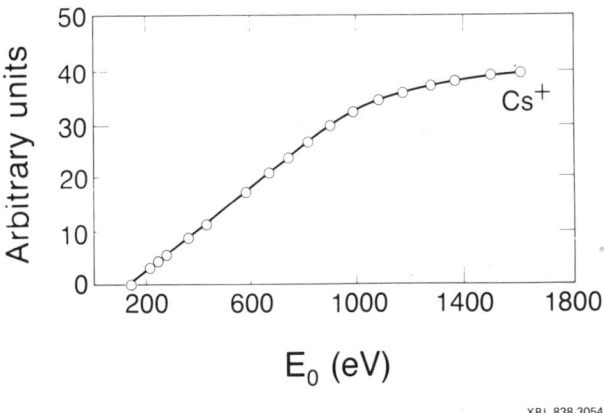

Fig. 7 Dependence of Cesium Ion Penetration of Molybdenum on Ion Energy

From this data, though admittedly incomplete, we have constructed the following model for the effects of cesium in the discharge: Approximately 60% of the cesium ions which strike the molybdenum converter electrode scatter as very low energy cesium ions. Because of their reduced energy, they are prevented from leaving the converter by the cathode sheath and are ultimately retained on the converter surface as neutral cesium atoms. The majority of the remaining 40% of impacting ions, leave the surface as low energy neutrals and they re-enter the plasma. If the sputtering ratio exceeds 0.6, additional cesium neutrals must arrive at the converter to maintain the desired cesium coverage. As the plasma density is increased, fewer cesium neutrals can penetrate the plasma and the converter surface becomes under-cesiated. If more cesium neutrals reach the converter than are required, one can easily maintain the desired coverage by increasing the converter potential which increases the cesium sputtering rate. If we are generally undercesiated as our test results indicate, it is probable that the sputtering ratio does exceed 0.6 even at the reduced converter potential at which we operate. An indication of this is shown in the photographed waveform shown in Fig. 8. H- ions were being extracted from the LBL surface production test source by a pulsed (3 sec.) 15 kV power supply. The lower flat trace is the three second pulse of the acceleration potential as read across a voltage divider. The self-extraction source was running dc, but prior to the high voltage pulse, the converter potential had been turned off. This allowed the converter to float electrically, and its floating potential was just 30 volts below anode potential. At this potential, cesium ions which strike the converter do very little sputtering of the cesium coverage, hence the cesium coverage increases. Just prior to the application of

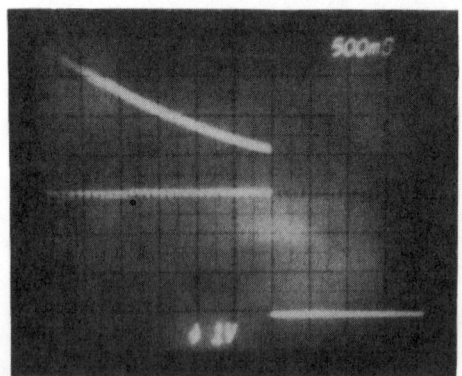

Fig. 8. Extracted H- output vs. time (See text)

the acceleration potential, the converter bias was raised to -150 volts, and one can see (on the top trace) that the H⁻ output is considerably higher at the beginning of the pulse. With time, the cesium coverage sputters away and the H⁻ output drops back to its original dc operating level.

SECONDARY ELECTRON PRODUCTION

By changing Faraday bias potentials, Arifov[9] was able to determine the secondary electron ratio for Cs^+ ions striking tantalum that contained an adsorbed film of cesium atoms. In order to change the film coverage, he determined the secondary electron ratio vs. the temperature of the tantalum target and this data is shown in Fig. 9.

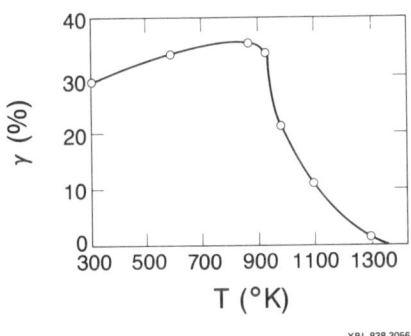

Fig. 9. Secondary Electron Ratio for 300 eV Cs^+ on Tantalum vs. Temperature

The secondary electron production ranges from 30 to 35% as the target is heated up to about 900°K. Above this temperature the secondary production drops as the cesium film is driven from the heated surface. At a temperature of about 1350°K, and for ion energies up to 1 keV, the secondary electron ratio drops to less than 1%. Thus the presence of the adsorbed film on the target is the source of the ion-electron emmission. This data would indicate that ~ 30% of the converter power supply drain is due to secondary electrons, which then enter the plasma with energy equal to the bias potential. The magnitude of photon produced secondary electrons from the converter surface is not known, but they would add to the drain of the power supply as well.

CONCLUSION

We have attempted to determine the approximate magnitude of the effects which can occur at the surface of a converter, when cesium is added to a hydrogen discharge. Due to the lack of experimental data, one can only make his best guess based on that data which does presently exist. Our best guesses may change with the addition of time, experience, and experimentally obtained physics data.

ACKNOWLEDGEMENTS

This work was supported by the Director, Office of Energy Research, Office of Fusion Energy, Development & Technology Division of the U.S. Department of Energy under Contract No. DE-AC03-76SF00098.

REFERENCES

1. L. B. Taylor and I. Langmuir, Phys. Rev. $\underline{44}$, 423, (1923).
2. G. I. Dimov, (Private Communication)
3. Atomic Data for Controlled Fusion Research, C. F. Barnett et. al. ORNL-5207, Feb. 1977.
4. J. B. Taylor and I. Langmuir, Phys. Rev. $\underline{44}$, 6, 423, (1933).
5. D. Rosenberg and G. K. Wehner, J. Appl. Phys. $\underline{33}$, 5, 1843, (1962).
6. H. L. Bay, J. Roth, and J. Bohdansky, J. Appl. Phys. $\underline{48}$, No. 11, 4723, (1977).
7. Thermionic Converters and Low Temperature Plasma. English edition edited by L. K. Hansen. Published by Technical Information Center/U.S. Dept. of Energy, DOE-tr-1, (1978).
8. W. Eckstein, (Private Communication)
9. U.A. Arifov, Interaction of Atomic Particles with a Solid Surface, Consultants Bureau, New York-London, (1969).

DISCUSSION

GRANNEMAN: You showed on a viewgraph that 65% of the cesium particles leaving the surface would leave as positive ions. At what value of the work function was that measurement done?
EHLERS: It is not a function of the work function at all.
GRANNEMAN: I guess that's a strong function of the work function of the surface. If you go to half a monolayer, the work function of the converter is at the minimum then the neutral fraction should increase.
EHLERS: What Arifov has done in this case is to essentially take a cold, clean molybdenum target, fire cesium ions at it, collect particles leaving, and he found that they are coming off as positive ions.
GRANNEMAN: But if you have half a monolayer of cesium, the work function decreases and the neutral fraction goes up considerably.
EHLERS: Conceivably those that are sputtered off that layer might come off positive too but I don't know that. If we had a sputtering ratio that was below 0.6, we'd have no problem. So it must be above 0.6 and just how much above 0.6, I don't know. But it looks like we don't have to supply an awful lot of additional cesium to the surface. We have to supply enough to get our yields back up. If we can supply more than the amount to where we run the converter at 100 volts or 150 volts, we'll crank the converter voltage up, the sputtering rate will go up and if we can still accord to that, we're still alright.
STIRLING: You showed a trace of the H^- current, falling during a 3 s pulse. What is your estimate of the H^- current density at the converter?
EHLERS: Well, I have no idea. This trace is the actual extracted current. We lose a factor of 2 or more inside the source just because of the ions that have angles such that they don't get out. On the curve I showed you, the current was about 750 mA at the peak and about 400 mA at the end. It means that at the exit slit, the current density is between 10 and 15 mA/cm^2. I can't chase that back to the converter. We have focusing, but those ions that come off with transverse energies above 2 or 2.5 eV never get out of the source and about half of our ions don't make it out.
MOSES: I didn't understand the last slide. You said that you would stop getting secondary electrons, but at those temperatures you will start getting electrons by thermal emission.
EHLERS: Not off molybdenum. Molybdenum has a pretty high work function and at the high temperature point the cesium is gone. With cesium on the surface you could get thermally emitted electrons, but cesium leaves as it becomes positively ionized.
PETERSON: It may be that one of the reasons molybdenum is better than tungsten is that molybdenum's mass is lighter than cesium's and you can't get directly that reflection from molybdenum whereas with tungsten you can. This allows the cesium to bury itself on the first collision. It can't be directly that reflected. This may add to the fact that you're not getting directly so much reflection as you might otherwise. As a matter of fact, it is sputtering but it affects the ability of the cesium to stick in the first place.

EHLERS: It would in turn say that with the single crystal where a higher energy of desorption is required that we should make a gain. Our work which we'll report on this afternoon didn't really show that but we weren't really in a situation where we were undercesiated because we were working with low densities. We could have been overcesiated at certain times. But in our TS-1 source we're out to get the maximum current and that means you're fighting right up to the limit and if you could extend the limit a little further, we'd run up there.

HERSHCOVITCH: I would like to add a remark that we basically saw something very similar in our hollow cathode discharge source which we ran steady state. What we did was to release a lot of cesium all of a sudden into the source and we were able to get 0.5 A and even a few shots at 800 mA. However, within 15 minutes we were back to the 0.25 A level which we could run steady state for hours. We did see the same effect.

SYSTEMATIC INVESTIGATION OF NEGATIVE ION PRODUCTION FROM LOW-WORK FUNCTION SURFACES

M. J. Coggiola and J. R. Peterson
SRI International, Menlo Park, CA. 94025

ABSTRACT

We have begun a systematic study of H^-/D^- ion production from low-work function metal surfaces. Initial experiments are focussing on low energy (50-500 eV) H_i^+/D_i^+ (i=1-3) impact on both clean polycrystalline Mo and minimum work-function Mo/Cs surfaces. A new, ultra-high vacuum ion-surface scattering apparatus has been constructed for this work, and is described here. Preliminary results are presented for the measured secondary electron yields from low energy He^+, Ar^+, H^+, H_2^+, H_3^+, and Cs^+ impact on polycrystalline molybdenum between 50 eV and 1500 eV.

INTRODUCTION

The LBL self-extraction negative ion source has become one of the most promising high current H^-/D^- sources designed for neutral beam heating[1,2]. A principle aspect of that source is a negatively biased converter surface which is exposed to a steady-state hydrogen plasma that also contains controlled amounts of cesium. The cesium is presumed to form a partial monolayer coverage on the converter, leading to a significant reduction in its work function. This in turn enhances greatly the production of H^-/D^- as the converter is bombarded by positive ions and perhaps energetic neutrals that emerge from the plasma.

Several experimental efforts have undertaken to provide the basic atomic physics data needed to understand and model the source operation. These experiments include, measurements on the actual source under operating conditions, and investigations of specific collision processes in an isolated and controlled environment. The source experiments have been done at LBL, and some recent results are presented elsewhere at this conference. Among the later type experiments is the work of Schneider et al.[3,4] on backscattered H^- yields from H_2^+/H_3^+ impact on targets of varying composition, the work of Seidl and coworkers[5-7] using Cs^+ projectiles, the thermal energy H-atom studies by Graham[8], the experiments by van Wunnik[9] on small-angle proton scattering from cesiated tungsten, and the work of Eckstein and coworkers[10] on backscattering from cesiated surfaces. Each of these studies has provided some additional insight into the actual operation of the converter type source, but each has been limited in scope. In view of this, we have undertaken a more comprehensive series of experiments of the later type: separate investigations of the various important collision processes under controlled conditions.

APPARATUS

The recently constructed apparatus used here consists of three main vacuum chambers, shown schematically in Figure 1, each of which will be briefly described in turn.

The ion source was designed to provide positive ion beams over a wide range of energies, with special emphasis on the low energy region. The source uses the commercially available Colutron components including; the hot filament discharge source, extraction and focussing lenses, an E x B velocity filter, a 2 stage deceleration lens, and a retractable Faraday cup current monitor. The modular source design also makes it easy to remove the discharge source which is used for H_i^+, and replace it with a surface ionization Cs^+ source. Typical operating pressures in the source are 3×10^{-6} torr with the ion beam running.

A stage of intermediate differential pumping is required to isolate the source from the target. The differential pumping region is a small-volume, all stainless steel chamber which sits directly on a 330 l/sec turbomolecular pump. A 2 mm diameter aperture is used between the

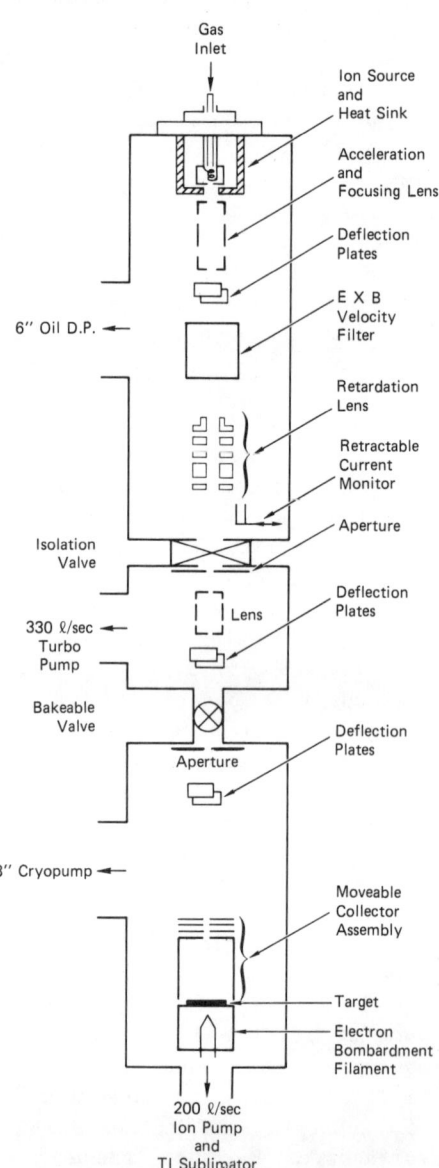

Fig. 1. Schematic diagram of the ion-surface apparatus.

source and differential chambers, both as a pumping restriction, and for beam collimation. Base pressure in the differential region is 1 x 10^{-8} torr, rising to 2-3 x 10^{-8} torr with the ion beam on. Located within this differential chamber is a small Einzel lens which can be used to refocus the ion beam. For studies involving neutral projectiles, a small charge-transfer gas cell can be added to the beam line within this chamber.

The interaction chamber is the largest of the three, with a volume of approxiamtely 120 ℓ. It is designed for ultra-high vacuum operation, being of all stainless steel construction and utilizing metal gaskets wherever possible. Pumping for this chamber is provided by a closed-cycle, liquid-He cryopump equipped with activated charcoal cryopanels. In addition a 200 1/sec ion pump is used to maintain a low pressure (10^{-8} torr) during regeneration of the cryopump. To date, an ultimate pressure of <2 x 10^{-9} torr is reached in this region without bakeout, and using a Viton o-ring to seal the top (24" diameter) flange.

The important components of the interaction region are shown in more detail in Figure 2. The target surface, located in this chamber, is clamped to a ceramic mounting block which also holds a small tungsten filament. The filament is positioned behind the target, and can be used to heat the surface by electron bombardment. A thermocouple attached to the surface monitors the temperature. The entire mounting assembly can be rotated about the surface normal to change the angle of impact with respect to the ion beam.

At present, the major diagnostic device installed on the chamber is a quadrupole mass spectrometer. The primary use of the mass spectrometer will be for secondary ion mass spectroscopy

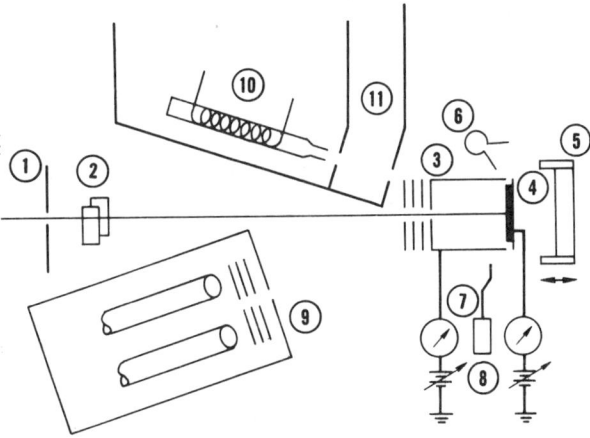

Fig. 2. Details of the interaction region.
1) Aperture (2 mm dia)
2) Deflection Plates
3) Moveable Collector Assembly 4) Target
5) Permanent Magnet Electron Suppressor
6) Cesium Dispenser
7) Kelvin Probe
8) Biased Electrometers 9) Quadrupole Mass Spectrometer
10) H-Atom Source
11) Differential Pumping Region.

(SIMS) in which a 2-3 keV Ar^+ beam from the source will be directed at the surface, and the sputtered positive ions observed. It is also important to correlate behavior of negative ion production to gross surface characteristics. One surface characteristic which is known to be very important in controlling the yield of negative ions is the work function (ϕ). Of the several methods available for work function measurement, we have chosen the contact potential (or Kelvin) method using a commercially available probe (Delta-Phi-Electronik) with a gold reference electrode.

The first experiments include measurements of the yield of secondary electrons from various converter surfaces under ion bombardment. In order to collect these secondary electrons, a detector assembly was designed in the form of a deep Faraday cup with a beam entrance aperture. A series of guard apertures were placed at the entrance to the cup to suppress any secondary production which could occur as the beam entered the detector. Because this detector system must be mounted concentric with the ion beam axis in front of the target surface, no other diagnostics could be used in this configuration (e.g. SIMS, Kelvin probe, etc.). This problem was overcome by constructing a large (3" x 6") translation stage which can hold the detector, the Kelvin probe, and several other devices. With the stage at its extreme retracted position, the quadrupole mass spectrometer has a clear view of the surface for SIMS work, and yet the secondary electron/ion detector and/or Kelvin probe can be quickly moved into place.

Secondary electron yield measurements require careful consideration of the effects due to backscattered and sputtered negative ions. For this reason, a small permanent magnet assembly can be lowered into position adjacent to the target. The resulting magnetic suppression of secondary electrons will allow these effects to be accounted for quantitatively.

For a majority of studies, it will be neceassary to cesiate the surface to some known extent. For this purpose, a small directly heated Cs dispenser will be the alkali source. Following cesium deposition, the Kelvin probe can be easily moved into position for an accurate measurement of ϕ. A further important capability designed into the interaction chamber (but not yet implimented) is the addition of a differentially pumped neutral hydrogen atom source to provide adsorbed atomic hydrogen on the surface in order to access its influence on negative ion production.

RESULTS AND DISCUSSION

The first experiments undertaken were measurements of the secondary electron yields produced by ion bombardment of "clean", molybdenum. Here, clean refers to the fact that the target has not been cesiated, and has had only minimal surface preparation. Specifically, the Mo target was fabricated of 1.25 mm thick sheet stock (Alfa Products) with a stated purity of 99.95%. The target was ultrasonically cleaned using successive baths of detergent, distilled water, ethanol, and a final distilled water rinse. After installation and pumpdown, some measurements were made on the surface without further preparations, however, all of the results given here were obtained following baking of the target for varying periods of time at 400-450 deg C. Initially upon heating the target, a noticeable pressure rise occured in the interaction chamber indicative of surface outgassing. Following several hours of operation at 450 deg C, the pressure returned to its base value of $< 2 \times 10^{-9}$ torr, and remained at this level even during successive heating cycles.

As a test of system operation, several ions were studied for comparison with previous work. The first measurements made were with He^+ and Ar^+ between 0.5 and 2.5 keV. The results for both He^+ and Ar^+ are shown as a function of impact energy in Figure 3 along with curves fitted to the previous results of Mahadevan et al.[11] It is clear from the figure that significant differences exist between the present work and the older data. What is most surprising about these discrepancies is that the Ar^+ data

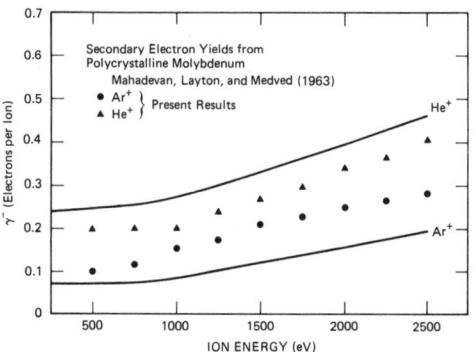

Fig. 3. Secondary electron yields for He^+ (▲) and Ar^+ (●) on bare Mo. Solid lines are to fits data from Ref. 11.

appear to differ in one direction, while the He^+ data differ in the opposite direction. It is unlikely that any type of systematic error can account for this behavior. Similarly, any surface contamination effects might be expected to produce equal errors for both projectiles. In fact, data taken prior to surface heating which most certainly involved a heavily gas covered target, did give results substantially larger than those in Figure 3. Following several baking cycles, the secondary yields were reduced to the reported values, with no further reduction produced even when measured with the target at 300-400 deg C.

A reasonable explanation for these differences lies in the details of the surface itself. Large and Whitlock[12] first showed that for a molybdenum target, significant changes in the secondary electron yields from keV proton bombardment could be seen depending on whether the surface was chemically etched, vacuum baked at 450 deg C, or flashed to 1750 deg C. Vance[13] later extended this work and showed that the effects were due to surface carbon contamination. He found, for example, that above 300 eV, Ar^+ on Mo that had been flashed to 2000 deg C gave essentially the same results as obtained by Mahadevan, while flashing the target to the same temperature in an oxygen atmosphere yielded a secondary coefficient nearly twice as large. He concluded from this and other evidence that flashing the target in oxygen caused the surface carbon to be chemically removed, thus producing an atomically clean

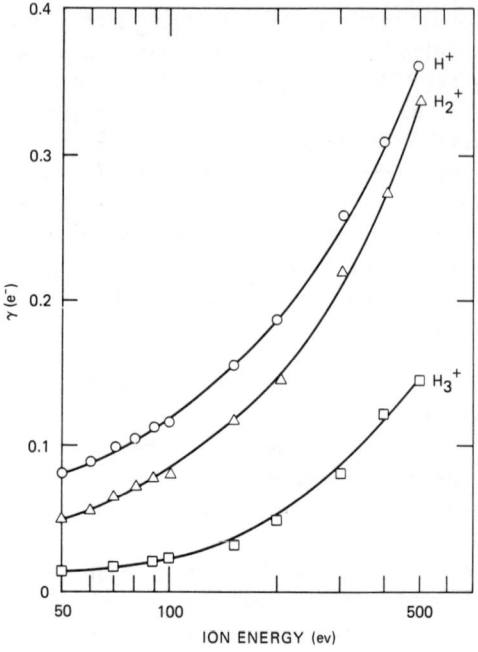

Fig. 4. Secondary electron yields for H^+ (○), H_2^+ (△), and H_3^+ (□) on clean Mo target.

surface. Interestingly, he found that the same treatment had little or no effect on the He^+ secondary yields. The degree of surface carbon contamination is apparently not predicatable or reproducible, and therefore results using different sources of molybdenum may not be directly comparable unless they have all been chemically depleted of their carbon. The results in Figure 3 for both Ar^+ and He^+ are within 20% of the values found by Vance for carbon depleted molybdenum. This agreement is most likely due to the particular nature of the molybdenum used in the present work (e.g. low surface carbon) rather than to any preparations (cleaning heating, etc.). The results do show, however, that the basic apparatus can yield reliable secondary electron emission coefficients that characterize the surface being studied.

Following these demonstration experiments, secondary electron yields were measured for H^+, H_2^+, and H_3^+ for ion energies between 50 eV and 500 eV (typical of the ion energies striking the converter in the LBL source). These results are shown in Figure 4. Here again, several previous workers have measured these quantities for reportedly clean molybdenum targets[12-15]. In view of the preceeding discussion, all of these results (including the present ones) must be assumed to reflect some degree of surface carbon contamination. In general, the values shown in Figure 4 are in reasonable agreement with some of the more recent experiments in the low energy region. However, the present data show a much faster rise with energy than do any of the other studies. This is true for each of the three projectiles, and is most noticable for H_2^+. Whereas Large and Whitlock[12] found that above 10 keV, the secondary yield seemed to scale with the number of atoms in the projectile ion, the present results do not. For example, at 100 eV, H^+ gives $\gamma = 0.116$, while H_2^+ at 200 eV yields $\gamma = 0.147$, and H_3^+ at 300 eV gives $\gamma = 0.083$. At higher energies, this scaling is assumed to arise from the breakup of the molecular ion prior to impact, with each

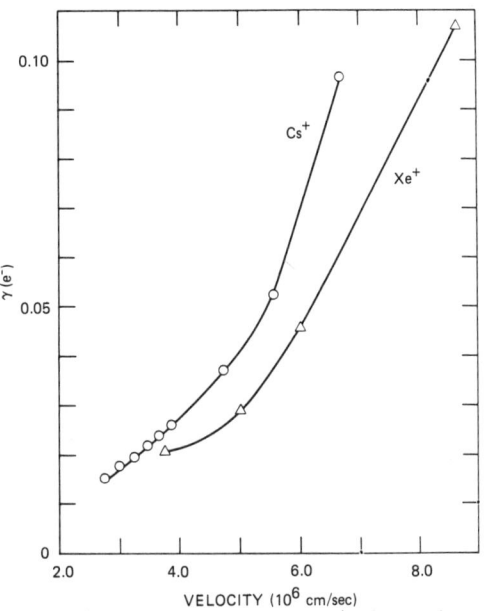

Fig. 5. Secondary electron yield for Cs^+ (○) impact on bare Mo as a function of velocity. Xe^+ data from Ref. 14.

fragment producing electron emission via a simple "kinetic" mechanism. At lower energies, where secondary ejection is more likely due to an Auger process, this same scaling is not expected to hold as rigorously.

It now appears reasonably certain that in an operating LBL source, substantial amounts of Cs^+ ions also bombard the converter surface. Thus, we have also made preliminary measurements of the secondary electron yields for this ion on our Mo target. The results are shown in Figure 5 plotted as a function of projectile velocity. No previous measurements have been made of γ for Cs^+ impact on molybdenum, however, Seidl[7] has measured γ for Cs^+ impact on a Cs/H/Mo surface and found comparable yields over this energy range. Ferron et. al.[14] have reported results for Xe^+ on Mo over this same range, and these data are included in Figure 5 for comparison with the present values. The Cs^+ and Xe^+ data are reasonably similar in shape and magnitude, perhaps reflecting the fact that only the projectile momentum is important here.

In any case, the strong variation of these secondary electron yields with surface preparation and hence surface carbon content make it difficult to apply these results to the situation of interest, namely the LBL self-extraction negative ion source. Since the converter in that source is never flashed to 2000 deg C in an oxygen atmosphere, it is clear that any surface carbon contamination cannot be removed in this manner. On the other hand, since the converter is exposed to a dc hydrogen plasma discharge, what is the fate of this surface carbon? To investigate this question, we plan to make additional secondary electron yield measurements following the in-situ exposure of our target to a similar dc hydrogen discharge. Changes in either the electron yields or the observed SIMS spectra may provide clues as to the disposition of this carbon contamination.

This work was supported by the U.S. Department of Energy, Division of Magnetic Fusion Energy.

REFERENCES

1. K. N. Leung and K. W. Ehlers, Rev. Sci. Instrum. 53, 803 (1982).
2. K. N. Leung and K. W. Ehlers, Proceedings of the Second International Symposium on the Production and Neutralization of Negative Hydrogen Ions and Beams, Brookhaven, N. Y. (1980) p. 65.
3. P. J. Schneider, K. H. Berkner, W. G. Graham, R. V. Pyle, and J. W. Stearns, Phys. Rev. B23, 941 (1981).
4. P. J. Schneider, Ph.D. thesis (University of California, Berkeley, 1981) (unpublished).
5. A. Pargellis and M. Seidl, Phys. Rev. B25, 4356 (1982).
6. M. Seidl and A. Pargellis, Phys. Rev. B26, 1 (1982).
7. M. Seidl, A. Pargellis, and J. Greer, Notas de fisica 5(1), 393 (1981).
8. W. G. Graham, Phys. Lett. 73A, 186 (1978).
9. J. N. M. van Wunnik, B. Rasser, and J. Los, Phys. Lett. 87A, 288 (1982).
10. W. Eckstein, H. Verbeck, and R. S. Bhattacharya, Surface Sci. 95, 380 (1980); 99, 356 (1980).
11. P. Mahadevan, J. K. Layton, and D. B. Medved, Phys. Rev. 129, 79 (1963).
12. L. N. Large and W. S. Whitlock, Proc. Phys. Soc. 79, 148 (1962).
13. D. W. Vance, Phys. Rev. 164, 372 (1967); 169, 252 (1968).
14. J. Ferron, E. V. Alonso, R. A. Baragiola, and A. Oliva-Florio, J. Phys. D14, 1707 (1981).
15. P. Mahadevan, G. D. Magnuson, J. K. Layton, and C. E. Carlston, Phys. Rev. 140, 1407 (1965).

WORK FUNCTION DEPENDENCE OF SURFACE PRODUCED H⁻ IN THE PRESENCE OF A PLASMA*

M. Wada,[†] R. V. Pyle, and J. W. Stearns

Lawrence Berkeley Laboratory
University of California
Berkeley, CA 94720

ABSTRACT

The maximum H⁻ flux from a negatively biased "converter cathode" occurs at the work function minimum. A cesiated hydrogen plasma produces a partially-cesiated surface at the converter. The cesium coverage can be controlled by the cesium partial pressure, the bias on the converter and the plasma density, while the work function of the converter surface is measured by the photo-electric effect, using a bright light source and a series of fliters. The angular dependence is measured by rotating the converter.

INTRODUCTION

A high-current negative-ion (H⁻) beam can be produced by immersing a negatively biased "converter" surface in a cesiated hydrogen plasma[1]. The electric field from the plasma to the converter accelerates plasma ions to the surface. Some part of the flux is converted to H⁻ ions which are accelerated back across the sheath and can be extracted from a slit placed opposite the converter.[2] The population of Cs in the discharge controls the work function of the converter and the H⁻ current, which is greatest at the lowest work function value.

Leung and Ehlers measured the energy distribution of the H⁻ beam from their surface-plasma ion source and observed two distinct groups in the range from the bias potential to about twice that potential.[3] The higher energy components were further resolved and identified as backscattered ions from H^+, H_2^+ and H_3^+ bombardment on the converter, while the lowest energy portion was identified with collisionally desorbed H^-.[4]

---

* This work was supported by the Director, Office of Energy Research, Office of Fusion Energy, Development and Technology Division of the U.S. Department of Energy under Contract No. DE-AC03-76SF00098.

[†]Present address, Hitachi Corp., Japan.

Leung and Ehlers found that their low energy component was substantially larger than the backscattered component. The ratio of the two major H⁻ components is very important to the H⁻ beam quality and the application of the source to an intense neutral beam system. This is because the higher energy (backscattered) portion also has a large angular distribution.

Hiskes[5] compared the backscattered and desorbed portions based on a theoretical model, and estimated the H⁻ production efficiency per incident hydrogen nucleus to be less than 1% through desorption, compared to some tens of percent by backscattering. Although Cs^+ ions in the plasma may enhance the desorption, backscattering seemed to be the dominant H⁻ production mechanism.[6]

The discrepancy might be explained by the difficulty in seeing all of the backscattered portion because of its large divergence.

This experiment was designed to measure the angular distribution as well as the energy distribution of H⁻ from the converter surface to help resolve the discrepancy.

EXPERIMENT

A schematic representation of the apparatus is shown in Fig. 1. A discharge between the filaments and the chamber walls

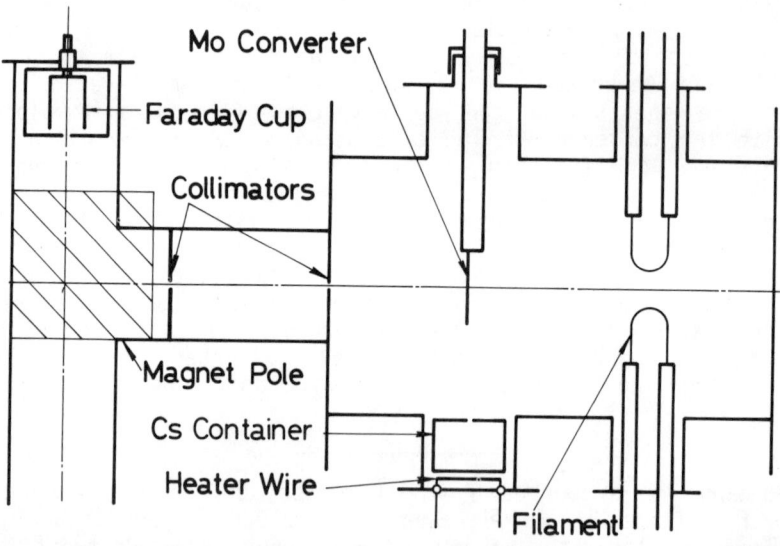

Fig. 1. Schematic of the apparatus showing the rotatable molybdenum converter for angular distribution measurements.

creates the plasma. A magnetic "bucket" geometry is created by six rows of samarium-cobalt permanent magnets placed around the 15 cm diameter chamber.[7] The filaments are maintained at -90 V with respect to the chamber walls and the discharge current is controlled by adjustments in the filament temperature as the H_2 gas and Cs vapor densities are varied. The hydrogen pressure is usually kept at ~1 mTorr while the Cs coverage on the Mo converter is increased by heating the Cs oven, which puts more Cs into the discharge. Once introduced, the Cs remains in the discharge with a long time constant. The rotatable converter is not cooled, so the Cs coverage is in dynamic equilibrium with the plasma and depends on plasma temperature, density, Cs fraction and bias on the converter.

The negative ions from the converter must traverse 7.5 cm of plasma before entering the first slit at the exit of the chamber. A second slit completes the collimation of the beam before it enters a small magnetic analyzer. A slitted Faraday cup at the focus of the analyzer is used to measure the partial current and provides a momentum resolution of about 1.5%.

Previous experiments[2-4] have established that the maximum H^- yield occurs when the surface work function is at a minimum. This corresponds, also, to a maximum yield in the low energy portion of the spectrum. Thus by monitoring the low energy H^- peak as Cs is introduced into the plasma, the minimum work function conditions can be found. Then, with small adjustments in the oven temperature and discharge parameters, the conditions can be stabilized so that the H^- yield is constant over several hours.

For the results reported here, the converter bias was maintained at -150 V. The plasma parameters were $n_e = 3 \times 10^{10}$ cm^{-3} and $T_e = 2.5$ eV, as measured with a small Langmuir probe.

The normal to the converter was initially aligned with the collimation to the analyzer. After the plasma had been stabilized, the analyzer magnet was swept and the H^- energy spectrum recorded. The converter was then rotated by 5° and the procedure repeated. This process continued until the converter had been rotated by 45°, and then repeated once more at the 0° position to confirm the constancy of the H^- yield which was generally within 2% of the original measurement. The entire proces required about 4 minutes to complete.

The angle could be set to about 0.5° while the collimation admitted ions over a range of about 1.40°. Repetition of the experiment while rotating the converter in the opposite direction gave essentially the same results.

A typical raw data result at 0°, with the Cs coverage near optimum is shown in Fig. 2. This result is very similar to ones measured by Leung and Ehlers.[3] Note the sharp rise at the lowest Hall probe values, and the shoulders at the larger Hall probe values. The width of the curve corresponds in energy with H^- ions originating at the surface with nearly zero-energy,

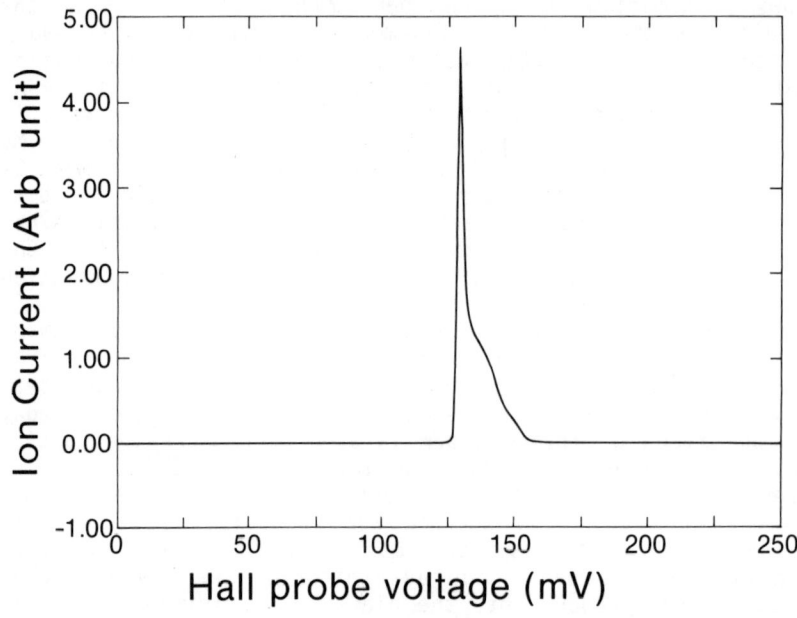

Fig. 2. Raw unscaled data showing the H⁻ spectrum. No other spectra appear over the range normally swept.

to backscattered H⁻ ions from H_3^+, H_2^+ and some H^+ that had been accelerated to the converter across the sheath.

Figure 3 is a plot of relative H⁻ yield vs energy and angle of the converter. The energy scale has been normalized so that 100% corresponds to the bias on the converter (100% = 150 eV).

The minimum energy H⁻ ion that can be seen at each angle is governed by ions that are scattered parallel to the converter surface and then accelerated normal to the surface across the sheath. This leads to an expression for the minimum energy,

$$E_{min} = |V_c - V_p|(1 + \tan^2 \theta_0) \qquad (1)$$

where V_c and V_p are the converter and plasma potentials and θ_0 is the angle from the surface normal to the collimation line. We can use this equation as a check to see that no H⁻ that did not originate at the converter surface entered the analyzer. (e.g. H⁻ ions that originated elsewhere could be reflected from the sheath.)

If we now make a linear interpolation of the H⁻ yield at each energy from the 0° measurement to each succeeding angle measured and ending with I = 0 at θ_0 + 1.4°, then we can obtain the total relative H⁻ yield at each energy.

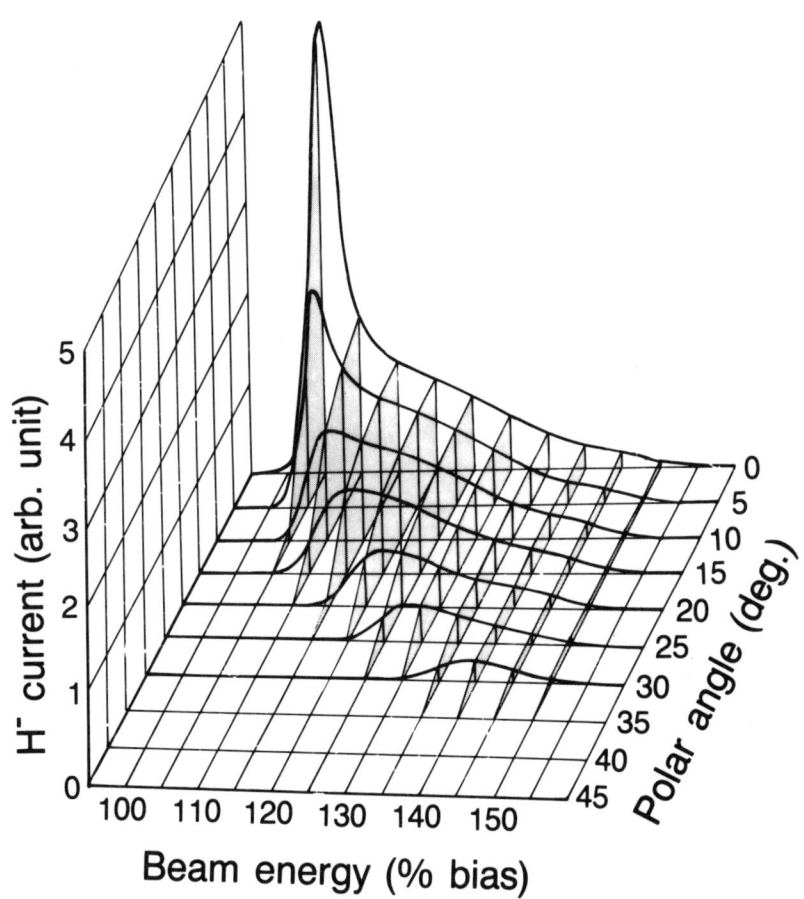

Fig. 3. H⁻ spectra vs energy, scaled to the converter bias voltage (-150 V), for ten orientations of the converter in 5° steps.

The results of this calculation are shown in Fig. 4 where the total relative H⁻ yield is plotted vs normalized energy.

If we call the low energy component that part of the curve in Fig. 3 which lies between 100% and 115%, then it becomes apparent that most of the H⁻ yield comes from backscattered energetic hydrogen. The peak at ~130% corresponds to backscattering from

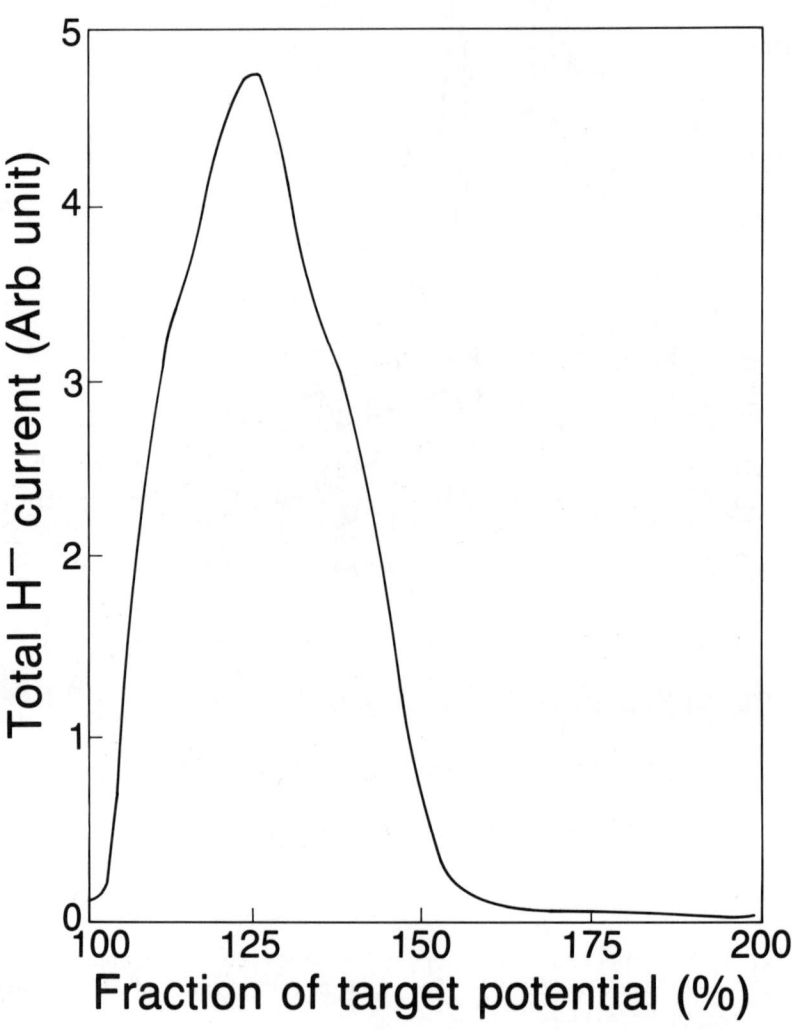

Fig. 4. Integrated relative total H⁻ flux from the converter vs H⁻ energy scaled as in Fig. 3, using data shown in Fig. 3.

H_3^+. The sharp rise from ~155% to lower energies is accounted for by backscattering from H_2^+ while the small higher energy H^- yield is from H^+. The large amount of the molecular hydrogen-ion species in the discharge compared to H^+ is probably caused by the large cesium concentration necessary to maintain the minimum work function on the converter.

The converter, in this case, was made of polycrystalline molybdenum. It has been suggested[8] that lower work functions and better cesium adhesion might be obtained from monocrystalline surfaces. Work is now in progress to test this idea with several single crystal materials in various orientations.

REFERENCES

1. Y. I. Belchenko, G. I. Dimov and V. G. Dudnikov, Izv. Akad. Nauk. SSR, Ser. Fiz. 37, 2537 (1973); Zh. Tech. Fiz. 45, 68 (1973), (Sov. Phys. Tech. Phys., 20, 40 (1975)).

2. M. Wada, Ph.D. Thesis, Univ. of California, Berkeley, CA (1983).

3. K. N. Leung and K. W. Ehlers, Proceedings of the Second Symposium on the Production and Neutralization of Negative Hydrogen Ions and Beams, BNL 5134 p. 65, Brookhaven National Laboratory, Upton, NY (1980).

4. K. W. Ehlers and K. N. Leung, Rev. Sci. Instrum. 51 (6), 721 (1982).

5. J. R. Hiskes, Jour. de Physique, Tome 40, Colloque C-7, Suppl. 7, Vol. 11, C-479, July (1979).

6. A. Kh. Ayukhamov and E. Turmashev, Sov. Phys. Tech. Phys, 22, 10, 1289 (1977).

7. R. Limpaecher and K. R. MacKenzie, Rev. Sci. Instrum., 44, 726 (1973).

8. J. R. Hiskes and P. J. Schneider, Proceedings of the Second Symposium on the Production and Neutralization of Negative Hydrogen Ions and Beams, BNL 5134 p. 15, Brookhaven National Laboratory, Upton, NY (1980).

PLASMA-SURFACE INTERACTION INVOLVED IN H⁻ GENERATION

H.-M. Katsch
Universität Essen, Essen, West-Germany

K. Wiesemann
Universität Bochum, Bochum, West-Germany

ABSTRACT

H⁻ ions were created on a ceasiated stainless steel converter surrounded by a hydrogen discharge. The energy of the H⁻ ions is equivalent to the difference between converter and plasma potential. The energy-half-width of the H⁻ ions was found to be about 5 eV. The maximum H⁻ current compared to the incoming positive ion current was 38%. With increasing converter potential there is an increase of the sputtering rate of Cs from the converter. Therefore after biasing the converter negative the production of H⁻ decreases in time to a much lower equilibrium rate. This rate is determined by the neutral Cs density.

INTRODUCTION

Ehlers and Leung[1] demonstrated the production of H⁻ ions at a converter in a magnetic box. While the production of H⁻ at a clean copper converter surface is very small a strong enhancement of H⁻ production could be found by introducing Cs. Energy analysis of the H⁻ ions showed the validity of the reflection model[2] in the case of clean converter surfaces. In the case of the ceasiated converters the energy of the H⁻ ions decreased. This can be explained by a change of the production mechanism as given by the desorption model[3,4].

EXPERIMENTAL ARRANGEMENT

The experimental arrangement is shown in Fig.1. A low pressure hydrogen plasma is produced in a magnetic box.

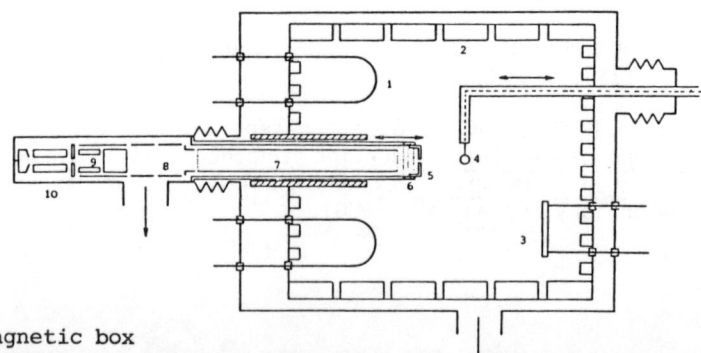

Fig.1. Magnetic box
1≙tungsten filament, 2≙permant magnets,
3≙Cs dispenser, 4≙spherical probe, 5≙converter with orifice (⌀=0.1 mm)
6≙energy analyzer, 7≙tube, 8≙electrostatic lens, 9≙deflection plates,
10≙quadrupol mass spectrometer.

Three different converter configurations were installed in the magnetic box (Fig.2). In the case a) the electron temperature Te and the total electron density n_e was measured by a spherical probe. Measurements of the density and the energy distribution of the plasma ion components were made by an orifice probe with an energy analyzer (four grids) and a quadrupol mass spectrometer (QPMS). In the case b) the energy of the H^- ions produced on the orifice probe used as converter was detected by a simple two grid ion energy analyzer (left side) using a special modulation technique[5].

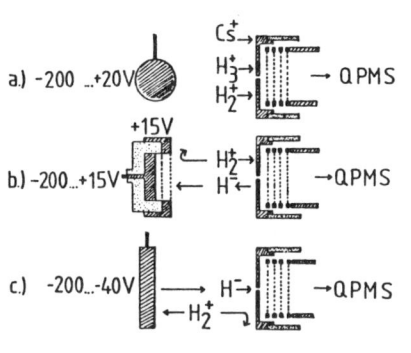

Fig.2. Converter configurations

To check the H^- density case c) is used. A plane probe (left side) is now used as the converter and ions produced are measured by the orifice probe. Typical parameters of the plasma are: Te ≈ 1 eV, ion temperature Ti < 0.1 eV, $n_e = 2 \cdot 10^9$ cm^{-3}, maximum density of Cs^+ $n_{Cs^+} \leq 2 \cdot 10^6$ cm^{-3}, neutral density of H_2 $n_{H2} \approx 10^{13}$ cm^{-3}, neutral density of Cs $n_{Cs} > 10^{10}$ cm^{-3}, maximum total pressure $p_t = 4 \cdot 10^{-4}$ Torr. The relativ density of H_2^+ and H_3^+ depends on the neutral pressure and the discharge current Id. Measurements are made in a parameter region of Id=270-470 mA and $p_t = 4 \cdot 10^{-4}$ Torr. Here we find 70% H_2^+ and 30% H_3^+. The H^+ concentration is below 3%.

MEASUREMENTS

The energy distribution of the H^- ions for different converter potentials Uc is obtained using the configuration b) (Fig.3) and the density of the H^- is checked using arrangement c). The total energy of the H^- ions corresponds to the converter potential. The energy-half-width of H^- increases with increasing converter potential. The shape of the energy distribution is shown in Fig.4. The distributions are asymmetric with a tail at higher energies. Time dependent measurements using arrangement b) show no enrichment of Cs^+ ions in the plasma-wall sheath of the converter. This may be due to

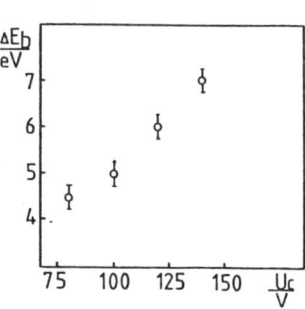

Fig.3. Half-width of the H^- energy distribution ΔE_b as a function of converter bias voltage U_c

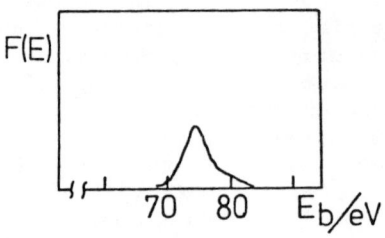

Fig.4. Energy distribution of H⁻ ions, converter potential, U_c=-80 V (energy resolution is about 0.1 eV).

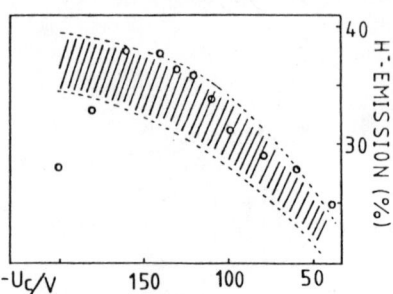

Fig.5. Maximum H̄ current per incoming total positive ion current on a caesiated converter in percent, shaded region gives results from the increase of ion current to the spherical probe.

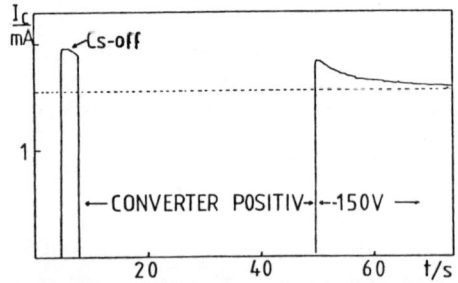

Fig.6. Converter current as a function of time for different converter potentials U_c (+15 V, -150 V), dotted line is the converter current due to positive ions from the plasma (clean surface). The increase of the current is due to H⁻ production on the caesiated converter surface.

the low Cs density and the low electron temperatures in the plasma. Comparing the measurements with configurations b) and c) it was found, that the increase of the total converter current Ic is only due to H⁻ production. In fig.5 the maximum H⁻ current is shown per incoming total positive ion current on a caesiated converter. A simple time dependent measurement of H⁻ production could be done by measuring the converter current with a transient digitizer. First the converter is biased positively with respect to the plasma potential U_p. Then the converter is chopped to -150 V. The H⁻ current saturates faster than within 50 μs. When the converter is switched positive for some minutes and then biased negatively again a higher saturation level of Ic is obtained for some milliseconds. Then the H⁻ current decreases again by sputtering of Cs atoms, till it reaches after some seconds the former level. The enrichment of Cs at the converter surface does not change with the following conditions: plasma off, or converter biased up to 15 V positive. When the converter is negative there is found a typical saturation level for each converter potential depending on the neutral Cs density. After switching on the negative converter bias and switching off the Cs supply the H⁻ production disappears within some seconds. The time of disappearance depends on the converter potential. This decrease of the H^- production can be slowed down by switching the converter positive. If the converter is biased negatively again the decrease of H^- current starts with the prior value and decreases again, see Fig.5.

REFERENCES

1. K.W. Ehlers, and K.N. Leung, Rev. Sci. Instrum. 51(6), 721(1980).
2. J.R. Hiskes, Journal de Physique, XIVe Conference Internationale sur les Phénomènes d'Ionisation dans les Gaz, Grenoble (France) 1979.
3. K. Wiesemann, Notes of Lectures given at Ecole Polytechnique, Ecole Polytechnique, Palaiseau · Cedex (France).
4. K. Wiesemann, K. Prelec, and Th. Sluyters, Journal of Appl. Phys. Vol. 48, No 7, 2668(1977).
5. M. Katsch, and K. Wiesemann, Plasma Phys. Vol. 22, 627(1980).

PRODUCTION OF H⁻ IONS FROM POLYCRYSTALLINE AND SINGLE CRYSTAL (110) TUNGSTEN AND MOLYBDENUM SURFACES

P.J.M. van Bommel[*], K.N. Leung and K.W. Ehlers
Lawrence Berkeley Laboratory, University of California,
Berkeley, California 94720, USA.

ABSTRACT

The H^- production on polycrystalline and monocrystalline (110) tungsten and molybdenum converters in a self-extraction cesiated surface plasma source has been compared. The H^- ions are mainly produced via ion impact desorption. No significant difference in H^- yield is measured between the four converter surfaces. The optimum converter voltage for H^- production, increases with the amount of cesium added to the plasma.

INTRODUCTION

There are several approaches for the production of negative ions [1,2]. At present the most evolved H^- sources are those based on plasma surface interactions [3,4].

In surface plasma type of sources an auxiliary electrode, the so-called converter, is introduced into the hydrogen discharge. This converter is biased negatively with respect to the source chamber wall (anode) and therefore to the plasma. Cesium is introduced into the discharge. As a result surfaces are (partially) covered by cesium. In general this results in a lowering of the work function. Positive hydrogen and cesium ions are accelerated towards the converter across the plasma sheath. Hydrogen atoms are either desorbed by ion impact or reflected from the converter surface. Part of these atoms are negatively charged because of the low work function of the cesiated converter. Any negative ions formed at or near the surface are accelerated back across the sheath. Subsequently these negative ions pass through the plasma and the extraction slit. In our case no external extraction field is present; this is called "selfextraction"[3].

Measurements [5] have shown that, in the case of positive hydrogen ions reflected specularly from cesiated monocrystalline (110) and polycrystalline tungsten surfaces, there is a difference of a factor of 3 in the H^- yield, in favour of the monocrystalline surface. Measurements of the fraction of negative hydrogen ions in the total number of particles (H^0 and H^-) leaving the surfaces, show a higher yield for monocrystals of a factor 2. The optimum H^+ to H^- conversion efficiency is reached at velocities perpendicular to the surface of 7×10^4 m/s [6,7]. These velocities correspond to "perpendicular" energies of around 5 eV. Hydrogen particles desorbed through ion impact from the converter, present in surface plasma sources, are believed to leave the surface with energies of the same order of magnitude [8].

---

[*] Permanent address: FOM-Institute for Atomic and Molecular Physics, Kruislaan 407, 1098 SJ Amsterdam, The Netherlands.

Because of the differences in H^- yield, mentioned above, it is worthwhile to do a comparative test on polycrystalline and monocrystalline (110) tungsten converter surfaces in a surface plasma H^- ion source. Poly- and monocrystalline (110) molybdenum surfaces are included in the test because traditionally molybdenum is the one most extensively used material in these types of sources.

APPARATUS

A schematic diagram of the experimental arrangement is shown in Figure 1. The device is a cylindrical ion source [3] of the bucket type, with a diameter of 20 cm and a length of 18 cm. The stainless steel chamber of this bucket source is surrounded externally by 10 columns of cobalt-samarium magnets ($B \approx 3.6$ kG). The extraction side of the chamber is closed by the plasma electrode, except for a slit with an area of 0.5 × 11 mm². Normal operation included an arc current of 4 A, an arc voltage of 70 V, a hydrogen pressure of $\approx 1.5 \times 10^{-3}$ torr, a converter voltage of -200 V and a converter current of ≈ 100 mA. During operation a continuous gasflow of hydrogen is maintained.

The water-cooled circular converter consists of an oxygen free copper plate with a diameter of 5.7 cm and a thickness of 1 cm. On thi s copper plate a polycrystalline molybdenum foil (thickness 0.25 mm) is brazed with Au-Ni brazing material.

A tungsten polycrystal, a tungsten monocrystal (110) and a molybdenum monocrystal (110) are brazed directly onto the copper through holes in the molybdenum foil. The dimensions of the crystals

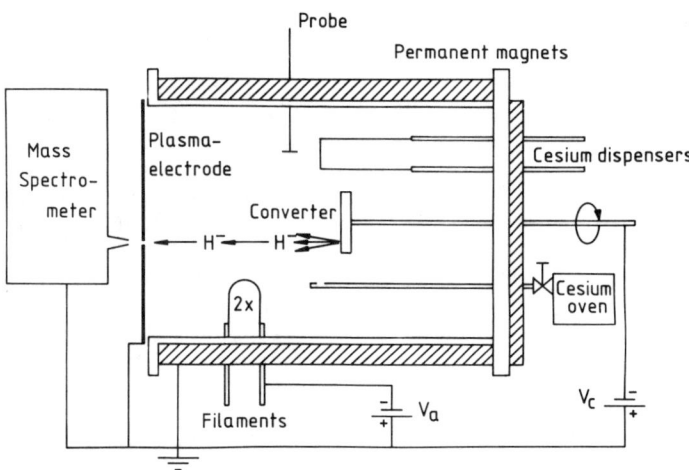

Fig. 1. Schematic diagram of the multicusp negative-ion source. The source is not drawn to scale. The diameter of the source is 20 cm, its length is 18 cm. The converter is covered with a polycrystalline foil. In holes in the foil three crystals (Mo(110), W(110) and W) are brazed.

are 16 × 19 × 1 mm³. The crystals are mechanically polished down to a grain size of 1 μm. Subsequently the crystals are electrolytically polished.

Only one type of converter material at a time is seen by the viewing system formed by the extraction slit and the entrance slit of the mass spectrometer (fig. 1). The latter is situated 3.8 mm behind the extraction slit. The distance from the converter to the extraction slit is approximately 9 cm.

Cesium is introduced into the discharge either via dispensers (SAES Getters, Italy) [9] or via a molybdenum pipe connected to a cesium oven. With cesium added to the discharge, Langmuir probe measurements indicate that the plasma density is 7×10^9 cm⁻³. The electron temperature is 0.8 eV. The plasma potential is ≈ 1.5 V positive with respect to the anode.

The mass spectrometer, consisting of an electromagnet and a Faraday cup, has an entrance slit with a width of 0.12 mm. The maximum mass number that can be detected is about 100. The half opening angle of the system formed by the extraction and the entrance slit is 2.8°. The mass spectrometer is operated at anode potential. For negative ions with the same mass over charge ratio, the mass spectrometer acts as an energy analyser.

EXPERIMENTAL RESULTS

During operation, the surfaces are cleaned due to sputtering by Cs^+ ions. This is confirmed by the mass spectrum, shown in figure 2. The heights of the O^-, OH^- and O_2^- peaks are small compared to the one of H^-.

Figure 3 shows energy spectra of H^- ions leaving the plasma. All detected H^- ions are generated on the surface, because electron capture in the plasma by fast neutrals, generated at the surface, is negligably small. Each of the four converter surfaces can have the highest H^- yield, depending on the cesium pressure in the source.

There are two mechanisms for the surface production of H^- ions. One mechanism is ion impact desorption [10]. The desorbed H^- ions leave the plasma with an energy eV_c, corresponding to the converter voltage V_c. The other mechanism is reflection of positive hydrogen ions from the surface, as negative ions [7]. Depending on the incident positive hydrogen ion species, the energy of

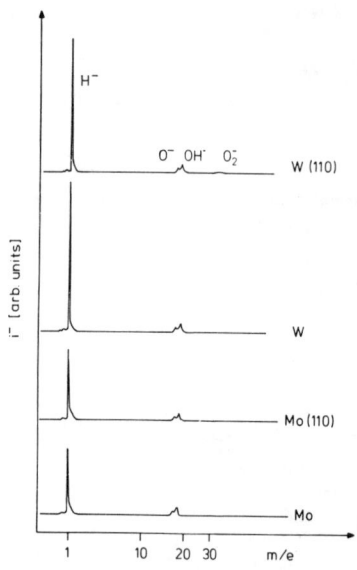

Fig. 2. Mass spectra of the negative ions leaving the four converter surfaces. The mass spectra correspond to the second row of energy spectra in figure 3.

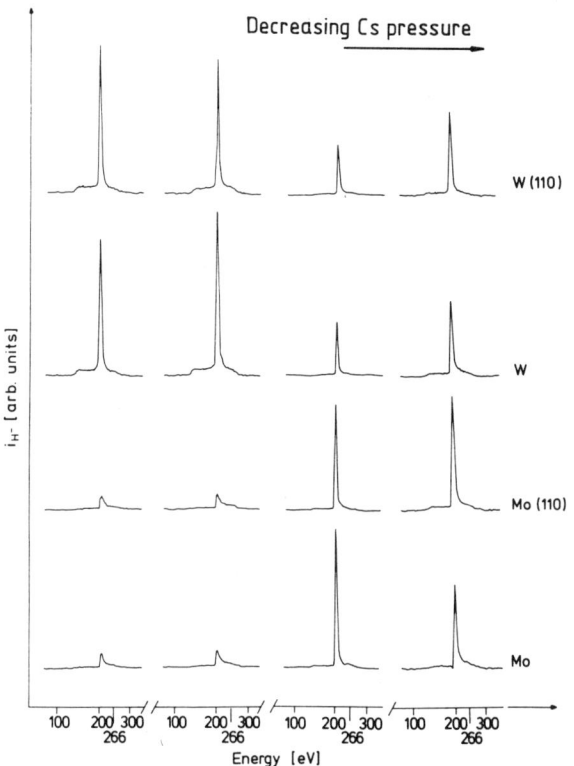

Fig. 3. The energy spectra of the H⁻ ions leaving the four converter surfaces at four cesium pressures in the source. The converter voltage is -200 V, the converter current 100 mA. The arc current is 4 A, the arc voltage 70 V. The pressure in the source is 1.5×10^{-3} torr.

the H⁻ ions reflected from the surface can be up to 2 times higher than eV_c.

It follows from the H⁻ spectra in figure 3 that the H⁻ ions are mainly desorbed (sputtered) from the surface [4]. The back-scattered fraction is small, less than 5%, compared to the desorbed fraction.

The desorbed H⁻ fraction is high because hydrogen is bound to the surface by the cesium [11,12]. A strong pumping effect is observed when cesium is introduced into the discharge [9].

A significant difference in H⁻ yield between the monocrystals and the polycrystals is not observed. Apparently, besides the negative ionization process another process is important. We suggest that this process is related to surface roughness. On a microscopic scale the polycrystals are more rough than the monocrystals and therefore their effective area for hydrogen adsorption and desorption is larger. Furthermore the incoming positive ions have a larger probability to impinge under an angle smaller than normal on a polycrystalline surface. At smaller angles the desorption coefficient is known to be larger [8].

Figure 4 shows energy spectra of the H^- ions leaving the polycrystalline molybdenum converter surface at various converter voltages between -150 V and -600 V. The cesium pressure in the source is high. It can be seen from figure 4 that the maximum H^- yield is obtained for a converter voltage of approximately -600 V.

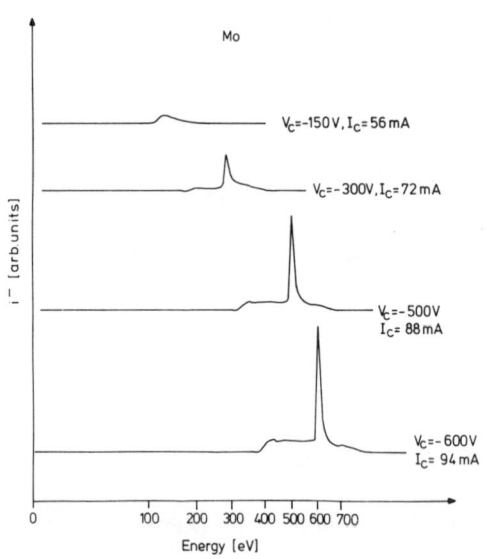

Fig. 4. Energy spectra of the H^- ions leaving the polycrystalline molybdenum converter surface at four different converter voltages. The cesium pressure in the source is relatively high. Except for the converter voltage, the discharge parameters are the same as in figure 3.

In figure 5 is shown the height of the H^- energy peaks as a function of the converter voltage. For this particular cesium inlet, the optimum H^- yield is found at a converter voltage of about -300 V, except for the polycrystalline molybdenum surface where the optimum is around -400 V, independent of the arc current. It is observed that these optima shift to higher (absolute) converter voltages when the amount of cesium added to the discharge increases. The highest optimum voltage we measured was -600 V. This value is appreciably higher than the voltage commonly applied in plasma surface sources [1].

An explanation can be found in the enhanced sputtering yield when the incident Cs^+ ion energy is increased. When the cesium pressure rises, more cesium is adsorbed. Therefore, to maintain an optimum coverage of half a monolayer [5,6,7], also the sputtering rate must increase, which is done by raising the converter voltage. Thus the maximum H^- yield is expected when the converter voltage at the particular value of the cesium pressure, is such that the cesium adsorption dynamically establish a coverage of about half a monolayer.

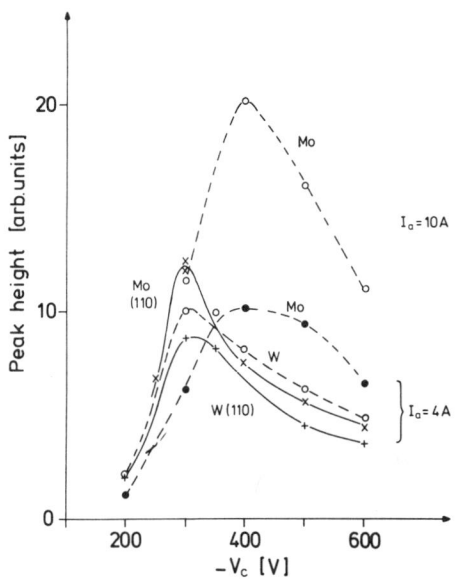

Fig. 5. The height of the ion impact desorption peaks as a function of converter voltage. The discharge parameters are the same as in figure 3. One run is done at an arc current of 10 A.

CONCLUSION

The dominant process for the H^- production is found to be ion impact desorption. The reflection peak height is below 5% of the desorption peak. No appreciable difference between mono- and polycrystals was observed. The higher H^- yield of monocrystals is counterbalanced by the higher surface roughness of polycrystals. For H^- production, the optimum converter voltage with respect to the anode increases with the amount of cesium added to the plasma.

ACKNOWLEDGEMENTS

This work is performed under US-DOE contract No. DE-AC03-76SF00098. One of the authors (PJMvB) was sponsored by the Stichting voor Fundamenteel Onderzoek der Materie (FOM) with financial support of the Nederlandse Organisatie voor Zuiver-Wetenschappelijk Onderzoek (ZWO) and EURATOM.

REFERENCES

1. K.W. Ehlers, J.Vac.Sci.Technol. A 1, 974 (1983).
2. M. Bacal, Physica Scripta 2, 467 (1982).
3. K.N. Leung and K.W. Ehlers, J.Appl.Phys. 52, 3905 (1981).
4. Yu.I. Bel'chenko, G.I. Dimov and V.G. Dudnikov, Proc.Symp.on the Production and Neutralization of Negative Hydrogen Ions and Beams, Brookhaven, USA, p. 79 (1977).
5. P.J.M. van Bommel, J.J.C. Geerlings, J.N.M. van Wunnik, P. Massmann, E.H.A. Granneman and J. Los, J.Appl.Phys. 54, 5676 (1983) and this proceedings.
6. H.J. Hopman, P.J.M. van Bommel, P. Massmann and E.H.A. Granneman, Proc.2nd Int.Symp.on the Production and Neutralization of Negative Hydrogen Ions and Beams, Brookhaven, USA, p. 233 (1980).
7. J.N.M. van Wunnik, J.J.C. Geerlings, E.H.A. Granneman and J. Los, Surf.Sci. 131, 17 (1983).

8. M. Kaminsky, Atomic and Ionic Impact Phenomena on Metal Surfaces, Springer-Verlag, Berlin, FRG (1965).
9. S.J. Hellier, Proc.4th Int.Symp.on Residual Gases in Electron Tubes, Florence, Italy, p. 2PP (1971).
10. E. Taglauer and U. Beitat, Proc.5th Int.Conf.on Plasma Surface Interactions in Controlled Fusion Devices, Galinburg, USA (1982); J.Nucl.Mat. $\underline{111}$ & $\underline{112}$, 800 (1982).
11. J.R. Hiskes, A. Karo and M. Gardner, J.Appl.Phys. $\underline{47}$, 3888 (1976).
12. M.L. Yu, Proc.Int.Symp.on the Production and Neutralization of Negative Hydrogen Ions and Beams, Brookhaven, USA, p.48 (1977).

H⁻ PRODUCTION FROM DIFFERENT METALLIC CONVERTER SURFACES*

K. N. Leung and K. W. Ehlers
Lawrence Berkeley Laboratory, Berkeley, Ca. 94720

ABSTRACT

The relative yield of H⁻ ions generated from various metallic surfaces (such as Mo, Ti, V, Nb, Pt, Pd, Rh, Cu, Ta, Al, Au and stainless steel) are compared in a multicusp source with and without the presence of cesium. Result of the investigation shows that one can optimize the H⁻ yield and formation process by choosing the proper converter material for a given range of source operating conditions.

INTRODUCTION

For heating plasmas and for current drive in some fusion reactors, high energy neutral beams will be required.[1] The high neutralization efficiency of H⁻ or D⁻ makes them favorable to form neutral atoms with energies in excess of 160 keV.[2] It has been shown that H⁻ ions can be generated by surface conversion processes[3] and a self-extraction negative ion source based on this principle has already been operated successfully to generate a steady-state H⁻ ion beam current greater than 1 A.[4]

In this experiment, we compare the negative ion yield generated by different converter materials in the pure hydrogen and cesium modes of operation. It is shown that H⁻ ions can be formed by both desorption and reflection processes even in the absence of cesium in the hydrogen discharge. The fraction of H⁻ ions formed by the desorption process can be high. We find that the number of H⁻ ions generated by this process is closely related to the amount of OH⁻ ions present in the self-extracted beam. When cesium is added to the discharge, the H⁻ yield can vary from one converter material to the other. Under optimum conditions, the majority of these H⁻ ions are formed by the desorption process.

EXPERIMENTAL SETUP

A schematic diagram of the experimental arrangement is shown in Fig. 1. The device is a cylindrical multicusp ion source (20 cm diam by 18 cm long) with the open end enclosed by a stainless-steel plate. The chamber is surrounded externally by 10 columns of samarium-cobalt magnets ($B_{max} \approx 3.6$ kG) to form a linecusp configuration for primary electron and plasma confinement.[5] A steady-state hydrogen plasma is produced by primary electrons emitted from two 0.05-cm-diam tungsten filaments which are biased at -70 V with respect to the anode (chamber wall). In normal operation, the source pressure is adjusted to 1.5×10^{-3} Torr. If needed, metallic cesium is introduced into the discharge through a molybdenum tube connected to a cesium oven.

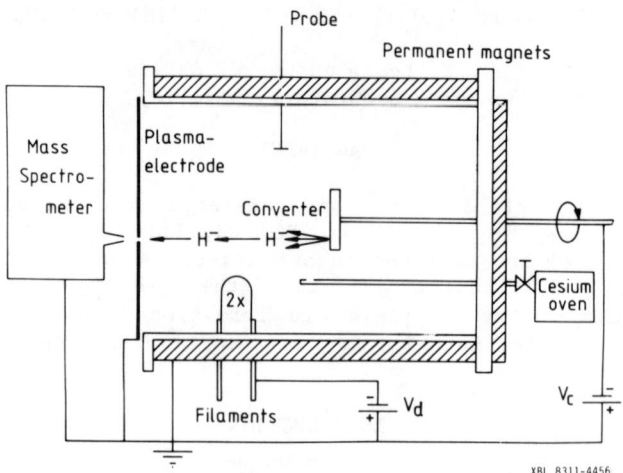

Fig. 1 Schematic diagram of the multicusp negative-ion source.

In order to investigate the negative ions produced by different converter materials, a rotable water-cooled copper disk with four different materials brazed on the four quadrants was employed. Negative ions formed at the converter surface will accelerate across the sheath, pass through the plasma and the exit aperture on the stainless steel plate, and then enter into a magnetic-deflection mass spectrometer. With this arrangement, only one type of converter material at a time is seen by the viewing system formed by the exit aperture and the entrance slit of the mass spectrometer (Fig. 1). The yield of H⁻ ions for each metal can be compared under identical conditions of gas pressure, discharge power, cesium environment, and converter potential. For negative ions with the same mass-to-charge ratio, the mass spectrometer also acts as an energy analyzer.

EXPERIMENTAL RESULTS

(a) Pure hydrogen mode operation

The source was first operated without cesium and with a discharge current of 4 A. Langmuir probe measurements indicate that the plasma density is approximately 10^{10} cm^{-3}. The electron temperature is about 1 eV and the plasma potential is ~1.5 V positive with respect to the anode.

Figure 2 shows the energy spectra of the H⁻ ions produced by Rh, Pt, Pd and Mo. Two distinct groups of H⁻ ions can be easily identified. The energy of one group is approximately equal to the sheath potential (~E), indicating that the ions leave the converter surface with very small energy. These H⁻ ions are desorbed (or sputtered) from the converter surface by bombardment with energetic hydrogen ions.[6] In this process, the average energy gained by the H⁻ ions is usually small (~5 ev), and their final energy

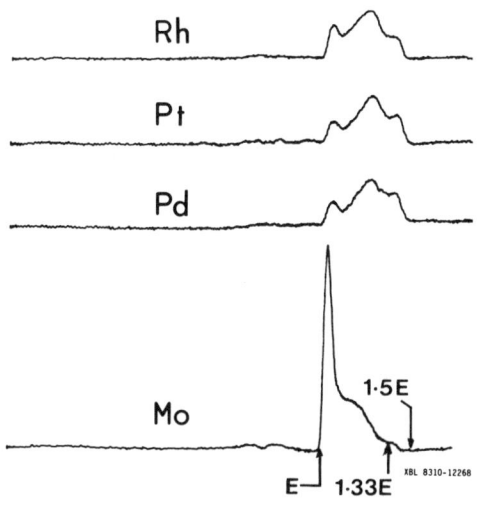

Fig. 2 Energy spectrum of the H⁻ ions produced on Rh, Pt, Pd and Mo at a converter voltage of 600 V.

should be approximately equal to the sheath potential.

The second group of H⁻ ions has a higher energy and they are produced by reflection (or back-scattering) of hydrogen ions from the converter.[7] All positive hydrogen ions (H^+, H_2^+, H_3^+) acquire the same amount of energy E as they cross the sheath. However, the molecular ions H_2^+ and H_3^+ are fragmented to form atomic particles with energy E/2 and E/3 respectively. If H⁻ ions are converted from the three groups of backscattered atomic hydrogen particles, their average energy at the detector should be close to the limiting values of 2E, 1.5E and 1.33E. Because of this species effect, three superimposed peaks should appear in the energy spectrum of the H⁻ ions formed by the reflection process. In this experiment, the species distribution H^+ : H_2^+ : H_3^+ = 4 : 35 : 61. Therefore only the peaks resulting from H_3^+ and H_2^+ ion reflection are visible in the energy spectra shown in Fig. 2.

Figure 2 also shows that the desorbed H⁻ ion peak is much higher for Mo than the other three metals. By comparing the mass spectrum of negative ions generated on these converter materials (Fig. 3), one finds that the peaks corresponding to mass 16 (O⁻) and mass 17 (OH⁻) are also much higher for Mo than Rh, Pt and Pd. In fact, for a gold converter, Fig. 4 shows that the O⁻ and OH⁻ peaks are not visible in the mass spectrum. Correspondingly, the H⁻ion energy spectrum in Fig. 5 illustrates the absence of the desorbed H⁻ ion peak. This result seems to suggest that hydrogen atoms are adsorbed on the converter surface mostly in the form of OH. Under ion impact, both the atomic ions, H⁻ and O⁻, and the molecular ion OH⁻ can be desorbed (or sputtered) from the converter.

The degree of forming the OH bond (and therefore the desorbed H⁻ ion yield) differs from one converter material to the other. Indeed, the mass spectra in Figs. 3, 4, 6 and 7 suggests that the OH bond is more likely to form on Mo, Al, Ta, and S.S. (stainless steel) surfaces. Notice that both Cu and Au produce mainly O_2^- ions, but this molecular ion forms only a very small fraction of the negative ion spectrum of the Ta and Al converters.

Figure 7 shows that the desorbed H⁻ ion peak is the highest when Al is used as the converter. In fact, the mass spectrum for

Al shows the desorbed H^- ion peak is even larger than the O^- and OH^- peaks. The reason for this high desorbed H^- ion yield is not yet understood, but it is known that oxide layers always exit on Al surfaces. As a result more hydrogen atoms may adsorb on the Al surface than other converter materials.

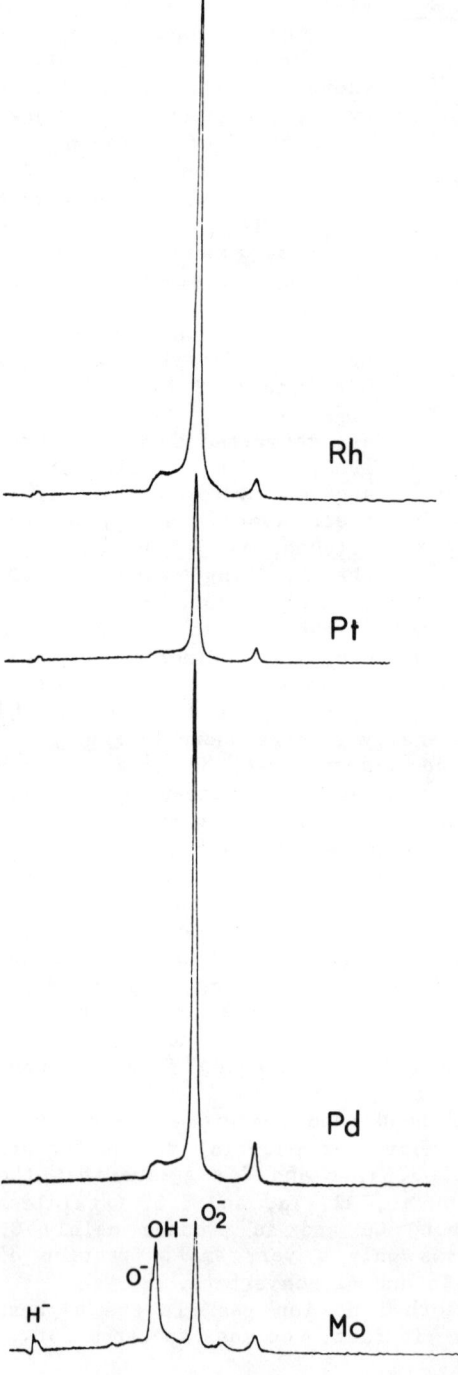

Fig. 3 The mass spectra of the negative ions produced on Rh, Pt, Pd and Mo at a converter bias voltage of 200 V.

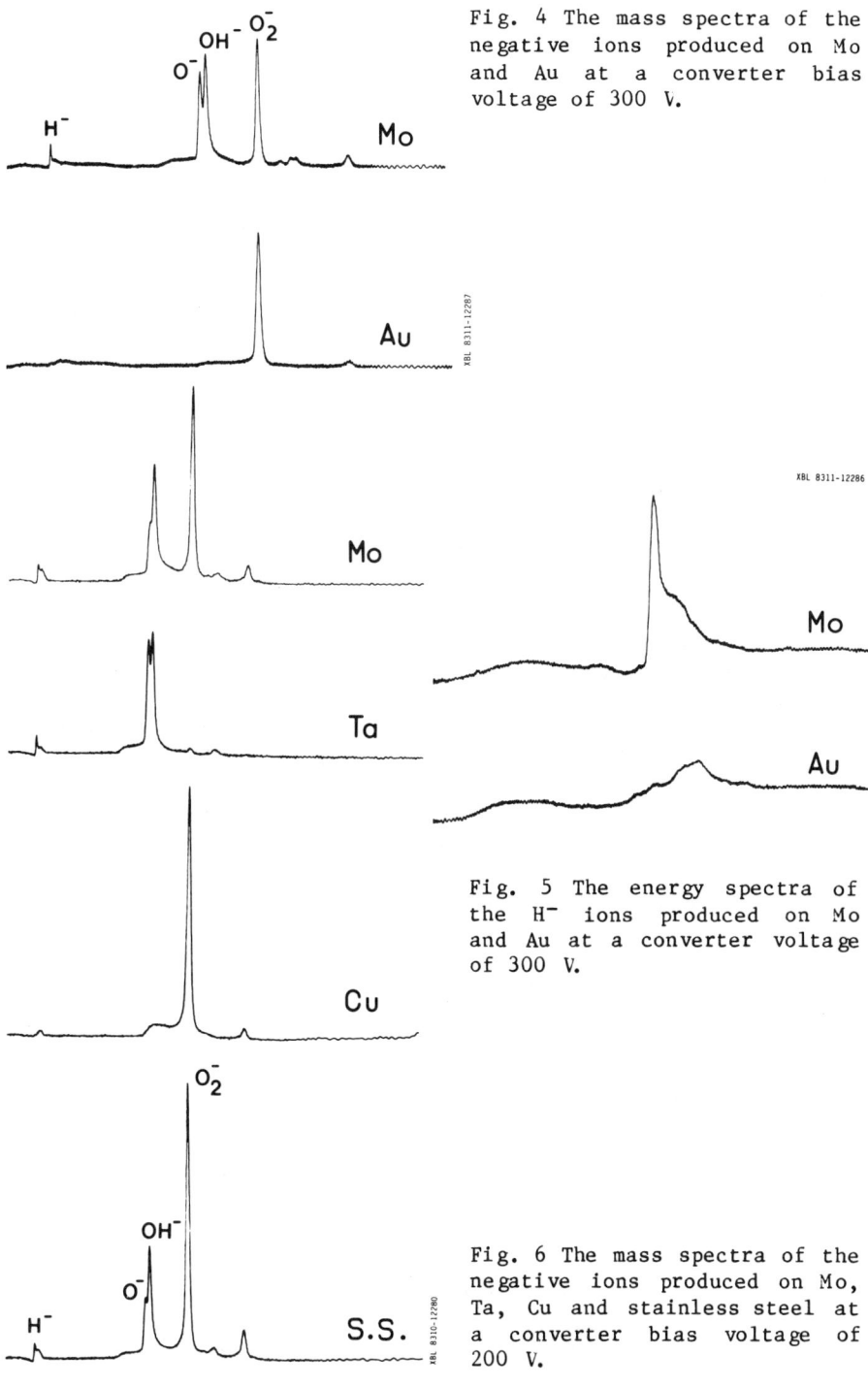

Fig. 4 The mass spectra of the negative ions produced on Mo and Au at a converter bias voltage of 300 V.

Fig. 5 The energy spectra of the H⁻ ions produced on Mo and Au at a converter voltage of 300 V.

Fig. 6 The mass spectra of the negative ions produced on Mo, Ta, Cu and stainless steel at a converter bias voltage of 200 V.

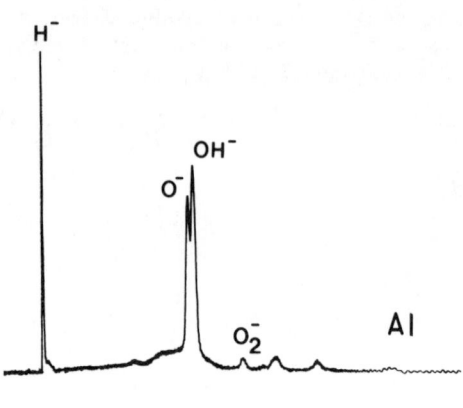

Fig. 7 The mass spectrum of the negative ions produced on aluminum at a converter bias voltage of 200 V.

(b) Cesium mode of operation

When cesium is added to the hydrogen discharge, the total H$^-$ ion yield in general is enhanced by more than two orders of magnitude. The impurity negative ions now form only a percent or two of the total self-extracted beam current. Figure 8 shows the energy spectra of the H$^-$ ions generated by Ta, Cu, stainless steel and Mo converters. The amount of H$^-$ ions produced differs from one converter material to the other. But for each of these four metals, the energy spectrum shows a very narrow peak near the converter potential. As in the case of the pure hydrogen mode, this group of H$^-$ ions are produced by a desorption

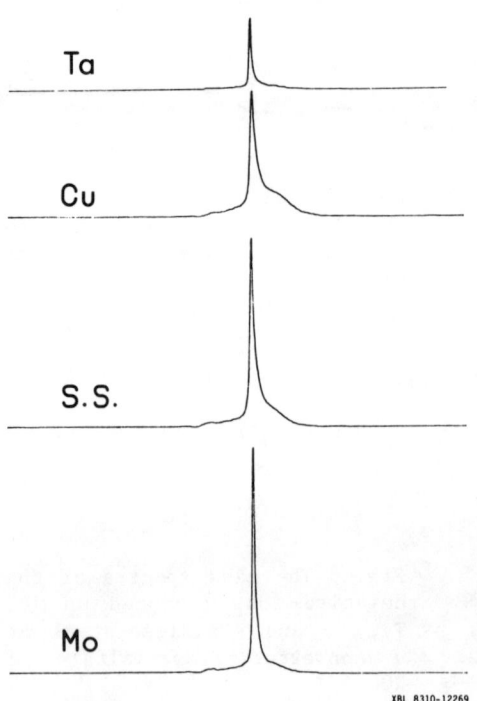

Fig. 8 The energy spectra of the H$^-$ ions produced on Ta, Cu, stainless steel and Mo at a converter voltage of 300 V.

process. With cesium added to the source, it is believed that the hydrogen atoms could adsorb on the converter surfaces in the form of CsH. Due to impact by positive hydrogen and cesium ions, H⁻ ions are desorbed from the converter surface. With cesium, H⁻ ion formation is much enhanced due to the lowering of the surface work function.[8,9] For Cu and stainless steel, the energy spectra in Fig. 8 show that a larger fraction of the H⁻ ions are produced by reflection of positive hydrogen ions.

When Rh, Pt, Pd, Al and Au are compared with Mo, we again find that Mo produces the highest desorbed H⁻ ion yield. However, when the source is operated with Ti, V, Nb and Mo converters, the energy spectra in Fig. 9 show that the majority of the H⁻ ions are formed by desorption and that Ti and V produce more desorbed H⁻ ions than Mo.

In the operation of the "self-extraction" negative ion source, the desorbed H⁻ ions are desired as they have a very small transverse energy component and therefore can leave normal to the converter surface. Thus these H⁻ ions can be better focused geometrically at the exit aperture of the source and then be accelerated to form a high energy beam. For this reason, Ti, V, and Mo are better candidates to be used as converter materials than the other metals we have tested in this experiment.

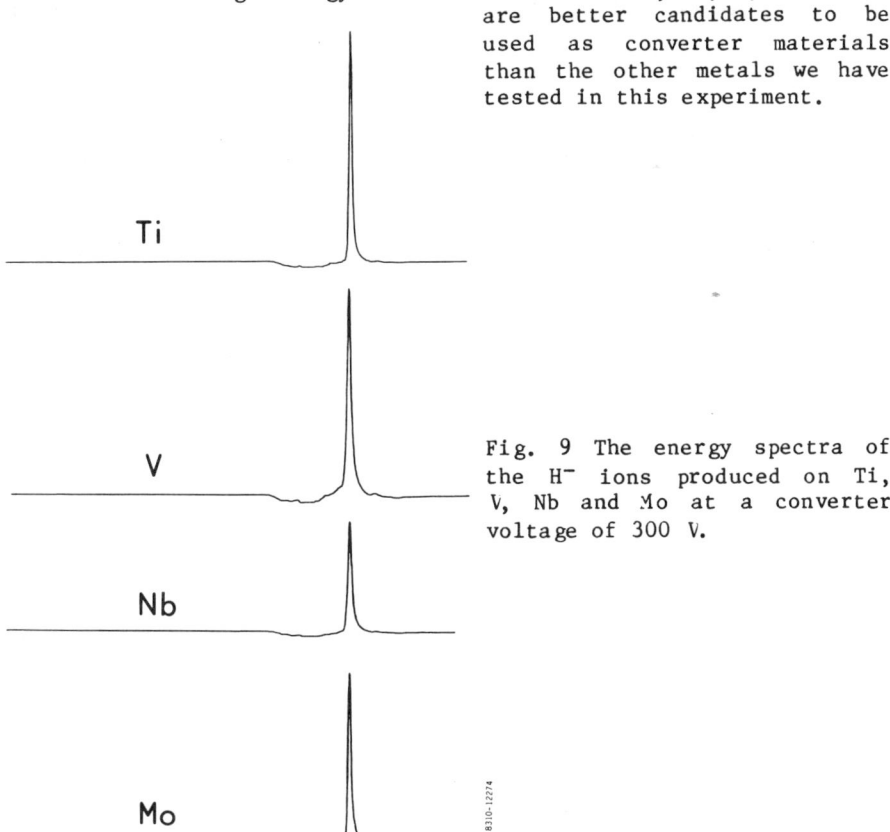

Fig. 9 The energy spectra of the H⁻ ions produced on Ti, V, Nb and Mo at a converter voltage of 300 V.

*This work is supported by U.S. DOE under contract number DE-AC03-76SF00098.

REFERENCES

1. L.D. Stewart, A.H. Boozer, H.P. Eubank, R.J. Goldston, D.L. Jassby, D.R. Mikkelsen, D.E. Post, B. Prichard, J.A. Schmidt, C.E. Singer, Proc. 2nd Int. Symp. on the Production and Neutralization of Negative Hydrogen Ions and Beams, Brookhaven, p. 321 (1980).
2. K.H. Berkner, R.V. Pyle, and J.W. Stearns, Nucl. Fusion $\underline{15}$, 249 (1975).
3. K. N. Leung and K. W. Ehlers, J. Appl. Phys. $\underline{52}$, 3905 (1981).
4. K.N. Leung and K.W. Ehlers, Rev. Sci. Instrum. $\underline{53}$, 803 (1982).
5. K.N. Leung, T. Samec and A. Lamm, Phys. Lett. A $\underline{51}$, 490 (1975).
6. M. Siedl and A. N. Paragellis, Bull. Am. Phys. Soc. $\underline{23}$, 804 (1978).
7. J. R. Hiskes, A. Karo, and M. Gardner, J. Appl. Phys. $\underline{47}$, 3888 (1976).
8. M.I. Yu, Phys. Rev. Lett. $\underline{40}$, 574 (1978).
9. P. Schnieder, Ph.D. Thesis, University of California, Berkeley (1980).

THE NEGATIVE FRACTION OF DEUTERIUM AND HELIUM SCATTERED FROM A SODIUM SURFACE

H. Verbeek, W. Eckstein, P.J. Schneider
Max-Planck-Institut für Plasmaphysik, EURATOM-Association,
D-8046 Garching/München, Federal Republic of Germany

ABSTRACT

Deuterium and helium were scattered from a Na surface at normal and glancing incidence and the reflected particles were measured in a glancing exit direction. The negative fraction of the reflected particles depends strongly upon the angle of incidence. Existing theories describing the production of negative ions cannot explain this behavior.

INTRODUCTION

A proposed method for detecting neutral hydrogen atoms with energies between a few eV and a few keV is by charge transfer through scattering from alkali surfaces to convert H^0 or D^0 to H^- or D^-. Negative fractions η^- of up to 50 % of the backscattered particles have been observed, depending upon the bombarding energy, scattering geometry, and the alkali coverage of the substrate[1-6]. For the detection and energy analysis of neutrals the scattering under grazing angles is favourable: In this case the scattering intensity is maximal and is centered around the direction of specular reflection. The energy distribution is a peak whose width and energy loss with respect to the primary energy is decreasing with decreasing incident energy. The unknown energy distribution of the H^0 or D^0 flux (from a fusion experiment for instance) can then be determined from the energy distribution of reflected $H^-(D^-)$ ions by unfolding a H^- yield per incident H^0 calibration curve. For the investigation of the feasibility of this detection method we consider the negative fraction η^- of all reflected particles rather than H^- and D^- yields. η^- describes the formation of the negative ions independently from the reflection coefficient which also depends on various parameters.

In earlier papers[5,6] we have presented measurements of η^- for Cs and Na targets bombarded with deuterium with an anlge of incidence of $\alpha = 85°$ and an exit angle $\beta = 85°$ (i.e. specular reflectance). η^- can be as large as 0.16, but below 1 keV it decreases rapidly with decreasing bombardment energy[6]. This is, however, the energy range of greatest interest. This result is in contrast to the fact that η^- increases with decreasing exit energy up to 0.5 at $\sim$ 300 eV (the lowest measurable energy in this experiment) when we scattered hydrogen from a Cs-target at normal incidence[2]. Since this exit energy dependence is so different, we made additional measurements with normal incidence and the same exit angle of $\beta = 85°$ as in the glancing experiments and found a trajectory dependence of the charged fraction[7]. The dependence upon the trajectory of the incoming particles is not well understood. To elucidate the processes involved, measurements of the negative fraction of He from Na were made[8]. When He is scattered from alkali targets a small fraction of He$^-$ ions is observed. There are, however, a few differences as compared to Hydrogen. The He$^-$ binding energy is only 0.076 eV, i.e. 1/10 that of the H$^-$ ion. The observed He$^-$ are most probable in the 1s 2s 2p ^{4}P state. This state can only be reached via an excitation of the neutral He atom into the ^{3}S state 19.8 above the ground state level[9]. Therefore an additional excitation process is necessary during the formation of the negative He$^-$ ion.

In this paper we compare the negative fractions as a function of the exit energy for normal and glancing incidence and glancing exit angles for D and He scattered from Na-targets.

EXPERIMENTAL RESULTS

The experimental procedure was described in detail in references 2, 4 and 5. We measured the energy distributions of reflected D$^+$, D$^-$ or He$^+$, He$^-$ ions and D^0 or He^0 neutrals and at each exit energy E, we determined the negative fraction
$\eta^- = N^-/(N^- + N^+ + N^0)$ where N^+, N^0 are the intensities of reflected positive or negative ions and neutrals respectively at these exit energies. This makes it possible to distinguish between particles which are scattered from the very surface of the target and those which penetrated and thus have lost more (or less) energy along their trajectory. Special care was taken to be sure that the Na-targets were clean. Slight contaminations of O and C had large effects on the measured negative fractions.

In Fig.1 the results are shown for D, in Figs.2 and 3 those for He. In the case of He a slightly different scattering geometry ($\alpha = \beta = 75^0$) was chosen for experimental reasons[8]. In Fig.1 the η^- data for normal incidence ($\alpha = 0$) are shown as dots for incident energies 3 and 6 keV. Crosses connected by lines represent the η^- data for glancing incidence ($\alpha = 85^0$) for the incident energy noted at each curve. At the bottom of Fig.1 the normalized energy distributions for $\alpha = 85^0$ are shown. For the larger incident energies the energy distributions show long tails on the low energy side indicating that an appreciable number of particles had penetrated into the bulk and thus had lost energy along their trajectory. The sharp peaks at the lowest incident energies are due to the fact that the scattering occurs at the surface with little penetration into the target. The spectra for $\alpha = 0$ (not shown here) extend from the highest possible exit energy (due to scattering from the surface atoms) down to the lowest detectable energies ($\sim$ 200 eV)[2].

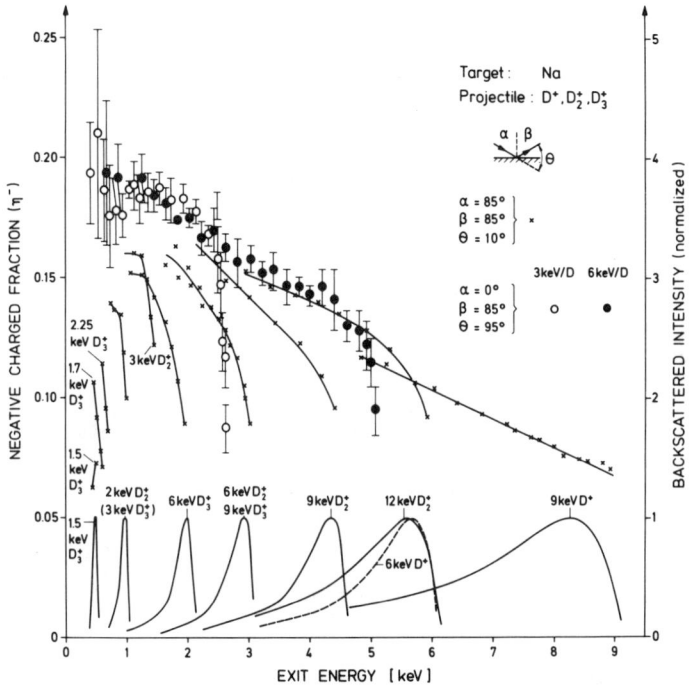

Fig.1: Negative fraction η^- vs exit energy. Dots o and ●.
$\alpha = 0$, $\beta = 85°$ for 3 keV and 6 keV incident D. Crosses
connected by lines: $\alpha = 85°$ for D incident at various
energies. At the bottom normalized energy distributions
$\alpha = 85°$, $\beta = 85°$ are shown.

Contrary to the case for glancing incidence, the netative fraction η^- for $\alpha = 0$ increases with decreasing exit energy and is independent of the incident energy over almost the entire exit energy range. Only those particles with exit energies corresponding to scattering from the surface atoms show much lower negative fractions than those scattered from deeper inside the target at the same exit energy. For both 3 keV and 6 keV η^- is low for particles scattered from the surface which gives rise to the apparent knee in the curves, but for particles which penetrated into the target the η^- data merge into a common curve. This is the same behavior as it was observed in reference 2. For glancing incidence ($\alpha = 85°$) and large incident energies ($E_o > 4.5$ keV) particles with energies corresponding to the high energy side of the spectra are scattered from the surface yield low η^- values. For those particles which have lost energy because they have penetrated into the target the η^- data are increasing and reach the common curve mentioned above.

For the highest incident energy (9 keV) no "surface effect" is discernible. But for decreasing incident energy the curves are much lower than for normal incidence and depend on the incident energy. For the lowest incident energies shown in Fig.1 only surface scattering is contributing to the reflected intensity. Therefore, it can be generalized: Scattering from surface atoms yields lower negative fraction η^- than scattering from the bulk. This effect increases with decreasing energy. (It should be mentioned here that we scattered from Na layers at least 1 μm thick, i.e. also bulk scattering is from the Na layer).

In Fig.2 the η^- data for normally incident He are shown. In this scattering geometry very few particles are reflected at ß = 75°. In addition the H$^-$ yield is very low. Therefore the statistical error was rather large. Also, some data were obtained using a He^3 beam which enabled measurments at higher velocities, within the fixed energy range of our accelerator. The data are plotted at energies equivalent to He^4 velocities. In this context He^3 and He^4 can be considered to be identical. Within the statistics the data points follow one common curve. Surface effects are not discernible, but this might be due to the poor statistics. The curve increases up to a maximum of $\eta^- = 1.3 \cdot 10^{-3}$ at ~ 10 keV, (which corresponds to a velocity of ~ 6.8 10^5 m/s) and decreases at higher energies. The solid line is drawn just to guide the eye.

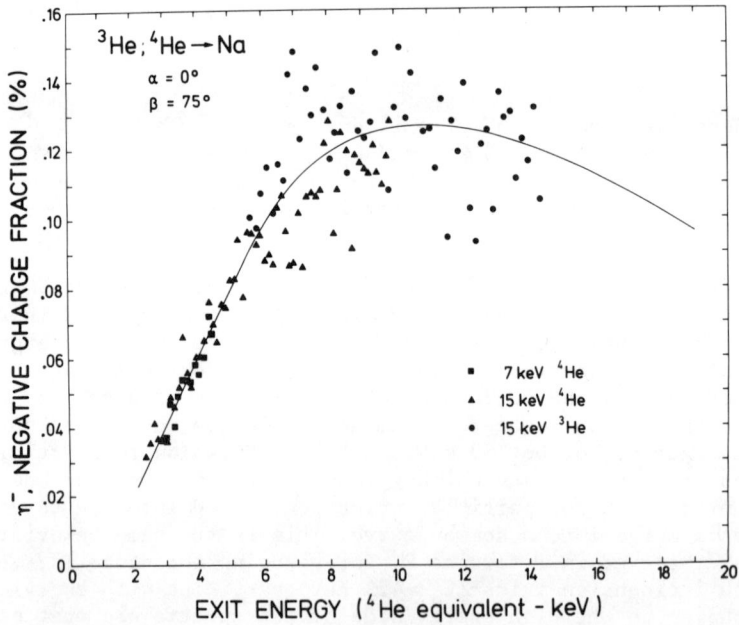

Fig.2: Negative fraction η^- vs. exit energy for He incident on Na, $\alpha = 0$, ß = 75°.

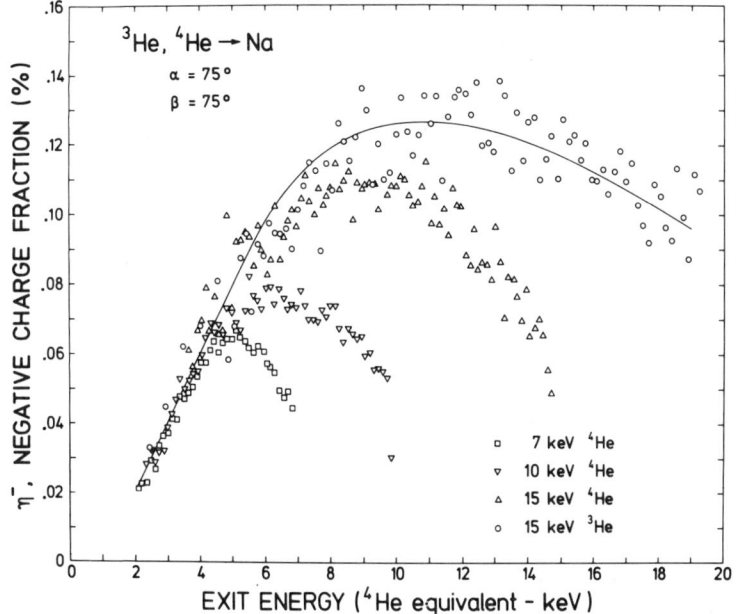

Fig.3: Negative fraction η^- vs. exit energy for He incident on Na $\alpha = 75°$, $\beta = 75°$.

In Fig.3 η^- data for glancing incidence ($\alpha = 75°$) are shown for 4 different (He^4-equivalent) incident energies. The 4 curves merge into a common curve at low energies. This curve is the same curve as observed for normal incidence in Fig.2. For high exit energies the η^- values are considerably lower than those of the common curve. Again, those particles, which have penetrated into the target and thus have lost more (or less) energy in a multiple scattering process show much larger negative fractions than those particles existing at equal energies but being scattered from surface regions.

DISCUSSION

Several theoretical approaches have been published to explain the H$^-$ or D$^-$ production during scattering at alkali surfaces[10, 11, 12]. As an incoming positive ion approaches a surface it is quite likely to undergo resonant neutralization followed by Auger de-excitation or direct Auger neutralization. This is supported by the results of Ref.5 were we scattered neutral D atoms and D$^+$ ions from a Cs surface and could not observe any significant differences in the charged fractions of the reflected particles. Because of the screening of the metal electrons negative H⁻ or D⁻ cannot exist inside the metal. The negative ion is formed outside the surface due to a resonant tunneling process. The affinity level of the H or D atoms is broadened and lowered down to the region of the Fermi level as the atom approaches the surface. Thus an electron can tunnel back and forth from the metal to the atom. Therefore there is a certain equilibrium negative fraction aquired adiabatically at a certain distance from the surface. From this the negative fraction of the particles leaving the surface can be determined as a function of the velocity perpendicular to the surface[11, 12]. In reference 10 Hiskes et al. describe the negative fraction by a formation probability and a survival probability during the way out. By v. Wunnik et al.[12] it has been shown that also the velocity parallel to the surface is important since only those metal electrons which have the right velocity component can be captured by the moving atom.

All these theories consider only the outgoing path of the negative ions. This is reasonable, since the negative ion is formed at some distance outside the metal surface and it should not depend upon the way the particles came to this point. Also in these theories it is assumed that the bulk electronic structure extends to the surface and ends here abruptly which is certainly a simplification.

The theories mentioned above cannot explain the larger differences in the negative fraction observed for normal and oblique incidence since the outgoing trajectories are the same in these experiments. As mentioned above for the D⁻ results, particles which have been scattered from the surface show much lower η^- values, than those scattered from deeper layers. The formation probability (Ref. 10) or the equilibrium charge state at some distance (Ref.11, 12) depend on the electron density of the metal. If one allows that this is not uniform but localized to the ion cores and especially at the surface it extends as the electron selvage into the vacuum, one can argue that particles which are scattered from the very surface for $\alpha = 0$ and along the surface for $\alpha = 85^0$ at low energies, have probed only regions of low electron density and thus the probability for electron capture was low resulting in a low η^- value. On the other hand, particles which have penetrated into the bulk (both for $\alpha = 0^0$ and $\alpha = 85^0$) probed the full bulk electron density causing a high probability for electron pick-up and a large negative fraction.

It should be mentioned here that this behaviour was observed for the alkali metal targets only. It seems that the very low electron density of alkalis as compared to transition metals is responsible for this effect.

It could be reasoned that this effect is due to incomplete neutralization of the incoming positive ions. Indeed, it was observed in Ref. 2 that hydrogen scattered from the very surface of a Cs target had a higher positive fraction than that scattered from the bulk. However, as the positive fraction is only a few percent, a very small effect can be expected in the neutral fraction (and nothing can be seen in our experiments). As the negative ions are created from neutrals, this cannot have any observable effect on the negative fraction.

Similar arguments should hold for the He results. There are, however, two large differences: 1. The deviation of η^- for $\alpha = 75°$ from the data obtained for $\alpha = 0$ extends to much lower exit energies, i.e. also for particles which have penetrated considerably. 2. The formation of the negative He$^-$ ion can only occur via an excited state of the He atom. The latter fact provides another reason for the differences of for $\alpha = 0°$ and $\alpha = 75°$. For $\alpha = 0°$ all particles are scattered by a large angle ($\theta = 105°$) and as computer simulation shows, have suffered among several small angle deflections at least one close encounter with a target atom. In this violant collision the neutral He atom can be put into the excited ^3S state which is necessary to pick up the extra electron to form a negative ion. For $\alpha = 75°$ the He atom suffers only a sequence of small angle deflections and no such violent collision. Therefore a reduced number of He$^-$ ions is observed. Once the particle has penetrated deep enough into the bulk, there is a chance to become excited in the many collisions it suffers in the target and hence the same η^- is observed as for normal incidence at the same exit energy. For the He$^-$ formation this effect seems to be much more important than the electron density effect, because in the case of glancing incidence also particles which have considerably penetrated into the bulk electron density show low η^- values.

It may be concluded that existing theories are still not sufficiently developed to explain all features of the production of negative ions during scattering.

REFERENCES

1. P.J. Schneider, K.H. Berkner, W.G. Graham, R.V. Pyle, and J.W. Stearns, Phys. Rev. B23 941 (1981)
2. W. Eckstein, H. Verbeek, and R.S. Bhattacharya, Surf. Sci. 99, 356, (1980)
3. W. Eckstein, H. Verbeek, P.J. Schneider, Proc. US-Mexico Joint Seminar on the Atomic Physics of Negative Ions, notas de fisica 5, 426 (1982)
4. J.N.M. van Wunnik, B. Rasser, and J.Los, Proc. 4th Int. Conf. on Solid Surfaces, Vol.II, p. 849 (1980)
5. P.J. Schneider, W. Eckstein, and H. Verbeek, Nucl. Instr. and Meth. 194, 387 (1982)
6. P.J. Schneider, W. Eckstein, and H. Verbeek, J.Nucl. Mat. 111 and 112, 795 (1982)
7. P.J. Schneider, W. Eckstein and H. Verbeek, Proc. 4th Int. Workshop Inelastic Ion Surface Collisions, Middelfart, Denmark (1982), Physica Scripta in the press
8. P.J. Schneider, W. Eckstein, and H. Verbeek, Proc. 10th Int. Conf. Atomic Collisions in Solids, Bad Iburg, Germany (1983) Nucl.Instr. and Meth. in the press
9. H.A. Massey: "Negative Ions" 3rd Edition, Cambridge 1976.
10. J.R. Hiskes and P.J. Schneider, Phys. Rev. B23, 449 (1981)
11. B. Rasser, J.N.M. van Wunnik, J. Los, Surf. Sci. 118, 697(1982)
12. J.N.M. van Wunnik, J.J.C. Geerlings, J. Los, Surf. Sci. 131, 1 (1983)

OBSERVANCE OF H⁻ BY SURFACE CHEMI-IONIZATION ON W(110)

R. L. Palmer
IRT Corporation, San Diego, CA 92138

ABSTRACT

During an investigation of secondary ion emission from W(110), spontaneous emission of H⁻ ions was observed in the presence of hydrogen and cesium vapors at surface temperatures between 600 and 1000 K. The interaction of neutral chemical species with the surface to produce H⁻ ions, i.e., surface chemi-ionization, compares favorably with production via ion-surface interactions in terms of yield and energy spread. Results for H⁻ produced by surface chemi-ionization are compared with secondary ion (sputtered) emission on clean and cesiated W(110).

INTRODUCTION

The generation of ions from surfaces is typically accomplished by the interaction of the surface with energetic ions, either as well defined beams or as a plasma. Surface production of ions can then occur via the ejection of surface species (sputtering), or by charge exchange. Plasma-surface interactions are most often used for generating practical currents of H⁻ and other ions although the interaction of positive ion beams with well defined energies and angles of incidence to produce negative ions via charge exchange is being increasingly studied.[1] The production of ions by surface chemi-ionization has, in contrast, not been previously studied as a means of producing useful currents of negative ions such as H⁻. Indeed, it was not the initial intent of the present investigation to study chemi-ionization. However, rather startling yields of H⁻ by chemi-ionization were fortuitously observed during the investigation of H⁻ production by ion-surface interactions. These results, although incomplete, encourage further investigation of the chemi-ionization mechanism as a practical source of negative ions.

EXPERIMENTAL

Figure 1 presents a schematic view of the apparatus used in this work showing the ion source, magnetic charge-to-mass ratio analyzer, and SIMS detector. The ion source is capable of giving stable ion currents of up to approximately 1 μa at energies from 300 eV to 75 keV. The ion beam spot size on the target is about 1-mm radius, allowing high collection efficiency and sharp focusing of the secondary ions. The target can be dosed with cesium using a bakeable valve located adjacent to the target. Various gases such as hydrogen may also be introduced from outside the vacuum chamber. Additionally, the target assembly includes a tungsten filament located directly behind the target so that it can be heated, either

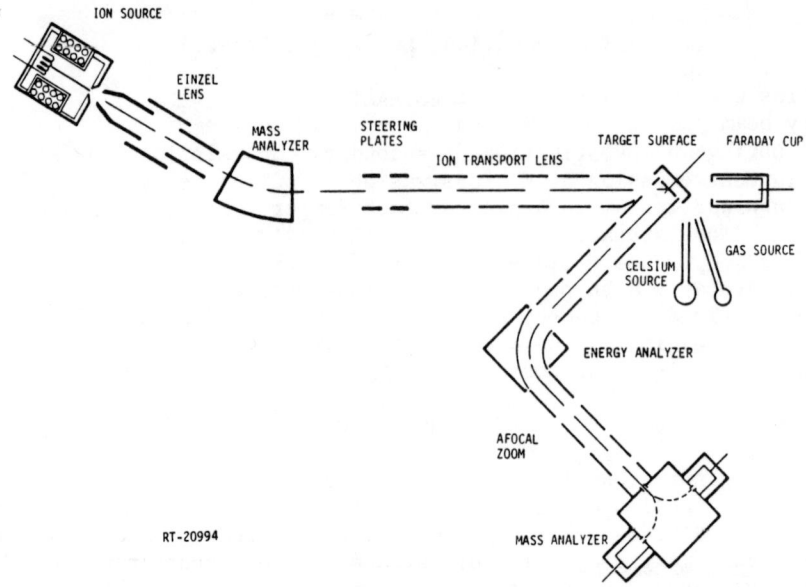

Fig. 1. Secondary ion mass spectrometer showing ion source, bending magnet, and target chamber

by radiation or to higher temperatures by electron bombardment. The capability of going to high temperatures is important for the cleaning of bulk refractory targets such as tungsten. In addition to high temperature heating, the target surface was also cleaned by sputter etching with 1-5 keV Ar^+ using the ion source described above.

The detector assembly resides in a stainless steel box which forms a secondary vacuum chamber that is pumped using the titanium filament. A nude ion gauge is used to monitor the pressure inside the box. After extensive bake-out, the pressure inside the box can be brought down to the low 10^{-9} torr regime, about a factor of ten lower than in the main chamber outside the box. Primary ions enter the box through a small hole. A combination of lenses and steering plates extending in a straight line from this hole focuses the ion beam through a hole on the side of the first extraction lens of the spectrometer. When the target is swung out of the way, the primary ion current can be measured directly using the Faraday cup located directly along the beam axis.

The extraction lens assembly is set at forty-five degrees to the incident ion beam and focuses the point of emission on the target surface at the entrance to the $90°$ spherical sector energy analyzer. The energy analyzed ions are bent through $90°$ into the mass analyzer. Positive ions are bent to the left and negative ions to the right where they are collected in Faraday cup detectors. The ion currents are measured directly using a Keithley Model 602 Electrometer. The entire detector assembly is nonmagnetic and the mass and energy analyzers are gold-plated to improve resolution.

The secondary ion energy and mass analyzer electrostatic lens optics were designed for maximum collection efficiency. The primary beam passes through a small hole in the first extraction lens and strikes the target at 45° when the target is in the measurement position at the entrance of the extraction lens system. The extraction lens produces an electric field pattern that pulls ions from the target and focuses them into a beam whose axis is normal to the target surface with a focal point at the entrance focal point of the electrostatic hemispherical energy analyzer. Since the usual electrostatic lens formulae are not applicable to the extraction region, electron models of this lens system were constructed and tested, the focal spot being observed on a glass screen coated with cathode ray tube phosphor. The target was simulated by a plate with an oxide-coated cathode mounted flush in a hole in its center. The cathode diameter was somewhat larger than the beam spot on the target in the finished apparatus. Several lens geometries were tested under a variety of focusing conditions until one was found which could consistently produce a small focal spot. Current measurements were performed which demonstrated that all of the electrons emitted from the cathode landed on the focal spot (none of them went astray, landing on electrode surfaces, etc.). The 90° spherical deflector energy analyzer has excellent transmission characteristics due to its inherently large acceptance angle. The exit focus of the analyzer is imaged at the analyzer exit slit by an afocal zoom electrostatic lens whose purpose is explained below. This slit is also the entrance focal point of a 90° deflection permanent magnet mass spectrometer. The mass spectrometer has two exit slits and two ion collectors on opposite sides so that it can collect equally well both positive and negative ions. The energy analyzer and mass spectrometer deflect the ion beams in orthogonal planes.

A permanent magnet-type mass spectrometer was used because of its compactness, simplicity, and ultra-high vacuum compatibility. Its relatively weak magnetic field strength does, however, limit its use to species of low mass such as the isotopes of hydrogen for which it was intended. The afocal electrostatic zoom lens has the favorable property of maintaining a constant distance between focal points and nearly constant magnification over a wide range of acceleration ratios. (The magnification, m varies as $R^{1/4}$ where R is the ratio of incoming to outgoing beam energy.)

The slit (1) at the exit of the energy analyzer and entrance of the mass spectrometer, and the slits (2) at the exits of the mass spectrometer are adjustable from outside the vacuum wall. This permits the energy resolution and mass resolution of the instrument to be varied independently during an experiment. With the slits fully opened, the collection efficiency approaches 100 percent, at least for thermal energy electrons. This was determin- using a heated cathode in place of the target as described above.

Ion energy spectra using this instrument are obtained in the following way. The potential between the hemispheres of the 90° analyzer is set to pass ions of a convenient energy, say 20 eV. (The design potential is 16/15 E_i, where E_i is the ion energy or 21.33 V in this case.) The target potential is then lowered, starting nominally from 20 V with respect to the analyzer to sweep the energy spectrum. In principle, for a target potential of 20 V with respect to the analyzer, only ions ejected with zero kinetic energy can pass through the 90° hemispheres. In practice, however, there is typically a contact potential different of up to 5 V between the target and the gold-plated analyzer. This contact potential difference, arising from the different work functions of the gold and clean or cesiated-molybdenum and tungsten surfaces, produces a corresponding shift in the measured ion spectra. The shift is toward higher measured energies for positive ions and toward lower energies for negative ions after lowering the W(110) work function by dosing with cesium. A comparison of positive and negative ion spectra, therefore, fortuitously provides a crude measure of the target surface contact potential relative to gold as well as indicating the true zero energy point on the energy scale.

RESULTS

The target for this study was a tungsten single crystal cut and polished by standard methods with the (110) face exposed. Prior to installing the crystal in the present apparatus it was put into another UHV apparatus where it was characterized using LEED and AES. After extensive heating and ion sputtering, all observable surface impurities were removed and a sharp LEED pattern characteristic of the W(110) surface was obtained. The crystal was then installed in the SIMS apparatus where it was bombarded with 900 eV Ar^+ ions and its temperature raised to 1600 K in order to remove surface contamination resulting from its brief exposure to air during the transfer.

An energy spectrum for H^- secondary ions produced by 600 eV He^+ striking the clean W(110) surface is shown in Figure 2 (solid curve). The primary ion current was in the 10^{-9} ampere range to avoid changing the surface condition while taking the data. At primary ion currents greater than 10^{-8} ampere the secondary yield was observed to change with time after the ion beam was turned on indicating a non-static surface condition. A single spectrum was typically obtained in 10-100 seconds during which time less than one percent of the surface is bombarded by ions at these currents. During and immediately following flashing the target to high temperatures no H^- (or H^+) signal is observable from the surface. After cooling to near room temperature the H^- signal builds to observable levels within a few minutes, perhaps as a result of the adsorption of hydrogen from the vacuum at 10^{-9} mbar. However, the addition of hydrogen gas had no apparent effect on the rate of recovery so hydrogen migration from the bulk may be suspected.

The effect of dosing the W(110) target with cesium is shown as the dashed curve in Figure 2. The most striking change is an

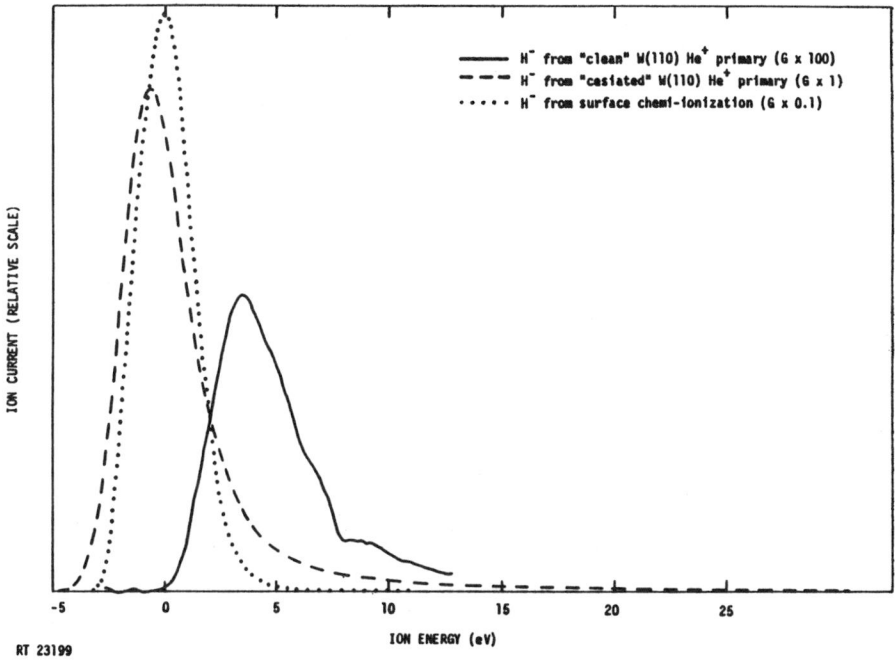

Fig. 2. H⁻ secondary ion spectra from 600 eV He⁺ on "clean" (—) and "cesiated" (---) W(110) and "thermal" H⁻ spectrum (···) from surface reaction with H_2 and cesium at 650 K

increase in the total secondary yield, defined as the ratio of secondary H⁻ ions to primary He⁺ ions, from less than 10^{-5} on the clean surface to about 10^{-3} on the cesiated surface. Additionally, the spectrum has shifted as a result of the lowered work function of the cesiated surface.

During the course of this investigation, several attempts were made to observe thermionic emission from the W(110) surface which would provide an experimental determination of the energy bandpass of the spectrometer. Heating the cesium free target in hydrogen produced a large H⁺ signal that originated from the tungsten heater filament rather than the W(110) surface. These ions had a very large energy spread due to the voltage drop along the length of the hot filament. On the cesiated surface, however, mild heating in the absence of an ion beam produced an easily observable H⁻ signal with a narrow energy spread. It was readily determined that this H⁻ current originated from the W(110) surface and not the heating filament or other possible sources such as the ion gauge or titanium getter filaments. The measured energy distribution of these thermally generated ions is also shown in Figure 2 (dotted curve). Heating the SIMS enclosure and releasing more cesium vapor effected an increase in the cesium pressure which was accompanied by a large increase in the H⁻ current. The addition of hydrogen gas also increased the H⁻ signal. None of these changes affected the shape of the energy of distribution.

The dependence of the H⁻ current on target temperature is shown in Figure 3. During heating from room temperature the ion current was dependent on the target history and the rate of heating and was not quantitatively reproducible. A typical curve obtained while heating the W(110) crystal at 10 Ksec^{-1} is presented as a qualitative indication of this behavior. Upon cooling the crystal from above 1200 K the results were highly reproducible and did not depend on the cooling rate up to 10 Ksec^{-1}. The crystal could, in fact, be maintained at any temperature in the range 950-625 K after cooling from above 1200 K with a resultant steady output of H⁻ at a rate characterized by the sharply peaked "cooling" curve of Figure 3.

DISCUSSION

While the initial motivation for this work was the study of secondary ion emission, attention is focused here on the interesting questions regarding the observance of spontaneous H⁻ emission

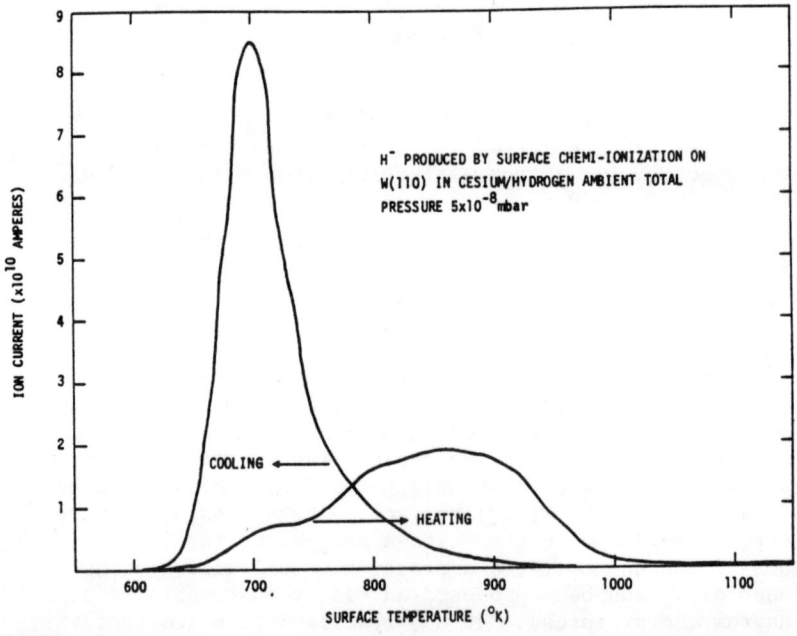

Fig. 3. Temperature dependence of spontaneous H⁻ emission upon heating from room temperature and during cooling from 1500 K

in the absence of ion bombardment. From the fact that a steady H⁻ current of 10^{-9} amperes was observed at a pressure of 5×10^{-8} mbar, one can estimate the conversion efficiency of the cesiated W(110) surface at 700 K from the impingement rate of H_2 at that pressure. A lower limit based on a collection area of 0.1 cm² and 100% transmission through the spectrometer is 10^{-3} for the conversion, per H_2 collision, to H⁻. It does not seem likely that simple thermionic emission can account for such a high relative ion current. Invoking the Saha-Langmuir equation gives a negative ion to neutral atom fraction of 10^{-6} for desorption from a 1.5 eV work function surface at 700 K, assuming unit sticking probabilities for all the species involved. Although large ion fraction enhancements over the Saha-Langmuir ratio have been reported,[2] a discrepancy of at least 10^3 seems unlikely and a more complicated chemical reaction path may be operative. It is known that cesium and hydrogen form a strongly interactive system with tungsten surfaces forming chemical bonds of the order of several electron volts.[3] Ions formed as the result of strong chemical interactions with the surface may be ejected with greater than thermal kinetic energies which would enhance the survival of the ion as it leaves the surface. Energetic neutral reaction products have, in fact, been observed in a number of studies of chemical reactions on surfaces, although the kinetic energies are less than one electron volt.[4]

The above arguments notwithstanding, the observed energy bandwidth of 3 eV (FWHM) for the spontaneous thermal H⁻ current is probably close to the limiting resolution of the analyzer, although some broadening may result from (1) a non-negligible variation of the local contact potential (work function) across the emitting surface, or (2) the larger emitting area of the thermal ions as compared with those generated by the 1 mm ion beam. However, the secondary ion spectra seen here are similar to those seen in earlier studies.[5] A comparison of the secondary and thermal ion spectra suggests that secondary ion emission from the cesiated W(110) surface is composed of a quasi-thermal component plus a higher energy tail that decreases exponentially. Whether this represents two distinct ion formation mechanisms is one of several questions that await further detailed study.

REFERENCES

1. c.f. J. N. M. van Wunnik, B. Rasser and J. Los, Phys. Lett. 6, 288 (1982); J. N. M. van Wunnik, R. Brako, K. Makoshi and D. M. Newns, Proceedings of the Fifth European Conference on Surface Science (ECOSS-5), University of Gent, 1982, pp 618-623.
2. c.f. Ralph Klein and Joseph Fine, Proceedings of the Fourth International Conference on Solid Surfaces and the Third European Conference on Surface Science, September 22-26, 1980, Cannes, France, Vol. II, pp 361-364.
3. C. A. Papageorgopoulos and J. M. Chen, Surface Sci. 39, 283 (1973).

4. R. L. Palmer and Joe N. Smith, Jr., J. Chem. Phys. 60, 1453 (1974); G. Comsa and R. David, Phys. Letters 49, 512 (1977); C. A. Becker, J. P. Cowin, L. Wharton, and D. J. Auerbach, J. Chem. Phys. 67, 3394 (1977); T. Matsushima, Surface Sci. 127, 403 (1983).
5. Sloane and Watt, Proc. Phys. Soc. 61, 217 (1948), Arnot, Proc. Roy. Soc. A 168, 284 (1938).

PRODUCTION OF HEAVY NEGATIVE IONS

EXPERIMENTAL INVESTIGATION OF VOLUME Li$^-$ PRODUCTION

M. W. McGeoch and R. E. Schlier
Avco Everett Research Laboratory, Inc.
2385 Revere Beach Parkway, Everett, MA. 02149

ABSTRACT

An experimental investigation of dissociative attachment in lithium is described. Using laser excitation of a supersonic lithium beam a population of metastable vibrationally-excited lithium molecules has been created in collision-free conditions. Electron-ion pairs have been created by three-step photoionization of lithium atoms in the beam, and the ion signal monitored by pulsed extraction of the Li$^+$ ions. By reversal of the extraction pulse lithium negative ions have been sought at the level of 1 in 2×10^4 electrons attaching to the vibrationally-pumped lithium molecules. The interaction time for attachment is limited by free plasma expansion at the ion acoustic velocity. We have set an upper bound on the attachment rate constant to lithium molecules in $\upsilon^*=11$ and $\upsilon^*=13$ states of $k \lesssim 1 \times 10^{-8}$ cm^3 sec^{-1}, corresponding to an attachment cross section $\sigma \lesssim 3 \times 10^{-16}$ cm^2.

INTRODUCTION

The present work was undertaken with the purpose of measuring the rate of dissociative attachment to Li$_2$ molecules in high vibrational states:

$$\text{Li}_2 \ (\upsilon^*) + e^- \rightarrow \text{Li} + \text{Li}^- \qquad (1)$$

Given a favorable disposition of the Li$_2^-$ potential curves the possibility exists for a strong enhancement of attachment rate with vibrational excitation, in a close analogy with attachment in H$_2$ [1] [2]. Using Li$_2$ the vibrational excitation can be performed with a visible photon (via the Li$_2$(B) state and radiative decay to Li$_2$ (υ^*)) whereas to perform the corresponding process in H$_2$ would require a vacuum ultraviolet photon.

At the commencement of this work in 1982 the Li$_2^-$ ($1^2\Sigma_g^+$) state, which is the prime candidate for a fast dissociative attachment channel, was thought [3] to have a repulsive potential curve, shown as a broken line in Fig. 1. More recently however a more accurate calculation [4] has given the Li$_2^-$ ($1^2\Sigma_g^+$) state a small degree of binding (Fig. 1), with crucial implications for the rate of dissociative attachment in Li$_2$.

The cross section for dissociative attachment may be viewed[5] as the product of two terms, one accounting for the trapping of the electron into an autodetaching state of the negative molecular ion,

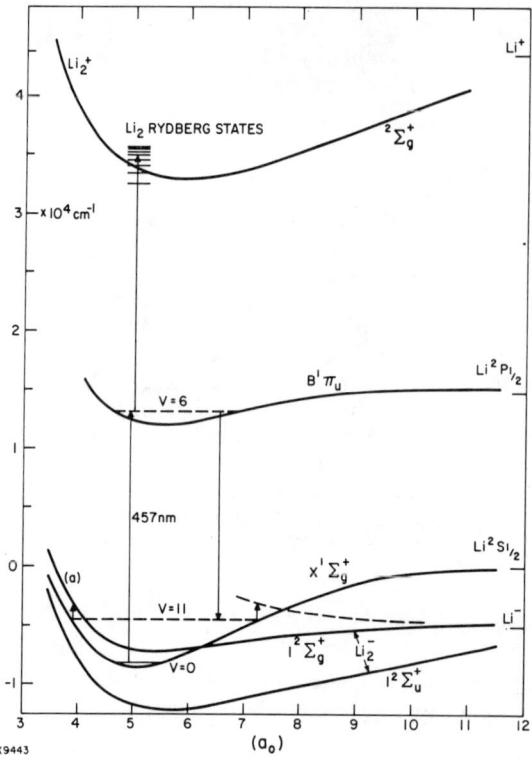

Fig. 1 Potential curves for Li_2, Li_2^-, Li_2^+ (ref. 4) and illustration of optical pumping to $\upsilon^* = 11$.

and the other accounting for the survival of the negative molecular ion to the point where it becomes stable to autodetachment and dissociates producing an atomic negative ion. In the case of a repulsive Li_2^- ($1^2\Sigma_g^+$) state, both terms can be optimized by preparing Li_2 (υ^*) (Fig. 1) so that low energy incident electrons form Li_2^- on resonance, and the time required to reach the stable internuclear distance is minimized. For such a case the cross section for dissociative attachment is estimated to range as high as 4×10^{-15} cm^2. However, for a weakly bound Li_2^- ($1^2\Sigma_g^+$) state, low energy electrons can only excite Li_2^- resonantly in region (a) (Fig. 1) from which the survival probability against autodetachment is small. For this case a reduction of cross section by several orders of magnitude can occur, depending upon the autodetachment rate (for Li_2^- ($1^2\Sigma_g^+$) $\rightarrow$ Li_2 ($X^1\Sigma_g^+$) + e^-).

The present experiments have successfully produced high Li_2 (υ^*) densities in collision-free conditions. Low energy electrons have been created in the same volume and Li^- ions looked for at moderate sensitivity. To date no Li^- production has definitely been observed, setting an upper bound of 1×10^{-8} cm^3 sec^{-1} for the dissociative attachment rate constant of Li_2 ($\upsilon^* = 13$). The preliminary conclusion is that the present result is consistent with recent theory predicting a weakly bound Li_2^- ($1^2\Sigma_g^+$) state which implies moderate rather than high attachment rates in Li_2.

LITHIUM BEAM APPARATUS

The optimum Li density for this type of volume Li source is $\approx 1 \times 10^{14}$ cm^{-3}, which is determined by the need to extract Li^- ions without collisional detachment. In order to create this density in

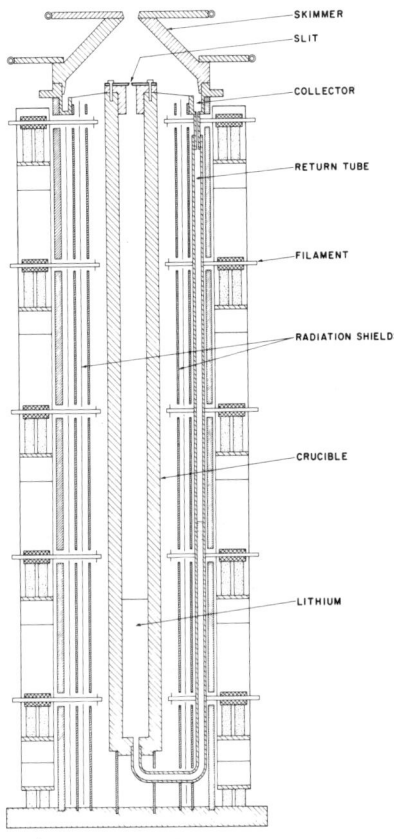

Fig. 2 Scale drawing of the recirculating Li beam source (crucible is 24 in. long).

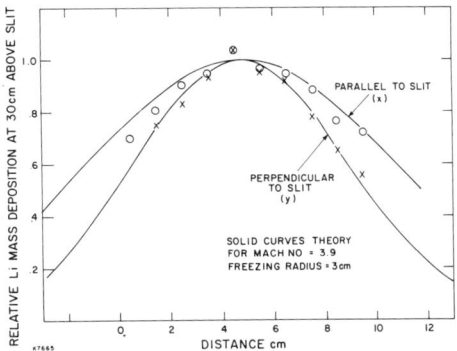

Fig. 3 Measured and fitted Li deposition at 30 cm above slit.

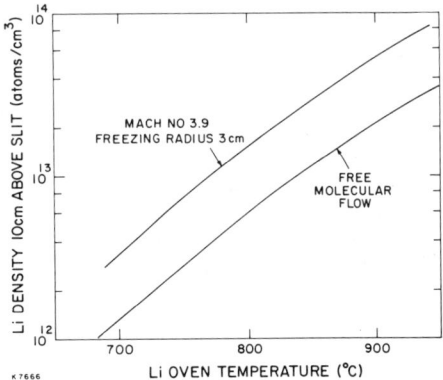

Fig. 4 Calculated Li density 10 cm above slit as a function of Li oven temperature.

a supersonic Li beam over a dimension of several cm a Li oven pressure of ≈ 10 torr and a slit area of ≈ 0.1 cm^2 were used, requiring a mass flow through the slit of the order of 100 g/hr. In order to provide reasonable experimentation times most of the lithium was recirculated by gravity.

The vapor source is shown in Fig. 2. The 900°C operating temperature is achieved using electron bombardment heating from 10 filaments operated in pairs, delivering up to 3KW at up to 6kV. The crucible was machined from a 2 in. diameter bar of 304 stainless steel with a 1 in. diameter hole gun-drilled to 1 in. from the bottom. The return tube was heliarc welded to the bottom of the crucible. Five heat shields are used, to keep the 900°C power requirement to ≈ 2.5kW. The slit consists of two stainless steel half-circles bolted to the cap of the crucible and typically spaced by 0.06 cm. The slit length is 1.2 cm. Above the slit is a conical condensing skimmer, water cooled at the apex and the rim via steel plates which maintain a temperature differential of ≈ 300°C between the condenser surface and the cooling water. The condenser surface must be maintained above 186°C, the melting point of Li, and below about 400°C, the temperature at which re-evaporation from the condenser impedes the supersonic Li beam. In operation the source can be run up to temperature, operated for 5 hours and cooled down approximately 10 times before it is necessary to replenish the lithium.

We have measured the Li mass flow by collecting the beam on a thin metal foil 30 cm above the slit and measuring its mass increase. The relative Li deposition is shown in Fig. 3. Also in Fig. 3 are shown the calculated deposition profiles for our geometry, assuming that the transition from collisional flow to free molecular flow occurs at a radius of 3 cm out from the slit, and that at that point the Mach number is 3.9. The measured total mass flow agrees with calculation to within ±50%. Using the same model we have calculated the lithium density as a function of oven temperature at the experimental location 10 cm above the slit (Fig. 4). Of this density, approximately 5% comprises Li_2 molecules.

At the experimental region the beam rotational temperature has been measured spectroscopically to be 150°K (±30°K), and its translational temperature is modeled to be the same. At this position the beam is completely collision-free, so that vibrational states Li_2 (υ*) prepared by optical pumping are truly metastable.

OPTICAL PREPARATION OF Li_2 (υ*)

Lithium molecules relax to >90% {υ=0} during the initial part of the supersonic expansion [6]. Using a tunable blue laser we can populate specific vibrational levels ($υ_B$) of the Li_2(B) state which subsequently radiate in 9 nsec to a distribution of Li_2(X) {υ*} levels determined by the appropriate Franck-Condon factors. It is a useful feature of Li_2 that 50% of the B→X radiative decay is into a close group of high vibrational levels whose centroid depends on $υ_B$. For example, with $υ_B$ = 6 the resultant decay is to υ* ≈ 11, as illustrated in Fig. 1. The range of $υ_B$ that is readily accessible extends up to $υ_B$ = 9, corresponding to υ* ≈ 16. For $υ_B$>9 the absorption cross section becomes too small.

Although the Li_2 molecules in the expansion are relaxed to a rotational temperature of 150°K, the Li_2(B-X) absorption bands still have a width of ≈ 10Å. Using our available narrow band lasers (0.25Å bandwidth) only a small fraction of the total X population may therefore be pumped into the B state. Typically the bandhead is the best place to pump because of its concentration of many superimposed rotational transitions. This sharp onset of the band is

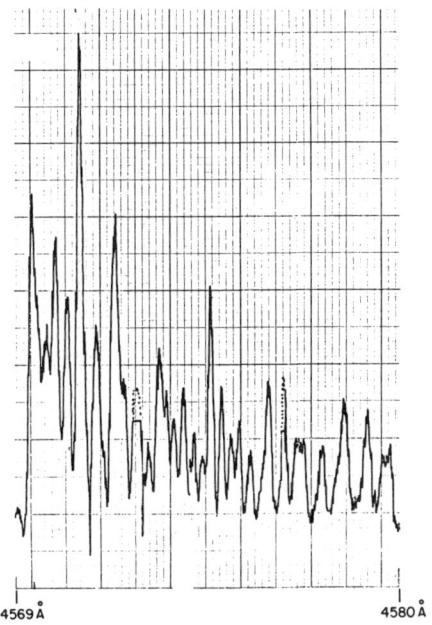

Fig. 5 Li_2^+ signal for excitation via the Li_2 (X-B), 0-6 Band, showing 150°K rotational distribution in Li_2(X) state.

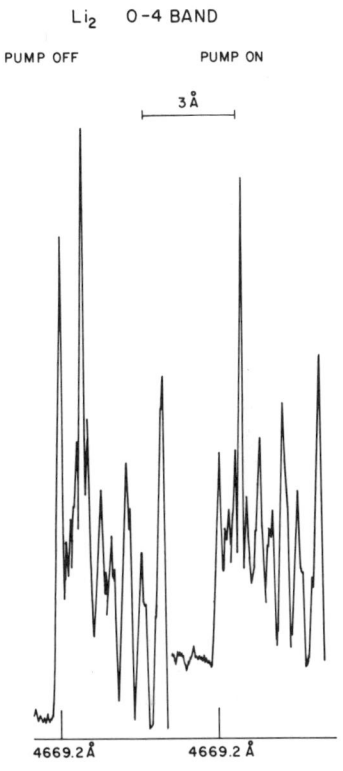

Fig. 6 Reduction of 0-4 bandhead line due to narrowband optical pumping, monitored by Li_2^+ signal.

illustrated in Fig. 5 for the 0-6 transition, which has its head at 4569.6Å. Pumping at this wavelength, for example, accesses ground state rotational levels J" = 0→5, within a 0.25Å bandwidth. By taking the ratio of the area of the 0-6 bandhead peak to the whole area we find that 8% of the Li_2 population is accessed by narrow band pumping at this beam temperature.

The degree of optical pumping into Li_2 (υ*) depends on the ratio of τ_p (the laser pulse duration) to τ_R (the B state radiative lifetime), and ranges from 0.25 for $\tau_p \ll \tau_R$ to 0.5 for $\tau_p \gg \tau_R$. For our experimental conditions τ_p = 5 nsec and the transfer efficiency is 0.31. When this is multiplied by the 8% of accessible population, the fraction of Li_2 molecules pumped into Li_2 (υ*) is 2.5%.

We have monitored the removal of bandhead population in a separate experiment in which optical pumping was performed on the 0-4 bandhead peak at 4669.2Å by a single 0.25Å, 1μJ pulse. After a delay of 100 nsec a more intense probe pulse was scanned through the 0-4 band to generate a Li_2^+ signal by two-step ionization (the

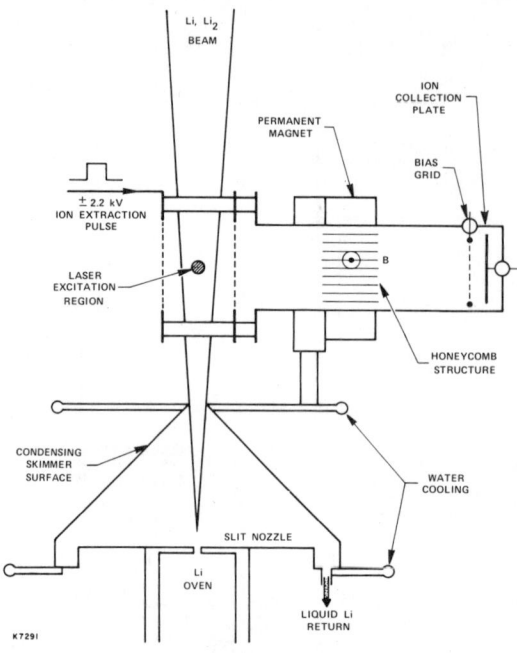

Fig. 7 Schematic of source slit, skimmer, extraction plates and ion drift tube.

second step being from $\upsilon_B=4$ to the Li_2^+ continuum). A reduction of bandhead intensity of 47% was measured (Fig. 6) after adjustment for a change in signal amplitude between the scans. It is probable that the difficulty of alignment in this experiment prevented us from observing all of the 62% reduction calculated to occur.

Using narrowband optical pumping we have demonstrated the production of 4×10^{10} cm^{-3} $Li_2(\upsilon*)$ molecules per pulse in a cylindrical volume ≈ 0.3 cm dia. x ≈ 3 cm long. The existence of a 'hole' in the rotational population after 100 nsec indicates that the beam is collision free over this timescale at the very least, as expected from the beam model.

ION COLLECTION

The laser excitation region is located between parallel mesh planes which carry an ion extraction pulse after a variable delay. This is shown schematically in Fig. 7, which also shows the 22 cm ion drift tube and collector plate, connected to a sensitive charge amplifier. Electrons are blocked by the presence of a 6 mm mesh honeycomb of length 3 cm in a crossed magnetic field of 200 G. The blocking factor is at least 2×10^4 in our experimental conditions as verified by comparing the noise level on negative ion extraction with the Li^+ signal on positive ion extraction (Li^+ creation is discussed below). Signal recovery is via either sample-and-hold circuits or a transient digitizer connected to a computer, which records 1000 pulse sequences in multiple channels reflecting the chosen experimental variables (such as the chopped application of optical pumping).

Li^+ CREATION BY TWO- AND THREE-STEP IONIZATION

In order to provide an electron density in the beam for attachment, the 3-step ionization of Li is used (Fig. 8). In our experimental conditions we compute a fractional ionization of 1×10^{-3}

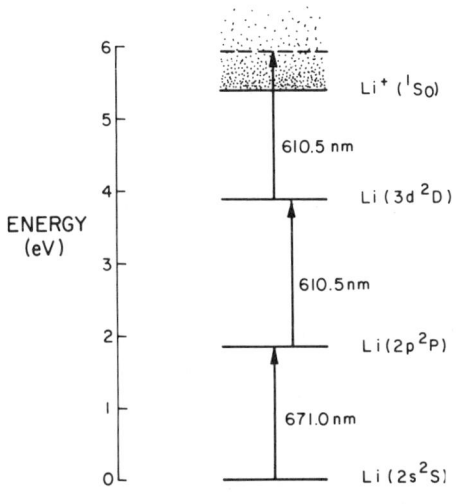

Fig. 8. Three-step photoionization of Li.

applying to a Li density of 3×10^{13} cm^{-3} at 850°C oven temperature. The calculated electron density is therefore 3×10^{10} cm^{-3}. We have monitored Li$^+$ creation in order to measure the electron density, but the collected charge is about 10 times less than the above calculation. At present it is believed that the discrepancy is due to space charge distortion of the extraction field, leading to extracted ion trajectories which miss the collector plate. When negative ions are looked for it is assumed that the Li$^-$ collection efficiency equals that for Li$^+$, and the absolute efficiency of collection is not used in the computation of the dissociative attachment rate constant.

An alternative method of creating electrons in the beam is via the two-step photoionization of Li$_2$. This is facilitated by the presence of many autoionizing Rydberg states of Li$_2$ which may be accessed from the Li$_2$ (B) state by the absorption of a further blue photon (Fig. 1). In the course of the present work we have discovered a new vibrational sequence of strongly autoionizing Rydberg series in Li$_2$ [7]. The Li$_2^+$ signal on the strongest of these resonances was as high as 0.4 of the Li$^+$ signal from three-step photoionization (all the lasers concerned having a similar pulse energy in the 5μJ to 10μJ range). The Li$_2$ fraction in the beam was 0.05, so that the two-step autoionization route was at least 10 times more efficient than the three-step process. For narrower bandwidths of laser than the 0.25Å used, the two-step process could be relatively even more efficient. In the present experiment the three-step atomic route was used, in order to avoid coupling with the optical pumping variable under study.

PLASMA EXPANSION

The cylindrical Li$^+$, e$^-$ plasma column created by photoionization immediately begins to expand due to the plasma pressure. The characteristic time for expansion is the ion acoustic transit time. It may be shown that electron-electron collisional relaxation occurs in $\approx 10^{-6}$ sec, but that electron-neutral and electron-ion relaxation are much slower. The only significant energy loss to the electrons in $\approx 10^{-6}$ sec is the work performed in expanding the plasma. The plasma expansion results in a limit to the interaction time of electrons with Li$_2$ (υ^*), which depends upon the initial plasma radius and the threshold energy for dissociative attachment ε_{DA}. The expansion may be shown to be self-similar, at least for a Gaussian radial density distribution at t = 0. The 1/e density radius at time t is

$$r_e(t) = r_{eo}(\gamma t + e^{-\gamma t}) \qquad (2)$$

in which r_{eo} is the initial radius and $\gamma = \dfrac{1}{r_{eo}}\sqrt{\dfrac{2\varepsilon_o}{m_i}}$,

where ε_o is the initial average electron energy and m_i is the mass of Li^+. In the same terms the average electron energy at a given time is

$$\varepsilon(t) = \varepsilon_o e^{-\gamma t}(2 - e^{-\gamma t}) \qquad (3)$$

and the time τ_{DA} available for attachment (for which $\varepsilon(\tau_{DA}) = \varepsilon_{DA}$) is

$$\tau_{DA} = -\frac{1}{\gamma}\ln\left[1 - \sqrt{1 - \frac{\varepsilon_{DA}}{\varepsilon_o}}\right] \qquad (4)$$

As an example, when the electron initial energy is 0.53 eV (as in the above photoionization process), and $r_{eo} = 0.1$ cm, a dissociative attachment threshold of 0.25 eV implies $\tau_{DA} = 0.34$ μsec.

Although there is an interaction time limit in the present experiment this consideration does not apply to scaled-up sources.

DISSOCIATIVE ATTACHMENT MEASUREMENTS

We have prepared $Li_2(\upsilon^*)$ $\{\bar{\upsilon}^* = 11,13\}$ at a density of 3×10^{10} cm^{-3} by optical pumping to $Li_2(B)$ $\{\upsilon_B = 6,7\}$ respectively. In these experiments the oven temperature was 850°C to 880°C and narrow band optical pumping was used, as described above. The fraction of electrons converted to Li^- that we were able to detect was noise-limited at 1 in 2×10^4. At this level we did not record consistent Li^- signals above the 2σ confidence level. If it is assumed that plasma expansion limits the interaction time to 0.3 μsec, then the upper bound on the dissociative attachment constant for these levels is $k_{DA} \lesssim 1 \times 10^{-8}$ cm^3 sec^{-1}. At the average electron energy of ≈ 0.4 eV this corresponds to $\sigma_{DA} \lesssim 3 \times 10^{-16}$ cm^2.

This upper bound is consistent with a weakly bound Li_2^- ($1^2\Sigma_g^+$) state rather than a gently repulsive state, which would be expected to show an order of magnitude larger σ_{DA} as discussed above.

A further enhancement in sensitivity of up to 100 times is achievable throught the use of (a) broad-band optical pumping, to use all the available Li_2, and (b) larger pumped volumes to allow a longer plasma expansion time. This combination leads to the requirement for a $\approx$ 100mJ pump laser of 10Å bandwidth, which we hope to employ in continued measurements.

ACKNOWLEDGEMENT

This work was supported by the Air Force Office of Scientific Research under contract no. F49620-82-C-0051.

REFERENCES

1. M. Allen and S. F. Wong, Phys. Rev. Lett. $\underline{41}$, 1791 (1978).
2. J. M. Wadehra and J. N. Bardsley, Phys. Rev. Lett. $\underline{41}$, 1795 (1978).
3. D. A. Dixon, J. L. Gole and K. D. Jordan, J. Chem. Phys. $\underline{66}$, 567 (1977).
4. D. D. Konowalow and J. L. Fish, private communication (1983).
5. An excellent discussion is given in 'Negative Ions' by B. M. Smirnov, Chapter 4 (McGraw-Hill Inc. 1982).
6. C. R. Wu, J. B. Crooks, S. Yang, D. R. Way and W. C. Stwalley, Rev. Sci. Instrum. $\underline{49}$, 380 (1978).
7. M. W. McGeoch and R. E. Schlier, Chem. Phys. Lett $\underline{99}$, 347 (1983).

FORMATION OF NEGATIVE IONS BY CHARGE TRANSFER: He$^-$ to Cl$^-$ *

Alfred S. Schlachter
Lawrence Berkeley Laboratory
University of California
Berkeley, CA 94720

ABSTRACT

Formation of energetic beams of negative ions of elements with atomic numbers 2-17 (helium to chlorine) by charge transfer in metal vapors is discussed.

INTRODUCTION

Negative ions are useful for atomic physics, for injection into accelerators, and for plasma physics. Energetic negative ions can be efficiently converted into neutral atoms, for which many uses are known or proposed relating to magnetically confined plasmas of fusion interest. Fast beams of H^0 and D^0 produced by electron detachment from H$^-$ or D$^-$ are presently being developed for heating of plasmas for fusion. Grisham and co-workers[1] have proposed using multi-MeV neutral beams of heavier atoms for plasma heating, made by neutralization of negative ions. The energy per atom is greater than that for H or D at the same velocity, so that less current would be needed to achieve a desired level of heating power. They also suggest that the injected beam could be used to drive current in a tokamak or for tandem-mirror-reactor end plugs.[2] Post and coworkers have discussed the use of a fast light-atom beam, e.g., multi-MeV Li^0, as a diagnostic for fast confined alpha particles resulting from deuterium-tritium reactions in a magnetically contained plasma: 2-electron transfer would neutralize alpha particles, allowing them to escape from the plasma. Afrosimov[4] has discussed neutral-particle diagnostics of plasmas.

Negative ions can be formed by several methods: a) direct formation by volume processes in a discharge; b) sputtering, backscattering, or desorption from a surface; and c) charge transfer of fast positive ions in an appropriate gas or vapor target. Method b) is used for high-current H$^-$ and D$^-$ sources,[5] in "universal" sources of heavy ions,[6] and is often used with tandem accelerators. Method c) has been used for production of intense beams[7,8] of H$^-$, D$^-$, and He$^-$, as well as heavier ions, and is the subject of this review, in which results of formation of negative ions heavier than H$^-$ or D$^-$ by charge transfer are summarized. The Aarhus group has made many of the measurements on heavy negative ion formation.[9] Tykesson has previously presented considerable data on this subject, and much of the data presented here is from that review or from papers by Heinemeier and Hvelplund. Binding energies of negative ions have been summarized by Hotop and Lineberger.[10]

Experimenters measure equilibrium charge-state fractions (equilibrium yields, F_i^∞) or optimum conversion efficiency (η_i^{opt}). The latter is dependent on the geometry of the experimental arrangement, and is a lower bound to the former.[11] Since data are sparse for formation of negative ions other than H- and D- by charge transfer, both are presented here. The reader is reminded that η_i^{opt} can be lower than F_-^∞ by an unknown amount.

Several systems considered here have more than 3 states, in which case charge-state fractions as a function of target thickness can exhibit complex behavior. An example is helium,[12] for which a minimum of 4 states must be considered: He+, He⁰(1s²)¹S, He⁰(1s2s)³S and He-; other states, e.g., He⁰ (1s2s)¹S or the P states must sometimes also be considered. The He- fraction exhibits an optimum fraction, F_{opt}, at a target thickness less than that for equilibrium (see discussion below).

He-, Ne-, Ar-

Donnally and Thoeming[13] showed in 1967 that He- is produced from He+ by a two-step process in cesium vapor, in which He triplet metastable atoms (1s2s)³S are produced in the first collision and He- in the second; Jorgensen et al.[14] had previously noted the role of the metastable He atom in He- formation. The process is

$$He^+ + Cs \longrightarrow He^0\,(1s2s)^3S + Cs^+$$

$$He^0\,(1s2s)^3S + Cs \longrightarrow He^-\,(1s2s2p)^4P + Cs^+.$$

This two-step process is necessary because He- is a quartet state, requiring all three electron spins to be aligned. Schlachter et al.[15] made a detailed study of this process, using a 4-state model to demonstrate the role of the He⁰ triplet metastable state in He- formation. Charge-state fractions for 25-keV He+ in cesium vapor are shown in Fig. 1a; the He- fraction is seen to reach a maximum at a target thickness of less than 1×10^{15} cm^{-2}. Singlet and triplet metastable atom fractions are shown in Fig. 1b, which were obtained from the data in Fig. 1a by use of a 4-state-component model; the triplet metastable fraction also has a maximum at less than 1×10^{15} cm^{-2}. Helium negative ions are created by electron attachment to triplet metastable atoms.

Schlachter et al. measured an optimum He⁻ fraction of 1.4% for 6-keV He⁺ in cesium vapor. The Belfast group studied similar systems.16 Formation of He⁻ by charge transfer has been studied in metal-vapor targets other then cesium[17]; results are shown in Fig. 2. A He⁻ beam of 70mA at 10.5 keV has been produced by charge transfer in sodium vapor.[8]

The He⁻ ion is believed to have only one bound state, the $(1s2s2p)^4P_J$ state (J = 5/2, 3/2, 1/2), with a binding energy of 0.078 eV and a lifetime of about 500μs (J = 5/2). Some experimenters have claimed the existence of a long-lived $(1s2p^2)^2P$ state of He⁻; recent calculations[19] and photodetachment[20] studies do not support the existence of this state. There is no bound state of Ne⁻, nor of the other rare gases (except He).

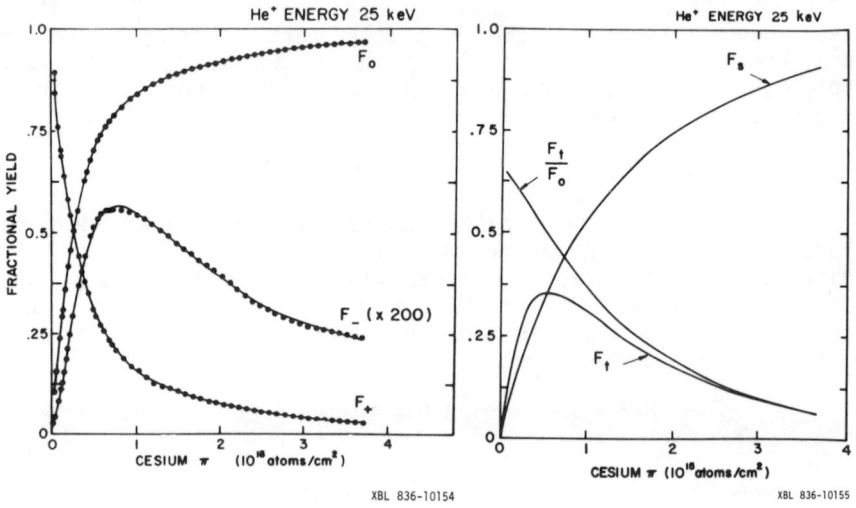

Fig. 1a Charge state fractions as a function of target thickness for 25-keV He⁺ in cesium vapor.[15]

Fig. 1b Computed fractions of He atoms in singlet and metastable triplet states for 25-keV He⁺ in cesium vapor.[15]

Li⁻, Na⁻

The Li⁻ ion is $(1s^2 2s^2)^1S$ and is bound by 0.62 eV. Equilibrium yields have been reported by the Aarhus group[9] for Na, K, and Cs vapor targets; conversion efficiencies at low energies have been reported by Steffens.[21] Results are shown in Fig. 3; the conversion efficiencies (1-20 keV) clearly lie below the equilibrium yields, as would be expected. The Na⁻ ion is $(3s^2)^1S$, with a binding energy of 0.55 eV. The only results for formation by charge transfer are shown in Fig. 4 (Aarhus group[9]).

Be⁻, Mg⁻

The Be⁻ ground state is not bound; the Be⁻ ion observed is metastable, probably $(1s^2 2s 2p^2)^4P$, with a binding energy of 0.24 eV. Results for Be⁻ formation from the Aarhus group[9] are shown in Fig. 5. The Mg negative ion is metastable, $(3s3p)^3P$, with a binding energy of 0.32 eV. Tykesson reports a conversion efficiency of less than 10^{-6} for Na and K targets at 20 keV.

B⁻, Al⁻

The B⁻ ion is $(2s^2 2p^2)^3P$, with a binding energy of 0.28 eV. Results from the Aarhus group are shown in Fig. 6. The Al⁻ ion yield is shown in Fig. 7 (measurements by the Aarhus group).[9] The binding energy of the ion is 0.46 eV for the $(3p^2)^3P$ state. There is also a metastable $(3p^2)^1D$ state.

C⁻, Si⁻

The C⁻ ion is $(2s^2 2p^3)^4S$, with a binding energy of 1.27 eV; there is also a metastable $(2s^2 2p^3)^2D$ state with a 0.035 eV binding energy. Formation by charge transfer has been measured by the Aarhus group,[9,10] by D'yachkov and Zinenko,[22] and by Nagata.[23] Conversion efficiencies (Nagata, 1-5 keV) lie below equilibrium yields (Tykesson, 3 and 4 keV to 70 keV)), for Na and Cs targets (Fig. 8). The Si⁻ ion ground state, $(3p^3)^4S$, has a binding energy of 1.385 eV. There are also $(3p^3)^2P$ metastable states with binding energies of 0.52 and 0.03 eV. The only reported results for formation by charge transfer are 24 conversion efficiency for 20 keV in a Na target.[9]

N⁻, P⁻

The negative ion of nitrogen, N⁻, has been reported[24] in a discharge, but has not been observed in charge transfer.[25] No results are known for formation of P⁻ whose states are $(3p^4)^3P$, 0.74 eV, and $(3p^4)^1D$, ä 0 eV.

O⁻, S⁻

The O⁻ ion is $(2s^2 2p^5)^2P$ with a binding energy of 1.46 eV. Results for formation by charge transfer are shown in Fig. 9. The results of D'yachkov et al[22] (2-8 keV) lie considerably above those of Nagata[23] (1-5 keV), probably indicating a larger angular acceptance in their apparatus. The measurements of the Aarhus group[9] (15 and 20 keV to 80 keV) are equilibrium yields. Large yields of O⁻ can be obtained by charge transfer in heavy noble gases.[26] Formation of S⁻ $(3p^5)^2P$, 2.08 eV) by charge transfer has been studied by Nagata[23] (Fig. 10), who measured conversion efficiencies.

XBL 8310-625

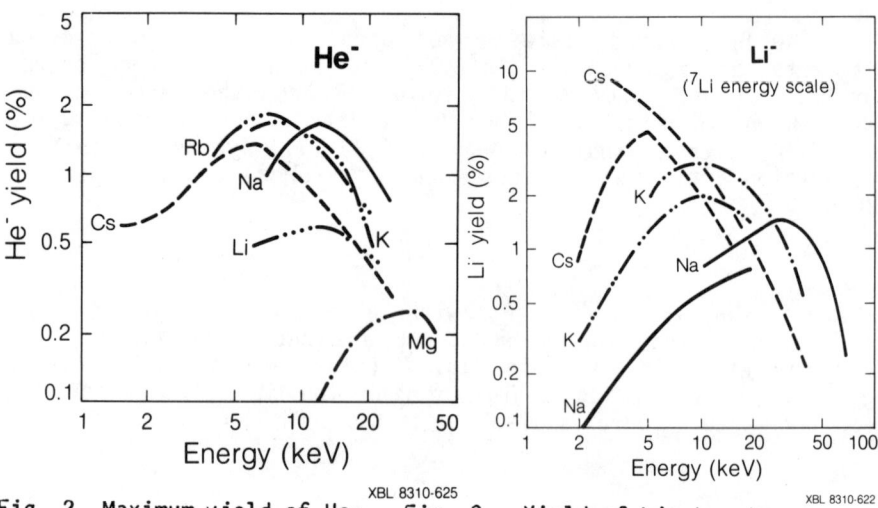

Fig. 2 Maximum yield of He- produced by charge transfer.[15, 17]

XBL 8310-622

Fig. 3 Yield of Li- by charge transfer in thick targets: equilibrium yields (3-40 keV, Cs; 5-40 keV, K; 10-70 keV, Na)[9] and conversion efficiencies (2-20 keV).[21]

XBL 8310-630

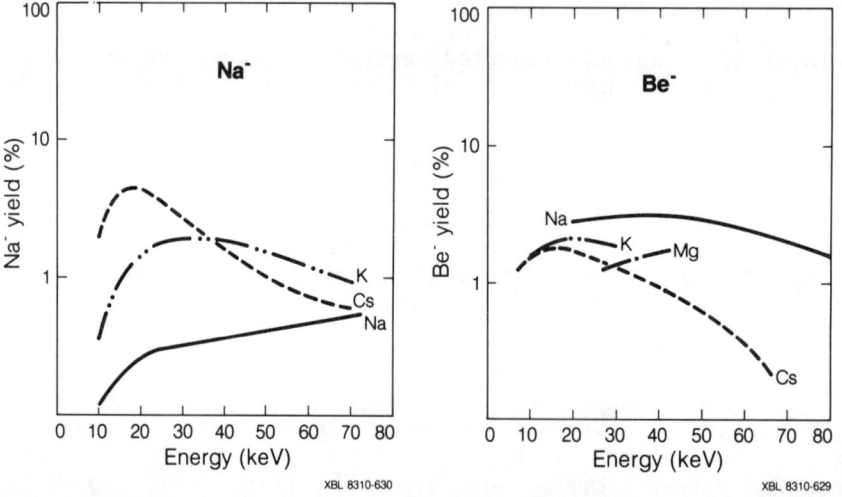

Fig. 4 Equilibrium yield of Na- by charge transfer in thick targets (from Tykesson).[9]

XBL 8310-629

Fig. 5 Equilibrium yield of Be- by charge transfer in thick targets (from Tykesson).[9]

305

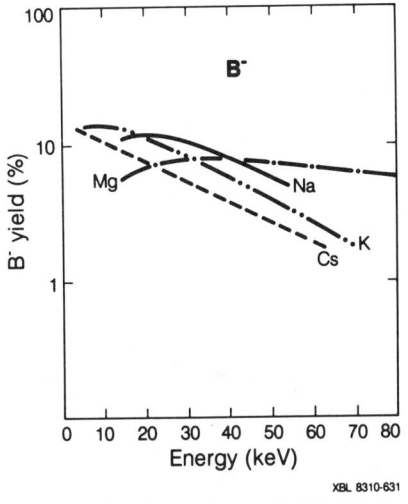

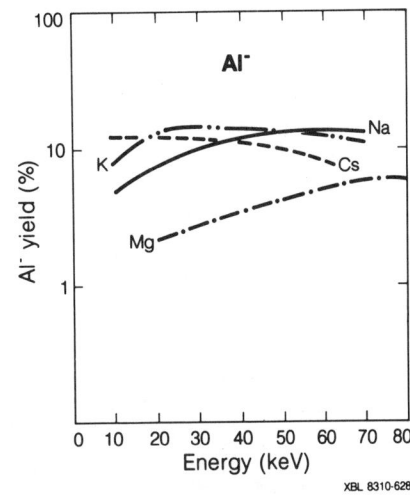

Fig. 6 Equilibrium yield of B⁻ by charge transfer in thick targets (from Tykesson).9

Fig. 7 Equilibrium yield of Al⁻ by charge transfer in thick targets (from Tykesson).9

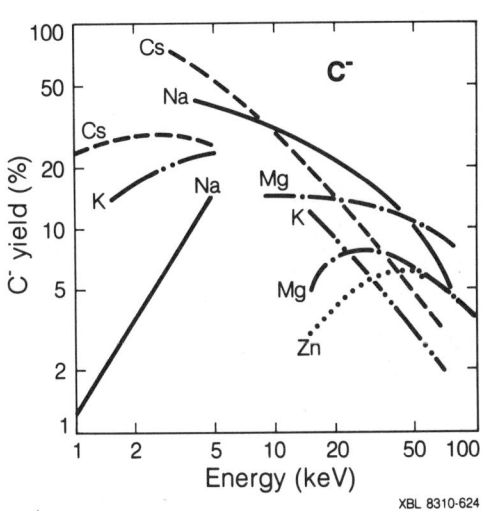

Fig. 8 Yield of C⁻ ions by charge transfer in thick targets: equilibrium yields (3-70 keV, Cs; 4-80 keV, Na; and 9-80 keV Mg)9 and conversion efficiencies.22, 23

F⁻, Cl⁻

There appear to be no results for formation of F⁻ (3.4 eV binding energy) by charge transfer. The Cl⁻ ion is $(3p^6)^1S$, binding energy 3.6 eV. Results (Fig. 11) by the Aarhus group[9] (Mg, 15-60 keV; Na, 20-80 keV) are in considerable disagreement with the D'yachkov and Zinenko results[22] (Mg, 15-100 keV; Zn, 12.5-100 keV) for Mg, which the former speculate could be due to insufficient target thickness and scattering losses in the latter's measurements.

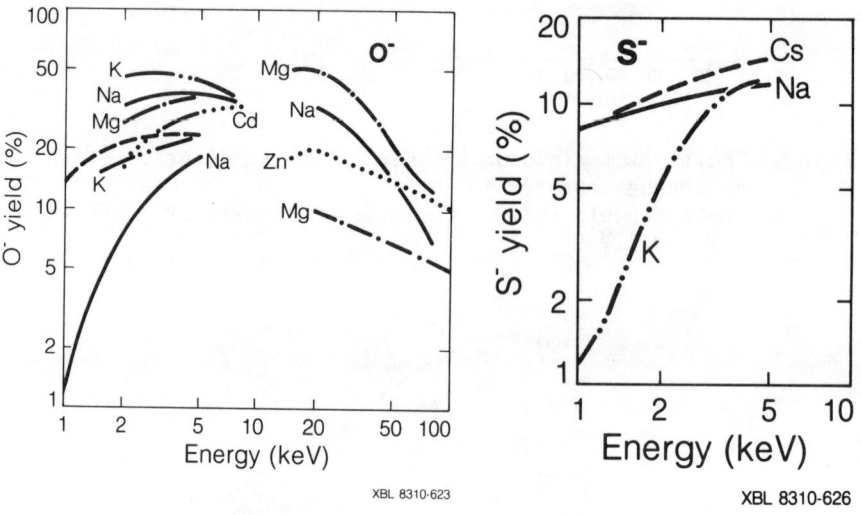

Fig. 9 Yield of O⁻ ions by charge transfer in thick targets: equilibrium yields (15-80 keV, Mg; 20-80 keV, Na)[9] and conversion efficiencies.[22]

Fig. 10 Yield of S⁻ ions by charge transfer in thick targets.[23]

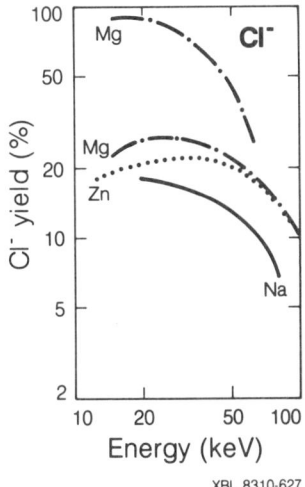

Fig. 11 Yield of Cl- ions by charge transfer in thick targets: equilibrium yields (15-60 keV, Mg; 20-80 keV, Na)[9] and conversion efficiencies.[22]

TRENDS

Heinemeier and Hvelplund[9] comment on trends observed in their measurements on negative-ion formation for a wide variety of projectiles in magnesium-vapor and sodium-vapor targets. The most important parameter is E_a, the projectile electron affinity. They find that F_-^∞ increases with increasing E_a, and that the velocity V_{max} at which the maximum negative fraction occurs decreases with increasing E_a. For low-electron-affinity projectiles, an alkali target is generally superior to Mg, while the Mg target is particularly useful for projectiles with large electron affinity. A major consideration for accelerator applications is that V_{max} be such that the projectile energy be greater than 20 keV; beam optics are better and scattering in the target is less at this energy that at lower energies. Angular scattering and energy straggling were found to depend only weakly on the atomic number of the projectile and target, but to depend strongly on the target thickness necessary for equilibrium. Heinemeier and Hvelplund's results are summarized in Fig. 12.

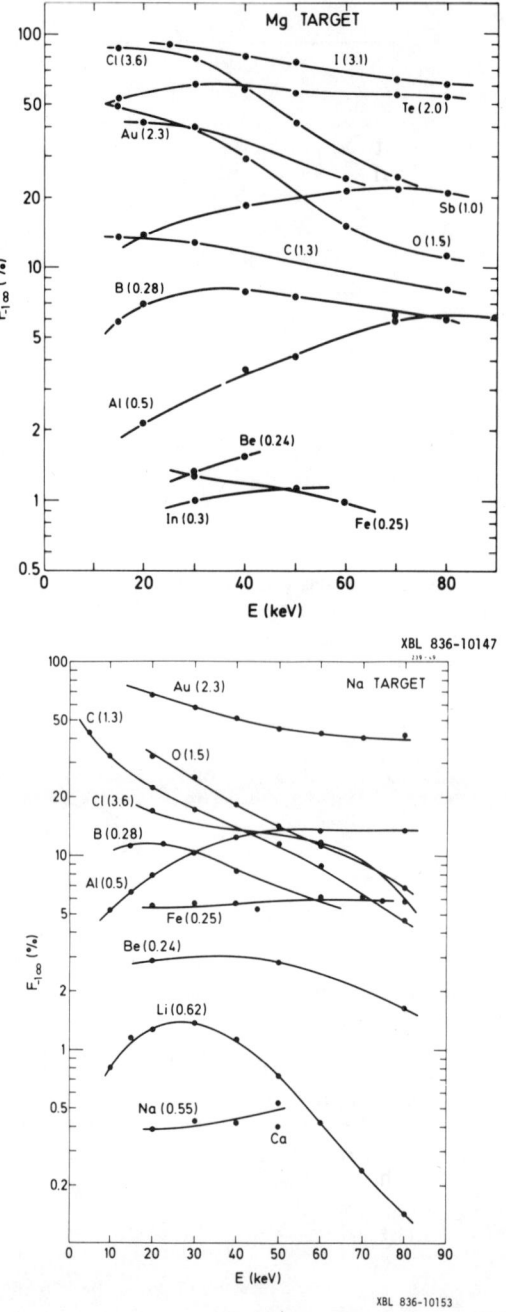

Fig. 12 Summary of the equilibrium-yield results of Heinemeir and Hvelplund (from Ref. 9), in Mg vapor (a) and Na vapor (b).

CONCLUSION

Formation by charge transfer of negative ions of species from He to Cl is reviewed in this paper. Negative ions of He, Be, and Mg are doubly excited autoionizing metastable states (Mg^- and N^- are not observed in charge transfer), and their optimal formation occurs for a target thickness less than that for equilibrium. Charge transfer is an efficient means of producing some negative ions, e.g. Cl^-, for which nearly 100% efficiency is obtained. Measurements are generally sparse; more experiments must be performed to find optimal charge-transfer media for most species.

This work was supported by the Director, Office of Energy Research, Office of Fusion Energy, Plasma Physics Division of the U.S. Department of Energy under Contract No. DE-AC03-76SF00098. The author would like to acknowledge the able assistance of Ms. Grace Yong in preparing this paper.

REFERENCES

1. L. R. Grisham, D. E. Post, D. R. Mikkelsen, and H. E. Eubank, Nuclear Technology/Fusion 2, 199 (1982); and proceeding of this conference.
2. D. E. Post, L. R. Grisham, J. F. Santarius, and G. A. Emmert, Nuclear Fusion 23, 3 (1983).
3. D. E. Post, D. R. Mikkelsen, R. A. Hulse, L. D. Stewart, and J. C. Weisheit, J. Fusion Energy 1, 129 (1981); L. R. Grisham, D. E. Post, and D. R. Mikkelsen, Nuclear Technology/Fusion 3, 121 (1983).
4. V. V. Afrosimov and A. I. Kislaykov, in Proceedings of the Course on Diagnostics for Fusion Reactor Conditions, Varenna, Italy (Sept. 6-17, 1982), EUR 8351-1 EN Vol 1, p. 289.
5. K. N. Leung, K. Ehlers, and others, proceedings of this conference.
6. See, e.g. R. Middleton, Nucl. Instrum. Methods 214, 139 (1983).
7. E. B. Hooper, Jr., P. Poulsen, and P. A. Pincosy, J. Appl. Phys. 52, 7027 (1981); R. Geller, B. Jacquot, and P. Sermet, Nucl. Instrum. Methods 175, 261 (1980).
8. E. B. Hooper, Jr., P. A. Pincosy, P. Poulsen, C. F. Burrell, L. R. Grisham, and D. E. Post, Rev. Sci. Instrum. 51, 1066 (1980).
9. P. Tykesson, presented at the 1978 Symposium of Northeastern Accelerator Personnel, Oak Ridge, Tennessee, October 23-25, 1978 (unpublished); J. Heinemeier and P. Hvelplund, Nucl. Instrum. Methods 148, 65 (1978); 148, 425 (1978); J. Heinemeier and P. Tykesson, Revue de Physique Appliquee 12, 1471 (1977).

10. H. Hotop and W. C. Lineberger, J. Phys. and Chem. Reference Data 4, 539 (1975).
11. A. S. Schlachter and T. J. Morgan, proceedings of this conference, and references therein.
12. See A. S. Schlachter, in Proceedings of the Workshop on Polarized Proton Ion Sources, University of British Columbia, Vancouver, May 23-28, 1983 (to be published in the AIP Conferences Proceedings series), and references therein.
13. B. L. Donnally and G. Thoeming, Phys. Rev. 159, 87 (1967).
14. T. Jorgensen, Jr., C. E. Kuyatt, W. W. Lang, D. C. Lorents, and C. A. Sautter, Phys. Rev. 140, A1481 (1965),
15. A. S. Schlachter, D. H. Loyd, P. J. Bjorkholm. L. W. Anderson, and W. Haeberli, Phys. Rev. 174, 201 (1968).
16. H. B. Gilbody, R. Browning, K. F. Dunn, and A. I. McIntosh, J. Phys. B 2, 465 (1969).
17. R. J. Girnius and L. W. Anderson, Nucl. Instrum. Methods 137, 373 (1976); B. A. D'yachkov and V. I. Zinenko, Sov. Phys. Tech. Phys. 16, 305 (1971); R. M. Ennis Jr., D. E. Schechter, G. Thoeming, D. B. Schlafke, and B. Donnally, IEEE Trans. Nucl. Sci. 14, 75 (1967); R. A. Baragiola, E. R. Salvatelli, and E. Alonso, Nucl. Instrum. Methods 110, 507 (1973).
18. K. F. Dunn, B. J. Gilmore, F. R. Simpson, and H. B. Gilbody, J. Phys. B 11, 1797 (1978).
19. A. V. Bunge and C. F. Bunge, Phys. Rev. A 19, 452 (1979).
20. R. N. Compton, G. D. Alton, and D. J. Pegg, J. Phys. B 13, L651 (1980). M. J. Coggiola, in Proceedings of the U.S. Mexico Joint Seminar on the Atomic Physics of Negative Ions, Notas de Fisica 5, 78 (1982); R. V. Hodges, M. J. Coggiola, and J. R. Peterson, Phys. Rev. A 23, 59 (1981); G. D. Alton, R. N. Compton, and D. J. Pegg, Phys. Rev. A 28, 1405 (1983).
21. E. Steffens, IEEE Trans. Nucl. Sci. NS-23, 1145 (1976); also shown in Ebinghaus et al, Proceedings of the Second International Conference on Ion Sources, Sept. 11-15, 1972, Vienna, p. 491.
22. B. A. D'yachkov and V. I. Zinenko, Sov. Phys. Tech. Phys. 18, 1087 (1974); B. A. D'yachkov, V. I. Zinenko, and A. V. Nasonov, Instrum. and Exper. Methods 1348 (1976) [Transl. of Priboryi Teknika Eksperimenta 5, 27 (1975)].
23. T. Nagata, J. Phys. Soc. Japan 46, 919 (1979).
24. H. Hiraoka, R. K. Nesbit, and L. W. Welsh Jr., Phys. Rev. Lett. 39, 130 (1977).
25. B. Hird and S. P. Ali, Phys. Rev. Lett. 41, 540 (1978).
26. A. B. Wittkower, P. H. Rose, R. P. Bastide, N. B. Brooks, and L. Hopwood, J. Can. Phys. 43, 404 (1965).

*This work was supported by the Director, Office of Energy Research, Office of Fusion Energy, Applied Plasma Physics Division of the U.S. Department of Energy under Contract No. DE-AC03-76SF00098.

DISCUSSION

MORGAN: Every time I hear N^- mentioned, it keeps jogging my memory. Don't I remember an article published by Hird in Canada in Physical Review Letters a few years back where he talked about seeing N^-?

SCHLACHTER: Yes, Hird and Ali reported in 1978 not seeing N^- in a charge-transfer experiment. This result is essentially in contradiction with the 1977 paper of Hiraoka and co-workers, who reported the presence of N^- in a discharge. I would think that metastable ions are more likely to be created by charge transfer than in a discharge.

MICHELS: N^- calculations have been done by several people and they all get the same result. The triplet p state is not stable but N^- is stable relative to the other two levels of nitrogen. Harris was the first one actually to do those calculations. The other comment I had is that it is not surprising that the ns np^2 combination is losing stability as you go from helium to berylium to magnesium because you're getting core polarization. That may explain why you lose it when you reach magnesium.

PETERSON: You're talking about the lifetime, not the stability?

MICHELS: The lifetime.

VERBEEK: Is it possible that you see these ions in SIMS experiments where you release them by sputtering? For instance magnesium and things like that.

SCHLACHTER: Metastable ions are not likely to be produced by sputtering. Middleton and Purser have reported not seeing Mg^- or N^- produced in a sputter ion source. I was, in fact, surprised when I heard that you have produced He^-, which is, of course, metastable, by reflection from a surface.

PANEL SESSION: FUNDAMENTAL PROCESSES

J.R. Hiskes (Moderator), M. Bacal, E.H.A. Grannemann,
A.M. Karo, M. Seidl, G. Wada, J.M. Wadehra

HISKES: We will discuss surface processes first, and after 45 minutes we will switch to a discussion of volume processes. I have asked three panel members to make a statement about some features of surface processes in order to get us started. After they have made their comments the audience will hopefully be stimulated to ask questions. The three panel members introducing the surface processes discussion are Ernst Granneman, Milos Seidl, and Gen Wada.
GRANNEMAN: Actually, I've prepared just general statements for both types of processes. I'm from the FOM Institute for Atomic & Molecular Physics which, as the name says, is doing fundamental atomic and molecular physics. The first two days of this conference already were quite a revelation to me in the sense that I have heard a lot of nice physics on the one hand. On the other hand, there is a lot of nice physics which is still unexplained in both, volume sources and plasma surface sources. A lot of topics have been only vaguely touched and people are not really sure of them, those I think should get some more attention. If I turn to the volume sources, then vibrationally excited H_2 molecules are considered to be important. They can be made either by the excitation by 20 eV electrons or they can also be made, as I understood from a poster paper, in wall collisions of H_2^+ ions. There is also wall relaxation of those vibrationally excited molecules. It is not really known what H_3^+ does. So there is a lot of physics here which in fact should be taken out of the source. People are trying to do measurements in sources, whereas in fact more of the efforts should be made outside the source. The same holds for plasma surface sources where the situation is slightly better, I guess, and where there have been done some experiments on well prepared surfaces. However, those experiments don't really cover all subjects. For instance, we still do not know what is the influence of hydrogen absorption in the cesiated converter layer, what is its influence upon the work function of the surface, how does the sputtering take place. Furthermore, another thing which is not clear to me is how does the cesium come off the converter? Does it come off positively charged? Does it come off neutral? So I think there are some more physics topics which should be covered outside the source and I hope that we at our laboratory can take some of those problems up and come up with some solutions.
SEIDL: There has been quite a progress in the understanding of the surface production in the sense that it is now clear that all processes really produce neutral hydrogen atoms that move away from the surface and that there is an electron transfer that produces the negative hydrogen ion. The actual intensity of the source depends on the product of these two terms, on the number of fast enough atoms produced and on the probability of electron transfer. The energy spread, which is the second important parameter for the sources, is obviously better if the velocity of the atoms leaving the surface

is small. Of course, then the probability for electron transfer is reduced too and there is some kind of an optimum that in fact has to be sought. Maybe one might even think of some kind of a process where one would use channeled hydrogen atoms that are directed exactly perpendicular to the surface and have enough energy so that they would have an optimum velocity for electron transfer. In actual sources there are probably three mechanisms at work, scattering, sputtering by hydrogen ions, and sputtering by cesium ions, but it would be good if somebody who has a surface conversion source would drill a hole in the converter and measure and analyze the ions hitting the converter to see how many are hydrogens and how many are cesiums. Also, if a high resolution energy analysis of the negative hydrogen ions is made, one could say how many of them are produced by cesium sputtering and how many are produced by hydrogen sputtering. The poster presented today by Wada et al. shows clearly that a large amount of negative ions is produced by scattering but they don't reach the extraction because the angle is in the wrong direction. So maybe one could somehow make use of them. It now appears that the optimum operation of the sources really depends on the cesium coverage and that is why they all operate the way they operate. More investigation should be devoted to understand the coverage of the converter by cesium ions alone and also basic measurements are needed for sputtering by hydrogen ions because I don't know about any detailed measurements of the yields of sputtering by hydrogen ions. As far as cesium sputtering is concerned, it is very clear that the energy dependence of the yield is such that at energies below 200 eV there is not much sputtering (or desorption) due to cesium if the only supply of hydrogen is the hydrogen adsorbed on the surface. Again, one should look into how important the implantation of hydrogen is. Such implantation could really change the concentration of the hydrogen in the metal considerably. One would like to have a much higher concentration than the concentration of hydrogen in the plasma. That might produce some kind of a continuous transition from the plasma into the metal. It wouldn't be such a clearly defined surface. In that context I don't know whether any charge transfer in atomic collisions between hydrogen and cesium would be of any importance if the concentration of hydrogen was very large close to the surface. There seems to be no data on cross sections for charge transfer at energies below 200 eV. Also, what would be the effect of alkaline earth metals if one had this large implantation of hydrogen on the surface. An interesting point is that in the sources there is no difference between monocrystalline surface and the polycrystalline surface. They give the same yield and maybe it is because the ion sputtering really perturbs the structure.

The last comment I would have is that there has been a large progress in volume production sources made simply by using the tandem design where you tailor your plasma to your needs. I wonder whether something like that wouldn't also be of importance for the surface produced ions, because I don't know how many negative hydrogen ions are lost as they go across the plasma sheet in front of that converter. Perhaps by reducing the electron temperature one could get some improvement.

WADA: I am principally concerned with the engineering applications of the sources rather than with the basic physics. I have built a small surface plasma source and have done an investigation of the work function correlation with the H^- produced. Basically, I do agree with the former two speakers that there is a self sputtering coefficient of cesium which should be measured. Sputtering of the cesium on the surface by the hydrogen particles should also be measured in order to fully understand the cesium recirculation inside of the source. But on the other hand, there are several other things that I don't understand about the surface plasma sources. Namely, sometimes if you operate the source at the identical condition you have a different value of the H^- yield. For instance, if you operate the source long enough then often you have a very large amount of H^-, in fact a very anomalous amount of H^-. The other point I really have interest in at the present moment is the surface plasma cascade proposed by J. Hiskes. I once tried to evaporate lots of cesium inside the chamber keeping the wall relatively cold and extract the H^- out of the source. The amount of H^- was really small but it seems to me that most of the H^- was produced by the backscattering mechanism rather than the cascading effect. I thought at that time that instead of increasing the converter potential, one might decrease the converter potential in order to have an efficient usage of cesium because with a lower converter potential one might expect less cesium to be removed due by sputtering. I tried that around a converter potential of 10 eV, but my source was not really designed to operate the converter at such a very low potential. I didn't have too much of H^- ions but I do believe that the experiment with converter operation at a relatively low voltage, like 30 eV or maybe even 20 eV, might be investigated in order to check the theory proposed by Hiskes and Karo and others. As for basic processes I do believe that the sputtering effect of cesium and hydrogen is very important for the more advanced surface H^- sources.

YORK: When running the converter at very low voltages in my operation of the cusp-field source, I find that when first starting the transfer of cesium I have run into such conditions. I could obtain a very bright beam, that is a beam with a very low ion temperature, but at the same time the total amount of beam is very small and as you add more cesium, raise the converter voltage, you get more beam. At the last meeting I remember there was a discussion on the creation of H^- ions on the surface and the survival rate as they leave the surface. It seems as though the voltage on the converter has something to do with that because at low voltages I can make a very bright beam but I can't make very much of it and I have to get more cesium on the converter and raise the voltage in order to create a higher intensity beam; there seems to be no way to get around that.

QUESTION: What is the spacing between your converter and source exit?

YORK: 12.97 centimeters. To determine the loss of beam as it exits, I've run studies by changing gas pressure, because if there is a lot of stripping of H^- ions, it should be very dependent on the gas pressure. Actually I have changed the gas pressure by a factor of about 2 to 3 and I don't see any effect.

VERBEEK: I would like to make a comment on the sputtering yields of hydrogen on cesium. I think at least an estimate should be able to be gotten from the work of Oen and Bohdansky who sputtered all kinds of materials with hydrogen and other light ions and found a universal curve. The single parameter that enters is the surface binding energy. One should be able to get an estimate from this work.

HISKES (to SEIDL): You have the universal curve of the negative ion energy which you published for cesium energies from 500 eV to about 2 keV. If you extend the energy down to 200 eV, is the shape of that curve still preserved?

SEIDL: Well, we haven't yet finished either the energy measurements or the angular distribution measurements of the beam and for that reason I didn't say a word about that.

HISKES: If the shape of the energy distribution curve is preserved for different incident cesium ion energies and if you knew the binding energy of the hydrogen on the surface then you could come up with a semi-empirical function which would be useful for extrapolating right down to the threshold.

SEIDL: We are just in the process of measuring the energy spread of the beam at these low energies. By the way, the binding energy of hydrogen on a cesiated surface is something that bothers me and it seems that there are all kinds of numbers in the literature. Some of them are very high, some of them are very low. Is there anybody who is familiar with these things?

GRANNEMAN: I think it shouldn't be too difficult to devise an experiment to measure it. I have found on several occasions at this conference that there have been developed very large theoretical models which have 14 coupled equations and in which all sorts of states are calculated. What I mean is that some of these parameters can be quite easily measured but that has never been suggested. I would like to make a point again that more of those things should be measured.

EHLERS: I have to agree with that. I don't like to measure the heat of sublimation in order to plug it into the formula to get my result. I'd much rather go directly to the result.

GRAHAM: Did anybody actually take the converter out of the ion source and look at its condition with an electron microscope to see whether it is damaged in any way, whether the surface is attacked or marked. In other words, is the converter surface you put in the same as the one you take out at the end of the run? For example, this business about the monocrystalline converter giving the same result as the polycrystalline converter, did somebody look to see if it's still a monocrystalline surface or has it been damaged by sputtering?

GRANNEMAN: Since the converter is still at Berkeley I haven't made an investigation. But I think under a microscope you should be able to get a good impression of what it looks like. But I don't believe you'll really ruin monocrystalline surface by sputtering a few 100 eV particles. I don't expect that. I just expect some sort of a regular wear. Maybe I could come back to the difference between monocrystalline and polycrystalline converters. We did fundamental measurements of the H^- production on polycrystals as well as monocrystals and we found that monocrystals should yield two times more H^- ions. Now in the real plasma source that wasn't apparently the case. You could optimize the source such that you get similar yields. Perhaps that could be explained by the fact that polycrystals have a much larger effective area. It is a bumpy structure. Everywhere you can have atoms, which can be hydrogen atoms, that can be sputtered off. I think from a fundamental point of view monocrystals should be better. I still believe that. But if you put it in a source, other things are very likely to be important too.

EHLERS: I think we've got to remember here that E. Granneman is using a completely different process. In his case the positive ion coming in is the negative ion going out. In our case the positive ion coming in is not the negative ion going out. It's a separate negative ion going out and we have two different particles. In other words, it's reflection versus adsorption or desorption and common rules may not apply.

HISKES: That depends on the angular distribution. While it has been shown by Wada that if you add up the contributions of negative ions from all angles up to 45° then most of the negative ions result from backscattering. However, if you look at the smaller angles, you have a smaller acceptance, the preference is then for hydrogen or cesium desorption. However, even at very small angles there is still some contribution from backscattering. If we look at angles within plus or minus 5 degrees, we still don't know how much of that distribution is backscattering versus cesium caused desorption. Some of these are the original ones that were incident and some are newly desorbed ones.

EHLERS: I think we know that. We have very, very small amounts of backscattered ions within our angular acceptance, it is mostly all desorption that we see. If there are large amounts of reflected ones we don't see what Gen Wada sees. That's a different story. They don't come out of our source. Secondly, you've got to be careful. If you were doing that work with a minimum work function cathode, in other words if you had no cesium at all we agree, 99% or thereabouts might be reflected so a lot depends on the cesiation of that surface.

LEUNG: I want to point out that Gen Wada's spectrum that he showed on his poster paper is quite different from ours. He showed a lot of reflected particles and a small amount of desorbed H^-. Now if your coverage is right and including hydrogen and cesium, then we normally see that the majority of H^- in the forward direction is made by desorption, very small amounts of it is made by reflection. It depends on what kind of condition you integrate all those reflected particles. If you start with a condition with a very small amount of the desorbed H^-, and you integrate all the reflected particles, the result is large. But if you start with very high desorbed part and very small reflected part and you integrate the reflected part, it may not be as big as the desorbed part. So you have to be very careful about those conditions when you compare these two sources.

HISKES: Even when you're looking at negative ions emitted at small angles to the surface normal, there are contributions from desorption and contributions from backscattering. If one could work with

different isotopes, like load the surface with deuterium, bombard with hydrogen, and alternatively bombard with deuterium and load the surface with hydrogen, one might be able to distinguish what is the principle mechanism for small angles.
LEUNG: Once you raise the gas pressure, the surface conditions are changed. Once you mix hydrogen with deuterium then the whole picture has changed.
HISKES: In what way?
LEUNG: In the spectrum when you have oxygen, you can see OH and OH^-. This is very complicated.
HISKES: When you compare hydrogen gas with deuterium gas, what differences do you see?
LEUNG: You mean the yield? Which yield, D^- or H^-?
HISKES: You imply you see differences with hydrogen and deuterium. What sort of differences do you see?
LEUNG: Well, you can see both contributions, a desorbed and a reflected part.
WADA: Because of the difference in mass, the optimum cesium coverage for a hydrogen discharge does not necessarily correspond to the optimum cesium conditions of the source for deuterium operation. At the same time, Leung's point is that if you prepare the surface in a nonconsistent way, then you will influence the H^- current; if you put in oxygen, usually what you see is a sudden increase in the H^-, at least that is what I've observed. If you introduce a small amount of xenon then there is a sudden decrease of H^- while at the same time you lose most of your photoelectric current required for measuring the work function. So it's a strong function of the surface conditions. Surface conditions are determined by the dynamics of discharge itself and I do understand Leung's point that if you're not careful about the preparation of the surface then what you're going to get is a totally different spectrum.
PETERSON: When you introduce oxygen and xenon, what quantity are you talking about?
WADA: A very small amount. Like 5×10^{-6} torr. Very small.
VERBEEK: What is the base pressure of H_2 or D_2?
WADA: Millitorr.
PETERSON: Then it's 5 parts per thousand.
WADA: I usually pulse it. I just open a small slot of a small cell inside of which I'm maintaining the 5×10^{-6} torr of oxygen. All of a sudden H^- goes up, then decays. Of course, I'm using up the oxygen. It seems to me that oxygen is recyling inside of the source. I've never done the study thoroughly so I cannot say anything more concrete.
VERBEEK: I'd like to come back to this suggestion to switch from hydrogen to deuterium because by this you should be able to learn something about the implantation. You could implant one isotope first or achieve some kind of implantation equilibrium, between implantation and desorption or sputtering; it is also known from the big plasma experiments that when you switch from hydrogen to deuterium it takes some time to get a deuterium discharge because of strong plasma-wall interactions. So just from the replacement of one isotope by the other you should be able to learn something.

HISKES: If there are no further questions on the surface subject, we can take up the subject of the physics of the volume sources. We have three persons who will present opening comments on volume processes: J. Wadehra, A. Karo, and M. Bacal.

WADEHRA: In the last two days I have had the opportunity to listen to some very interesting and very exciting results. Most of the results have raised a lot of questions for me and I'm sure they must have raised a lot of questions for you as well. On the first morning Marthe Bacal discussed the volume processes and efforts to identify them. In order to enhance the rate of mutual neutralization, the pressure was increased, and that resulted in the expected decrease in H^- density; the question remains if there is some simple physical picture for that. Is there some kind of a three body problem which is contributing also or something like Bates' model? That is the question I would like to know the answer to. Another question that came up in Marthe's talk was about mutual neutralization of H_3^+. What can we say about the cross sections and the rates of mutual neutralization of H_3^+ with H^-? There is a related question, namely what about the dependence of rotational and vibrational states of both H_3^+ and H_2^+ on mutual neutralization? What do we know about that? Again I don't have the answer to that, but that is a question which I am posing. Then we come to the discussion of Hiskes in which he mentioned the two chamber tandem source, the optimized source, which fortunately scales very nicely with the system scale length. Subsequently Leung described a three chamber tandem system with an increased ion density and reduced electron density at the same time and it would be interesting to see experimentally what the results are. I think that the three chamber system was presented in a very conceptual fashion. Then there was a talk by Srivastava on polar dissociation as another significant source of negative ions, namely what role does polar dissociation really play in the negative ion production. Srivastava's feeling is that since the resonance formation of H^- occurs only over a very limited range of energy, whereas polar dissociation occurs over a very large range, perhaps polar dissociation may be comparable source to the resonance formation of H^-. But I think there should be a warning there also, namely that polar dissociation is occuring at higher energies compared to resonance formation and therefore if one calculates the rate, which is the ultimate thing of interest, the question is how much does polar dissociation contribute to the rate and there, I guess, only a detailed calculation or perhaps an experiment can provide an answer. Finally, there was also a discussion by Tom Morgan about excitation associated with neutralization, the so-called NINE that Tom mentioned. We notice that there are almost no cross sections for that process at low energies and there are nine experiments for NINE and five theoretical calculations and almost all the information is on H^-. There is one calculation for lithium and one for sodium and nothing else is available. Maybe we should get some more information for excitation associated with neutralization processes at low energies, particularly at energies below 1 keV/amu. Then this morning we all saw a beautiful movie that Arnold Karo showed on the two-dimensional calculations. I would love to see that movie in 3-D, and I think we have to convince

Arnold to produce that in 3-D if he can. Another very interesting discovery was shown yesterday by Jim Peterson, and that is the discovery of a new negative ion, an ion of He_2^- which is rather unexpected because there is no theoretical calculation and no tradition whatsoever of the ion. I think that we should let the theorists face up to that challenge of predicting the lifetime that Jim has indicated.

Finally, in the poster session this afternoon I saw a poster by Malcolm McGeoch on Li^- production by dissociative attachment. The interesting thing he showed is that the maximum rate for that process, and that is experimental, is about 10^{-8} cm^3/s. That is also roughly the maximum rate for the hydrogen so I would like to urge him if he or someone else could really look at hydrogen problem once again and see if indeed that is also the rate for hydrogen.

KARO: The point at which I think we are now is one where we have a combination of theory and calculational tools (I am referring to computers), together with a variety of very, very interesting problems to which we can apply such tools. I would agree completely with an earlier comment that if an experiment can be done rapidly, efficiently, and with sufficient accuracy, one would not want to step in with a lengthy and costly calculation. So we want to judge very carefully those areas in which we would like to apply the calculational tools which we have. Certainly there are some areas where the experiments are very difficult and where perhaps a calculation can lend some support to one of several different results. We all know that there are occasions when experimentalists will disagree among themselves and where a calculation can in fact set bounds in such a way as to permit a choice between different experimental results. The other aspect of a calculational approach is that parameters can be readily varied, perhaps far more easily than they can be varied in the laboratory.

Often there are simple models which have been useful over a long period of time but which may have become, say, antiquated and where more realistic models are actually needed. I'll refer a little bit to that in just a minute. So we have to examine very carefully models that are in use, to see whether they are really applicable and whether they should be made more sophisticated. I'll use the opportunity right now to say that I would in fact like to produce a 3-dimensional movie. The codes that we've used are in fact 3-dimensional and we can apply them either to a 2 or 3-dimensional situation. But there are problems in trying to obtain 3-D graphics which would not be too confusing to the eye. We certainly are working in that direction, and in some of the other areas to which we've applied molecular dynamics we are in fact producing 3-dimensional graphics.

I'll say a few words about where we stand with computer molecular dynamics. This is a very powerful technique but it has to be used with discretion. One does not want to apply this technique to an area where hydrodynamics techniques would be more appropriate. However, molecular dynamics allows us to examine the microscopic details of both homogeneous and heterogeneous systems. One of the restrictions that we have, at least today, is that we can't really accept the tremendous cost of examining more than, say, a thousand or perhaps even ten thousand particles. There are special purpose microcomputers or minicomputers which can handle more particles, so I think we're approaching a time when we will be using special purpose computers to examine even more particles. But with the codes that we have on the machines we're using, we are restricted to this particular range. Another restriction is that we generally will follow the dynamics of an ensemble of particles up to, for example, about 100 picoseconds. If we have to extend either the number of particles or the length of time, then very special techniques or special purpose computers must be used. The other side of the coin is that we now have a very general ability to handle the particles of the ensemble so that they need not be just a set of atomic particles; they can be molecular, they can be clusters. We can essentially create whatever ensemble would be useful for the particular case we're looking at. We can treat solids, liquids, and gases, and mixed phases. This occurs because we now use a "neighborhood lookup" procedure where each particle knows what its neighborhood is, who's in it, how it's supposed to react with others in that neighborhood, whether it is itself bound to a molecule, whether it's free, whether it's going to be rebound to another molecule, and so on. So we have a very flexible code, but it has to be used within certain constraints. Of course, when we're treating surfaces or the interaction of some subspecies with a surface, the surface needs not be perfect. It can be regular, it can be composite, and it can include impurities. In other words, we can essentially compose whatever scenario we would like for the particular case we're looking at. Thus in molecular dynamics we monitor the time evolution of our system, and we monitor the energy flux within the system or within a portion of the system. For example, we can compute the energy content of the hydrogen molecule as it interacts with the surface and watch the way in which the various components of energy change as a function of time.

For the near future we expect to continue the vibrational relaxation calculations. We're beginning to look now at higher energy translational impacts of the molecules with the surface, and we may very well vary the surface depending upon the needs of the experimental community. The general backscattering calculations that we can carry out in the low energy, the 1-10 eV range, obviously are complementary to the very good calculations which are being carried out with the MARLOW code. For our purposes we feel that we can look at the low energy range of 0-10 eV. Here we need accurate interatomic potentials, and these can be obtained from a variety of sources. We have in the past examined some ways in which we can look at both light and heavy systems by using a model pseudopotential approach, and there are a variety of other ways in which one can obtain those potentials. We've seen also that we have very powerful ways of looking in extreme detail where the need exists at the electronic structure of composite systems. I am referring here to the slab type calculations that are being carried out for surfaces and for overlayers on the surfaces by Art Freeman using a rigorous full-potential self-consistent linearized APW method. Such calculations can also be carried out for finite clusters and this helps us in some of the molecular dynamics areas. What do we see when we carry out these calculations? Well, a variety of simple models has always been proposed for such overlayers, but in fact it turns out that what's really going on is very complex and very interesting. There are multiple polarizations within the cesium overlayer that lead to the lowering of the work function. We note also that the substrate is in fact not changed to as great a degree as the overlayer. One would want to look at a variety of overlayers and perhaps then be able to make some suggestions as to what would be good overlayers on good substrates. The result of the cesium on tungsten calculations is that we are dealing with polarized metallic rather than ionic overlayers, and certainly this is a new bit of information which I think will be useful in the long run.

BACAL: Well I must say I felt some emotion these two days since we started this meeting, because I saw that from this kind of fantasy, like believing in volume production, we can get something which may be useful. I think we have to appreciate the confidence and enthusiasm and the spirit of scientific adventure of all those who spent some time in trying to make a success of volume production, and there are quite a few of them here. Let me come back to what should we do. We should first try to understand the volume production. I would like to come back to what we were talking about three years ago, in trying to see what new diagnostics we would need, and I would say we didn't make big progress in studying vibrational excitation in plasmas from an experimental point of view. I think it is an important goal to verify finally the hypothesis that vibrational excitation is at the origin of volume production. We have no real proof of that and since it is becoming an area of technology, I think it should have a good, solid scientific basis. Therefore, couldn't we make an effort and get the support we need to make real big progress in detecting vibrationally excited molecules. Well, I can say we did our best in France. We have used the CARS diagnostics but which probably will be limited to levels lower than $v = 5$ if we succeed to do that. We will try a different approach in the following years, analyzing vibrationally excited molecules flowing in the gas from the discharge by a method based on the dissociative attachments. It means that we will use the dissociative attachment for diagnostic purposes, but that will be a measurement not in situ, it will be in the gas flowing outside. There are now lasers which would allow other diagnostics, like laser induced fluorescence. We have people who are enthusiastic to do these experiments. Couldn't we find the support to really start this kind of work? I think it's an important work because we shouldn't have to live with some illusions about what is important and what is not. Now, the same diagnostics would help us to measure the concentration of hydrogen atoms in the discharge and that is another important goal as was explained in the talk by J. Hiskes. So I think we could with a single effort get a lot of information which would be extremely useful.

   I would like to address another question to A. Karo, since we are talking about atoms. Couldn't the molecular dynamics also calculate the sticking probability of hydrogen? His talk explained that that is possible. Maybe that would be a good thing to do because I have read an excellent paper to appear by W. Cooper and Chun Fai Chan and others explaining how poorly we know the sticking probabilities of hydrogen and deuterium, and they appear to be different. So, there is something which is very intriguing in this problem by itself. It is of much larger interest than volume production, it goes on also in the equilibrium of positive ions sources and probably in other devices too.

Well, now I would say what is new. We have some new problems now which we didn't know about three years ago. I would say that they are related to the importance for the plasma potential profile of the small fields which exist in the plasmas and which may have some very important effects; we can only guess how they interfere in our models. So, I think that until now our models were related rather to an idealized plasma, like having a flat plasma potential, and with ions whose energy isn't dependent on any plasma parameter. Now we have to approach the problem in a new way, in just trying to understand the effect of ion velocities, how they change with pressure, how they change with profiles of different plasma configurations, and how finally they affect ion-ion recombination which is one of our dominant loss processes. The latter we should try to handle by introducing the rate of mutual neutralization. All the parameters we know are changing like pressure, plasma profile, and so on. In connection with that I'll say I found an area where our knowledge of cross sections is very poor and that is the cross section for H^- neutral elastic collisions. What is the cross section for elastic scattering of H^- on H_2? I also found, related to positive ions, that we don't know very well the same cross section for H_3^+ on H_2 in the range of energies of interest. What we could do was done by interpolating some data relevant to energies which were very different from the ones which are important and maybe we can find among our friends people who could help in that respect.

Finally the point which W. Cooper raised about H^-, about the heating of negative ions in collisions with positive ions, I didn't address it but it should be addressed in the future. Well, I would say that this will be all on the subject of fundamental processes. Let us turn now to the ion source question. I would like to express my feeling that we shouldn't overestimate the importance of the work which has been done, but should keep our minds open to other possibilities. We shouldn't think we have found the best, we have only started to look for the best. The thing which puzzles me, and I would like to point that out, is that we get the same number of negative ions in many configurations, in similar conditions which means that we shouldn't direct our attention to a certain configuration, to some complicated scheme when apparently the experiment indicates that the simplest is as good or even better. I would say that because we have developed a good technology for positive ions we should not say, let's use it for negative ions. We just were discussing this point during the poster session, that a good positive ion source may be very bad for negative ions and we heard several examples. Let's keep our minds open and look for the conditions which are important to produce negative ions in volume which are certainly different from the conditions required to produce positive ions. Let's not get trapped into one scheme but try to evaluate the real important things instead and not some technical accessories which may mask some less important aspect of the thing. Let me come back to the statement I made yesterday about this plasma electrode. You see, just a choice of area of the plasma electrode, which we didn't know before that it was important, could make us think that a certain configuration is good and another one is bad. The truth is that they are all good. We have just to choose a suitable area. I think we don't have to precipitate ourselves into a certain scheme. We have to be quiet and look into every parameter that is important, whether it is a simple thing to change it and get some simple and flexible and scalable configuration which should be easy and inexpensive to use.

PETERSON (to WADEHRA): I didn't understand your first question about ion-ion neutralization.

WADEHRA: I think two or three years ago Bates had shown that if you have three body collisions then the two body rate is enhanced in the presence of the third body. The model that he has suggested is that if the ions are for the most part inactive and the third body, a neutral body, collides with one of the ions, then you can get a bound state of the original two body system and the probability of a Landau-Zehner crossing is effectively enhanced in the presence of the third body. Therefore, the question that I was raising was, since M. Bacal has noticed in her experiment that on increasing the pressure she sees a higher mutual neutralization rate, if that was because of this three body effect.

BACAL: I don't know. Maybe J. Wadehra is the expert on the three body ion-ion recombination which I am not but my understanding is that we deal with pure two body effects which is mutual neutralization and not three body ion-ion recombination. He has in mind that the third body sticks there and helps some neutralization to occur but I think it's a matter of collision frequency. You can't keep this body for a very long time alive. So I would like that J. Wadehra would analyze the question and tell us at what collision frequency the mechanisms become active. My knowledge says that it is not.

PETERSON: By how much do you see an increase occurring in the rates?

BACAL: The change in H^- is at least by a factor of 10 but to this factor of 10 a part is contributed by other factors like electron temperature, speed of particles, and so on. I think we need a factor of 3 in the ion-ion recombination. My explanation was that due to the change in velocities of the positive and negative ions, this rate just changes. To prove that, we have to calculate the velocities correctly as a function of pressure, which we are about to finish and we're just writing a paper on that. But, J. Wadehra suggested that that could be a three body effect. I don't think so, but maybe somebody else knows. I think for that effect we would need much higher collision frequencies. Our pressure is 10^{-2} torr. I have asked this question last year at the gaseous electronics meeting. Prof. Bates was there but we asked somebody else and he told me no. Your case isn't that strange but maybe there is something new so, please, Joe (Wadehra) look into that.

WADEHRA: I just suggested that because Bates has indicated that two body rates are enhanced by three body collisions.

BACAL: I didn't ask Bates, this is true.

PETERSON: I don't think it will change very much with the internal energies of the molecules.
BACAL: This is another part of the question.
WADEHRA: Well, that was in fact the second question. How does the internal energy of the molecular ion affect the mutual neutralization, both of H_2^+ and H_3^+?
COOPER: I'd like to ask A. Karo to answer Marthe's question because I was going to ask him exactly the same thing. Could you use your molecular dynamics code to throw atoms at a surface and keep track of them as they wander around and see how many come off as molecules, that is to say to calculate the recombination rate?
KARO: Yes, in principle, if the time scales are reasonable there's no reason why that wouldn't be a very suitable problem for molecular dynamics. I haven't looked at the details, but I see no reason why it shouldn't be possible. I think there again we have to ask the question, whether the experimental procedure is going to be cheaper and easier to do than, say, carrying out of a calculation of this type. I judge this is an area where the calculation could be very useful.
COOPER: The reason why I would be interested in seeing you do it is that we derived numbers in our modeling of positive ion sources but this was the single fitting parameter which necessarily would cover all stems of the model. And we really have no idea whether it is correct or not. At least the number fell between 0 and 1, for the value of the recombination coefficient for atoms recombining to molecules on the wall while in the presence of the plasma.
HISKES: That's a little different from the case considered by Karo. In the source you already have atoms on the wall. In Karo's calculation there were no atoms initially on the wall.
KARO: We could put atoms on the wall. The code is in fact general enough to start with some suitable initial configuration, sprinkle atoms on the wall, and then have other atoms impact the surface. I could see a variety of possible starting points for this calculation and I think it's one of these cases where the easiest thing to do is to try it. The code could handle something of this type.
HISKES: If you recall the movie this morning, there was the one molecule that dissociated and one atom subsequently went off and bounced off the wall for one more time.
COOPER: It didn't recombine.
KARO: If we ran the movie backwards, it would.
POST: There's another community of people, the people who were trying to understand the behavior of plasmas at the edge of tokamaks and mirrors for impurity control and where conditions are quite similar to ion sources, densities in the 10^{12} to 10^{14} range, and temperatures in the 1 to 20 eV range. All of these fundamental processes are very crucial there, particularly when ions come in, hit the surface, recombine, then come off as molecules or as atoms, continue to bounce around on the walls, and sometimes we have to pump these. All these fundamental processes are extremely important in understanding these plasmas and we'd very much appreciate to find out about the results and also we're very interested in having data on reflection properties and questions such as W. Cooper was raising. What is the mix of atoms versus molecules and what is the

energy distribution? What are the excited states and the vibrational states of the molecules that come off the surface, what are volume recombination mechanisms? All of these make a tremendous difference in the behavior of the edge plasma in the energy balance. There exist situations in plasmas where one sees recombination but the ordinary two and three body rates are too small. Do negative ions play a large role in providing one with a recombination mechanism?

HISKES: I would like to ask the question: In the wall recombination, what is the principal energy we're concerned with? Something like 1 eV to 10 eV or 1 eV to a 1 keV.

POST: Probably 1 eV to 100 eV. It depends on the sheet potential of the electrostatic sheet. There are large modeling codes. There is a fair bit of experimental data. Getting basic data on the reflection probabilities, reflection characteristics of neutrals and ions in the few eV to tens of eV range is extremely difficult experimentally because it's my understanding there's no easy way to really detect and measure the energy of a 15 eV neutral. It is really very difficult and calculations may be all that's available. The calculations using the Trim and Marlow codes at the 10 and 20 eV level are beginning to be suspect at these low energies because of the purely repulsive potentials utilized in these codes.

KARO: Just a general comment: this obviously looks like an area where the type of programs and the calculations that we're able to carry out with these programs might well bear fruit.

BARNETT: In the talks there's been no discussion of fluctuations and oscillations, turbulence, as it affects the observations and as you try to explain the observations. Does this imply that no turbulence oscillations exist or is this not important?

HISKES: Are you referring to the electron energy distribution?

BARNETT: Well, just in general. Nobody has said anything about turbulence or fluctuations when they make their observations and I am just wondering if the observations take this into account or not?

HISKES: I wonder if K. Leung would like to comment? Do you see any fluctuations in your volume sources, Kao?

LEUNG: The only thing I know is that if in the bucket source you put your filaments way into the cusp field then you get fluctuations in the plasma density. That's the reason why we have to be very careful when we install the filaments. I think I did talk to P. Allison about fluctuations this afternoon. Maybe you could ask Paul.

ALLISON: It would depend on what you're trying to measure. I'd say yes, it does make a big difference.

SRIVASTAVA: I would like to comment on three points on volume processes which I noticed during the meeting today. There has been a lot of discussion about production of H^- via vibrationally excited H_2 molecules. Why only H_2? There are so many molecules that contain hydrogen and produce H^-, why not consider NH_3, CH_4? These have higher cross sections for dissociative attachment. Cross section for dissociative attachment for H^- production from H_2 is of the order of 10^{-21} cm^2. A small fragment of H_2 you could vibrationally excite, and vibrational cross sections are of the order of 10^{-19} cm^2, but it's a very small fraction of the total H$_2$ population which you'll be able to excite vibrationally and then produce H$^-$. Why have we not considered any other molecule as a source of H$^-$ for dissociative attachment even from vibrationally excited states? In my experiments I have seen that if you excite vibrationally, some of these molecules, like CH$_4$, the H$^-$ production goes beyond the limit that I can measure, because something saturates. It is not known what is going on right now from experiments on this vibrationally excited state. There was only one other group doing this work and they also have uncertain results. We don't have any experimental information. The effort is very meager. I think we should try some other molecules instead of H$_2$ as a source of H$^-$. The second thing I want to comment on is the dissociation. About polar dissociation, J. Wadehra mentioned we should be cautious because in case of H$_2$, the polar dissociation onset takes place at around 13 eV, while for other molecules where H$^-$ is produced or any other heavier atom is produced, the threshold is very high. In a normal plasma which we use for negative ion sources, the distribution peaks around 1 or 2 or 3 eV so we are talking about the tail which may have too few electrons to cause any polar dissociation. I agree there. What if we find a molecule which had an onset for dissociation around 5 eV or so and if the cross sections for polar dissociation are much much higher than for dissociative attachment, then that molecule can become a very important source of H$^-$ production or any heavier negative ion production. The third point I want to comment about is that I see two groups here. One is the volume processes, the other one is the surface production of the negative ions. Why not the combination of the two because as Sluyters and Prelec have shown in the past in their duoplasmatron source, they have produced H$^-$ by polar dissociation or dissociative attachment but then there are positive ions also and when they added cesium into it, they found enhancement of H$^-$. Why not develop a source which combines H$^-$ with both principles, the surface conversion because there are positive ions produced and either polar dissociation or dissociative attachment or any other plasma chemistry going on, and at the same time so combine the two together, the volume source and the surface conversion and try to produce a plasma. It would be interesting for the community to join hands together and see if a source can be built which can combine both the principles. Well, in all these systems the one thing I have been thinking lately is that in all these negative ion sources they are designed around some discharge, some filament or electron source or something like that. In order to produce a negative ion, we had to supply electrons from outside to produce negative ions. The polar dissociation which I had pointed out in case we can find some molecules like alkali hydrides or cesium hydride or some other molecule which may have an onset around 5 eV or 6 eV, then with these one can use the combination of a light source like a flash from a dye laser or something like that or a flash lamp to produce polar dissociation or positive or negative ion and then positive ions combining with surface conversion producing more negative ions. This is my comment about this negative ion production. I would like to see somebody comment or say something about some other molecules which have been tried and why they have not been used before, except H$_2$.

HISKES: I think the answer is, we're open to any specific suggestion. We have the tradition of working with cesium which is the worst conceivable material to work with. So I think we would be willing to accept almost anything as a candidate provided of course it's competitive with those things we have already explored.

PETERSON: You don't want to inject other ions than H^- into the plasma, basically. That's one thing that answers the first question. To answer your second question, if you want to produce from both volume and surface, then you are going to have two energy groups in the beam which is not exactly what you want. I think this is one of the reasons, it could generate an ion-optical problem.

SRIVASTAVA: By biasing the plasma with respect to the wall, we produce positive ions, for example, if CsH dissociates into Cs^+ and H^-. Now Cs^+ can be accelerated to the wall and converted into some more H^- from the wall. You have more H^- inside the plasma and you can accelerate it.

LEUNG: You've got here a very interesting question. Actually, a year ago when we had the positive ion workshop at Gatlinburg, one expert from Culham suggested putting some ammonic acid in the hydrogen discharge. He told me that when he was a graduate student he saw a lot of H^- when he added this to hydrogen. So immediately I bought a bottle of ammonia, a special kind of a bottle and then leaked the gas into the hydrogen. But there was no increase, actually there was a drop of H^-. If anybody has a good suggestion, I would like to try it.

BACAL: We have done two more experiments which seemed as if they should have been successful but were not. K. Leung put some cesium into the volume of a multipole and what we saw was a bit of gettering effect which was all. But if you put in more cesium, you've got a loss of the H^- current, so adding cesium for the volume source is not good. I don't know why but that's it. Now, about water vapor. One would say that water vapor is a good negative ion production molecule. So, when you add water vapor into some hydrogen, you will get OH^- and that's a very big danger. We have seen that clean hydrogen is still an academic system but for some reason it works. We had also thought about lithium hydride. Somebody in France has tried to measure cross sections. They are very low to produce H^-. Preferentially it will produce H^- not Li^-, but they are very low and it's impossible to make a guess of the lithium hydride. It's very difficult. It makes hydrogen. Every plasma chemical system which is more complex leads you to something else. Imagine HCl. You will have negative chlorine ions, but you don't want that. I mean, good molecules which have a big dissociative attachment don't produce hydrogen, they produce the other one, Cl^-, and other things. That's another aspect, if we want those ions these molecules are very good, but not for hydrogen. Let's find something, but we didn't find. We tried.

SRIVASTAVA: It has to be designed properly, not just putting something haphazardly and trying, one has to work it out on paper and which we know, create that kind of a situation and condition in the plasma and then we can really understand it. But then one has to try it and see if something happens because there are certain regimes of pressure and discharge conditions where things prefer to work.

BACAL: Well, I would like to ask another question which was posed to me by Prof. Capitelli, who is now working with us. Maybe we don't need another molecule, but some additive could be good. The idea is to add something, a small amount of nitrogen for example or a small amount of something else which would give a good shape to the electron energy distribution, like making a hole in the energy distribution in the range from 10 to 20 volts. That would be something. I mean, let's design an electron energy distribution in the suitable way. Maybe that is the way to have a success and perhaps adding a small amount of some impurity would help from that point of view.
PETERSON: As long as the impurity does not make negative ions itself.
BACAL: I mean, nitrogen would be good. I have added nitrogen but I didn't make it yet systematically. But I will do it. That sounds reasonable and maybe we can also simulate it. That is something we could do on the computer seeing how much nitrogen we need to get a good energy distribution.
EHLERS: I investigated about 20 years ago a completely different source geometry. I tried 10 or 12 different things including water vapor, cesium, nitrogen, methane, everything you could think of. Under no conditions did I see any improvement but I saw many times a sizeable decrease in negative ion production.
SRIVASTAVA: But there should be some reason for it, because in the situation using just H_2, it is just dissociative attachment. On the basis of the basic physics the dissociative attachment cross sections in other gases that you mentioned are much higher than what's happening.
EHLERS: Maybe it's not dissociative attachment.
PETERSON: I wanted to go back to a comment about the neutralization. The internal energy in molecules in fact can increase the rate by a factor of 3 or 4 or 5 and we have evidence of that from rates that have been measured in merged beams where they found rates that were of the order of 3×10^{-7} and when they were actually measured finally in systems, where the molecules were cold, they were around 6×10^{-8}. The ions were cold in the second case and hot in the beams.
SRIVASTAVA: And this is only a factor of 2 or 3?
PETERSON: It's about a factor of 4, I would guess, 4 or 5.
BACAL: Well, so translational energy reduces the rate and internal energy increases it.
PETERSON: Yes.
BACAL: Interesting. It means that we would be interested to have faster particles because translational energy reduces the ion neutralization rate, but not excited particles.
BARNETT: You can't say that, that translation always reduces the rate.
PETERSON: It doesn't change very much because in fact, the cross section goes as $1/E$, and the rate would not change very much.
BARNETT: It depends on the reaction.
PETERSON: The ion-ion data is pretty much the same at low energies. You're not talking about very high energies here.
BACAL: The slope changes. It's not everywhere $1/E$.
PETERSON: Eventually it does, when you get to higher energies.
BACAL: Finally, we don't know anything about $H_3^+ + H^-$. We talk

about a system which nobody has studied. If our ideas that H_3^+ is the dominant ion is true, which still has to be verified in all the good systems we have now, we do not know anything about how the rate varies with energy and this is a subject which really should be studied sometime.

HISKES: I want to thank the audience and thank the panel and this session is adjourned.

McFARLAND: At the time (fall 1980) that W. Stearns and I made the 150-1500 eV/AMU F_∞^- measurements on the alkaline earths, we were naturally impressed with the 50% fraction observed for Sr. Unfortunately, this was under conditions which were not optimal for negative D^- ion production for fusion. Scattering at equilibrium target thickness and at 250 eV/AMU was such that the D^- survival represented 10% or less of the original D^+ beam. Optimal D^- production of possibly 30% may be possible at a somewhat higher energy and lesser target thickness, but this remains to be proven.

Other factors including the difficulties of producing a satisfactory alkaline earth target and impurity possibilities has since caused OFE to discontinue for now efforts to use double charge exchange for the production of D^-.

NEGATIVE HYDROGEN ION SOURCES

REPORT ON THE BNL H⁻ ION SOURCE DEVELOPMENT*

Krsto Prelec
Brookhaven National Laboratory
Associated Universities, Inc., Upton, N.Y. 11973

ABSTRACT

This paper is a report on H⁻ ion source development at BNL over the past ten years, with most emphasis on the selected approach for the design of a steady state operating source. Sources of this kind will be used in neutral beam lines of future fusion devices, for plasma heating and toroidal current drive. In addition to the steady state operation, H⁻ ion sources for this application have to show very high gas and power efficiencies.

INTRODUCTION

The work on H⁻ ion sources began at Brookhaven National Laboratory in 1972 with a proposal[1] to replace the existing proton injection into the Alternating Gradient Synchrotron with a more efficient H⁻ injection, via double electron stripping. Very soon afterwards it was realized that negative hydrogen ions could find an application in neutral beam lines for fusion devices and the BNL program began to receive the support from the Office for Magnetic Fusion Energy. Except at the very beginning of the program, all BNL ion source models have been based on surface production of negative hydrogen ions using cesiated molybdenum as the converter. Initial tests have been made with modifications of negative hydrogen ion sources previously developed in USSR, where a lot of pioneering work in this field has been done; later on new approaches have been studied at BNL, promising substantially improved devices.

The first source studied at BNL was a hollow discharge duoplasmatron; when operating with hydrogen gas only, the H⁻ yield was initially about 5 mA[2], later to be increased to 9 mA[3]. Injection of cesium through the center tube resulted in an increase of the yield to 18 mA. Subsequently a parametric study was done and the yield in the hydrogen-cesium mode increased to 60 mA[4] (current density in the extraction aperture: 1.27 A/cm^2). Energy analysis of extracted H⁻ ions has shown[5] that the large increase in H⁻ yield as a result of cesium injection is due to surface produced H⁻ ions, mostly from the center tube. The hollow discharge duoplasmatron has been running in pulses of less than 1 ms long, which would limit its application to pulsed accelerators.

A small magnetron source, similar to the one developed at Novosibirsk, was the next to be studied. The first model was very simple, but also very fragile. Still, an H⁻ yield of 7 mA was obtained from a device that had a volume of a few cm^3 only. The next versions were more rugged, with more slits, so that eventually up to 1A of H⁻ ions could be obtained in 10 ms pulses.[6] The first studies of electrode heating followed and scaling laws were established;

---
*Work performed under the auspices of the U.S. Department of Energy.

however, it was determined that in order to match in a steady state
the source performance when operating with pulses of 10 ms duration,
removal of heat from electrodes at a rate of several kW/cm^2 would
be necessary. A program was therefore initiated to study methods
for efficient electrode cooling and up to 0.5 kW/cm^2 was removed
from an electrode by nucleated boiling of water.

A similar performance was achieved with Penning sources. The
basic design, two cathodes and the anode, was modified to include an
independently biased converter[7], placed opposite the extraction
slits. A maximum H^- current of 0.44 A was obtained from slits with
an area of 1 cm^2, in 3 ms pulses. As it was the case with the
standard magnetron, steady state operation at an extracted current
density of 0.1-0.2 A/cm^2 would require removal of several kW/cm^2
from the cathodes. A larger model was designed and fabricated[8],
incorporating nucleated boiling of water as the method for heat removal. This source was tested with the discharge only; many problems have been encountered when cesium was injected (nonuniform coverage, breakdowns) and tests have been discontinued.

A real breakthrough in the design of H^- sources was the introduction[9-12] of geometrical focusing of surface produced negative
ions and widening of the electrode gap in the magnetron
source.[12] In a pulsed mode of operation (10-25 ms), with 1 A of
H^- ions extracted[12], the gas efficiency was improved to 6% and
power efficiency to 8 kW/A. Based on these very promising results,
a larger magnetron was designed, fabricated and initially tested.
Although its design value for the H^- yield was 1 A in a steady state
operation, not more than 0.12 A was achieved due to reasons to be
described in more detail later. After initial tests, studies of
this model were also discontinued.

Even with the best results for the gas efficiency (6%), the gas
flow from a scaled-up source could be prohibitively high. Late in
1979 it was felt that any substantial improvement in source performance would require radical changes in the design. Several options
were considered and the hollow cathode discharge was chosen as the
best candidate for a good H^- source. The design was based on plasma
generation by hollow cathode discharges and H^- production on an independently biased converter with geometrical focusing. Many experiments have been performed with the objective to optimize the shape
of the cathodes, configuration of the magnetic field, and creation
and maintaining of the Cs layer on the surface of the converter. Up
to 0.5 A of H^- ions was achieved, with the source operating steady
state and only extractor pulsed. Details of the design and experimental results will be described later in the paper, but the conclusion is that a steady state H^- or D^- source with an excellent gas
efficiency can be designed, using hollow cathode discharges as a
plasma source and a porous converter with liquid cesium transpiration for production of H^- ions.

A STEADY STATE MAGNETRON SOURCE (MARK V)

Results with the improved design of a pulsed magnetron source
were very encouraging and justified the decision to proceed with
scaling-up of magnetrons to higher H^- currents and steady state

operation. Figure 1 shows the difference between the standard design (as basically developed at Novosibirsk) and the improved version[12] which incorporated the geometrical focusing and a larger anode-cathode gap. Comparison of the H^- yield from the two geometries is shown on Figure 2; it is evident that the efficiency

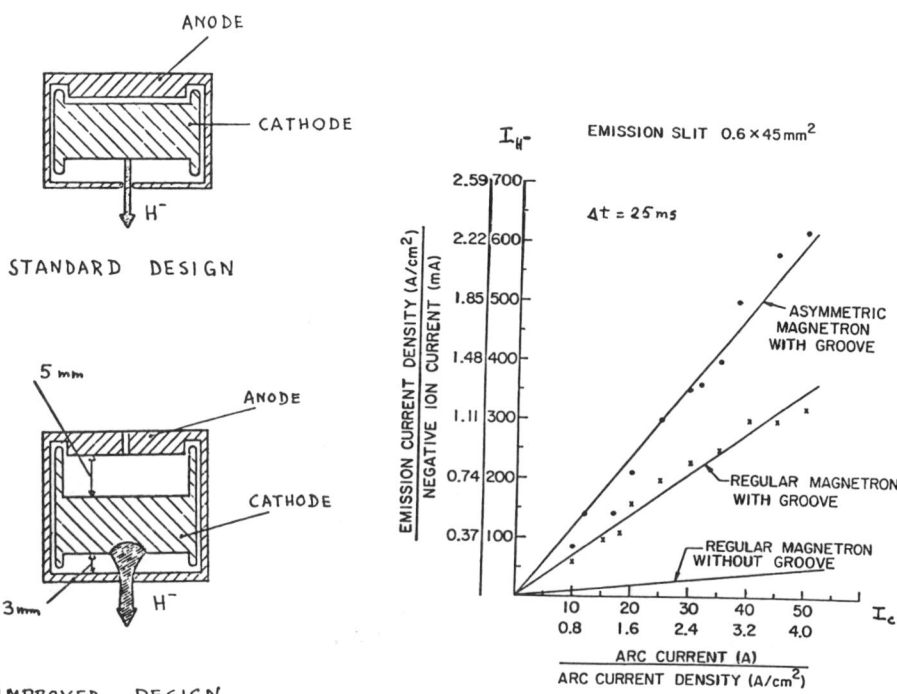

Fig. 1. Comparison between a standard magnetron and a magnetron with geometrical focusing and a wide chamber.

Fig. 2. H^- yield for different magnetron source geometries.

of H^- ion production has been greatly increased. Also, it has been found that for the same H^- current density the improved source geometry requires a much lower neutral gas pressure, which results in a higher gas efficiency. The improved power efficiency of the source has simplified the electrode cooling system: instead of nucleated boiling heat removal, standard water cooling became sufficient.

Table I shows the design parameters of the water cooled, steady state operating, Mark V magnetron source, while Figure 3 shows a cross section of the source.[13] The source was first tested with hydrogen only and runs of several days duration have been achieved at power levels of 60% to 90% of the design value. The cooling system performed well during these tests. For the H^- production tests a simple extractor was designed, with no cooling, so that the extractor voltage had to be pulsed (100 ms, 0.1 Hz). Maximum H^- yield was not higher than 0.12 A (source operating steady state, extractor

TABLE I

Design Parameters of Mark V Magnetron

Basic geometry	5 grooves, 5 slits
Cathode surface area	60 cm^2
Cathode current density	0.8 A/cm^2
Cathode power density	100 W/cm^2
Extraction slit area	2 cm^2
Extracted H$^-$ current	1 A
Extracted H$^-$ current density (in slits)	0.5 A/cm^2
Average H$^-$ current density	40 mA/cm^2
Power efficiency	8 kW/A
Gas efficiency	6%

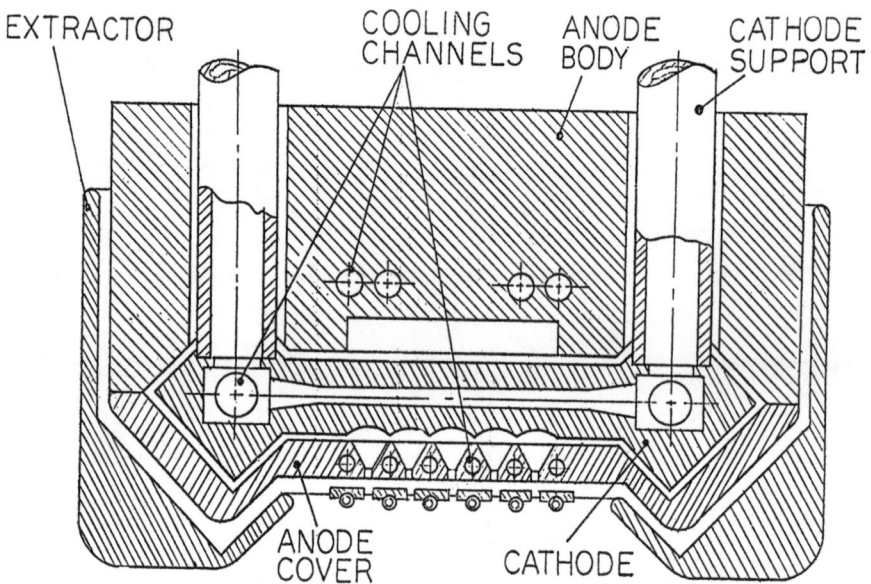

Figure 3. Cross section of a steady state magnetron source

voltage pulsed). Although the experiments with this source were discontinued, the main reason for the poor performance was found to be the nonuniformity of the cesium layer on the cathode. The channels for cesium injection were machined in the anode body of the source and cesium injected into the back part of the discharge chamber, away from the active surface of the cathode. For cesium it was very difficult to diffuse around the bends of the discharge chamber and produce the required layer on the front face of the cathode; instead, a much thicker layer would form on the anode (very efficiently cooled!) and on the back side of the cathode. This nonuniform distribution of cesium resulted in a nonuniform distribution of cathode current density, and eventually in the appearance of unstable operation and sparking. As the first step to ameliorate the situation, the cooling system was upgraded to operate at 150° C (pressurized water) in order to keep the surface temperature above this value. It was felt that in this way a higher density of cesium vapors would be achieved, condensation of cesium on anode wall reduced, and the uniformity of cesium layer on the cathode surface improved. Second, the focusing ratio (width of the groove to the width of the slit) was chosen too high, 10:1, and a large portion of H^- ions was hitting the anode wall in the vicinity of the slits. A value of 5:1 or even lower would be more appropriate, but the source operation was not studied after these modifications were incorporated.

A HOLLOW CATHODE DISCHARGE (HCD) SOURCE OF H^- IONS

Experience with all the models of H^- ion sources studied at BNL has been that they may operate very well with pulses even several tens of milliseconds long, but that problems would appear with longer pulses or in the steady state mode of operation. While a source may have yielded H^- ions with a current density of 1 A/cm^2 or more in pulses of 10 ms, not more than 0.2 A/cm^2 could be achieved with 100 ms pulses or only 0.06 A/cm^2 in the steady state. It has been also shown that the gas efficiency improves with the H^- current density in the extraction slits, indicating that it would not be easy to match the performance ($\approx 6\%$) of a small model when scaled up for higher currents and steady state operation. In 1979, studies began with the objective to design a source that would approach as much as possible an ideal device. (An ideal H^- source, based on surface-plasma method, would have in the discharge chamber a pure plasma with no neutral gas; a plasma dense enough to supply the required current density on the converter and thin enough for a safe transport of H^- ions out of the source; an independently biased converter so that the production of H^- ions can be optimized; a converter surface without cesium, if possible; if cesium is required, then its coverage should be uniform over the surface and maintained constant in time.) For plasma generation the best candidate appeared to be a hollow cathode discharge, a device that can operate with many different gases (hydrogen, helium, argon, etc.), at very high values of the gas efficiency (approaching in some cases 90%) and producing plasmas with densities $10^{13}-10^{14}$ cm^{-3}. First tests with a 3 mm tantalum hollow cathode producing a steady state plasma and a cesium covered molybdenum converter[14] have shown that it is possible to produce H^- ions by this approach, with converter current densities as high or higher than in other sources, but at background gas pressures of 10^{-3} torr or lower.

The existing Mark V magnetron structure was subsequently modified for tests with the HCD approach (Figure 4). The active part of the converter electrode has five grooves for geometrical focusing of H^- ions into the anode slits (focusing factor 5:1); the area of this part of the converter is 5 x 5 = 25 cm^2. Many different shapes of the cathode were studied and, finally, two rectangular cathodes (inside dimensions approximately 9 mm x 0.75 mm) were chosen to produce

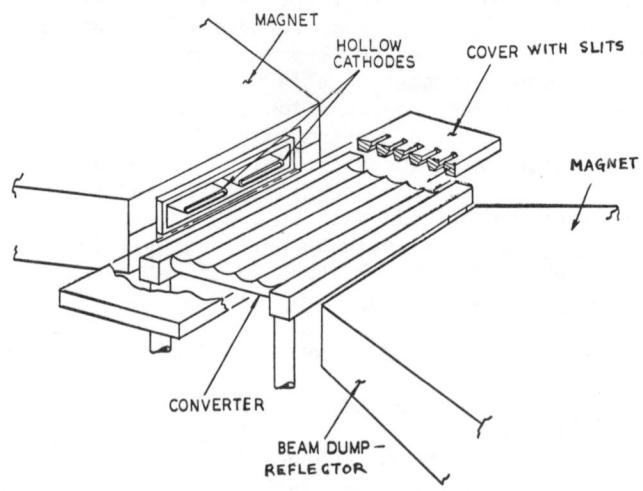

Figure 4. Cross section of the HCD source.

a flat plasma sheet in the vicinity of the converter. Magnetic field (100-200 G), required for the operation of hollow cathodes, was produced between the poles of an electromagnet. The initiation of the discharge was done by producing a Penning discharge first between the cathode and a hot tantalum filament,[15] placed in the vicinity of the opposite pole; after the discharge transferred from the Penning mode into the HCD mode (usually after less than one minute), the filament can be switched off. Plasma studies have shown that the density is uniform within 10% (Figure 5), high enough to reach a primary, positive ion current density on the converter of 0.5 A/cm^2, with a neutral background gas pressure of $(1-2) \times 10^{-3}$ torr. The required gas flow depends on the shape of the magnetic field and, as a general rule, with air-core systems the background pressure can be reduced to below 5×10^{-4} torr. Steady state operation was achieved (3-4 days continuous running), limited by factors not related to the cathode itself (Figure 6).

The converter was fabricated from solid molybdenum, with channels for water cooling. This was the simplest design, but it also limited the options for cesium injection. The highest H$^-$ yield was obtained when cesium was injected into the discharge from one or two heated perforated tubes, placed along the surface of the converter. Cesium injection through the hollow cathode itself, as an admixture to hydrogen gas, was also tried, resulting first in a lower arc voltage, but the H$^-$ yield was not as high as with the former method. The extractor was not cooled (solid tungsten wires) and its voltage

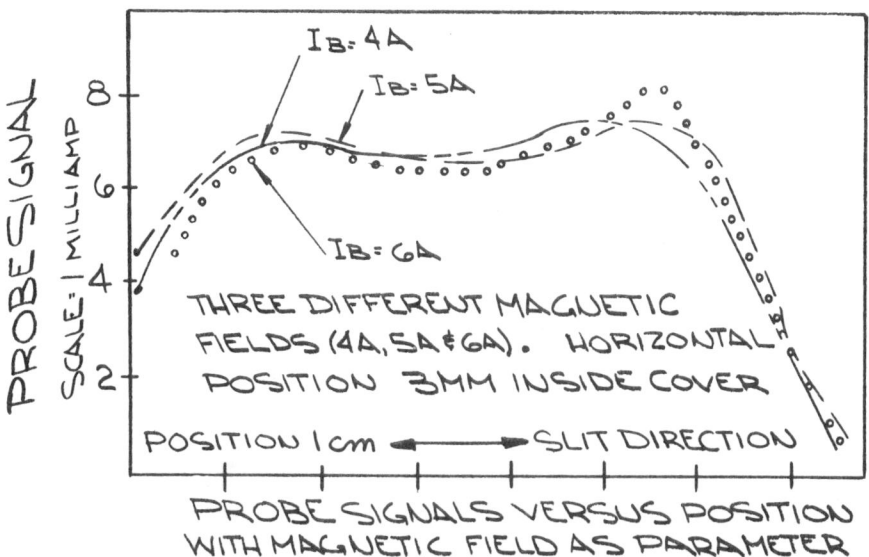

Figure 5. Plasma density distribution in the HCD source.

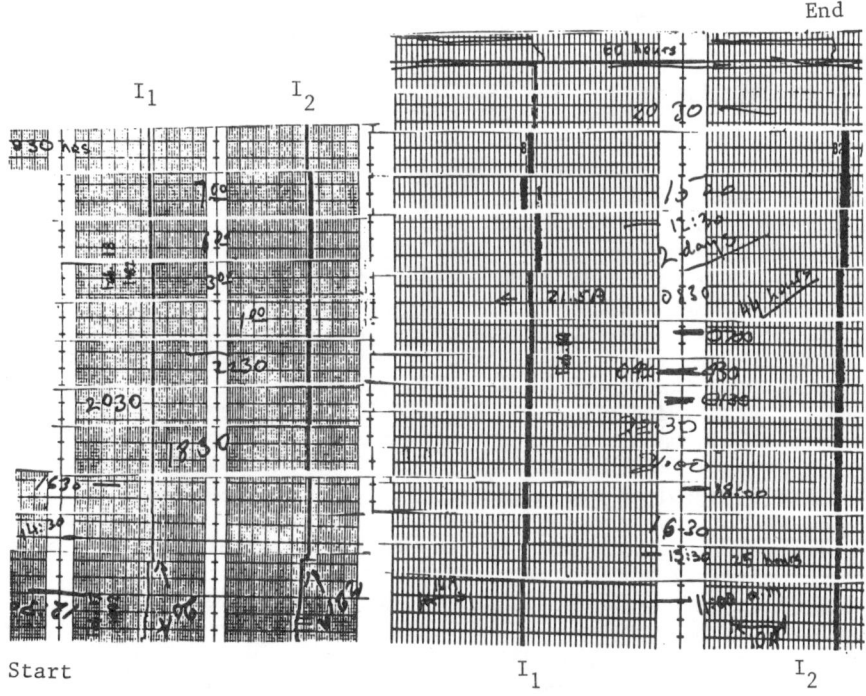

Figure 6. Chart recording of the HCD current.

had to be pulsed in order to avoid the overheating (pulse duration up to 1 s). With the plasma generator and converter operating steady state (converter bias: -100 to -150 V) and 7.5 kV extraction voltage pulsed, stable H^- currents of 0.2-0.3 A have been maintained over periods of several hours (Figure 7) with shorter peaks up to

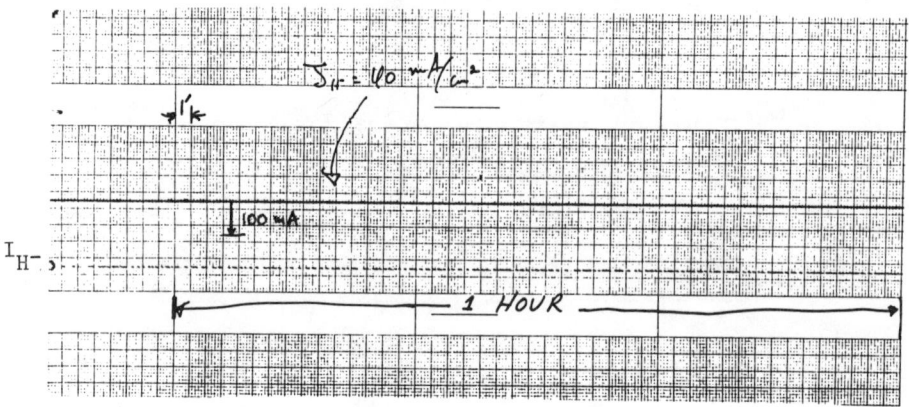

Figure 7. Chart recording of the extracted H^- current from the HCD source.

0.5 A. This corresponds to the extracted current density in the slits of 0.04 to 0.1 A/cm^2. As it was the case with Mark V magnetron source, the cesium coverage of the converter surface was not uniform and close to the required optimum.

There are two ways to try to improve the performance of the converter, either by developing a low work function surface that does not require cesium or by diffusing cesium through pores in the converter. The first approach was tried with some success[14] by using a cesium doped LaB_6 converter; no further tests have been done since then. In order to test the second method, several small-scale experiments have been done[16] at BNL. In the first, H^- yields from a solid and from a porous molybdenum surface were compared; in the latter case the yield was reduced by less than 20%. The second test was even more important; it is shown schematically on Figure 8. In front of a molybdenum piece, which could be either solid or porous, a sheet of plasma was produced using a hollow cathode. Negative hydrogen ions, formed on the surface of the negatively biased converter, were bent 90° in the magnetic field to reach a Faraday cup. No signal was observed on the cup in pure hydrogen or argon plasma nor in the cesium-hydrogen mode unless the converter bias (i.e., the energy of H^- ions) and the magnetic field were adjusted for the 90° bend. As a reference, solid molybdenum converter was used with the standard way of cesium supply through the discharge. The best value of the H^- current, collected on the cup, was 5 mA with the solid converter. With the porous converter, but no discharge and no pressure on the liquid cesium in the chamber, there was no cesium on the surface as determined by the absence of the photoelectric current

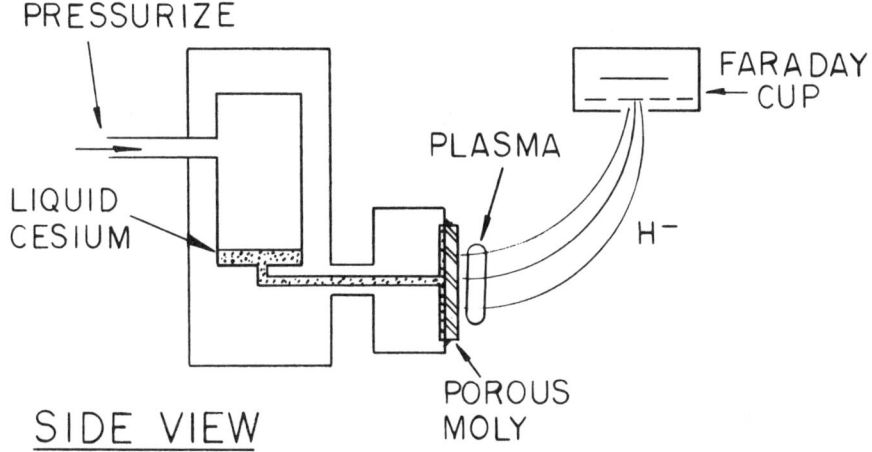

Figure 8. Experiment to compare the H⁻ yield for different methods of cesium supply.

under He-Ne laser illumination. Once the pressure on cesium was increased and plasma in front of the converter established, the H⁻ signal on the cup increased to 25 mA. This unexpected result can be explained by the fact that the cesium layer created by transpiration on the surface of the porous converter suffers much less sputtering of Cs^+ ions and, therefore, can be maintained closer to optimum conditions, than the layer that has to be maintained by deposition of cesium particles (Cs^0, Cs^+) from the discharge.

CONCLUSION

Experiments with hollow cathode discharges have shown that they can be used as steady state plasma generators with parameters as required for applications in H⁻ ion sources (plasma density, uniformity). For the converter, transpiration of liquid cesium through porous molybdenum seems to be the best choice at present, eliminating the need for cesium in the discharge. These features and very low values of the background gas pressure (10^{-4} to 10^{-3} torr) promise that an H⁻ source could eventually be developed with a performance equal to that of the best positive ion sources. Such a source could be scaled up to currents required for neutral beam lines as well as adapted easily to different types of accelerators.

REFERENCES

1. A. Maschke, K. Prelec, and Th. Sluyters, BNL Informal Report Nov. 1972 (unpublished).
2. Th. Sluyters and K. Prelec, Nucl. Instr. Meth. 113, 299 (1973).
3. K. Prelec and Th. Sluyters, in Proc. II Symp. on Ion Sources and Formation of Ion Beams, October 1974 (Lawrence Berkeley Laboratory Rep. LBL-3399), p. viii-6.

4. M. Kobayashi, K. Prelec, and Th. Sluyters, Rev. Sci. Instr. $\underline{47}$, 1425 (1976).
5. K. Wiesemann, K. Prelec, and Th. Sluyters, Jour. Appl. Physics $\underline{48}$, 2668 (1977).
6. K. Prelec, in Proc. Symp. on the Production and Neutralization of Negative Hydrogen Ions and Beams, September 1977 (Brookhaven National Laboratory Rep. BNL 50727), p. 111.
7. K. Prelec, Nucl. Instr. Meth. $\underline{144}$, 413 (1977).
8. K. Prelec, in The Physics of Ionized Gases, Inv. Lectures and Progress Reports of SPIG-78 (Institute of Physics, Beograd, Yugoslavia, 1978), p. 697.
9. T. Green, private communication.
10. F. Sternberg, Brookhaven National Laboratory Int. Rep. AGS 77-4 (1977).
11. Yu. I. Bel'chenko and V. G. Dudnikov, Preprint IYaF 78-95 (Novosibirsk, 1978).
12. J. G. Alessi and Th. Sluyters, Rev. Sci. Instr. $\underline{51}$, 1630 (1980).
13. K. Prelec, in Proc. II. Symp. on the Production and Neutralization of Negative Hydrogen Ions and Beams, October 1980 (Brookhaven National Laboratory Rep. BNL 51304), p. 145.
14. A. Hershcovitch and K. Prelec, Rev. Sci. Instr. $\underline{52}$, 1459 (1981).
15. V. J. Kovarik, A. I. Hershcovitch, and K. Prelec, Rev. Sci. Instr. $\underline{53}$, 819 (1982).
16. J. G. Alessi, A. I. Hershcovitch, and Th. Sluyters, Rev. Sci. Instr., to be published.

DISCUSSION

COOPER: When you ran your porous molybdenum converter with cesium transpiration and got 25 mA versus 5 mA with the solid converter does that mean that the current density of negative ions is 5 times higher than you can achieve with a solid converter?

PRELEC: Yes, I think so. Jim Alessi can describe the details of this test.

ALESSI: The negative ion current density was that much higher, but basically that was because the solid surface, when you try to deposit cesium vapor on it is just working so poorly compared to the porous surface. I think the porous surface was working similar to what we used to see in pulsed magnetrons. When we tried to deposit cesium vapor on the solid converter in the steady state, we just were getting a much poorer coverage than in the pulsed operation.

PRELEC: I would say that the increase in total H^- current is due to a higher current density and to a more uniform distribution of H^- ions.

BARNETT: Do you have any evidence that this is a surface effect and not a volume effect where you might be getting hydrogen interacting with cesium vapor rather than the surface?

PRELEC: This is a surface effect, because it you want to register the current on the Faraday cup, the voltage on the converter and the magnetic field have to be adjusted for the proper radius of curvature. Had those ions been produced in the volume, they would not have the energy they gain when they are produced on the surface.

MILEY: On the air core coils, does the better operation have something to do with the uniformity of the field or do you have any other guess as to why they are so much better than the other?

PRELEC: It certainly has to do with the shape of the field because with coils we have much more freedom than by using iron pole pieces. We were using iron pole pieces because that was the fastest way to start hollow cathode discharge experiments: we just removed the magnetron source and used its magnetic circuit with iron poles.

HENKES: What material are the hollow cathodes?

PRELEC: Almost exclusively tantalum tubes, either round or squashed. There was just one test made with rhenium.

THE STATUS OF ≳1 AMPERE H⁻ ION SOURCE DEVELOPMENT AT THE LAWRENCE BERKELEY LABORATORY

A.F. Lietzke, K.W. Ehlers and K.N. Leung
Lawrence Berkeley Laboratory
University of California
Berkeley, CA 94720

INTRODUCTION

This paper summarizes the effort to improve the operation of the ~1 A surface-production H⁻ ion source developed by K.W. Ehlers and K.N. Leung. The plasma chamber consists of a large magnetic bucket of oval cross section (Fig. 1). A concave cylindrical "converter" surface is suspended in the plasma chamber to direct any surface-produced negative ions through the exit aperture. The ion source has been mated to a tetrode accelerator for the "proof-of-principle" tests. Most of the problems discovered in the tests were associated with difficulties in controlling the production process. This paper describes the plasma chamber in greater detail and illustrates the quality of the present ion production. The acceleration difficulties have been deferred until a better test-stand is completed.

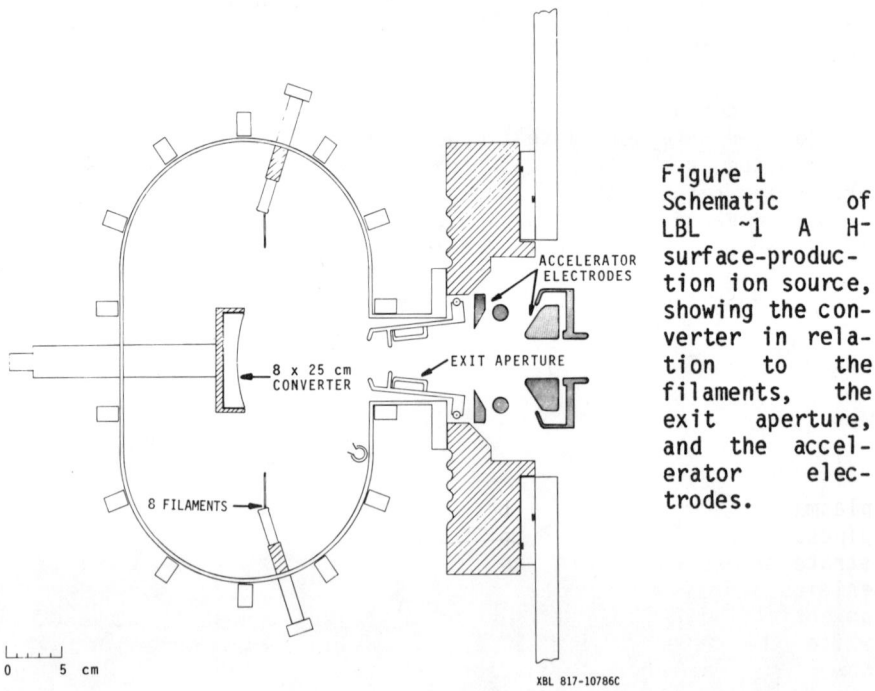

Figure 1 Schematic of LBL ~1 A H⁻ surface-production ion source, showing the converter in relation to the filaments, the exit aperture, and the accelerator electrodes.

0094-243X/84/1110344-09 $3.00 Copyright 1984 American Institute of Physic

PLASMA CHAMBER DESCRIPTION

The plasma chamber cathode consists of eight filaments (1.5 mm diameter tungsten) having a total emission area of ~55 cm^2. They are distributed uniformly in two magnetic field nulls (Fig. 2) which run parallel to the cylindrical axis of the chamber's anode wall. The anode is shielded by line cusps (total length = 950 cm) in the form of six hoops and two isolated lines of samarium-cobalt magnets. A weak filter (~50 G-cm) separates the filament null from the converter nulls, which are separated by ~300 G-cm from the aperture null and ~1200 G-cm from the wall nulls.

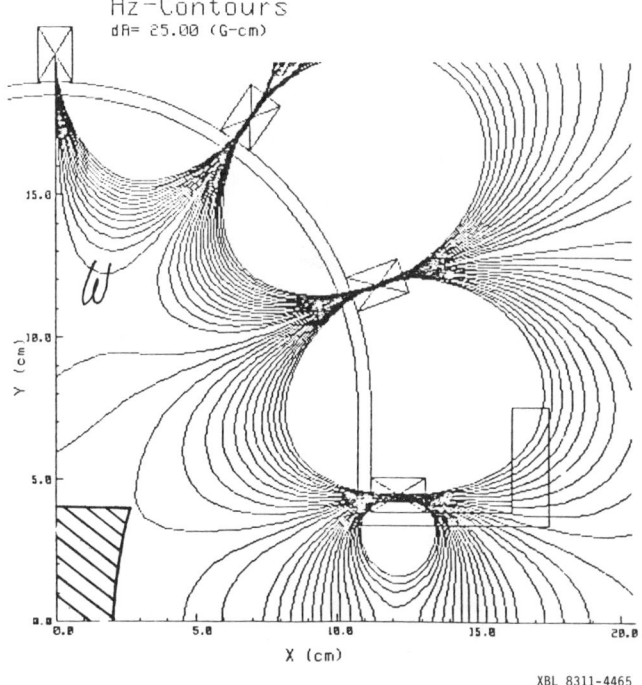

Figure 2 Calculated magnetic field structure in one quadrant of the plasma chamber showing the filament filter and exit filter in relation to the filaments and the converter.

XBL 8311-4465

The negative ions created on the converter (-130 V relative to the anode) are accelerated in the plasma sheath and penetrate the exit filter with a small displacement (~2 cm) parallel to the cylindrical axis. The converter is roughly in the center of the plasma chamber (Fig. 3), being suspended by its water cooling pipes. The exit surface (molybdenum) is brazed onto a copper substrate which runs cold but is shielded by two hot molybdenum shields (Figs. 4 and 5). The inner shield is at the converter potential (with the intention of being hot enough to stay clean) while the outer shield floats (thereby reducing the converter power supply requirement) and protects the intermediate boron nitride insulator from plasma bombardment and tungsten deposition.

CBB830-8753

Figure 3 End photo showing the converter and its shields in relation to the filaments and the large epoxy accelerator insulator.

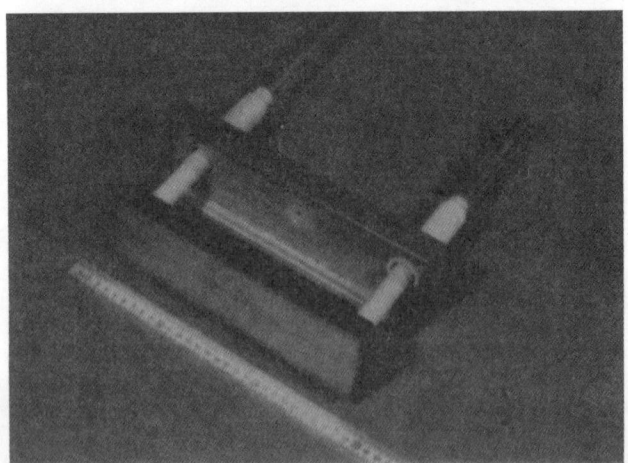

CBB 830-8741

Figure 4 Photo of the converter and its insulated box (separated for detailed observation).

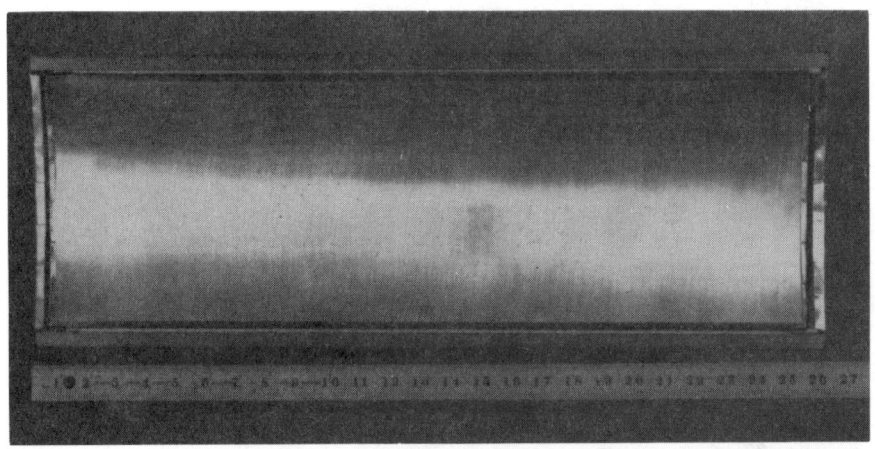

CBB 830-8743

Figure 5   Exit view of the 8 x 25 cm converter showing the outer shields in operating position.

Both shields are creased and constrained in a manner to minimize warpage while hot.

Cesium has been injected at three locations. At this writing, it is being injected behind the converter, from one end-plate where the injector (Fig. 6) mounts onto a standard flange. The injector has a high temperature needle-valve which separates the cesium reservoir from the main vacuum. Both the valve and the reservoir operate inside a temperature-controlled oven (250 - 270 °C). The cesium vapor is conducted to the plasma chamber by an ohmically heated tube.

An inter-cusp cage is being used to test theories of cesium transport. It enhances the H⁻ output only when it is negative (~50 V, 4 A) and the walls are somewhat loaded with cesium. The H⁻ uniformity is diagnosed by a movable, screened, faraday cup operated with the screen 12 V below anode and the cup 24 V above anode (sufficient for the cup current to be composed of only fast H⁻ and secondary electrons from the screen.

OPERATION

At the standard "no cesium" operating point, there is 9 kW of arc power (90 V, 100 A), 1.2 kW of converter power (150 V, 8A) and 6 kW of filament power (8 filaments x 7.5 V, 100 A). This produces an output current in the neighborhood of 20-50 mA. Optimum injection of cesium (without other changes) nearly doubles the converter current (14 A), raises the arc current 20% and delivers over 1 A through the exit throat. Fig. 7 is raw data illustrating typical operation after the plasma system is well-conditioned. This data was collected after five hours of continuous operation at the 1 A output level, so when the arc is turned ON, the floating converter produces some H⁻ energetically

Figure 6 The cesium injector consists of a high temperature regulating valve which separates the reservoir from the ohmically-heated tube leading to the main plasma chamber.

CBB 830-8747

capable of penetrating the exit filter. Biasing the converter to its previous optimum produces a rapidly rising, but short-lived, surge of H⁻ (as the accumulated cesium is depleted). Biasing the cage to its previous optimum yields another 100 mA increase. The big increase in H⁻ occurs ~1 sec after the cesium valve is opened. Steady state is attained in ~1 minute and maintained for 10 minutes by slowly closing the cesium supply valve. Shutdown in this history was initiated by first floating the cage (to check its effect under steady Cs input conditions). Closing the cesium valve immediately starts the decay of the H⁻ output (the slope is utilized to estimate the cesium "pumping" time). "Refloating" the converter produces another (but smaller) surge in H⁻ (as the converter passes through its optimum coverage again). The following negative converter current results from electron collection caused by a "load" resistor to anode. It disappears when the arc is extinguished.

Fig. 8 shows some profiles taken under optimum cesium conditions. These are not "typical" profiles (which suffer from non-uniform cesiation) but they illustrate the vignetting and attenuation the H⁻ beam suffers before arrival at the accelerator aperture ($z \simeq 18$ cm). The non-uniform cesiation is inversely related to the positive ion variation (Sputtering and cesium ionization are believed to be the mechanisms.) The focal plane non-uniformity in the focussed direction depends only upon the transverse

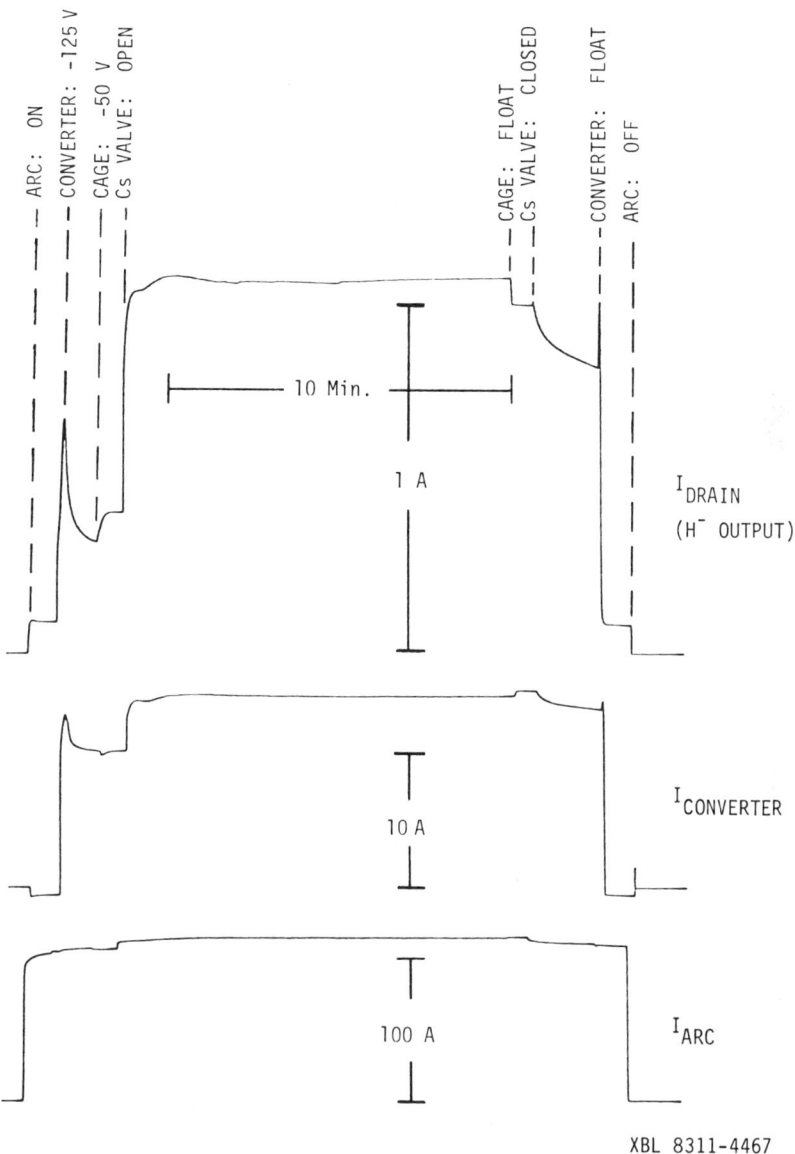

Figure 7 Standard operating sequence: 18 minutes from arc on to arc off, including converter on, cage on, cesium valve open. ~10 minutes where the cesium valve was slowly closed to maintain a steady output, cage off, cesium valve closed, converter off, arc off.

H^- temperature (a good fit is obtained by assuming a 6 eV Gaussian velocity distribution) and the vignetting of the exit aperture.

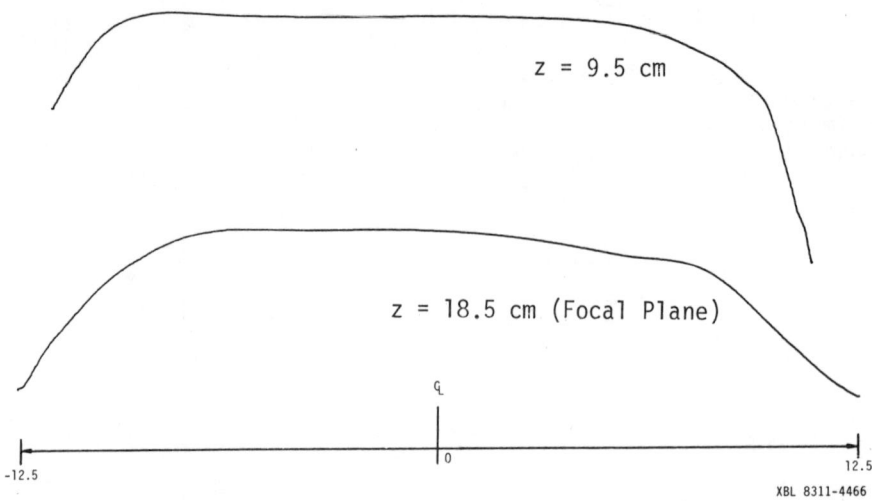

Figure 8 The H⁻ intensity, parallel to the converter axis, suffers from vignetting and attenuation (in spite of geometric focusing). Profiles were taken at $I_{Drain} \gtrsim 1$ ampere within four minutes of each other under very steady conditions. These are not typical profiles (which also suffer from non-uniform cesiation).

ENGINEERING EFFORT

Hardware modifications consumed much of the past year, but have recently permitted controlled and reproducible behavior (of the type shown in Fig. 7, 8). Modifications were indicated by the types of problems identified in the "proof-of-principle" tests:

1. Lack of H⁻ steadiness (caused by strong coupling between a poorly controlled cesium feed system and the filament emission).
2. Lack of H⁻ steadiness (caused by frequent cathode spotting from the converter and filaments, which produced large cesium transients on the converter).
3. Heavy cesium consumption (necessitating frequent clean-up and frequent spotting).

PHYSICS EFFORT

Some phenomena are not yet "explained"; some are not even reproducible enough to study, but some, like the profiles, seem well understood. Others, like the high cesium consumption rate, the anomalously high cesium reservoir temperatures, the causes of the H⁻ attenuation, the mechanism(s) for electron leakage, variations in the optimum converter voltage, and the non-linear input/output relations, remain subjects of speculation and investigation.

CONCLUSIONS

The system, as it stands, works very well up to 1.3 A. Indeed, the hardware is recently capable of controlled experimentation. Investigations are believed to be necessary to examine engineering trade-offs, and push-out (or remove) fundamental limitations. Some hardware improvements already await testing and upcoming beam tests may reveal new problems.

ACKNOWLEDGEMENTS

Bill Yant and Carolyn Wong deserve special attention for working so well under adverse conditions: Bill for his help with the modifications and data collection, and Carolyn, for her help with this manuscript.

This work was supported by the Director, Office of Energy Research, Office of Fusion Energy, Development and Technology Division of the U.S. Department of Energy under Contract No. DE-AC03-76SF00098.

DISCUSSION

HISKES: Do you have any idea how much of the converter current is negative ions?
LIETZKE: I don't think I have a good idea. When cesium is injected we get about a factor of 1.3 to 1.4 increase in the converter current. How much of this is electrons or cesium ions is not clear.
HENKES: What was the residence time of cesium if you don't switch off the discharge? Do you have any estimate of the half time, until your H^- current really is decreasing due to the loss of cesium?
LIETZKE: If I don't switch off the discharge and maintain the converter voltage? After I switch off the cesium supply, I have measured values from half a minute up to 12 minutes but anything over a minute is a recent result that followed the installation of hot molybdenum shields on the bucket wall. I never saw anything longer than a minute before I added those shields. At the present time roughly one quarter of the anode area is covered by hot molybdenum. The cesium forms a dark brown coating on the cold walls.
DAGENHART: What extraction voltage were you using? You had a very nice flat drain current there. What voltage was that at?
LIETZKE: That was just self extraction at 135 V. That is just the negative current that is able to somehow find its way to the outside world. Over 95% is believed to be energetic negative ions.
DAGENHART: Ken Ehlers showed a decaying H^- curve yesterday. Is it possible that that cesium problem comes from backstreaming positive ions?
LIETZKE: The decaying curve that Ken Ehlers showed is the change in the cesium level on the converter and it is a result of not operating at the optimum converter voltage. The converter can be turned off, loaded up with cesium, and for a short time with a higher voltage on the converter, the negative ion yield would be higher. However, the higher voltage on the converter will slowly erode cesium on the converter to a level lower than optimum, so you end up with a yield lower than the optimum. What I have shown you here is what happens at a steady state optimum.
STATEN: How much more current can you obtain initially when you load the converter? It would seem that that might indicate the kind of currents you might be able to obtain if you were better able to control cesium on the surface, either by using the porous surface or some other method.
LIETZKE: I think I've seen under conditions where the steady state is in the neighborhood 1 A excursions up to 1.4 A but I don't normally look for that so I haven't tried to optimize that.

SHORT-PULSE OPERATION WITH THE SITEX NEGATIVE ION SOURCE[*]

W. K. Dagenhart, W. L. Stirling, G. M. Banic,
G. C. Barber, N. S. Ponte, and J. H. Whealton
Oak Ridge National Laboratory, Oak Ridge, TN 37831

ABSTRACT

The successful high current, long-pulse SITEX source experiments with H$^-$ beams have been extended to D$^-$ operation and to short-pulse H$^-$ and D$^-$ beam acceleration. Extracted D$^-$ beam current densities of 100 mA/cm^2 at the plasma grid have been achieved for 3 s at a total beam of 260 mA and 10 keV. The extracted electron-to-D$^-$ ratio is <5% with ≈100% of the extracted electrons collected at 2 keV. Short-pulse beams down to 10 ms in length have been accelerated with the minimum pulse length determined by the power supply time constants. Optics of the accelerated beam is $\theta(1/e)_\perp = \pm 0.48°$, which corresponds to a normalized emittance perpendicular to the slot of 0.0007π cm-mrad. These data place an upper limit on the surface conversion D$^-$ ion temperature of 0.7 eV.

INTRODUCTION

The SITEX negative ion source has been used successfully for the generation of long-pulse H$^-$ beams.[1-6] H$^-$ beams of 650 mA, 18 keV, and 10 s have been produced at extracted current densities of j_{H^-} = 130 mA/cm^2. The extracted electron-to-H$^-$ current ratios for H$^-$ are I_e/I_{H^-} = 0.15. These electrons are recovered on the source structure at 10% of the first accelerator gap potential difference. Plasma generator pressures are 2–4 mtorr for reliable operation, giving a nucleon gas efficiency of 3%. The arc efficiency is now quite good at 6 kW of total ion source power per ampere of accelerated H$^-$ beam (not including accelerator supply power). Plasma generator reliability is 100%. A degassing phase of a few hours is necessary before the heavy negative ion impurities fall from ∼10% to ∼1%. The principal remaining problem is one of getting good cesium control to get flat beam pulses, to get good pulse current control, and to minimize cesium consumption.

Papers presented at the Third Neutral Beam Heating Workshop in Gatlinburg, Tennessee,[5,6] and at the Fifth Topical Meeting on the Technology of Fusion Energy at Knoxville, Tennessee,[7] show how the

0094-243X/84/1110353-10 $3.00 Copyright 1984 American Institute of Physics

---

[*] Research sponsored by the Office of Fusion Energy, U.S. Department of Energy, under Contract No. W-7405-eng-26 with the Union Carbide Corporation.

SITEX ion source can be the basis of a reliable neutral beam heating module. The Knoxville paper[7] presented a 1-MW beam line conceptual design for a 200-keV beam. Ion optics calculations with the Oak Ridge National Laboratory (ORNL) code indicate a total root mean square beam divergence of 0.4° for the tetrode accelerator employed.

The present investigations were initiated to determine the performance with D^- operation and to determine if the accelerated beam could be chopped or put out in bursts of short pulses. D^- operation was very successful, and the beam can be modulated as reported in the following sections. The emittance measurements have resulted in a normalized emittance perpendicular to the slot of 0.0007π cm-mrad.[8] A paper in these proceedings by Stirling et al.[9] gives the details of emittance measurements. A paper in these proceedings by Whealton et al.[10] gives the details on the accelerator design using the ORNL optics code.

ION SOURCE OPERATION

Figure 1 shows the basic operational principles of the SITEX source that have previously been described. The hot-filament-initiated reflex arc discharge strikes with 100% reliability and requires only 2-4 mtorr of H_2 or D_2 gas pressure to operate. Gas is admitted to the cathode cavity to initiate the arc discharge, and more gas is admitted to the anode cavity to improve the uniformity of the plasma density. Cesium is admitted to the anode cavity from a stainless steel oven whose flow is valve and temperature controlled. Cesium provides the partial monolayer coverage of the molybdenum converter to lower the surface work function to the minimum of about 1.5 eV. In SITEX, source cesium also provides the bulk of the converter current in the form of Cs^+ bombardment. With pure H_2 or D_2 arc operation, the converter current is limited to about 3 A from H^+/D^+ ions. When cesium is admitted to the discharge, this converter current rises to the 8- to 15-A level. Calorimetry of converter

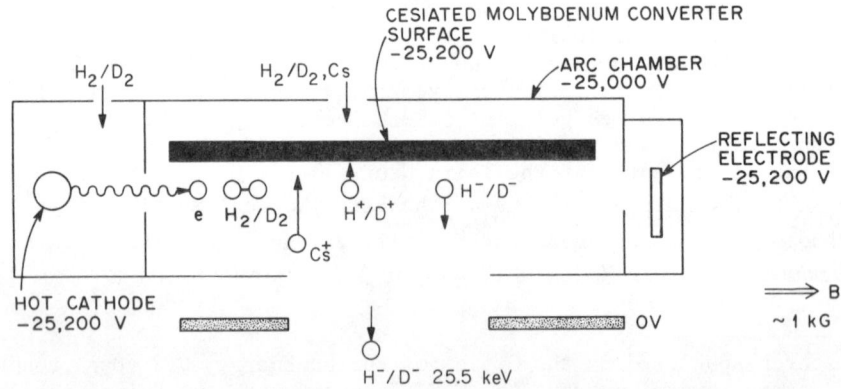

Fig. 1. SITEX negative ion source plasma generator principles.

water cooling during earlier experiments verified that most of the converter current is from positive ion bombardment, rather than from secondary or emitted electrons. Output current increases with cesium density. To obtain optimum cesium coverage of the converter, the converter temperature must also increase. At the higher cesium densities, the converter operating temperatures must be increased up to 350–450° C. Thus, the majority of our H^-/D^- output is from the Cs^+ sputter mechanism. No significant difference in source operation has been noticed when switching from H_2 to D_2 source gas. Figure 2 shows the power supply hookup for the source.

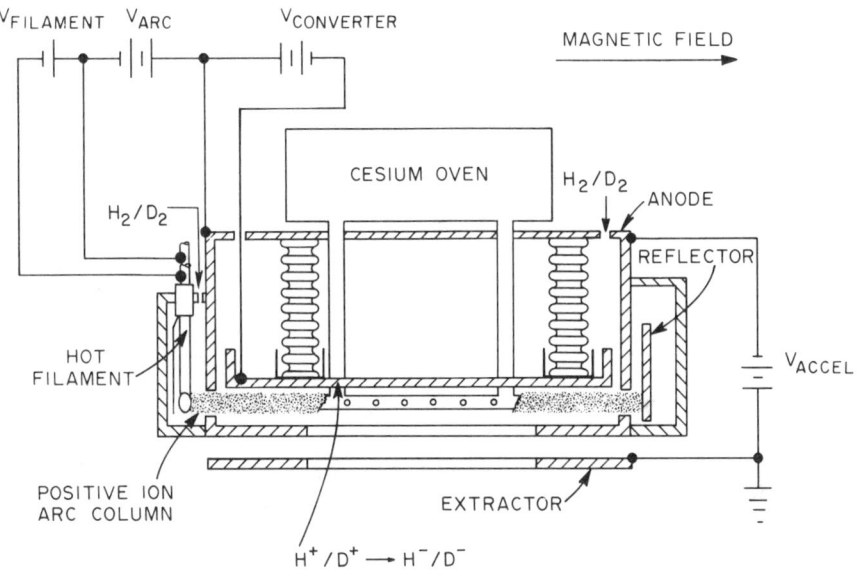

Fig. 2. SITEX negative ion source electrical supply hookup.

The converter emitting surface is cylindrical and has been positioned according to the results of the ORNL optics code. The plasma electrode and extraction electrode geometry also resulted from this calculation. The calculated angular divergence perpendicular to the slit was $\theta_{rms\perp} = 0.28°$ (total angle). Figure 3 shows the segmented Faraday cup used to measure the beam current and the angular divergence of the beam.

Figure 4 shows a perspective of the single-slit accelerator and the momentum dispersion properties of the source magnetic field. The $\vec{E} \times \vec{H}$ field in the accelerator gap separates the 15%/5% extracted electron fraction for H^-/D^- operation and carries the electrons on cycloidal trajectories up into the electron recovery electrode region. Once inside the electron recovery electrode region, an electric field component is created parallel to $\vec{B}$, which forces the electrons to be collected onto the electron recovery electrode. The electron

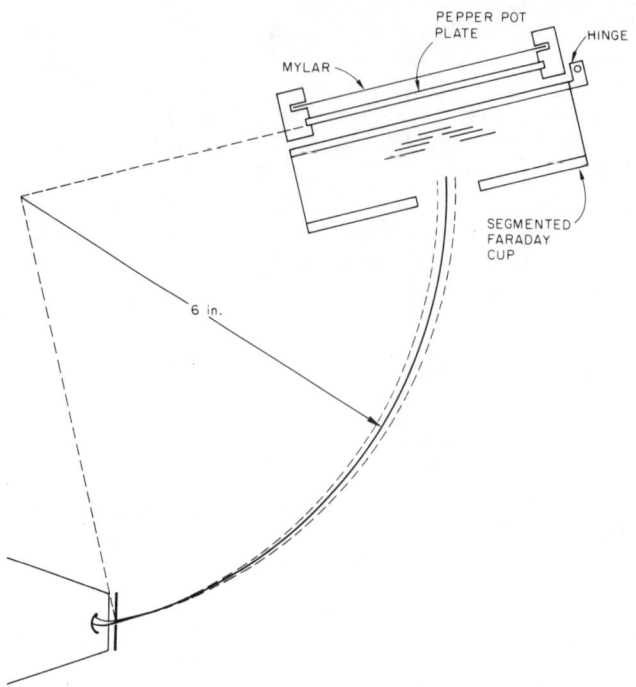

Fig. 3. Faraday cup geometry for measuring the beam emittance.

recovery energy must be about 10% of the first gap potential energy difference. Heavy ions accelerated through the extraction gap are separated according to their momentum. The ~1% of heavy negative ion impurities is separated from the H^-/D^- beam, resulting in a 100% pure H^-/D^- beam.

A long-pulse H^- beam current trace is shown in Fig. 5 for extraction slits of 5-cm^2 total area. Also shown in this photo are the other power supply signals. The loss of beam current from 650 mA down to ~200 mA after 23 cm of path length can be explained by charge exchange using the measured tank pressure and published cross sections. The Faraday cup current at 90° has been verified calorimetrically. A higher pumping speed system would greatly lower this loss. The data for this beam are contained in Table I. The details of these experiments have been reported elsewhere.[1-6] Details of the present D^- experiments are contained in the second column of Table I, where the extraction is now from a single slit of only 2.5 cm^2. The purpose of the present experiment was to determine the best optics we could achieve in the direction perpendicular to the slot by calculating the optimum geometry with the ORNL optics code. Since the geometrical optics abberrations had been minimized, this would also set an upper limit on the effective ion temperature of surface-conversion prepared H^-/D^- ions. Previous experiments have

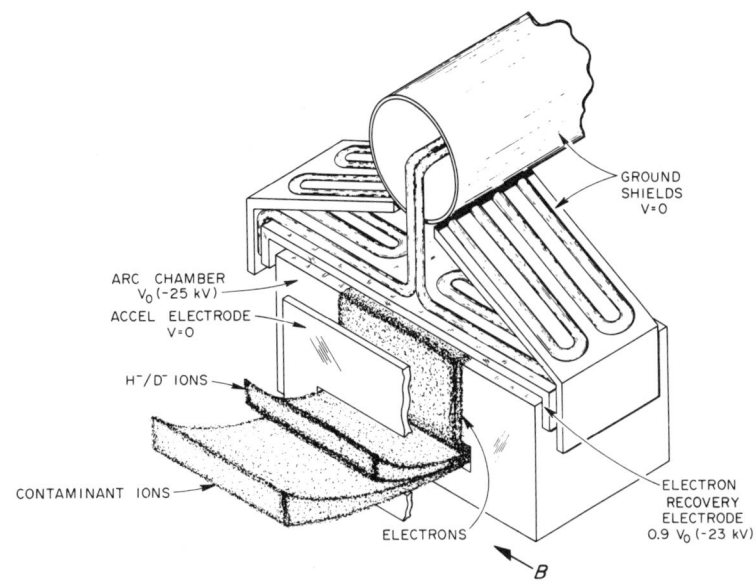

Fig. 4. Perspective of the SITEX negative ion source extraction geometry showing the momentum dispersion of the source magnetic field.

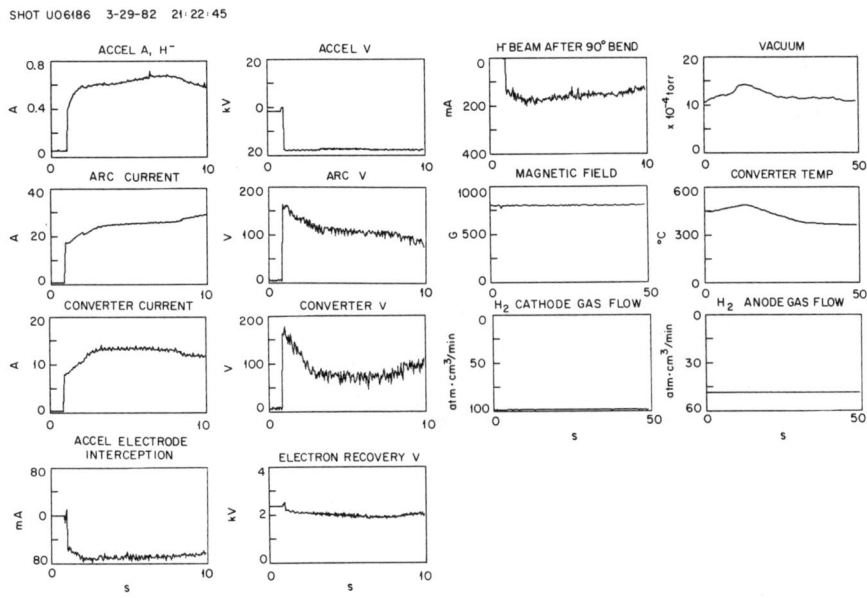

Fig. 5. Long-pulse H^- beam from the SITEX negative ion source of 650 mA, 18 keV, and 10 s.

Table I H⁻/D⁻ performance status for the SITEX source

Parameter	Achieved	
	H⁻	D⁻
$I_{accel}{}^a$ (mA)	650	260
$j_{extracted}$ (mA/cm²)	130	100
Extraction area (cm²)	5	2.5
V_{accel} (kV)	18	10
Pulse length (s)	10	5
Source discharge pressure (mtorr)	4	4
Extracted $I_e/I_{H^-,D^-}$ (%)	15	5
Electron recovery at 10% V_{accel}	yes	yes
Arc efficiency (kW/A of beam)	6	8
Beam divergence (degrees)	2[b]	(1/e) = ±0.48°

[a] Contains no electrons.
[b] Unoptimized accelerator.

shown that source beam current is directly proportional to source area.

Figure 6 shows a single 10-ms beam pulse. The effect of the power supply cable capacitance can be seen in the slow rise and fall

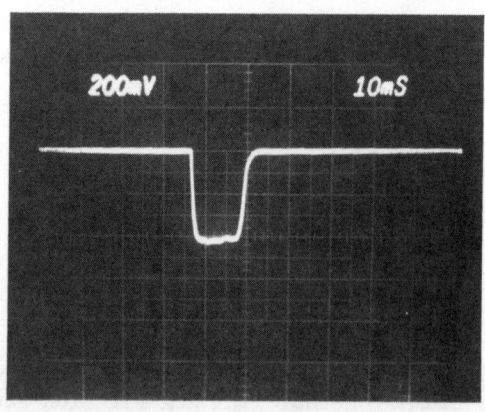

FARADAY CUP
PROBE 1 (20 mA/div)
(RESISTOR)

P43 — 9/22/83 DATA

Fig. 6. Short 10-ms D⁻ pulse waveform.

time of the I_{D^-} beam pulse. This power supply rise and fall time has also been measured with a resistive dummy load. The slight ripple in the peak amplitude is due to ripple on the converter power supply. Figure 7 shows a short-pulse train of 10-ms I_{D^-} pulses to the Faraday cup. The small drift in amplitude is again due to changes in the cesium density in the plasma generator.

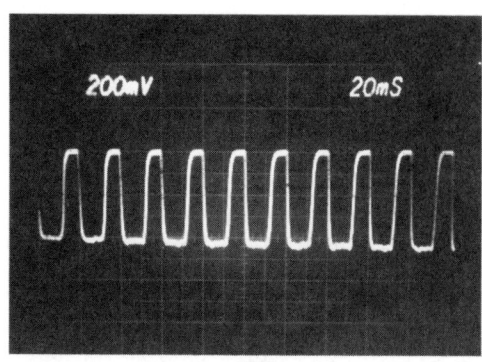

FP1 (20 mA/div)
(RESISTOR)

Fig. 7. Burst of 10-ms D^- pulse achieved with a dc arc operation by modulating V_{accel}.

SUMMARY

The successful H^- long-pulse beam experiments have been extended to D^- beams for both short- and long-pulse operation. Extraction of D^- beams was at about the same current density as for H^- extraction, with the 3-s beam time being dictated by beam target considerations. Reasonably flat beam current out to 10 s should be possible (20% amplitude drift) and has been done for H^- beams. The 20% amplitude drift during the 10-s beam pulse has been attributed to cesium distribution problems in the plasma generator, and new developments are expected to remove this limitation. The plasma generator can be run steady state.

Limitations on the shortest beams of 10 ms produced in these experiments were placed by the time constants of the power supply system. Much shorter pulses should be available from the ion source.

The emphasis of these experiments was to determine the effective ion temperature of the surface-produced H^-/D^- ions. This limit has been reduced to 0.7 eV. Further improvements in the geometrical design and alignment of the source parts may reduce this limit further.

REFERENCES

1. W. K. Dagenhart, W. L. Stirling, H. H. Haselton, G. G. Kelley, J. Kim, C. C. Tsai, and J. H. Whealton, in Proceedings of the Second International Symposium on the Production and Neutralization of Negative Ions and Beams, BNL-51304, Brookhaven National Laboratory, Upton, New York (1980).
2. W. K. Dagenhart, C. D. Croessmann, and W. L. Stirling, Bull. Am. Phys. Soc. 26, 987 (1981).
3. W. K. Dagenhart, W. L. Stirling, and J. Kim, Negative Ion Beam Generation with the ORNL SITEX Source, ORNL/TM-7895, Oak Ridge National Laboratory (May 1982).
4. W. K. Dagenhart and W. L. Stirling, Bull. Am. Phys. Soc. 27, 1136 (1982).
5. W. K. Dagenhart, W. L. Gardner, G. G. Kelley, W. L. Stirling, and J. H. Whealton, in Proceedings of the Third Neutral Beam Heating Workshop, CONF-811018, Oak Ridge National Laboratory (1982).
6. W. L. Gardner, W. K. Dagenhart, G. G. Kelley, W. L. Stirling, and J. H. Whealton, in Proceedings of the Third Neutral Beam Heating Workshop, CONF-811018, Oak Ridge National Laboratory (1982).
7. W. K. Dagenhart, W. L. Gardner, W. L. Stirling, and J. H. Whealton, Nucl. Technol./Fusion 4, 1430 (1983).
8. W. K. Dagenhart, W. L. Stirling, J. H. Whealton, and J. J. Donaghy, Bull. Am. Phys. Soc. 28 (8), 1150 (1983).
9. W. L. Stirling, W. K. Dagenhart, J. H. Whealton, and J. J. Donaghy, "Normalized Emittance of SITEX Negative Ion Source," in these proceedings.
10. J. H. Whealton, R. J. Raridon, R. W. McGaffey, D. H. McCollough, W. L. Stirling, and W. K. Dagenhart, "2-D Accelerator Design for SITEX Negative Ion Source," in these proceedings.

DISCUSSION

SCHILLING: It seems that you and Berkeley spend a fair amount of your experimental time fighting the cesiation problem. Do you have any plans to look into a porous type converter like K. Prelec described this morning? It sounds very attractive.

DAGENHART: We have had thoughts on it but we haven't had the funds to go into that. That is a very interesting development. One of the important things is to maintain high current densities, around 100-130 mA/cm^2, and we feel that we have to maintain some cesium in the discharge itself. It is not just the problem of the cesium coverage on the converter because Cs^+ ions produce a considerable number of the H^- or D^- ions. We must therefore maintain the cesium pressure in the discharge. That doesn't say that you cannot use a porous dispenser but you may wind up putting too much cesium on a converter. You may need to use two dispensers, one somewhere else in the discharge chamber and one in the converter to maintain optimum coverage.

ALLISON: What would you say the level of high frequency fluctuations was?

DAGENHART: We really had no time to optimize or minimize that. I have some traces that show relatively little noise, perhaps 5%, and some of them that have a considerable amount of noise, up to 100% modulation. We have not made a study and I would hesitate to quote what the discharge noise was. We plan to do that but that is something we haven't done yet.

BARNETT: What is the frequency range that you see?

DAGENHART: Of course, I showed you 120 Hz from the power supplies. From the plasma generator we've seen on the Faraday cup frequencies to several hundred kHz.

LEUNG: How do you know how many electrons you extract and how do you estimate what part of them you recover?

DAGENHART: I have to give you two answers to this. One is that the electrons circulate around in our electron recovery system and in a sense they never leave the source. The power supply that recovers the electrons has one of its terminals on the ion source at anode potential and the other is a couple of kilovolts higher where the electrons are actually collected; electrons simply circulate around in that supply. If there is a breakdown to ground or if there is some sort of a corona discharge, then those electrons would go through the accelerator supply and would be recorded as H^- ions. Now we have current shields which we monitor calorimetrically as well as electrically, downstream from the source, where we can pick up any of these electrons and typically their current is quite low, it is almost an immeasurable current. Of course, when we start up the operation, we have arcs in those electrodes, but during normal operation virtually 100% of the electrons are recovered as long as the geometry is correct and the electron collector is in the system. We can also look at the source and if there are many milliamperes of electrons going down field you can actually see a glow where they impact because we have a magnetic field and they would be fairly concentrated. When we first started this source development we had electrons going to the ground. We drilled holes in copper and in stainless steel water cooled jackets. It is quite easy to see tens of milliamperes of electrons on some uncooled parts that we had in there. Now it is just absolutely no problem to recover electrons without any cesium if we run this the source steady state. We operated the plasma generator in a mode to generate up to 1 amp of electrons coming out the front. We had the converter all the way up to the extraction slit, as close as we could get the discharge to the extraction slit, with the purpose to generate the maximum number of electrons and we recovered virtually 100% of them. This is not the way we operate now, the converter is 4 mm back now, and from our initial results when we had roughly three electrons for every negative ion we reduced the electron load to 5-15% of the H^- current. I don't know if I've answered your question but we've verified the electrical measurements calorimetrically on the side shields. We have an electrical monitor on the side plates and then we have roughly a particle accountability through all the power supplies. We can roughly account for all the particles. I don't know what the accuracy of that is. I would have to look at that to see. But certainly the electrical and calorimetrical measurements put a quite low error bar on those two.

PULSED MULTIAMPERE SOURCE OF NEGATIVE HYDROGEN IONS

Yu. I. Bel'chenko and G. I. Dimov
Institute of Nuclear Physics
Novosibirsk 90, U.S.S.R.

ABSTRACT

A description is given of a pulsed model of a multiampere honeycomb surface plasma source of negative hydrogen ions (H^-). The H^- emitter is the cesium-coated cathode surface in a high-current, glow discharge; the surface consists of a large number of spherically concave indentations. The concavity of the emitting surface of the indentation and of the adjacent cathode sheath, where ions are accelerated, ensures two-dimensional geometrical focusing and a 20-fold compression of the H^- stream from a larger part of the cathode surface into the emission slits of the multiaperture extraction system. A pulsed (0.2 to 0.8 ms) H^- beam carrying a current in excess of 11 A was produced from 600 indentations with an emitting surface area of ~60 cm^2 and accelerated to 25kV. The gas efficiency of the source was $\gtrsim 20\%$. The thermal loads on the electrodes were <1 kW/cm^2.

INTRODUCTION

The injection of fast hydrogen atoms is one of the main methods of storing, heating, and controlling the distribution of particles in the plasma of thermonuclear devices. In existing injection systems, atoms having an energy of ~100 keV/nucleon are produced through neutralization of positive ions. As the particle energy is increased to the range of 200 keV to 1 MeV in injectors of the next generation, neutralization by stripping of negative ions must be used. In contrast to positive ions, such negative ions have at these energies a high neutralization efficiency (80%-100%) in a plasma target[1] or a laser target.[2]

In recent years work has been done to develop high current negative-ion sources as well as efficient neutralizers for such high energy injectors. In addition to traditional charge exchange sources[3,4] and volume sources of negative ions,[5,6] which are experiencing their second birth, intensive studies are under way on new types of sources: surface conversion sources[7] and surface plasma sources.[8-16] Surface-plasma sources (SPS) have already been developed that allow the production of pulsed H^- (D^-) beams with a current of several amperes;[11-13] negative beams with a current of 0.7-1.1 A have been produced and accelerated from quasi-steady-state SPS.[15,16]

In an SPS the plasma sheath adjacent to the emitter of secondary ions (the cathode) delivers to the emitter surface intense fluxes of ~100 eV ions and atoms (Fig. 1). Upon collision with the surface, primary ions and atoms are partially reflected and sputter hydrogen particles adsorbed on the emitter. Due to a surface work function reduced to ~1.5 eV by cesium adsorption, a significant fraction of fast particles leaving the emitter escapes beyond the surface potential barrier in the form of negative ions.[17] These negative ions are accelerated in the sheath of potential next to the electrode and after passing through the plasma, enter the beam-forming region.

SPS with geometric focusing are promising for use in controlled thermonuclear fusion. In these SPS the negative ion emitting surface has the form of concave semicylindrical grooves or spherical indentations (Fig. 1B).

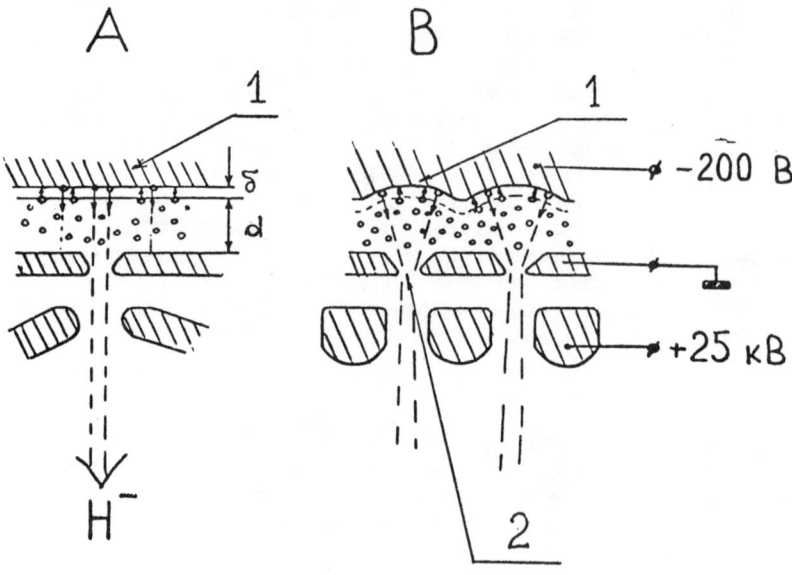

Fig. 1. Scheme of the surface-plasma method of generating negative ions. (A) with plane electrode geometry (a planotron). (B) with geometrical focusing: [1] cylindrically or spherically concave cathode indentations, [2] slit-type or round emission apertures.

As a result, the negative ions accelerated in the sheath are focused from the concave surface of the emitter into relatively small emission slits or apertures. Because of a high compression of the negative-ion stream into the emission holes, up to 80% of the working surface of the cathode can be used in SPS with geometric focusing. The reduction of the relative area of the emission slits (by a factor of 3-5 in SPS with grooves and by a factor of 10-20 in honeycomb SPS) makes possible a significant decrease in hydrogen and cesium consumption and a 10-20% gas efficiency for reduced discharge power and thermal loads of ~1 kW/cm^2 on the electrodes.

One dimensional geometric focusing was first implemented in semiplanotron sources[10,11] and is being used successfully in other SPS modifications: in a ring source,[12] in a magnetron with grooves,[13] and a multicusp source.[15] Two-dimensional geometric focusing has been tested in a small honeycomb SPS[14] and also is used in the multiampere honeycomb source (MHS) described below, which has a much larger emitter surface.

DESIGN OF THE SOURCE

A diagram of the multiampere honeycomb source (MHS) is presented in Fig. 2. The high current glow discharge, with an open electron drift, is localized in the gap between the cathode emitting surface 1 and the anode cover 2 with emission holes. At the edges the discharge space is limited by side projections 1a and 1b of the cathode, by the ignition gap 3, and by the electron-dump region (not shown in Fig. 2).

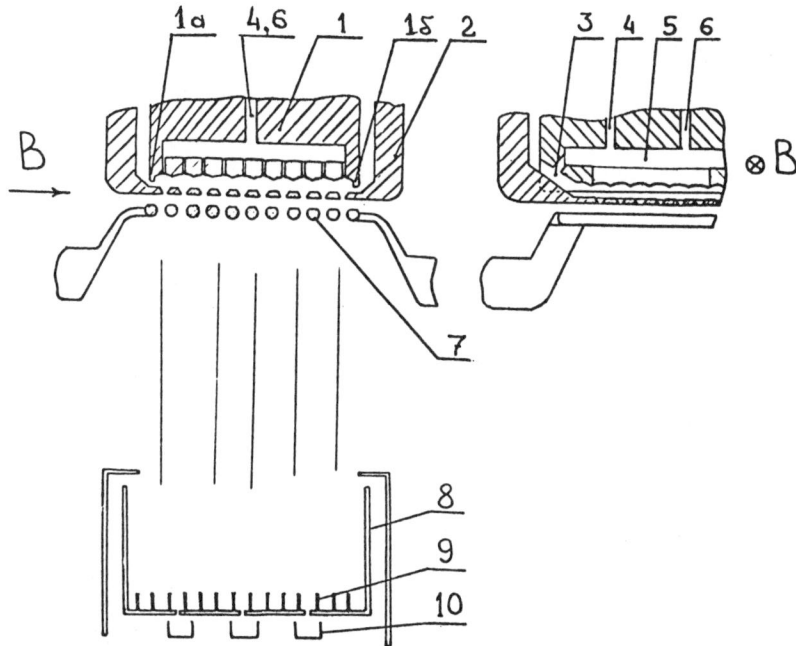

Fig. 2. Diagram of a honeycomb source. At left - view along the magnetic field; at right - across the magnetic field. (1) cathode; (1a and 1b) side projections of cathode; (2) anode body; (3) ignition gap; (4,6) H_2 and Cs supply channels; (5) distributor cavity in cathode; (7) extractor electrode; (8) Faraday cup for measuring the total negative ion current; (9) fins for intercepting secondary electrons (10); Faraday cup for measuring current density over the beam cross section.

A magnetic field of 0.5-1.5 kG is generated parallel to the plane of the discharge gap by means of an electromagnet and special pole tips. The magnetic field causes oscillations of the electrons in the space between the side projections of the cathode, and also attenuates and removes from the beam electrons extracted together with the negative ions. The width of the zone of electron oscillations between the side projections of the cathode is 3 cm, and the thickness is 1-2 mm.

In the ignition gap of the cathode, the height of the side projections of the cathode has been increased to 4-5 mm, thereby improving electron confinement in the oscillation zone and lowering the initial hydrogen density in the ignition gap that is required to maintain a self-sustained discharge. The loss of plasma from the ignition gap due to electron drift in the crossed E x B fields of the electrode gap helps the propagation of the discharge over the entire electrode gap, which is 20 cm long. The discharge current was varied in the range of 0-700 A, the discharge voltage was in the range 150-200 V under operating conditions, and the discharge pulse length was varied from 300 µs to 2 ms.

Hydrogen is supplied to the discharge chamber by means of pulsed electromagnetic valves through special channels (4) and the cavity (5) inside the cathode, which is connected to the main chamber by narrow (~0.05 mm) slits. A distributor plate with a variable number of bypasses is installed in the inside cavity of the cathode to ensure the required hydrogen density profile over the length of the discharge gap (the plate is not shown in the figure).

Cesium is supplied to the ignition gap and to the cathode surface from independently heated containers holding a mixture of cesium chromate and titanium and located outside the source. Supply is accomplished via channels 6 and through the inside cavity of the cathode.

The emitting surface of the cathode has dimensions of 3 x 18 cm^2 and consists of ~600 spherical, concave indentations arranged on the cathode surface in orthogonal or hexagonal matrices (Fig. 3). To ensure more complete utilization of the cathode surface, the indentations overlapped, i.e., their diameter was greater than the distance between the centers of adjacent rows and the boundaries between neighboring cells created either a square or honeycomb structure (Fig. 3). The radius of curvature of indentations within a square structure was 3 mm, and the distance between the centers of rows of adjacent craters was 3 mm. Indentations with a 3.5 mm radius of curvature were used in the honeycomb structure.

The cathodes were made from plates of monocyrstalline molybdenum. This material sputters lightly, has a low probability of forming arc spots, and, in contrast to tungsten, has a better adsorption of cesium from the flux of cesium ions and atoms incident on the cathode. Forced cooling of the electrodes was not used in the pulsed source described here. Conditioning and outgassing of the electrodes at the start of the working cycle were accomplished with pulsed low current glow discharges. Switching the source to a high current glow discharge operation was speeded up significantly when the cathode was heated to 400°C by the heater built into its cavity (the heater then was shut off).

Conical emission holes with a 0.9 mm diameter were drilled at the negative ion focusing points on the anode cover (allowing for the displacement due to the magnetic field of the source). The accuracy of the alignment of cathode indentations and emission apertures was checked by the sputtering marks that appeared on the anode cover at negative ion focusing points after the source was run.

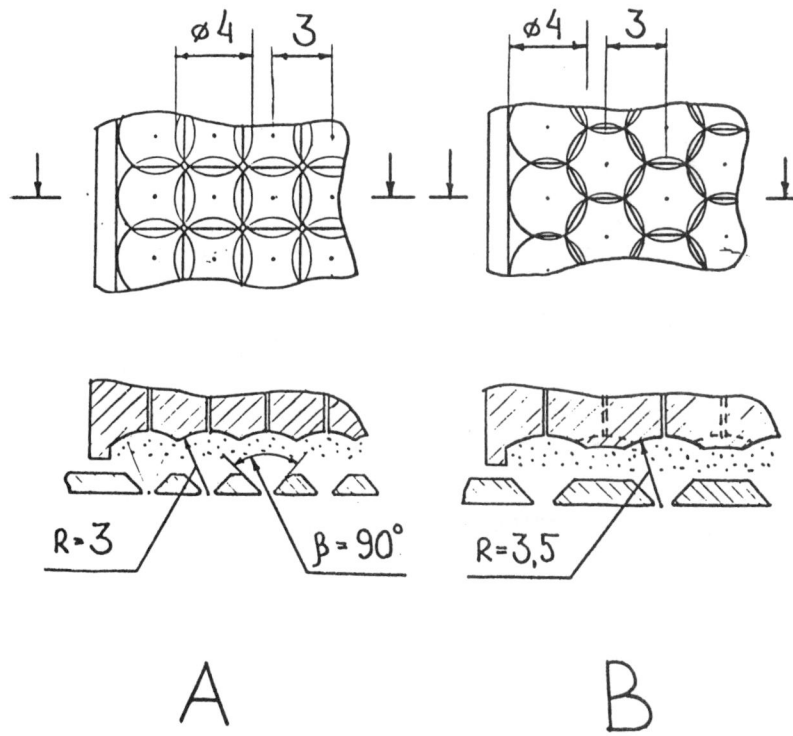

Fig. 3. Arrangements of spherically concave indentations on the cathode surface. (A) orthogonal arrangement; (B) hexagonal arrangement.

The negative ions emerging through the emission holes are extracted by supplying a pulsed voltage (25 kV, 0.2-0.8 ms) to the body of the source. The electrodes of the multiaperture extraction system (7 in Fig. 2) are grounded through the current measuring resistor. To increase the dielectric strength of the extraction gap (1.3 mm long, with an area of 3.3 x 20 cm^2), the grid of the extraction electrode has been fabricated from profiled molybdenum and is heated by heaters built into the grid wires.

In the source being described, the electrons that enter the extraction gap as a result of diffusion from the plasma sheet as well as those formed on the cesium-coated walls of the emission slits and those formed by destruction of negative ions are accelerated to the full extraction voltage. The height of the trochoid of electrons drifting in the crossed E x B fields of the extraction gap exceeds the width of the extraction gap; therefore, the stream of accelerated electrons is intercepted by the extraction electrode and by the pole tips of the magnet.

The negative ion beam current and its density distribution were measured with Faraday cup collectors (8 and 10 in Fig. 2) located at a distance of 20-25 cm from the extraction region. Fins (9), intercepting secondary particles moving in the stray field of the source, serve to suppress the departure of secondary electrons from the large collector. Separate measurements were made of the current on the extractor grid and of the total current in the extractor circuit. In addition to the H⁻ beam current, the total current included the flux of electrons and other negative particles. A mass analysis of the extracted negative ion beam was performed by means of an additional magnetic analyzer with an entrance slit 20 mm² in area, that could be moved across the negative ion beam. Measurements of the source gas efficiency were made by using noise-immune fast ionization detectors that register both the total flux of hydrogen particles and the slow thermal component of the hydrogen flux emerging from the source.

The cesium density in the source and the cesium consumption were determined by means of a surface ionization type cesium detector.

CHARACTERISTICS OF THE SOURCE

The main factors that determine the efficiency of negative ion production in high current sources are the thickness and uniformity of the cesium coating of the cathode, the hydrogen density profile over the discharge gap, and the uniformity of the distribution of the discharge current (especially the ion current) over the cathode emitting surface.

When these factors were optimized, a pulsed beam of H⁻ ions with a current in excess of 11 A was produced from the source and accelerated to 25 keV. The total negative ion current recorded (allowing for heavy negative ions) reached 12 A, and the current in the extraction circuit was ≤ 25A. The H⁻ yield was proportional to the discharge current (curve I in Fig. 4a). For comparison, Fig. 4a shows the analogous relationship (curve II) for a small honeycomb source.[14] Figure 4b shows the dependence of the H⁻ current density, averaged over the beam cross section, on the average cathode current density for the MHS and for a small honeycomb source. Conversion efficiency of the discharge current into an H⁻ beam, $\alpha = j-/\langle j_d \rangle$, of about 2% was achieved with discharge current densities of 8 A/cm² on the cathode of the MHS.

A typical negative ion current density distribution along the source is shown in Fig. 5. The drop in the negative ion emission at the edges of the source was due to the reduced density of the plasma in those regions of the discharge. Over the flat part of the H⁻ current density distributions in Fig. 5, the conversion efficiency of the cathode current into the H⁻ beam current ($j-/\langle j_d \rangle$) reaches a value of 2.7%, which is approximately half the value attained in a honeycomb source with a relatively small beam cross section.[14] This may be due to a higher current density in the regions of the cathode surfaces from which H⁻ ions are not extracted.

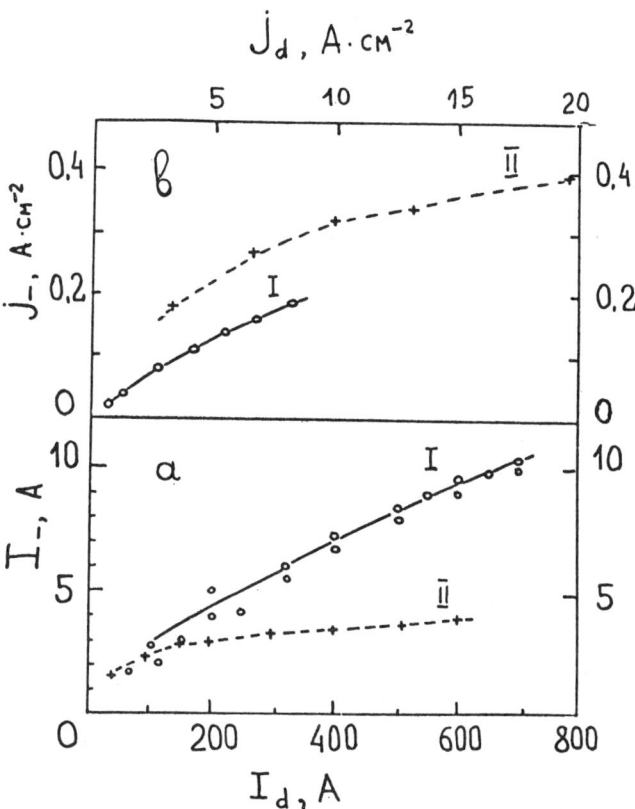

Fig. 4. (a) Dependence of H⁻ current I_- on discharge current I_d; (b) dependence of average H⁻ current density j_- in the beam on discharge current density j_d on the cathode (I - multiampere honeycomb source; II - small honeycomb source).

A magnetic analysis of the composition of the accelerated negative ion beam showed that heavy (O⁻, etc.) impurity ions comprise 10-20% of the total beam current for discharge currents of ≤200A. During the beginning phase of the heating of the cesium chromate, the heavy ion component may amount to 40-50% of the beam current.

In the range of discharge currents 400-700 A, the extracted beam is 95-99% H⁻ ions (averaged over the pulse); the heavy ion component is substantially lower at the end of the discharge pulse than at the beginning. This fact may be explained by the removal of impurities from the cathode surface during ion bombardment. The cesium chromate (C_2CrO_4) used in the source was the main source of oxygen.

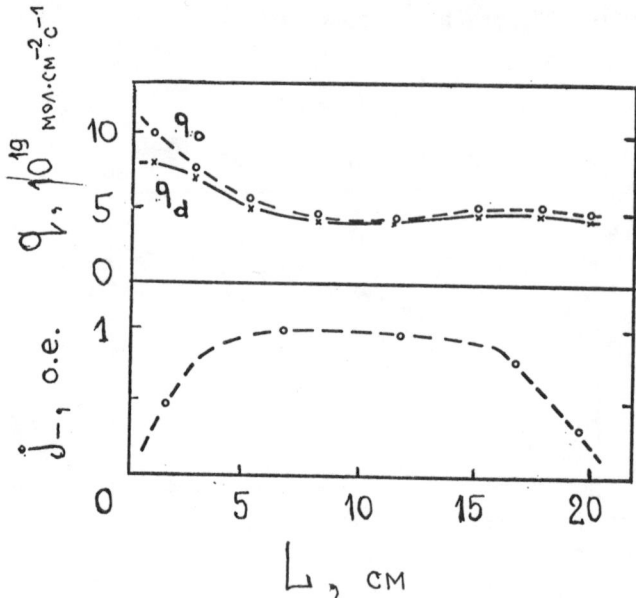

Fig. 5. Distribution of the density of the hydrogen flow from the source emission apertures (q_0 - before triggering of discharge, q_d - during discharge) and of the negative ion current density j_- over the length of the source across the magnetic field; L - distance from the ignition gap.

In the investigated range of discharge currents (<700 A), the loss of H⁻ ions in the plasma sheet, which was due mainly to collisions with plasma electrons, was small. The thickness of the plasma sheet of the high current source did not exceed 5×10^{12} cm⁻²; accordingly, the transmission coefficient for negative ions during motion through the source plasma had the following value:

$$\mu_e = \frac{j_-}{j_-^0} \simeq \exp\left(-n_e d \frac{\langle \sigma_e v_e \rangle}{v_{H^-}}\right) \gtrsim 0.9$$

(σ_e is the cross section for H⁻ ion loss in collisions with electrons; the other notations are standard).

In the MHS the hydrogen density required to initiate a high current discharge depended on the magnetic field, the cesium concentration, the height of the side projections of the cathode, and other factors. Under the operating conditions of the source the initial hydrogen density in the ignition gap was ~3 x 10^{15} molecules/cm^3; the discharge propagated into the main part of the gap at a gas density of $\gtrsim 6$ x 10^{14} molecules/cm^3. A typical density distribution of the stream of hydrogen particles from the emission slits along the source before and during discharge is shown in Fig. 5. In contrast to SPS with a dense plasma,[18] in this source the total flux of hydrogen particles from the emission holes decreased insignificantly when the discharge was initiated (Fig. 5). The flux of hydrogen particles Q_d emerging from the source during the discharge pulse consisted of H$^-$ ions, fast H^0 atoms, and a slow component: thermal hydrogen molecules and atoms. When the H$^-$ ion beam current was $I_- = 11$ A, the total flux of hydrogen particles from the source was (converted to atoms) $Q_d = 3$ x 10^{20} atoms/s; accordingly, the pulsed gas efficiency of the source was $\eta = I_-/Q_d \gtrsim 20\%$. Because of the pumping effect of the plasma sheet,[18] the slow component of the hydrogen flux from the source decreased 30-40% from the initial value when a high current discharge was triggered. Model experiments using a scaled-up discharge chamber and a geometry similar to a cell of the high current source showed that when the thickness $n_e d$ of the sheet of the discharge plasma increased in the range 10^{12} to 2 x 10^{13} cm^{-2}, the flux of thermal hydrogen from the chamber decreases exponentially with a constant $n_e \lambda \cong 8$ x 10^{12} cm^{-2}. Reduction of the density of H$_2$ molecules in the discharge chamber and in the extraction region is important for eliminating the additional loss of negative ions upon collisions with H$_2$ molecules. Such losses were insignificant in this source because of the reduced H$_2$ density in the main portion of the discharge gap and a small flow of H$_2$ molecules into the extraction region. (This flow was small because of the small area of the emission holes.) The attenuation of the H$^-$ ion stream due to ion losses on hydrogen molecules during the transit from the source cathode to the collector (~20 cm) was:

$$\mu_{H_2} = \exp - \langle \sigma_{H_2} n_{H_2} Z \rangle \simeq \exp - (4 \times 10^{-16} \cdot n^\circ_{H_2}) \simeq 0.8$$

(σ_{H_2} is the H$^-$ loss cross section for collisions with H$_2$; n_{H_2} is the hydrogen density in the discharge chamber; and Z is the coordinate in the direction of motion of an H$^-$ ion). Let us note that as the number of cells increases, so too do the total flux and the effective thickness of the "cloud" of hydrogen molecules that destroys negative ions, $\int_0^\infty n_{H_2} dZ$. This cloud emerges from the emission holes and is pumped through the extraction and beam forming regions. When a ribbon beam with an emission cross section a x b is extracted, the effective thickness increases linearly with an increase in the smaller transverse dimension b and logarithmically with an increase in the large dimension a:

$$\int_0^\infty n_{H_2} dZ = n°_{H_2} \cdot d_{H_2} + \left(n°_{H_2} \frac{S_{em}}{a \cdot b}\right) \cdot b \frac{2}{\pi} \ln\left(\frac{\pi}{2} \frac{a}{b}\right)$$

where d_{H_2} is the total effective thickness of molecular hydrogen in the discharge chamber and emission holes.

The cesium coverage of the cathode is $2-4 \times 10^{14}$ atoms/cm^2, which ensures high values of the coefficients of secondary negative ion emission $K^- \cong 0.2-0.7$ and of secondary electron emission $\gamma = 3-6$ in SPS[19]; it is regulated by the rate of supply of cesium to the discharge chamber and depends on the electrode temperature, the cathode current density, and other factors. The thickness N of the cesium coating is determined by the relationship between the rate j_{Cs} of arrival of particles on the surface and by the length of time τ the cesium spends on the surface:

$$\frac{dN}{dt} = \langle \alpha j_{Cs} \rangle - \frac{N}{\tau}$$

where α is the sticking coefficient of cesium.

In the intervals between discharge pulses the value of N depends on the thermal flux of cesium from the discharge chamber volume: $j_{Cs} = \frac{n_{Cs} \cdot \bar{v}}{4}$ (actually on the thermodesorption flux from the opposite surface) and on the rate of cesium desorption from the electrode: $N/\tau = N\omega_0 \exp(-W/kT)$. Optimum cesium coverage on molybdenum is maintained with a relatively low volume density of cesium, since the sticking coefficient α of cesium for the optimal coverage is close to 1, and the heat of desorption from an optimal layer on molybdenum is high: $W \sim 2$ eV. Because of the low volume density of cesium ($\lesssim 10^{12}$ atoms/cm^3) and the small area S_{em} of the emission holes, the cesium flux from the source in the intervals between discharge pulses is small: $Q_{Cs} = \frac{n_{Cs}}{4} \times S_{em} \cong 0.5-1 \times 10^{16}$ s^{-1}; hence the rate of required cesium supply compensating for this "thermodesorption" consumption is low.

When discharge is triggered, the cesium coverage of the source electrodes changes. The cesium adsorbed on the electrodes is sputtered by fast discharge particles; the coefficient of sputtering of cesium from an optimal coating on molybdenum increases with the energy and mass of the particles bombarding the surface, having a value of order 10^{-3} for 100 eV hydrogen ions. The cesium desorbed from the electrodes is ionized in the discharge chamber with a mean free path of $\lambda_{Cs} = \nu_{Cs}/n_e \langle \sigma_{Cs}\nu_e \rangle$ (where σ_{Cs} is the cross section of ionization of cesium by electrons) and is returned to the cathode by the electric field of the discharge.[21] For plasma densities $n_e = 2 \times 10^{12}$ to 10^{13} cm^{-3}, the mean free path is $\lambda_{Cs} = 1-0.2$ mm. As measurements have showed, the total cesium flux onto the cathode during discharge is 2-3 orders of magnitude larger than the stream onto the anode, and it consists basically of Cs$^+$ ions. The thickness of the cesium layer on the cathode is determined by the sticking efficiency of cesium from the incident Cs$^+$ flux, which in turn depends on the cathode material, the thickness and structure of the cesium layer, the presence of surface contamination, and other factors. Because of a favorable atomic mass ratio, molybdenum adsorbs cesium well from the flux of ions. Thus, when a clean molybdenum surface is bombarded with 100-200 eV ions, only around 5% of the incident ions are reflected. The remaining 95% temporarily stick to the surface and fly off (~60%) in the form of slow cesium ions.[20] The cesium sputtering increases with increasing cesium coverage. This contributes to the rapid transport of cesium to cesium-depleted sections of the cathode both during startup of the source and during the discharge pulse.

In the MHS it is possible to obtain both, stable regimes with an optimal cathode coverage throughout the entire discharge pulse, and regimes with an increasing (or decreasing) cesium coverage of the cathode, by controlling the electrode temperature, the cathode current density, and the cesium supply rate. Because of the fast ionization of cesium in the discharge and its transport to the cathode, the loss of cesium through the emission holes of surface plasma sources during the discharge pulse is insignificant.

Let us present a summary of the basic parameters attained in the high current H$^-$ source:

H$^-$ beam current	– over 11 A
Heavy impurities (average over a pulse)	– <3%
Beam energy	– 25 keV
Beam pulse length	– 200-800 μs
Average emission current density of H$^-$ ions in beam	– 180 mA/cm^2
Same, in the flat part of the distribution	– 220 mA/cm^2
Area of beam emission cross section	– 60 cm^2
Total current in extraction circuit	– $\leq$25 A
Number of indentations	– 600
Total operating surface of cathode	– 74 cm^2
Emitting surface of cathode	– 63.6 cm^2
Total area of emission apertures	– 3.86 cm^2
Gas efficiency (pulsed)	– at least 20%
Cesium consumption (due to loss between pulses)	– no more than 20 mg/hr
Discharge voltage	– 150-200 V

Discharge current — 700 A
Average current density on cathode — 9.5 A/cm^2
Thermal load on electrodes — 1 kW/cm^2
Magnetic field in the discharge chamber — 700-900 G

In conclusion let us take note of the properties of the MHS that are promising from the standpoint of its use in high energy neutral injectors for controlled thermonuclear fusion.

1. Low consumption of both the primary working material (hydrogen) and the secondary emission catalyst (cesium).

2. Shielding of 90% of the working surface of the cathode from direct impact of surface-damaging external fluxes of fast particles (positive ions accelerated in the extraction gap, etc.) through the emission holes.

3. A moderate level of power deposited on source electrodes; the possibility of heat removal by cooling in the steady state case.

4. The possibility of using a single source to obtain negative hydrogen ion beams of unlimited current with a high average emission density ($\sim$100 mA/cm^2) close to the average emission current density of high current positive ion beams.

Based on the pulsed source described here, we propose the creation of a quasi-steady-state version of a high current honeycomb source of H$^-$ ions.

REFERENCES

1. G.I. Dimov, A.A. Ivanov, and G.V. Roslyakov. Fizika Plazmy, 6, 933 (1980).
2. J.H. Fink. Preprint UCRL-87301, Livermore, 1982.
3. N.N. Semashko, A.N. Vladimirov, V.V. Kuznetsov, et al. Inzhektory bystrykh atomov vodoroda (Fast Hydrogen Atom Injectors). Moscow, Energoizdat Publishers, 1981, p. 142.
4. E.B. Hooper, Jr., P. Poulsen, and P.A. Pincosy. J. Appl. Phys., 52, 7027 (1981).
5. K.N. Leung, L.W. Ehlers, and M. Bacal. Rev. Sci. Instrum., 1, 56 (1983).
6. T.S. Green, A.J. Holmes, and A. Walker. Bull. Am. Phys. Soci., 27, 1055 (1982).
7. J. Los, E.A. Overbosch, and J. Van Wunnik. Proc. Sec. Int. Symp. on Production and Neutralization of Hydrogen Ions and Beams, Brookhaven, N.Y., BNL-51304, 1980, p. 23.
8. G.I. Dimov, X European Conference on Contr. Fusion and Plasma Phys., Moscow, 1981, vol. II, p. 35.
9. Yu. I. Bel'chenko, G.I. Dimov, and V.G. Dudnikov. In Proc. Symp. on Production and Neutralization of Negative Hydrogen Ions and Beams, Brookhaven, NY, 1977, BNL-50727, p. 79 (1977). (See also Preprint No. 77-56, Institute of Nuclear Physics, Novosibirsk, 1977.)

10. Yu. I Bel'chenko and V.G. Dudnikov. Preprint 78-95, Institute of Nuclear Physics, Novosibirsk, 1978. Also Journal de Phys., Colloque C7, No. 7, 40, p. C7-501 (1979).
11. Yu. I. Bel'chenko and V.G. Dudnikov. Trudy XV Mezhdunarodnoi konferentsii po yavleniyam v ionizovannykh gazakh (Transactions of the 15th International Conference on Phenomena in Ionized Gases.) Minsk, 1981, part II, p. P-1504.
12. V.L. Komarov and A.P. Strokach. Zhurnal tekhnicheskoi fiziki, 49, 75 (1979).
13. J.G. Alessi and Th. Sluyters. Rev. Sci. Instrum., 51, 1631 (1980).
14. Yu. I Bel'chenko. Preprint 82-54, Institute of Nuclear Physics, Novosibirsk, 1982. Also Fizika Plazmy, 9 (1983), 1219.
15. K.N. Leung and K.W. Ehlers. Rev. Sci. Instrum., 53, 803 (1982).
16. W.K. Dagenhart and W.L. Stirling. Bull. Am. Phys. Soc., 27, 1136 (1982).
17. M.E. Kishinevskii. Zhurnal Tekhnicheskoi Fiziki, 48, 773 (1978).
18. A.N. Apolonskii, Yu. I Bel'chenko, G.I. Dimov, and V.G. Dudnikov. Pis'ma v Zhurnal Tekhnicheskoi Fiziki, 6, 86 (1980).
19. V.G. Dudnikov and G.I. Fiksel'. Fizika Plazmy, 7, 283 (1981).
20. U.A. Arifov. Vzaimodeistvie Atomnykh Chastits s Poverkhnost'yu Tverdogo Tela (Interaction of Atomic Particles with the Surface of a Solid). Moscow, Nauka Publishers, 1968, p. 102.
21. Yu. I. Bel'chenko, V.I. Davydenko, G.E. Derevyankin, et al. Pis'ma v Zhurnal Tekhnicheskoi Fiziki, 3, 693 (1977).

PRODUCTION OF HIGH BRIGHTNESS H⁻ BEAMS IN SURFACE PLASMA SOURCES

G. E. Derevyankin and V. G. Dudnikov
Institute of Nuclear Physics, Siberian Branch,
USSR Academy of Sciences, Novosibirsk

ABSTRACT

This paper examines topics related to the production of H⁻ beams of maximum brightness in surface-plasma sources. It is shown that a maximum brightness of $\cong 2 \times 10^8$ A/cm^2 rad^2 can be obtained in Penning geometry sources in which efficient generation of negative ions with a reduced energy of ion thermal motion in the plasma of <0.5 eV is possible under optimized conditions. In sources with slit optics, the brightness basically is limited by supplementary increase in temperature of ions during beam extraction and formation due to fluctuations of beam intensity. In sources with a round emission aperture, a beam with a brightness of $\cong 10^7$ A/cm^2 rad^2 can be produced even when there are significant fluctuations of intensity ($\cong$ 10-15%).

INTRODUCTION

In recent years interest in the acceleration of negative ions of hydrogen and deuterium has grown, as evidenced in the increased number of accelerators and laboratories where work on H⁻ ion sources has been undertaken. This development is due to two factors. On the one hand, by the start of the 1970's the pressing need to utilize in accelerators all the advantages offered by charge exchange technology had become fully evident[1-3]. (These advantages include charge exchange injection of protons into ring accelerators, the acceleration of H⁻ and D⁻ ions in powerful accelerators at meson and neutron factories, etc.) This generated a demand for appropriate ion sources, which long went unsatisfied. The best sources of the charge exchange type at the time allowed the production of H⁻ ion beams having an intensity of 20-50 mA.[4] This turned out to suffice for converting the proton synchrotron of the Argonne Laboratory to charge-exchange injection[5], but a number of shortcomings inherent in sources of this type limited their extensive use in other accelerators.
A radical solution to the problem proved possible on the basis of a new surface-plasma mechanism of negative ion formation in gas discharges that was discovered at the Institute of Nuclear Physics in Novosibirsk[6,7]. A complete understanding has now been gained of the basic principles underlying the formation of negative ions in surface-plasma sources (SPS). The physical foundations of this method have been covered in detail in Refs. 8 and 9 and in a number of other materials in these collections. This has made possible the design of SPS for accelerators and a comparatively early production of H⁻ beams of the required intensity. As a result, in addition to the Brookhaven and Los Alamos groups, which undertook such work almost immediately after the first positive results were obtained in Novosibirsk, work on SPS also is being done at Batavia and in the

Argonne, Rutherford, and a number of other laboratories[10-15]. Thus, obtaining the requisite ion beam intensity has become a solvable technical problem. As far as obtaining H⁻ beams with good ion-optical characteristics is concerned, these problems are now under intensive study. It seems useful in this regard to attempt to summarize the experience gained to date in the area of increasing the brightness of H⁻ beams formed in SPS, since this topic is the most important one for accelerator sources.

GENERAL

The brightness is a generally accepted characteristic of the quality of an ion beam:

$$B = \frac{2I^-}{\pi^2 \cdot E_X \cdot E_Y} \left(\frac{A}{cm^2 \, rad^2}\right) \quad (1)$$

where I^- is the beam intensity, and E_X and E_Y are the normalized emittances with respect to one of the transverse degrees of freedom.

$$E_{X,Y} \simeq \beta \cdot \gamma \cdot a_{X,Y} \cdot \Delta\alpha_{X,Y} \quad (2)$$

where β and γ are relativistic factors, and $a_{X,Y}$ and $\Delta\alpha_{X,Y}$ respectively, are the half-dimension and angular half-spread of the ion paths in one of the transverse directions (Fig. 1). If we assume that the ions in the source plasma have a Maxwellian distribution with temperature T_i (in energy units), and if the conditions of Liouville's theorem are fulfilled during beam acceleration and focusing, then the following expression is valid for the emittance[16,17]:

$$E_X \cdot E_Y \simeq 4X_0 Y_0 \frac{T_i}{Mc^2} \quad (3)$$

where $4X_0 Y_0$ is the beam cross section at the plasma emission surface. Substitution in eq. (1) gives

$$B \simeq \frac{2j_0^- \cdot Mc^2}{\pi^2 \cdot T_i} \quad (4)$$

Hence we can see that to obtain a high brightness, the emission current density j_0^- during ion extraction from the source plasma must be increased where possible, and the ion temperature in the plasma must be quite low. Expression (4) defines the beam brightness that can be obtained under ideal conditions. In fact, some increase in temperature of the ions will always take place, so the question is how much can it be reduced. Experience shows that the degree of additional increase in temperature of the ions depends on the design of the source, the conditions of beam extraction and primary formation in the extraction gap, the operating conditions of the source, and the conditions of subsequent beam transport and focusing.

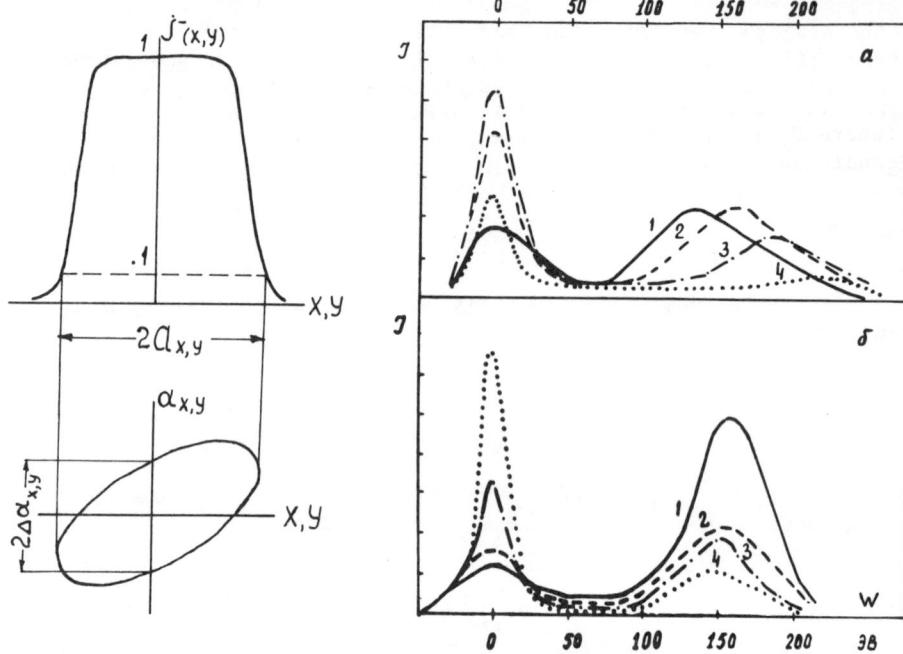

Fig. 1. Phase diagram of the beam for determining the emittance.

Fig. 2. Energy spectra of H⁻ ions produced in a planotron: a - for different voltage drops $\overline{U_p}$ across the discharge:
1 - U_p = 120 V; 2 - 150 V; 3 - 160 V; 4 - 210 V. b - for different flow rates Q of hydrogen molecules per pulse:
1 - Q = 10^{16}; 2 - 1.2 x 10^{16}; 3 - 1.7 x 10^{16}; 4 - 2.2 x 10^{16}.

GEOMETRY OF THE SOURCE

The maximum emission current density attained to date in SPS is $\cong$ 8 A/cm², [18] but, as will be shown below, an emission current density of $\cong$ 2 A/cm² is realistic when intense beams are extracted.

The effective ion temperature T_i in the source plasma is determined by the physical processes responsible for the formation of negative ions[8,9], and it therefore depends on the design and operating conditions of the source.

The maximum efficiency of H⁻ generation is achieved in sources in which the beam is formed from ions emitted by the surface of an emitter that has an optimal cesium coating. These include sources with a planotron (magnetron) geometry and some modifications thereof[18]. Two such sources have found application in accelerators[19,20]. In this case the effective temperature will be determined by the energy spread and angular spread of the ions leaving the emitter. Exhaustive data are not available for these distributions in SPS.

A number of indirect findings enable us to conclude[21,22] that the angular distribution of the ions has a width $\cong \pm(0.2-0.3)$ rad. The energy spectra of ions extracted from a source with a planotron geometry are shown in Fig. 2[7].

It can be seen that the total width of the spectrum is $\simeq eU_p$ (where U_p is the voltage drop across the discharge). Under these conditions it is difficult to anticipate that the effective ion temperature would be less than 10 eV.

Therefore, V. G. Dudnikov proposed for accelerators a surface-plasma source with a Penning geometry[23]. Later, a greatly improved version of this source was specially developed for accelerators, with charge exchange injection and for powerful linear accelerators[24] (Fig. 3). Research on sources of this kind subsequently started at a number of laboratories[14,25-27]. In these sources the fast ions

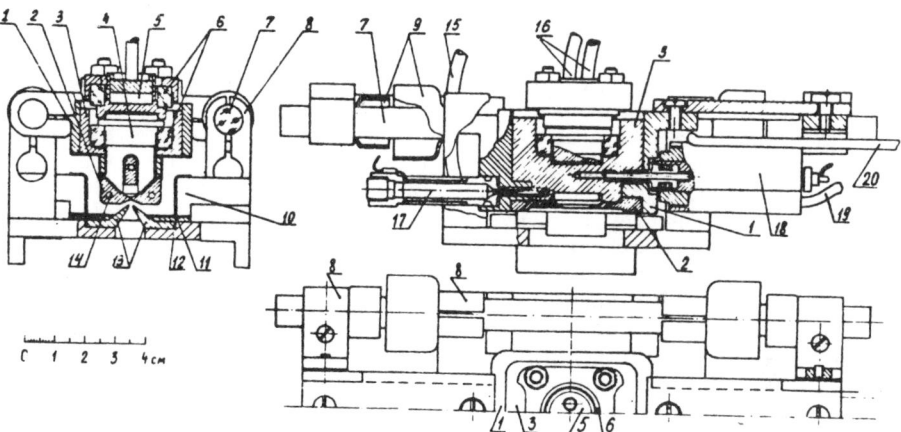

Fig. 3. Design of a Penning geometry source. 1 - body of the discharge chamber; 2 - chamber cover with emission slit; 3 - anode; 4 cathode; 5 - cathode cooling; 6 - cathode insulator; 7 - high-voltage insulators; 8 - brackets; 9 - shields; 10 - magnetic poles; 11 - projections of poles; 12- base; 13 - extractor electrodes; 14 - anode cooling channels; 15 - anode cooling pipes; 16 - cathode cooling lines; 17 - cesium feed; 18 - hydrogen valve; 19 - hydrogen supply; 20 - valve cooling pipes.

emitted by the cathode cannot pass into the emission slit of the source. The beam is formed of slow H⁻ ions produced in the recess adjacent to the emission slit, as a result of resonance charge exchange (on hydrogen atoms) of the fast ions emitted from the cathode and accelerated in the cathode layer of the discharge. The principal source of atomic hydrogen in the discharge is the dissociation of molecules by electron impact. According to the Franck-Condon principle, the atoms of dissociated molecules should have an energy of $\cong$ 2.2 eV if the molecule dissociates from the ground state. In the discharge plasma of SPS, the probability of stepwise dissociation of molecules from excited states may be high; then the energy of the atoms may be significantly lower. Furthermore, the wall recombination coefficient (atoms to molecules) is $\simeq 0.2$[16]. The atoms therefore will effectively exchange their energy in collisions with neutral molecules and the walls. We thus may expect that the average energy, and hence the H^- ion temperature in the recess, will be <1 eV. Hence it follows that SPS with a Penning geometry are preferable for purposes of shaping intense beam of H^- ions. Subsequently, an analogous conclusion was formulated in Refs. 28 and 29 following a comparison of the brightness characteristics of SPS with planotron and Penning geometries.

BEHAVIOR OF THE DISCHARGE

Oscillations of the gas-discharge plasma are the main source of the temperature increase of ions in the beam that ultimately determines beam brightness. By influencing the motion of ions in the plasma, the electric fields of these oscillations directly increase their temperature.

The fluctuations of plasma density and electron temperature lead to rapid changes in the location and shape of the plasma emission boundary, giving rise to oscillations in the ion beam being formed and thus causing subsequent heating of the ions in the beam.

As in the case of positive ion sources, the stability of the discharge with respect to the development of oscillations depends on the geometry of the discharge chamber, magnetic field configuration, and the design of the electric power supply and gas supply. These aspects have been examined previously on numerous occasions[30,31].

Here we shall examine only the specific features of the behavior of the discharge in SPS that are associated with the presence of cesium.

Relaxation Oscillations. The main purpose of having cesium in the discharge[9] is to reduce the work function of the surface of the molybdenum cathode. This decrease is effected when the optimal cesium coverage in Fig. 4 is used, since this results in a sharp increase in the rate of H^- generation. The change in the work function of the cathode surface will entail a change in the overall secondary electron emission coefficient $\gamma = j_e/j^+$ (where j_e and j^+ are the densities of the electron current from the cathode and of the ion current to the cathode, respectively). An increase in γ leads to a decrease in the voltage drop across the discharge (Fig. 5). Thus, in a pure hydrogen discharge $\gamma \simeq 0.1$-0.2 and $U_p \simeq 400$-600 V, whereas $\gamma \simeq 5$-6 and $U_p \simeq 100$ V when an optimal cesium coverage is used. In order for an optimal equilibrium coverage to exist, the cesium flux from the surface due to thermal desorption and bombardment by ions and fast neutrals must equal the cesium flux onto the surface. In other words, for an equilibrium cesium concentration to exist on the surface, a corresponding optimal cesium density must be maintained in the discharge. If the amount of cesium in the discharge is suboptimal, then equilibrium is only possible at some point on the left branch of the curve in Fig. 4. But because of the

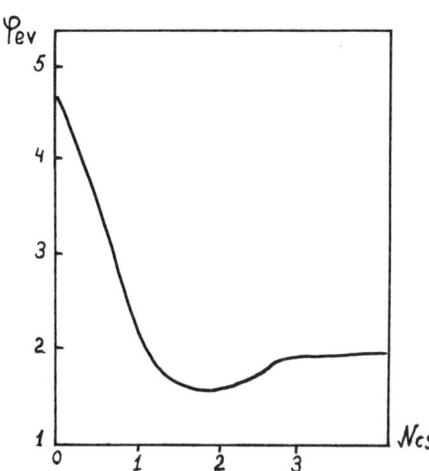
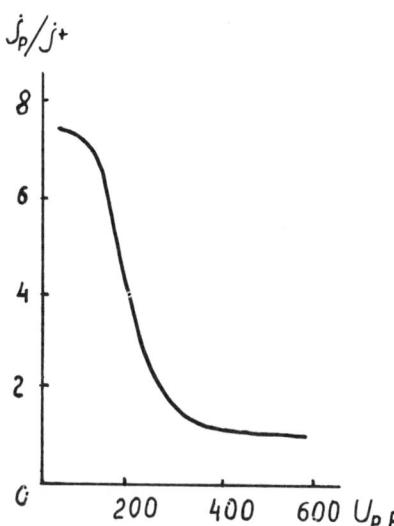

Fig. 4. Diagram of the dependence of the work function φ of the cesium-coated molybdenum cathode on the concentration N_{Cs} on the surface.

Fig. 5. Dependence of the ratio of the cathode current density to the ion current density (j_p/j^+) on the discharge voltage.

pumping effect of the discharge[32-34] and the increase in the desorption rate with increasing U_p, this equilibrium turns out to be unstable and low frequency oscillations arise in the discharge with a period of 5-10 μs. This period is determined by the amount of time required for cesium transport between the electrodes (the velocity of cesium atoms is $\cong 10^4$ cm/s, and the characteristic transport distance is 1-2 mm). This regime is a transient one, from a high voltage discharge with no cesium ($U_p \simeq 400$-600 V) to a low voltage discharge with cesium ($U_p \simeq 100$ V). Here the voltage across the discharge may undergo fluctuations over the maximum range when there is a minimal supply of hydrogen. This regime is not of interest from the standpoint of generating negative hydrogen ions.

The Noisy Regime. As the source heats up and the hydrogen and cesium densities in the discharge rise, the level of oscillations can be reduced comparatively easily to 20-30%, the oscillations become more regular in character, and their spectrum shifts into the RF region, extending from 1-2 to 10-15 MHz (Fig. 6a). In this regime stable operation of the source is achieved with a high rate of negative ion generation if the beam formed has satisfactory quality (the characteristic value of the normalized emittance is 2-5×10^{-4} cm rad for a 0.1 A current).

Attempts have been made to study these oscillations[35], but the results are extremely modest. The nature of the oscillations remains obscure; nonetheless it has been established that the amplitude of the

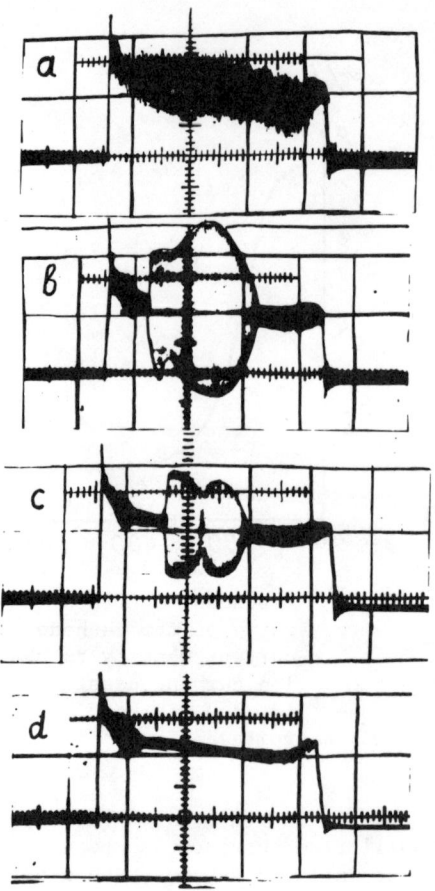

Fig. 6. Oscillograms of the discharge voltage under different conditions: a - noisy regime; b, c - regime of RF oscillations; d - noiseless regime.

fluctuations in U_p and J_p has a definite tendency to decrease with diminishing magnetic field and increasing hydrogen density. Moreover, by proceeding in this direction one generally can produce a discharge without observable fluctuations in the frequency band up to 50 MHz. Because of the wide spectrum, no definite dependence of the characteristic frequencies of the oscillations on the magnetic field, hydrogen density, or discharge current can be found.

The high rate of H^- generation in this regime attests to the fact that a coverage somewhat above the optimal or optimal is maintained on the cathode, i.e., the working point in the diagram in Fig. 4 is located on the right branch. It follows from the considerations presented above that this situation should be stable, but this does not eliminate the possibility of fluctuations of the cesium concentration on the surface around the mean equilibrium state. The fluctuations in the cathode emissivity due to these variations lead to sharp changes in the voltage drop across the discharge, and hence in the electron energy distribution. This is the cause of the oscillations in the plasma column, and is evidenced in the observed dependence of the level of the oscillations on the magnetic field, hydrogen density, and other factors.

Thus processes that occur on the cathode surface and that lead to changes in cathode emissivity are a source of noise in the discharge. Hence it follows that if we work on the plateau in the diagram in Fig. 4 where the work function is practically independent of the cesium concentration on the surface, then we may expect stabilization of the regime. Indeed, experience has showed that by increasing the cesium density in the discharge, beyond the density that is optimal from the standpoint of negative ion yield, it is possible to obtain a regime with no observable fluctuations by using a moderate hydrogen supply and not too weak a magnetic field (Fig. 6) and to maintain it for a long period of time[36,29].

The Regime of RF Oscillations of discharge voltage and current is characterized by regular oscillations at a frequency of 10-15 MHz and by a level of up to 30-50%, with a characteristic oscillation rise time of 10 to 100 μs (Fig. 6b and 6c). This regime becomes established either from a regime with a low noise level (a few percent) or from a noiseless regime after some amount of time has elapsed (from a few minutes to a few hours). Cases have been observed in which the discharge began to generate RF after one or two days of operation under noiseless conditions. It should be noted that RF components with a relative amplitude of 5-10% always can be seen in the spectrum of a noisy discharge. No definite dependence on the magnetic field is observed for oscillations of this type, and the dependence on the hydrogen density in the discharge is weak and unclear. Therefore their nature remained obscure until recently. The beam emittance generally is much lower in this regime than in the noisy regime, but is higher by a factor of 2-3 than in the noiseless regime.

In Ref. 36 attention is drawn to the fact that the frequency of the RF oscillations is close to the oscillation frequency of the H^+ ion in a parabolic potential well having a depth on the order of the ionization potential and a width equal to the characteristic transverse dimension of the plasma column. A 1/L dependence of the frequency on the distance L between the cathodes was found in Ref. 27.

On these grounds the oscillations are identified as ion-acoustic oscillations. It was recently established[36] that stable generation always can be either eliminated or significantly reduced, at least for a short period of time (5-10 min during operation at 50 Hz), by lowering the discharge current by 10-20% or by slightly decreasing the hydrogen supply. This suggests that the generation of the RF oscillations is associated with a disruption of equilibrium between the production and loss of charged particles in the plasma column. In particular, the plasma in clean discharges usually acquires a small positive potential with respect to the anode as a result of the higher mobility of the electrons. In a surface plasma source a cesium coating also reduces the work function of the anode. Under these conditions the anode may become an effective emitter of electrons into the plasma. If the electron stream from the anode exceeds the electron current onto the anode from the plasma, then the anode potential drop is reversed[37]. This leads to a fast build-up of ions in the plasma and to sharp expulsion of electrons, giving rise to oscillations in the anode plasma sheath.

The consideration presented above should be viewed only as an attempt to establish a relationship between the processes that occur on the cathode and anode of an SPS discharge with oscillations, since such a relationship now seems to be the most natural one and is not inconsistent with available empirical data. Therefore, on the whole the problem of producing intense beams of H⁻ ions requires detailed investigation and a search for methods of effective suppression of plasma oscillations in surface-plasma sources.

ION EXTRACTION

The implementation of a noiseless discharge regime creates conditions favorable to the establishment of a stable plasma emission surface that does not experience perturbations. In this case an increase in beam brightness is possible only by reducing the average ion energy on the plasma emission surface, increasing the emission current density, and minimizing ion heating due to aberrations during acceleration in the extraction gap.

It was demonstrated above that in a Penning source we may expect to obtain an ion temperature of less than 1 eV in the plasma adjacent to the emission slit. In the case of SPS, a further decrease in ion temperature during adiabatic expansion of the plasma escaping into the vacuum (which is standard practice in positive ion sources[38]) is difficult because of the short lifetime of H⁻ ions in the plasma.

When ions are extracted from the plasma, the emission current density is known to be

$$j_0^- \sim E_0^{3/2}/D^{1/2},$$

where E_0 is the electric field strength in the plane of the extractor electrode and D is the length of the extraction gap. The figure $E_0 \cong 10\text{-}15$ kV/mm that has been attained in SPS (corresponding to an extraction voltage of 20-25 kV for $D \cong 2$ mm) is limited by the dielectric strength of the gap. As D is reduced, the scattering effect of the slit in the extractor electrode increases; hence the aberrations also grow. Obviously, the level of aberrations should be considered acceptable if the increase induced by them in the transverse energy of ions in the beam does not exceed the ion energy on the emitter.

In the overwhelming majority of existing SPS for accelerators, the ions are extracted through a narrow slit oriented across the magnetic field, generally with a cross section of 0.5 x 10 mm². For convenience, hereinafter we will designate the direction across the emission slit (along the magnetic field) by the X coordinate, the direction along the emission slit (across the magnetic field) by the Y coordinate, the direction of beam travel by the Z coordinate, and the angular slope of the paths in the planes (XZ) and (YZ) by α_x and α_y, respectively. Numerous tests and a number of design considerations indicate the expediency of using a quasi-Pierce geometry of the extracting gap[24] (Fig. 3). Numerical simulation of beam formation in such a geometry[39] has demonstrated that for an emission current $j^- \cong 3$ A/cm² equal to the calculated j^- according to the 3/2 law, the beam formed acquires an emittance $E \cong 3 \times 10^{-7}$ cm rad due to aberrations. This corresponds to an ion temperature of $\cong 0.07$ eV on the emitting surface. If the calculated current density j^- is increased to 5 A/cm^2 and the emission current density is left as before, then the emittance increases to 7×10^{-7} cm rad, and hence the ion temperature rises to 0.4 eV. The calculated phase diagram is shown in Fig. 7. The emission boundary takes on a concave shape,

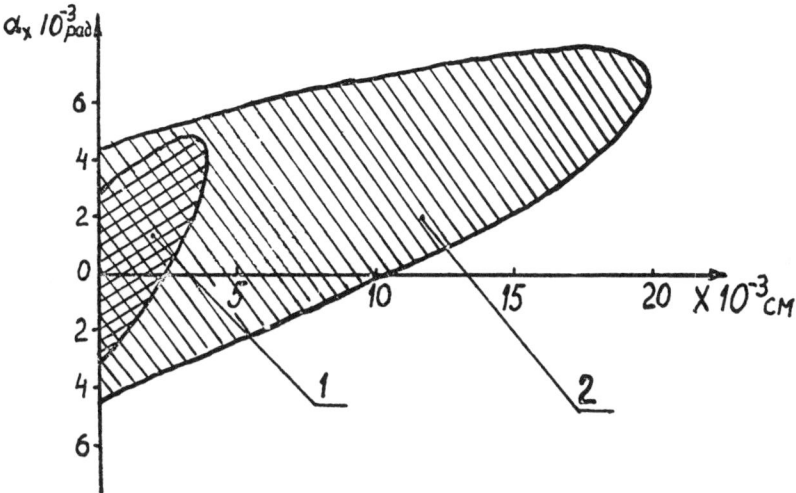

Fig. 7. Calculated phase diagram of the beam in a surface-plasma source due to aberrations in the extraction gap (1) and to charge exchange (2).

and the increase in emittance basically occurs during the initial forming segment, over a distance on the order of the width of the emission slit. Hence we can see that according to the foregoing, at the anticipated ion temperature $\lesssim 1$ eV it is not effective to increase the emission current density beyond 3 A/cm^2.

Similar findings also are presented in Ref. 28. It was assumed in the calculations that the plasma emits 0.4 eV ions normal to the emission boundary; no allowance was made for the magnetic field or the effect of accompanying electrons. Therefore, the working current density of 1.5-2 A/cm^2 that has been attained in existing SPS seems to be close to the optimum concerning the brightness.

The optimal regime for ion extraction with minimal ion temperature increase due to aberrations is realized when the density j of the ion current from the plasma to the plasma boundary is equal to the emission current density defined by the 3/2 law.

In this case an emission boundary of optimal shape is formed. In SPS this regular pattern of behavior is evident only when the fluctuations of the total ion current are of a quite low level ($<1\%$) and the magnetic field strength is not very high ($\lesssim 1$ kG). This manifests itself in the fact that for a specified gap geometry and extraction voltage, there exists an optimal value of the discharge current (ion density) for which a maximum current density is observed in the beam, while the total beam current increases with the discharge current. This is illustrated in Figs. 8 and 9, which show

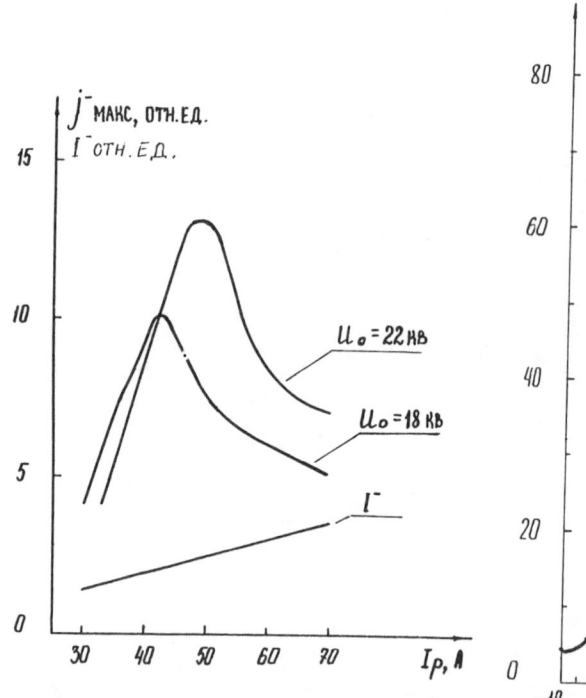

Fig. 8. Dependence of the maximum current density in the beam (j_{max}^-) and of the total current (I^-) on the discharge current for different values of the extraction voltage U_0.

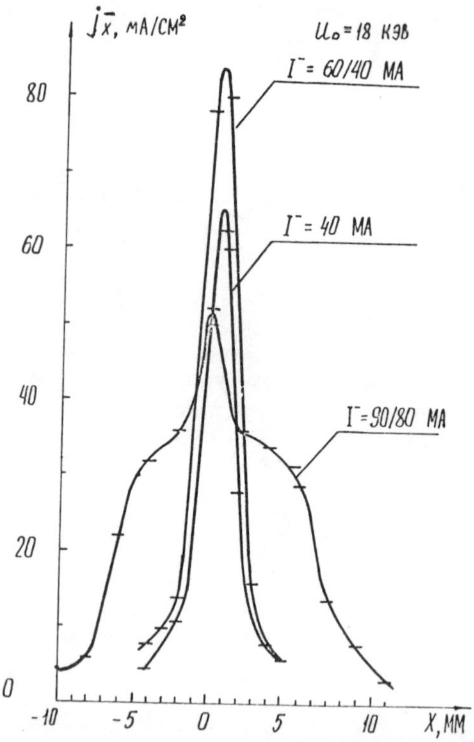

Fig. 9. Beam current density distribution as function of the beam current.

the dependence of the maximum current density at beam center on the discharge current and the transformation of the current density distribution for different values of the total current (discharge current).

It should be noted that despite the fact that oscillations are absent in the discharge or their level is very low, in some cases significant fluctuations may appear in the beam, originating in the extraction gap during ion extraction[36].

BEAM FOCUSING AND TRANSPORT

The quality of beam forming can be estimated conveniently from the average energy of ion thermal motion in the beam, referred to the emission slit. This energy is determined by using expression (4) and the measured values of the beam emittance in each direction.

$$W_X \simeq \frac{Mc^2 \cdot E_X^2}{2X_0^2} \; ; \; W_Y \simeq \frac{Mc^2 \cdot E_Y^2}{2Y_0^2} \qquad (5)$$

By comparing the values obtained in this manner with the expected plasma ion temperature T_i, it is possible to estimate the degree of ion heating during focusing for each direction separately.

In the overwhelming majority of existing SPS, H⁻ ions are extracted through a narrow emission slit with a ratio of $\cong 20$ or more between the sides. Therefore, a ribbon beam with a regular divergence of 0.5-1×10^{-2} rad with respect to X is formed immediately after the extracting electrode. Hence one-dimensional focusing is needed to obtain a quasi-parallel beam in both transverse directions. To this end, the proposal was made in Ref. 23 to use a 90° bending magnet with a field index $n \cong 1$ (this scheme is now the one commonly used; see Fig. 10). When the beam moves in such a field, its size with respect to Y undergoes practically no change, but increases in the X direction by a factor of 10-20 depending on the initial divergence. The radius of the mean ion path in the magnet usually is 7-10 cm. In this case, during focusing of a beam carrying a 0.1 A current at an energy of $\cong 20$ keV, the strength of the space charge exceeds the focusing power of the magnetic field by a factor of 3-4. Therefore, effective focusing is possible only if the negative space charge of H⁻ ions in the beam is largely compensated by positive ions of the residual gas. The gas pressure required for this to happen ($\cong 10^{-4}$ torr) is maintained in the magnet by the hydrogen stream from the source, and additional hydrogen supply generally is not needed. As a result, a high coefficient of beam transmission through the magnet is ensured ($\cong 90\%$).

The degree of compensation of the beam space charge, as determined by the ratio of the density of positive ions (n^+) to the density of negative ions (n^-) in the beam, is governed by the balance between the rate of positive ion production in the beam and the rate of positive ion loss from the beam. It is clear from general considerations that the production rate is proportional to n_0 (the gas density), and the loss rate decreases with increasing mass of the positive ions. If we assume that the rate of positive ion loss is determined solely by diffusion, as numerical simulation shows[40], a state in which $n^+ \simeq n^- \gg n_e$ (where n_e is the electron density) may occur even when the gas density is low. This has been confirmed experimentally in the case of an electron beam[41]. In reality, the loss of compensating ions from the beam is enhanced by Coulomb collisions with beam ions, by recombination, and by other effects. Furthermore, anomalous loss of ions is possible. For example, one mechanism that leads to a higher ion loss is the driving of ion oscillations in the potential well by fluctuations in the intensity of the H⁻ ion beam, the transport of compensating ions along the beam, etc.[42] As a result, beginning at low pressures ($\cong 10^{-5}$ torr) the degree of compensation of the beam space charge increases with

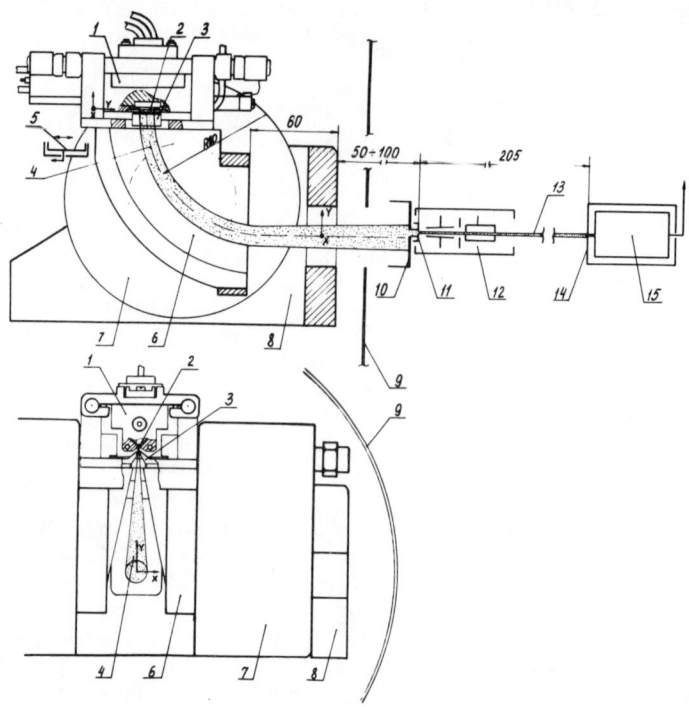

Fig. 10. Diagram of beam formation and of the investigation of the beam ion-optical characteristics. 1 - discharge chamber of source; 2 - emission slit; 3 - plates of extraction electrode; 4 - negative ion beam, 5 - movable collector for measuring the extracted beam; 6 - poles of bending magnet; 7 - sealed coils of magnet; 8 - yoke of magnet; 9 - shielding of ion source; 10 - collector for measuring the final beam; 11 - collector for measuring the current density; 12 - deflection system; 13 - ion beamlet; 14 - shield with analysing aperture; 15 - collector for measuring the ion distribution in the beamlet by using a current amplifier.

increasing residual gas density. A state in which $n_e \ll n^+ \simeq n^-$ occurs is realized because of the fact that electrons are expelled from the beam by the negative space charge. When the residual gas density is close to the critical[43],

$$n_c \simeq \frac{2}{R_0 \cdot \sigma_i \cdot v} \left(\frac{T_i}{M}\right)^{1/2}$$

complete compensation of the space charge is achieved: $n^+ \simeq n^- + n_e$.

A further increase in gas density only leads to an increase in the beam plasma density--an overcompensated state.

The action of the beam space charge is known to lead to an increase in the regular beam angular divergence while the beam is moving in a drift space. This means that as a first approximation, the space charge behaves like a linear diverging lens, and therefore should not lead to a significant increase in beam emittance. The situation changes fundamentally if the beam has fluctuations of intensity. Then the corresponding fluctuations of the space charge lead to rapid changes in the orientation of the phase ellipse in the phase plane; on average, this is equivalent to a significant increase in beam emittance.

In this regard the overcompensated regime is interesting in that if the characteristic times of the fluctuations in the beam are shorter than the loss time of compensating ions, the change in ion density in the beam will be compensated by the redistribution of electrons. The experimental investigation of the brightness characteristics of a beam with a compensated space charge[20,28,35] has showed that even when the intensity fluctuations of an ion beam are significant ($\cong$10-15%), its emittance decreases by a factor of 2-3 as the residual gas pressure increases. However, even in the case of a clean noiseless discharge, beams with a reduced temperature W_Y less than 5 eV and W_X less than 27 eV cannot be obtained in the source in this manner[35].

Detailed studies of the conditions of formation of an intense beam in SPS[36] have made it possible to determine that significant additional heating of ions in the beam occurs when it is focused and transported in a bending magnet. To minimize this heating, the beam extracted directly from the source must have intensity fluctuations as small as possible ($\lesssim$ 0.2-0.3%, in any event no more than 1%). The ions are most sensitive to heating in the X direction. This is explained by the fact that when a beam is focused in a bending magnet, its size with respect to X increases by a factor of 10-20; hence the value of the ion temperature in this direction is reduced by a factor of 100-400 compared to the initial value. At the same time, the ion temperature with respect to Y differs little from the initial temperature. Naturally, under such strongly nonequilibrium conditions the "cooled" degree of freedom turns out to be highly susceptible to additional heating. At least two sources of such heating can be pointed out.

As a result of the nonlinearity of focusing in the magnetic field, there is a coupling between the X and Y degrees of freedom, both to each other and to the longitudinal motion. This may cause aberrational heating of the ions in the X direction. As was demonstrated in Ref. 44, the aberrations can be reduced considerably if the magnetic field topography is properly selected. However, this is confirmed experimentally only for low beam currents (1-2 mA).

The collective processes that may develop in the plasma of a compensated beam obviously will be a more powerful heating source when an intense beam is focused[45,46]. It follows from the analysis given in Ref. 47 that the three-component plasma of a compensated beam always is unstable if the beam ion velocity exceeds the average plasma electron velocity. In SPS with an ion energy of $\cong$ 20 keV, this condition probably is met when the gas density is $n_0 \gtrsim n_e$. Figure 11 shows the spectra of electrons escaping the beam. We can see that the average electron energy is less than 10 eV.

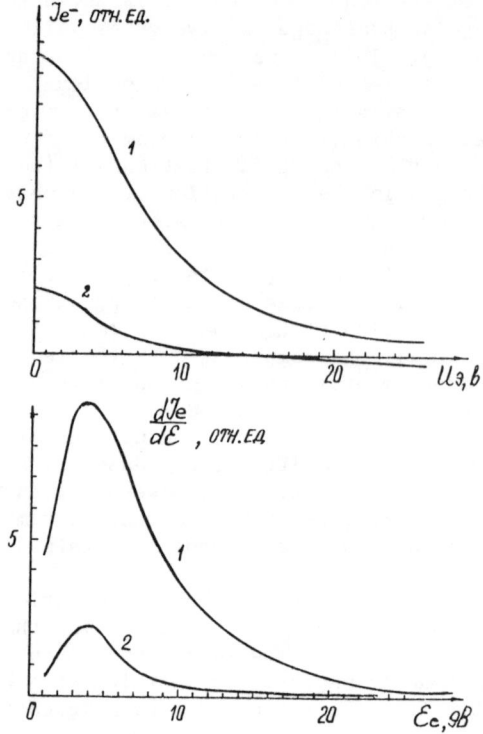

Fig. 11. Spectra of secondary electrons escaping from the beam.
$\underline{1}$ - secondary emission from collector is not suppressed;
$\underline{2}$ - secondary emission is suppressed.

Experiments under optimal conditions using an 80 mA beam at a fluctuation level below 0.5%[36] have showed that when such a beam passes through a bending magnet, ion heating does indeed take place in the X direction. The beam emittance triples from the initial $E_X \cong 2 \times 10^{-6}$ cm rad, whereas E_Y remains unchanged at the value $E_Y \simeq 2 \times 10^{-5}$ cm rad.

It also was established in these experiments that supplementary heating of the ions in the beam may increase significantly if secondary electrons are dumped into the beam from the surfaces surrounding the beam and exposed to it. Analogous effects were observed previously in compensated electron beams[48,49].

It has not been determined what mechanism is responsible for the observed heating. It most likely will depend on the experimental conditions in each specific case. For example, the authors of Refs. 42 and 50 believe that the driving of Langmuir ion oscillations by the beam may be the mechanism. Nor has it been excluded that a mechanism analogous to the one suggested in Ref. 51 may be at work here.

Density fluctuations in the beam are a source that causes the development of a beam instability that leads to the heating of ions in the beam. The observation of these fluctuations has made it possible to establish the following properties[36]:

1. Current density noise at a level of at lest 3-5% always is observed in the beam, even when the total beam current noise is at a very low level ($\cong 0.1\%$).

2. The current density noise level increases with decreasing size of the measuring slit (which ranged from 0.8 to 0.2 mm in experiments).

3. The noise level increases with increasing distance from the source to the detection plane.

4. Measurements of the linear current density using a narrow slit (0.2 mm wide) show that the noise level of j(Y) with the slit oriented along X always is higher than for j(X), with the slit oriented with respect to Y. The anisotropy of the noise level, which is 2-3 at a distance of 3 cm from the extracting electrode, increases to 10-20 at the exit from the bending magnet.

The first two properties indicate the statistical nature of the current density fluctuations, whereas their relative level is proportional to $1/\sqrt{j}$ (j is the measured current through the slit).

The first two properties indicate that the initial statistical fluctuations are subsequently aggravated during beam transport on fine-structured potential distributions in the plasma of the compensated beam. The perturbations of this potential are quickly leveled out by the motion of electrons along the magnetic field lines; as a result, anisotropy of the current density noise level is observed. Of course, this is possible when $n_e \gtrsim n^-$, i.e., the beam must be overcompensated. The gas density in the magnet probably is a fortiori above-critical, since the beam traverses a significant fraction of its path in the magnet in the gas stream flowing from the source.

This conclusion has been confirmed by experiments using a small bending magnet with an average path radius of 3.5 cm and an aperture of 1.5 cm. In this magnet the beam is quickly extracted from the gas jet; this cannot be expected in a large magnet with a path radius of 8 cm. The best result was obtained with a beam carrying 40 mA of current at an energy of 20 keV. The phase diagram of this beam is shown in Fig. 12. Emittances $E_X \cong 7 \times 10^{-7}$ cm rad, $E_Y \cong 1.4 \times 10^{-5}$ cm rad and $W_Y \cong 0.3$ eV, $W_X \cong 0.4$ eV correspond to this diagram.

AN AXISYMMETRIC BEAM

In the initial stage of research into SPS, the resultant beam emittance was approximately the same in the X and Y directions because of the noise in the source plasma and beam heating during beam formation, despite the fact that the ions were extracted from a narrow slit with a ratio of sides of $\cong 20$. When bright beams are generated in sources with slit optics, E_X turns out to be much smaller than E_Y. Subsequently, during the acceleration, focusing, and transport of such a beam, the direction with a lower ion temperature will suffer an increase in the temperature because of the unavoidable coupling between the degrees of freedom, and the high beam brightness attained in the source will be lost. Therefore, the

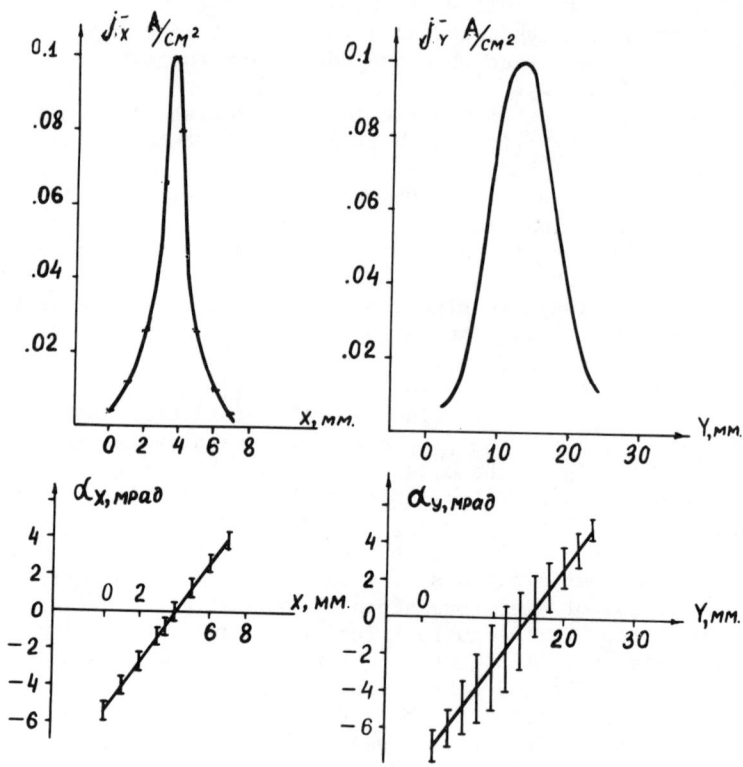

Fig. 12. Phase diagrams of the beam under optimized conditions: $I^- = 0.04$ A; $W_0 \cong 22$ keV; $E_X \cong 7 \times 10^{-7}$ cm rad; $E_Y \cong 1.4 \times 10^{-5}$ cm rad.

question of the shaping of an axisymmetric beam in SPS becomes extremely important. By virtue of the considerations presented above regarding aberrations in the extraction gap, the question of the acceptable intensity of an axisymmetric beam remains open. The first results from experiments on ion extraction from a round aperture[52] allow us to hope that in any case the production of quite bright axisymmetric beams with currents of 20-30 mA is possible in SPS.

In Penning geometry SPS the emission current density is proportional to the cathode current density up to values of $\cong 400$ A/cm^2. Therefore, in sources with a round emission aperture it is reasonable to reduce the cross section of the discharge channel so as not to reduce the overall energy efficiency. Experiments have demonstrated that the stability of the discharge is retained as the channel cross section is reduced all the way down to a value of 3×4 mm^2. As follows from Fig. 13, the emission properties of the source are practically unchanged.

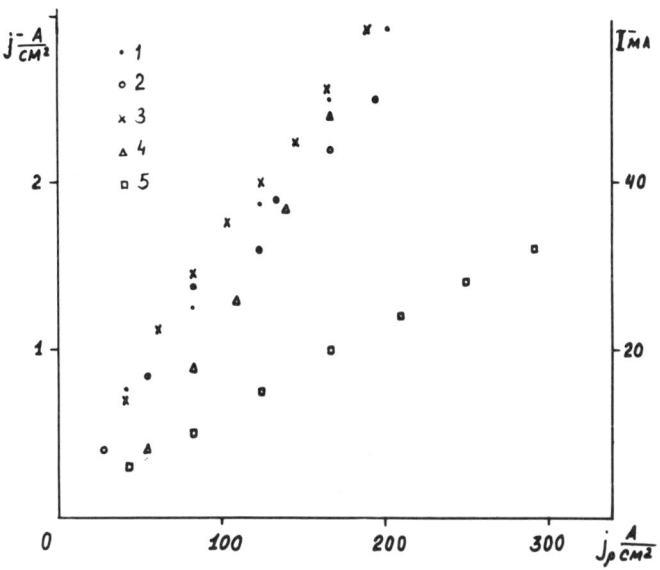

Fig. 13. Dependence of emission current density on discharge current density for various cross sections of the discharge channel (1, 3 x 15 mm²; 2, 3 x 10 mm²; 3, 3 x 6 mm²; 4, 3 x 4 mm²) and on the total current from an aperture 1 mm in diameter for a channel cross section of 3 x 4 mm² (5).

This fact was taken into account during the development of a source with an emission aperture of 1 mm diameter; the design of this source is shown in Fig. 14.

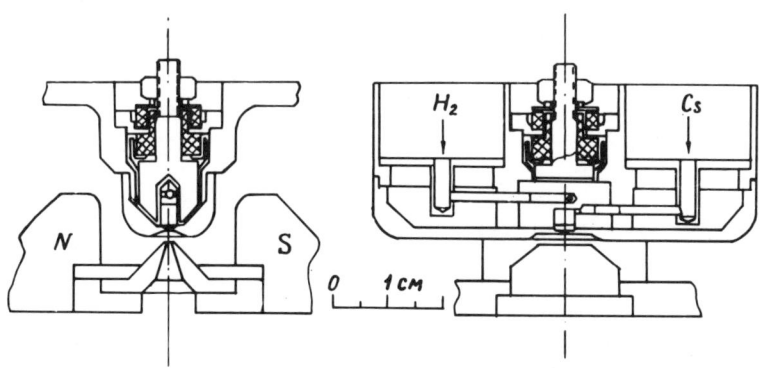

Fig. 14. Design of a source for producing an axisymmetric beam.

Fundamentally, this source differs little from the one shown in Fig. 3[24]. The discharge channel cross section in the source is 3 x 4 mm^2. A reduction in the average power of the discharge made it possible to not use forced cooling, and thus to greatly simplify the design. Cylindrical quasi-Pierce optics were used in the source. The characteristic operating conditions of the source were: discharge current, $\cong$ 30 A; voltage across discharge, $\cong$ 100 V; pulse length, $\cong$ 100 μs; pulse repetition rate, 50 Hz; extracting voltage, $\cong$ 20 kV; current in the beam, 20 mA. As yet it has not been possible completely to eliminate the oscillations in the discharge; the level of fluctuations in the beam generally is $\cong$ 10%. Despite this, a beam with relatively good characteristics as shown in Fig. 15, is produced from the source without supplementary focusing. As

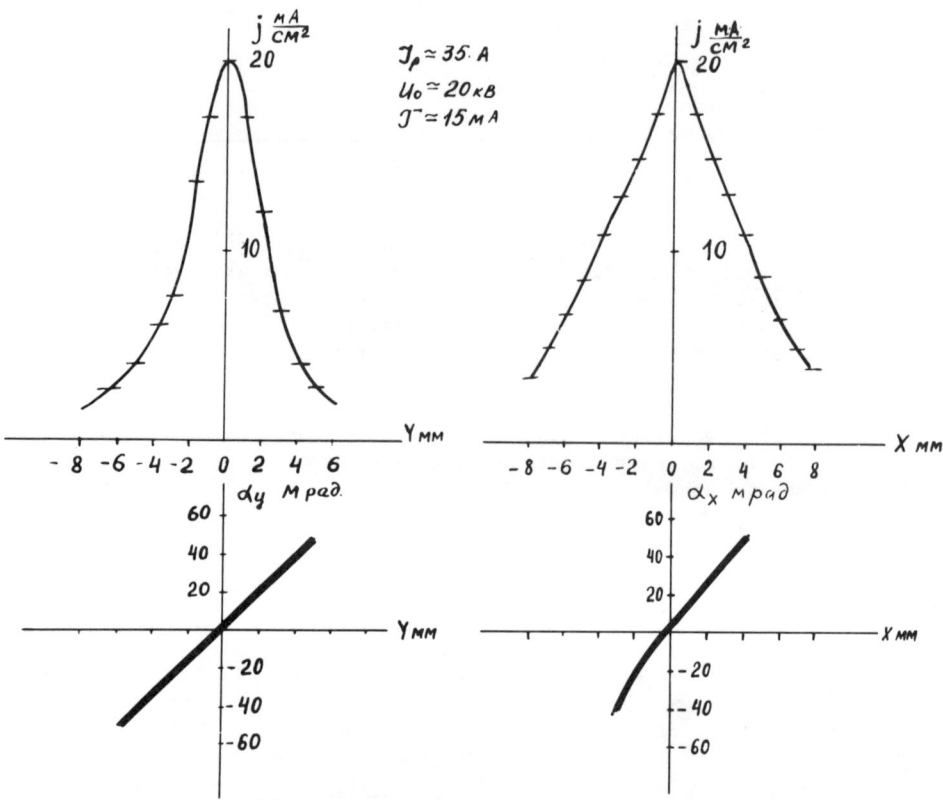

Fig. 15. Phase diagrams of an axisymmetric beam.

we can see from the figure, within the accuracy of the measurement an axisymmetric beam is formed in the source. The beam emittance is somewhat higher than in Ref. 52. This is due to the fact that in the measurements the emittance is determined at the level of 0.1 of the maximum in the distribution, which includes 90% of the beam in the four-dimensional phase volume, whereas in Ref. 52 this fraction is 67%. A brightness of $\cong 2 \times 10^7$ A/cm^2 rad^2 corresponds to the measured value of the emittance: $E_X \cong E_Y \cong 1-1.5 \times 10^{-5}$ cm rad.

REFERENCES

1. L. W. Alvarez, Rev. Sci. Instr., 22, 705 (1951).
2. G. I. Budker, G. I. Dimov, and V. G. Dudnikov, Atomnaya Energiya, 22, 348 (1967).
3. G. I. Dimov and V. G. Dudnikov, Fizika Plazmy, 4, 701 (1978).
4. G. I. Dimov and G. V. Roslyakov, Pribory i Tekhnika Eksperimenta, No. 1, 29 (1974).
5. J. D. Simpson, IEEE Trans. Nucl. Sci., NS-20, No. 3, 198 (1973).
6. Yu. I. Bel'chenko, G. I. Dimov, and V. G. Dudnikov, Doklady Akademii Nauk SSSR, 213, 1283 (1973).
7. Yu. I. Bel'chenko, G. I. Dimov, and V. G. Dudnikov, Izvestiya Akademii Nauk SSSR, Seriya Fizicheskaya, 37, 2573 (1973).
8. Yu. I. Bel'chenko, G. I. Dimov, and V. G. Dudnikov, Fizicheskie osnovy poverkhnostno-plazmennogo metoda polucheniya puchkov otritsatel'nykh ionov [Physical Principles of the Surface-Plasma Method of Producing Negative Ion Beams], Preprint 77-57, Institute of Nuclear Physics, Novosibirsk, 1977.
9. V. G. Dudnikov, Proc. 2nd International Symposium on the Production and Neutralization of Negative Hydrogen Ions and Beams, BNL-51304, New York, 1980, p. 137.
10. K. Prelec and Th. Sluyters, IEEE Trans. Nucl. Sci., NS-22, No. 3, 1662 (1975).
11. P. W. Allison, IEEE Trans. Nucl. Sci., NS-24, No. 3, 1594 (1977).
12. C. Schmidt and C. D. Curtis, Proc. 1976 Proton Linear Accel., AECL, Ontario, 1976, p. 402.
13. C. Ankenbrandt et al., Proc. 11th International Conference on High Energy Accelerators, Geneva, 1980, p. 260.
14. P. E. Gear and R. Sidlow, Low Energy Ion Beams (London, 1980), p. 281.
15. K. Matsuo et al., Proc. Symp. on Ion Sources and Appl. Technology, Tokyo, 1981, p. 43.
16. M. D. Gabovich, Fizika i Tekhnika Plazmennykh Istochnikov Ionov [Physics and Technology of Plasma Ion Sources], (Atomizdat Publishers, Moscow, 1972).
17. I. D. Lawson, The Physics of Charged Particle Beams (Oxford, 1977).
18. G. I. Dimov, Poluchenie Intensivnykh Puchkov Otritsatel'nykh Ionov Vodoroda [Production of Intense Beams of Negative Hydrogen Ions], Preprint 81-98, Institute of Nuclear Physics, Novosibirsk (1981).
19. C. W. Schmidt and C. D. Curtis, IEEE Trans. Nucl. Sci., NS-26, No. 3, 4120 (1979).
20. D. C. Barton and R. L. Witkover, IEEE Trans. Nucl. Sci., NS-28, No. 3, 2681 (1981).
21. W. G. Graham, See ref. 8, p. 53.
22. J. N. M. van Wunnik, Electron Transfer in Gas Surface Collisions (Amsterdam, 1983).

23. V. G. Dudnikov, Trudy IV Vsesoyuznogo Soveshchaniya po Uskoritelyam Zaryazhennykh Chastits [Transactions of the Fourth All-Union Conference on Charged Particle Accelerators] (Nauka Publishers, Moscow, 1975), vol. 1, p. 323.
24. G. I. Dimov, G. E. Derevjankin, and V. G. Dudnikov, IEEE Trans. Nucl. Sci., NS-24, No. 3, 1545 (1977).
25. P. W. Allison, IEEE Trans. Nucl. Sci., NS-24, No. 3, 1594 (1977).
26. D. R. Moffet and R. L. Barnet, IEEE Trans. Nucl. Sci., NS-28, No. 3, 2678 (1981).
27. E. A. Wadlinger, J. A. Farrel, and H. D. Dogliani, IEEE Trans. Nucl. Sci., NS-30, No. 2, 1408 (1983).
28. Th. Sluyters and V. Kovarik, in Proc. 1979 Linear Accel. Conf., BNL-51134, New York, 1979, p. 428.
29. H. V. Smith and P. W. Allison, Rev. Sci. Instr., 53, 406 (1982).
30. G. Lejeune, Proc. 2nd Symp. on Ion Sources and Formation of Ion Beams, LBL-3399, Berkeley, 1974, I-1.
31. T. S. Green, See ref. 30, I-2.
32. Yu. I. Bel'chenko et al., Pis'ma v Zhurnal Tekhnicheskoi Fiziki, 3, 693 (1977).
33. H. V. Smith and P. W. Allison, IEEE Trans. Nucl. Sci., NS-26, No. 3, 4006 (1979).
34. A. V. Zharinov and Yu. V. Sanochkin, Fizika Plazmy, 9, 397 (1983).
35. G. E. Derevyankin and V. G. Dudnikov, Formirovanie puchkov ionov H⁻ Dlya Uskoritelei v Poverkhnostno-Plazmennykh Istochnikakh [Formation of H⁻ Ion Beams for Accelerators Using Surface-Plasma Sources], Preprint 79-17, Institute of Nuclear Physics, Novosibirsk, 1979.
36. G.E. Derevyankin, V. G. Dudnikov, and M. L. Troshkov, Osobennosti formirovaniya puchkov ionov H⁻ v Poverkhnostno-Plazmennykh Istochnikakh Dlya Uskoritelei [Characteristics of the Formation of H⁻ Ion Beams in Surface-Plasma Sources for Accelerators], Preprint 82-110, Institute of Nuclear Physics, Novosibirsk, 1982.
37. V. L. Sizonenko, Zhurnal Tekhnicheskoi Fiziki, 51, 2283 (1981).
38. Th. Sluyters, Proc. 2nd Intern. Conf. on Ion Sources, Vienna, 1972, p. 190.
39. V. I. Davydenko and N. G. Khavin, Chislennoe Modelirovanie Formirovaniya Puchkov Ionov H⁻ v Poverkhnostno-Plazmennykh Istochnikakh [Numerical Simulation of H⁻ Ion Beams in Surface-Plasma Sources], Preprint 79-27, Institute of Nuclear Physics, Novosibirsk, 1979.
40. O. A. Anderson and E. B. Hooper, See ref. 8, p. 205.
41. V. I. Kudelainen, V. V. Parkhomchuk, and A. V. Pestrikov, Zhurnal Tekhnicheskoi Fiziki, 53, 691 (1983).
42. D. G. Dzhabbarov and A. P. Naida, Zhurnal Eksperimental'noi i Teoreticheskoi Fiziki, 78, 2259 (1980).
43. M. D. Gabovich et al., Zhurnal Tekhnicheskoi Fiziki, 44, 861 (1974).

44. J. D. Sherman and P. W. Allison, IEEE Trans. Nucl. Sci., NS-26, No. 3, 3916 (1979).
45. M. D. Gabovich, Uspekhi Fizicheskikh Nauk, 121, 259 (1977).
46. J. M. Dolique, See ref. 8, p. 215.
47. R. J. Turnbull and E. B. Hooper, See ref. 8, p. 236.
48. M. V. Nezlin et al., Zhurnal Eksperimental'noi i Teoreticheskoi Fiziki, 53, 1180 (1967).
49. M. V. Nezlin et al., Zhurnal Eksperimental'noi i Teoreticheskoi Fiziki, 55, 397 (1968).
50. D. G. Dzhabbarov and N. P. Naida, Fizika Plazmy, 6, 577 (1980).
51. M. V. Nezlin, Plasma Phys., 10, 337 (1968).
52. J. D. Sherman, P. W. Allison, and H. V. Smith, See ref. 9, p. 184.

OPERATIONAL EXPERIENCE WITH THE BNL MAGNETRON H⁻ SOURCE*

Richard L. Witkover
Accelerator Department, Brookhaven National Laboratory
Associated Universities, Inc., Upton, N.Y. 11973

ABSTRACT

A magnetron H⁻ source with a grooved cathode has been in operation at the BNL Linac for over 18 months. The source has run at 5 pps with a 600 μsec pulse width for periods as long as 5 months. Its development and performance will be discussed.

INTRODUCTION

The H⁻ source which was installed in September 1982 in Pit II of the BNL linac was based on the original magnetron source reported by Dimov et al.[1] and further developed by Sluyters[2] and Prelec at BNL. This design served as a starting point for Schmidt[3] at FNAL who brought it to full operational capability for accelerator use. While retaining the basic physical parameters the FNAL design provided a more durable configuration. A 90° bending magnet was added to select the H⁻ beam and focus in both planes. A cold-box cryopumped the cesium which left the source, preventing plating of the column components. The cesium delivery system was redesigned using a boiler to allow long term operation. The result was a truly operational source which ran at 15 pps with a 0.1% duty factor and 40-50 ma H⁻ current. This source was installed in the FNAL preinjector in late 1977. A second preinjector was adapted to H⁻ operation a year or so later.

Based on the success with H⁻ injection at Fermilab, Brookhaven started planning for a similar modification[4] using the magnetron source which had performed so well at FNAL. It was hoped that little modification and no development would be required since both machines operated at the same duty factor.

The actual mating of the source to the accelerating column was an obvious problem. The FNAL high gradient column used SF_6 gas in a liner to provide insulation between external electrodes, allowing close spacing. The BNL column design operates in air so the external electrode separation had to be much wider. Inside the BNL column, long conical sections were used to connect the actual accelerating electrodes with the external portions. The shallow cones in the Fermilab column allowed the source to be placed close to the first electrode. The source sits above the beam line, with the beam bent onto the axis by the 90° magnet, requiring considerable radial clearance. To get sufficient voltage hold-off, the BNL source had to be moved to the back-plate of the column. At this location the beam had to travel about 0.5 M from the 90° bending magnet to the first electrode.

---

*Work performed under the auspices of the U.S. Government of Energy

An estimate of the emittance at the extractor was made from FNAL data and used to redesign the bending magnet to reduce aberration terms and increase the focussing normal to the bend plane. This higher magnet gradient (n = 1.35) when coupled with a downstream quadrupole doublet allowed the production of a double waist at the aperture of the first electrode. To keep these pulsed quadrupoles at dome reference potential, a quartz cylinder was used to insulate them from a drift tube which is at extractor voltage. The calculated beam transport is shown in Figure 1. A schematic of the complete source assembly is presented in Figure 2. The difficulty in moving the source forward is clearly evident in this drawing. Note also the location of a beam toroid at the end of the extraction tube.

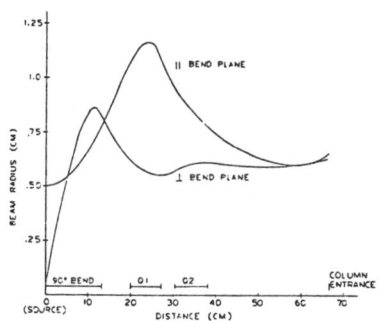

Fig. 1. 20 keV transport plot.

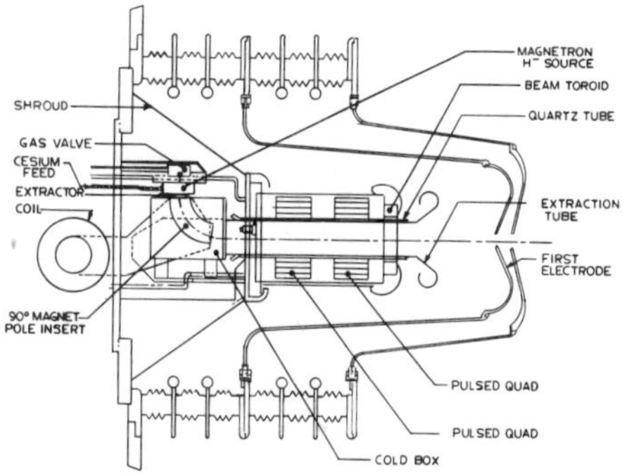

Fig. 2. Source assembly in BNL column.

SOURCE EVALUATION

Tests of the FNAL design source showed it could operate at the 5 pps and 200 sec pulse width required by the BNL linac. The discharge parameters were the same as reported by Schmidt[3], but the source was being stressed thermally. After several weeks of running, the source was disassembled and inspected. The MACOR insulators appeared to be melting away at the front of the source. The source temperatures were higher than that at Fermilab but seemed to be below the manufacturers limit. The thermocouples read only the average temperature of anode and of the cathode. Even though the duty factor for both sources is the same, the longer pulse at BNL probably caused the peak temperature to rise higher than at FNAL, reaching the point where the MACOR would break down and melt away. To counter this problem, a set of alumina insulators was installed. After two weeks of around-the-clock running, there was no melting or physical damage evident.

The emittance of the magnetron source was measured after the 90° bending magnet.[5] The results are shown in Table I for beams of 40 mA.

TABLE I
Emittance of the flat cathode magnetron source
for two bending magnet designs

Magnet Type	Narrow Plane $(\varepsilon\beta\gamma)_{90}$	Wide Plane $(\varepsilon\beta\gamma)_{90}$
FNAL (n=1.0)	0.07π cm-mrad	0.20π cm-mrad
BNL (n=1.35)	0.035π cm-mrad	0.14π cm-mrad

The improvement made by the new magnet is clearly seen. In addition, for the same source parameters, 30-50% more beam was observed after the magnet. During the course of these tests, improvements were made in some of the electronic support equipment. One area of improvement was in the operation of the pulsed gas valve, which uses a piezoelectric crystal to allow a burst of gas into the source. A voltage pulse of fixed amplitude (~ 100 V) was varied in width to control the gas setting. The long-term stability was found to be poor, with the output varying as the source heated. An adjustable bias voltage had been used to compensate for this variation. Short term drift was also observed. These problems (and the bias adjustment) were eliminated by redesigning the gas valve pulser to provide higher voltage pulses (> 150 V) and designing a regulator circuit which automatically adjusted the pulse width to keep the average vacuum readout constant.

Controlling the cesium boiler temperature was another problem area. The cesium delivery system consisted of a copper boiler topped by a mechanical valve. This was connected to a feedtube which delivered the cesium vapor to the source. The valve and feedtube are heated to prevent cesium condensation. Only the heater on the boiler is adjustable. It was found that ambient temperature variations and drift of the valve and feedtube temperatures affected the boiler temperature. A local servo-loop was put around the boiler heater and thermocouple, which eliminated this problem. Variations of tens of degrees in the valve or ambient temperature produced only a fraction of a degree change in the boiler.

INSTALLATION IN PIT II

The source was installed in the second preinjector pit (Pit II) at the linac for testing at 750 kV. Figure 3 shows a photograph of the source assembly with the vacuum baffle removed. A beam transformer at the end of the quadrupole can measure the net current leaving the drift tube. Figure 4 shows the column back plate with the source and turbomolecular pump installed in pit II.

Fig. 3. Source assembly mounted on column back plate with vacuum shroud removed.

Fig. 4. View of column back plate mounted on Pit-II column. Insulated cesium boiler is angled item in center. Dome-turbomolecular pump is in the foreground.

Some initial problems with column conditioning were overcome by cleaning the ceramics and electrodes. The commissioning of the remaining support hardware went smoothly. However, after 6 months of trying, it was never possible to reproduce the stable operation of the test box for more than 8-10 hours. In the test box the source had run at 150 V and 150 amperes discharge, in the column it would run at 220 V, 70 Amperes. The extracted current was only about 18 mA, barely enough to meet the minimum AGS needs.

In an effort to get stable source operation, many substitutions of source components were made but without success. The alumina ceramics were replaced by the original MACOR insulators but no improvement was seen. Most suspect was the cesium delivery hardware. Cesium was introduced by breaking a 5 gm glass cesium ampule within the evacuated copper boiler by externally crushing the boiler tube and glass ampule. Occasionally, it was necessary to re-crush the boiler to prevent glass pieces from capping the cesium. Several times a copper colored film was found over the cesium when the boiler was removed and inspected.

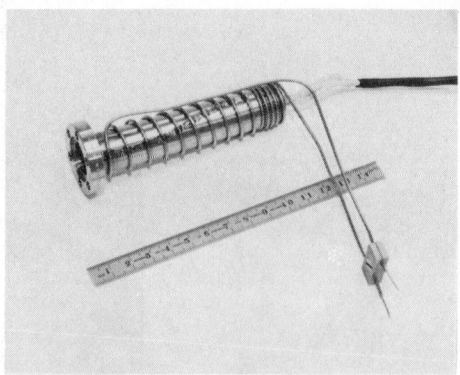

Fig. 5. Stainless steel cesium boiler with armored heater wire in place. Temperature is monitored by a thermocouple mounted at the bottom.

Because of the problems with the copper boilers, a re-usable stainless steel boiler was developed. This unit features a carefully placed heater winding which keeps the temperature constant to within 1 degree Centigrade throughout the volume of the cesium pool. The temperature is then graded uniformly up to that of the valve (Figure 5). The cesium is put into the boiler in an Argon atmosphere glove box. A special fixture breaks the ampule letting the cesium filter through a heated mesh into the boiler. About 70-80% of the cesium reaches the boiler with this technique. The new boiler was much more convenient to use but made no improvement to the ailing discharge.

The feedtube was suspected of being blocked because the internal heater element occluded much of the tube. This was redesigned using a larger diameter tube to greatly increase the conductance. The net result was a lowering of the required boiler temperature from 190°C to ~ 140°C, to maintain the same cesium in the source.

After all the changes and refinements to the source and its support hardware, the behavior was the same. Returning the source to the same configuration and components used in the test box still gave poor results. It was suspected that some contaminant existed in the column which had not been present in the test box. The early difficulties in voltage conditioning of column had been traced to oil contamination from a prior vacuum pump failure. This had been cleaned up well enough to hold high voltage on the column but perhaps was still the cause of the poor source operation.

THE GROOVED CATHODE

It had been suggested by Sluyters and Alessi[6] that a groove in the magnetron cathode could produce more H⁻ current at lower discharge current density. With the poorly operating source a 70 A discharge produced only 15-18 ma of accelerated H⁻ beam. Since the groove in the cathode would increase the surface area contributing to the extracted beam, it was felt that such a modification might result in acceptable current even with the poorly operating source.

In April of 1982 a groove was machined into a molybdenum cathode. The arc length was 4 times the height of the anode slit. The cathode was put into a source body and installed in the column. Conditioning and performance were the same as for prior flat cathode designs: dying out after about 8 hours of high discharge current operation (120-150 A). Beam current was similar to that observed for the flat cathode.

At this point the discharge power supply voltage and the pulsed gas were reduced. The beam current soon rose above its original value. This procedure was repeated until the H^- arc conditions peaked at ~ 150 V, 50 A. The pulse width, which in the high discharge current mode was thermally limited to 200 μsec, had to be continually increased to keep the source from getting too cold. It soon reached the 320 μsec limit set by the PFN in the Discharge Pulser. Under these conditions 40-45 mA was observed in the Dome of which 90% could be transported through the column to the Low Energy Beam Transport (LEBT). Up to 85 mA could be obtained in the Dome by lowering the gas, but the LEBT beam current never exceeded 50 mA.

Emittance measurements were made at 760 KeV in a viewing box about 5 M from the column. The normalized values at 90% for a 35 mA beam are:

Narrow plane - 0.19π cm-mrad   Wide Plane - 0.26π cm-mrad

These values are for the grooved cathode, but the flat cathode gave about the same result when extrapolated to this current. These data are about twice that reported by FNAL[7], possibly due to space charge growth in the 20 KeV transport.

The PFN was modified to allow a pulse width of nearly 600 μsec at a 3-Ohm impedance. The source was run at 5 pps at a duty factor of 0.3% around the clock for 2 and one-half months. When it was inspected no damage was found on the MACOR ceramics. Some small black flakes of molybdenum were present, but less than after a 2 week run in the high current mode. These flakes, which were also found on the extractor, probably caused the observed 10% current drop-off by obscuring the slit.

The cathode was eroded by about .020" on the side nearest the cesium inlet, and half that amount on the other side. A sharp image of the anode slit .010" deep was found in the center of the groove, caused by high energy protons which had been stripped in the extractor region.

Due to the good lifetime and reliability of the source and successful acceleration of 25 ma of H^- to 200 MeV, the AGS was shut down for complete conversion to H^- injection. During this period the source was cleaned and reinstalled with a new grooved cathode but with the original boiler and cesium. Work was begun to convert the main preinjector (Pit I) to an H^- source.

The second grooved cathode conditioned much like the first. With this source new peak intensity and integrated beam records were set for the AGS in a short time. The source ran a little over 2 months until a short circuit in its heater required replacement of the feedtube. This was done without disassembling the source by keeping it in a plastic argon-filled bag while the feedtube was replaced.

This same source was kept running for another 3 months while Pit I conversion was completed and commissioned. In total, it operated for 5 months without disassembly. The source had performed well during this period, showing no deterioration in its output at the end of the 5 months. This is to be contrasted to the flat cathode source which drops to half its output in 2 months.[8]

Figure 6 shows the source as the anode cover plate was removed. Some of the molybdenum flaking can be seen. Part of the image of the anode slit is visible. Figure 7 compares the cathode with a new grooved one. The ablation on the cesium inlet side is clearly displayed.

 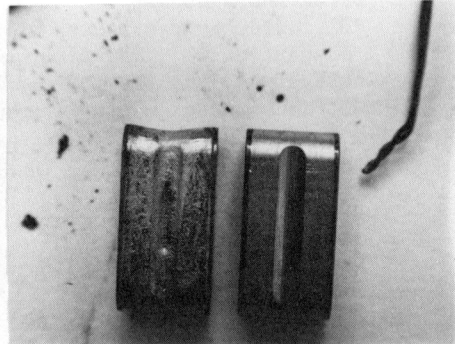

Fig. 6. Opening grooved cathode source after 5 months operation.

Fig. 7. Grooved cathode after 5 months (left) vs. new one.

The cesium boiler was found to be nearly empty when it was opened. There was no residue of any kind observed. This cesium had lasted for over 8 months of full time operation for an average rate of 0.6 mg/hr. After cleaning and charging with a new load of cesium, the boiler was re-installed on the source, which had been cleaned and fitted with a new grooved cathode.

The temperature at which the cesium pool is maintained has varied as the source and cesium delivery system changed. After increasing the conductance of the feedtube, the boiler temperature had to be lowered from 190°C to 140°C. When the stainless steel boiler was installed, the temperature went up to 160°C. This was partly because the stainless is an inferior thermal conductor to copper and because the thermocouple had been placed too close to a heater wire. When the grooved cathode was installed, the temperature was still measured in this manner, but now had to be set at 120°C. Later the thermocouple was moved to a recess in the bottom of the boiler where the temperature was measured to be 8°C cooler than the cesium pool. The temperatures read by the thermocouple indicated a cesium temperature of 105°C.

PIT I--OPERATION

Operation from Pit I begin in late April 1983. Table II shows typical parameters for the source.

Table II
Source parameters

Function	Name	Readback Nominal	Range	Units
Discharge PS Voltage	DISP	255	250→270	V
PFN Voltage	PFNV	190	185→205	V
Extractor PS Voltage	EXTV	17.5	17→18	kV
Dome Vacuum	DVAC	8	7→10	μTorr
Cesium Boiler Temp.	CSTP	99	98→100	°C
Cesium Boiler Current	CSBI	480	450→520	mA
90° Magnet Current*	90DI	7.50	7.00→8.00	A
Discharge Current	DISI	40	38→50	A
Discharge Voltage	DISV	135	125→150	V
Cesium Valve Temp.	CSVT	300	280→330	°C
Cesium Feedtube Temp.	CSFT	330	300→340	°C
Source Anode Temp.**	SANT	200	160→220	°C
Source Cathode Temp.**	SCAT	380	350→400	°C
Cold Box Temp.	CBXT	−30°C	−40→−15	°C
Extractor Delay Time	EXTD	1830	1830	μsec
Extractor Pulse Width	EXTW	400	300→500	μsec
Gas Delay Time***	GASD	930	700→1130	μsec
Discharge Delay Time***	DISD	900	1130→700	μsec
End of Discharge	EDIS	500	400→600	μsec
Dome Beam Current	DMBI	40	35→60	mA

* Current shown for Pit I. Magnet in Pit II 1 A less.
** Depends on many conditions, especially EDIS.
*** Sum of these must equal EXTD.

Figure 8 shows the discharge voltage and current in the Dome and the beam current measured in the first LEBT transformer. The beam current is always quiet, in contrast to the high current mode in which high frequency noise was often seen. The ripples in the beam current may be due to mismatch problems in the PFN.

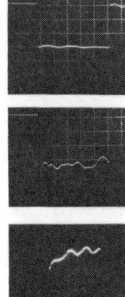

Fig. 8. Source discharge voltage 130 V, 500 μsec.
Source discharge current 40 A, 500 μsec.

Figure 9 shows a long time plot of source parameters. The data is taken every 20 minutes by the AGS PDP-10. The microprocessor in the Dome sends the readbacks of all parameters, which are then logged on a disk file. Operating personnel can then print the history of 10 parameters for a period of up to 1 week. Using this tool, long term drifts with time constants of hours to days become apparent, as does the effect of parameter change.

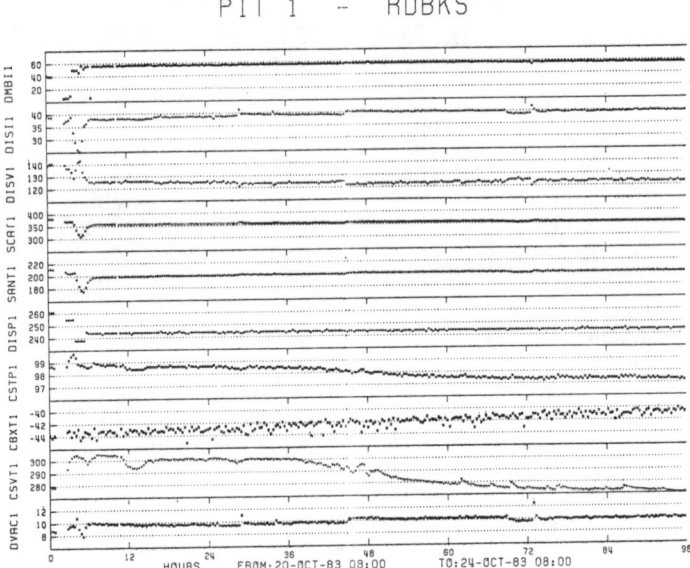

Fig. 9. Source parameters over a 4-day period.

During the summer 1983 shutdown, the source was cleaned and re-assembled with the original cathode. A new design pulsed gas valve (Figure 10) was used which greatly simplifed the valve calibration process by using a 10:1 lever arm to extend the adjustment range. The new valve appears to be stable and reliable in operation.

Fig. 10. New pulsed gas valve. Levers give 10:1 mechanical gain in settability.

The source start-up was interrupted by a direct lightning strike during a storm. One of the pulsed quadrupoles developed a short circuit to ground. Floating the quad pulser allowed the source conditioning to continue but the source showed only very slow improvement. After several weeks of effort the Dome beam current was only 25-35 mA and the source impedance about 4 Ohms. Most disturbing was the very limited range of stability. Changing the gas by 0.1 µTorr or the Discharge Power supply by 2 volts, or the boiler temperature by 0.5°C would cause the current to drop below 20 mA and the impedance to go to 5 Ohms or more.

Because of the quadrupole failure, poisoning of the source by Freon 113 coolant was suspected. A mass spectrometer installed at the ground side of the column showed a Freon 113 signature. The source was removed and given a minor cleanup. The quadrupole short was repaired and all Freon connections redone. The mass spectrometer, now installed in the Dome on the pumpout port, indicated that the Freon-113 had been significantly reduced, but a trace was still discernable.

The source was started up and conditioned very well. The beam current soon increased until over 50 mA was obtained in the Dome. The ranges of parameters which could be tolerated were back to their previous points.

NEUTRALIZATION PHENOMENA

Emittance measurements in the 20 KeV Test Box had shown distinct space charge influence on the early portions of the beam.[9] For the first 50-100 µsec, changes were seen in the phase space area and orientation of the beam normal to the 90° bend plane. This was shown to be greatly reduced by the increase of the background gas pressure, until stripping occurred. Similar effects have been seen by others.[10]

When the source was moved to the column, the same effect was observed. In the column, however, increasing the background gas is not a desirable option. An alternate method is to make the beam as late as possible after the gas pulse, letting the neutrals flow into the beam transport channel. The effect upon the rise time of at 760 KeV is quite apparent in Figure 11. The benefit of improved rise time must be balanced against the loss of peak current due to stripping at later times. As the beam pulse width is made longer, the start must occur earlier to assure sufficient gas in the source for proper discharge.

The transport line from Pit II to the Linac is about 15M long. The vacuum in the line is normally in the low 10^{-7} Torr range. A chopper located 3 M from the column produces fast rise and fall times or completely inhibits the beam in case of a malfunction. It was observed that beam chopped from a flat portion of the pulse acquired a tilt by the end of this line. To test if this was a neutralization effect, various vacuum pumps along the line were turned off, changing the background pressure. The results are shown in Figure 12. It is clear that the slope is reduced, then reversed, and finally the whole beam is lowered in intensity as the background ion concentration increased.

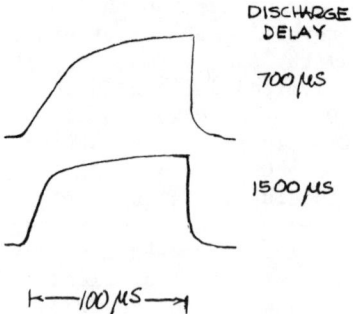

Fig. 11. Effect of background gas on source current rise time.

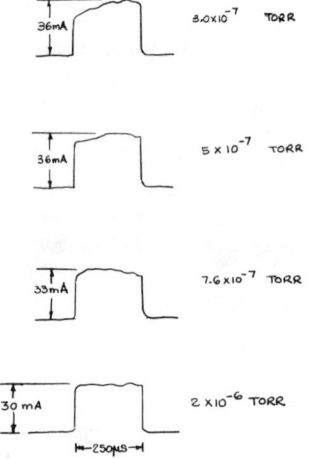

Fig. 12. Variation of beam current in LEBT with background pressure.

SUMMARY

The use of a curved cathode magnetron H^- source has been very successful at BNL. The lower dissipation requirements have allowed extension of the duty factor by at least a factor of 3 beyond the flat cathode source. By lowering the stress on the source, the stability seems improved as well.

The durability of the source is such that it can easily run 3 months with little or no loss of intensity before maintenance is performed, and has been run up to 5 months.

The use of a 20 KeV beam transport line in the column has been successful although it appears emittance growth may have occurred.

ACKNOWLEDGEMENTS

Coworkers on this project were D. Barton and K. Reece. Mechanical engineering was provided by W. Van Zweinan, E. Grove, R. Alforque, and J. Sheblein. Long hours by the BNL Linac group helped make the project succeed.

The many helpful discussions and suggestions from T. Sluyters, J. Alessi, and K. Prelec of BNL and C. Schmidt of FNAL are gratefully acknowledged.

REFERENCES

1. Yu. I. Belchenko, G. I. Dimov, V. G. Dudnikov, Proc. 2nd Sym. on Ion Sources and Formation of Ion Beams, Berkeley, CA, VIII-1, LBL-3399 (1974).
2. K. Prelec and Th. Sluyters, Proc. 1975 Particle Accel. Conf., IEEE Trans. Nucl. Sci. NS-22, 3 (1975) 1662.
3. C. W. Schmidt and C. D. Curtis, Proc. 1979 Particle Accel. Conf., IEEE Trans. Nucl. Sci., NS-26, 3 (1979) 4120.
4. D. S. Barton and R. L. Witkover, Proc. 1979 Linear Accel. Conf., BNL, Upton, NY, BNL-51134 (1979) 47.
5. D. S. Barton and R. L. Witkover, Proc. 1981 Particle Accel. Conf., IEEE Trans. Nucl. Sci., NS-28, 3 (1981) 2681.
6. J. G. Alessi and Th. Sluyters, Rev. Sci. Instrum., 51, 12 (1980) 1630.
7. Loc. cit., Ref. 3, p. 4121.
8. Loc. cit., Ref. 3, p. 4121.
9. Loc. cit., Ref. 5, p. 2683.
10. V. Stipp, A. DeWitt, and J. Madsen, Proc. 1983 Particle Accel. Conf., IEEE Trans. Nucl. Sci., NS-30, 4 (1983) 2743.

DISCUSSION

PETERSON: Why does the discharge run at lower cathode current intensity with curved cathodes?

WITKOVER: That is probably better answered by the people who developed that. It wants to run actually at a much lower gas density and at a much lower voltage. It seems almost as if it's a different mode of operation but perhaps I should refer that question to Jim Alessi.

ALESSI: I think, basically, that the charge exchange between the cathode and the exit slit, when the source is running at high current densities, reduces the focusing effect, so that at lower current densities the focusing works better.

DEVELOPMENT OF A MULTICUSP H⁻ ION SOURCE FOR ACCELERATOR APPLICATIONS

R. L. York and Ralph R. Stevens, Jr.
Los Alamos National Laboratory*
Los Alamos, New Mexico 87545

ABSTRACT

The development of a multicusp surface-production H⁻ ion source (Berkeley concept) designed specifically for accelerator use is described. The goal of this development effort has been to provide a suitable H⁻ ion source for the Proton Storage Ring now being constructed at LAMPF. The ion source that has been developed is now capable of long-term operation at 20 mA of H⁻ current at 10% duty factor and with normalized beam emittance of 0.13 cm-mrad (95% beam fraction). The development program will be described with particular emphasis on beam emittance measurements.

INTRODUCTION

A multicusp surface-production negative ion source has been developed at LAMPF to satisfy the operation of the Proton Storage Ring as well as existing LAMPF requirements. This source must provide an H⁻ beam with a peak intensity of 20 mA and sufficient quality to match the acceptance of the accelerator. To insure an accelerator quality H⁻ ion beam, the geometrical admittance of the extracted beam was collimated in the ion source and thus the geometrical admittance of the ion source was used to determine the emittance of the extracted beam. In fact, it has been found that the measured emittance values are somewhat higher than the geometrical admittance predictions.[1] With this restriction, the operating parameters of the ion source were studied in an attempt to increase the beam brightness.

EXPERIMENTAL APPARATUS

The H⁻ ion source development program has been carried out at the high-voltage test stand in the LAMPF injector complex.[2] The test stand provides the capability of intensity and emittance measurements for both unanalyzed and mass-analyzed beams. Beam currents are measured using both beam-current toroids and biased Faraday cups. The beam-current monitors have been calibrated using the LAMPF H⁺ duoplasmatron and the absolute accuracy of the beam current measurements is better than 3%. The emittance scanners are conventional slit and collector systems with spatial resolution of 0.2 mm and angular resolution of 1.5 mrad. The emittance data are processed by the SEL 840 LAMPF control computer and on-line reanalyzed emittance scans are provided to the test stand.

---

*Work supported by the US Department of Energy.

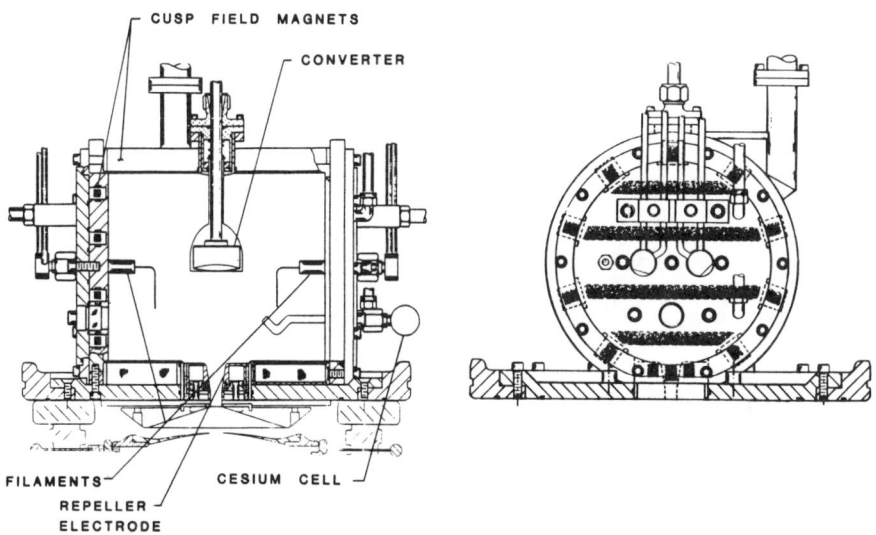

(a)　　　　　　　　　　　　　　(b)

Fig. 1. (a) A top view of the cylindrical source design.
(b) A side view of the cylindrical source design where the shaded areas represent the permanent magnets.

The prototype ion source used in these experiments is similar to the Berkeley source.[3] The source employs a cylindrical geometry with ten magnets around the cylinder and four magnets in each endplate as shown in Fig. 1. The cylindrical design is large enough to accommodate a 5 cm diameter converter in the magnetic field-free region in the center of the source. A 0.154 cm tungsten filament, approximately 18 cm in length, is mounted in each endplate. Water cooling is provided to the converter, the filaments, the endplates, the repeller electrode, and each of the ten magnets. The source housing is only cooled indirectly through contact with the individual magnet holders. As shown in Fig. 1(a), the geometrical admittance of the source is defined by the diameter of the 5 cm converter and particle flight path of 12.9 cm to the 1.27 cm diameter channel in the repeller electrode. The ion beam is extracted through a break in the cusp-field geometry. The magnets are positioned symmetrically around the source housing, extending the length of the cylinder except along the beam axis. There are breaks in the bar magnets along the beam axis to allow for installation of the converter and for beam extraction. As shown in Fig. 1(b) magnets are positioned above and below the extraction aperture in a symmetric manner in the repeller electrode to minimize plasma loss area and to provide an essentially magnetic-field-free region for beam extraction. All of the magnets around the cylinder and in the endplates are samarium-cobalt magnets except for the three bar magnets on the extraction side of the source and those in the repeller electrode. It was necessary to make these magnets Alnico-8 magnets to insure a magnetic-field-free extraction region.

EXPERIMENTAL RESULTS

Cesium Transfer

Originally cesium was transferred into the source using a method similar to that used by the Berkeley group. First, the source was conditioned by running relatively high continuous arc currents (40 to 50 amps) for several hours. The ratio of H^- current to heavy ion impurity current was monitored to measure the condition of the source. When the ratio of H^- ion current to impurity current (primarily O^- and OH^-) was unity, cesium was transferred by heating the cesium oven to 270° C and opening the valve of the oven for approximately one minute. With this method of transfer, the source was found to operate stably for up to 24 hours. However, the period of stable operation varied from one transfer to another.

To achieve greater long term operational stability, a method of continuous cesium transfer was adopted. A temperature controller was used to regulate the temperature of the cesium oven. Thus, after the source was conditioned the cesium valve was left open and the flow of cesium into the source was determined by the temperature of the oven.

This method of cesium transfer gives us the capability of studying the various source parameters under controlled cesium conditions. For an arc current of 40 A, a pulse length of 800 μsec,

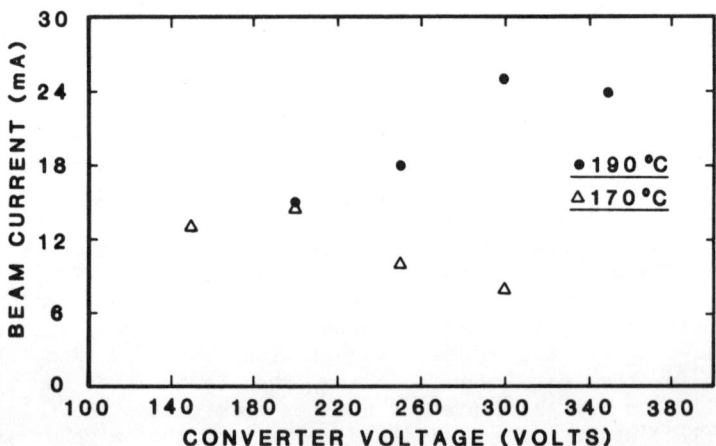

Fig. 2. Beam current vs converter voltage for two different cesium oven temperatures.

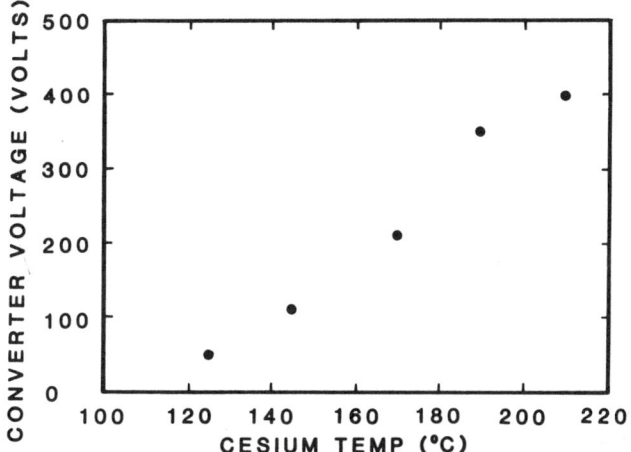

Fig. 3. Converter voltage for maximum beam current vs cesium-oven temperature.

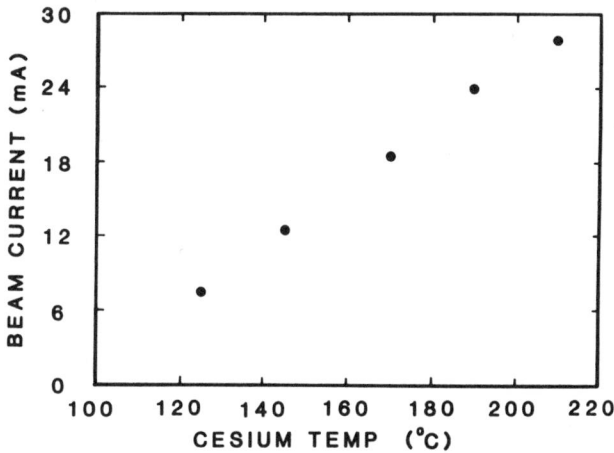

Fig. 4. Maximum beam current as a function of cesium oven temperature.

a repetition rate of 120 Hz, and a molybdenum converter, a study of converter voltage and cesium temperature was performed. At each cesium-oven temperature, we found a converter potential which yielded the maximum beam current. Two typical converter voltage versus beam-current scans are shown in Fig. 2. In general, the greater the cesium transfer rate, the higher the optimum converter voltage and the larger the maximum beam current. Fig. 3 shows the variation of the optimum converter voltage as a function of cesium-oven temperature. In Fig. 4, the beam currents corresponding to the converter voltages shown in Fig. 3, are plotted as a function of the cesium-oven temperature.

Thus, Fig. 4 indicates that higher beam currents can be obtained by simply transferring cesium at a higher rate. However, since the measured emittance values are determined by the geometrical admittance of the source, these higher beam currents at higher converter voltages have higher emittance values. All of the emittance values measured at the various cesium temperatures are plotted in Fig. 5 as a function of beam current. This plot shows that over the range of cesium conditions investigated, the observed emittance values increased with beam current.

To further investigate how the beam current yield depends on cesium-transfer rate the source was operated at various duty factors. When operated at low duty factor (1%), the source performance was equivalent to that at higher duty factor (8%), but the source required a lower rate of cesium transfer. The measured emittance values at lower duty factors exhibit the same relationship to beam current shown in Fig. 5.

Converter Material Study

In the development program, a molybdenum converter has been used in most of the testing. However, titanium and niobium converters have also been tested. The titanium converter was extremely difficult to operate. The converter surface sputtered severely and this sputtering caused sparking that made it very difficult to maintain the voltage on the converter. There was also evidence of the titanium converter pumping a significant amount of hydrogen, especially when the converter was allowed to float at

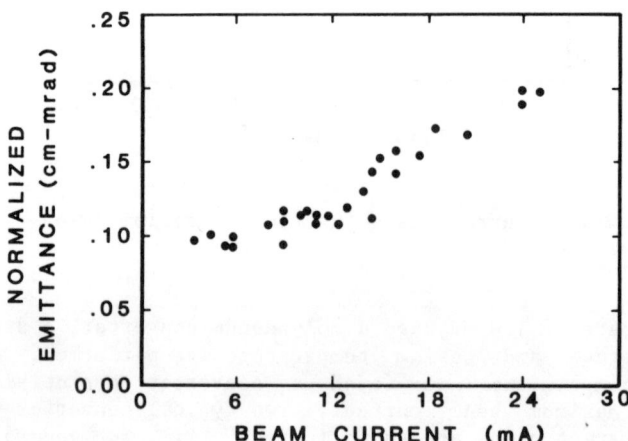

Fig. 5. Normalized emittance vs beam current.

plasma potential. The small amount of data that was obtained indicated this converter did not yield significantly more beam current even though it drew approximately twice as much converter current as the molybdenum.

A niobium converter was also installed in the same geometry as the molybdenum converter. Its operation was very similar to the molybdenum in terms of sparking and converter current. However, the niobium yielded both a brighter and higher-intensity beam than the molybdenum when operated under the same conditions as shown in Fig. 6. At an intensity of 20 mA, the niobium converter produced a beam that is 1.5 times as bright as that of molybdenum. Although the beam quality is improved, a niobium converter may have some long-term operational problems. When the source was disassembled, the surface of the converter was badly spalled and cracked.

Lifetime Tests

Several lifetime tests were carried out to evaluate the long-term performance of this ion source for accelerator service. The source was operated for extended periods up to 200 hours and ion beam measurements were taken continuously during the tests at periodic intervals. The tests were terminated either by accelerator failure or by choice but not because of ion source failure. In general, the beam-current and emittance values were constant to

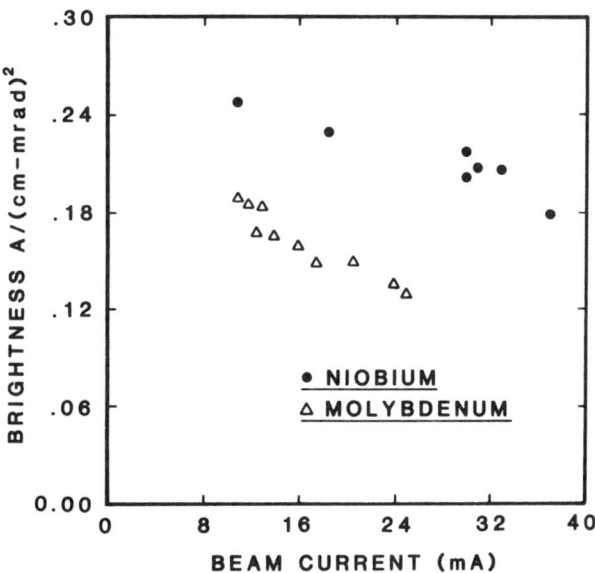

Fig. 6. Beam brightness vs beam current for identical converters made of niobium and molybdenum.

within ±5% of setpoint values after equilibrium had been reached. During these tests, neither the arc nor the gas flow had closed-loop control, so the above variations reflect the open-loop behavior of the source. A hydrogen-gas controller has now been installed and the present performance is deemed adquate for LAMPF operation.

At the end of each test, the tungsten filaments were removed and measured. On the average, the diameter of the filaments decreased 0.001 inches/100 hours of operation for 20 mA beam operation. This sputtering rate implies a 600 hour filament lifetime (10% reduction in diameter). However, some regions of the filaments near the cesium transfer tube did exhibit a greater reduction in diameter which would entail a shorter lifetime. This transfer tube has been moved farther from the filaments and a more uniform reduction in filament diameter is now expected. The cesium consumption for 20mA operation was 0.012 grams/hour. Thus, the present design is expected to have an operating lifetime in excess of 500 hours; with suitable improvements, a 1000 hour lifetime can be achieved.

CONCLUSION

The multicusp surface-production negative ion source can provide moderate intensity beams at high duty factor suitable for long-term accelerator operation. Peak beam currents of 38 mA have been obtained with normalized emittances less than 0.20 cm-mrad and ion source lifetimes in excess of 500 hours can be expected. Although development work is continuing to obtain brighter beams at the 20 mA level, the present performance of this source is deemed adquate for operation of the Proton Storage Ring at LAMPF. Installation of the source into the 750 kV Cockcroft-Walton dome has begun.

REFERENCES

1. R. L. York and Ralph R. Stevens, Jr. IEEE Trans. on Nucl. Sci., Ns-30, No. 4, 2705 (1983).
2. Ralph R. Stevens, Jr., John R. McConnell, E. P. Chamberlin, R. W. Hamm, and R. L. York, Proceedings 1979 Linear Acc. Conf., Brookhaven, BNL 51134, 405 (1979).
3. K. W. Ehlers and K. N. Leung, Rev. Sci. Instrum. 51, 721 (1980).

DISCUSSION

SCHILLING: I wonder how much the effective increase of the converter surface area due to this cracking and sputtering has increased the apparent brightness by giving you more current.

YORK: I thought about that and that is certainly possible. The one thing I can say is that we studied molybdenum, niobium and titanium, and all the surfaces seem to be sputtered, even molybdenum, although it is not sputtered as much as niobium. After we took the molybdenum converter out, it was roughened, not quite as bad as the niobium, but it was roughened.

PROGRESS IN DEVELOPING A 'VOLUME' HYDROGEN NEGATIVE ION SOURCE

M. Bacal, F. Hillion, M. Nachman* and W. Steckelmacher**
Laboratoire de Physique des Milieux Ionisés,
Groupe de Recherche N° 29 du C.N.R.S.,
Ecole Polytechnique, 91128 Palaiseau Cedex, France

ABSTRACT

The H^- ion extraction from several configurations of the multicusp plasma generator (hybrid, tandem, conventional) is investigated and the effect of the area of the plasma electrode of the extractor is discussed. The relationship between the extracted H^- current and the negative ion density in the plasma is studied in the hybrid multicusp generator.

INTRODUCTION

Several experiments have been conducted recently to extract volume produced H^- ions from multicusp plasma sources. Leung et al[1] have shown that the addition of a magnetic filter to a classical multicusp source enhanced the H^- ion yield and reduced the extracted electron component. In the presence of the magnetic filter the ratio of I^-/I_e appeared to be very sensitive to the positive (with respect to the anode) bias of the plasma electrode (PE). An optimum I^-/I_e of 1/120 was observed by biasing the PE at + 2.5 V in the presence of the filter, but no improvement was obtained when biasing the plasma grid in a classical multicusp source. Holmes et al[2] have studied several transverse magnetic field arrangements : permanent filter fields were formed by the permanent magnets on the anode, while an electromagnetic filter field was produced by a coil enclosed in the rods running through the discharge.

Previous work at Ecole Polytechnique using the photodetachment technique to measure the density of H^- in the body of the plasma has shown that certain hybrid magnetic multipole configurations contain a high density and a high proportion of H^- ions[3]. The study by this technique of the 'target' plasma of the tandem multicusp plasma generator[4], formed by the introduction of a magnetic filter in a multicusp source, indicated comparable characteristics.

It is the purpose of this work to study the relationship between the extracted H^- and electron currents and the respective densities in the multicusp plasma ; to analyze the effect of biasing positive the PE of the extractor ; to compare the performance of various multicusp ion source configurations.

*   Permanent address : Ecole Polytechnique, Montreal, Canada.
** Permanent address : University of Sussex, Brighton, England.

EXPERIMENTAL SET-UP

Fig. 1 is a schematic diagram of the hybrid multicusp plasma generator. The side wall of the cylindrical stainless steel chamber (25.4 cm diameter and 23.6 cm high) is surrounded by ten columns of ceramic or samarium-cobalt magnets. There are no magnets on the end plates. Ten hairpin-shaped filaments are arranged on a circle (19.2 cm in diameter) ; the center of each filament is located between two magnet columns. Because the filaments are situated in the multi-cusp magnetic field and because there are no magnets on the end plates, we denote the multicusp configuration shown on Fig. 1 as "hybrid". A more detailed description of the plasma source is given in Ref. 3. One end of the chamber is enclosed in part by an annular grid (G), and in part by the PE of the extractor (9 cm diameter), which contains a small extraction slit. G can float, or be connected either to the anode, or to the PE. G separates the plasma from an annular liquid nitrogen trap.

In order to form a conventional multicusp ion source configuration four extra rows of magnets are applied on the top flange. In this case two groups of four thoriated tungsten filaments are placed close to the chamber axis. A samarium-cobalt magnet filter can be added to produce a tandem multicusp plasma generator (Figure 1 in Ref. 4). In these experiments the discharge voltage was 50 V. The plasma was produced under continuous pumping and gas flow conditions. The pumping occurred through a total area of 1.5 cm^2 : the slit in the plasma electrode and five additional openings, 0.6 cm diameter, in the wall located behind the LN$_2$ trap. The latter provided additional cryogenic pumping. The background pressure was 10^{-5} Torr.

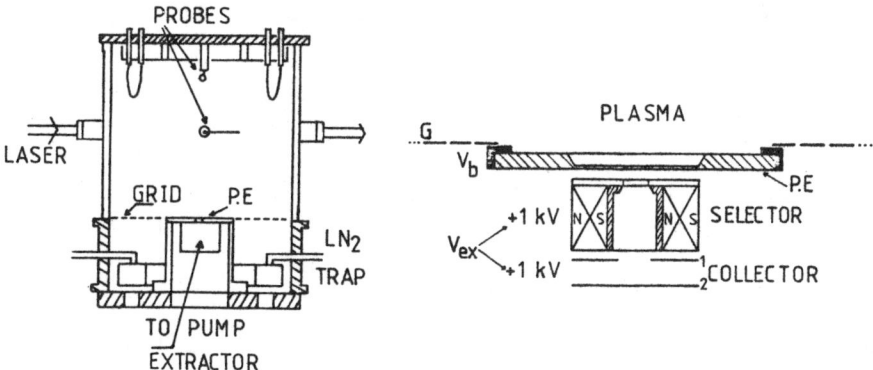

Fig. 1. Diagram of the hybrid multicusp plasma generator with extraction system.

Fig. 2. Diagram of the $\overline{H}$-electron separator. G-grid, PE-plasma electrode.

Electron separation. In experiments on H^- extraction from a plasma it is important to separate the electrons from the negative ions and to have a clear measure of the H^- and electron fluxes separately. This problem was solved by using an extraction system formed of four electrodes (Fig. 2). The first electrode, in contact with the plasma, denoted as plasma electrode, PE, contains a relatively narrow extraction slit (0.2 x 0.6 cm^2). The second electrode, denoted as 'separator', is located at 3 mm from the PE and contains a slit of 0.3 x 0.8 cm^2. A pair of ceramic magnets located in the separator just behind the slit create a transverse magnetic field ($B_d \simeq 300$ G) strong enough to deflect the accelerated electrons onto a grooved graphite lining inside the separator. The H^- ions are little affected by the presence of this field, which should cause only a small lateral displacement of the H^- ion flux reaching the collector. The collector consists of two sections (Fig. 2) : the first section contains a slit (0.5 x 1.0 cm^2) and is placed at 1.5 mm from the selector ; the second section is located at 8 mm from the first one. This enables us to evaluate the divergence of the H^- ion beam. Unless specified otherwise, the total current, I_c on both sections of the collector is reported.

A series of tests were conducted to make sure that, with the magnets in place in the separator, I_c was due solely to H^- ions. The hybrid multicusp source (Fig. 1) was first operated with an argon plasma and the currents on the selector, I_s, and on the collector, I_c, were measured in the absence and in the presence of the pair of magnets in the selector, with a positive extraction potential, V_{ex}. With the magnets in place, there is no current on the collector, but a current of 10 mA on the selector. (Table 1). With a negative V_{ex}, most of the extracted positive ions are found on the collector. Subsequently argon was replaced by hydrogen and I_c and I_s were measured in the presence of B_d with a positive or negative potential applied to the selector and collector (Table I). A positive bias on the plasma electrode, V_b, leads to the characteristic rise of the negative ion current to the collector described in Ref. 1.

Gas	P	I_d	V_b	B_d	V_{ex}	I_s	I_c
	mTorr	A	V	G	kV	µA	µA
Ar	3	0.5	0	300	+1.0	10^4	0
	3	2.25	0	300	-1.0	4.5	72.5
H_2	2.9	2.0	0	300	+1.0	2.85×10^3	2
	2.9	2.0	+1.4	300	+1.0	700	15.3

Table I. Data illustrating the tests of the e-H^- separator.

The test in argon has also indicated that the effect of secondary electrons emitted from the selector is negligible. In order to evaluate the effect of secondary emission from the collector, we compared the values for I_c when the collector was biased positive by 19 V with respect to the selector, with those measured when the selector and the collector were at the same potential. The results are shown on Fig. 5 in Ref. 4. Note that due to secondary electron emission the negative ion current is minorated by 20 % when the measurement is made with the collector and selector at the same potential, as it is the case in the present paper and in Ref. 4.

Focussing of H^- ion beams. The H^- ion currents to the two sections of the collector are shown in Fig. 3 as a function of V_{ex}. The selector and the two sections of the collector are at the same potential. This measurement was made at constant current through the first gap ; this was achieved by maintaining I_s = 1.2 mA by suitably varying the discharge current I_d. Note that the total H^- current increases smoothly from 300 V and reaches a maximum at 1200 V. It can be noted that the H^- current to the second section of the collector attains a plateau at 1700 V.

The measurement of the negative ion density in the center of the plasma was made by the photodetachment technique[5], while the plasma parameters were measured with a tantalum disc probe (4 mm diam)[3].

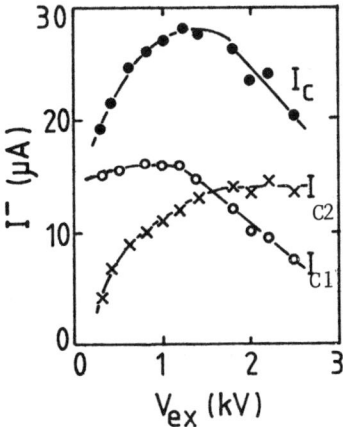

Fig. 3. Variation of the H^- currents to the two sections of the collector versus the positive extraction potential. The total current to the collector is also shown. The current through the acceleration gap was maintained constant (I_s = 1.2 mA). P = 2.8 mTorr.

EFFECT OF THE AREA OF THE PLASMA ELECTRODE

The apparatus shown on Fig. 1 provides the possibility of testing the effect of the area of the plasma electrode, PE, of the extractor, since this area can be increased by connecting G to the PE. In this way the diameter of the PE can be increased from 9 cm to 25 cm.

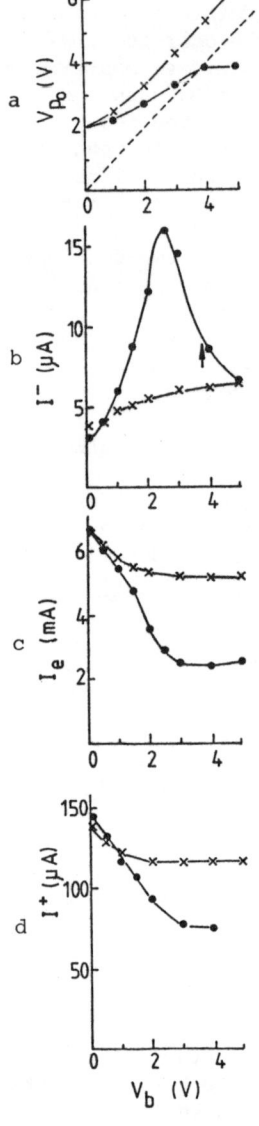

In order to provide the possibility of comparing the performances of the different multicusp configurations, the results presented in Figures 4 - 7 correspond to the optimum gas pressure (3 mTorr) and to the same discharge parameters (50 V, 2 A).

1. <u>Conventional multicusp plasma generator with samarium-cobalt magnets</u>. Fig. 4a shows the variation of the plasma potential in the center of the plasma chamber, V_{po}, versus V_b. When $V_b = 0$, V_{po} is 2 V more positive than the chamber wall or anode, and 0.9 V more positive than the plasma potential near the extractor (not shown). When G is maintained at the anode potential ($V_g = 0$), V_{po} first increases smoothly with V_b and attains a plateau value of + 3.8 V. Note that when $V_b = + 3.8$ V, there is no potential difference between the PE and the plasma center i.e. $V_b = V_{po}$. When G is connected to the PE ($V_g = V_b$) V_{po} follows the increase of V_b and, for $V_b > 2$ V, V_{po} stays continuously 1.25 V more positive than the PE. This situation, similar to that described in Ref. 1, corresponds to a large area PE covering the entire extraction end of the source.

Thus it appears that it is possible to flatten the axial plasma potential profile in a conventional multicusp source without magnetic filter by reducing the area of the PE. This is also true in a hybrid multicusp configuration

Figures 4 b, c, d, show respectively the variation versus V_b of the extracted negative ion current, I-, electron current, I_e, ($I^- = I_c$ and $I_e = I_s$ when $V_{ex} = +1$ kV), as well as that of the positive ion current ($I^+ = I_c$ when $V_{ex} = -1$ kV). When G is connected to the anode, I- increases from 3 to 16 µa as V_b changes from 0 to + 2.5 V. The extracted electron current is 6.6 mA when $V_b = 0$ and decreases to 2.9 mA at $V_b = +2.5$V.

Fig. 4. Conventional multicusp source : effect of the PE area. Variation versus the bias of the PE of : a) plasma potential in the center of the plasma ; b) extracted negative ion current ; c) extracted electron current ; d) extracted positive ion current.
•• 9 cm diam PE, $V_g = 0$. xx 25 cm diam PE, obtained by connecting the PE with the grid. P = 3 mTorr, $I_d = 2$ A, $V_d = 50$ V. Samarium-cobalt magnets. The arrows in (b) indicate the PE bias for which $V_{po} = V_b$.

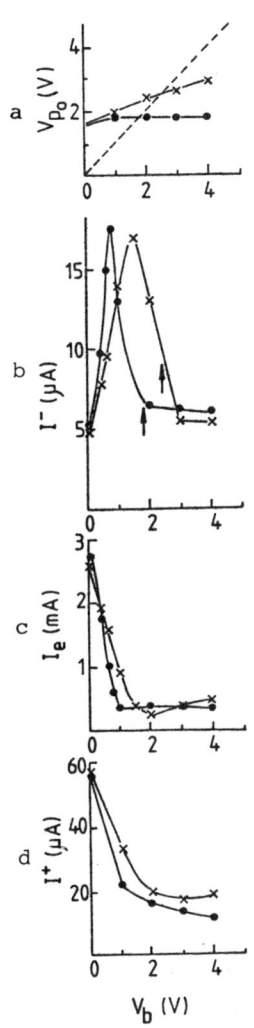

Fig. 5. Tandem multicusp source : same dependences and conditions as in Fig. 4.

The ratio I^-/I_e is thus improved from 1/2200 to 1/180. There is no maximum in I^- and the ratio I^-/I_e is only slightly improved when a large area PE is used, i.e. when G is connected to the PE.

2. Tandem multicusp plasma generator with samarium-cobalt magnets. Fig. 5 presents data similar to those of Fig. 4, but relevant to the tandem multicusp plasma generator (Fig. 1 in Ref. 4). I^- exhibits a maximum when the PE is biased positive with any of the two tested values of the PE area. The increase in the PE area shifts positive by 0.7 V both the PE bias for which $V_{po} = V_b$ and that for which I^- is maximum. The arrows on Fig. 5b indicate the PE bias for which $V_{po} = V_b$. As in Fig. 4b, this value is 1 V more positive than the one for which I^- is maximum. The maximum value of I^- is the same as in Fig. 4b ; however I_e at optimum V_b is 7.6 times lower in the tandem generator with the largest PE area (which is the best) than in the conventional multicusp generator. Biasing positive the PE improves here the ratio I^-/I_e from 1/520 at $V_b = 0$ to 1/22 at $V_b = +1.5$ V.

Note that the positive ion currents generated by the conventional multicusp configuration are about 2.5 times larger than those generated by the tandem configuration. This suggests that an excellent positive ion source is not necessarily the best volume H^- ion source, as discussed in Ref. 3.

The effect of the PE area upon V_{po} in the conventional multicusp plasma generator can be explained qualitatively by applying the requirement of plasma neutrality : a higher electron loss to a large area positively biased PE will produce a larger variation of V_{po}, which will thus follow V_b. A more detailed analysis will have to explain why the tandem multicusp source behaves differently. The lower average mass of the positive ions in the target section of this configuration can contribute to this effect.

HYBRID MULTICUSP PLASMA GENERATOR : EFFECT OF THE STRENGTH OF THE MULTICUSP MAGNETIC FIELD

The effect of the PE area is similar

here to that shown in Fig. 4a : when the PE is 25 cm diam ($V_g = V_b$), there is no flattening of the plasma potential profile and no maximum of I^-. We will present data taken with a 9 cm diam PE and a floating grid ; the results obtained with a floating grid are very close to those obtained when the grid is connected to the anode. Figure 6 and 7 present the variation of the plasma parameters versus the positive bias of the PE ; they illustrate the effect of the strength of the multicusp magnetic field since Fig. 6 shows the data obtained with 10 columns of ceramic magnets (2 x 1 x 20 cm^3), and Fig. 7 shows data obtained with 10 columns of stronger, samarium-cobalt magnets (0.9 x 1.3 x 20 cm^3). We report here the plasma parameters measured in the center of the plasma and near the PE. The plasma potential and the electron temperature go down when the multicusp magnetic field goes up. The electron temperature at the plasma center stays constant when V_b is increased, but that measured next to the PE exhibits a sudden rise when V_b varies between 1.5 and 2 V. The plasma electron density, n_e, is higher at the plasma center than at the plasma border, and goes down when V_b goes up. The negative ion density, n_-, and relative density, n_-/n_e, measured in the center of the plasma, are also shown. Both n_e and n_- are higher when the magnetic field is stronger, but n_-/n_e is lower in this case (Fig. 7b). Both n_- and n_-/n_e exhibit a maximum at V_b = + 1 V in the case of ceramic magnets (Fig. 6b) ; in the case of the stronger magnets, n_- goes slightly down with V_b, while n_-/n_e goes up. Note that n_-/n_e varies in the range 6.5-9.4 %.

Figure 8 presents the variation of I^- and I_e versus V_b. I^- is higher and I_e is lower at the optimum V_b when stronger magnets are used. The ratio I^-/I_e is improved when biasing positive the PE from 1/663, at $V_b = 0$, to 1/9, when $V_b = + 1.5$ V. The negative ion current is 22.5 µA at optimum V_b, i.e. 30 % higher than in the cases illustrated by Figs. 4 and 5, while the electron current is 0.2 mA, i.e. 50 % of that shown on Fig. 5.

DEPENDENCE OF EXTRACTED CURRENT VS PARTICLE DENSITY AND DISCHARGE CURRENT IN HYBRID MULTICUSP PLASMA

Figure 9 presents the variation versus the electron density in the center of the plasma of the measured electron and negative ion currents extracted when the PE and G are connected to the anode. The negative ion density in the center of the plasma is also shown. The dotted line represents the thermal electron flux through the extraction slit (of area A = .12 cm^2) ; this flux is calculated using the values for n_e and kT_e measured in the center of the plasma. The measured I_e varies somewhat slower ($I_e \propto n_e^{0.75}$) than the calculated electron thermal flux, but the two values are very close in the studied density range. This indicates that there is practically no potential barrier for the electrons at the plasma border in front of the extraction slit.

Figure 9 shows that in the studied density range the negative ion density in the center of the plasma varies in proportion to n_e, as predicted by our theoretical model for the "high density regime"[5,7].

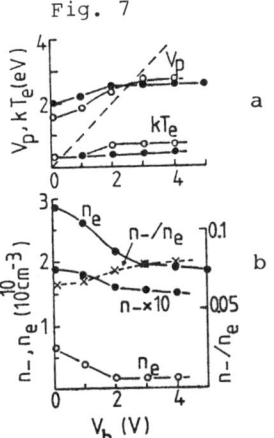

Fig. 6. Hybrid multicusp plasma generator with ten ceramic magnet columns. Variation versus the bias of the PE of a) plasma potential and electron temperature ; b) electron and negative ion density. 3 mTorr, 50V-2A.
●● center of the plasma
○○ plasma border, next to PE (11.3 cm from plasma center)

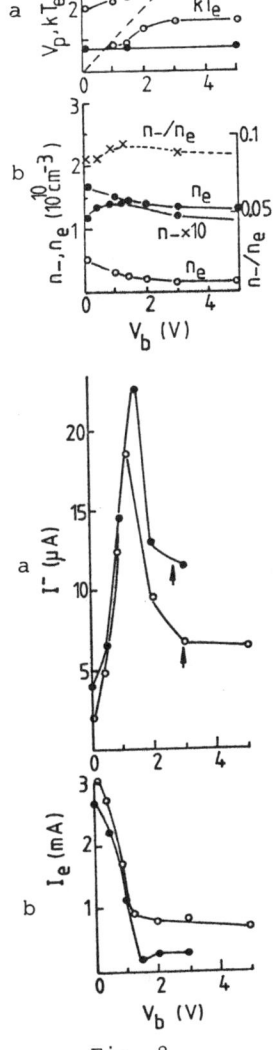

Fig. 7. Hybrid multicusp plasma generator with ten Sm-Co$_5$ magnet columns. Same dependences and conditions as Figure 6.

Fig. 8. Hybrid multicusp plasma generator Variation versus the bias of the PE of
(a) extracted negative ion current
(b) extracted electron current
●● Sm-Co$_5$ magnets
○○ ceramic magnets
The arrows in (a) indicate the PE bias for which $V_{po} = V_b$. 3 mTorr, 50 V - 2 A.

The extracted negative ion current increases more rapidly at low density and then varies approximately in proportion to n_e.
We have plotted on Figure 10 the dependence of I^- versus n_-, both for $V_b = 0$ and for $V_b = +1.25$ V, (close to optimum). The dotted lines represent the negative ion thermal flux through the extraction slit, calculated using Eq.(1) with different values for the negative ion temperature, and assuming that n_- in front of the slit is equal to the value measured in the plasma center.

$$I^- = A\, e\, n_-\sqrt{\frac{kT_-}{2\pi m_-}} = 7.5 \times 10^{-15}\, n_-\sqrt{kT_-} \quad (A) \qquad (1)$$

where n_- is in cm^{-3} and kT_- in eV.

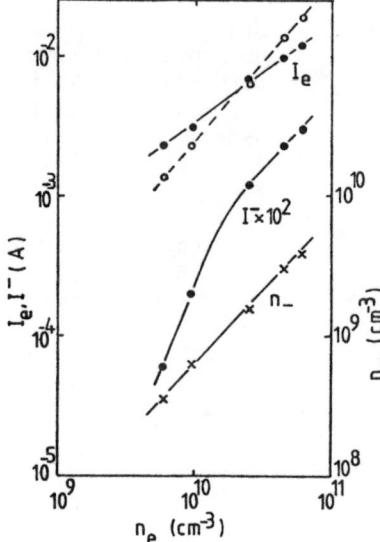
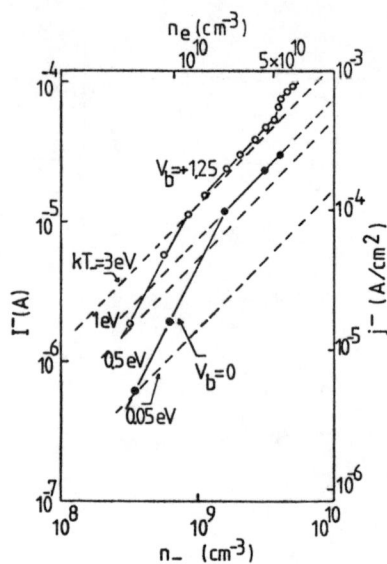

Fig.9. Variation versus the electron density in the center of the hybrid multicusp plasma generator with ceramic magnets of the extracted electron and negative ion currents, with $V_b = V_g = 0$. The dotted line represents the calculated thermal electron flux through the extraction slit. The negative ion density measured in the center of the plasma is also shown. 3 mTorr, 50 V.

Fig.10. Dependence of extracted negative ion current, I^-, and negative ion current density, j^-, upon the negative ion density in the center of the hybrid multicusp plasma with ceramic magnets, at $V_b = 0$ (●●) and $V_b = +1.25$ V (○○). $P = 3$ mTorr, $n_-/n_e = 0.08-0.09$. The dotted lines represent the calculated values (Eq. 2) for constant kT_-.

Note the existence of two regimes both when $V_b = 0$ and when V_b is optimum : at low density I^- varies in proportion to n_-^2, while at high density I^- varies in proportion to n_-. Note that at high density the negative ion current at optimum V_b is approximately two times larger than at $V_b = 0$. When $V_b = 0$, at the lowest density studied, I^- is close to the thermal flux value for $kT_- = 0.05$ eV, while it is five times larger than this flux in the high density regime. The question of negative ion temperature in plasma is discussed in Ref. 7 and it is shown that for the conditions of the studied plasma, the negative ion temperature should be close to gas temperature which should not exceed 0.05 eV. Since I^- is larger than the H^- thermal flux, it is reasonable to conclude that, as in the case of the electron extraction, there is no

Fig. 11. Variation of the extracted negative ion and electron currents at optimum V_b, versus the discharge current in the hybrid multicusp plasma generator with Sm-Co$_5$ magnets at various hydrogen pressures. □ - 1.8 mTorr ; ● - 2.9 mTorr ; o - 3.6 mTorr ; x - 4.9 mTorr, + 8.5 Torr. The actual I^- current is 20 % higher, due to the effect of the secondary electron emission from the collector.

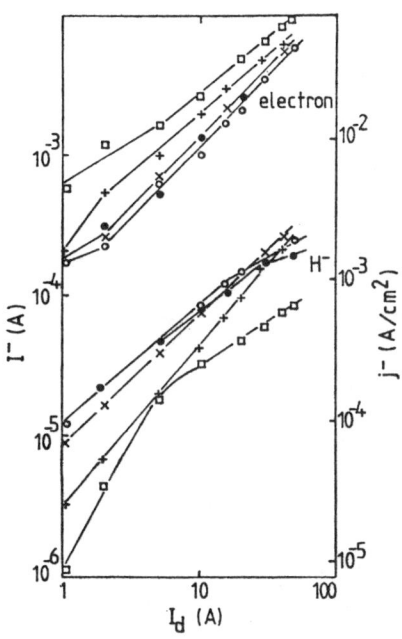

potential barrier and furthermore, a region of maximum density for the negative ions in front of the slit. Due to the application of the positive extraction voltage, the plasma region in front of the slit may become the most positive part of the plasma, in which a large density of negative ions bluids up. Fig. 11 presents the variation versus I_d of I^- and I_e at optimum V_b, for different pressures in the range 1.8 - 8.5 mTorr. The highest I^- and lowest I_e are observed at 3.6 mTorr. At this pressure for $I_d = 50$ A, the negative ion current density is 2.4 mA/cm^2 and I^-/I_e is 1/24.

CONCLUSION

In all the multicusp plasma configurations we have studied (conventional, hybrid, tandem) the extracted negative ion current can be optimized and the electron current reduced, by a suitable choice of the area and of the positive bias of the plasma electrode. This effect is a general one, related to applying a positive bias to an electrode in contact with the plasma and is not due to the presence of the magnetic filter. However in the presence of the magnetic filter a larger area PE can be used.

The reduction of the extracted electron current at optimum V_b is a consequence of the reduction of the plasma electron density. However the maximum of I^- at optimum V_b is not always associated with a maximum of n_- in the center of the plasma : in Fig. 7b n_- drops monotonously when V_b is increased. This may indicate that varying V_b redistributes the negative ions inside the plasma and possibly leads to the formation of a region of maximum H^- density in front of the extraction slit in the PE.

The choice of the PE area which gives the possibility to obtain the characteristic I^- maximum versus V_b, leads also to a reduction of the axial potential gradient across the extraction side of the plasma. One is tempted to conclude[1] that the role of the positive bias applied to the PE is to help the negative ions to escape from the positive plasma potential well. However the present work shows that I^- attains its maximum when V_b is approximately 1 V lower than the plasma potential in the center of the plasma. It seems impossible that the cold negative ions could escape across this 1 V barrier. It is not clear why I^- would drop dramatically when a more positive bias is applied to the PE.

It is shown that V_b affects the electron temperature in the proximity of the PE, but not in the center of the plasma. This may optimize the negative ion formation by dissociative attachement or other mechanisms next to PE.

It is shown that a strong multicusp magnetic field improves the I^-/I_e ratio, especially through a reduction in the electron drain.

The comparison of various configuration of the multicusp plasma generator from the point of view of high negative ion yield, indicates that at the optimum pressure (which is the same ∼ 3 mTorr) the best performance is obtained with the hybrid multicusp generator with samarium-cobalt magnets. In this configuration, we found that at optimum V_b the extracted negative ion current is larger by a factor of 8 compared to the thermal negative ion flux calculated for $kT_- = 0.05$ eV. The extracted electron current at $V_b = 0$ is close to the corresponding thermal flux. It follows that in front of the extraction slit, when the extraction voltage is applied,
1. there is no potential barrier for the plasma negative particles,
2. there is a region of maximum negative ion density which is larger than the value measured in the center of the plasma.

ACKNOWLEDGEMENTS

This work was supported by Ecole Polytechnique of Palaiseau, France, and Ecole Polytechnique of Montreal, Canada. The authors are grateful to H.J. Doucet and K.N. Leung for many enlightening discussions.

REFERENCES

1. K.N. Leung, K.W. Ehlers and M. Bacal, Rev. Sci. Instrum., 54, 56 (1983).
2. A.J.T. Holmes, G. Dammertz, T.S. Green and A.R. Walker, Proc. Intern. Ion Engineering Congress, ISIAT'83 & IPAT'83, Kyoto (1983).
3. M. Bacal, A.M. Bruneteau and M. Nachman, J. Appl. Phys., 54, (1983), to be published.
4. M. Bacal and K.N. Leung, Proceed. of this Symposium.
5. M. Bacal, Physica Scripta, T2/2, 467 (1982).
6. K.W. Ehlers and K.N. Leung, Rev. Sci. Instrum., 52, 1452 (1981)
7. M. Bacal and A.M. Bruneteau, Proceed. of this Symposium.

EXTRACTION AND ACCELERATION OF H⁻ IONS FROM A MAGNETIC MULTIPOLE SOURCE

A. J. T. Holmes and T. S. Green
Euratom/UKAEA Fusion Association, Culham Laboratory,
Abingdon, OX14 3DB, England.

ABSTRACT

Experimental measurements of an H⁻ accelerator are described, based on extraction of H⁻ ions and electrons from a magnetic multipole volume H⁻ source. The accelerator includes a trap for the extracted electrons, which gives an overall accelerator power efficiency of 81%. Measurements of the focussing properties of the accelerator are described and compared with a simple analytic model for the ion trajectories. This H⁻ accelerator has several important focussing advantages for neutral beam injection over comparable positive ion accelerator columns.

INTRODUCTION

The past few years have seen the development of plasma sources which produce H⁻ ions in the plasma by dissociative attachment. This process was first examined by Bacal and co-workers[1,2], and later taken to the stage of ion beam extraction by Holmes et al[3,4] and Leung et al[5]. The development of a neutral beam injector based on H⁻ ion sources using dissociative attachment will require the development of efficient H⁻ plasma sources and also the ability to accelerate and form highly collimated ion beams.

In this paper we describe experiments made on accelerating H⁻ ions in a triode extraction system using the H⁻ ion source described in the companion paper in these proceedings by the authors. These experiments will demonstrate that H⁻ ions produced by dissociative attachment can be easily focussed to give a highly collimated beam under all conditions.

EXPERIMENTAL APPARATUS

The source and accelerator are shown in Fig. 1. The plasma generator is of the magnetic multipole type where the permanent magnets on the outside of the source confine the plasma. The low pressure gas in the source is ionised by fast electrons, which are emitted from the hot wire filaments and accelerated to around 90 volts by the cathode sheath. The source contains a magnetic filter, which is formed by an anti-symmetric magnet pattern as seen in Fig. 1. This filter has the effect of creating a low temperature plasma in the vicinity of the extraction aperture. The electron temperature of around 0.5 eV and the apparent existence of vibrationally excited hydrogen molecules leads to the production of H⁻ ions which can then be accelerated to form a beam.

Fig. 2 shows a simple diagram of the triode accelerator with potentials and gap distances. Although the accelerator has three electrodes it is unlike a triode positive ion accelerator, as the two inter-electrode gaps both accelerate the ions and there is no suppression or decel gap. The second electrode contains magnets which create equal and opposite magnetic fields, as seen in Fig. 2. The first field maximum deflects electrons into the second electrode, and the second opposing field maximum straightens out the H⁻ beam so that it suffers only a slight sideways deflection, but is unsteered.

The principle diagnostics used in the following experiments are calorimeters, which are also electrically isolated to enable current and power measurements to be made. One calorimeter takes the form of a tube at 30 cms from the source, and the other is a plate at the end of the beam line which collects the particles which pass through the tube. There is an array of faraday cups mounted in the plane of the end calorimeter which can be used to examine the beam profile.

The beam profile measured on the faraday cups is near to gaussian in shape. In this instance we can estimate the overall beam divergence, Ω, from the ratio of the power loadings on the two calorimeters. For a gaussian profile the divergence is:

$$\Omega = \left\{ a \left[\ln(1 + T_i/T_0) \right]^{-\frac{1}{2}} - \rho \right\} \cdot L^{-1}$$

where a is the inner radius of the outer calorimeter (11 mm), L is the beam line length to the outer calorimeter (300 mm), and ρ is the extraction aperture radius (5 mm). T_i and T_0 are the inner and outer power loadings respectively.

A long range magnetic field can be applied in the beam line to deflect the particles reaching the end calorimeter and faraday cup array. The deflection observed is in good agreement with that expected for H⁻ rather than electrons, substantiating our assumption that the flux of particles exiting from the magnetic trap is dominated by H⁻ ions.

EXPERIMENTAL RESULTS

A simple method of focussing the H⁻ beam is obtained by operating at constant arc current and beam energy and varying the potential, V_1, across the first gap, until the minimum divergence is obtained. Fig. 3 shows the variation of the total current density with V_1 at constant beam energy, V_b, as well as the fractional current densities which are transmitted to the two parts of the calorimeter. Two distinct operating regions are observed. Firstly, when V_1 is less than 0.05 V_b, essentially no H⁻ ions are extracted. When V_1 exceeds this limit a rapid increase in J_- is initially observed followed by a slow increase. The current to the central part of the calorimeter shows a pronounced maximum at about the point corresponding to focussing of the beam.

An alternative method of displaying this focussing effect is to plot the beam divergence, derived from the ratio of the two calorimeter powers versus V_b/V_1 (simplified as U.) The result is seen in Fig. 4, where the beam comes into focus when U is approximately 7. A similar result has been obtained in tetrode accelerators of positive ions[6] with a decel gap.

However, the H⁻ accelerator does show a major difference to the positive ion accelerator in the relationship between divergence, beam current and beam energy. For tetrode positive ion accelerators the minimum beam divergence is obtained at a unique beam perveance which is determined by the accelerator geometry. In the case of the negative ion accelerator, minimum beam divergence occurs at a fixed value of U for a given beam current over a wide range of beam energies (V_b). This is seen clearly in Fig. 5. At low beam energies, however – less than 6 keV – the beam perveance becomes too great to be easily transported through the gap. Fig. 5 shows that the initial response of the accelerator is to require an increase in the value of U. The corollary to this experiment is to fix the beam energy and vary the beam current. The value of U needed to focus the beam in this case is shown in Fig. 6. In all cases U is close to 6.5, which is the minimum value seen in Fig. 4.

This is an unexpected result which implies that the plasma boundary is essentially rigid (as in an electron gun) and is only weakly affected by the electrostatic force. Positive accelerators, in contrast, have a flexible plasma boundary which constrains them to operate on a perveance versus divergence curve which has a "V" shape (Green[7]). The negative ion accelerator can hence operate over a wide range of beam energy and current and have a collimated beam. The minimum beam divergence itself is a very weak function of beam current, and decreases slowly with increasing current, as seen in Fig. 7. This effect may indicate the presence of an electro-static pinch created by the need for a positive beam potential to expel positive ions created by ionisation.

The electrons extracted from the plasma obey different source scaling parameter laws to the H⁻ ions and are confined within the accelerator. The magnetic field of the electron trap has two equal and opposite field maxima, which allows passage of the H⁻ ions with a lateral displacement of the order of 1 mm, whereas the electrons are deflected by the first field maximum into the second electrode. However, not all the electrons are trapped in this way, as secondary electron emission can allow some of them to get through the fields. These transmitted electrons constitute 10% of the total electron flux at present and are highly aberrated, with the result that they strike the third electrode, which is at earth potential. Hence the overall accelerator efficiency, ε, which is defined as:

$$\varepsilon = H^- \text{ beam power/Total drain power}$$
$$= V_b I_-/(V_b I_- + 0.9 V_1 I_e + 0.1 V_b I_e)$$
$$= (1 + 0.9b/U + 0.1b)^{-1} \qquad (1)$$

where $b = I_e/I_-$.

Fig. 7 in the companion paper indicates that I_e/I_- can be significantly less than unity when the beam-forming electrode is biased in excess of 5 volts positive to the anode box. If b is equal to 1, for example, then ε is 81% when U is 7.

DISCUSSION
H⁻ ION OPTICS

Simple analytic models of the optics of double gap accelerators have been developed by Holmes[6], Kim[8], and Green[7]. These models are based on the thin lens approximation of Davison & Calbick[9]. These models relate the mean beam divergence, Ω, to the plasma boundary curvature parameter, k, and V_1, d_1 and d_2 shown in Fig. 2. We define

$$k = d_1 \Theta / \rho$$

where Θ is the initial angle the outermost ion trajectory makes to the beam axis (negative values equal convergence).

$$U = V_b/V_1$$

$$\gamma = d_2/d_1$$

Then, using Ref. 6, we have:

$$\frac{3\Omega}{2} = \Theta U^{-\frac{1}{2}} - U^{-\frac{1}{2}} \rho \frac{(1+k)}{f_1} - \frac{\rho(1+k)}{f_2} - \frac{2\Theta d_2 (U^{\frac{1}{2}} - 1)}{f_2 (U-1)} \quad (2)$$

$$f_1^{-1} = \frac{V_2/d_2 - 4V_1/3d_1}{4V_1}$$

$$f_2^{-1} = - V_2/4V_T d_2$$

from which we have:

$$\frac{3d_2\Omega}{2\rho} = \frac{(U-1)}{4U} - U^{-\frac{1}{2}}(\frac{(U-1)}{4} - \frac{\gamma}{3}) + k[\gamma\frac{(11 \cdot U^{\frac{1}{2}} - 3)}{6U} - \frac{1}{4}(\frac{(U-1)}{U^{\frac{1}{2}}} - \frac{(U-1)}{U})] \quad (3)$$

This equation predicts that the beam divergence is independent of beam current and energy, and only depends on U and the position of the electrodes, if k is constant.

Experimentally we know that when $\gamma = 1.9$ and U is 6.6, Ω has a minimum value which can be approximated to zero if we neglect the residual divergence. This indicates that the value of k is + 0.10, corresponding to a fixed convex boundary with a radius of curvature of 110 mm. It is probable that the plasma boundary may be very close to being truely flat within experimental errors, as the aperture radius is only 5 mm. Using this value of k, Equation 2 predicts a divergence versus U curve, which is shown in Fig. 4 by the dashed line. Apart from the fact that the experimental divergence does not go to zero, due to space charge and aberrations, the theory above is in reasonable agreement with experiment.

Equation 3 also applies to positive ion beams, except here k is related to the extracted beam current via the Langmuir-Blodgett equation. This effectively means that only one beam current will have the right value of k to give a zero value for Ω, otherwise known as the "perveance match condition".

Equation 3 also predicts that if we wish to increase the value of U in order to raise the accelerator efficiency we must increase the value of γ. This modification is also now being investigated.

ELECTRON TRAPPING

The magnets in the second electrode create a zero net flux electron trap with 340 gauss cm of transverse flux in each of the two lobes. This flux is sufficient to trap all electrons whose energy is less than 8 keV. An illustration of the effect of electron trapping is shown in Fig. 8. Here the variation of the extracted total currents of electrons and H^- is shown as a function of the bias potential of the beam-forming electrode relative to the anode of the plasma source. The electrons and H^- ions have different parametric dependences on bias potential, indicating that the trap separates the particles. The effectiveness of the electron trap is seen by comparing the current to the earth or third electrode, I_3, to the total electron current which flows to the second and third electrodes only. Only 10% transmission is measured. The data indicates that these electrons are not stripped from H^-, as they have a very different scaling with source parameters to the H^- ions, and the transmission fraction is independent of all source parameters.

The transmitted H^- ions suffer no increase in divergence and a slight transverse displacement but no deflection in transiting the second electrode. This displacement is estimated to be of the order of 1 mm. However this displacement will lead to beam steering by aperture offset effect[6] at the third electrode, and this has been observed. This effect can be minimised by reducing the focal length of this electrode lens by increasing d_2/d_1. This modification will also result in a larger value of U needed to create a collimated beam - a result which will increase the accelerator efficiency and possibly reduce electron transmission.

CONCLUSION

Focussed H^- ion beams have been produced by a triode accelerator. The accelerator operates in a manner which indicates that it has a rigid plasma boundary which may be held in place by the transverse fields of the source. This effect has the useful advantage that a collimated H^- beam of any current can be extracted over a wide range of beam energy. This result is contra-distinction to the perveance match condition needed to focus positive beams in d.c. accelerators.

The electron component of the beam can be removed from the beam by the second electrode trap without significantly affecting the H^- beam, with a result that the accelerator efficiency has reached 81%.

REFERENCES

1. M. Bacal, E. Nicolopoulou, H. J. Doucet. Proc. Int. Symp. on Production & Neutralisation of H⁻ Ions & Beams, Brookhaven. BNL 50727, p.26, 1977.
2. M. Bacal. Phys. Scripta, T2/2, p.467, 1982.
3. A. J. T. Holmes, T. S. Green, M. Inman, A. R. Walker, N. Hampton. 3rd Varenna-Grenoble Symp. on Heating in Toroidal Plasmas, p.95, Grenoble 1982.
4. A. J. T. Holmes, G. Dammertz, T. S. Green, A. R. Walker. Proc. Int. Ion Engineering Congress, p.71, Kyoto 1983.
5. K. N. Leung, K. W. Ehlers, M. Bacal. Rev. Sci. Instr., 54, p.56, 1983.
6. A. J. T. Holmes and E. Thompson. Rev. Sci. Instr. 52 p.172, 1981.
7. T. S. Green. J. Phys. D., 9, p.1165, 1976.
8. J. Kim, J. H. Whealton, G. Schilling. J. Appl. Phys., 49, p.517, 1978.
9. G. J. Davisson, C. J. Calbick. Phy. Rev., 38, p.585, 1931.

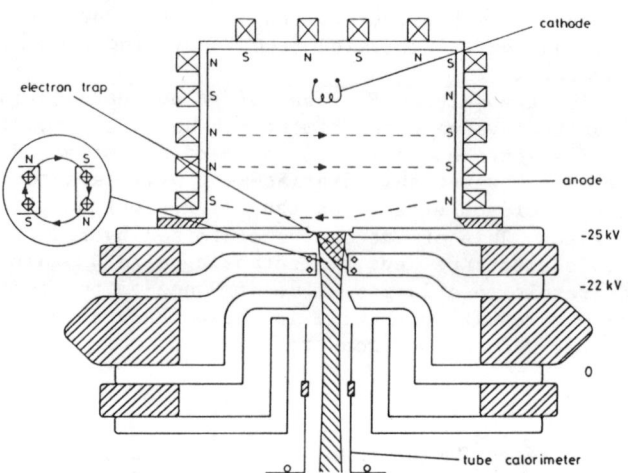

FIG. 1. Cross-section view of plasma source and accelerator. Inset is a view of the electron trap in the second electrode.

FIG. 2. Schematic drawing of the accelerator electrodes showing the zero net magnetic flux trap for electrons.

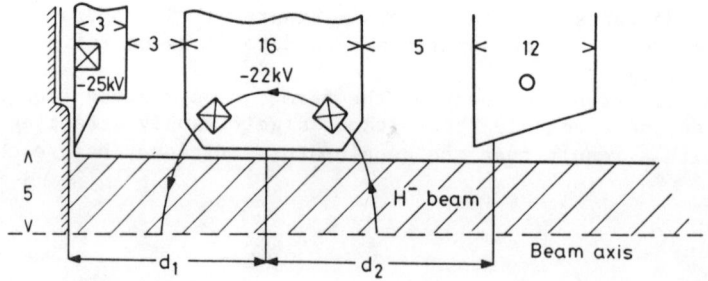

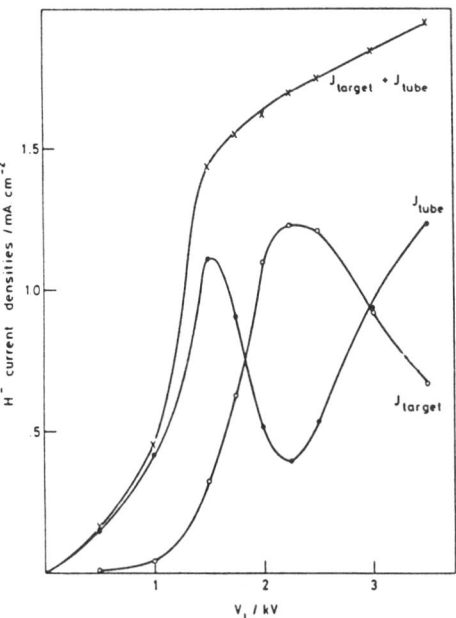

FIG. 3. H^- currents to the inner and outer calorimeter (labelled "target" and "tube" respectively) at I_{Arc} = 10 A, V_b = 15 keV.

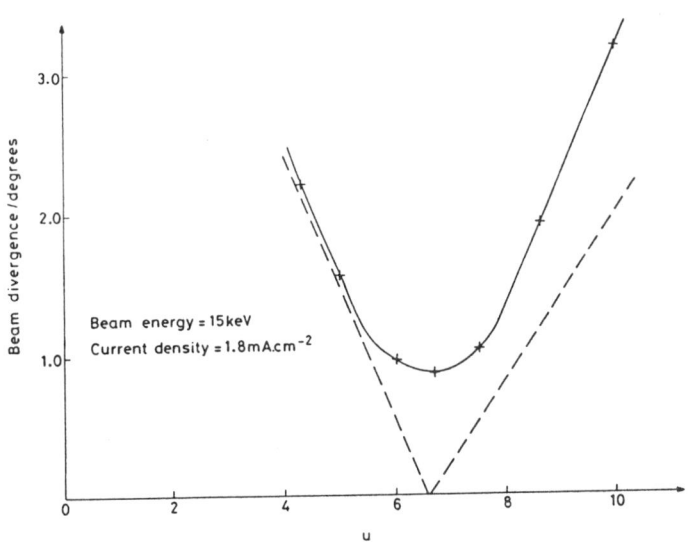

FIG. 4. Beam divergence plotted as a function of U for the data presented in Fig. 3.

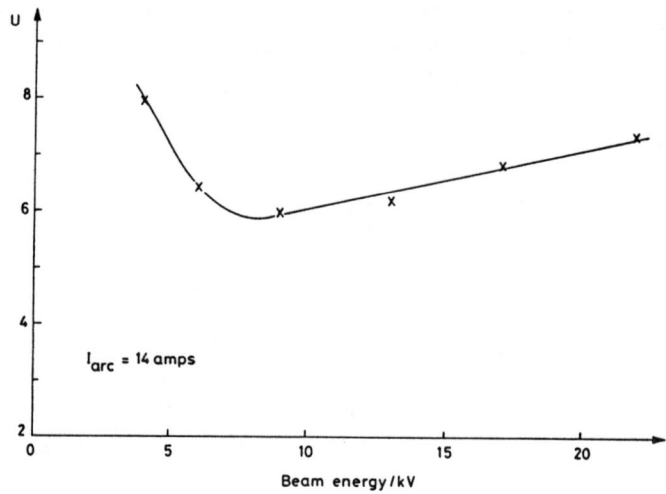

FIG. 5. The value of U needed to focus the H$^-$ beam plotted as a function of beam energy at constant beam current.

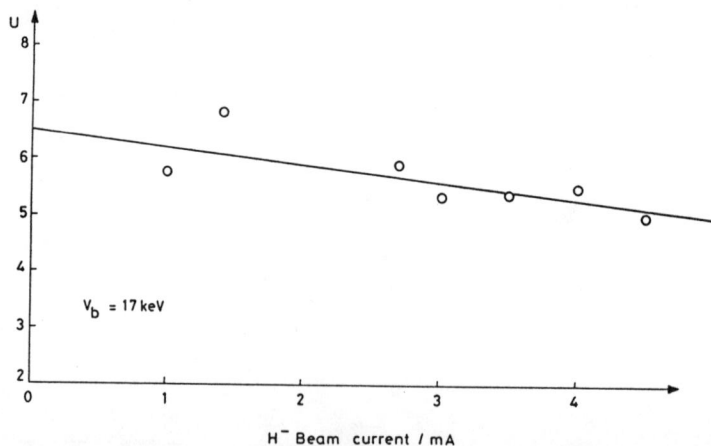

FIG. 6. The value of U needed to focus the H$^-$ beam plotted as a function of beam current when V_b = 17 keV.

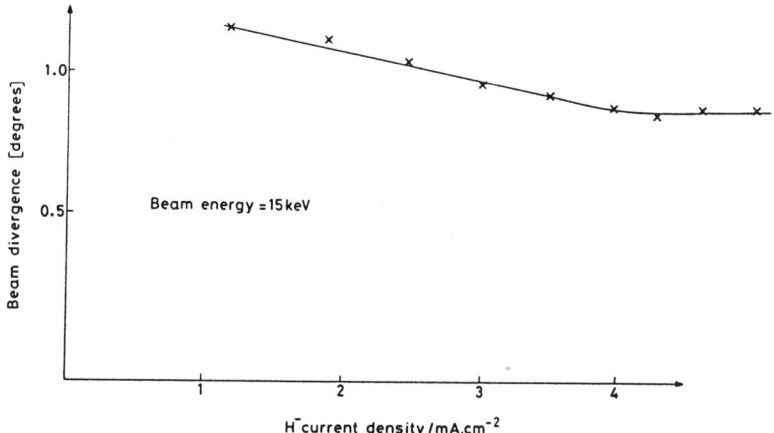

FIG. 7. Beam divergence plotted as a function of beam current at 15 keV beam energy.

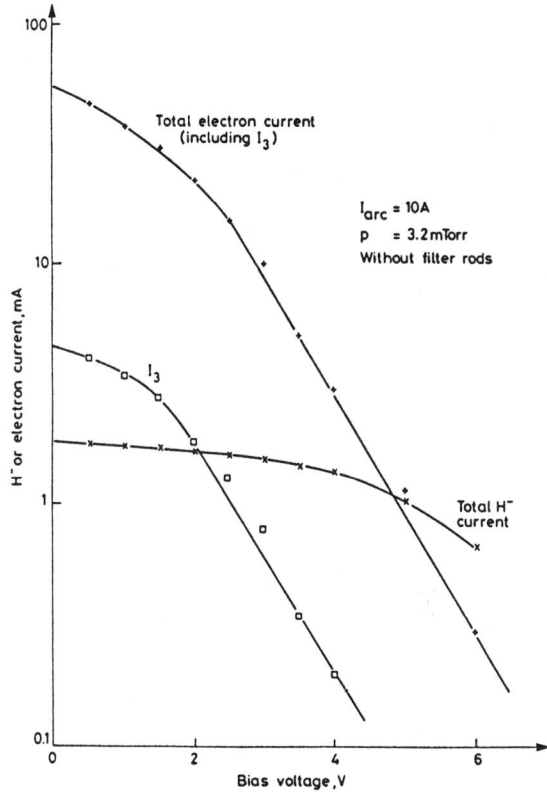

FIG. 8

The separate scaling of electron and H^- currents is shown as a function of the bias of the beam-forming electrode. The current I_3 is the electron current which gets through the electron trap to the third electrode.

LARGE NEGATIVE IONS SOURCE FOR ENERGETIC NEUTRAL BEAMS

M. Delaunay, R. Geller, C. Jacquot, P. Ludwig, P. Sermet,
J.C. Rocco, F. Zadworny
Association EURATOM-CEA
Département de Recherches sur la Fusion Contrôlée
Service d'Ionique Générale, 85 X, 38041 Grenoble (France)

J.B. Bergström[*], G. Hellblom[*], R. Pauli[*], H. Wilhelmsson[**]
[*]EURATOM-NE Association, Studsvik-Energiteknik,
S-61182 Nyköping (Sweden)
[**]Chalmers University of Technology Fak, 40220 Göteborg 5 (Sweden)

ABSTRACT

A negative ion source using the double electron capture can be extrapolated to a 10 A H⁻ ion source in d.c. state. Examined designs can handle 5 A/m of H⁻. These designs associate an electron cyclotron positive ion source with a monogrid to extract the beam and a magnetic compression to increase the positive current density in the cesium exchanger. After the H⁻ production in the cesium, the magnetic field is cancelled to eliminate the electrons before the H⁻ accelerator. The electrostatic accelerator is magnetic field free.

INTRODUCTION

The program at the Grenoble DRFC Laboratory through a French-Swedish collaboration to develop multiampere long-pulse negative ion sources has as its long term goal their proof of principle on a one megawatt beam line at large energies [1]. This choice is not entirely arbitrary because more recently, this requirement was suggested as a guide line by the office of Fusion Energy (U.S. Department of Energy) [2]. Such a demonstration would not have to show that the source could be scaled to 10 A, but that a practical beam line could be constructed based on it.

Since several years, the double charge exchange method using the production of negative ions by the interaction of a proton beam with a supersonic jet of Cesium has shown that it was possible to produce and accelerate a lot of tens of mA to 30 KV without troubles if we respect a electric field strength in the electrostatic accelerator of 25 KV/cm maximum for a D.C. operation [3]. The conversion efficiency in the Cesium cell was in agreement with the different cross sections datas and 160 mA of D⁻ has been achieved with a efficiency of 25 %. The difficulty with the double charge exchange methods was the limitation in D⁻ current density (5 mA/cm^2) due to the low positive ions current density consistent with the low voltage extraction.

Consequently, because a minimum of D⁻ current density of 25 mA/cm^2 at the entrance of the accelerator should be achieved for an optimal design of a neutral beam injector we have to develop a new D⁻ source which permits the extrapolation to a large current. Others specifications for this source should be the possibility to associate its to a strong-focussing of electrostatic accelerator and to limit the fraction of electrons in front of the accelerator. We will suppose that the more suitable neutralizer to be use with a D⁻ accelerator is the Laser Photo Detachment Neutralizer. The components of such a system would represent extrapolations, although reasonable ones [3]. Nonetheless, the advantages which appear possible with such a system appears sufficiently attractive to be considered in a future prototype. Nevertheless the main difficulty with a D⁻ source using the double charge exchange method is to get more than 25 mA/cm^2 of negative ions after the Cesium cell at low energy (500 V). The minimum positive current interacting with the Cesium should be 0.1 A/cm^2. In conventional two grid ion optics widely used, lower voltages (< 1 KV) can result in seriously reduced performance. This is due to the $V^{3/2}$ variation of current. The net to total voltage ratio should be compatible by the need for a barrier to prevent electron backstreaming and the minimum divergence of the beam. This condition is difficult to satisfy in the range of few tens of mA/cm^2 in the range of 0.5 KV extraction and the gap ratio between the electrodes very small in this case (< 1 mm) does not permit an easy mechanical extrapolation.

By using a three grids system (accel.decel. system) the advantage over the two grids system is the ability to operate at low ion energies without the extreme penalties in current density. But in our case (100 mA/cm^2 to be extracted) the voltage ratio between the acceleration voltage and the extraction voltage is greater than one and a supplementary problem is created with the addition of Cesium ions falling on the acceleration electrode. A lot of energetic secondary electrons are accelerated towards the accelerator and are responsable of parasite effects (breakdowns).

As discussed in connection with Child's law, the ion current density varies inversely as d^2 where d is the acceleration distance. For ion optics employing multiple grids, this acceleration distance is closely related to the spacing between grids. The one grid approach omits the screen grid and used only an accelerator grid biased relatively to the plasma source at the extraction voltage. The acceleration distance is simply the thickness of the plasma sheath next to the accelerator grid. It also means that the accelerator grid impingement increases greatly over that of the two grid approaches. The sputtering damage associated with this increased impingement can still be acceptable if operation is restricted to low voltages and low plasma densities. A possibility to solve this difficult problem in the double charge exchange problem is to associate an electron cyclotron resonance source with a large beam adiabatic compression from the extraction electrode to the Cesium charge exchange collision.

OPTIMISATION OF THE POSITIVE PROTON SOURCE (Fig. 1)

A new electron cyclotron resonance source has been developed using one magnetic field gradient instead of the classical magnetic mirror configuration. The radio frequency power is matched to the plasma source through a waveguide with a ceramic window inside a multimode cavity large in comparison with the RF wavelength. The RF wave is introduced towards the strongest magnetic field and the resonance condition $\omega_{RF} = \omega_{ce}$ must be satisfied inside the radio frequency cavity. A plasma is created and a strong absorption mechanism is observed. This absorption can be related to transformation of the injected electromagnetic wave to a slow fastly damped plasma wave. The theory of linear transformation for a one dimensional plasma indicates that the transformation should occur in a region where the cold refractive index becomes infinite for a wave propagating along the density gradient. This density gradient is perpendicular to the magnetic field and the position of the transformation region is defined by the equality between the RF frequency and the upper hybrid frequency ω_u where : $\omega_u = \omega_{RF} = (\omega_{pe}^2 + \omega_{ce}^2)^{1/2}$. ω_{pe} is the plasma frequency. The region of plasma wave propagation is bounded on one side by the cyclotron resonance surface $\omega = \omega_{ce}$ and on the other by the surface $\omega = \omega_u$. The RF energy is absorbed in the gap between these boundaries. The metal of the box returns the wave reflected from the opacity to the plasma so the opacity does not affect the integral result of the absorption.

With an RF generator of 10 KW maximum at a frequency of 8.27 GHz the ionic plasma current density in front of the extractor can be changed from 10 mA/cm^2 to 150 mA/cm^2 versus the RF power (Fig. 2). The optimum gas pressure inside the cavity is 1. mTorr (Fig. 3). The plasma volume is 5 liters and the total drain current is 25 A. The plasma is uniform on a total of 10-15 cm of diameter. The magnetic field value at the resonance surface is 2.8 KG. The current plasma density J_p depends of the value of the B field magnitude at the extractor system. Pratically the ratio $J_p(B_{ex})^{-1}$ is constant to : 2 10^{-4} A/cm^2 G. in a range of density 10 to 150 mA/cm^2. The J_p current density is roughly proportionnal to the cubic density of RF power and the source can operate in a range of B_{ext} field as low as 200 Gauss. The density of the plasma source is close to the cut off plasma density (ne = 6 10^{11} cm^{-3}) and the estimated electronic temperature is 10 eV. In these conditions, a large fraction of protons has been detected with a classical magnetic filter immersed in the hydrogen beam and 85 % of the total positive beam are protons. A theoretical model has been developed [4] to explain the high H$^+$ fraction ratio. In an E.C.R. source with a confinement merit factor of neτ = 10^7 cm^{-3}s and a moderate plasma density, the high proton ratio cannot be explained by the classical dissociation of hydrogen molecules in the ground state. A lot of new phenomena can be included in this model and among them, the RF dissociation of molecules into vibrationnal states can explain the high atom fraction. These vibratory states of molecules can be due to a RF wave interaction or to a fraction of energetic electrons in the discharge. This part of vibrationnal states of molecules can be estimated to be 20 % of the total number of molecules and consequently the classical dissociation of H$_2$(V) into two atoms can be enhanced. With a lifetime of 50 µs to 100 µs for $H_2(V)$ and an electronic temperature of 10 eV, 80 to 85 % of protons are obtained comparable to the experimental data.

THE BEAM OPTIC - CHOICE OF THE ONE GRID SOLUTION

The advantage of the one grid optic is its capability to operate at combinations of moderate current density and low ion energy that are not practical for any other ion acceleration approach using grids. On the other hand the heating limit and the various sputtering are serious problems despite the refractory metals using. A typical range of parameters for a definition of a one grid system is the extraction of 10 mA/cm² of H^+ or D^+ in a range of 0.3-0.5 KeV energy.

Generally, if we consider the geometry of a one grid system the geometric transparency T_G is 50 % and the thickness of the grid is equal to the maximum aperture dimension in the grid (Fig. 4). At this point the beam is still reasonably collimated and the impigement does not greatly exceed that which would be expected from the projected blockage of the grid. In the first approximation, for a low current plasma density (J_p = 20 mA/cm²) and a typical aperture dimension (a = 0.5 mm) the plasma sheath dimensions d is given by the Child's law is greater than a/2 and in a first approximation the effect of the aperture size can be neglected in the calculation of the real transparency of the grid. The transparency of the extractor T is given by :

$$T = T_g \left(1 + 2.87 \ 10^3 a \ \frac{J_p^{1/2}}{V^{3/4}} \right)^{-1} \quad \text{MKSA}$$

V is the extraction potential, a is aperture size (slit geometry), J_p is the Bohm ionic plasma current and T_g is the geometric transparency. Different extracted current density are indicated on fig. 5 versus V and J_p confirmed by experimental datas.

A one grid optic system is always divergent due to the negative focal length of the one slit aperture. The maximum perpendicular energy for the hydrogen ions can be estimate by an analytical formula taking into account the effect of the slit aperture size a given by :

$$(eV_\perp)_{eV} = \left(\frac{a}{7 \ 10^{-4} \ \frac{V^{3/4}}{J_p^{1/2}} - a} \right)^2 eV_{//} \quad \text{.MKSA}$$

where $eV_{//}$ is the extraction energy for the ions. Different perpendienergies are shown on fig. 6 versus V and J_p. Consequently with a one grid system, the best result is to try to use low plasma current density J_p with the highest extraction potential compatible with a good efficiency for the double charge exchange mechanism. Typically with

$eV_{//}$ = 400 eV), the transverse energy spread of 6.5 eV obtained with J_p = 20 mA/cm² must produced adequatly good beams at 200 to 300 KeV energies. We indicate on the fig. 6 the maximum perpendicular energy compatible with different exit angles for large voltage (200 KV).
The application of the criteria to minimize the perpendicular energy spread for the ions should lead to a very small ionic current extracted compatible with a minimum H⁻ density of 25 mA/cm². A direct H⁻ production of 25 mA/cm² obliges to interact 0.1 A/cm² of proton with a Cesium supersonic jet and a magnetic adiabatic compression of the beam is necessary from the extractor to the Cesium cell.

H⁻ CURRENT DENSITY PRODUCTION

Because low positive current density can be got at low magnetic field value in the vicinity of the extractor with low divergence and high transparency, the extracted beam can be compressed at moderate magnetic field value (2 to 3 KG) with large compression factor γ :

$$\gamma = \frac{B_{on\ Cesium}}{B_{ext.}} = \frac{J_{cs}^+}{J_{ex}} \qquad J_{ex} = J_p \times T.$$

Because the compression of the proton beam should be adiabatic in order to cancel the velocity along the large dimension, the final perpendicular energy is increased proportionally to the magnitude of the magnetic field. That is a consequence of the magnetic momentum invariant. Then after the H⁺, H⁻ conversion the magnetic field should be decreased to a value close to the initial value to fulfill two conditions. The first one is to minimize the perpendicular velocity spread compatible with the final divergence. The second condition is the ability to cancel very fastly the magnetic field in front of the accelerator to eliminate the electrons and to minimize the effect of the cancellation of the field on the H⁻ beam.

Finally, the magnetic field configuration is more or less symetric on the Cesium cell and minimize the width of the Cesium curtain and consequently the Cesium pollution. A typical schematic view of a double capture H⁻ source is shown on fig. 1. The B field on the positive ion extractor is between 100 and 500 G. The maximum B field on the Cesium cell is 2500 Gauss and the maximum width tolerable for the Cesium curtain is 2 to 4 cm. The half length L between the positive source and the Cesium or between the Cesium and the accelerator entrance should be greater than :

$$L > \frac{M\ V_{//}\ \gamma}{eB}$$

Classically this half length should be of the order of 0.5 m to satisfy the adiabatic criteria. Consequently the pressure along the beam transportation should be less than 10^{-4} Torr and a strong pumping system should be associate with the H⁻ source.

The electron elimination is obtained by a strong non adiabatic filter (high gradient : $\frac{dB}{dz} > 500$ G/cm). The magnitude of the magnetic field before the cancellation is small (100 G to 500 G) in order to minimize the spread velocity of the H⁻ after the electron separation. The spread velocity along the slit direction is proportionnal to the B field value and to the aperture size of the entrance electron of the accelerator. This effect is negligeable in comparison with the positive ion spread if B is less than 500 Gauss and the aperture size dimensions less than 1 cm.

The accelerator is magnetic field free and the electrostatic compression of the beam depends of the magnitude of the electric field that can be tolerate in the acceleration region. Preliminary calculations including a electrostatic compression of 3 to 5 with a parallel beam at the exit of the accelerator using a slit geometry shown the maximum value of electric field versus the current density (fig. 7).

A 10 A H⁻ SOURCE SCHEMATIC VIEW (FIG. 8)

In order to build a reasonable size negative ion source, a typical negative ion source must be operated with an output current of negative ions of 5 A/m. If the most attractive alternative is to use photodetachment to convert the negative ions to neutrals, it is necessary to minimize the cross sectional area of the laser beam to maximize system efficiency. These considerations lead naturally to the thin sheet beams with a maximum width of 2 cm. Consequently the minimum current density for the negative ions at the exit of the accelerator should be 25 mA/cm² at 200 KV. This electrostatic preaccelerator is magnetic field free. The magnetic field is cancelled at the accelerator entrance and contributes to the separation of the electrons from the negative ions beam of the beam.

For a minimum divergence beam of 0.5° at 200 KV a typical positive source with a current plasma density of 40 mA/cm² can be used with an extraction voltage of 400 V. An average position current density of 15 mA/cm² at 400 V with a magnetic field value on the extraction region of 250 G is extracted. A maximum divergence for the positive ions corresponding to a perpendicular energy of 10 eV is tolerated. The total positive current extracted is 36 A/m corresponding to a total slit width of 24 cm. At the extraction exit we use an adiabatic magnetic compression with a compression factor of 6. To fulfill the adiabatic criteria the minimum distance from the positive source to the Cesium curtain is 50 cm and with a background pressure lower that 10^{-4} Torr. The positive ion losses are less than 5 %. With a proton ratio of 90 % 31 A interacts with a Cesium curtain of 4 cm width. The choice of the maximum width for the Cesium exchanger is compatible with a good definition of the Cesium supersonic jet. The minimum conversion efficiency for the double charge exchange mechanism is 25 % and 8 A/m of negative ions are produced. The maximum H⁻ current density is 20 mA/cm² in a magnetic field of 1.5 KG. By decreasing adiabatically the magnetic field from 1.5 KG to 250 G, the final divergence of the H⁻ ions is conserved (10 eV) and the current density is 3.5 mA/cm². The goal of the accelerator is to accelerate the ions at 200 KV with an electrostatic compression ratio of 3 to 5. The accelerator employs a maximum transverse electric field and the maximum parallel electric field should be of the order of 20 KV/cm compatible with the accelerated H⁻ current density. Because the magnetic field is totally cancelled on the accelerating part of the H⁻ beam, a strong decreasing of the B field from 250 G to zero with a magnetic gradient of 500 G/cm is built by using an iron plate with aperture size (1 cm × 1 m) and a total transparency of 66 %. If the background pressure is less than 10^{-4} Torr. The H⁻ beam transmitted to the accelerator is 0.q and the total accelerated H⁻ current is of the order of 5 A/m with a final divergence of 0.5° at 200 KV. Adequate electron control (5 %) with the sharply decreasing of the B field is under investigation. The effect of the cancellation of a low magnetic field of 250 G to zero with an half width slity of 0.5 cm contributes to give a supplementary velocity along the great dimension of the slit given by :

$$V_{\perp (v)} = \frac{e}{8m} B^2 y^2$$

In our case the supplementary perpendicular velocity is 0.2 eV and is negligeable in comparison with the 10 eV given by the monogrid optic.

CONCLUSION

Designs that we have examined can handle 5 A/m of H⁻ ions by using a double charge exchange mechanism to produce negative ions. The E.C.R. positive source is operated in a range of frequency corresponding to the same range of frequency that the low hybrid frequency heating (3-4 GHz). The using of these wavelength can be envisaged because we use a moderate positive current density in the source (40 mA/cm^2). The magnetic field in the source is typically 1.5 KG for the resonance and 0.25 KGauss on the monogrid. The monogrid with a realistic transparency of 35 % and a final divergence equivalent to 10 V can support an ionic flux of 16 W/cm^2 ($\Delta T = 1500°K$). Because we used a very strong supersonic Cesium jet to minimize the neutrals lifetime, the Cesium curtain dimension transverse to the beam is limited to few centimeters. Consequently a magnetic compression of the positive beam and a magnetic decompression of the negative beam are used to increase the number of A/m. After the magnetic decompression of the beam, a very strong B cancellation serves to eliminate the electrons from the beam. The electrostatic accelerator is classical, magnetic field free. It increases the negative ion current density to 25 mA/cm^2 compatible with 5 A/m and a good efficiency for the photodetachment neutralizer. The final divergence of the neutrals can be expected inside an half width angle of 0.5° at 200 KV. The pollution due to the Cesium is minimized by using a supersonic Cesium curtain. If a new feature of conceptual negative ion-based-neutral-beam systems is that the transverse field focusing can be used to transport beam of negative ions to highest energies, it eliminates direct streaming of particles through the transport system and the Cesium problem can be solved easily.

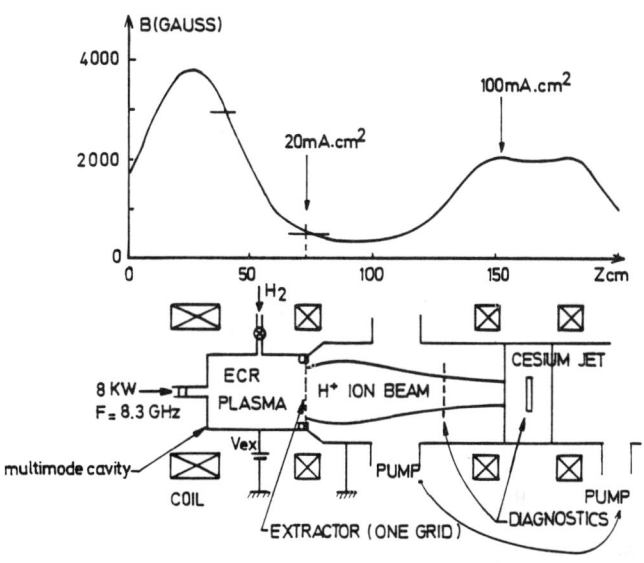

Fig 1
E.C.R. ION SOURCE LAMPION 3

MAIN CHARACTERISTICS

- hydrogen pressure = $2 \times 10^{-4} - 3 \times 10^{-3}$ m Bar
- microwave frequency = 8.3 GHz
- resonance magnetic Field = 2950 Gauss.
- extraction : one grid $S = (14 \times 14) cm^2$ $e = 5 \times 10^{-2}$ cm
- H^+ current (near the extraction system) = 4 A. (20 mA.cm^{-2})
- H^+ energy : 0,3-0,6 keV.
- Proton ratio : 85% (15% $H_2^+ - H_3^+$)

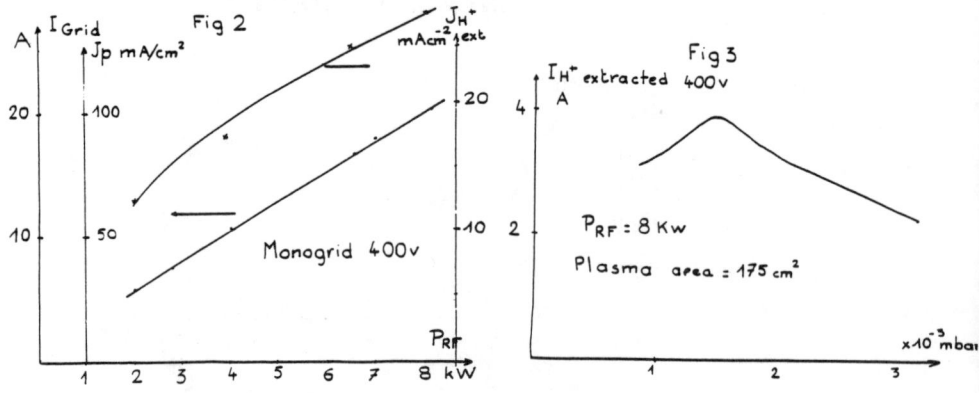

Fig 2

Fig 3

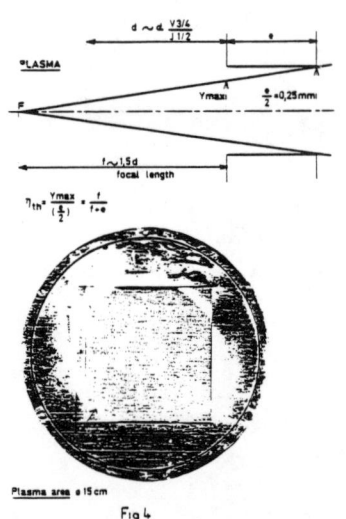

Fig 4

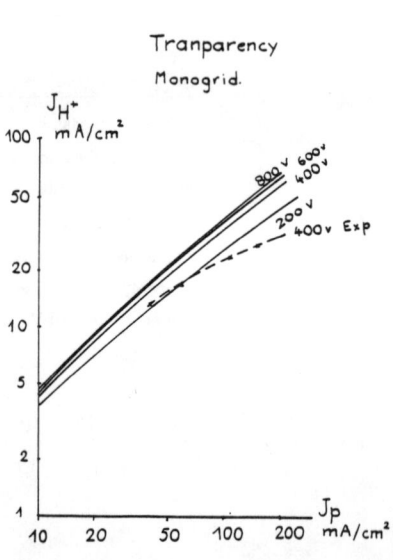

Fig 5

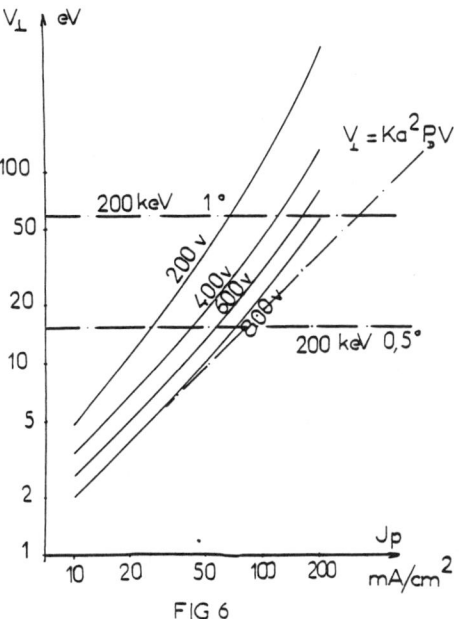

FIG 6

Slit accelerator 80kV
Compression ratio 3

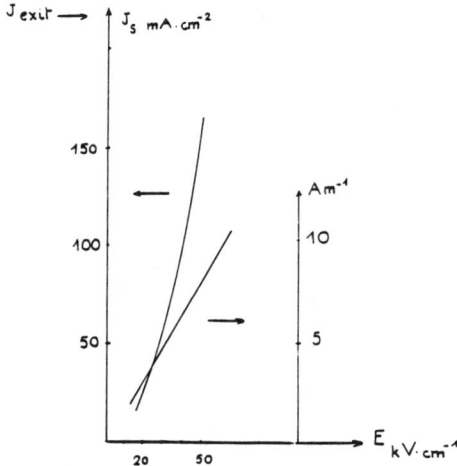

Fig 7

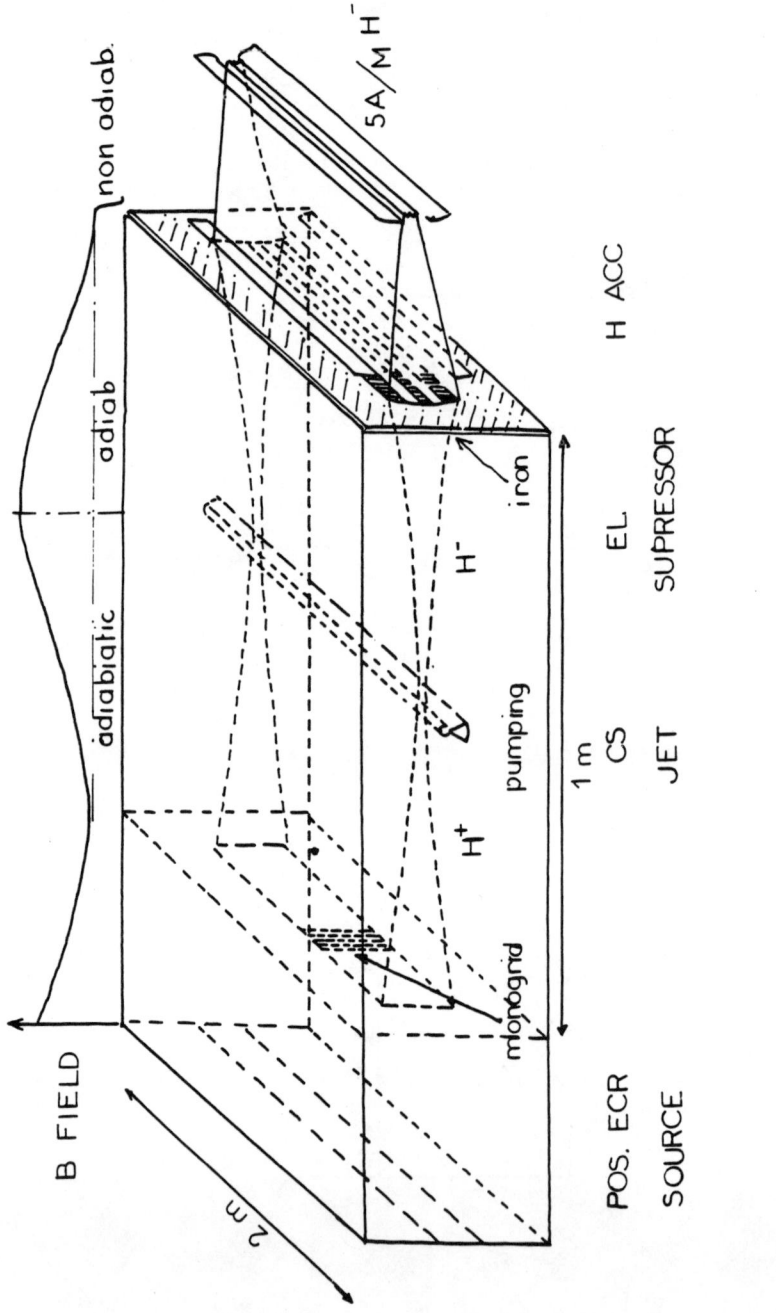

FIG 8

REFERENCES

1. M. Delaunay and al., Proceedings of the 2nd Grenoble-Varenna International Symposium, 3-12 sept. 1980, Como (Italy), "a new type of neutral injector based on the production of D⁻ by double electron capture".
2. D.A. Goldberg and al., Journal of Fusion Energy 3, 1 (1983), "conceptual design of a 200 KeV, 1MW negative ion-based".
3. C. Jacquot, 12th Symposium on Fusion Technology, 1982, Jülich (F.R.G.), p. 125.
4. M. Delaunay and al., to be published.

DISCUSSION

HENKES: Did you make any measurement of the beam scattering on the cesium jet?

JACQUOT: No, not in this experiment, but in previous experiments other causes of divergence were large in comparison with the scattering due to the collisions in the cesium jet and we could not see any change in the divergence due to scattering.

NORMALIZED EMITTANCE OF SITEX NEGATIVE ION SOURCE

W. L. Stirling, W. K. Dagenhart, and J. H. Whealton
Oak Ridge National Laboratory
Oak Ridge, Tennessee 37830

J. J. Donaghy
Washington and Lee University
Lexington, Virginia 24450

ABSTRACT

An emittance measurement employing two techniques are being made on SITEX. To this end, a 2-D calculation was performed to design the accelerator in order to reduce electric field aberrations. The calculated normalized emittance is 6×10^{-4} πcm mrad for an angular divergence $\theta_{RMS} \approx 0.28°$. Status of the experimental findings are presented and a comparison made to the calculated value which will yield the ion sputter energy.

INTRODUCTION

An experimental determination has been made of the normalized emittance of the SITEX negative ion source.[1] To this end, an extraction system has been designed in conjunction with a newly designed, focussing converter system for minimum electric field aberration of the extracted ions. The converter geometry was determined self consistently with the accelerator geometry through an exact 2-D solution of the coupled Poisson-Vlasov equations.[2,3] The geometry of the system and the computer determined beam envelope is shown in Fig. 1 for a converter bias of -150 V relative to the discharge chamber, a negative ion current density of 50 mA/cm^2 at the converter surface, a magnetic field of 1300 gauss directed out of the paper and an accelerating potential of 18 kV. Fig. 2 shows a plan view of the ion source. For the conditions stated above, the mean RMS divergence angle of the beam is calculated to be 0.28°. The beam is 0.5 mm wide perpendicular to the exit slit with a current density of 400 mA/cm^2 at the extraction grid. The aberrations of the beam in the direction parallel to the magnetic field (long slit dimension) were not important, because the transverse emittance was measured only over a short region of the beam extracted from the center of the slit. Assuming a sucessful reduction in electrostatic aberrations on beam divergence, one can determine an accurate estimate of the ion sputter energy originating at the converter and thereby determine a lower limit to the expected beam emittance.

Figure 1
2-D OPTIMIZED SITEX ACCELERATOR

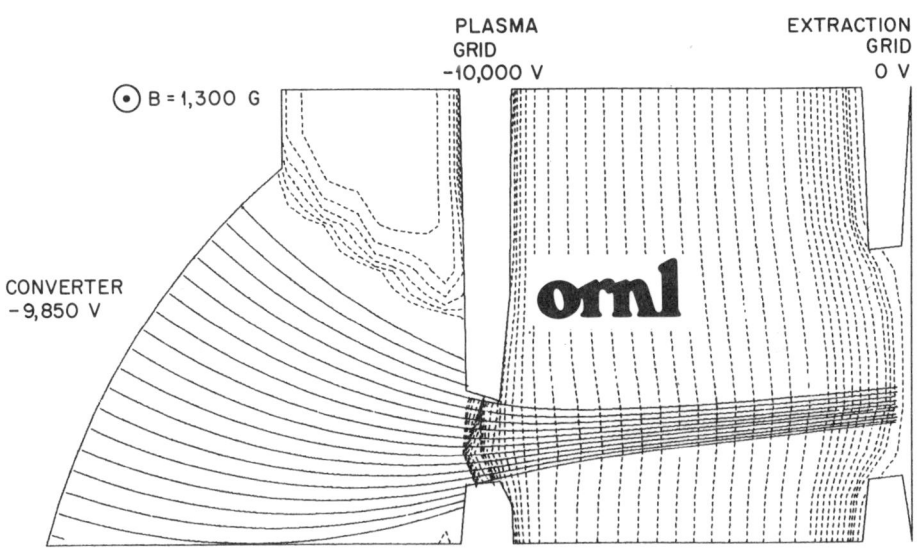

SITEX ELEVATION

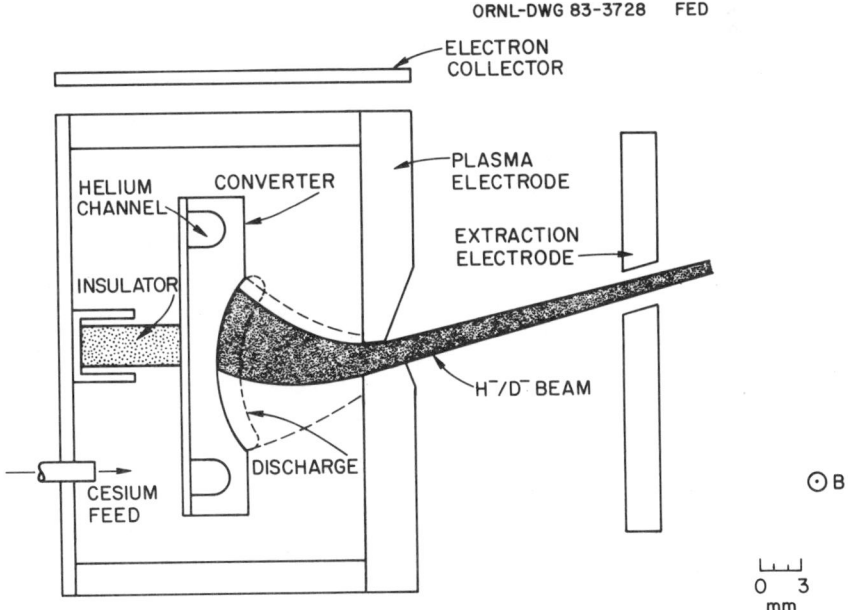

Figure 2
EXPERIMENTAL LAYOUT

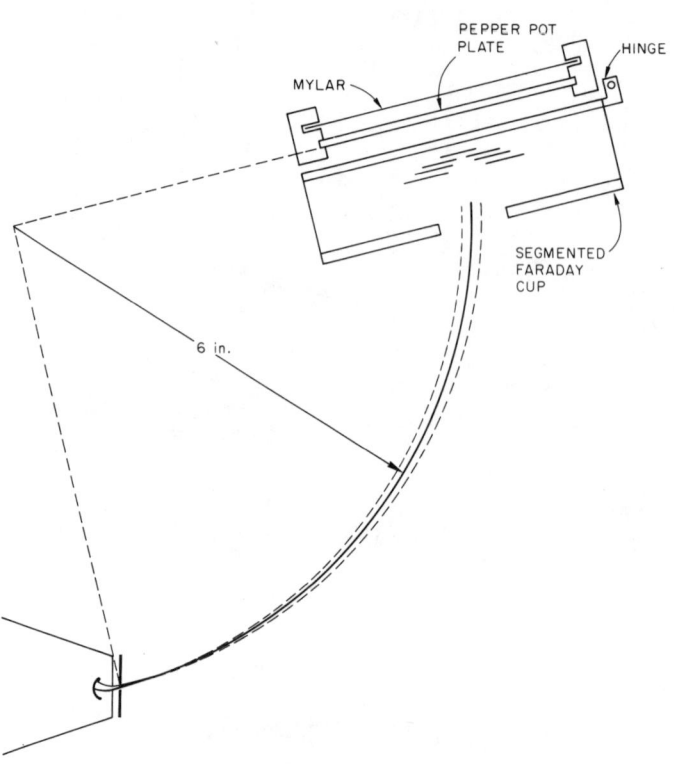

Figure 3

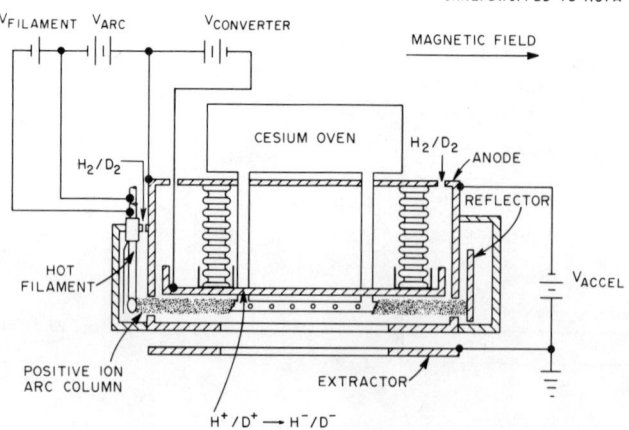

EXPERIMENTAL DESCRIPTION

An experimental layout is shown in Fig. 3. A segmented, shielded Faraday cup is used to locate the beam as well as estimate the beam profile width and angular divergence. The Faraday cup can be rotated out of the beam to permit a pepper pot scan[4] of the beam to be made from which the beam emittance can also be determined. Since the Faraday cup, Fig. 4, is a current measurement, each segment of the cup was fabricated with upturned "lips" or "edges" on the ends to ensure secondary particles emitted from a plate would follow a magnetic field line back into the plate of origin. Various combinations of positive and negative biases on adjacent segments revealed no current leakage from one plate to another. The pepper pot geometry used is shown in Fig. 5. Mylar was chosen as a "film" due to experience with a pinhole camera diagnostic used in positive ion source development. A roll of mylar sheet, 0.15" thick, could be fed into the exposure position of the pepper pot as needed in order to make multiple exposures without breaking vacuum.

RESULTS

From the Faraday cup readings one can calculate the beam profile. The beam profile width yields the angular divergence at the extraction grid, which coupled with the calculated beam size at the extraction grid, permits a determination of beam emittance.

Typical operating conditions are summarized in Table 1 and a typical beam profile is shown in Fig. 6. The profile is a Gaussian peak riding on top of a rather broad, low intensity distribution. This profile distribution is always seen in a careful beam analysis. The Gaussian peak carries the ion temperature information and the low intensity spread results from electrostatic field aberrations. Fig. 7 is a semi-logarithmic plot of several profile scans for both hydrogen and deuterium. There is no significant difference in source operation or beam output between hydrogen and deuterium. Also plotted in Fig. 7 are the lines for $T_i = 0.5$ eV and $T_i = 1$ eV under the present operating conditions. The variation in T_i is from 0.7 eV to 1.25 eV corresponding to beam divergences of $\pm 0.48°$ and $\pm 0.64°$ respectively. The angular divergence of the beam is related to the profile width as shown in Fig. 8 through the expression

$$R \sin \theta = \text{profile width}.$$

The normalized emittance ε is now calculated from the expression

$$\varepsilon = \beta \times \text{angular divergence} \times \text{beam size}$$

$$\beta = v/c$$

where v = beam velocity and c = velocity of light.

$$\varepsilon = 10^{-3} \pi \text{ cm mrads}$$

Figure 4
FARADAY CUP

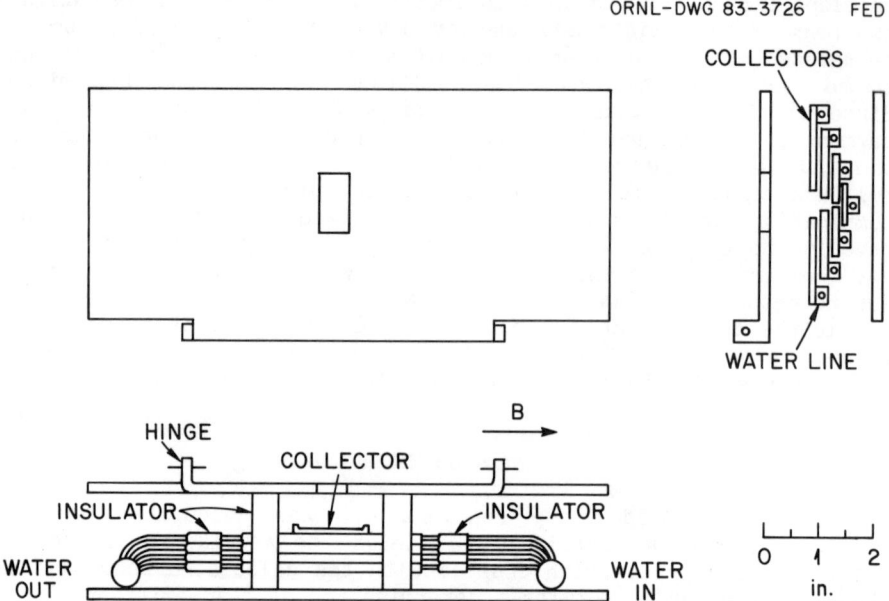

Figure 5
PEPPER POT GEOMETRY

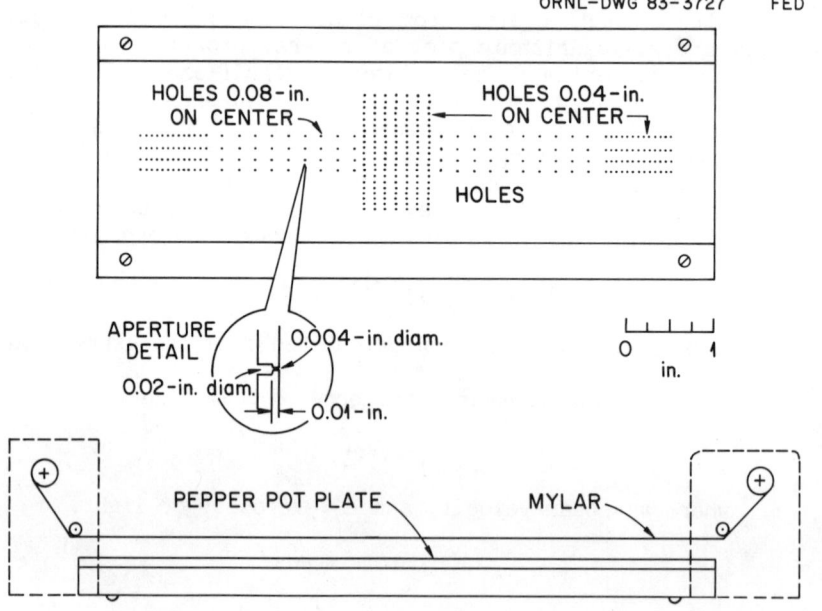

TABLE 1

PARAMETER	H⁻	D⁻
I_{accel} (mA)	550	250
V_{accel} (kV)	18	10
$j_{extracted}$ (mA/cm²)	110	100
Pulse Length (s)	10	5
Normalized Emittance $E_{N\perp}$ (π cm mrad)	not measured	10^{-3}
Source Pressure (mT)	4	4
Electron/Ion Ratio (%)	15	5
Electron Recovery at 10% V_{accel}	yes	yes
Arc Efficiency (kW/A)	5	5

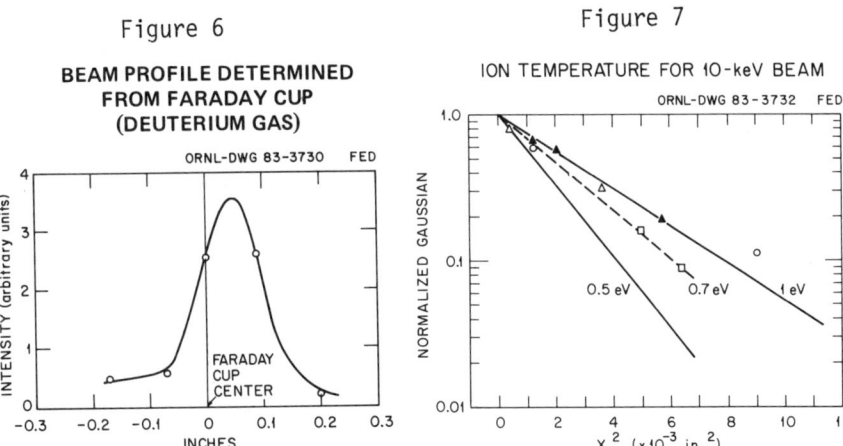

Figure 6

BEAM PROFILE DETERMINED FROM FARADAY CUP (DEUTERIUM GAS)

Figure 7

ION TEMPERATURE FOR 10-keV BEAM

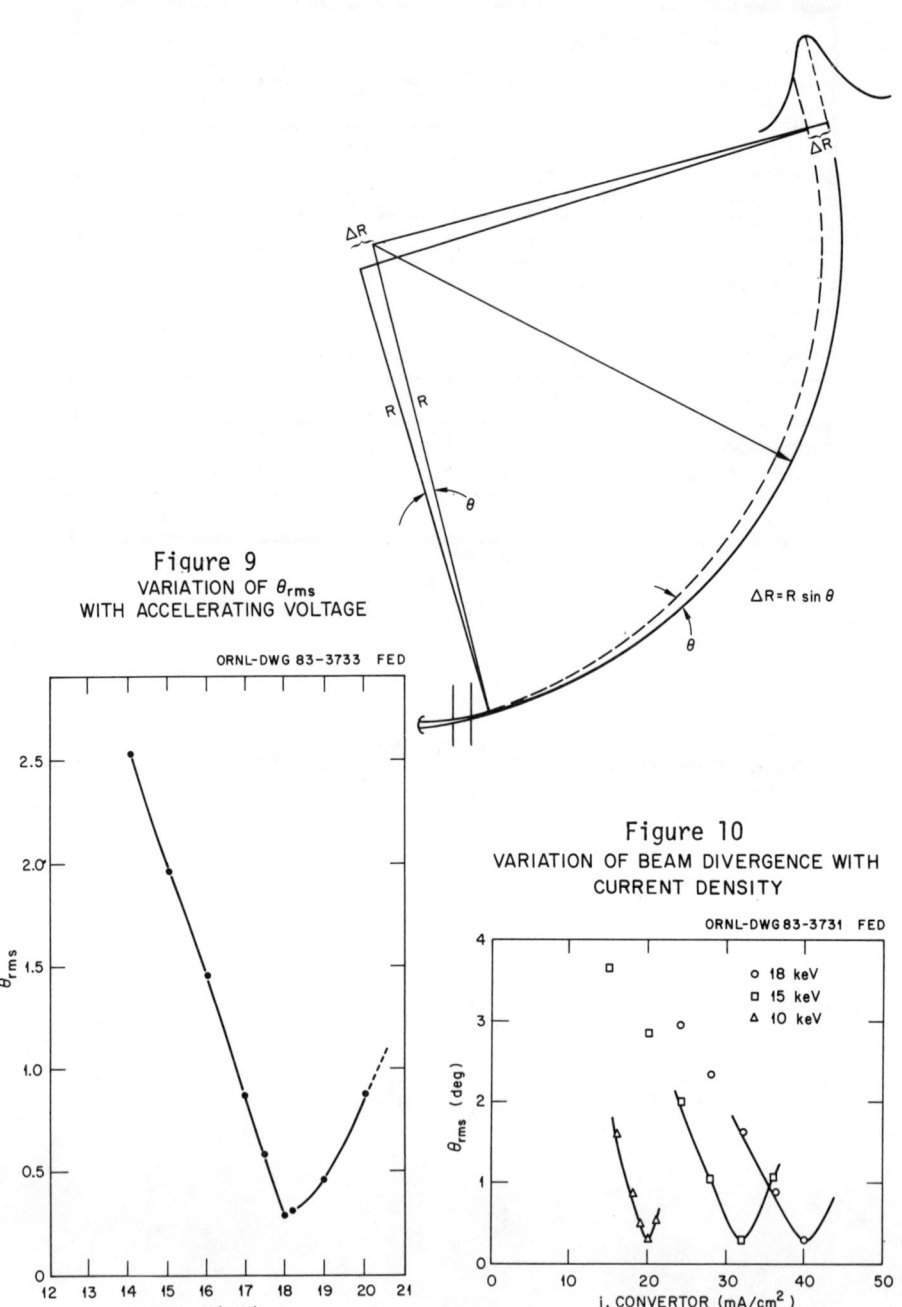

Figure 8
DETERMINATION OF BEAM DIVERGENCE FROM PROFILE WIDTH

Figure 9
VARIATION OF θ_{rms} WITH ACCELERATING VOLTAGE

Figure 10
VARIATION OF BEAM DIVERGENCE WITH CURRENT DENSITY

A pepper pot scan requires 200 to 300 identical pulses to make an exposure. Calculations of the beam profile reveal that the beam divergence is a strong function of several parameters. Fig. 9 shows a variation of the divergence with accelerating potential. The variation with beam current density at the converter is shown in Fig. 10 for three different beam voltages.

These two effects are coupled through the internal impedance of the high voltage supply thus requiring some means of voltage regulation. In the runs thus far, the regulation has been poor with the voltage varying ±2% from pulse to pulse. From Fig. 9, a 2% voltage swing produces an angular divergence increase from 0.28° to ~.5°. In addition the beam moves radially at the pepper pot by the same percentage as the voltage shift of ±2% or 0.12" which is about equal to the calculated beam size. Voltage regulation of about 0.1% must be achieved in order to make a meaningful pepper pot scan.

DISCUSSION OF RESULTS

M. Seidl and A. Pargellis[5] have recently reported the measurement of the energy distribution of negative hydrogen ions for Cs^+ ion energies varying from 500 eV to 2000 eV. The energy distribution is for ions traveling perpendicular to the surface. They found that a universal relationship exists between the energy distribution and the ratio of the H^- ion sputter energy, E, to the incident Cs^+ ion energy, U, independent of incident Cs^+ ion energy. The peak in the energy distribution occurs for $E/U \simeq 0.013$ with a full width half maximum (FWHM) of $E/U \simeq 0.005$. Extending these results to a Cs^+ ion energy of 150 eV as used in this experiment, the peak of the distribution would be 1.95 eV with a FWHM of 0.75 eV. Our value of 0.7 eV is in very good agreement with their FWHM value of 0.75 eV.

REFERENCES

1. W. K. Dagenhart, W. L. Gardner, W. L. Stirling, and J. H. Whealton, Nucl. Tech./Fusion, 4, pp. 1430-1435, September 1983.
2. J. H. Whealton, paper presented in these proceedings.
3. J. H. Whealton, Nucl. Instrum. Methods, 189, 55-70 (1981).
4. J. J. Donaghy, private communication.
5. M. Seidl and A. Pargellis, Phys. Rev. B, 26, pp. 1-9, July 1982.

A SCALED, CIRCULAR-EMITTER PENNING SPS FOR INTENSE H⁻ BEAMS*

H. Vernon Smith, Jr., Paul Allison, and Joseph D. Sherman
Los Alamos National Laboratory, MS-H818, Los Alamos, NM 87545

ABSTRACT

The Los Alamos versions of the Penning Surface-Plasma Source (SPS) routinely generate H⁻ ion beams with pulsed currents over 100 mA. However, these sources employ geometries that result in the extraction of slit beams (0.5 x 10 mm^2). Our modeling with the SNOW code indicates that the beam from a 5.4-mm-diam circular emitter will have lower emittance and divergence for transport to and injection into our radio-frequency quadrupole (RFQ) accelerator. This paper describes a newly constructed Penning SPS that has most of its discharge chamber dimensions scaled up by a factor of 4 to accommodate this circular emitter.

INTRODUCTION

As part of an ongoing effort to study the acceleration of H⁻ ions in a RFQ accelerator[1], we studied[2,3] several H⁻ SPS sources and built an injector[4] incorporating a Penning SPS.[5] We find that after extracting, accelerating, and transporting the slit beam from the injector source to the emittance scanners (∼30 cm total distance), coupling effects cause the transverse plane emittances to be nearly equal,[4] even though the initial transverse-plane-emittance ratios almost reflect the 0.05:1 ratio in slit dimensions. Using this finding, it is straightforward to show that use of a circular emitter having the same H⁻ emission current density and total current as the slit emitter results in a lower H⁻ beam emittance. For a slit emitter of area 4ab (20a = b) and total current I, the emission current density is $5I/b^2$. For the aperture emitter of area πR^2 and total current I, the emission current density is $I/\pi R^2 = 5I/b^2$, so R = 0.25b. Since the two-dimensional, normalized emittance ε_y is proportional to the emitter dimensions whether ion temperature or aberrations dominate the optics, a circular emitter is expected to produce a lower emittance H⁻ beam than the slit emitter.

Even if the 0.05:1 ratio in transverse beam emittance at extraction could be preserved in transporting the slit beam to the RFQ, coupling of the two transverse and the longitudinal plane emittances, mostly by space-charge effects in the RFQ, will cause the ratio of the transverse plane emittances to be nearly 1:1 at the RFQ exit[6]. Therefore, we built a new Penning SPS incorporating a circular emitter and a spherical extractor. See Ref. 7 for an account of our previous Penning SPS circular aperture work.

---

*Work supported by the US Air Force Office of Scientific Research.

SOURCE DESIGN

We used the SNOW code[8] to study the ion-extraction and beam-formation optics. The extraction system is designed to provide a total H⁻ current of 160 mA. The electrode design resulting from the SNOW calculations is shown in Fig. 1A. The emission aperture is 5.4-mm diam; the extraction electrode, 3.4-mm diam; the extraction gap, 4.7 mm; and the gap voltage, 29 kV. We succeeded in keeping the designed extraction gap electric field below 120 kV/cm. The emittance predicted by the SNOW code for the H⁻ beam at the exit of the extraction electrode ($z = 12.5$ mm in Fig. 1A.) is 0.009π cm·mrad. This number is derived in the following manner. A total of 670 rays were launched from the injection plane, located at $Z = 0$ in Fig.1A. The rays were given a distribution of angles with respect to the Z-axis appropriate for an H⁻ ion temperature of 4 eV, the average of our previous estimates for this parameter (3 eV in Ref. 4, 5 eV in Ref. 3). SNOW self-consistently calculated each ray through the extraction optics to the extractor exit. The distribution in phase space of the surviving 519 rays at $Z = 12.5$ mm is shown in Fig. 1B. The two-dimensional normalized rms emittance is calculated from this distribution according to the formula

$$\varepsilon_{x,y}^{\text{aperture}} = (\beta\gamma/\sqrt{2}) \left[\overline{r^2}\ \overline{r'^2} - \overline{r\ r'}^2 \right]^{1/2}, \qquad (1)$$

where β and γ are the usual relativistic parameters. If the H⁻ beam current fluctuates no more than $\pm 20\%$ about the 160-mA design value, the SNOW code predicts the time-averaged emittance will increase by a factor of 1.9 because of variations in the phase-space orientation after extraction. Thus, it may be possible to keep $\varepsilon_{x,y}$ below our previously recorded lowest value of 0.02π cm·mrad.[3,4] The SNOW calculations assume an injected ion energy of 100 eV to avoid nonuniform current density build-up, thereby probably underestimating the H⁻ beam emittance.

Since the SNOW design calls for an H⁻ emission density of 700 mA/cm² compared to our previously measured values of 2-3 A/cm², we decided to scale the source size up by a factor of 4 and reduce the plasma density and the H⁻ emission current density by a factor of 4 according to the law of similarity. This reduction of the plasma density results in a similar decrease in the cathode-power loading, thus allowing a substantial increase in the arc duty factor. The enlarged source, shown in Fig. 2, has a cathode-cathode gap of 17 mm, large enough to accommodate the emission aperture. We refer to this enlarged Penning SPS as the 4X source. A comparison of the source dimensions and anticipated operating parameters of the 4X source with the values for 100-keV injector source is given in Table I.

Unlike the 100-keV injector source whose arc magnetic field is driven by a permanent magnet circuit, the 4X source magnet circuit is driven by an electromagnet coil, allowing the arc magnetic field to be varied. The lower arc field, coupled with the higher gap voltage, results in a bend angle of 4.6° for the 4X source, compared to 8.1° for the injector source. The low bend angle will allow the 4X source to be close-coupled to the 100-keV injector column as is the present injector source, shown in Fig. 7 of Ref. 4.

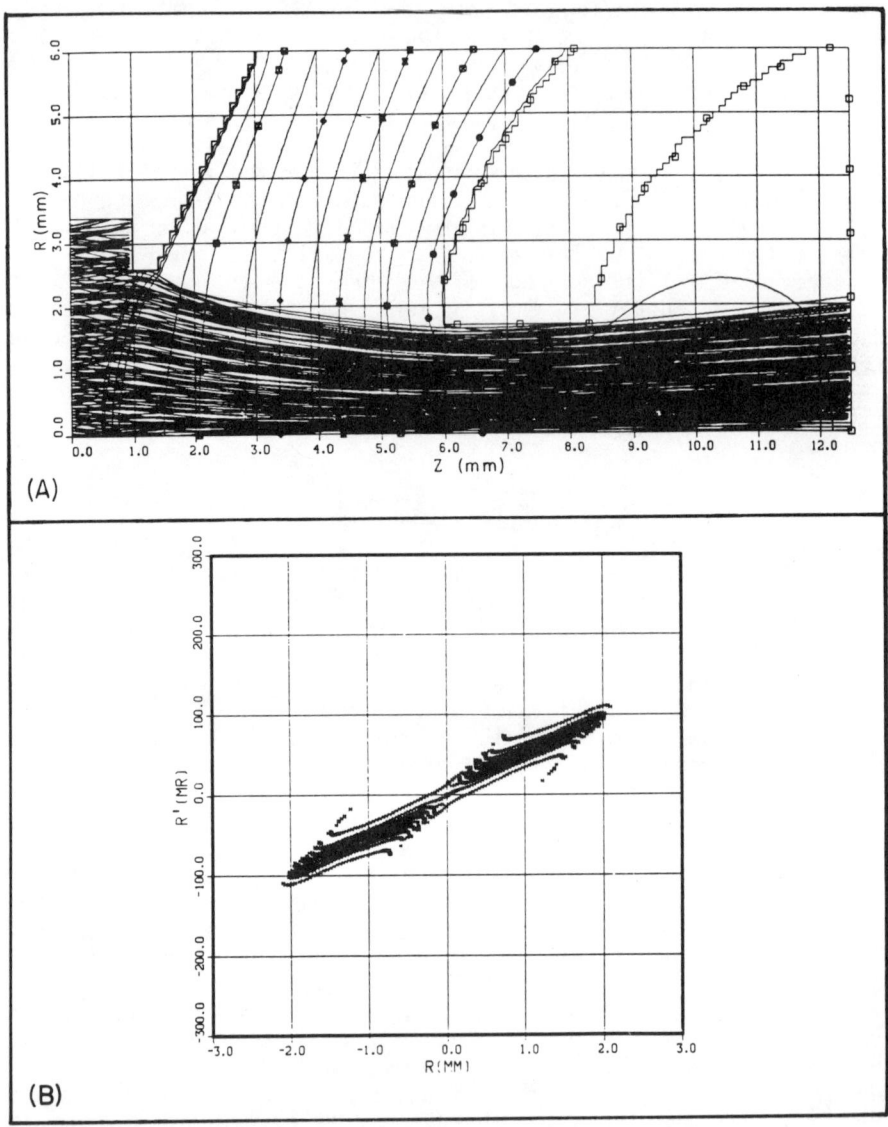

Fig. 1. (A) SNOW-code calculation of the H⁻ ion trajectories for the 4X source extraction optics; (B) Phase-space diagram at Z = 12.5 mm calculated by the SNOW code.

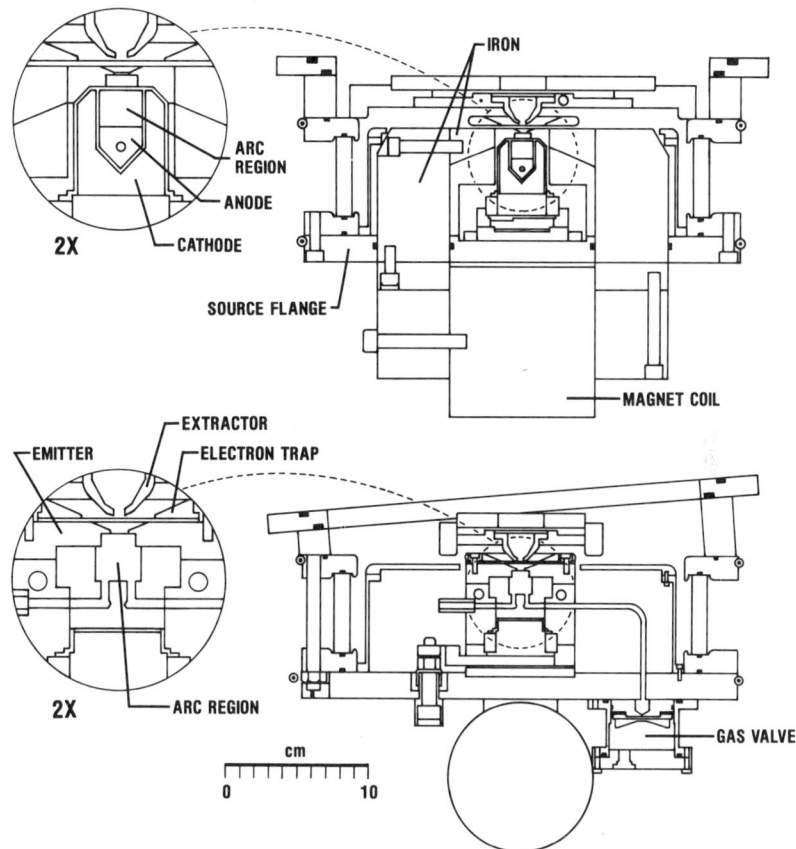

Fig. 2. The 4X scaled source. Top view: arc magnetic field in the plane of the paper; lower view: arc magnetic field direction (x) out of the paper. At the left are 2X blowups of the arc region.

SOURCE STATUS

A photograph of the assembled 4X source is shown in Fig. 3. We will study the arc discharge, H⁻ beam extraction, and discharge oscillations, as well as the H⁻ beam emittance. The enlarged arc volume may allow some discharge plasma measurements. We also plan to study the electron loading of the extraction electrode and ways to control and/or alleviate this problem.

Fig. 3. The 4X source.

Table I. Comparison of 4X Source with 100-keV Injector Source

DIMENSION/PARAMETER	4X SOURCE	100-keV INJECTOR SOURCE[a]
Cathode-cathode gap, mm	17	4.3
Arc slot width, mm	12	3
Arc slot length, mm	16	12
Arc magnetic field, T	0.05[b]	0.22
Emitter dimensions, mm	5.4 diam	0.5 x 10
Extraction gap, mm	4.7	2.5
Extraction voltage, kV	29	22
Arc voltage, V	100	100
Arc current, A	210	180
H⁻ current, mA	160	160
Cathode power load, kW/cm^2	1.5-4	7-16
Duty factor, %	~5	0.5

a) Ref. 4.
b) 0.14 T if magnetic field suppresses electrons in extraction gap.

REFERENCES

1. F. O. Purser, E. A. Wadlinger, O. R. Sander, J. M. Potter, and K. R. Crandall, IEEE Trans. on Nucl. Sci. NS-30, 3582 (1983).
2. P. Allison, H. V. Smith, Jr., and J. D. Sherman, Proc. 2nd Int. Symp. on the Production and Neutralization of Negative Hydrogen Ions and Beams, Upton, New York, October 6-10, 1980. Brookhaven National Laboratory report BNL-51304 (1980), p. 171.
3. H. V. Smith and P. Allison, Rev. Sci. Instrum. 53, 405 (1982).
4. P. Allison and J. D. Sherman, "Operating Experience with a 100-keV, 100-mA H⁻ Injector," Proc. this Conf.
5. V. G. Dudnikov, "Surface-Plasma Source of Negative Ions with Penning Geometry," Proc. of the IV USSR Nat. Conf. on Particle Accelerators, Moscow, 1975, Vol. I, p. 323.
6. E. A. Wadlinger, Los Alamos National Laboratory, private communication, October 1983.
7. J. D. Sherman, P. Allison, and H. V. Smith in Ref. 2, p. 184.
8. J.E. Boers, "SNOW-A Digital Computer Program for the Simulation of Ion Beam Devices," Sandia National Laboratory report SAND79-1027, 1980.

THE PLASMA FOCUS AS A SOURCE OF COLLIMATED BEAMS OF NEGATIVE ION CLUSTERS AND OF NEUTRAL DEUTERIUM ATOMS*

V. Nardi, C. Powell
Stevens Institute of Technology, Hoboken, NJ 07030

ABSTRACT

We report the space anistropy and brightness B_4 (i.e., the momentum normalized density in four dimensional transverse phase space) of a high-intensity pulsed source of neutral-atom and negative-ion-cluster beams with energy/atom $E \gtrsim 0.2$ Mev, ion clusters with m/Z (a.u.) > 200. The source is formed in an 0.5 MA plasma focus-PF-discharge. The energy spectrum of different particle species is obtained from a 12.2 kG magnetic analyzer, energy filters and time resolved detectors. Collimated particle beams are ejected within a < 6° cone along the discharge axis inside a < 3 mm diameter plasma channel (neutral atoms, ion clusters, impurity heavy ions at 0°, electron beams, ion clusters and negatively-charged ion clumps at 180°). Pulsed kA currents of ions (and neutral fluence of comparable intensity at 180°) are detected in the 6° cone at 0° with $B_4 \sim 10^7$ (mA/cm^{2}rad^2) for particle energies $E \gtrsim 200$ KeV. In the 180° direction the source ejects multiple pulses of electron and ion beams in alternating sequency (typical pulse duration $\sim$ 10 ns) with a net negative charge which provide charge neutralization for ion and ion cluster beams. The source which can operate - in principle - at a high repetition rate has a scaling law in which the particle-beam intensity increases without a detectable increase of the angular dispersion.

INTRODUCTION

It is well established[1,2] that PF discharges (powered from capacitor banks of $\sim$ 1 kJ to > 100 kJ at a voltage $V_0 \sim$ 10-200 kV, with peak current in the axial pinch from $\sim$ 0.1 MA to > 2 MA) are intense sources of highly collimated beams of MeV positive ions (at 0°) and of electrons (at 180°, along the discharge/electrode axis). Each of these kA pulsed beams with duration $\sim$ 1-100 ns has a broad energy spectrum with several peaks[3] (in a single discharge) for different values of the particle energy E (up to $E \gtrsim 500$ V_0 for V_0 = 15 kV) and carry about 10% of the bank energy. New data[3] indicate that the 0° beam in deuterium discharges has a component of neutral particles and includes large clusters with m/Z > $10^2 - 10^3$ (mass m, charge Z of particles in atomic units). From the tests at 0° we report here some of the characteristics of this component (m/Z > 200) which has a number of particles $N_0(m) \sim$ 1-10% of the D^+- ion population $\int N(E)dE$ (integration on the interval 0.3 MeV < E < 7.5 MeV of the D^+-ion spectrum) inside

---

*Work supported in part by ONR, Arlington, VA.

the same solid angle dω at 0°. A brief description of the system and of the methods of observation are included. The particle beam at 180° - with a negative net charge - has an intensity which is several orders of magnitude higher than the m/Z > 200 component at 0°. We report some of the characteristics of this 180° beam (which includes ion clumps with m > 10^{10}-10^{12}, the probable mass limit for large m values of the population of accelerated particle clusters) and the methods of observation of the tests at 180°. Filamentation due to space charge is related with the large value of the emittance of clusters at 180°. Our data are obtained from single PF discharges. A repetitive mode of operation with a repetition rate ≳ 1 pulse/μs has been tested in experiments of other laboratories[4]. On the basis of the feasibility of the repetitive mode the PF can be considered - consistently - as a source of neutral and negatively-charged particle beam with outstanding capabilities in many applications for which negative-ions and neutral beam sources are presently considered.

BEAM SOURCE - EMISSION AT 0°

The PF with coaxial electrodes (hollow anode dia. 3.6 cm, cathode dia. 10 cm; Fig. 1) is operated with 15 kV (∼ 5 kJ) unless it is otherwise indicated. Details on construction and method of operation are extensively reported in the literature [5,1,2,3]. We report here methods of observation and some of the results.

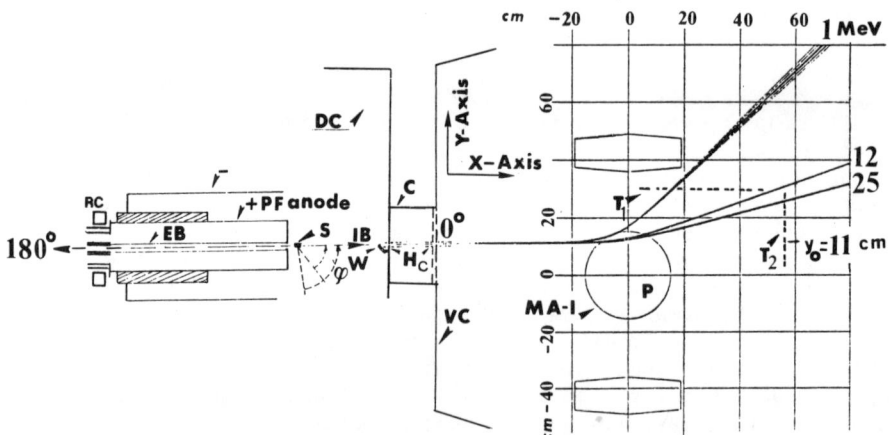

Fig. 1. Schematic of: PF electrodes with magnetic analyzer (MA-I, at right) in vacuum chamber (VC); magnet poles (P) dia. 30 cm for tests on ion beam at 0°. Fast opening valve (W) seals collimator apertures H_c. Targets (T_1,T_2) and computer-generated trajectories of 1, 12 and 25 MeV D^+ ions for B=10 kG. At left Rogowski coils (RC) monitor negative current of EB inside dielectric-wall drift chamber.

The electron beam (EB) with an energy peak at $\sim$ 400 keV is extracted via a dielectric pipe at 180° attached at the front end of the anode. EB propagates in a drift chamber with the same pressure p of the discharge chamber (DC). In other experiments[3] we have observed the ion emission at other angles $\varphi \neq 0°$, 180° and a variety of magnetic analyzers (including a Thomson spectrometer) has been used.

The analyzer vacuum chamber (VC) holds target strips (T_1 and T_2) and maintains a vacuum of approximately 10^{-5} torr during the length of the experiment. The fast opening valve (W) fires the PF via a fiber optic link within 10 millsec of uncovering the entrance aperture of the magnetic analyzer, thus maintaining a high vacuum environment within the analyzer. The collimator assembly is formed of a plasma shield, vacuum valve W, and a collimator (two 0.5 mm dia. holes separated by a 10 cm drift region) collimates the incoming beam and transports it to the analyzer chamber. The distance from the W entrance is 10 cm from the front end of the anode. The analyzer is calibrated with a 15μ Curie; 5.4-MeV α-source (Am^{241}); α-trajectories and landing points on T_1, T_2 are numerically calculated and experimentally verified for each of the chosen values of B. All ion tracks are observed after 3 hours of etching in a 6.25% solution of NaOH at 75°C. Different values B (7.5-12.25 kG) of the analyzer magnetic field are used with filter (mylar foils) of different thickness convering different parts of the target T_1 - on which tracks of D^+ ions and impurity ions (C^+, N^+, O^+, etc.) are observed - and of T_2 used for detection of neutral and ion cluster tracks. The density profile $(dN+N_o)/dA$ in Fig. 2 is obtained from a target T_2 of CR-39 exposed to four consecutive shots with a total neutron yield $\Sigma n_i = 4\bar{n}$ ($\bar{n} = 2 \times 10^8$, n_i = yield in i-th shot, $\bar{n} = 2 \times 10^8$ is the typical mean value of the yield/shot on hundreds of shots) at a pressure p = 4.5 Torr in the discharge chamber (DC). In this case the undeflected component at 0° has a low intensity and represents the lower limit of tracks density for beam detectability with $dN = 1.7 \times 10^3$ $cm^{-2} \gtrsim N_o$; $N_o/dA = 1.3 \times 10^3$ cm^{-2} is the background count. The observed fluctuations in $dN + N_o$, N_o are $\lesssim (dN + N_o)^{1/2}$, $N_o^{1/2}$ (Poisson's standard deviation) respectively. The track density is reported as a function of the scanning coordinate z (z is transversal to beam and parallel to magnetic-analyzer field B). Each data curve is generated from the scanning of a 0.7 mm wide strip with center at a distance Δy from the center of the undeflected (neutral component) ion beam in z=0, y_o=11 cm; the value of Δy (cm) of each scanned strip labels the corresponding curve. Half of the target surface - from center to left edge (z<0 side) - is covered with a mylar foil 2.5μm thick: Note sharp drop of the track density on screened area of target (z<0). The observed track density dN/dA in a single shot at 0° (B $\gtrsim$ 10 kG) can be higher by a factor 10^3 (three orders of magnitude) than the maximum value of $dN (\Delta y=0)/dA$ ($\sim 1.7 \times 10^3$ cm^{-2}) reported in Fig.2 for the same electrode voltage and pressure (4.5 Torr) in DC. An example, a relatively-high intensity beam with $Z/m \cong 0$,

$dN\ (\Delta y=0)/dA = 1.7 \times 10^6\ cm^2$ at 0° (B = 10 kG, single shot) is reported in Fig. 3. This large fluctuation from shot to shot can be explained from a high-intensity (I_o) narrow beam (of angular spread << 6°) which can be randomly oriented within a 6° full-width cone and is embedded in a "peripheral" beam of peak intensity $I_p \sim 10^{-3}\ I_o$ with an almost-uniform distribution within the 6° cone. In some shot ($\lesssim$ 10%, from our observations) the I_o beam enters the collimator.

In Fig. 4 the observed maximum value of $dN/dA_o d\omega_o$ on T_2 from Fig. 2 is plotted as a function of $y = y_o + \Delta y$; dA^o is the area of the collimator pinhole; $d\omega = d\omega_o\ dA/\Delta A$, $d\omega_o = 7.85 \times 10^{-5}$ steradians is the geometric angular width of the collimator aperture; $\Delta A(y)$ is the elliptically-shaped projection on T_2 of the collimator aperture $[\Delta A(y=0) = \pi\ (0.65)^2\ cm^2]$; the vertical error bar is $(dN + N_o)^{\frac{1}{2}} + dN_o^{\frac{1}{2}}$. As a reference the energy E = E(y) of a singly-charged (Z=1) particle is also plotted as a function of the particle landing point y on T_2 for m = 200 and B = 12.25 kG.

In Table I is reported the maximum value of the energy E_m for different ion species (possible contaminants of the gas filling in DC) which are screened out from the 2.5 μm thick mylar filter.

Table I Energy of particles with 2.5 μm range in mylar.

m (mass in a.m.u.)	E_m (Mev)
2 (D)	0.25
4 (^{4}He)	0.65
12 (C)	2.15
14 (N)	2.40
63 (Cu)	4.50

The filter screening effect and a comparison of the plots E = E(y) with the observed fluence $dN/dAd\omega$ for different values of y on T_2 indicate that the tracks on T_2 are produced from ion clusters with m >> 200. (No ion with E_m < 4.5 Mev, - so that the ion can be stopped from the filter - Z=1, can land at a point Δy < 8 cm on T_2 if B = 12 kG; the same argument indicates that m >> 10^4 for $\Delta y \lesssim$ 1 cm). Characterizations of etched particle tracks from track morphology (as well as other observations on PF particle emission[3]) indicate that many tracks are produced from composite objects which can split in several fragments on impact on the target surface. In some case the presence of recoiling ions from the target material can be inferred from the presence of several small cavities inside a larger track pit[6] as in Fig. 5. Alternative explanations (e.g., the emission of a few high-energy

particles from an exploding particle cluster) in which the cluster
energy is not equally distributed among the fragments can also be
considered. In all cases the composite nature of the primary
particle reaching the target seems an essential element of the
track-producing process.

As a reference the typical energy spectrum (fluence $dN/dEd\omega$)
of the D^+ ion emission is reported in Fig. 6 from T_1 target
data[3].

BRIGHTNESS

Pulsed ion currents of $\sim 10^5$ A and neutral, $Z/m \sim 0$ current
of comparable value with a pulse duration of $\sim 10-20$ ns are
obtained from each discharge of a plasma focus (PF). The value of
the particle current I which is significant for practical applications (e.g. continuous mode of operation) can be assessed from
$I = N/\Delta t$ (N = number of particle per pulse; Δt = time interval
between pulses in a repetitive mode of operation). The data presented in this paper have been obtained from the usual (pulsed)
mode of operation of Mather-geometry plasma focus (PF) fed from
5 kJ (at 15 kV) capacitor bank. A repetitive mode of operation
with a high repetition rate (~ 1 pulse/μs) has been proposed
(Ref.7) and successfully tested ($\sim$ one pulse/3 - 5 μs) as reported
in Ref. 8. In the repetitive mode the PF is fed with a millisec
current pulse of 50-100 kA; the PF operates as self-opening and
closing switch with a closing-opening rate = $1/\Delta t$= generation rate
of particle beams. The duration of the pulse train and the pulse
repetition rate have limitations set from the power source and from
the cooling system for the PF electrodes. On a "fine" time scale
(10-100 ns) each pulse of the pulse train is considered to have a
composite structure quite similar to the time structure of the
beam generated in a single shot of the PF described in this paper.
From the definition of source brightness B_4 in the literature[9]
and from an assumed PF repetition rate $R = 10^5$ sec^{-1} we estimate
an equivalent brightness (in terms of an identical flow of charged
particles with Z=1) $B_4 = eR\ dN/dA_o d\omega \sim 240$ mA (sr^{-1} cm^{-2}) of
neutral atoms and of $Z/m \sim 0$ clusters at 0° at the exit of the
collimator aperture (e = 1.6 x 10^{-19} coul; dN from Fig. 3).

The peak neutral brightness during the 10-20 ns duration
of a single pulse has the value $B_{4p} \gtrsim 10^5$ mA/cm^{2}sr. In this case we
can estimate an energy $Em_D/m \sim 200$ keV/deuteron mass in neutral
and $Z/m \sim 0$ beam from the energy limit of etchable track formation,
consistently with test with formvar filters 0.1 μm thick.

A brightness $B_4^+ \gtrsim 10^5$ mA/cm^{2}sr for the D^+ emission in the
forward direction in a repetitive mode ($R = 10^5$ sec^{-1}) is similarly estimated from T_1-ion tracks ($E \gtrsim 0.3$ MeV) and from a value
$dN(0.2\ MeV \leq E \leq 0.3\ MeV) \sim 10\ dN(E \gtrsim 0.3\ MeV)$. During a single
pulse we have a peak value $B_{4p}^+ \gtrsim 10^7 - 10^8$ mA/cm^{2}sr.

The total D^+ current $I_i(6°)$ in the 6° cone before entering the
collimator (if we include the low energy component $E \gtrsim 20$ keV) has I_{ip}
$\gtrsim 10^5$A in a single pulse[10] and ~ 100 A in the repetitive mode. We
can asses an equivalent (with Z = 1 particles) current $I_{op} \gtrsim 100$ A of
neutral and $Z/m \sim 0$ beam by taking before the collimator the same
proportionality factor (10^{-3}) between $Z/m \sim 0$ particle and D^+-ion
population as we have determined after collimator-MA system.

We observe: (i) That the brightness for neutral and $Z/m \sim 0$ component is usually increasing with the brightness of the low-energy population component (E < 1 MeV) of the D^+ spectrum.
(ii) A strong dependence of the neutral-component brightness at 0° on the pressure of the discharge (beam-source) chamber and on the applied votage. An increase of the pressure from p = 4.5 Torr to 7.5 Torr decreases the brightness B_4 by a factor 10^{-2} (15 kV on the electrodes). An equal reduction is observed by decreasing the electrode voltage from 15kV to 12.5 kV.

180° EMISSION

Ion and electron beam at 180° have an angular dispersion < 1°-0.1°. The observed EB filamentation focalizes also the ions accelerated and trapped in the filaments[11]. Since the filling pressure with 180° drift chamber is p, we determine the typical values of E (with a resolution not matching the 0.5-5% resolution of MA-I at 0°) from nuclear activation (of C^{12}, B^{10}, etc.), etchable tracks and filtering, penetration range and damage on targets for ions; from time of flight via RC signals, dendrite lengths, and limiting frequency of x-ray bremsstrahlung for EB. The EB source is located at $\sim$ 1 cm from the front end of the hollow anode. EB and ion beam at 180° are extracted from the fully open anode or from a circular aperture (usual dia. 2r = 3.7mm) at the center of a metallic disc which partially closes the anode front. The transversal dimension $2r_o$ of the EB source (and/or that of the axial region where the EB source is located in a sequence of shots) is determined as the maximum value ($r = r_o$) for which we can still detect a dependence of the EB current I_e on the aperture radius r $[\partial I_e(r)/\partial r \cong 0$ for $r \gtrsim r_o]^{11}$. The result $r_o \lesssim 1.5$ mm fits the x-ray pinhole photograph data. In each PF shot the number of peaks observed in the $|\dot{I}_e|$ signals is the same as the number of peaks of $|\dot{I}_{PF}|$. The ion beam intensity at 180°, I_i (180°) = N/Δt is estimated: (A) from target damage (as it is done for I_i (0°)) and/or (B) from $I_e - I_{e-plasma} - I_i$ (180°) with a determination of $I_{e-plasma}$ (the return current carried from background - plasma electrons) and of I_i (180°) from two RC's and a mylar filter $\sim$50μm thick; $I_{e-plasma} + I_i$ (180°) = I_e (after filter) - I_e (before filter; I_e (after filter) $\sim$ 5 I_e (before filter); see Ref. 11. Two components contribute to I_i (180°) = I_{ica} (180°) + I_{iPF} (180°). I_{iPF} (180°) [= αI_{iPF} (0°); $\alpha = 10^{-1} - 10^{-3}$ depending on the value of E that we consider and on the pressure p] is ejected from the same plasma region (within 1-2 mm) of the exploding PF pinch where the EB source is located. I_{ica} (180°) is the component which is generated from EB in the drift-chamber background gas and is accelerated by EB collective fields. Tests to determine the relative values of I_{ica} (180°) and I_{iPF} (180°) for different values of E are in progress. From the morphology of target damage and from etched track density we conclude tentatively that I_{iPF} (180°) $\sim 10^{-2} - 10^{-3} I_{ica}$ (180°) for E $\sim$ 1-3 MeV deuteron, the dominant

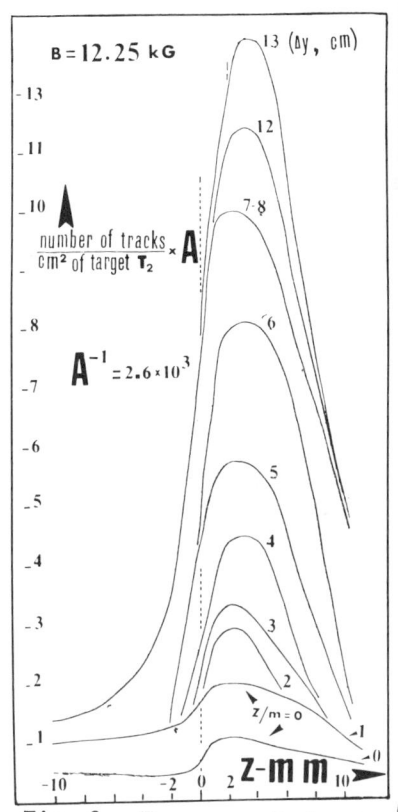

Fig. 2. Transversal density profile from T_2. N_O from $\Delta y=0$ line for $z < 0$.

Fig. 4.

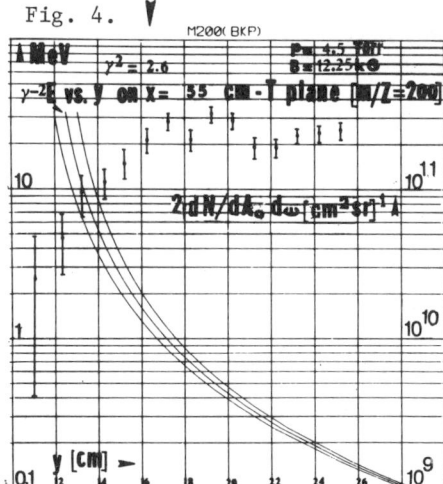

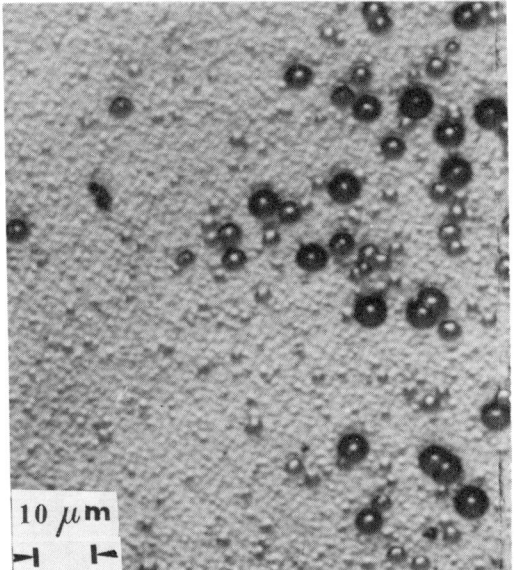

Fig. 3. Optical microscope (OM) photograph of T ($y=y_o$, $z=0$) filter covers left side.

Fig. 5. SEM microphotograph of CR-39 etched tracks.

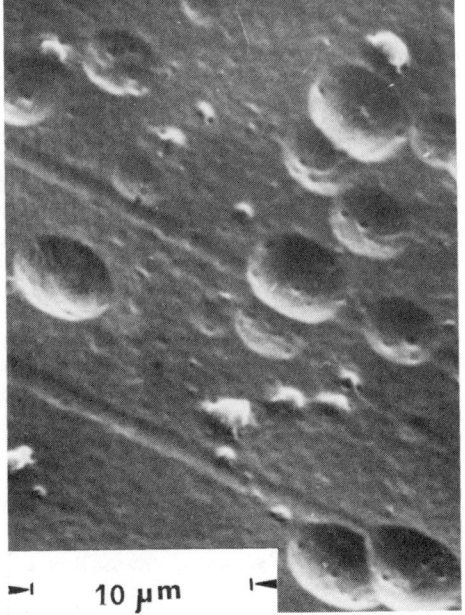

values in ion clumps with $10^{10} - 10^{12}$ ions in each clump (which probably represent the upper limit of the cluster mass spectrum at 180°). The formation of ion clumps and their direct kinetic energy is controlled from EB which collectively accelerates the ions during the clump formation process. This is a picosec process which occur at any distance from the source, depending on local conditions. Charge neutrality or a net charge $Z/m \leq 0$ is usually established in the clump. The 180° brightness for deuterons in clumps with $Z/m \leq 0$ is $B_4 \sim 4 \times 10^7$ mA/sr cm^2, with $d\omega \sim 4 \times 10^{-6}$sr (from filament focalization of ions), $dA \sim 10^{-2}$ cm^2, $R = 10^5$ s^{-1} and 10^{10} (~ 1 MeV) deuterons in each clump (conservatively, 10/shot).

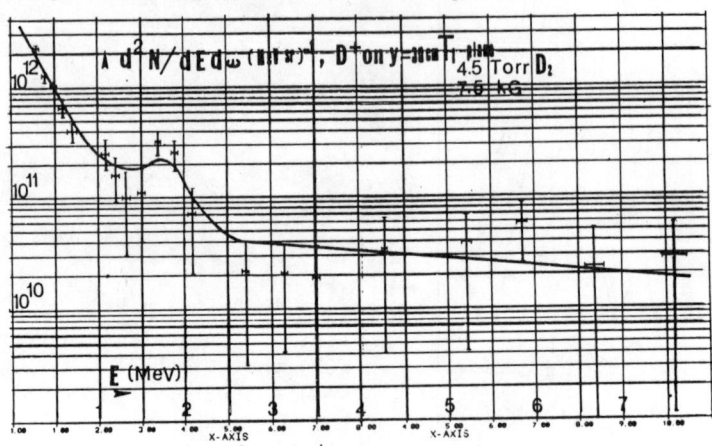

Fig. 6. $dN/d\omega dA$ vs. E from T_1.

References

1. W. H. Bostick, V. Nardi, W. Prior, J. Plasma Phys. 8,7 (1972) and Nuclear Fusion, Suppl., 3 497 (1977, 2, 143 (1978).
2. R. L. Gullickson et al., in Energy Storage, Compression and Switching, Vol. 2, (Plenum, N.Y., 1983) p. 579.
   V. A. Gribkov, Ibid, Vol. 1 (Plenum, N.Y., 1976) p. 271.
3. V. Nardi et al. in Controlled Fusion and Plasma Physics, Vol. 7D-I (European Phys. Soc., Aachen, 1983) p. 489.
4 J. Salge et al., same as Ref. 2, Vol. 2 (1983) p. 75.
   H. Conrad, J. Salge, Ibid, p. 99.
5. J. Mather, Methods of Experimental Phys. Vol. 9-B (Academic Press, N.Y., 1971) p. 187.
6. R. L. Fleisher, (General Electric Res. Lab., Schenectady, N.Y.), private communication, July 1983.
7. V. Nardi, Proc. Int. Conf., Radiation Test Facilities, ANL, 1975. (P.J. Persiani edit., Aug. 1975) p. 527.
8. J. Salge, et al., Nuclear Fusion 18, 972 (1978).
9. A. Van Steenbergen, IEEE Trans. Nucl. Sci., NS-12, 746 (1965) and Nucl. Inst. Meth. 51, 245 (1967).
10. W. H. Bostick, et al., J. Nuclear Mat. 63, 356 (1976).
11. V. Nardi, et al., Phys. Rev. 22A, 2211 (1980).

ACCELERATORS AND ACCELERATED BEAMS

473

TRANSVERSE-FIELD FOCUSING ACCELERATOR

O.A. Anderson
Lawrence Berkeley Laboratory
University of California
Berkeley, CA 94720

A. INTRODUCTION

The fact that ribbon beams can be focused by transverse electric fields at the beam axis has been known for a long time. Such focusing, in its rudimentary forms, is so easy to analyze that it was used by Pierce in his classic book[1] to introduce the subject of beam focusing (Chapters 3 and 4); he reserved the more difficult case of a straight-axis beam for later chapters. The illustrations which accompanied his discussion of two simple transverse-field examples are reproduced in Fig. 1. A fairly thorough analysis (though neglecting emittance effects which we will include in this paper) of a number of electrode configurations for transverse-field beam transport was given in the book by Kirstein, et al.[2] Of those configurations analyzed, only one seems to have been built and tested. This was a prototype electron transporter built in England in 1958 by Hogg.[3] He moved to the United States before completing the project, but the device apparently performed as expected. So, historically, it seems that the only important application of transverse electric-field focusing has been its well-known usage in electrostatic analyzers alluded to by Pierce.

Recently, however, the author realized that transverse-field focusing (TFF) can be useful in producing powerful D- ribbon beams for magnetic fusion.[4-8] In a TFF accelerator, the beam gains energy in a succession of isolated low gradient stages. The maximum

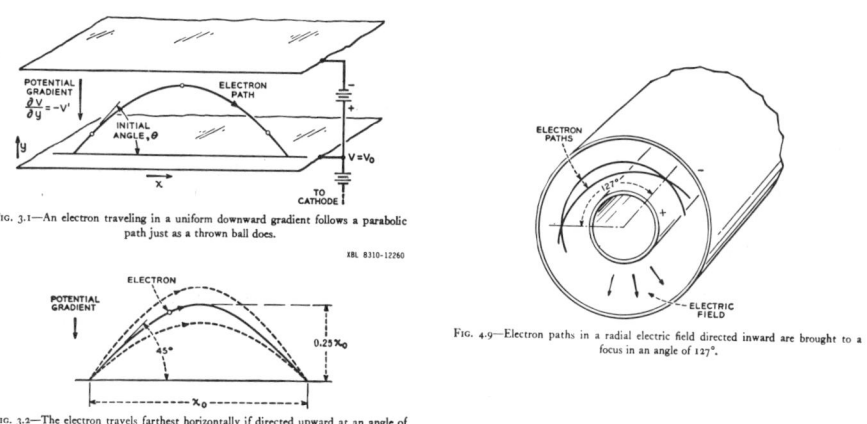

Fig. 1. Illustrations from Pierce's book.

0094-243X/84/1110473-16 $3.00 Copyright 1984 American Institute of Physics

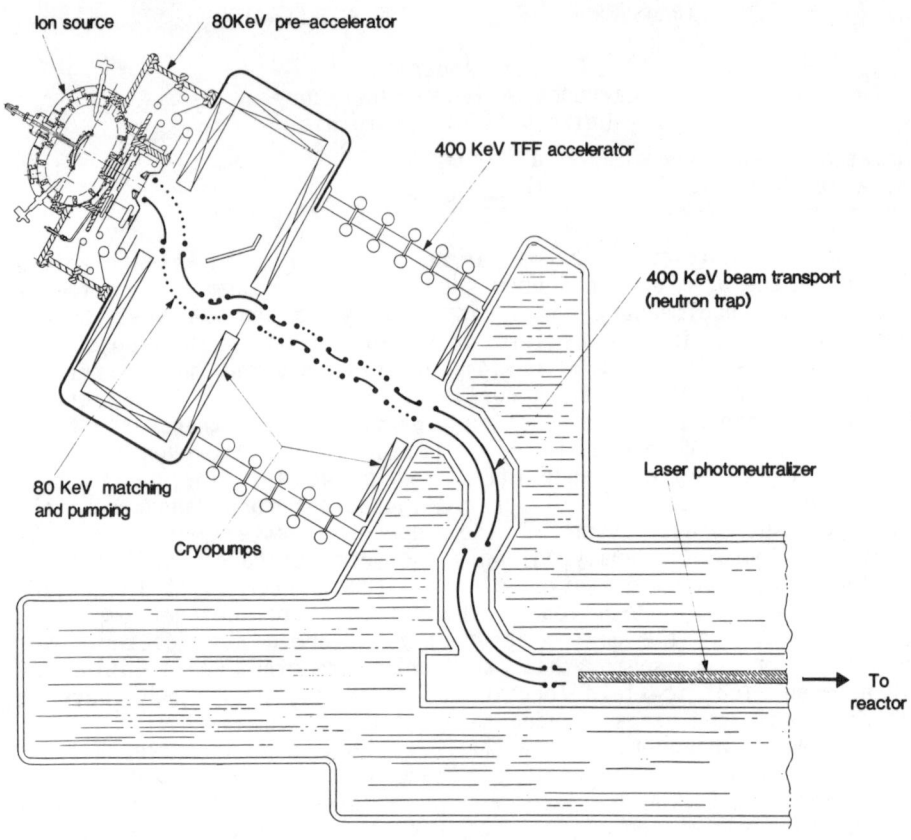

Fig. 2. Plan view of 400 keV TFF-based beam line.

field is lower than in a Pierce gun of the same perveance, and this fact, along with basic TFF properties, should make such an accelerator resistant to breakdown.

The TFF principle lends itself to a negative-ion based neutral beamline in a number of other ways. A typical conceptual design for a 400 KeV, 2 to 6 MW beamline is shown in Fig. 2. It starts with a surface-conversion D- source[9], which produces a ribbon beam one or more meters wide, and ends with a ribbon-shaped photoneutralizer.[10,11] (Gas-cell neutralizers which might be used in the short term also work best with ribbon beams.[12]) Thus, the TFF ribbon accelerator fits well into the beamline. Furthermore, the differential pump and neutron trap require bends (as shown in Fig. 2) to avoid line-of-sight flow of gas or neutrons, and TFF transport is ideal for this purpose. (A comparison of TFF with the alternative concepts has been presented by Cooper.[13])

The purpose of the present paper is to state the essential facts about TFF transport and acceleration as simply as possible and yet include all the information needed to design a system like the one in Fig. 2. In section B, we summarize our previous analysis[5] of basic TFF physics, this time keeping the beam-line designer in mind rather than the theoretician. Section C presents specific calculations for a 400 KeV conceptual accelerator and compares the predicted performance with computer simulations. Oscillations in unmatched beams are discussed in section D and applied to the design of a matching section. Finally, in section E, we put all the parts together in a single beam simulation that starts inside the D-source and ends at the photoneutralizer, thus covering all the elements of Fig. 2.

Many topics are omitted from the present report for lack of space. These include the analysis of systems for confining the beam edges and for removing residual high-energy electrons. We also omit discussion of beam stripping, production and behavior of secondary particles, calculation of the resulting power loads and the engineering and design work on the water-cooled electrodes and on the differential pump. Our aim here is to concentrate on the essential analytic results that have allowed us to design the large, complex system modeled in section E.

B. SIMPLIFIED THEORY

1. Introduction

The basic equations governing TFF transport and acceleration were derived previously (Ref. 5). Here we only summarize the results but will discuss their physical interpretation and practical application much more extensively. Our heuristic approach here ignores some fine points which would lead to small corrections (of the order of 1%) and concentrates on the information needed to design a practical TFF system, such as the accelerator discussed in section C.

2. Curvature and Current Density

Figure 3 shows schematically a TFF transport section and introduces our notation. The arrangement shown produces a constant potential ϕ_m along the midline between the various pairs of circular-arc electrodes; this midline is the equilibrium orbit for ion trajectories with energy $q\phi_m$. At the midline radius r_m, the electric field required to balance centrifugal force is

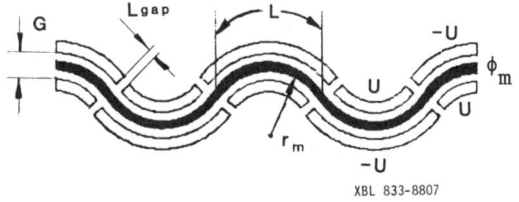

Fig. 3. TFF transport system, with notation used in this paper.

$$E_m = \frac{2\phi_m}{r_m} . \qquad (1)$$

The current density that can be transported by TFF is affected by beam emittance and by the field reduction in the gaps. If these effects are neglected, the current density is[5] $J_{max} = 9J_{CL}$, where

$$J_{CL} = \frac{C_{CL}\phi_m^{3/2}}{r_m^2} \qquad (2)$$

and C_{CL} is the coefficient in the Child-Langmuir formula for a plane diode. The TFF current scales the same as in a diode but the characteristic length has a different significance and there is a new factor of 9 in the resulting equilibrium. We will see later that when gaps and emittance are considered, a typical value is

$$J_{typ} = 5J_{CL}, \qquad (3)$$

a useful rule of thumb.

In practice, one usually deals with a given total current I_L (per unit length) and uses equation (3) to estimate the required beam thickness corresponding to a given ϕ_m and r_m. If the beam is launched with a different thickness, it will adjust itself by expanding or contracting in the appropriate manner. It will overshoot and finally oscillate about the correct value. This effect is utilized in beam matching, discussed in section D. In the remainder of the section, however, we assume the parameters have been properly adjusted for constant beam thickness.

3. Correction for Gap

Electrostatic focusing forces are proportional to the square of the field. Thus, in TFF transport, although the field direction reverses from section to section, the focusing force remains constant except in the gaps. When gaps are considered, the mean force is reduced according to the factor

$$f = \frac{\langle E^2(s) \rangle}{E_m^2} \qquad (4)$$

where the average is taken along the midline over the total section length L (see Fig. 3); E_m is a constant, determined from Eq. (1). The transportable current, still neglecting emittance, is

$$J = fJ_{max} = 9fJ_{CL} . \qquad (5)$$

A simple estimate of f comes from the hard-edge approximation in which E(s) equals 0 in the gaps and equals E_m elsewhere. Then, clearly, $(1 - f) = L_{gap}/L$. Normalizing with the gage G (Fig. 3), we have the gap factor

$$(1 - f)\frac{L}{G} = \frac{L_{gap}}{G}$$

This is plotted as the dotted line in Fig. 4. Much better accuracy can be obtained by modeling the electrodes as shown in Fig. 5. (It is assumed that $L > 3G$.) The field can then be obtained by conformal mapping,[14] which yields the gap factor shown in the solid line (Fig. 4). As expected on physical grounds, the gap factor remains finite even with vanishing gap length.

In the useful range $0.35 < L_{gap}/G < 1.45$, a good approximation for the focusing factor is

$$f = 1 - 0.65 \frac{L_{gap}}{L} - 0.45 \frac{G}{L} . \qquad (6)$$

This is used for the TFF accelerator design discussed in section C.

4. Effect of Emittance

Besides the gap factor, Eq. (5) needs another correction to take account of finite beam emittance. The emittance ε (phase plane area over π) is proportional to the transverse random velocity, and there is a corresponding beam pressure proportional to ε^2. The gradient of this pressure must be added to the space charge electric force to balance the average focusing force. Thus, Eq. (5) becomes

$$\frac{J}{9J_{CL}} + \frac{r_m^2 \varepsilon^2}{2u^4} = f \qquad (7)$$

where u is the mean half thickness of the beam and $J = I_L/2u$. Eq. (7) may be derived using the beam pressure approach[15] or the Floquet method.[5]

TFF transport may be used at low energy (for differential

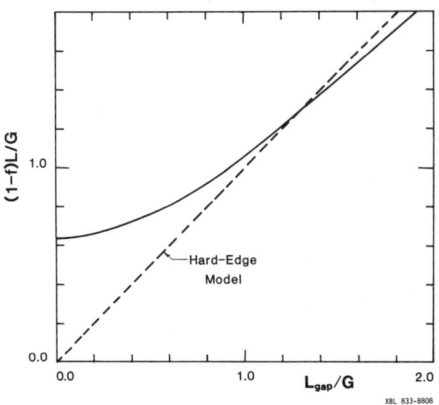

Fig. 4. Result of electrode gap calculation.

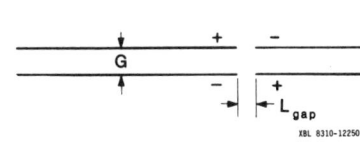

Fig. 5. Model for calculating gap factor f(s).

pumping) or at high energy (in a neutron trap). In the latter case, one may wish to minimize the electrode potential differences. This can be done by adjusting the beam thickness with a matching section (see section D). One finds that for a given safety factor G/u, the relative potentials are minimized when the pressure term is one-half the space charge term, i.e.,

$$\boxed{\frac{J_{max}}{9J_{CL}} = \frac{r_{opt}^2 \varepsilon^2}{4 u_{opt}} = \frac{2}{3} f} . \qquad (8)$$

Taking a typical value $f \sim 0.85$, one gets the rule of thumb, Eq. (3). The boxed equations (1), (6), and (8) are sufficient to design an optimized, high-energy transport section (or the final stage of the TFF accelerator).

To illustrate, we have chosen a beam-clearance safety factor $G/u = 6$. Also, we have defined the relative potential U, i.e., $\phi_{electrode} = \phi_m \pm U$. (For simplicity, we ignore a small logarithmic correction.) Then the electric field E_m is $2U/G = U/3u$, and Eq. (1) becomes $U/3u = 2\phi_m/r_m$. Thus, for our chosen safety factor, we have

$$\frac{u}{r_m} = \frac{U}{6\phi_m} \equiv a . \qquad (9)$$

If we are designing a 400 keV transport system and wish to specify an electrode potential U of 50 kV, then we have $a = 1/48$. From Eqs. (8) and (9),

$$r_{opt} = \sqrt{\frac{3}{2f}} \frac{\varepsilon}{a^2} . \qquad (10)$$

A typical unnormalized emittance at 400 keV for a D- beam produced by surface conversion is $\varepsilon = 0.011$ rad-cm and a typical f is 0.85, so we find $r_{opt} = 35$ cm.

The line current per cm at this optimum radius, from Eqs. (8), (2) and (9) is

$$I_{max} = 2uJ_{max} = \frac{2U\phi_m^{1/2} C_{CL} f}{r_{opt}} . \qquad (11)$$

For D- with the above parameters, we get $I_{max} = 0.059$ A/cm = 5.9 A/m. If H- were being used, then $I_{max} = 8$ A/m, which is more than is normally produced at present by the surface source, so that our design allows for future improvement. The example discussed in section C is based on the above current.

Different optimizations may be required in other cases. For example, volume production of D- could possibly reduce the emittance and at the same time deliver more current. If we reduce ε by a factor of 3, the above approach gives r_{opt} = 12 cm, I_{max} = 17.7 A/m, and E_{max} = 67 kV/cm. If the latter were unacceptable, we would have to constrain the field and accept a somewhat lower current.

5. Acceleration

In a TFF acclerator, the beam energy is increased by a longitudinal field in the gaps, where the equilibrium orbit has an inflection point. The beam axis is approximately straight there and thus the acceleration and transverse-focusing effects are effectively decoupled -- a great simplification in the analysis.

In a typical design, we increase the beam energy 30% in each gap. The radius r_m increases at each stage and so does the potential difference 2U between the electrodes; thus, one designs the system to minimize U for the last stage. If we use symbols r_0, ϕ_0, and f_0 for quantities in the optimized stage, we find for any other stage

$$\frac{r}{r_0} = g^{1/2} \left(\frac{\phi}{\phi_0}\right)^{3/4} \tag{12}$$

where

$$g = \frac{3f/f_0}{2 + (\phi/\phi_0)^{1/2}} .$$

In deriving this result, we assumed the beam thickness is held constant. Therefore in Eq. (6), G will also be constant in order to maintain the safety factor G/u. However, it is desirable to keep L_{gap} as short as the design field (e.g., 40 kV/cm) permits so that L_{gap} changes from gap to gap. There are various approaches to choosing the section lengths. In the design shown in the following section, L was varied in a way that kept f approximately constant.

C. TFF ACCELERATOR

The parameters chosen for the last stage of a conceptual 80 - 400 keV TFF accelerator were the same as for the transport section discussed above. In addition, the maximum field in the last accelerating gap was specified to be 40 kV/cm. The model used six accelerating gaps with 30% energy gain per stage.

Fig. 6a shows the electrode shapes and potentials used for a computer model of the accelerator designed according to the formulas of section B. In this preliminary model, some realistic features, such as rounded edges and electrode openings for gas removal, have been omitted, but will be included later. Fig. 6b shows the equipotential contours for this structure. Numerically computed trajectories of 60 beamlets in the self-consistent field appear as the solid black band in Fig. 6c; there are too many beamlets for any of them to be resolved. Figs. 6c1-4 show the phase-space locations of the 60 beamlets at 80, 180, 300, and 400 keV. In this example the initial distribution is rectangular. Fig. 6d shows the equipotential contours in the presence of the space charge; note these contours are barely changed from the vacuum field contours of Fig. 6b. This is characteristic of strong focusing and explains the extremely wide range of currents that can be transported by TFF.

The complete system (Fig. 2) includes a preaccelerator, based on the Pierce gun principle, which is highly sensitive to changes in current and dominates the overall system performance. But it is interesting to consider the behavior of the TFF accelerator by itself as the current or initial beam divergence is varied. Figs. 7a, 7b, and 7c illustrate the effect of varying current. (To simplify interpretation, an elliptical emittance is used here.) Notice that the beam thickness initially contracts or expands if the current is lowered or raised, respectively. But the damping action of the TFF acceleration leads to nearly identical beam envelopes in the final sections. This fact is made especially clear by the phase space plots. In this particular design, it happens that a horizontal elliptical phase envelope at the entrance maps into another horizontal ellipse at the center of the penultimate section, so that if the system were truncated there (at 300 keV energy) the exit beam envelope would be parallel in each case. However, note the variation with current of the beamlet pattern within the ellipse.

Cases with converging or diverging entrance beams were also studied, and again a surprising tolerance for variation was observed. Other studies have shown that allowable variations in voltages or mechanical alignment also fall within reasonable bounds.

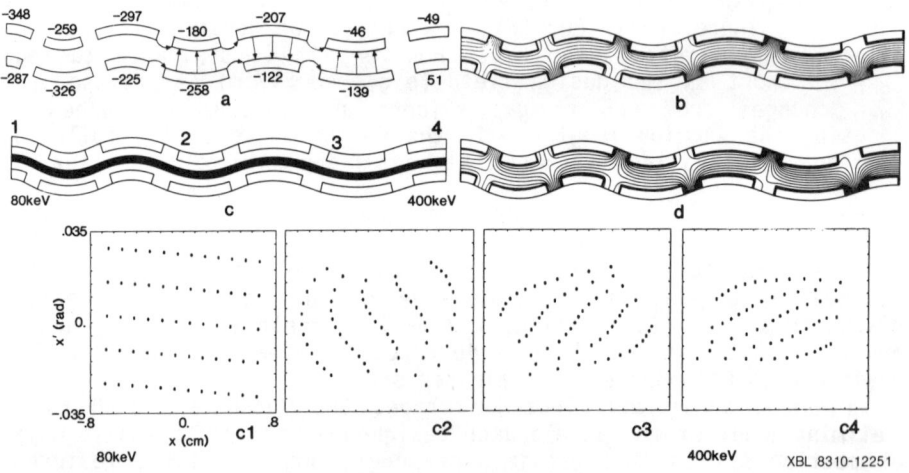

Fig. 6. TFF 400 keV, 8A/m H- accelerator. (a) Computational model with voltages in kV. Total length is 92 cm. Arrows are in direction of electric force. (b) Equipotentials for vacuum field. (c, d) Beam envelope, phase plots and equipotentials for 8A/m.

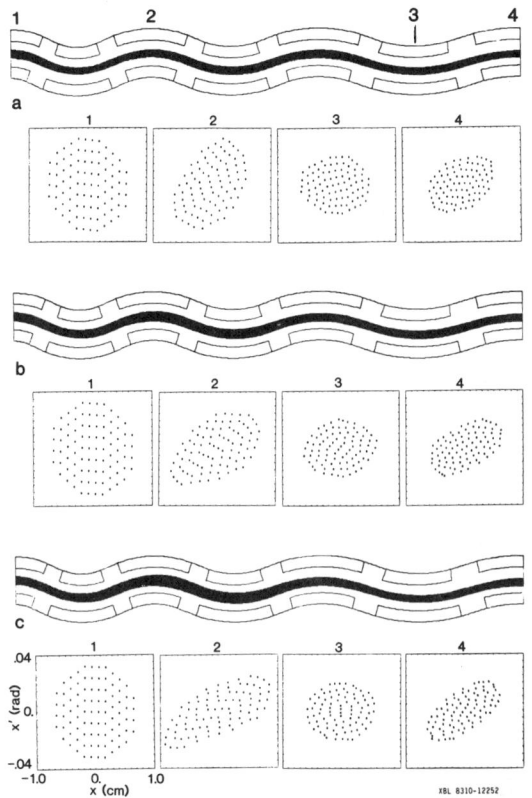

Fig. 7. Same parameters as Fig. 6c except: (a) 5A/m, (b) 7A/m, (c) 9A/m.

D. TFF TRANSPORT AND MATCHING

1. Oscillations in a Transport System

In section B, we discussed TFF transport for the case of constant beam thickness. The present section deals with cases where there may be large thickness variations, either produced deliberately in order to optimize potentials (section B) or inadvertently by, e.g., an unexpected change of input current. Here we also mention a complication: even with initial conditions chosen to eliminate such variations, another oscillation with a different wave number persists.

Both effects are governed by the basic equation[5,15]

$$\frac{d^2w}{ds^2} + g(s)w - P - \frac{\varepsilon^2}{w^3} = 0 \qquad (13)$$

where, for convenience, we have replaced the symbol x_0 in Ref. 5

by w to stand for the beam half-thickness. We have also simplified the notation here by introducing the normalized line perveance

$$P = \frac{1}{9C_{CL}} \frac{I_L}{\phi_m^{3/2}}$$

and the scaled focusing factor

$$g(s) = \frac{2}{r_m^2} f(s) = \frac{E^2(s)}{2\phi_m^2} .$$

The gap periodicity appears in g(s). For the special case of a matched beam, the solution of Eq. (13) has Floquet oscillations with the same periodicity. In TFF these oscillations are very inconspicuous and have been neglected up to this point.

For other initial conditions, a different sort of oscillation is excited, with arbitrarily large amplitude, up to the limit set by the electrode spacing G. We will call these conspicuous oscillations "envelope oscillations." Their wave number depends on the beam emittance and, in the absence of gaps, lies approximately in the range from $\sqrt{2}/r_m$ to $2\sqrt{2}/r_m$.

2. Floquet Oscillations

We digress here to compare the magnitude of Floquet oscillations for TFF and AG (alternating gradient) types of strong focusing. In accelerator terminology, the basic configurations are FOFO and FODO respectively, where F and D stand for the focus and defocus sections and O stands for the gaps. In AG systems, the D and F sections are the same length, which causes substantial ripple, e.g., 30%. In TFF systems, there is defocusing in the gaps because of space charge, but since the gap lengths are short, the ripple is quite small, typically 1-2%. This can simplify the analysis, as we shall see.

3. Envelope Oscillations

In the absence of gaps (f = 1), Eq. (13) is a straightforward, non-linear, but autonomous equation with periodic solutions which can be found by perturbation methods, or even (in principle) expressed in terms of known functions.[15] A complete treatment of the non-periodic case (both frequencies present) probably can be done only computationally. But approximate analytic results are useful, and so we use an approach here based on the fact that in TFF the gaps have a large (10 to 15%) effect on the average focusing force f, and a similar effect on J [see Eq. (7)], and yet they produce a negligible ripple.

Our approach also assumes that there are a sufficient number of gaps in the system.[15] Under these conditons, it is justifiable to replace g(s) in the envelope equation, Eq. (13), by its mean value g and write

$$\frac{d^2w}{ds^2} + gw(s) - P - \frac{\varepsilon^2}{w^3(s)} = 0 \tag{14}$$

which is now autonomous. The equilibrium value of w in this approximation, which we call u as in section B, is found by setting w" = 0, and gives the force-balance condition, Eq. (7).

When Eq. (7) is not satisfied, the non-linear Eq. (14) has oscillatory solutions. The wave number depends on amplitude, except in the special cases when either P or ε vanishes.[15] The amplitude dependence of the wave number has been analysed by a perturbation method[15] and the variation turns out to be only a few percent in practical cases. Therefore, since we are neglecting small corrections, we simply quote the linearized wave number obtained from Eq. (14). Its square is

$$\Omega_{env}^2 = g + \frac{3\varepsilon^2}{u^4} \tag{15}$$

where u is again the equilibrium half-thickness. In practice, u is unknown and must be obtained from Eq. (14) by solving the fourth degree equation obtained by setting w" = 0. Newton's method gives the result shown in Fig. 8 where we define the normalized variables for thickness and pressure:

$$y = \frac{g}{P} u \qquad \text{and} \qquad p = \frac{g^3 \varepsilon^2}{P^4} . \tag{16}$$

Usually p is a small parameter and one could solve by simple iteration, obtaining

$$y = 1 + p - 3p^2 \ldots \tag{17}$$

An interpolated formula accurate to 5% for $0 < p < \infty$ is

$$y \cong (1 + p^{3/4} + p)^{1/4} . \tag{18}$$

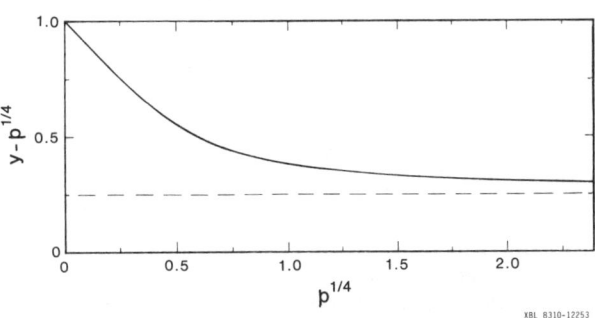

Fig. 8. Universal curve for calculating thickness of matched beam.

Fig. 8 or Eq. (17) or (18) may be used with Eq. (15) to express the envelope wave number of a transport system in terms of known quantities r_m, ε, P, f.

4. Matching Section

The purpose of a matching section (Fig. 9) is to make a transition between two transport sections of different beam thickness. For example, in the optimized system shown in Fig. 2, the beam is compressed in the matching section by a factor of 3 or 4. This is done by designing for a large-amplitude envelope oscillation which is allowed to continue for only a half-period. First we discuss a design for a hypothetical system which, unlike Fig. 9, has several gaps so that Eq. (14) applies.

The quantities of interest are w_1 and w_2, the maximum and minimum values of w in the oscillation, and the half-period π/Ω_{env}. For $w_1/w_2 = 4$, the linearized result for Ω_{env}, Eq. (15), is surprisingly accurate. But a linearized approach gives large errors for w_1 and w_2, and in this regime the non-linear Eq. (14) is very difficult to use as it stands. Luckily, complete information about the solution of (14) is not needed, and a different approach is available. We specify the maximum and minimum values w_1 and w_2, along with the beam perveance P and emittance ε, and only need to solve for the appropriate radius r_m, contained in the definition of g. Integrating Eq. (14),

$$\frac{1}{2}\dot{w}^2 + \frac{g}{2}w^2 - Pw + \frac{\varepsilon^2}{2w^2} = \text{const.}$$

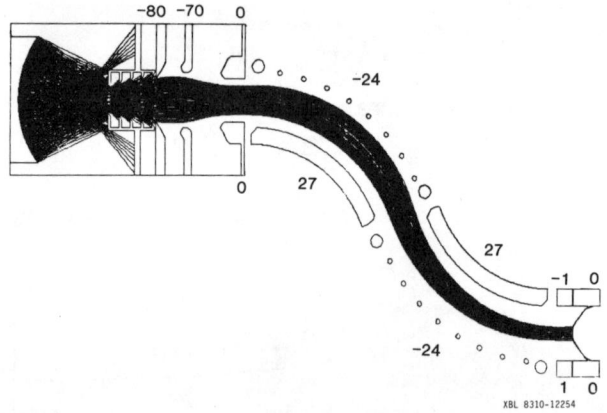

Fig. 9. TFF matching section, with beam exiting to grounded terminal electrode and downstream plasma. Voltages are in kV. The model for this simulation includes the open rod structure needed for differential pumping (see Section A). The rod voltages were 1 kV higher than for the solid electrode model (Fig. 10).

At the extrema w_1 and w_2, $\dot{w} = 0$, so that

$$\frac{g}{2} w_1^2 - Pw_1 + \frac{\varepsilon^2}{2w_1^2} = \frac{g}{2} w_2^2 - Pw_2 + \frac{\varepsilon^2}{2w_2^2}$$

Solving,

$$g = \frac{2f}{r_m^2} = \frac{P}{\bar{u}} + \frac{\varepsilon^2}{\langle u \rangle^4} \qquad (19)$$

where $\bar{u} = (u_1 + u_2)/2$, $\langle u \rangle = \sqrt{u_1 u_2}$. This result gives the required radius under the assumption of several gaps in the matching section. If there is only a single gap, as in Fig. 9, one must start from Eq. (13), which has been done in Ref. 15. It turns out that Eq. (19) can still be used with a revised value for f. For the geometry of Fig. 9, $f \approx 0.86$. This value, along with the other parameters, yields $r_m \cong 21$ cm, the radius used in the computer simulation of Fig. 9.

E. COMPLETE D- BEAM SYSTEM

Fig. 10 shows the D- beam computational result for the complete system of Fig. 2. The beamlet trajectories begin at the converter plate, follow straight paths through the source chamber and collimator, cross the plasma sheath and enter the 80 keV preaccelerator. From this point on, space charge is included in the computation. After preacceleration, the beam is compressed by the TFF matching section and passes through a short 80 keV transporter. The TFF accelerator takes the beam up to 400 keV. At this energy, the beam winds through the TFF neutron trap and exits (through a short termination) to the neutralizer.

There are several things to notice about Fig. 10. (1) In setting up the computation, the voltages needed no adjustments. All the TFF electrode voltages were calculated from the vacuum-potential formula

$$\phi_{electrode} = \phi_m - 2\phi_m \ln(1 \pm G/2r_m) .$$

A voltage correction for space charge was neglected. (2) The TFF accelerator and neutron trap were designed first, as described earlier, on the basis of a future availability of 8 A/m of H-. Later, the preaccelerator and matching section were designed and optimized for the existing production of 5.2 A/m. As the figure shows, this discrepancy is easily tolerated by the strong-focus system. (3) The high-perveance preaccelerator produces some beam aberrations which become particularly noticeable after the beam is compressed by the matching section. (We have observed a similar effect when beams are compressed by a straight-axis focusing system.) However, the aberration shows no growth through the remainder of the TFF system. For example, the phase plots at

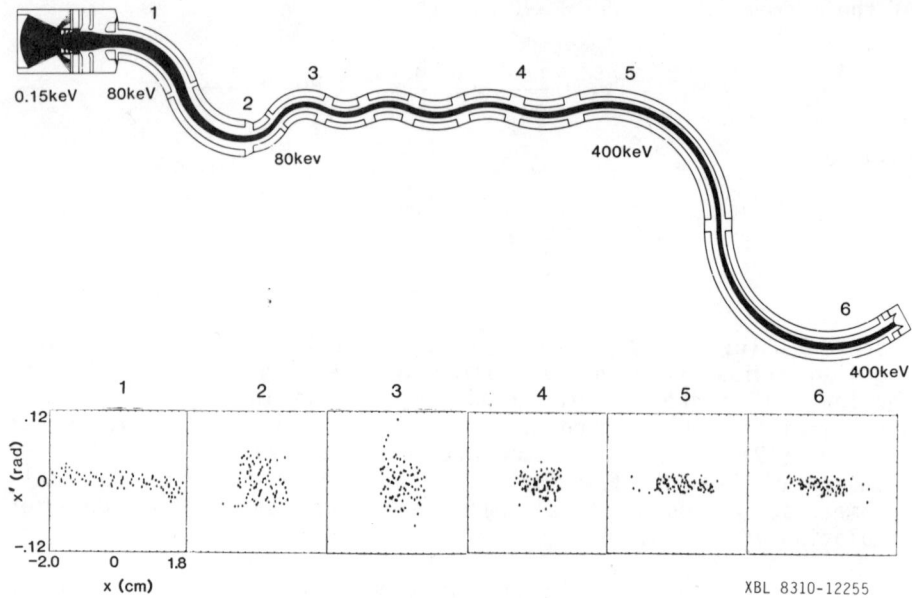

Fig. 10. Simulation of complete D⁻ beam system. See text.

the entrance and exit of the neutron trap are identical except for a 180° rotation in the phase plane. (4) The halo of aberrated beam at the exit contains only three beamlets (out of 88), i.e., about 3% of the total current. The total beam width including the halo is only half the gage width, giving adequate electrode clearance. The three halo beamlets were deleted from the top portion of the figure to give a clearer picture of the main beam.

ACKNOWLEDGEMENT

The author thanks all the people who have made important contributions to the TFF project since its inception, especially Ludmilla Soroka and Bill Cooper. More recently, Chun Fai Chan, Joe Kwan and Deon Vogel have helped in various ways. Above all, special thanks to David Goldberg, who taught me about ESQ's and such and was there at every step during the early exciting synergetic days when we developed TFF theory.

The author thanks Van Nostrand Rheinhold Company for permission to reproduce the illustrations from Ref. 1.

This work was supported by the Director, Office of Energy Research, Office of Fusion Energy, Development and Technology Division of the U.S. Department of Energy under Contract No. DE-AC0376SF00098.

REFERENCES

1. J.R. Pierce, Theory and Design of Electron Beams, 2nd Ed. (D. Van Nostrand, N.Y., 1954).

2. P.T. Kirstein, G.S. Kino, W.E. Waters, Space Charge Flow, (McGraw-Hill, N.Y., 1967).

3. H.A.C. Hogg, "Periodic Electrostatic Beam Focussing," Proc. IEE, Part B, Suppl. 12-13, p. 1016 (1958).

4. O.A. Anderson, "The TFF Accelerator for Fusion Reactor Injection," Bull. Am. Phys. Soc., 27, 1141 (1982). See also four other TFF-related papers in the same session.

5. O.A. Anderson, D.A. Goldberg, W.S. Cooper, and L. Soroka, "A Transverse Field Focusing (TFF) Accelerator for Intense Ribbon Beams," 1983 Particle Accelerator Conference (Santa Fe, NM), published in IEEE Transactions on Nuclear Science, Vol. NS-30, No. 4, p. 3215 (1983).

6. O.A. Anderson, W.S. Cooper, J.A. Fink, D.A. Goldberg, L. Ruby, L. Soroka, and J. Tanabe, "Efficient, Radiation-Hardened, 400 and 800-keV Neutral Beam Injection Systems," Nuclear Technology/Fusion, 4, p. 1418 (1983).

7. O.A. Anderson, W.S. Cooper, D.A. Goldberg, L. Ruby, and L. Soroka, "Negative-Ion-Based Neutral Beam System for FED-A Current Drive," LBL-14880, August, 1982, published in "Fed-A, An Advanced Performance FED Based On Low Safety Factor and Current Drive," Oak Ridge National Laboratory Report, ORNL/FEDC-83/1, p. 4-82 (1983).

8. W.S. Cooper and R.V. Pyle, Scientific Editors, The National Negative-Ion Based Neutral Beam Development Plan, Lawrence Berkeley Laboratory, PUB-464 (1983).

9. K.W. Ehlers and K.N. Leung, "Self-Extraction H- or D- Ion Source," Bull. Am. Phys. Soc., 27, 1057 (1982).

10. M.W. McGeoch, "Laser Neutralization of Negative Ion Beams for Fusion," Proc. of 2nd Int. Symp. on Production and Neutralization of Negative Hydrogen Ions and Beams, Brookhaven National Laboratory, Upton, NY (1980).

11. J.H. Fink, "Photodetachment Technology," Proc. of 3rd Int. Symp. on the Production and Neutralization of Negative Ion and Beams, Brookhaven National Laboratory, Upton, NY (1983).

12. J.H. Fink, Personal Communication (1983).

13. W.S. Cooper, "Summary of the Status of Negative-Ion Based Neutral Beams," Nuclear Technology/Fusion, 4, p. 632 (1983).

14. D.A. Goldberg and K. Halbach, to be published.

15. O.A. Anderson, in preparation.

DISCUSSION

HISKES: What is the height of the beam, about 30 cm?

O.A. ANDERSON: On the test stand TS-IIIA the beam height is 25 cm, but since the beam is only a little over a centimeter thick in the transport and accelerator sections, it's a pretty good ratio, something like 25:1. I think we can learn a lot from that. If you are talking about amperes per meter, we have a quarter of a meter and all these current values (on the viewgraphs) have to be divided by 4.

GRANNEMAN: You said nothing about confinement in the direction perpendicular to the board. There is no focusing in that direction, so what happens there?

O.A. ANDERSON: Electric force is uniform in the central region but it is tilted at the edges (by the shaped electrodes). There is a component of the electric field directed toward the center of the strip beam. We calculated that 10% field tilt is sufficient. That means the restoring field is 1/10 of the transverse field. The space charge field is also about 1/10 of the transverse field, that is why it is strong focusing. A small amount of tilt compensates for the space charge force and emittance pressure.

DAGENHART: In one of your viewgraphs you had a .028 π rad cm emittance. Is that normalized?

O.A. ANDERSON: No, that's not normalized.

DESIGN DESIDERATA FOR A LAMINAR FLOW QUADRUPOLE-FOCUSED
ACCELERATION COLUMN*

A.W. Maschke

Accelerator Department
Brookhaven National Laboratory
Associated Universities, Inc.
Upton, New York 11973

The "Pierce" design acceleration column has been widely used to accelerate high curent beams. It operates well in the space charge limited condition, and will produce beams with a temperature comparable with that of the source. It is restricted in current density, however, by the Child-Langmuir relation. If the ion source itself is not the limiting constraint, then the achievable current density is limited by the electric field at which sparking occurs.

$$J = 5.44 \times 10^{-8} \frac{E^2}{V^{1/2}} \quad (1)$$

where J is current density in amps/m^2, E is the electric field in the column and V is the terminal potential. One sees clearly that the achievable current density decreases as one goes to higher voltages. This can be easily overcome by using electrostatic quadrupole focusing in the acceleration column.

Now it can be shown[1] that the space charge limited current density in a constant energy quadrupole transport channel is greater than that given by Eq. 1, if one assumes that the electric fields on the quadrupoles can be as high in the ion source extraction electric fields. In practice, this is a conservative assumption. It follows that if the beam can be transported a large distance at the C-L current density limit, it can surely be accelerated as it goes from quadrupole to quadrupole. Hence, the necessity of having a high gradient acceleration column goes away.

The design of a laminar flow quadrupole-focused acceleration column can be broken down into two parts. One is to match the ion source into the beginning of the acceleration column. We did this by building a transport system consisting of about twenty electrostatic quadrupoles. The beam can then be analyzed at the exit. In this way one can empirically determine when a space charge limited laminar flow condition has been achieved.

With this having been done, we know the proper geometry and voltage for the first quadrupole of the acceleration column, when the beam has only the energy of the extraction voltage. Now it only remains to know how to change the quadrupole parameters as

---

*Work performed under the auspices of the U.S. Department of Energy.

one goes up in voltage. This is given by the laminar flow condition:[2]

$$\left[\frac{r_b}{r_q}\right]^2 \frac{E_q^2 \ell_q^2}{V^{1/2}} = \text{constant} \qquad (2)$$

Here r_b is the beam radius, r_q the quadrupole half-aperture, E_q is the electric field in the quad, ℓ_q the quadrupole length and V is the energy of the beam in the quadrupole of interest. The only other constraint is to make sure the optical changes from quadrupole to quadrupole are gradual. This translates into a design which does not change the beam energy by a large amount in a single gap. For the column that we tested here, the change was limited to 15% per quad., i.e., a factor of four in voltage in the space of 10 quads.

The usual situation would be to keep the beam radius and quadrupole radius constant. Also, if the electric field was high at the beginning, one might chose to keep it constant also. In this case the only variable is the quadrupole length. One has then that $\ell_q \propto V^{1/4}$. In the acceleration column that we tested here, we used ten identical quadrupoles (because we had them) and let the quadrupole voltages increase as $V^{1/4}$. This was no problem because the electric fields were low. Notice that the spacing between the quadrupoles does not explicity enter the relationship for laminar flow. However, the phase advance/cell increases as the quad separation increases. Above 90° the thin lens approximation breaks down. This is not a practical limitation, since stable laminar flow is usually seen around 70-80° phase advance. In the column we tested, we let the cell spacing vary as $V^{1/4}$ also. This gives a phase advance/cell which goes as $V^{-1/2}$.

References

1. Formulary for MEQALAC Design, BNL 51119 (1979).
2. Space Charge Limits for Linear Accelerators, BNL 51022 (1979) and Design and Operation of a Laminar Flow Electrostatic Quadrupole Focused Acceleration Column, BNL 51692 (1983).

DISCUSSION

LEUNG: Can you use these quadrupoles to turn the beam around corners?
MASCHKE: Well, you'd have to put the electrostatic bending elements between the quadrupoles or, as it has been suggested, you could actually put electrostatic quadrupoles in a bending magnet. In this case the magnetic field will bend the beam and the electrostatic field will do the focussing. How well you can hold voltages under those circumstances is something you have to look at.
BARNETT: What is the maximum practical energy you can accelerate particles up to, using this technique? Is it 1 MeV?
MASCHKE: No, I think the only limit is how high a voltage you want to live with, but certainly because the voltage on the quadrupoles doesn't get worse, you could go on for a very long time. For instance, in the case we've done we had ten quadrupoles and we increased the voltage a factor of 4 and so we are pretty confident that we can maintain laminar flow for 50 quadrupoles which means you could go up by a factor of something like 4^5 which is a giant factor. Since in a normal situation the extraction voltage would be 30 or 40 kV, something on that order, the system we're building with 20 quadrupoles will take us up by a factor of 16, from essentially 30 keV to 0.5 MeV. But that is only 20 quadrupoles and I don't think there is in principle any problem in going to 10 or 20 MeV with a system if you're willing to live with such a big high voltage terminal.
HENKES: What is the typical energy gain, how many keV per centimeter?
MASCHKE: It can be very low. In the system built, we are going up in steps of 15%. At some point in the system you can't keep going up 15% per gap because if you're at MeV level, 15% is getting to be a lot of voltage and so you would trim that down and lower your rate. But unlike a Pierce column, there is no pressure to have high accelerating fields, so you don't pay a penalty for having a very long acceleration column, except to the extent that that gives you more quadrupoles and at some point you are going to worry how long you can stay in this system. So far we have exhibited 50 quads. It might be longer. It also depends on the source temperature. I mean the laminar flow condition implies that you have a zero emittance beam and of course you don't. You have a finite ion temperature in the source but if the temperature is low enough so the beam can get through the system, it certainly can go through 50 quadrupoles.

THE AMSTERDAM "MEQALAC" RF ACCELERATION SYSTEM

E.H.A. Granneman, R.W. Thomae, F. Siebenlist, P.W. van Amersfoort,
H.J. Hopman, J. Kistemaker
FOM-Institute for Atomic and Molecular Physics,
Amsterdam, The Netherlands

H. Klein, A. Schempp, T. Weis
Institut für Angewandte Physik, University Frankfurt,
Frankfurt, Federal Republic of Germany

ABSTRACT

An experiment is described which is designed to investigate the possibilities of RF acceleration of intense negative ion beams to energies $\lesssim$ 1 MeV/nucleon. In a MEQALAC system a large number of small, parallel beamlets is accelerated simultaneously. (Multiple Electrostatic Quadrupole Array Linear ACcelerator.) Calculations have been performed to find conditions for stable transport through the quadrupole periodic focusing system. In the first stage of the project four 40 keV He^+ beams are to be accelerated to 106 keV. The theoretical (average) space charge limited current is $\sim$ 3 mA/channel for this case. The status of this accelerator is discussed. Also parameters are given for a 100 mA, 6 MeV Li^- accelerator. Such an accelerator can be used for plasma diagnostics.

1. INTRODUCTION

In present day fusion experiments H°, D° heating beams with particle energies in the range of 50 - 120 keV are used. For the next generation of devices such as INTOR, 175 keV neutral beams are planned. Even higher particle energies are desirable for fusion reactors. Grisham et al. [1,2] and Post et al. [3] have shown that the most ideal energy of light atoms (H, D, Li, C, O) is of the order of 1 MeV/nucleon.

It is highly unlikely, that multi-ampere beams can be accelerated to these energies by means of the commonly used DC acceleration systems. For that reason the use of RF acceleration techniques has to be tested. With RF acceleration the technological difficulties are in principle independent of the final energy and therefore it is also possible to use other types of light negative ions. The higher mass has the advantage that the current requirements on the negative ion source are drastically reduced: in case of e.g. 12 MeV C° beams the $\sim$ 75 MW neutral beam power believed to be necessary for ignition of the INTOR plasma is obtained with a current of only $\sim$ 6 A.

RF acceleration techniques can also be used to produce diagnostic beams with high particle energies. Grisham et al. [4] showed that with a Li° beam with an energy $\lesssim$ 5.2 MeV and a current of $\sim$ 50 mA the energy distribution of fusion alpha particles can be determined.

Other applications of RF accelerators can be found in ion implantation of semiconductors and metal surfaces.

RF acceleration has a number of distinct advantages above DC acceleration:

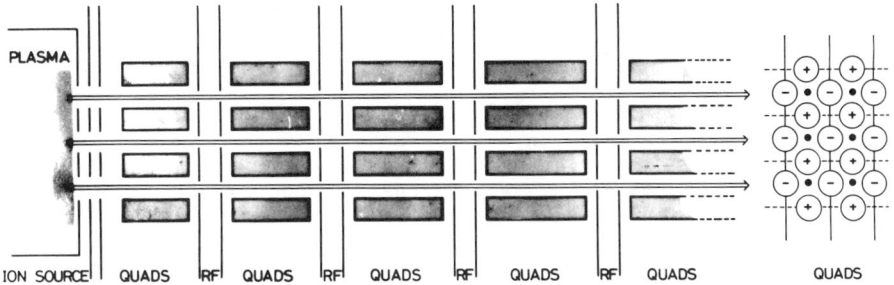

Fig. 1. The MEQALAC structure. A large number of beams is injected into an RF gap/quadrupole structure and accelerated to high energies. The special arrangement of the quadrupole elements makes it possible to stack many beams within a small area. The length of the quadrupole elements increases in proportion to the particle velocity.

- Only particles with one particular value of M/Z are accelerated; a strong discrimination against electrons and (negative) impurity ions takes place.
- The highest voltages present are those on the source extraction grids and in the RF gaps ($\sim 20 - 100$ kV), independent of the final particle energy. Consequently the voltage hold-off problems are strongly reduced.

The major disadvantage of these systems is the inherently lower acceleration efficiency (see section 3). Recently the FOM-Institute for Atomic and Molecular Physics, Amsterdam and the Institut für Angewandte Physik, Frankfurt started a joint project in Amsterdam to investigate the possibilities of accelerating intense negative ion beams by means of RF fields. It was decided to build a so-called MEQALAC system (Multiple Electrostatic Quadrupole Array Linear ACcelerator, after Maschke [5]). A MEQALAC system consists of the following components (see figure 1): From a multi-aperture ion source a large number of beamlets are extracted and subsequently injected into a multichannel RF structure. RF voltages are applied on a (usually) large number of acceleration gaps. In each gap a fixed amount of energy is gained by the particle. In between the acceleration gaps strong focusing quadrupole lenses are placed to ensure stable transport. These quadrupoles provide the restoring forces necessary to prevent beam space charge blow-up and to correct for the defocusing action of the accelerating gaps. The advantage of a system like this is that in principle an arbitrarily large current can be transported and accelerated, just by increasing the number of beamlets. The space charge of the total beam is of no importance since the individual beamlets are shielded from one another by the quadrupole elements. By stacking the channels as shown in figure 1 a dense packing of beamlets is obtained. This system was invented by Maschke [5]. A first accelerator of this type was built by his group in 1979 [6]. With a 4 MHz RF system 9 Xe^+ beams were accelerated from 15 to 75 keV (total average current 2.8 mA). A design of a 200 MHz H^- accelerator is described by Keane and Brodowski [7].

Because limited experimental information is available on the acceleration of space charge loaded beams in MEQALAC structures it was decided to start with an experiment with positive instead of negative ions: we developed a 40 MHz Interdigital H resonator for the acceleration of four He$^+$ beams with average currents of $\sim$3 mA (each) from 40 to $\gtrsim$ 100 keV. The main objectives are:
- to test the principle of IH-MEQALAC systems
- to study the transport and acceleration of space charge loaded beams
- to determine requirements on the emittance of the injected beams
- to find scaling laws of important parameters, such as resonator dimensions, resonator frequency and acceleration efficiency.

In section 2 the theoretical limits on the space charge loading of the beamlets are discussed. Section 3 gives experimental details on the FOM-MEQALAC system. Finally section 4 gives the parameters of a high power system: a 100 mA, 6 MeV Li$^-$ accelerator.

2. SPACE CHARGE LIMITED TRANSPORT AND ACCELERATION

In linear accelerators the forces on the beam must be such that stable transport is guaranteed. This is accomplished by applying periodic focusing forces in the transverse as well as in the longitudinal direction. In the transverse plane this force is generated by the electric field of the quadrupoles; it drives the particles back toward the beam axis. In the longitudinal direction it is the net force being the difference between the forces exerted by the RF field on the synchronous particle (i.e. the one moving with the average velocity) and on the particles at other (longitudinal) positions in the bunch. This net force keeps the bunch together in axial direction.

The stability of the focusing system is determined by the parameter μ, the phase advance per cell, for zero space charge: $\mu_o = 2\pi L / \lambda_o$. λ_o is the "wavelength" of the radial particle motion (see figure 2); L is the cell length.

When the current increases the phase advance per cell is reduced ("depressed") because of the space charge fields opposing the external fields. In the absence of space charge the transport is stable when $\mu_o < 180°$.

The transport (and acceleration) of realistic, non homogeneous, space charge loaded beams can only be studied by means of particle simulation techniques. From thos types of studies [8,9] it is concluded that the zero current phase advance μ_o should not exceed 90°, preferably $\mu_o \lesssim 60°$. Only in case of a beam homogeneously filled with space charge the radial transport can be calculated by solving the (analytic) Kapchinsky-Vladimirsky (KV) equations [10].

Maschke [11] and Reiser [12] showed that if the modulation in the beam envelope is small an approximated solution can be derived for the space charge limited current. Reisers [12] so-called smooth approximation gives a solution of the KV equations; it results in the following expression for the space charge limited current for bunched beams:

$$I_{MAX} = \tfrac{1}{2} I_o \beta^3 \ G(\bar{a},\bar{b}) \ (\frac{\mu_o \alpha}{L}) \ (1 - \frac{\varepsilon^2}{\alpha^2}) \tag{1}$$

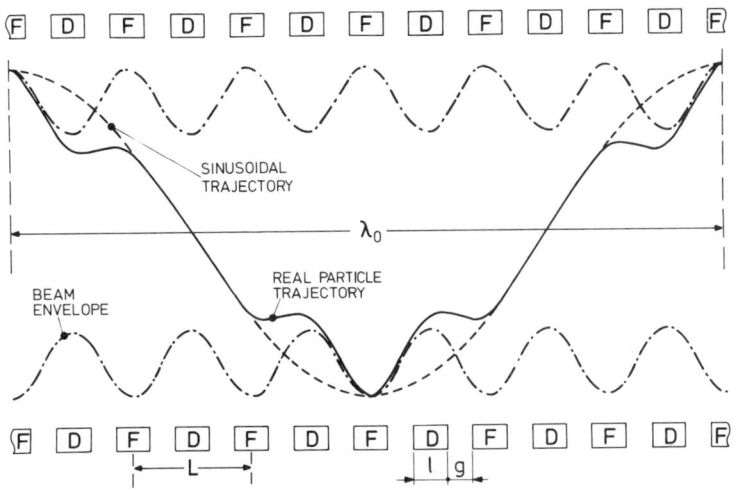

Fig. 2. The quadrupole periodic focusing system. A FODO focusing grid is applied (foc.quad. - drift space - defoc.quad. - drift space - focu.quad - ...). Each transverse unit cell (L) contains two quadrupole singlet lenses. The net effect of many cells is an oscillatory particle motion which has two characteristic frequencies, a high frequency oscillation with a period equal to the length of one cell and a low frequency oscillation with a period of many cells. In an infinitely long quadrupole channel the transport is stable when the beam envelope reaches maximum and minimum values in the centers of the focusing (F) and defocusing (D) planes, respectively.

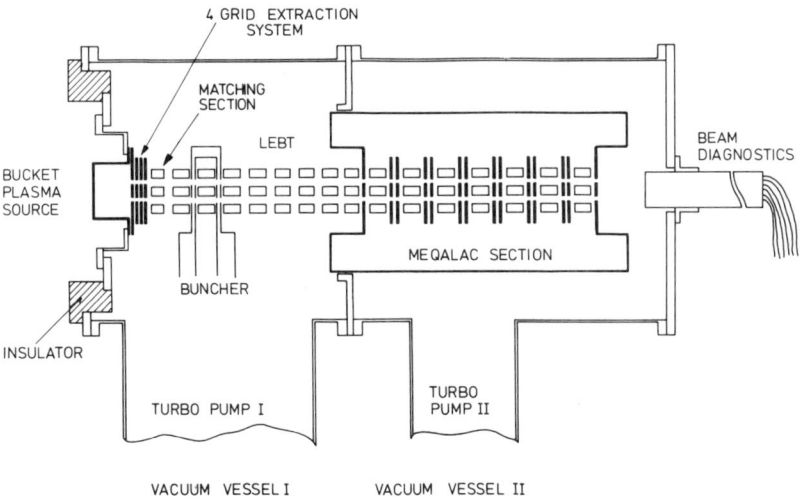

Fig. 3. The experimental set-up. For details, see text.

This relation predicts the space charge limited current in the transverse (T) as well as in the longitudinal (L) direction. μ_o, ε, α, L and $G(\bar{a},\bar{b})$ generally have different values in both dimensions. μ_o is the zero-current phase advance per cell, α and ε are the (unnormalized) channel acceptance and beam emittance, respectively.
$I_o = 4\pi\varepsilon_o mc^3/q = 1.24 \times 10^7$ A for He$^+$ ions. $G(\bar{a},\bar{b})$ is a bunch shape factor: $G_T = (1 - \bar{a}/3\bar{b})^{-1}$; $G_L = 3\bar{a}/2\bar{b}$ where $\bar{a}$ and $\bar{b}$ are the average values of the beam envelopes in the transverse and longitudinal dimensions, respectively. For DC beams $G_T = 1$. α_T is a function of the quadrupole channel parameters (ℓ, g, a_o, voltage); α_L is a function of the RF frequency, the RF field strength and the synchronous phase ϕ_s.

It has been shown by Struckmeier and Reiser [9] that exact numerical integration of the KV equations as well as computer simulations for both KV and Gauss distributions give results which agree well with relation (1) for $\mu_o \lesssim 90°$.

3. THE FOM-MEQALAC SYSTEM

A schematic picture of the FOM experiment is shown in figure 3. From an ion source four He$^+$ beams are extracted. A low energy beam transport section (LEBT) transports the ions from the high pressure ion source region ($p \gtrsim 10^{-3}$ torr) to the low pressure RF acceleration region ($p \lesssim 10^{-6}$ torr). The DC beams are bunched by means of a two-gap buncher and subsequently injected into the MEQALAC accelerator which features a FODO (transverse) focusing grid. After acceleration the beam quality, the beam energy and current are monitored by means of various types of diagnostics. For further information the reader is referred to Van Amersfoort et al. [13] and Siebenlist et al. [14]. We now discuss the most important parts of the system.

SOURCE

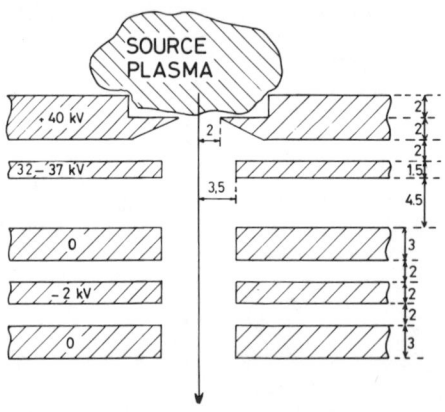

Fig. 4. The 40 keV extraction system of the bucket source. By varying the voltage on the second electrode (between 32 - 37 kV) the shape of the emittance diagram can be changed. All dimensions are in mm.

The source is a bucket type ion source (diameter 14 cm, depth 11 cm) which runs DC [14,15]. Typical arc voltages and currents are 100 - 140 V and 5 - 20 A, respectively. The extraction system, see figure 4, is a modified version of the four-grid extraction system described by Holmes [15]. With this set-up 40 keV beams are extracted with currents variable between 2 - 8 mA and (unnormalized, RMS) emittances between (10 - 15) .π mm . mrad. The duty cycle of the extraction system can be varied between 1% and 100%. An example of the emittance of a 6 mA beam is shown in fig. 5.

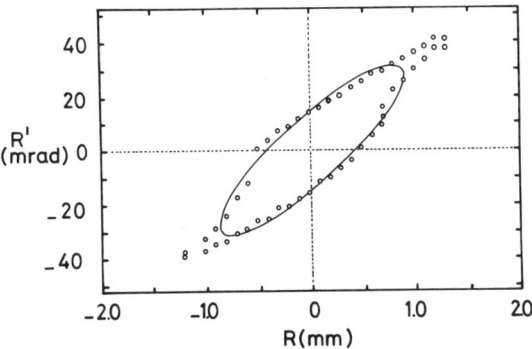

Fig. 5. The emittance of a 6 mA beam as measured 42.5 mm behind the last source extraction electrode. The second electrode is kept at 34 kV. The dots indicate points where the current has dropped to 10% of the maximum value (in the center of the beam). The ellips gives the RMS emittance: $\varepsilon = 13.23 \; \pi$ mm.mrad. (area ellips).

LOW ENERGY BEAM TRANSPORT SECTION (LEBT) AND BUNCHER

The low energy beam transport section transports the beams from the ion source to the MEQALAC system. It consists of a series of 34 quadrupole singlet lenses, the first of which is located 17.5 mm behind the last grid of the source extraction system. All LEBT singlets are held at the same potential except for the first five which can be excited independently. The first five quads match the cylindrical-

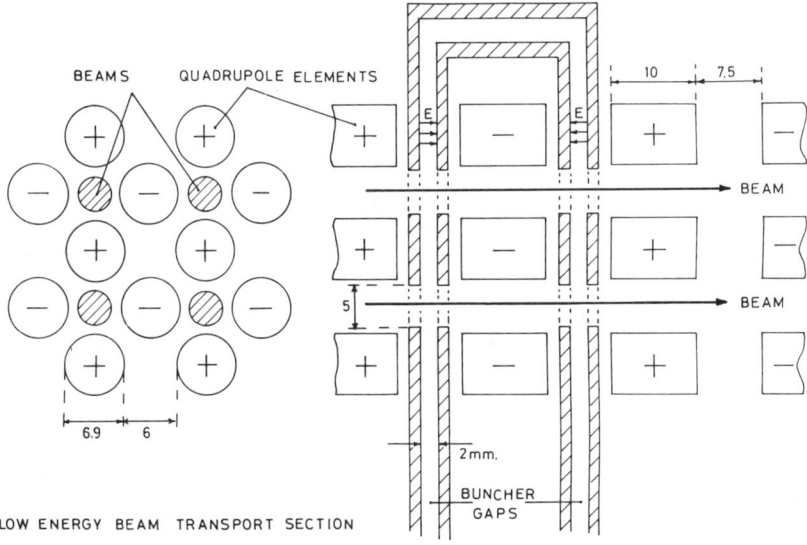

Fig. 6. Characteristic dimensions (in mm) of the LEBT section. Also the two-gap buncher is shown.

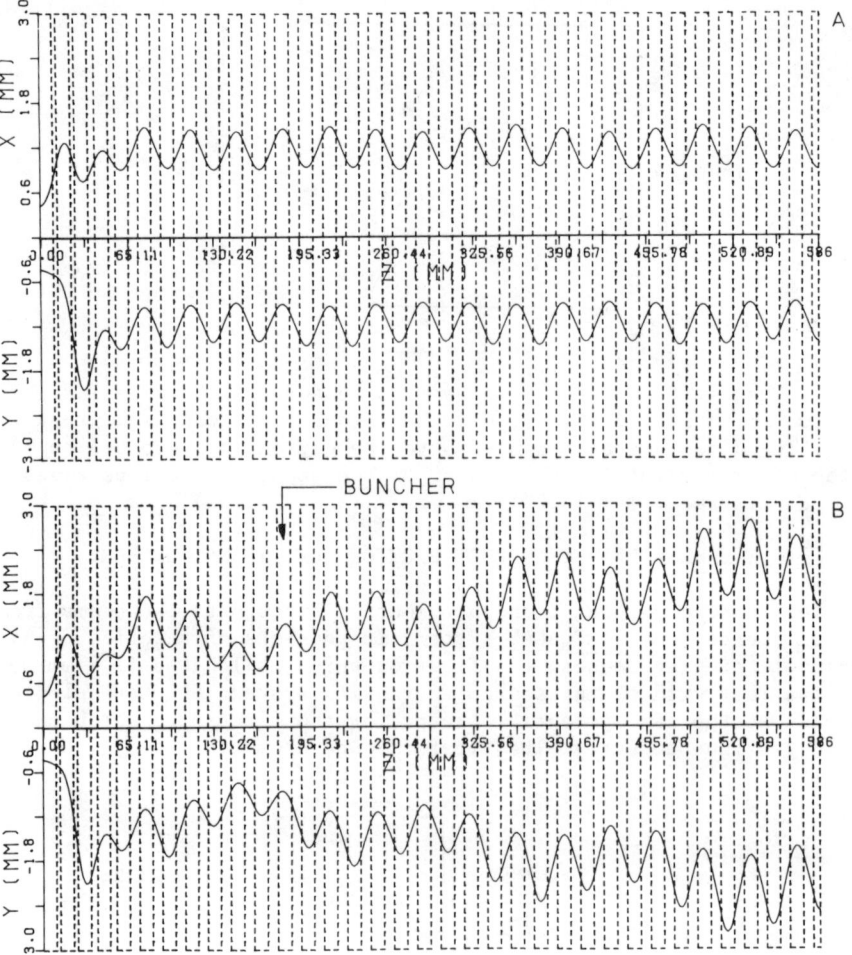

Fig. 7. KV beam envelope calculations for the LEBT section. In the top and bottom halves of each of the two pictures the envelopes in the XZ and YZ planes, respectively, are shown. The 6 mA beam of fig. 5 is injected. The quad lengths are 10 mm; the gaps are 7.5 mm, except for the first four, which are 3,3,5 and 5 mm. The voltage on all LEBT quads is ± 2.62 kV. The total number of quads is 34.
a. The beam is matched to the periodic focusing system of the LEBT; the buncher is not excited; the voltages on the first five matching section quads are: ± 2.62 kV; $\mp$ 4.67 kV; ± 4.15 kV; $\mp$ 2.72 kV; ± 2.21 kV.
b. The buncher is excited such that the current increases a factor 4, linearly with z. The beam is ideally matched to the periodic focusing system of the MEQALAC accelerator (not shown). The voltages on the matching section quads are: ± 2.62 kV; $\mp$ 4.76 kV; ± 3.88 kV; $\mp$ 1.99 kV; ± 1.87 kV.

ly symmetric beam extracted from the source to the periodic focusing system of the LEBT and the MEQALAC sections. Figure 6 gives the characteristic dimensions of the LEBT section. It also shows the two-gap buncher which matches (bunches) the incoming DC beam to the longitudinal acceptance of the RF accelerator with a calculated efficiency $\eta \sim 60\%$ [13]. The RF voltage required in each of the two gaps is ~ 800 V. For 40 keV He$^+$ beams to be transported through this periodic focusing system, with $\mu_0 = 60°$, quadrupole voltages of ± 2.62 kV are needed. In figure 7 beam envelopes, calculated with the kV equations [9,10], are shown for two situations: In figure 7a the 6 mA beam shown in figure 5 is matched to the periodic focusing system of the LEBT; the buncher is not excited. In figure 7b the increase in beam space charge caused by the buncher, is taken into account. The matching section is excited such that the beam extracted from the source is matched to the periodic focusing system present in the MEQALAC accelerator.

MEQALAC SECTION

The RF resonator must satisfy the following requirements: resonance frequency 40 MHz ($\beta\lambda/2 = 17.5$ mm for 40 keV He$^+$); He$^+$ ions must be accelerated from 40 to $\gtrsim 100$ keV with high efficiency. It was decided to use a modified version of the interdigital H resonator (TE 111); see figure 8: On opposite sides of a cylindrical resonator two hollow rectangular boxes (1,2) are mounted. These have a rectangular, fingerlike, structure on one side. The fingers meet on the axis of the cylinder. The quadrupole lenses (3) are mounted inside the fingers. The RF current flows as indicated in figure 8. The RF potential difference present between top and bottom of the cylinder is thus used to generate a longitudinal electric field on axis. The current produces a magnetic field (H) flowing around the boxes. Because the magnetic field must have space to turn around, the end plates (4,5) closing the resonator are located at a distance of ~ 10 cm from the sides of the boxes. The beams enter and leave the resonator through two inserts (8,9) which allow them to propagate in a (RF) field free space up to the first accelerator gap, and from the last acceleration gap onwards. In this way the LEBT section can be continued up to the first RF gap.

The properties of this resonator were investigated by varying the gap widths (i.e. the capacitance of the structure) and the distance between subsequent gaps. The desired resonance frequency was obtained with gap widths of 2.0 mm. The total capacitance is ~ 350 pF; the gap aperture diameter 5.0 mm.

The axial distribution of the electric field is obtained by pulling a small ($\emptyset = 2$ mm) perturbation ball through the gaps and measuring the resulting frequency shift. This measurement yields $E^2(z)$, see figure 9. When the capacitance per unit length is constant the structure is expected to have a relatively flat electric field distribution between gaps 2 and 19, with a slight decrease in field strength at both ends. Since the capacitance per unit length decreases with increasing $\beta\lambda/2$ this decrease is enhanced at the high energy end of the resonator. The field strength in gaps 1 and 20 is

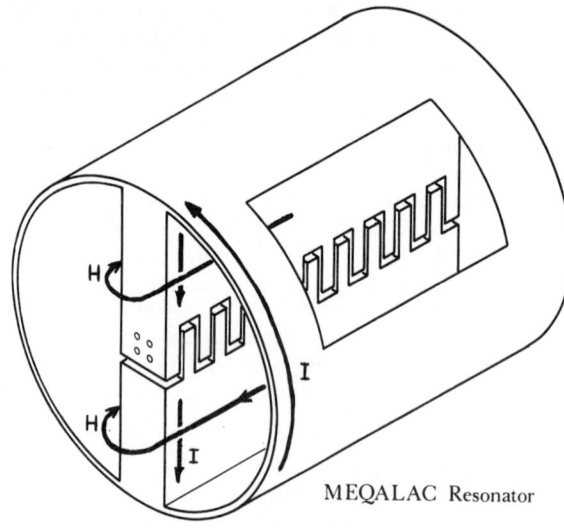

Fig. 8. The MEQALAC accelerator structure. The resonator is a modified interdigital H resonator (TE 111). The number of gaps is 20. For simplicity fewer gaps are shown in figure b. The gaps are water cooled (not shown in picture). All dimensions are in cm. For more details, see text.

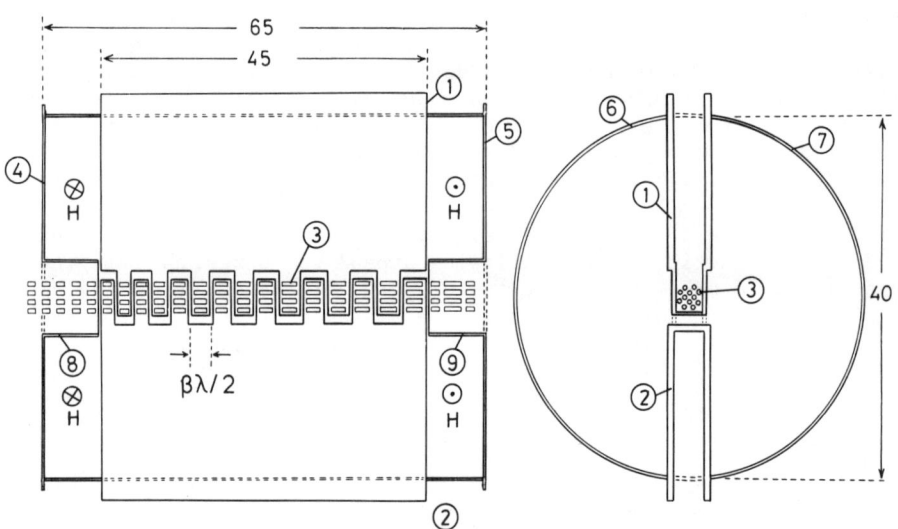

roughly half of that in the rest of the gaps. For this structure a shunt impedance of 12 MΩ was measured. In the final structure in which the finger structure is machined from one peace a shunt impedance ≳ 16 MΩ is expected. This means that four 40 keV beams with an average current of 3 mA/channel can be accelerated to 100 keV with an efficiency ≈ 60% (conversion of RF power into particle power). In table 1 a list of characteristic parameters of the accelerator is presented.

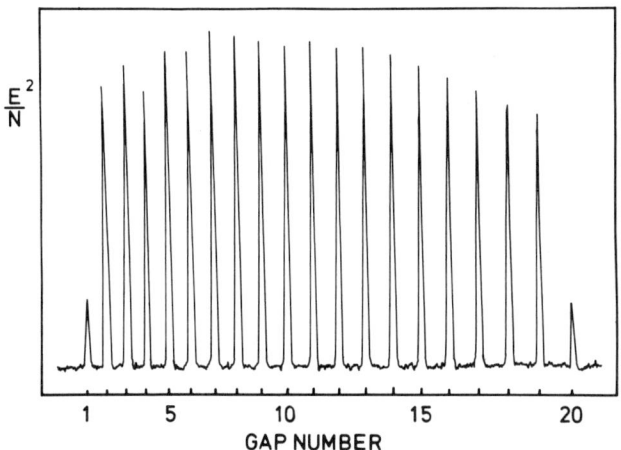

Fig. 9. The square of the electric field E along the accelerator axis, as obtained with a perturbation ball measurement. Plotted is E^2/N; N is the RF power loss in the resonator structure. The small gap-to-gap variation is caused by minor deviations in the gap widths.

4. A 6 MeV Li⁻ ACCELERATOR

Based on our prototype experience we can scale the present structure up to a high energy, high power, system. As an illustration we present a design of a 6 MeV, 100 mA Li⁻ accelerator; see table 1. An accelerator of this type can in principle be used for diagnosing α-particle energy distributions in large fusion experiments [4]. If we take into account a neutralization efficiency [16] of 50% this yields a neutral beam with an equivalent current of 50 mA. The accelerator is made up of two stages. In the second stage the resonance frequency is doubled. The acceleration efficiency is 67%. The total length of the system is 6.2 m. By increasing the injection energy or the number of channels the total accelerated current can be scaled up to higher values (with similar efficiencies). Furthermore it is in principle also possible to design a system which consists of a larger number of sub-sections, thereby providing the option to vary the particle energy in a number of discrete steps. By adding a final section which contains many small deflection plates each individual beamlet (⌀ = 5 mm) can be steered to hit the same (small size) particle detector.

ACKNOWLEDGEMENTS

This work is sponsored by F.O.M. with financial support by Z.W.O./ EURATOM.

Table 1: The characteristic parameters of two MEQALAC accelerators. In the first column the present He$^+$ system is given. In the second and third column a 100 mA 6 MeV Li$^-$ accelerator is presented. This accelerator consists of two stages. In the first stage acceleration takes place from 90 to 1210 keV; in the second stage to 6 MeV. The subscripts L and T denote the longitudinal and transverse dimensions, respectively; the subscripts MAX and AV stand for peak and average, respectively. μ is the phase advance per cell. The acceleration efficiency is defined as the fraction of the total RF power which is converted into particle power.

Parameter	Present exp.	Li accelerator Stage 1	Li accelerator Stage 2	DIM.
Particle	He$^+$	^{6}Li$^-$	^{6}Li$^-$	
Injection energy	40	90	1210	keV
Exit energy	106.5	1210	6010	keV
RF frequency	40	40	80	MHz
Synchronous phase ϕ_s	-38	-33	-16	°
Transit time factor	0.9	0.9	0.9	
Gap electric field ampl.	2.6	9.3	8.0	MV/m
Averaged acc.el.field	0.1	0.4	1.35	MV/m
Number of gaps	20	41	49	
Number of channels	4	16	16	
Overall beam dimensions	4	25	25	cm²
Length resonator	65	260	355	cm
Diameter resonator	40	65	65	cm
Capacitive load (Av.)	7.0	2.6	0.7	pF/cm
Quality factor Q	1800	2900	4100	
Shunt impedance R_{po}	16	44	120	MΩ
R_{po},eff	9	26	91	MΩ
RF power losses N	0.5	49	253	kW
$\beta\lambda/2$ first cell	1.75	2.28	3.96	cm
$\beta\lambda/2$ last cell	2.80	7.73	8.62	cm
Width RF gaps	0.20	0.40	1.30	cm
Quad spacing/length; g/ℓ	0.75	0.75	0.80	
Diameter quad channel	0.60	0.60	0.60	cm
Useful diameter	0.50	0.50	0.50	cm
Quadrupole voltage	±2.62	±4.09	±5.97	kV
Zero current μ_{OT}	60	80	20	°
Zero current μ_{OL}	20.0	34.5	10.6	°
Depressed μ_T	7.3	24.0	4.8	°
Depressed μ_L	8.0	10.4	6.7	°
Channel acceptance α_T	108.π	83.π	26.π	mm.mrad
Channel acceptance α_L	240.π	343.π	44.π	mm.mrad
$I_{T,MAX}$	24.1	59.8	151.0	mA
$I_{L,MAX}$	22.6	142.0	130.5	mA
$I_{T,AV}$	3.0	6.3	7.8	mA
$I_{L,AV}$	2.8	15.0	6.7	mA
Total current $I_{TOT,AV}$	11.2	101	101	mA
Acceleration efficiency	60	70	66	%

REFERENCES

1. L.R. Grisham, D.E. Post, D.R. Mikkelsen and H.P. Eubank, Nucl. Techn./Fusion $\underline{2}$, 199 (1982).
2. L.R. Grisham, D.E. Post, D.R. Mikkelsen, these proceedings.
3. D.E. Post, L.R. Grisham and R.J. Fonck, Physica Scripta $\underline{T3}$, 135 (1983).
4. L.R. Grisham, D.E. Post, D.R. Mikkelsen, Nucl.Techn./Fusion, to be published; Princeton report: PPPL-1886.
5. A.W. Maschke, Brookhaven Natl.Lab. report BNL-51209 (1979).
6. G. Gammel, J. Brodowski, J. Keane, A. Maschke, E. Meier, R. Mobley, R. Sanders, 1981, Part.Acc.Conf.Washington DC. IEEE Trans. Nucl.Sci. $\underline{NS-28}$, 3482 (1981).
7. J. Keane, J. Brodowski, 1981, Part.Acc.Conf.Washington DC. IEEE Trans.Nucl.Sci. $\underline{NS-28}$, 3485 (1981).
8. I. Hoffmann, L.J. Laslett, L. Smith, I. Haber, Part.Acc., to be published.
9. J. Struckmeier, M. Reiser, Part.Acc., to be published. GSI (Darmstadt) preprint GSI-83-11.
10. I.M. Kapchinsky, V.V. Vladimirsky, Proc.Int.Conf.on High Energy Acc., CERN, Geneva, 1959, page 274.
11. A.W. Maschke, Brookhaven Natl.Lab.report BNL-51022 (1979).
12. M. Reiser, Part.Acc. $\underline{8}$, 167 (1978); J.Appl.Phys. $\underline{52}$, 555 (1981).
13. P.W. van Amersfoort, E.H.A. Granneman, J. Kistemaker, F. Siebenlist, R.W. Thomae, H. Klein, A. Schempp. FOM-report nr. 55.120, November 1982.
14. F. Siebenlist, R.W. Thomae, P.W. van Amersfoort, E.H.A. Granneman, J. Kistemaker, H. Klein, A. Schempp. FOM-report nr. 56.298, May 1983.
15. A.J.T. Holmes, E. Thompson, F. Watters, J.Phys. $\underline{E14}$, 856 (1981).
16. L.R. Grisham, D. Post, B. Johnson, K. Jones, J. Barette, T. Kruse, I. Tserruya, Wang Da-Hai, Rev.Sci.Instr. $\underline{53}$, 281 (1982).

TvdV

DISCUSSION

EHLERS: Including your buncher, what will your effective acceptance be in reference to a DC beam?
GRANNEMAN: This particular buncher accepts 60% of the particles. If you really want to get more than that, you have to do some more gentle bunching and it should be possible to go up to 80 or 90%. However, in this first experiment we're not aiming for that.
JACQUOT: At the beginning of your talk you have mentioned that the power of the oscillator is about 1.5 kW. As you accelerate 11 mA with an energy gain of about 60 keV, this corresponds to about 600 W of the beam power, so that the power efficiency is not 60%. How do you define your efficiency?
GRANNEMAN: I define the power efficiency as the fraction of the RF power which is converted into particle power.
JACQUOT: The fraction of the RF power in the cavity?
GRANNEMAN: I think I know what the problem is. The slide you are referring to is a list I made about a week before we left. We subsequently finished the measurement, all RF level measurements on the accelerator, and the value of 60% is based on that. This is really based on measured values so this should be realistic.

THE BNL RFQ*

Salvatore T. Giordano
Accelerator Department, Brookhaven National Laboratory
Associated Universities, Inc., Upton, N.Y. 11973

ABSTRACT

A brief outline of the theoretical design has been presented at the Symposium. Particular emphasis has been placed on the RF design of the main cavity, end cells, and power feed. The relevance of the design has been considered with respect to the tuning and operating characteristics of the structure.

RFQ

An RFQ is presently being constructed here at BNL for the acceleration of polarized protons. I was asked to present a verbal report on the status of the machine. Our present schedule calls for the acceleration of beams by mid February, 1984. A complete detailed report will be presented at the International Accelerator Conference at GSI in May, 1984.

The RFQ is a relatively new device which offers some interesting future possibilities. It was decided that we develop in-house, for future use, the technology for designing and fabricating RFQ structures. Programs for beam dynamics, RF characteristics, vane profile design and machining techniques now exist here at BNL. We would like to thank the RFQ group at Los Alamos and Saclay for their help.

RFQ DESIGN PARAMETERS

Frequency	201.25 MHz
Ion	H^-
Number of Cells (in the vane)	144
Length	130.28 cm
Intervane Voltage	63 kv
Peak Surface Field	20.9 MV/m
Average Radius, r_o	0.4638 cm
Final Radius, a_f	0.299 cm
Final Modulation, m_f	1.969
Initial Synchronous Phase, ϕ_i	$-90°$
Final Synchronous Phase, ϕ_f	$-30°$
Estimated Peak RF power	60 kW
Nominal Current Limit	56 mA
Nominal Acceptance	0.27π cm-mr (normalized)
Initial Energy	20 keV
Final Energy	760 keV

---

*Work performed under the auspices of the U.S. Department of Energy.

506

THE STAFF

Full time: Salvatore Giordano, Ray McKenzie-Wilson, Philip Warner
Part time: Hugh Brown, Thomas Clifford, Mario Puglisi

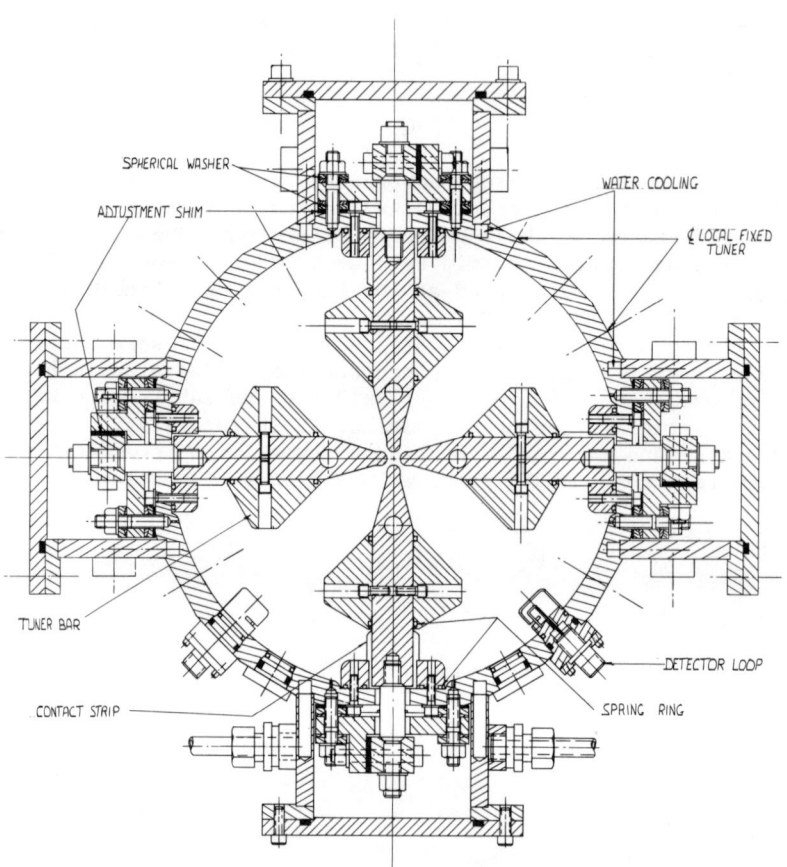

Fig. 1. Mechanical cross section.

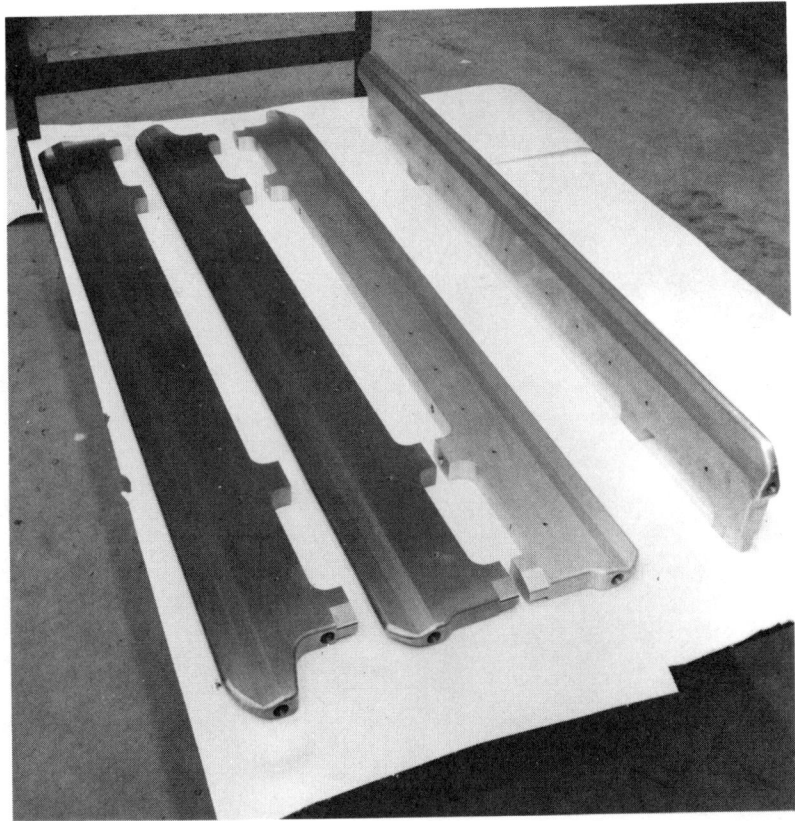

Fig. 2. Vanes prior to final machining and copper plating.

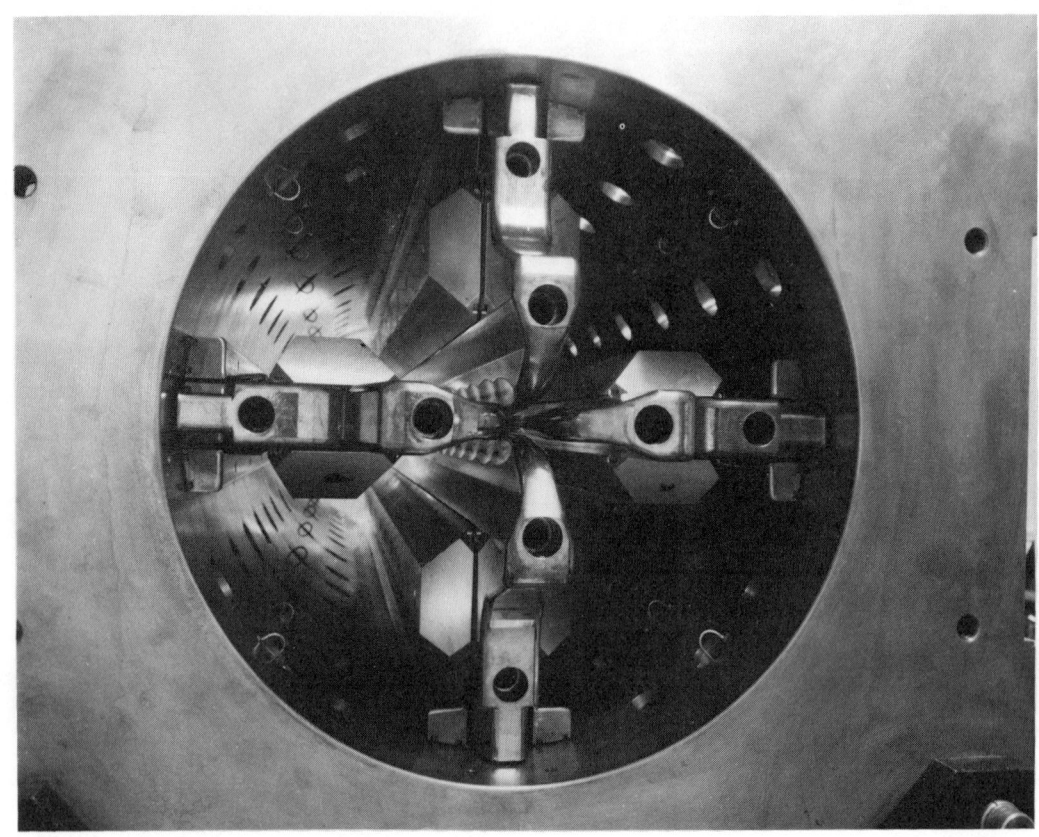

Fig. 3. Vanes and tuner bars installed in cavity.

Fig. 4. Preliminary RF test with end sections removed.

DISCUSSION

HENKES: If you couple the RFQ to a DC ion source, how much of the current of the source will be trapped by the accelerating field?
GIORDANO: Some of our calculations indicate that it is something like 90% to 95% capture.
BARNETT: This design is for 760 kV?
GIORDANO: 760 kV, with 20 kV input.
COOPER: What current and what ion?
GIORDANO: Actually, there have been a few successful RFQ's. The latest one we heard about is at CERN, it has accelerated a current of 30 mA but in our particular case we are not interested in a high current. Our current requirement is about 20 µA of polarized H^- ions, but I think that a machine like this can go to 50 mA.
MASCHKE: I would like to correct something you said regarding the current. First RFQ was built by the Soviets and in 1975 they reported currents up to 200 mA with 100 kV injection.
GIORDANO: The machine has, I think, potential for high currents but in our particular case we have not investigated that aspect.
GRANNEMAN: Could you tell me what the shunt impedance is?
GIORDANO: I am hoping that the Q turns out to be something of the order like 8,000 or 10,000. The shunt impedance does not have the same significance for RFQ's as it has for conventional linacs. A significant number is that the RF power required is about 70 kW to achieve the 760 kV of acceleration. This power can go up or down depending on what the final Q of the cavity is.

OPERATING EXPERIENCE WITH A 100-keV, 100-mA H⁻ INJECTOR*

Paul Allison and Joseph D. Sherman, AT-2
Los Alamos National Laboratory, Los Alamos, NM 87545

ABSTRACT

According to beam dynamics calculations it should be possible to accelerate a high-perveance beam in a radio-frequency quadrupole (RFQ) accelerator with low emittance growth and nearly 100% capture efficiency. A 100-mA, 100-keV H⁻ ion injector with a 5-Hz, 1-ms duty factor was built for use with this accelerator, but the beam emittance at 100 keV was found to be two to four times the value previously determined at 20 keV. This emittance growth was traced to the 20-keV beam transport, where an instability occurred in the background plasma created by beam ionization of the residual gas. The injector has been rebuilt with a shorter transport length, resulting in greatly reduced emittance growth.

ORIGINAL INJECTOR

The first version of the injector was intended to provide a 250-keV, 20-mA beam for studies with a drift-tube linac. Beam was extracted from a Penning source with a circular aperture[1], focused with a 90°, n = 0.5 dipole magnet and with a quadrupole doublet before reaching an emittance station in front of the column, 60 cm from the source (Fig. 1). With the successful test of an RFQ at Los Alamos[2] the injector was modified to produce a 100-mA, 100-keV beam. A 10- by 0.5-mm slit was used as the beam emitter, the magnet was changed to n = 0.9, and the column was shortened to provide a higher gradient. The Y-axis is along the 10-mm direction of the slit. Calculations still indicated favorable beam transport, assuming complete space-charge neutralization of the beam.

Emittances were measured with an electric-sweep scanner,[3] and the results quoted here are rms normalized values, as calculated in the computer analysis according to the formula[4]

$$\epsilon_x = \beta\gamma \sqrt{\overline{x^2}\,\overline{x'^2} - \overline{xx'}^2} \quad .$$

The first measurements showed that only about half of the extracted current was transmitted to the accelerating column entrance (Fig. 1). Experiments with various commonly available gases, ranging from hydrogen to xenon, showed that beam transmission could be dramatically increased by flooding the transport system with xenon, with lighter gases giving poorer results. Measured transmissions at

---

*Work supported by the US Dept. of Defense, Defense Advanced Research Projects Agency, and Ballistic Missile Defense Advanced Technology Center.

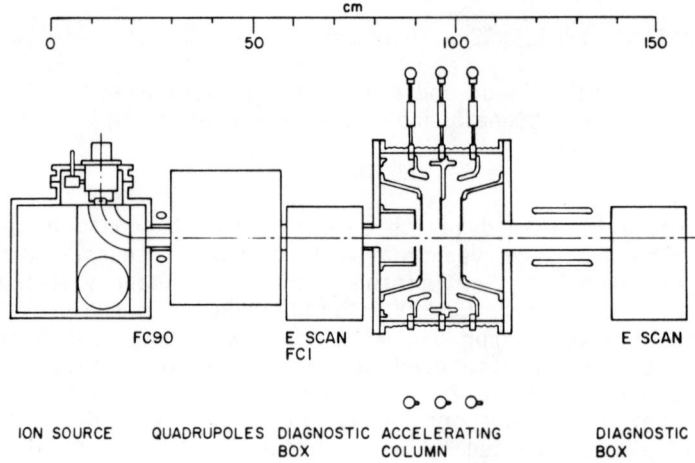

Fig. 1. Layout of original 250-keV, 20-mA injector, modified for use at 100-keV, 100-mA operation.

the 90° magnet exit (FC90) and at the column entrance (FC1) are shown in Fig. 2 as a function of neon and of xenon densities. Also shown are the calculated currents, assuming full transmission from extraction except for stripping losses. These results can be partially understood by comparing the stripping losses at the Gabovitch critical density[5] for the various gases (Table I). At lower densities a negative ion beam will be underneutralized, according to his theory. It is seen that stripping losses for light gases are substantially higher than for heavy gases. Improved beam transmission then presumably results from a balance between the beam's stripping losses and reduced effective space charge. Much more troublesome was the emittance growth. Whereas the emittance of the beam from the 10- by 0.5-mm slit was found to be ∼0.02 by 0.01 (π cm·mrad) 20 cm from the source, it was typically 0.07 by 0.03 at the emittance scanner 60 cm from the source. An interesting observation had been that the emittance scanner currents (proportional to the phase-space

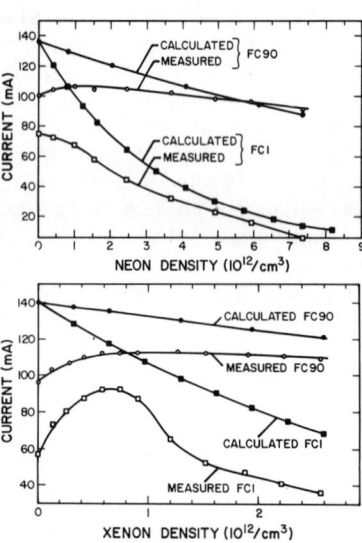

Fig. 2. Measured and calculated beam transmission from extraction to the 90° magnet exit (FC90) and the column entrance (FC1) versus densities of neon and xenon gases.

TABLE I

PROPERTIES OF GASES FOR H⁻ BEAM NEUTRALIZATION AT 20 keV

Gas/amu	n_{ao} (10^{12}/cm^3)	Losses $n_{ao}(\sigma_{-10}+\sigma_{-11})$ (%/cm)	σ_i (10^{-16} cm^2)
H$_2$/2	23.6	2.12	1.5
He/4	66.4	3.14	0.38
Ne/20	16	0.88	0.7
N$_2$/28	2.38	0.38	4
O$_2$/32	2.7	0.34	3.3
Ar/40	1.7	0.29	4.7
Kr/83.8	1.02	0.32	5.4
Xe/131.8	0.59	0.23	7.6

n_{ao} = gas density for exact neutralization (Gabovitch critical density)

= $2\sqrt{8kT_i/\pi M_i}/Rv_-\sigma_i$ where

T_i = ion temperature ~ 0.1 eV
M_i = mass of neutralizing gas
R = beam radius - 1 cm
σ_i = cross section for ionization of gas by H⁻ ion

densities), had much greater fluctuation than the beam current itself, and that the amplitude of fluctuations was progressively larger farther along the transport toward the RFQ. Dzhabbarov[6] has reported beam-plasma oscillations produced by an H⁻ beam with conditions similar to those in our injector, and he showed that oscillations in current and in beam potential were damped at high argon density in the transport line. In accordance with Dzhabbarov's results, we examined the effects of much higher density of xenon.

The 20-keV emittance scanner (Fig. 1) was positioned in the beam center, and the scanner current, proportional to $d^2i/dxdx'$ as a function of x', was recorded for 50 beams pulses for low and for high xenon density. The extracted current (Fig. 3a) did not change significantly and had ±5-10% fluctuation; however, the phase space rotated and shrank at high density (Fig. 3b). The most interesting change was a substantial reduction in the pulse-to-pulse fluctuations in the phase-space density, (Fig. 3c). We found that the emittance decreased markedly in the two planes as a function of xenon density (Fig. 4). Excessive stripping losses with high xenon density made this method of emittance control unacceptable. It may also be that the beam plasma cannot adjust fast enough to track the variations in current. We calculated that if the effective current varied by ±3 mA about zero, then the resulting phase-space orientations, averaged in time, would account for the observed emittance growth. The neutralization time constant varied from about

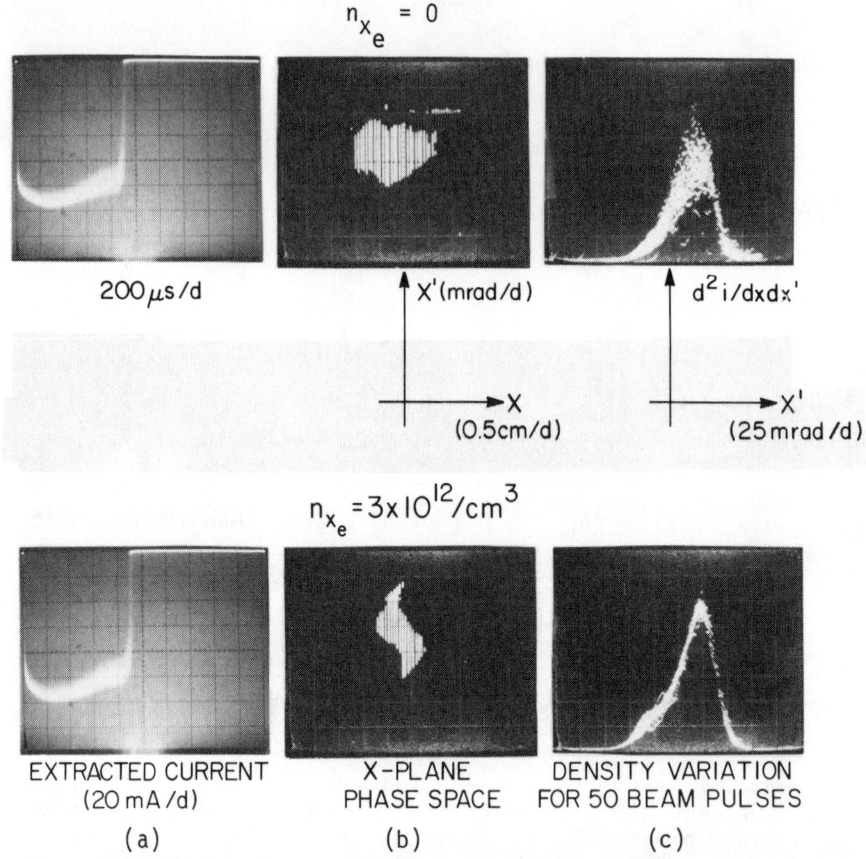

$n_{x_e} = 0$

$200 \mu s/d$

X'(mrad/d)

$d^2i/dxdx'$

X
(0.5cm/d)

X'
(25 mrad/d)

$n_{x_e} = 3 \times 10^{12}/cm^3$

EXTRACTED CURRENT
(20 mA/d)
(a)

X-PLANE
PHASE SPACE
(b)

DENSITY VARIATION
FOR 50 BEAM PULSES
(c)

Fig. 3. (a) Waveform of extracted current, (b) phase-space contour plot in X-plane, (c) phase-space density variation for 50 consecutive beam pulse in the X-plane.

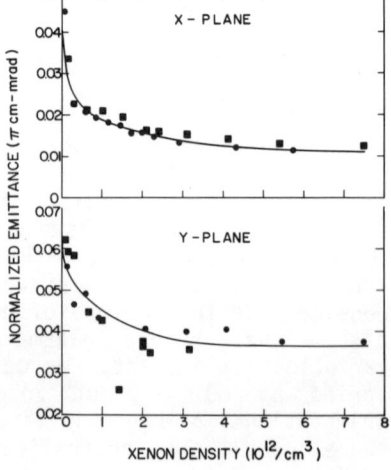

Fig. 4. Emittance variation at column entrance vs xenon density. Circles and squares are for slightly different source operation.

2 µs at $4 \times 10^{12}/cm^3$ with xenon to about 30 µs with only residual hydrogen--about the right range for this effect because the observed beam fluctuations are mostly below 1 MHz. Considerations[7] of the likely beam-plasma densities, however, suggest that there will be adequate electrons present to accommodate the oscillations. The cause of the emittance growth is therefore unclear from our measurements; however, a solution to the problem seemed clear: reduce the length of the beam-transport line.

REVISED INJECTOR

The primary objective of rebuilding the injector was to shorten the 20-keV transport as much as possible. With a shortened transport line, a high density of xenon does not lead to excessive beam stripping. We believed that direct extraction at 100 keV was not desirable for our slit emitter source, so a new 20-keV source was built, using samarium-cobalt permanent magnets. In the configuration chosen (Fig. 5), the physical length of the magnet structure extends only 35 mm from the emission slit, and the bend angle at 22-keV energy is 8.1°. This allows refocusing or further accelerating with a minimum of drift, and the second-order aberrations of the magnet are negligible. The effect of energy dispersion is greatly reduced, and it is possible to measure beam emittance much closer to extraction than before.

The source performance was investigated on a separate test stand at 20-keV extraction energy 92 mm from the emission slit, about twice the length of the new injector 22-keV transport, so instabilities in the neutralizing gas presumably would be more severe than on the injector. Within the ±25% scatter of our measurements, there was no effect on emittance of xenon density from $0-6 \times 10^{12}/cm^3$. There was a definite effect on the phase-space orientation, however, indicating that the beam is generally underneutralized with only residual gas. The emittance at 150 mA is 0.027π cm·mrad in the Y-plane by less than 0.004 in the X-plane. Over 180-mA current has been extracted, and the noise in a 1-MHz bandwidth has been as low as ±2% (Fig. 6a). The phase-space densities in the two planes near the beam center line are also shown

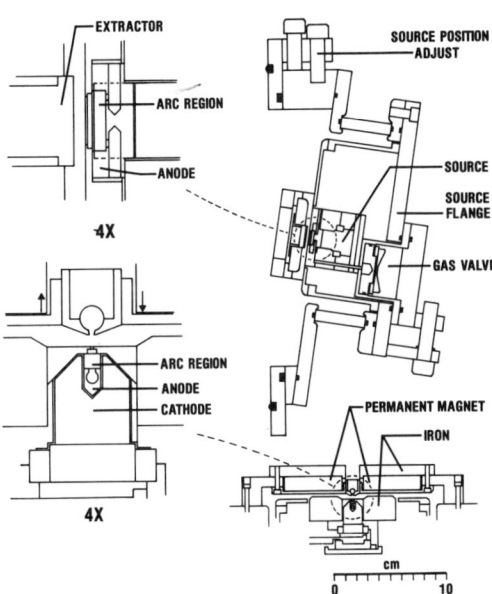

Fig. 5. Schematic of small angle source (SAS).

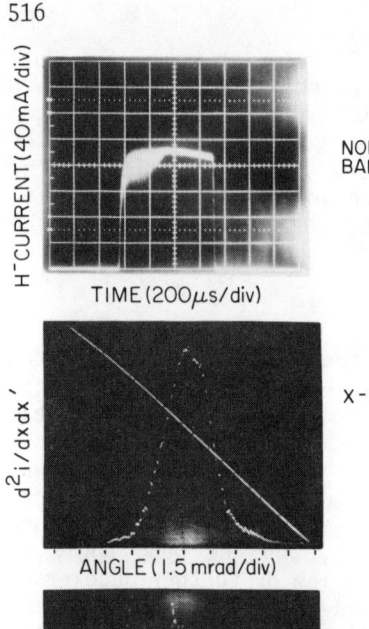

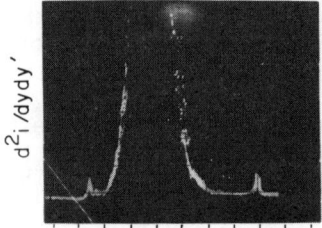

Fig. 6. (a) Waveform of 180-mA beam pulse from the SAS, (b) angular spread in the X-plane, (c) angular spread in the Y-plane.

(Fig. 6b and 6c). In the narrow and diverging plane, $\Delta x'_{rms} = 1.2$ mrad is typical. In the large plane, the beam is nearly parallel (with a size about equal to the slit width) so that the angular spread at the scanner is close to that at the plasma. The value $\Delta y'_{rms} = 8.3$ mrad then can be used to calculate for the H⁻ ions that $kT_{eff} = 3$ eV from $kT = 2\phi(\Delta y'_{rms})^2$, where ϕ is the beam voltage. The small peaks in the y' distribution (Fig. 6c) are from ions coming from the ends of the 10-mm slit; thus, the width of the main peak is seen to be unaffected by end effects.

The original accelerating column was replaced with a shorter system. Favorable results from optical calculations and the desire to minimize emittance growth in the 20-keV transport in the column led us to couple the source as closely as possible to the column. An electron suppressor electrode at the entrance is followed by the accelerating gap (Fig. 7). Beam current, as measured at the Faraday cup 46 cm from the source, is about as expected. At the design arc current of 200 A, we measured over 140 mA of current at 100-keV energy.

At an arc current of 160 A, a substantial set of 100-keV emittance measurements were made. Emittance was found to increase with electron suppressor voltage in approximate agreement with emittance-growth calculations made with the SNOW[8] code. Current transmitted

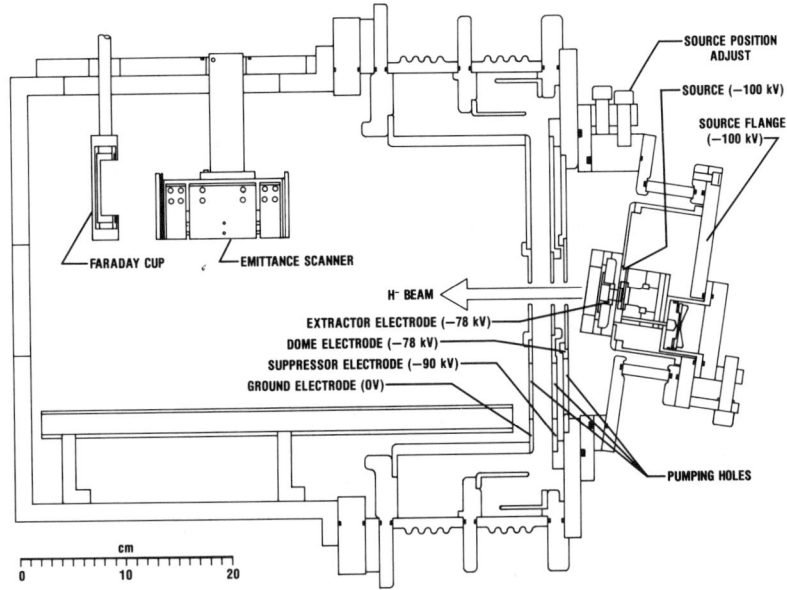

Fig. 7. Layout of the revised 100-keV injector.

to the Faraday cup increased by 20-100% when xenon was used in the beamline at a density of $\sim 10^{12}/cm^3$; however, emittance in both planes grew by 20-90%, probably because of the change in the first-order transport optics, as observed by substantial rotations of the X- and Y-plane emittances. This change has evidently led to increased third-order aberrations in the column. The Y-plane emittance was about equal to the 20-keV measurement, but the X-plane value was substantially larger, never being less than 0.019 π cm·mrad. With zero xenon density, we measured emittances of 0.021 by 0.023 (π cm·mrad) in the Y- and X-planes having a measured current of slightly over 100 mA.

CONCLUSION

A Penning source with a small bend angle has been used to produce a 100-keV beam of high brightness. The nature of beam-transport instabilities leading to deneutralization and emittance growth is not fully understood and is under further investigation.

REFERENCES

1. Joseph D. Sherman, Paul Allison, H. Vernon Smith, Jr., "H⁻ Beam Formation From a Penning Surface Plasma Source Using Circular Emission-Extractor Electrodes," Proc. of the Second International Symposium on the Production and Neutralization of Negative Hydrogen Ions and Beams, BNL-51304, p. 184 (October 1980).
2. R. H. Stokes, T. P. Wangler, and K. R. Crandall, "The Radio-Frequency Quadrupole - A New Linear Accelerator," IEEE Transactions on Nuclear Science, Vol. NS-28, No. 3, p. 1999 (June 1981).
3. Paul W. Allison, Joseph D. Sherman, and David B. Holtkamp, "An Emittance Scanner for Intense Low-Energy Ion Beams," IEEE Transactions on Nuclear Science, Vol. NS-30, No. 4, p-2204 (August 1983).
4. Frank J. Sacherer, "RMS Envelope Equations with Space Charge," IEEE Transactions on Nuclear Science, Vol. NS-18, No. 3, p. 1105 (June 1971). Note that some authors define ε_{rms} as four times this value.
5. M. D. Gabovich, L. S. Simonenko, and I. A. Soloshenko, "Space Charge Neutralization of an Intense Negative-Ion Beam," Sov. Phys. Tech. Phys. 23(7), p. 783 (July 1978).
6. D. G. Dzhabbarov and A. Naida, "Spatial Development of the Instability of a Dense Beam of Negative Ions in a Rarefied Gas," Institute of Physics, Zh. Eksp. Teor. Fiz. 78, 2259-2265 (June 1980).
7. Michael E. Jones, Huan Lee, Don S. Lemons, Los Alamos National Laboratories report, to be published.
8. Jack E. Boers, "SNOW - A Digital Computer Program for the Simulation of Ion Beam Devices," Sandia National Laboratory report SAND79-1027.

DISCUSSION

HERSHCOVITCH: A while ago you sent me some papers by Gabovitch where he reported that by examining various gases the best results were obtained with krypton. Have you tried it or do you have any plans to try it?

ALLISON: We used a variety of gases including krypton. Basically the light gases were bad and the heavy ones were good, but I'm not sure we could tell the difference between krypton and xenon. Krypton has higher stripping loss at its critical pressure and this loss is an important consideration.

WHEALTON: What is the effect on the emittance of varying the source magnetic field?

ALLISON: We can't vary it, it is a permanent magnet source.

WHEALTON: It used to be a variable one, right?

ALLISON: Previously, with the 90° bending magnet it was not easy to vary the magnetic field because it was coupled to the extracting voltage, but it would be interesting if the small angle source had a variable magnetic field to make that measurement. In the source that was described by Smith in the poster session there will be a variable magnet.

SLUYTERS: What is the acceptance of the RFQ?

ALLISON: The acceptance is about $\varepsilon_{rms} = 0.02 \ \pi$ cm-mrad.

EHLERS: Because you have a magnetic field in there, do you have to essentially run that first gap at a fixed voltage to have the right amount of bend?

ALLISON: Yes.

EHLERS: This means you have to juggle the plasma density in order to get the proper meniscus shape, is that true?

ALLISON: Yes. We take a very large divergence because of that.

EHLERS: Do you see effects when you change the plasma density?

ALLISON: We've measured the emittance at 1 mA and at 100 mA and it is about the same, but I can't say we've explored everything in between. By the time the beam has gone through the accelerating column, the emittance in the plane perpendicular to the slit has increased to approximately the value parallel to the slit. This emittance growth masks the effects of meniscus shape. One could probably prevent this growth with some work but the planes are going to be coupled in the RFQ and quadrupoles anyway so it's really not very important.

ACCELERATED BEAM FROM CUSP H⁻ ION SOURCE

A. Takagi, Y. Mori, Z. Igarashi, K. Ikegami, C. Kubota
and S. Fukumoto
National Laboratory for High Energy Physics

ABSTRACT

A multi-cusp H⁻ ion source was built for the charge-exchange multiturn injection project in the KEK 500 MeV booster proton synchrotron. More than 20 mA H⁻ ion beam was extracted from the ion source, 18 mA was injected in the linac and 8.5 mA was accelerated up to 20 MeV. Beam was successfully injected in the booster and a new intensity record of 500 MeV proton beam was obtained.

INTRODUCTION

Requirement to get higher beam intensity for the KEK 12 GeV synchrotron has been strongly emphasized by the users in these recent years. As a beam intensity increases, radiation and residual activity problems become serious because of the growing beam loss at injection or during acceleration in the machine. In order to increase beam intensity without these prblems, charge-exchange multi-turn injection scheme has been currently used in the various laboratories and at KEK it was also proposed three years ago[1]. It is favorable for the dual mode operation of ordinary and polarized beams in the synchrotron complex.

In this injection scheme, H⁻ ions are injected in the same phase space area of the circulating protons in the synchrotron so that accumulated protons can be increased compared with an ordinary multi-turn injection scheme.

A multi-cusp H⁻ ion source was constructed for this purpose and it is the same type that LAMPF has developed for the storage ring[2].

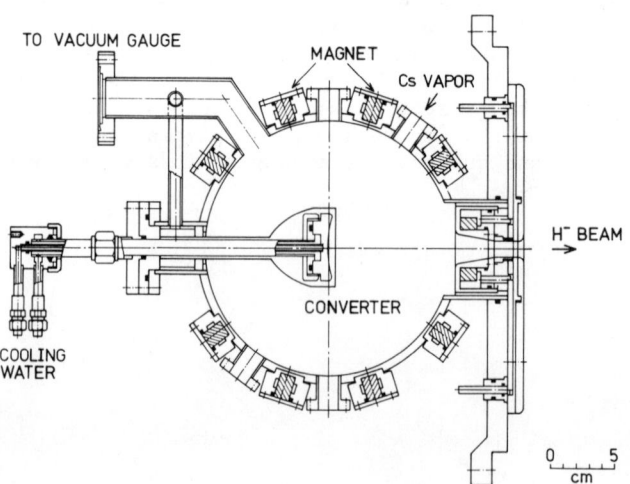

Fig. 1   Schematic view of multi-cusp H⁻ ion source.

The ion source was directly installed in the accelerating column of 750 kV preaccelerator just like a duoplasmatron. In the high voltage dome, it is not necessary to use any analysing or focusing magnet because the components other than the H^- ions included in the beam, electrons are main, are the same order of H^- ion current and beam loading due to these components is not so serious for the Cockcroft-Walton preaccelerator.

A first beam acceleration test was performed in last September and the beam was successfully injected in the booster synchrotron.

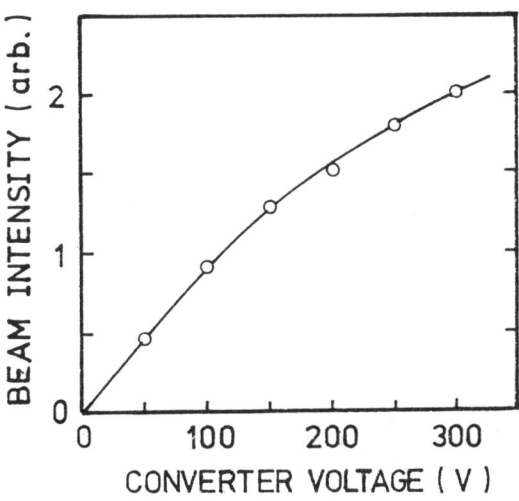

Fig. 2 Variations of H^- ion current as a function of converter voltage.

H^- ION SOURCE AND BEAM ACCELERATION

Schematic view of the multi-cusp H^- ion source is shown in Fig. 1. The ion source consists of a cylindrical plasma chamber, a molybdenum converter and a couple of tungsten filament. The plasma chamber is surrounded by Alnico permanent magnets which form a multi-cusp magnetic field to confine a plasma efficiently. The converter is

Fig. 3 Photograph of ion source mounted in 750 kV accelerating column.

cooled by water and a negative voltage of - 300 ∿ 450 V is applied to it. Cesium is supplied through a small (6 mmφ) heated stainless steel pipe from a reservoir which is placed at the outside of the chamber. The operating temperature of the reservior is about 160°C.

Typical arc conditions are as follows: arc current is 60 ∿ 80 A and arc voltage is - 100 ∿ - 110 V. The pulse duration of the arc is about 100 μsec and the repetition rate is 20 Hz at the maximum. Operating hydrogen gas pressure is 3 ∿ 7 × 10^{-4} Torr and the gas consumption rate is almost one-tenth of that of the duoplasmatron.

The extracted H^- ion current depends largely on the converter voltage. Fig. 2 shows the variations of the H^- ion current as a function of the converter voltage. The beam current increases gradually by increasing the converter voltage. It seems that this ion source needs at least 5 or 6 hours conditioning by hydrogen mode discharge before introducing the cesium in the source. We observed small decrease of O^- ions from the source after the conditioning above mentioned. By this procedure, the various contamina-

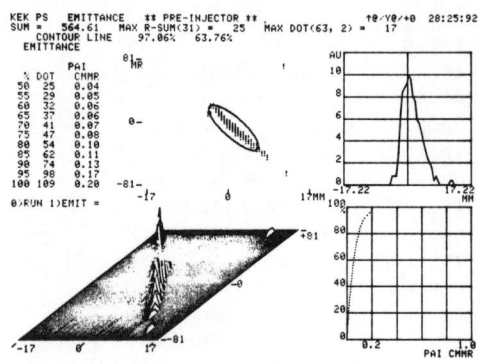

Fig. 4   Beam emittance of H^- ion beam.

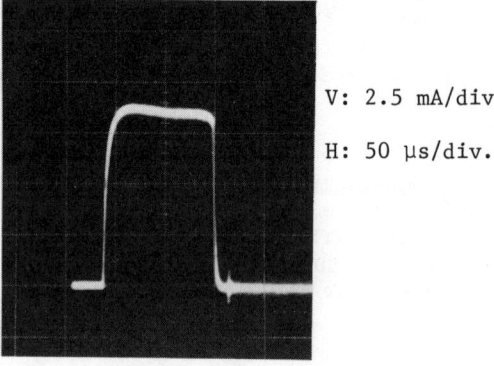

V: 2.5 mA/div.

H: 50 μs/div.

Fig. 5   Beam shape of 20 MeV H^- ion current.

tions in the source, especially attached on the converter surface might be effectively removed. As for the cathode filament, it was noticed that its life time was short and the beams were not stable for 50 Hz ac heating. Although no systematic test was carried out, the lives of the 1 mm diameter filament were 30 hours or less when heated by the ac current of 55 A whereas the 1.2 mm diameter one lasted more than 100 hours and was not broken during the test run when heated by a dc current of 78 A.

The ion source was directly mounted in the 750 kV accelerating column as shown in Fig. 3. The H^- ion beam was accelerated up to 750 keV by a three electrodes system, whose total gap length is 76 cm, focused by quadrupole magnets and injected to the 20 MeV linac. It was hard to measure the H^- beam current only just after the acceleration of 750 keV because of the electrons in the beam. However, after passing through four quadrupole magnets, the measured beam current was about 20 mA at the maximum, so it was conceived that more than 20 mA H$^-$ ion beam was extracted from the ion source. The beam emittances were measured at the entrance of the linac. A typical emittance configuration, when the beam current was 11 mA, is shown in Fig. 4. The normalized (= phase space area × $\beta\gamma/\pi$) 95 % emittance was 0.17 cm·mrad. After the acceleration by the linac, we have obtained 8.5 mA H$^-$ ion current at the maximum so far. Fig. 5 shows the beam shape of 20 MeV H$^-$ ions measured by a current transformer. As can be seen from this figure, the beam is surprisingly stable and noiseless. By changing the duty foctor, the beam current changed slightly, however, the operational condition did not change drastically. The emittance of 20 MeV beam was almost same as that of 750 keV beam.

The beam was successfully injected in the booster synchrotron and a new intensity of 500 MeV protons, 7.13×10^{11} ppp, was recorded.

CONCLUSION

A multi-cusp H$^-$ ion source was built which produces more than 20 mA H$^-$ current with a normalized emittance of 0.17 cm·mrad and the beam was accelerated up to 20 MeV by the linac. A first beam test was successfully finished, however, further development work on the ion source is still needed. Especially, the life time of the tungsten filament must be much longer for the long period routine operation.

We appreciate very much Dr. R.L. York at LAMPF of his useful advices on the source designing and operations and members of Accelerator Division, directed by Prof. T. Kamei, for their operation and tuning of the synchrotron.

REFERENCES

(1) T. Kawakubo et al.: Proc. 3rd Symposium on Accelerator Science and Technology, Osaka, 31, 1980.
(2) R.L. York and R. Stenvens, JR.: IEEE, NS-30, (1982), 2705.

2D ACCELERATOR DESIGN FOR SITEX NEGATIVE ION SOURCE*

J. H. Whealton, R. J. Raridon, R. W. McGaffey,
D. H. McCollough, W. L. Stirling, and W. K. Dagenhart
Oak Ridge National Laboratory
Oak Ridge, Tennessee 37830

ABSTRACT

Solving the Poisson-Vlasov equations

$$\nabla^2 \phi = \int f d\mathbf{v} - e^{-\phi} \quad (1)$$

$$(\mathbf{v} \times \mathbf{B} + \nabla \phi) \cdot \nabla_v f + \mathbf{v} \cdot \nabla f = 0 \quad (2)$$

where the magnetic field, $\mathbf{B}$, is assumed constant, we optimize the optical system of a SITEX negative ion source in ∞ slot geometry. Algorithms designed to solve the above equations were modified to include the curved emitter boundary data appropriate to a negative ion source. Other configurations relevant to negative ion sources are examined.

BODY

First we will consider the SITEX negative ion source (Ref. 2) which is shown in Fig. 1. On the left is a curved converter surface where ions are ejected out via a plasma sheath due to the adjacent source plasma at an energy of approximately 150 volts. There is a magnetic field perpendicular to the paper of approximately 1300 gauss. Upon entering the accelerator, the negative ions leave the positive source plasma ions behind and accelerate to the desired potential. The optics is determined by the ion initial ejection velocity, the shape of the plasma extraction double sheath, and the applied electrostatic fields as manipulated by the geometry of the electrode shape. The plasma model above assumes that the excess source plasma positive ion density is representable by a Boltzmann distribution; otherwise it is assumed to exactly cancel out the electron density. This matter should really be modeled by solving three Vlasov equations coupling into a Poisson equation instead of one Vlasov equation and a Boltzmann equation coupling into a Poisson equation. The modeling of the source plasma is important: for example, if the Boltzmann term in Eq. (2) were not present then we would get a result like Fig. 2 instead of Fig. 1 which is a solution to Eq. (1) and (2) as they stand for an optimum geometry at optimum perveance. The RMS beam divergence is 0.28° due entirely to electrostatic aberrations. The variation of divergence with beam current or perveance is shown in Fig. 3; a situation of lower than optimum perveance is shown in Fig. 4 and a case of higher than optimum perveance is shown in Fig. 5. The phase space emittance as a function of acceleration potential is shown in Fig. 6. The accuracy of construction of this negative ion accelerator is hinted at by the sharp variations shown in Fig. 3 and is elucidated in Fig. 7. Fig. 7 shows how a variation in the focal point in either the beam direction $f(x)$, or the transverse direction $f(y)$, effects the RMS beam divergence.

Another negative ion source (Ref. 3) incorporates a long region (thousands of Debye lengths long) when the negative ions are transported through a plasma before being accelerated. The plasma model is very important in modeling this configuration as illustrated in Fig. 8 for five different plasma temperatures. We are proposing to study this device using a triple Vlasov model.

Several accelerators (Ref. 4) using a Soviet source were examined.

A study is underway to determine if there are aberrations in the LBL-TFF accelerator (Ref. 5) shown in Fig. 9 for some nonoptimized cases. Shown in Fig. 10 is a preliminary finding of one accelerator gap from which the aberrations can be deduced.

REFERENCES

1. J. H. Whealton, R. W. McGaffey, and E. F. Jaeger, Appl. Phys. Lett. 36, 91 (1980); J. H. Whealton, Jour. Comput. Phys. 40 491 (1981); J. H. Whealton, Nuc. Instrum. Meth. 189, 55 (1981).
2. W. L. Stirling, proceedings this conference.
3. Lietzke/Ehlers/Leung, proceedings this conference.
4. Smith/Allison/Sherman, proceedings this conference.
5. O. Anderson, proceedings this conference.

---

*Research sponsored by Union Carbide Corporation under contract No. W-7405-eng-26 for the U.S. Department of Energy, DOE.

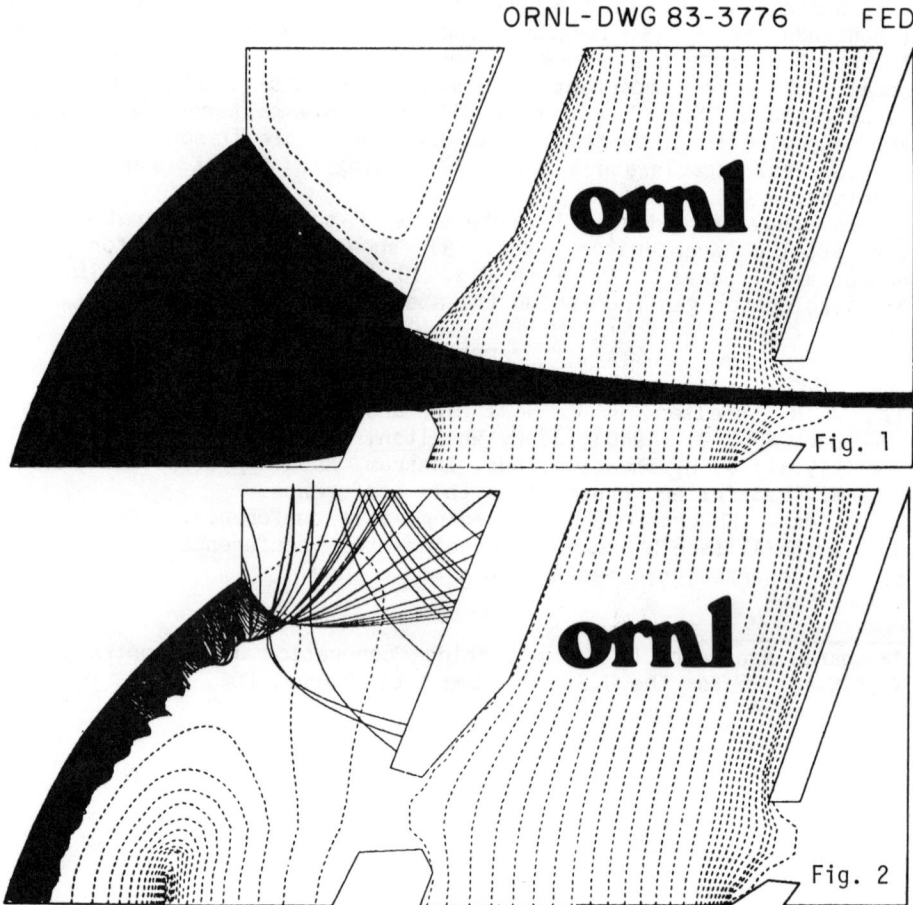

Figure 3

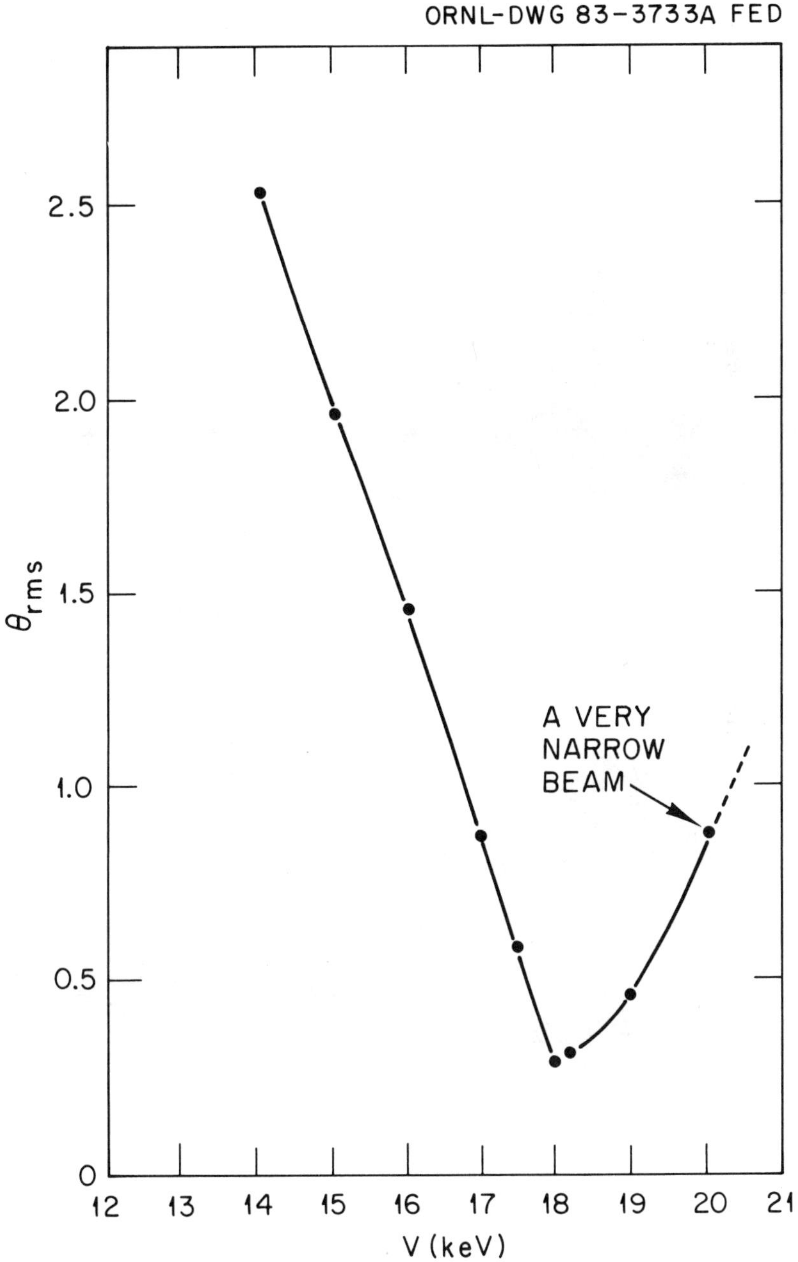

ORNL-DWG 83-3733A FED

A VERY NARROW BEAM

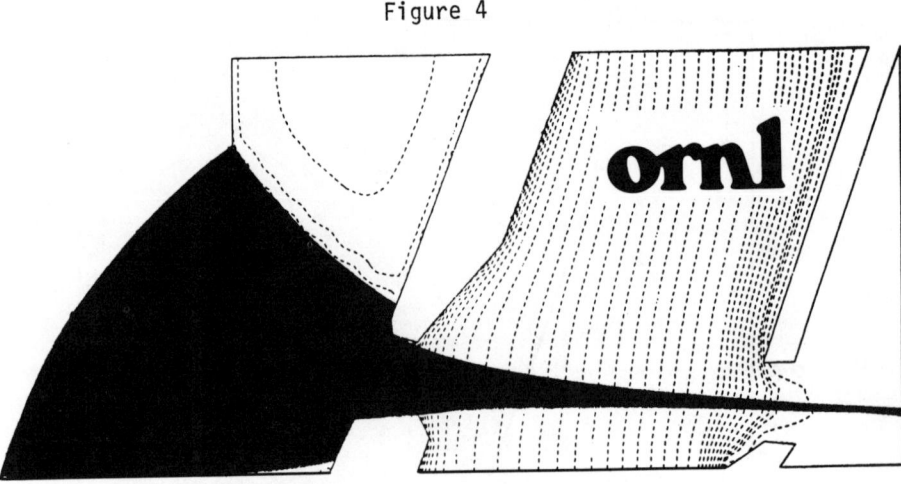

Figure 4

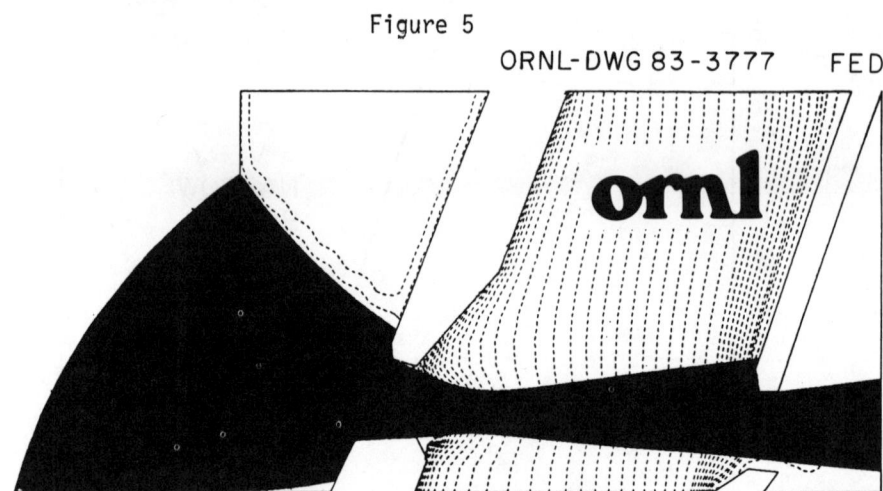

Figure 5

ORNL-DWG 83-3777   FED

529

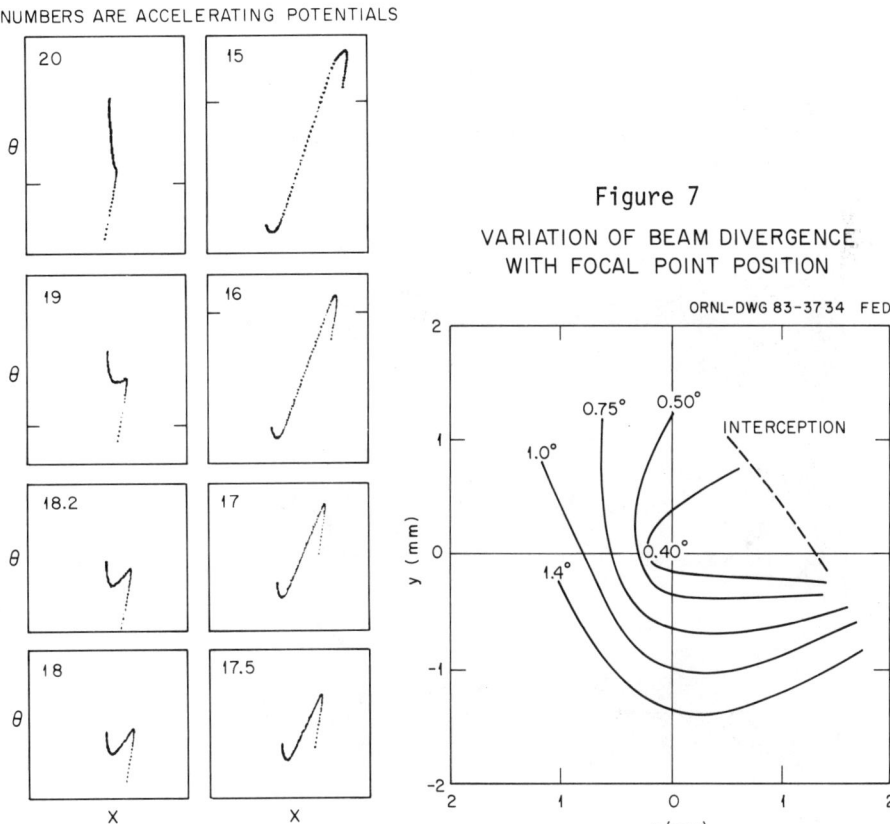

Figure 6
ORNL-DWG 83-3778 FED
NUMBERS ARE ACCELERATING POTENTIALS

Figure 7
VARIATION OF BEAM DIVERGENCE
WITH FOCAL POINT POSITION

Figure 8

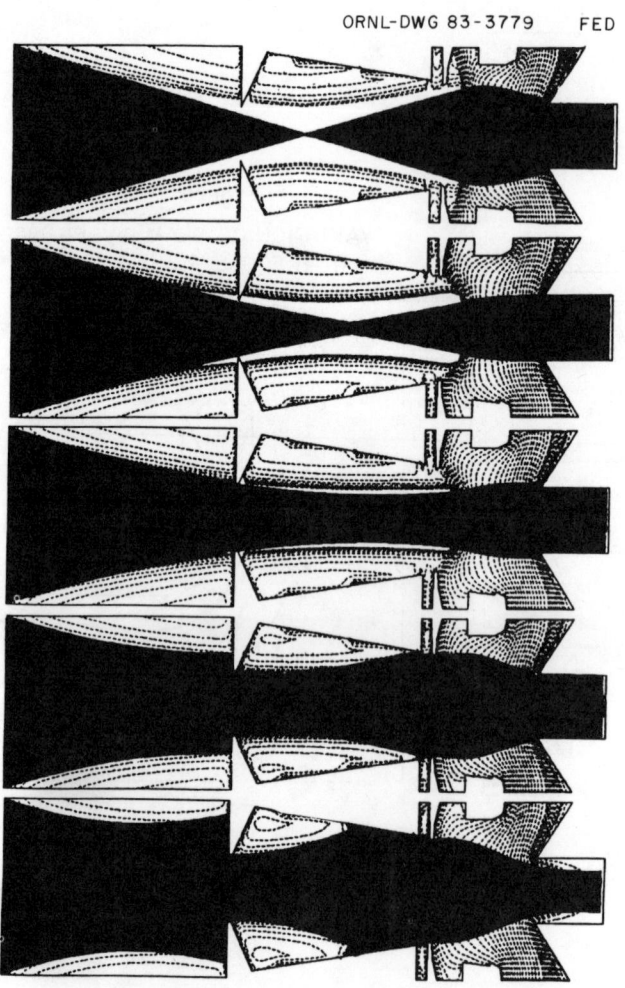

Figure 9

LBL's TFF ACCELERATOR

ORNL-DWG 83-3781   FED

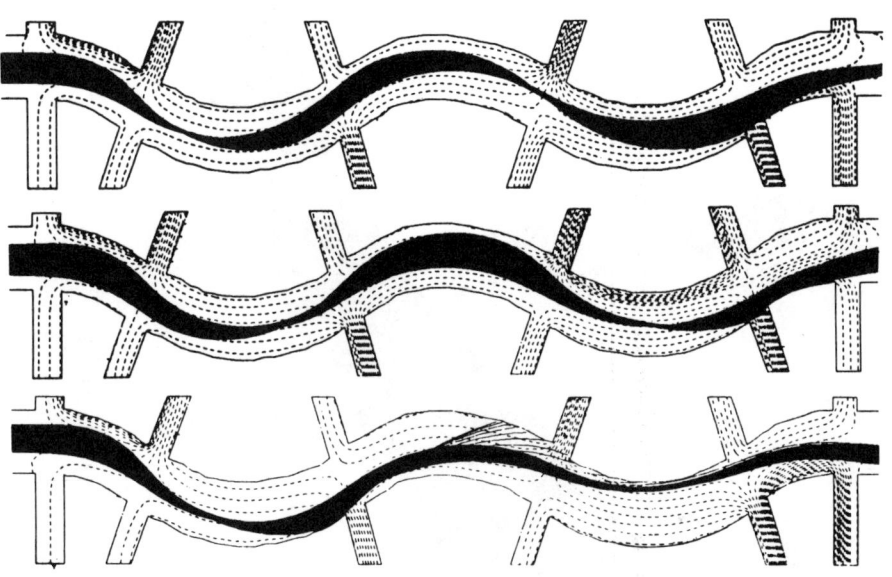

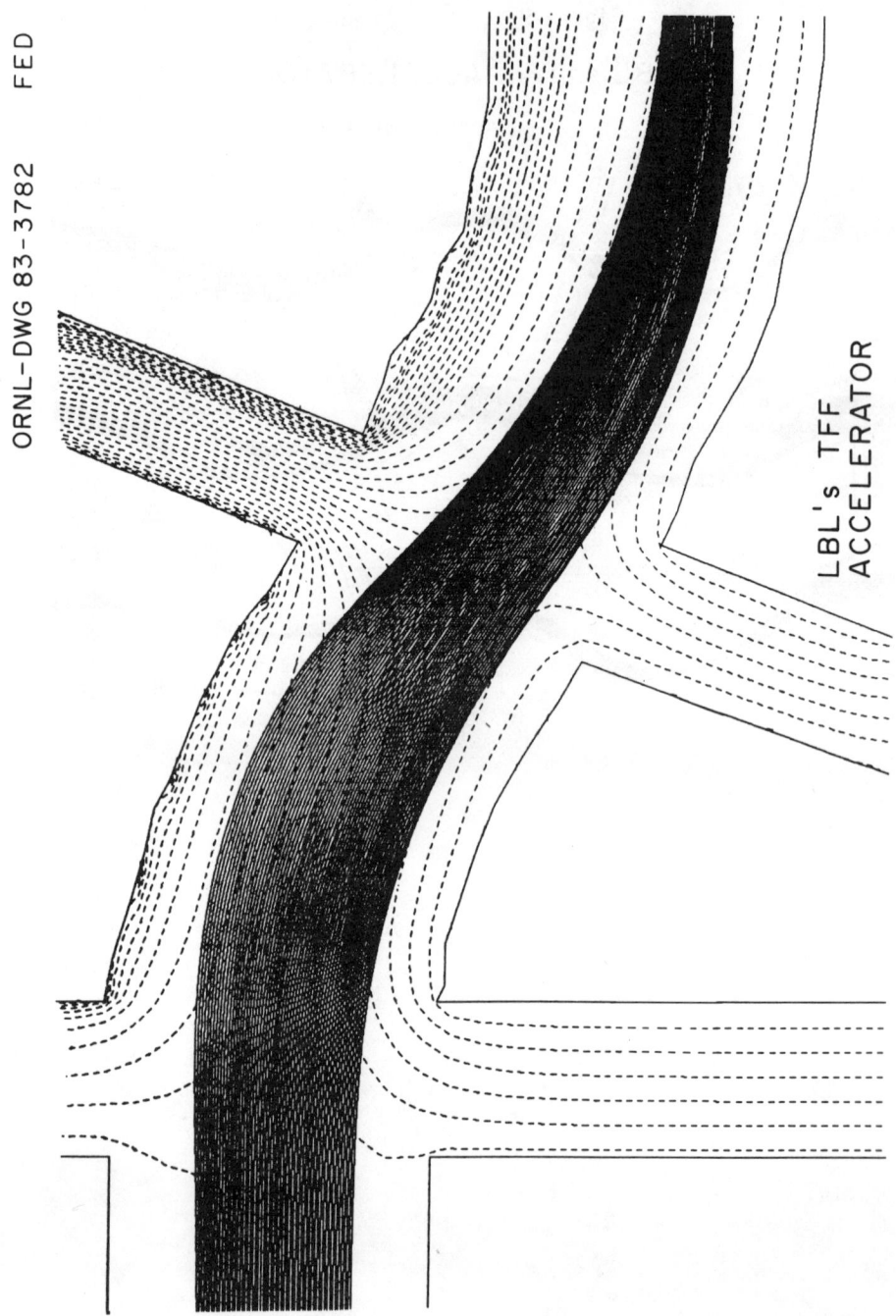

Figure 10

END EFFECTS IN SLOT EXTRACTION FOR A SITEX NEGATIVE ION SOURCE*

R. W. McGaffey, P. S. Meszaros, J. H. Whealton,
R. J. Raridon, and D. H. McCollough
Oak Ridge National Laboratory
Oak Ridge, Tennessee 37830

ABSTRACT

A 3D solution to the Poisson Vlasov equation of the form

$$\nabla^2 \phi = \int f d\mathbf{v} - e^{-\phi} \quad (1)$$

$$(\mathbf{v} \times \mathbf{B} + \nabla \phi) \cdot \nabla_v f + \mathbf{v} \cdot \nabla f = 0 \quad (2)$$

is obtained in the extraction region using the technique of Ref. 1. A constant magnetic field $\mathbf{B}$ is imposed. Algorithms designed to solve the above equations were modified to include the curved emitter boundary data appropriate to a negative ion source. The end of the slot was optimized to minimize aberrations.

BODY

Using a 2D electrode shape as shown in Fig. 1, the end was optimized by examining the configurations shown in Fig. 2. The optimized design is shown in Fig. 3C. Less optimized designs are shown in Figs. 3A-B as indicated by the corresponding symbols in Fig. 2. The parallel and transverse divergence is shown in Fig. 4 for the configuration A and C of Fig. 3. The dependence of RMS beam divergence (totally due to electrostatic aberrations) as a function of current density is shown in Fig. 5 for the configuration A of Fig. 3. In Fig. 6 is shown a 3D plot of the solution to the Poisson-Vlasov equation for the configuration C of Fig. 3. In Figs. 7-9 are shown different viewpoints of Fig. 6 which are easier to interpret. In Fig. 7 the ions actually are injected normally to the converter; but since the scale of the Z and Y axis are different, they appear to be oblique. An emittance in V_Z, V_X phase space is shown in Fig. 10.

REFERENCES

1. J. H. Whealton, R. W. McGaffey, and P. S. Meszaros, IEEE Conference on Plasma Science, San Diego, California, 1983; J. H. Whealton, Nuc. Instrum. Meth. 189, 55-70 (1981).

---

*Research sponsored by Union Carbide Corporation under contract No. W-7405-eng-26 for the U.S. Department of Energy, OFE.

Figure 1

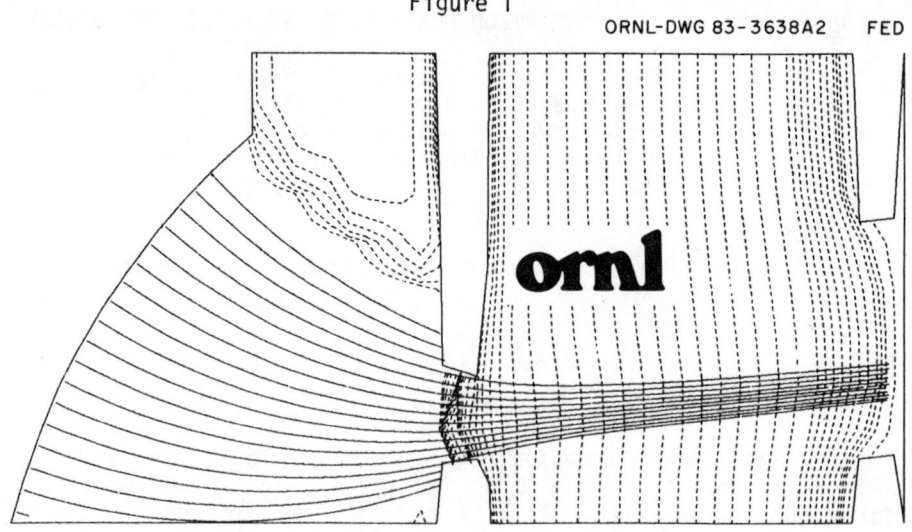

ORNL-DWG 83-3638A2   FED

Figure 2

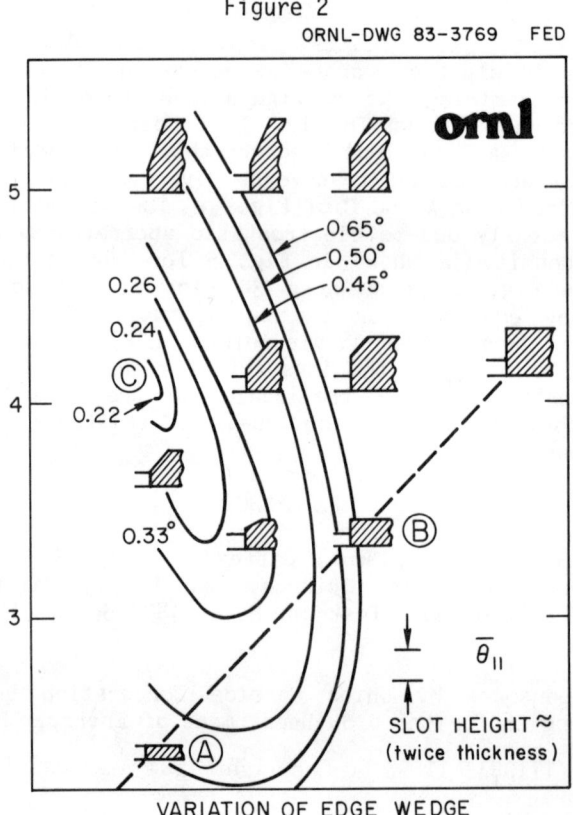

ORNL-DWG 83-3769   FED

VARIATION OF EDGE WEDGE

Figure 3
ORNL-DWG 83-3770 FED

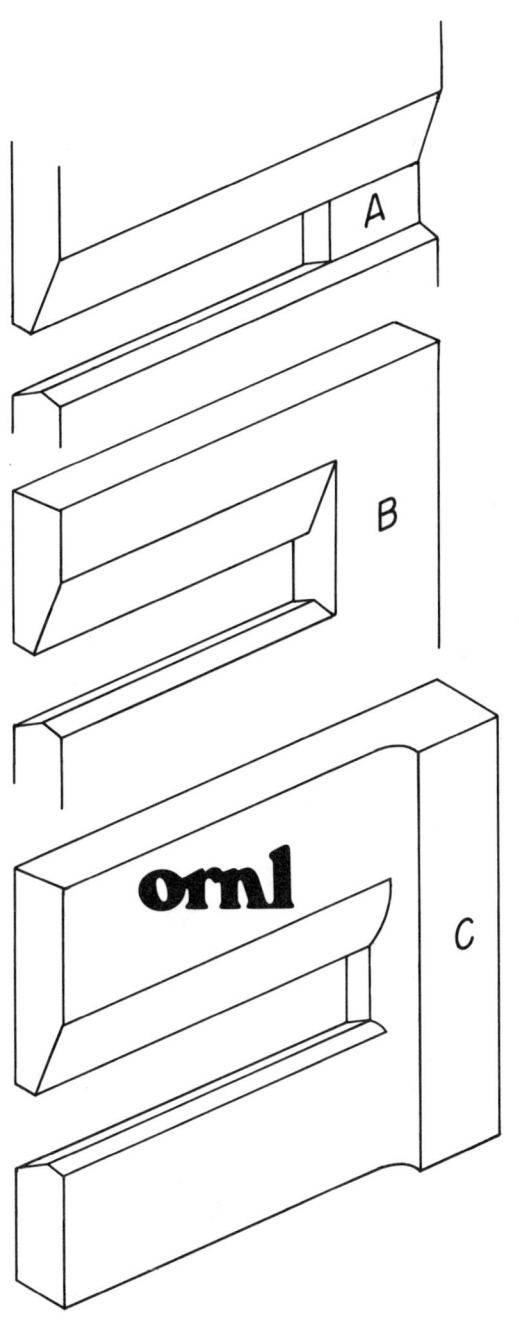

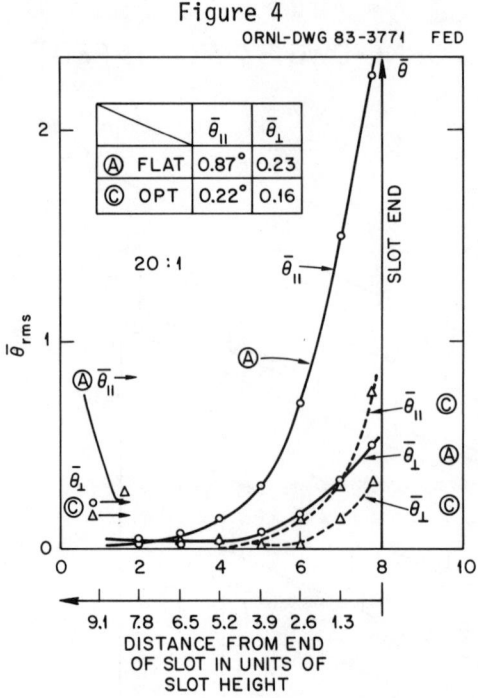

Figure 4

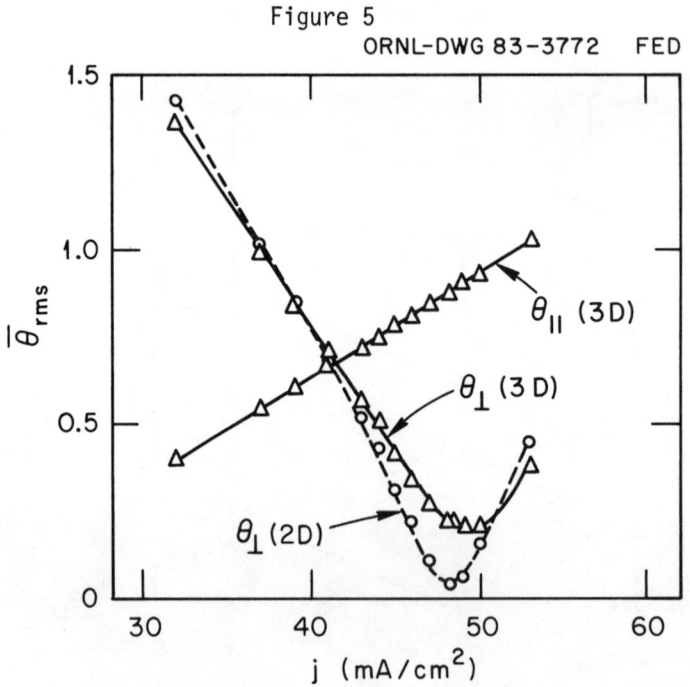

Figure 5

Figure 6

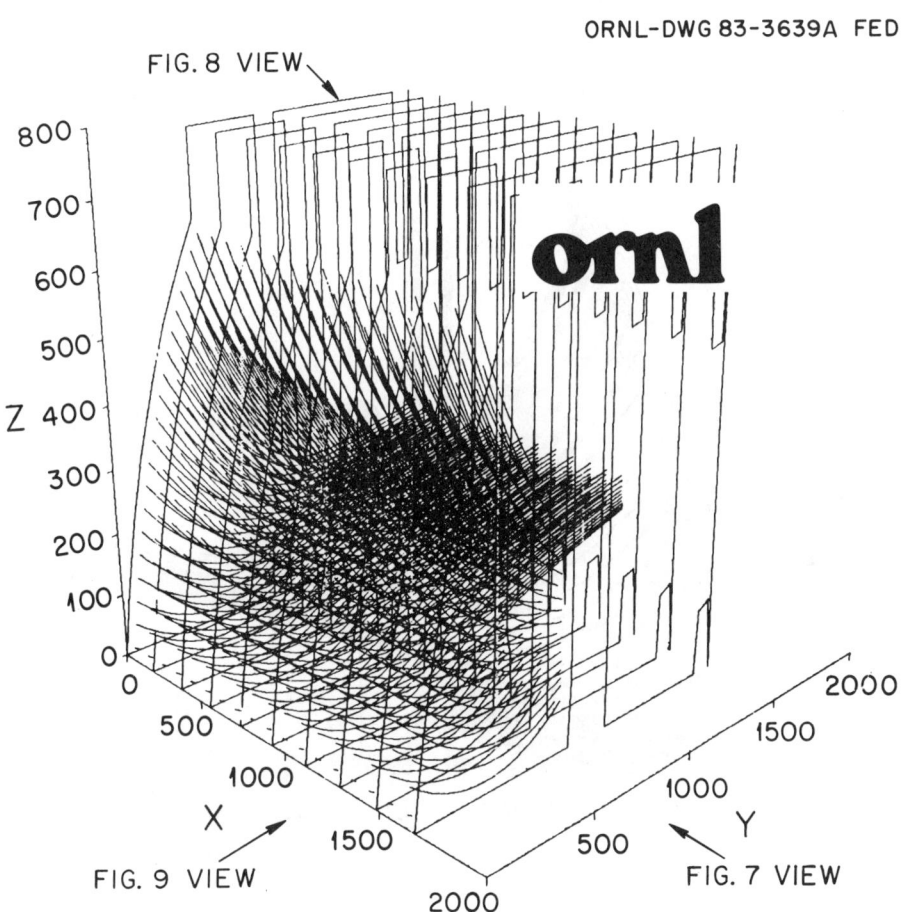

Figure 7

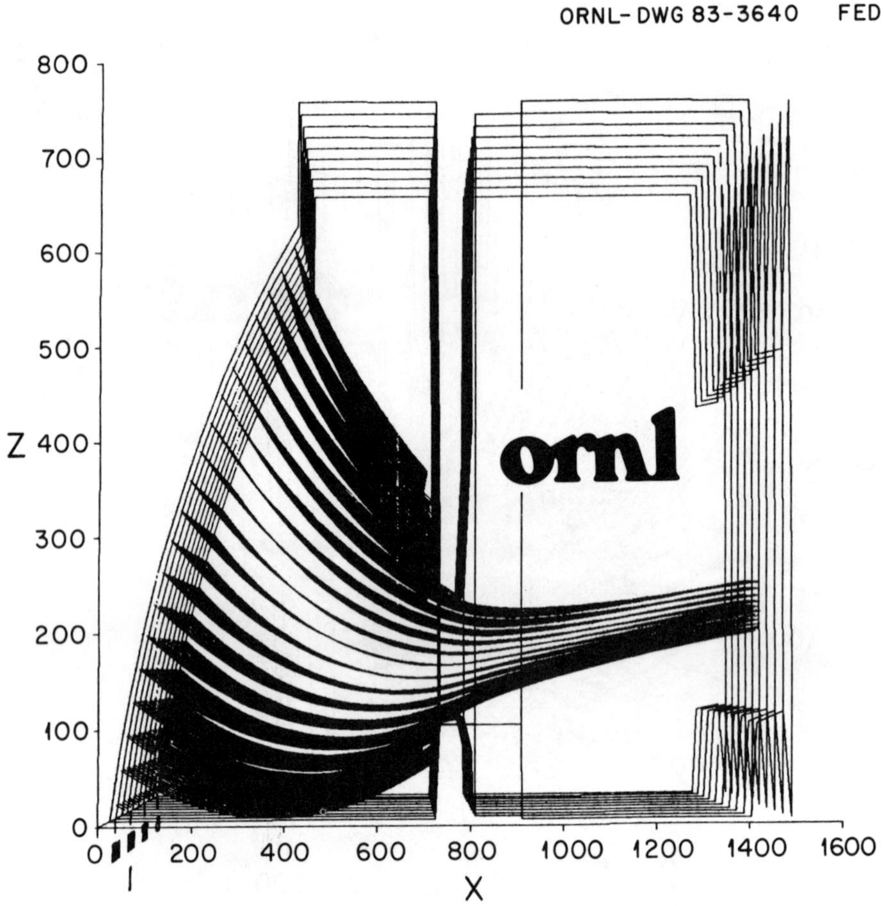

Figure 8

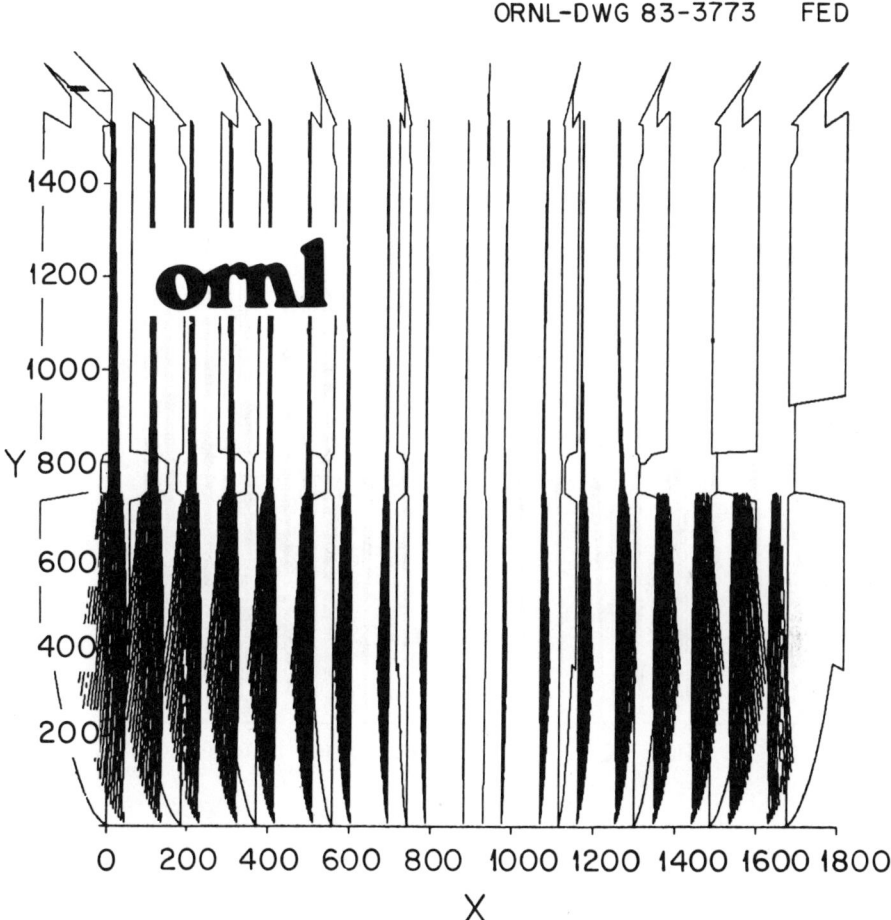

Figure 9

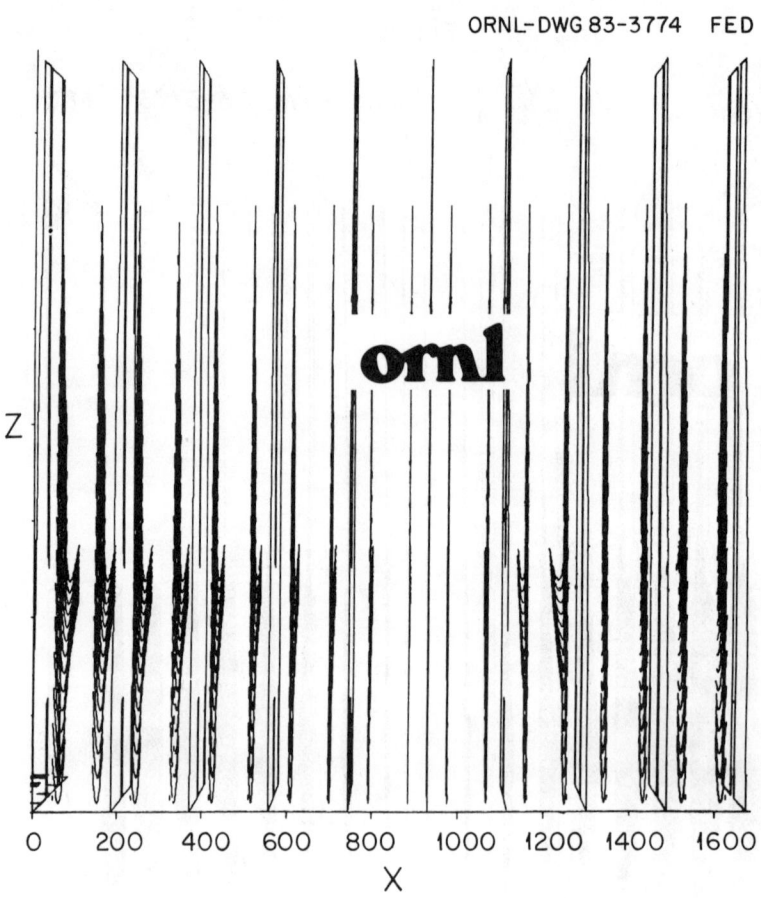

ORNL-DWG 83-3774 FED
ornl
Z
0
200
400
600
800
1000
1200
1400
1600
X

Figure 10

ORNL-DWG 83-3775 FED
θ_z
z

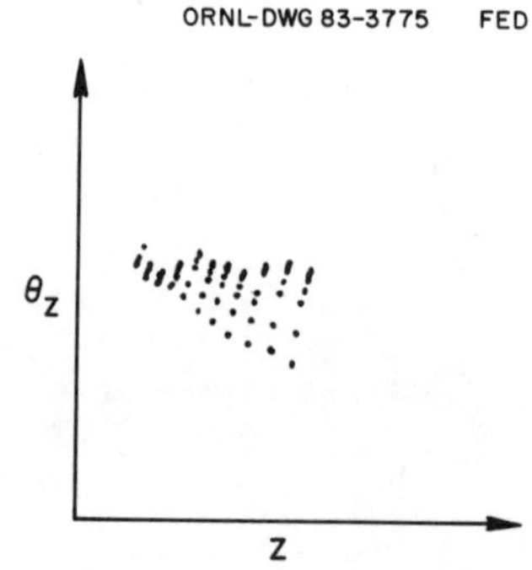

DISCUSSION

EHLERS: What converter potential do you use for these considerations?
WHEALTON: 150 V.
PETERSON: Is it at all possible to deal with time dependent space charge effects such as might affect the emittance of pulsed beams Paul Allison was talking about.
WHEALTON: Well, we haven't really done that. We can certainly put in the time dependent potentials and we are considering expansion of the code in that dimension to deal with gyrotrons, but right now, as it stands, we can't.
ALLISON: In your calculations for the Berkeley source, was the parameter you varied the electron temperature in order to get the beam blow up?
WHEALTON: No, it would be the positive ion temperature in this case. We compensated charge.
ALLISON: What was the range of variation?
WHEALTON: It was very large, it was a hideous variation, something like 0-50 V. It was not the point to say that the observed blowup is due to a high temperature plasma, the point was to say that if there is a little bit of space charge blow up, the beam will be going everywhere. I did not even want to conjecture that the beam blowup was due to a high temperature plasma rather than it was due to an imperfect plasma although a very high temperature plasma is a form of an imperfect one at least from the concept of the device.
LEUNG: Your slot is very small but the converter is very large and it looks like a bit of beam is being lost to the walls. Why do you use a big converter?
WHEALTON: The object of this study, which did not have an excess of funding, was to demonstrate what the temperature of the ions coming out of the converter surface was and not to build a source with as large a current as we could or to even be particularly astute in designing it for maximum converter usage. The objective was to elaborate on what is the random transverse energy at the converter. In order to do that we had to have a very good optical system or else the electrostatic aberrations will hide the ejection transverse energy. In fact, W. Stirling reported that we found the transverse energy to be 0.7 eV, which is rather low.
LEUNG: Do you assume in your calculations a zero transverse energy to start with?
WHEALTON: That's correct.
EHLERS: 0.7 eV, you say?
WHEALTON: That is correct, for W. Stirling's measurements.
EHLERS: Is that geometry defined? That is much lower than other measurements indicate. I mean, you can't get that low an energy. Do you read it in both planes?
WHEALTON: No. We're reading it in the transverse plane.
EHLERS: Then it is a function of geometry as well.
WHEALTON: Yes, it's a function of geometry. The aberration can only get worse. You can't easily argue that we can reduce it. However, you can argue that the ion temperature should be lower than measured

and the object of this study was to get an upper bound on the ion temperature.
EHLERS: You should look at it in the other dimension too. There it would not be geometry defined.
WHEALTON: Well, I agree with you.
COOPER: To elaborate on Ehlers' point, the apparent temperature you measure is a function of the slot width. You can be reading a geometrically reduced temperature.
WHEALTON: That is not true. These ions come out with a certain temperature and the emittance is a function of the slot width but the temperature isn't.
EHLERS: Those that have an additional temperature can't come out. That's what Cooper is saying.
WHEALTON: You're saying we are filtering out the high temperature component. Well, maybe you're having a key for having a zero temperature ion source. In our case we filter out ions with over 4.2 eV.
COMMENT: Are you building this?
WHEALTON: Well, yes, this is how we got 0.7 eV.
COMMENT: With the shaped ends against the slot?
WHEALTON: The ends were not shaped in the experiment we are quoting. That's one of the reasons why the experiment that Ehlers proposed was not performed. The slot ends were not optimized. This accelerator was built before the structure was designed for the third dimension. We were getting electrostatic aberration which we didn't want to bother with at that time.
JACQUOT: Please explain to me what the input data are in your model, the shape of the electrodes, the voltage, and the plasma. What are the input data to calculate your model?
WHEALTON: As you say, the voltage on the electrodes, the shape of the converter surface, the current density at the converter, magnetic field, which is 1300 gauss, plasma temperature, plasma density.
JACQUOT: Plasma density?
WHEALTON: In this case I was alluding to the nature of the model which has a Boltzmann distribution which is cancelling out the negative ion distribution. The positive ion space charge that cancels out the electron space charge is neglected and that's one of the defects of the model that I was alluding to. I think these things should be more reliable.
LEUNG: I would like to ask a question about the experimental measurement of the divergence. Could it be that you are extracting volume produced H^- too in your source, in addition to the surface produced?
WHEALTON: The mean free path is rather large. The neutral gas density is quite small.
QUESTION: What fraction of negative ions is volume produced?
STIRLING: We don't know whether we have any excitation up to higher states leading to dissociative attachment.
DAGENHART: I would like to point out that the opening of the front is 2 mm and the converter surface is 6 mm back so the angular acceptance for transverse components coming off the converter is roughly 18°.
EHLERS: What is this in eV?

WHEALTON: I think Dagenhart is arguing that significant volume negative ion extraction is inordinately improbable since the beam free path is large compared with the apparatus scale and the fact that newly borne ions would miss the accelerator. I think we assume that that's a negligible fraction at this time.
DAGENHART: Plus or minus 15 eV.
EHLERS: Oh no, no way.
BACAL: What is the gas pressure?
WHEALTON: The gas is neglected here other than in forming a plasma. That is, volume processes that are occurring here are neglected in this calculation.
STIRLING: To answer the question by Leung, the converter voltage must be on or you don't see anything and there has to be cesium in the source as well.
LEUNG: If you look at the energy spectrum of the beam you should be able to identify two groups in the negative ion beam coming out of the source but he is claiming that one is due to charge exchange. In your case I don't know what is the source pressure so you can identify those two groups.
WHEALTON: Unless one group is very small.
LEUNG: But if you make a scan then you should be able to tell how large is one group in respect to the other.
STIRLING: You're talking about the charge exchange or what's coming off the converter?
WHEALTON: That's about 4 to 6 mm distance and the pressure is a few millitorr.
LEUNG: Either charge exchange or volume produced ions at the extraction region will give you a very low temperature.
WHEALTON: But then it wouldn't depend on the converter voltage.
LEUNG: It won't depend on the converter voltage.
WHEALTON: I think W. Stirling just said it was very sharply dependent on the converter voltage and therefore most of the negative ions are produced on the converter and are not volume produced.

NEUTRALIZERS

PHOTODETACHMENT TECHNOLOGY*

Joel H. Fink[†]
Lawrence Livermore National Laboratory, Livermore, CA 94550

ABSTRACT

This report gives an analysis of a neutral beam line formed of negative ions and stripped in a photoneutralizer. Estimates are made of its performance when neutralized by an atomic-iodine laser.

INTRODUCTION

Photodetachment, in a high-power neutral beam injector, is an effective method of neutralizing high-energy negative ions. It can neutralize a large fraction of a negative ion beam, while leaving many of the negative impurity ions unneutralized. A well-designed system requires the addition of neither gas, vapor, nor plasma.

Although the physics of photodetachment is well-known,[1] some technological advances must be made before we can design and build a large photoneutralizer. As presently conceived, we need an efficient, high-power laser of suitable wavelength, along with an optical resonator of high gain in which the negative ion beam is neutralized. Suitable mirrors with high reflectivity and low scattering have been developed;[2] they are capable of handling high-power irradiances. In addition, however, a highly transparent window is needed to separate the laser gain medium from the evacuated resonator through which the ion beam passes.

Preliminary studies of laser windows and resonator designs are in process.[2] Currently under development is a continuously operating laser that can be used to neutralize negative hydrogen, deuterium, and tritium beams,[3] i.e., the supersonic chemical oxygen-iodine laser.[4] Meanwhile, a study contract to evaluate the efficiency and cost of a chemical recycling plant is in process.[5]

In the following I derive a series of equations to determine the operating efficiency and fraction of neutrals obtainable with a photoneutralized negative ion beam line, and discuss the significance of these results with respect to the design of various neutral beams.

---

*Work performed under the auspices of the U.S. Department of Energy by the Lawrence Livermore National Laboratory under contract number W-7405-ENG-48.

[†]On assignment from Negion, Inc., Hayward, CA.

0094-243X/84/1110547-14 $3.00 Copyright 1984 American Institute of Physics

Figure 1 is a block diagram of a deuterium neutral beam injector. Negative ions are extracted from their source, accelerated to the desired energy, and injected into the laser resonator. A large fraction of the negative ions is neutralized as they travel through the resonator toward their target. Subsequently, the small fraction of the beam that remains ionized is deflected out of the neutral beam path, to be collected at the ion dump.

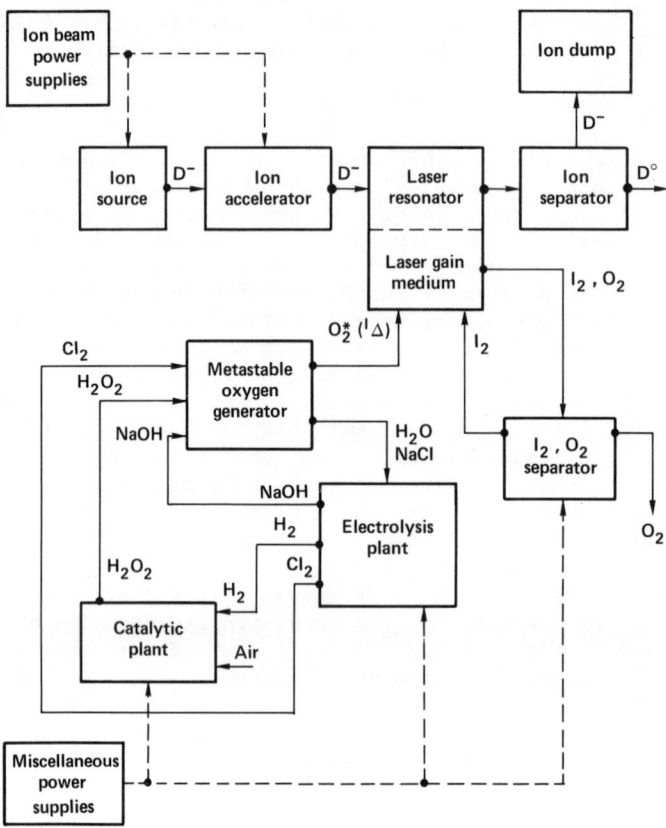

Fig. 1. Block diagram of a deuterium neutral beam injector that uses photoneutralization.

The laser is fueled by metastable oxygen molecules $O_2^*(^1\Delta)$ and iodine vapor. Through a series of involved reactions, the metastable oxygen dissociates the iodine molecules and raises the newly liberated iodine atoms to an excited spin-orbit $I^*(^2P_{1/2})$ state. The concentration of excited iodine soon becomes sufficient for lasing, and stimulated emission occurs at a wavelength of 1.315 μm, with the transition from the excited state $I^*(^2P_{1/2})$ to the ground state $I(^2P_{3/2})$. Metastable oxygen molecules re-excite the ground state iodine atoms, and the cycle repeats many times as the gases flow through the laser at supersonic velocities. Lasing stops at the far side of the laser where the supply of metastable oxygen is insufficient to maintain the necessary population inversion of excited iodine.

Metastable oxygen molecules are generated by bubbling chlorine gas through a solution of hydrogen peroxide and sodium hydroxide whereby

$$Cl_2 + H_2O_2 + 2NaOH \rightarrow O_2^*(^1\Delta) + 2NaCl + 2H_2 . \tag{1}$$

For an experimental installation chlorine, peroxide, and sodium hydroxide could be purchased and supplied in large tanks, but in an actual reactor they would most probably be continuously recycled at less cost. The sodium chloride and water that remain after the production of metastable oxygen can be reconverted into sodium hydroxide and chlorine.

The chemical recycling method indicated in Fig. 1 is a commercial process using an electrolyic diaphragm cell[6] in which

$$2NaCl + 2H_2O \rightarrow 2NaOH + Cl_2 + H_2 . \tag{2}$$

Hydrogen produced by this reaction is subsequently used to make hydrogen peroxide by means of the Quinone catalytic process,[6] another commercial procedure. The power required to operate the laser is the sum of the power needed to operate the catalytic and electrolysis plants, various gas compressors, filters, etc., plus the power needed for the boiler that separates the spent iodine from the oxygen.

MAXIMUM BEAM-LINE EFFICIENCY

Neglecting any loss of neutrals betwen the neutralizer and the neutral beam target, we define the efficiency of a beam line that uses a gas neutralizer as ε_B, the efficiency with which the high-energy negative ion beam was formed, multiplied by η, the fraction of negative ions that is neutralized. This assumes that (a) the power needed to operate the gas cell and to pump away the gas introduced by the gas cell is negligible; (b) the background gas pressure in the accelerator and beyond the neutralizer is low; and (c) the beam path from the neutralizer to the beam target is of reasonable length.

In contrast to a gas cell, which requires negligible power, the power P_N needed to operate a photoneutralizer is significant. Hence, the efficiency of such a neutral beam line is

$$\varepsilon_P = \eta/(\varepsilon_B^{-1} + P_N/P_B) , \tag{3}$$

in which P_B is the power of the negative ion beam as it enters the neutralizer.

The maximum fraction of neutrals obtainable from a negative ion beam passing through a gas cell is determined by the cross sections of the various interactions between the beam and the background gas.[7] On the other hand, the neutral fraction obtainable with a photoneutralizer is a function of its design and of the power used to operate it. A more powerful photoneutralizer can always be used to make the neutral fraction larger than that produced by a gas (or vapor) cell.

The efficiency of a photoneutralized beam line can exceed that of a gas cell if the power saved (by forming a smaller negative ion beam with a larger neutral fraction) is greater than the power spent to operate the photoneutralizer. For this reason, a photoneutralizer can be designed to produce an optimum neutral fraction that corresponds to a maximum beam-line efficiency. The condition for the maximum is established by setting the derivative of the beam-line efficiency with respect to the neutral fraction, equal to zero. The result is

$$\varepsilon_B^{-1} = \eta_o \, d(P_N/P_B)/d\eta - P_N/P_B \, , \tag{4}$$

in which η_o is the optimum neutral fraction at which the neutral beam-line efficiency is maximum. When we introduce Eq. (4) into Eq. (3), we get an expression for the maximum beam-line efficiency:

$$\varepsilon_P = [d(P_N/P_B)/d\eta]^{-1} \, . \tag{5}$$

NEUTRALIZER THICKNESS

Assume that a negative ion current of I_o enters the photoneutralizer and a current I_z leaves there unneutralized as shown in Fig. 2. With the beam uniformly exposed to light of an average irradiance of ϕ_w (W/m^2), electrons are stripped from the negative ions at a rate of

$$dI/dt = -\rho \, I \, , \tag{6}$$

where

$$\rho = \sigma \, \phi_w/(h\nu) \, , \tag{7}$$

and σ is the photodetachment cross section, while $(h\nu)$ is the energy of the photons irradiating the beam.

When we introduce the ion velocity $v = dz/dt$ into Eq. (6) and integrate over the length of the neutralizer Z, we get the neutral fraction of the output beam as

$$\eta = 1 - \exp\left[-\sigma/(h\nu) \, v^{-1} \int_0^Z \phi_w dz\right] \tag{8}$$

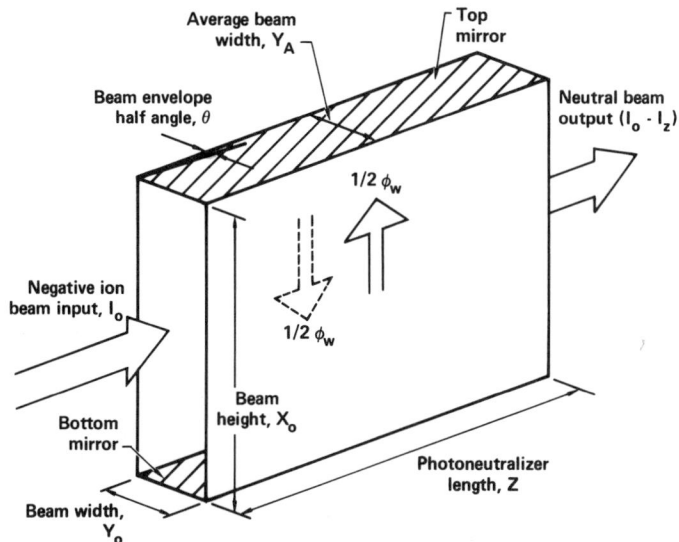

Fig. 2. Geometry of a photoneutralizer.

where

$$\eta = 1 - I_z/I_o . \tag{9}$$

If S (W/m) is defined as

$$S = \int_0^Z \phi_w \, dz , \tag{10}$$

i.e., the thickness of the radiant flux in the photoneutralizer, then the length of the neutralizer must be approximately

$$Z = S/\phi_w , \tag{11}$$

and from Eq. (8),

$$S = (2eV/M)^{1/2} (h\nu/\sigma) \ln [1/(1 - \eta)]. \tag{12}$$

In Fig. 2, the irradiance of the top and bottom mirrors, ϕ_M, is half of ϕ_w, i.e., $S/(2Z)$ (W/m^2), because ϕ_w equals the radiant flux going up toward the top mirror plus that going down toward the bottom mirror. Meanwhile, the ion beam passing through the neutralizer has an average width Y_A so that it is exposed to a total irradiance of

$$P_o = S \, Y_A . \tag{13}$$

Figure 3 shows the fraction of neutrals obtainable from a neutralizer of thickness S.

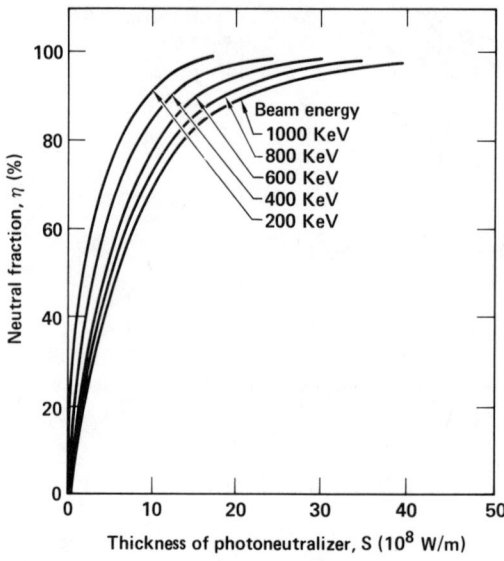

Fig. 3. Neutral fractions obtainable from negative deuterium ion beams as a function of irradiance thickness in a photoneutralizer.

EVALUATION OF P_N/P_B

The gain of the resonator G can be defined as the ratio of the radiant power in the resonator P_o to the power needed to excite it P_L. If ε_L is the efficiency with which the laser produces the light that excites the resonator, the total power required to operate the photoneutralizer must be

$$P_N = P_L/\varepsilon_L , \qquad (14)$$

whereby

$$P_N = P_o/(G\ \varepsilon_L) . \qquad (15)$$

As the power in the negative ion beam is

$$P_B = I_o V , \qquad (16)$$

the ratio P_N/P_B becomes

$$P_N/P_B = \Gamma^{-1} \ln[1/(1 - \eta)] , \qquad (17)$$

in which

$$\Gamma = (M/2e)^{1/2} V^{1/2} (I_o/Y_A)\ \sigma/(h\nu)\ G\ \varepsilon_L . \qquad (18)$$

We introduce Eq. (17) into Eq. (4) to get the condition for maximum beam-line efficiency:

$$\Gamma = \varepsilon_B [\eta_o/(1 - \eta_o) + \ln(1 - \eta_o)] , \qquad (19)$$

while from Eq. (5) the maximum efficiency is found to be

$$\varepsilon_P = \Gamma (1 - \eta_o) . \qquad (20)$$

Figures 4 and 5 show the optimum neutral fraction and the maximum beam-line efficiency, respectively, as functions of parameter Γ. Note that the larger the Γ, the higher the neutral beam-line efficiency and the greater the neutral fraction. As Γ gets even larger, the optimum neutral fraction nears 100%, while the maximum beam-line efficiency approaches the efficiency with which the negative ion beam was formed. Figure 6 shows the maximum beam-line efficiency as a function of the associated optimum neutral fraction.

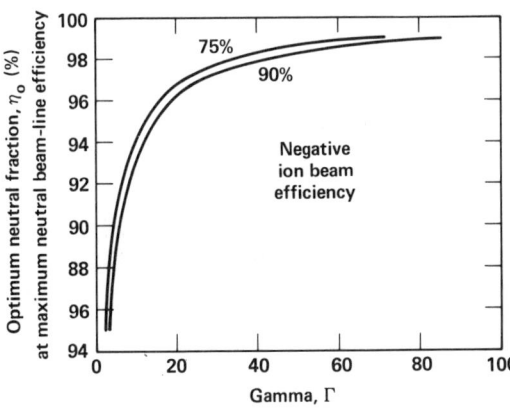

Fig. 4. Optimum neutral fraction as a function of Γ.

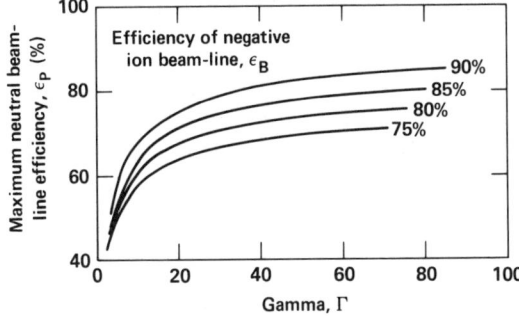

Fig. 5. Maximum neutral beam-line efficiency as a function of Γ.

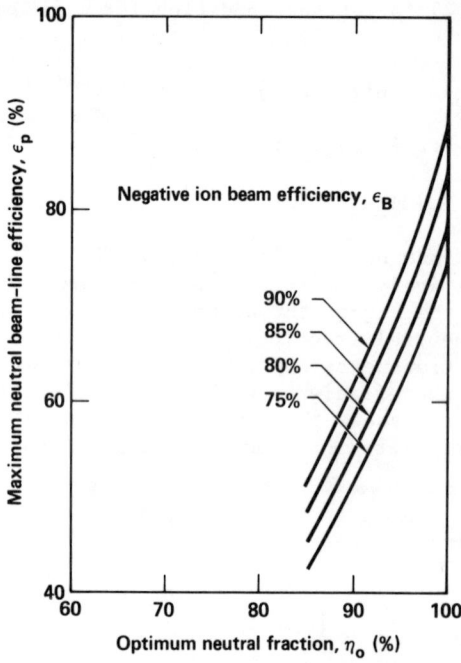

Fig. 6. Maximum neutral beam-line efficiency as a function of the optimum neutral fraction.

EVALUATION OF Γ

The development, at this time, of a photoneutralizer for a deuterium beam is not sufficiently advanced to evaluate Γ with any assurance. Specifically, we can only estimate the gain that might be attained in a suitable laser resonator, while the ultimate efficiency of the laser is uncertain. Nevertheless, it is evident that an efficient negative ion beam and a large Γ are essential to form a neutral beam injector of optimum performance.

As defined by Eq. (18), Γ is composed of six factors. The first, $[M/(2e)]^{1/2}$, is a function of beam composition; for a deuterium beam, it is approximately equal to 10^{-4} ($v^{1/2}$ s/m). The second, $v^{1/2}$, relates to beam energy, and the third, I_o/Y_A, is a function of beam geometry.

Having originated from a slit in an ion source similar to the Lawrence Berkeley Laboratory (LBL) self-extracting negative ion source,[8] the ion beam is very narrow (see Fig. 2). It enters the neutralizer with a cross section of Y_o by X_o (m^2), where $Y_o \ll X_o$.

For the neutralizer to function efficiently, its width must closely match that of the ion beam. Thus, the average width of the neutralizer is the same as the average width of the ion beam traveling through it. If the neutralizer length is Z, then

$$Y_A = Y_o + \theta Z , \qquad (21)$$

in which θ (rad) is the half angle of the beam envelope (Fig. 7). To irradiate the most ions with the least light, the ion beam cross section should be rectangular (Fig. 2), and is approximately such when the neutralizer is located close to the ion source where the initial beam width Y_o is sharply defined.

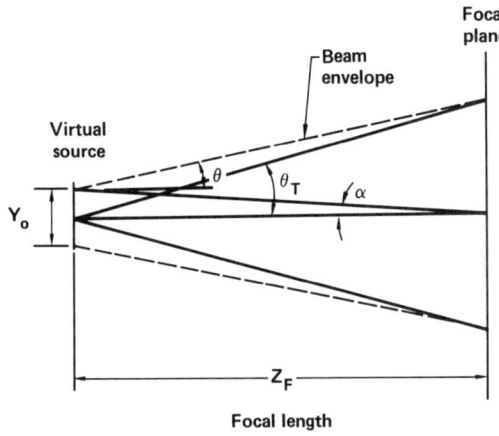

Fig. 7. Schematic of the high-energy beam from the virtual source to the focal plane.

As the negative ion beam passes through the neutralizer, it must be space-charge neutralized to prevent the half angle of the beam envelope from becoming unnecessarily large. This is accomplished by adjusting the background pressure. However, the pressure must not be so high as to strip an excessive fraction of the negative ion beam in front of the neutralizer, or to ionize a significant fraction of the high-energy neutrals beyond the neutralizer.

If we assume the space-charge forces are neutralized, the average width of the neutralizer can be minimized by means of suitable beam optics. Let θ_T represent the divergence angle corresponding to that of an ion emitted anywhere over the surface of the source with a transverse velocity equal to the most probable transverse velocity, and a longitudinal velocity equal to $(2eV/M)^{1/2}$. Then

$$\theta = \theta_T - \alpha , \qquad (22)$$

in which α is the angle formed by the beam geometry as shown in Fig. 7. Because the emittance is an invariant of a beam of given energy,

$$\theta_T = C/Y_o , \qquad (23)$$

in which C is a constant related to the beam emittance.

From Eq. (21), then,

$$Y_A = Y_o + CZ/Y_o - \alpha Z \qquad (24)$$

and minimum Y_A is obtained when

$$Y_O = (CZ)^{1/2}. \qquad (25)$$

Neglecting α,

$$\text{minimum } Y_A \leq 2(CZ)^{1/2}. \qquad (26)$$

From the dimensions of a negative ion source currently under development at LBL,[9] we estimate C at 0.034 mm-rads. Thus, for a neutralizer 2.0 m long, minimum Y_A is about 0.05 m, and

$$I_O/Y_A = 20 \, I_N/\eta, \qquad (27)$$

in which the negative ion beam current I_O is taken to be equal to the neutral beam current I_N, divided by the neutral fraction η.

The fourth term of gamma equals the photodetachment cross section σ at the wavelength of the laser, divided by the corresponding photon energy $h\nu$. With an atomic iodine laser, $\sigma/(h\nu)$ = 0.012 m^2/Joule, as shown in Fig. 8. Photons from an atomic iodine laser do not have sufficient energy to strip C^-, O^- or OH^- ions.[10] On the other hand, we can see from Eq. (12) that the mass and photodetachment cross section of O_2^- are such that a photoneutralizer capable of neutralizing 95% of a 200-keV D^- beam will strip 75% of the O_2^- impurity in the beam. Obviously, O_2^- is undesirable.

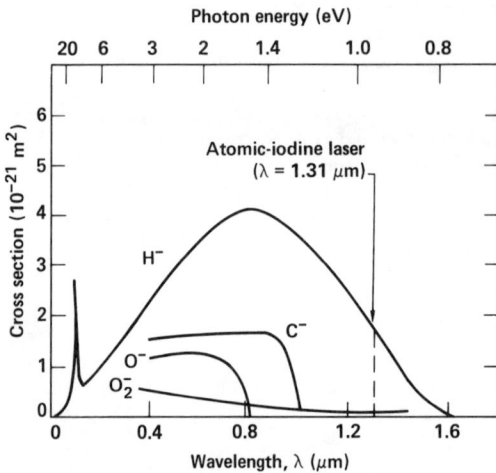

Fig. 8. Photodetachment cross section of negative hydrogen, carbon, oxygen, and oxygen molecules. (Adapted from Ref. 11.)

The fifth term G is the gain of the laser resonator. It is inversely proportional to the optical losses resulting from scattering and absorption in the laser window, mirrors, and gain medium, plus the losses resulting from optical diffraction within the resonator itself. I will not pursue this subject here because it is discussed elsewhere in these proceedings.[2] However, I believe that the development of low-loss mirrors of very high reflectivity makes a resonator with a G of 500 possible.

The sixth and last term is laser efficiency ε_L. If ε_C is the efficiency with which the metastable oxygen is recycled; $(\varepsilon_{in} - \varepsilon_{out})$, the fraction of the metastable oxygen that is consumed in the laser; and ε_U, the fraction of metastable oxygen that excites the iodine atoms; then

$$\varepsilon_L = 0.96 \ \varepsilon_C (\varepsilon_{in} - \varepsilon_{out}) \ \varepsilon_U \ , \tag{28}$$

in which 0.96 is the ratio of the energy of metastable iodine to metastable oxygen molecules. Although it is impossible to know at this time what the efficiency of a fully developed supersonic atomic iodine laser will be, my estimates fall within 3 to 12%.

NEUTRAL BEAM CURRENT

We can determine the neutral beam current needed to obtain a maximum beam-line efficiency by introducing the six factors of Γ, previously discussed, into Eqs. (18) and (20). The result is

$$I_N = 83.3 \ (\varepsilon_P/\varepsilon_L) \ V^{-1/2} \ \eta_o/(1 - \eta_o) \ . \tag{29}$$

This equation can be solved for any maximum beam-line efficiency ε_P by using the corresponding optimum neutral fraction η_o shown in Fig. 6. Figure 9 shows the result for a neutral beam line based upon a negative ion beam whose efficiency ε_B is 85%.

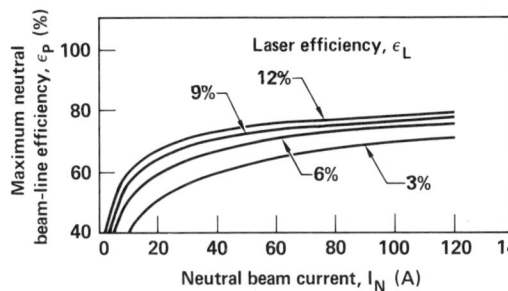

Fig. 9. Maximum efficiency of a 200-keV neutral beam line obtainable with a neutral beam current, I_N. The negative ion beam was formed at an efficiency of $\varepsilon_B = 85\%$, while the average width of the resonator was 0.05 m, and its length, 2.0 m.

Because so much energy is lost to the resonator mirrors, in contrast to the energy used for photodetachment, the larger the beam current neutralized in a resonator of given mirror area, the more efficient the beam line. However, there is a limit to the neutral current that can be carried in a beam of reasonable size. Ion sources deliver only so many amperes of negative ions per meter of beam height (X_o per Fig. 2). The present LBL surface conversion source is expected to deliver 6 A/m, but this may be larger in other sources now under development. With a 10% loss of negative ions during acceleration, the present source will provide about 5.4 A/m of neutrals.

LITHIUM BEAMS

Of interest is a comparison of the photodetachment of a Li^- beam with a D^- beam, when both use the chemical iodine laser previously discussed.[11,12] Assume both beams operate with the same neutral fraction, whereby the ratio of the thickness of their respective neutralizers, per Eq. (12), is

$$S_{Li}/S_D = (M_D/M_{Li})^{1/2}(V_{Li}/V_D)^{1/2}(\sigma_D/\sigma_{Li}) . \qquad (30)$$

If the two beams are of identical height and width,

$$\Gamma_{Li}/\Gamma_D = (M_{Li}/M_D)^{1/2}(V_{Li}/V_D)^{1/2}(J_{Li}/J_D)(\sigma_{Li}/\sigma_D) , \qquad (31)$$

in which J_{Li} and J_D are the current densities of the lithium (Li) and deuterium (D) ion beams, respectively.

At the laser wavelength 1.315 μm, the photodetachment cross section of D^- is about 1.8×10^{-21} m^2, while that of Li^- is roughly 1.4×10^{-20} m^2 (Figs. 8 and 10).[13] For beams of the same energy, $S_{Li}/S_D = 0.069$, and $\Gamma_{Li}/\Gamma_D = 14.5 (J_{Li}/J_D)$. For beams of the same velocity (where $V_{Li}/V_D = M_{Li}/M_D$), $S_{Li}/S_D = 0.129$, and $\Gamma_{Li}/\Gamma_D = 27.0 (J_{Li}/J_D)$. In both cases, the thickness of the Li neutralizer is of the order of 10% of that of D. Furthermore, if the current density of the Li^- negative lithium ion beam could be made to approach that of today's D^- sources, Γ_{Li} would be quite large, whereby the efficiency of the neutral Li beam line would be about that with which the Li^- ion beam was formed. Consequently, an efficient negative ion source and accelerator are essential to form an efficient Li neutral beam.

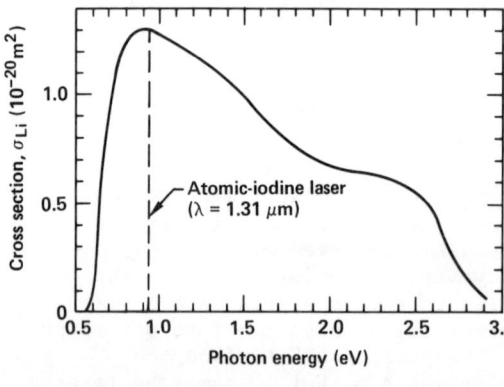

Fig. 10. Photodetachment cross section of negative lithium ions. (Adapted from Ref. 13.)

CONCLUSION

This analysis shows that the maximum efficiency of a neutral beam requires the optimization of both the negative ion beam and the photoneutralizer. The efficiency of the neutral beam line can not be greater than that of the negative ion beam from which it was formed; therefore, the performance of the ion source, accelerator, and ion transport must be optimized.

To effectively use a photoresonator, the negative ions must be formed into a thin ribbon beam. In addition, the beam current density should be high, while the negative ion beam is space-charge neutralized.

Although the chemical iodine laser is the preferred choice for photoneutralizing hydrogen, deuterium, tritium, and lithium beams, experimental verification of the laser performance at full power is required. While some performance data of the supersonic laser are available, its ultimate overall efficiency is obviously not known. To help estimate this, a study of the efficiency of the laser chemical recycling system has been initiated. In addition, computer studies are planned to design the laser resonator and to optimize the resonator gain, while advanced concepts of the window are verified. I believe that components of a full-scale, multi-ampere demonstration of photoneutralized negative ion beams could be available in about 4 years.

REFERENCES

1. E. W. McDaniel, <u>Collision Phenomena in Ionized Gases</u> (John Wiley & Sons, N.Y., 1964).
2. V. Vanek, D. Goebel, T. Hursman, and D. Copeland, "Technology of a Laser Resonator for a Photodetachment Neutralizer," prepared for the Third International Symposium on the Production and Neutralization of Negative Ions and Beams, Brookhaven National Laboratory, Upton, N.Y., Nov. 14-18, 1983.
3. TRW, Inc., Redondo Beach, Calif., under contract to the Air Force Weapons Laboratory, Kirtland Air Force Base, Albuquerque, N. Mex.
4. W. E. McDermott, N. R. Pchelkin, D. J. Bernard, and R. R. Bousek, "Electronic Transition Chemical Laser," Appl. Phys. Lett. <u>32</u>, 469 (1978).
5. Tera Corporation, Berkeley, Calif., under a small Business Innovation Research contract to the Department of Energy.
6. M. Sittig, <u>Inorganic Chemical and Metallurgical Process Encyclopedia</u> (Noyes Development Corp., Park Ridge, N. J., 1968).
7. K. H. Berkner, R. V. Pyle, and J. W. Sterns, "Intense Mixed Energy Hydrogen Beams for CTR Injection," Nucl. Fusion <u>15</u>, 249 (1975).
8. W. Ehlers, "Negative Ion Sources for Neutral Beam Systems," J. Vac. Sci. Technol. <u>A</u>, 974 (1983).

9. W. S. Cooper, O. A. Anderson, K. W. Ehlers, J. W. Kwan, A. F. Lietzke, H. M. Owren, J. A. Paterson, and W. F. Steele, <u>A New 2 A, 160 keV, H$^-$ TFF Test Facility</u>, Lawrence Berkeley Laboratory, Berkeley, Calif., LBL-16289(15), submitted to the 25th Annu. Mtg. of the Am. Phys. Soc., Division of Plasma Physics, Los Angeles, Calif., Nov. 7-11, 1983.
10. C. F. Barnett, "Atomic Data for Controlled Fusion Research," Oak Ridge National Laboratory, Oak Ridge, Tenn., ORNL-5206, 1977.
11. L. R. Grisham, D. E. Post, D. R. Mikkelsen, and H. P. Eubank, "Plasma Heating with Multi-MeV Neutral Atom Beams," Nucl. Technol./Fusion, $\underline{2}$, 199 (1982).
12. L. R. Grisham, D. E. Post, and D. R. Mikkelsen, "Multi-MeV Li Beams as a Diagnostic for Fast Confined Alpha Particles," Nucl. Technol./Fusion $\underline{3}$, 121 (1983).
13. H. J. Kaiser, E. Heinicke, R. Rackwitz, and D. Feldmann, "Photodetachment Measurements of Alkalai Negative Ions," Z. Phys. $\underline{270}$, 259 (1974).

GLE/rp/mm/4151t

NEUTRALIZATION EFFICIENCY OF PLASMA TARGETS FOR HIGH ENERGY
NEGATIVE IONS*

A. I. Hershcovitch, B. M. Johnson, V. J. Kovarik, M. Meron,
K. W. Jones, K. Prelec
Brookhaven National Laboratory, Upton, New York 11973

L. R. Grisham
Princeton Plasma Physics Laboratory, Princeton, New Jersey 08544

ABSTRACT

Plasma targets generated by hollow cathode discharges are used to neutralize a variety of multi-MeV negative ions as heavy as 28 a.m.u. Our plan is to determine the neutralization efficiency of hydrogen and argon plasma targets for D^-, Li^-, C^-, O^-, and Si^- at beam energies of 2-8 MeV. The experiment is still in progress. Encouraging initial results are reported.

INTRODUCTION

Tokamak and mirror fusion reactors will require some method of auxiliary heating. In addition, tokamaks will need a noninductive method for current drive, while TMX devices require maintenance of a potential barrier at the end plugs. For heating, deuterium beams with energies in the hundreds of KeV will be required for these reactors, while for current drive or mirror plugging, the deuterium beams need energies of about 1 MeV. As an alternative to the use of deuterium beams for auxiliary heating, mirror plugging and current drive, neutral beams of heavier atoms such as O^0 with energies of about 1 MeV/nucleon have been proposed.[1]

At present, experiments, designed to determine the neutralization efficiency by argon and hydrogen plasma targets of multi-MeV D^-, Li^-, C^-, O^-, and Si^- are in progress at the Brookhaven National Laboratory Tandem Van de Graaff Facility. These experiments are similar to a previous study[2] in which the neutralization efficiency of a differentially pumped gas cell was measured for Li^-, C^-, O^-, and Si^- at 2-7 MeV with target gases of N_2, CO_2, and Ar.

Neutral and charged particle fractions produced by passing negative ion beams through plasma targets have been measured. From this data neutralization efficiencies have been computed. Since no sufficient data for the interaction of these multi-MeV ions with plasma ions and electrons exists, it was impossible to design the best plasma neutralizer. Therefore, a plasma neutralizer suitable for 200-500 KeV H^-/D^- beams has been used. The plasma neutralization utilizes plasmas injected from hollow cathode discharges.

EXPERIMENTAL ARRANGEMENT

The negative ions were produced and accelerated by the Tandem Van de Graaff Accelerator Facility at Brookhaven National Laboratory. Figure 1 shows the beam line which has been used to make

---

* Work performed under the auspices of the U.S. Department of Energy.

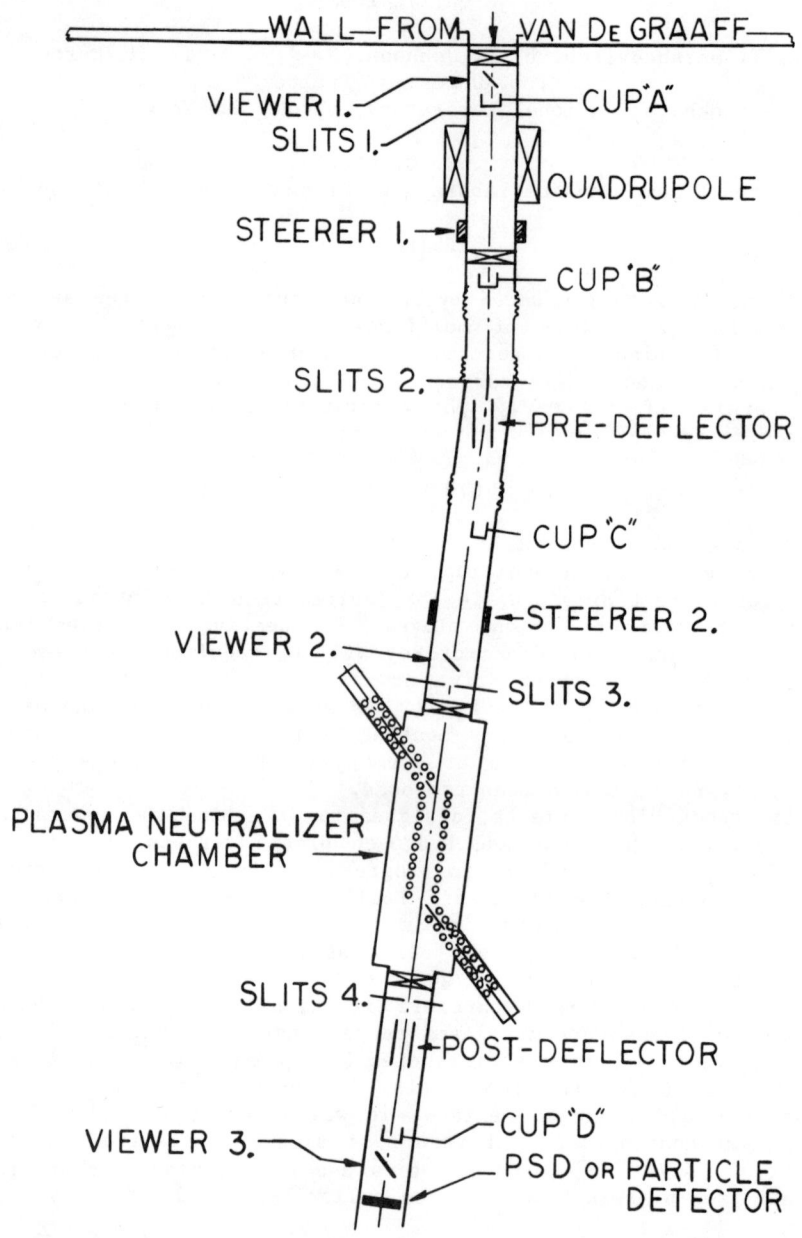

Fig. 1. Diagram of the experimental arrangement

the measurements. In this beam line, the foredeflectors which are located 5 m ahead of the entrance to the neutralizer are used to deflect the negative ions through 2° to purify the beam of other charge states which are formed during transport. This geometry prevented the achievement of maximum neutralization efficiencies due to some formation of other charge states before the beam entered the neutralizer. At a distance of 20 cm downstream from the neutralizer exit, postdeflector plates separate the resultant charge states. A surface-barrier particle detector is translated across the end of the beam line to measure the final charge-state distribution.

Figure 2 is a more detailed diagram of the plasma neutralizer, which is fed by two hollow cathode discharges. The arrangement is similar to a configuration[3] which was used to estimate the power efficiency of plasma targets produced by hollow cathode discharges. But unlike that previous scaled-down version, this neutralizer has a plasma target which is 1 m long, and it is differentially pumped. The plasma is injected from two hollow cathodes through collimators into the main chamber at 30° to the plasma target axis where the plasma is confined by a solenoidal magnetic field of about 200 Gauss. During the experiments, additional partitions with apertures were added on both sides of the neutralizer to reduce the gas pressure in the rest of the beam line. Thus, the neutralizer target is differentially pumped from the small chambers which house the cathodes, and the beam line is differentially pumped from the plasma target.

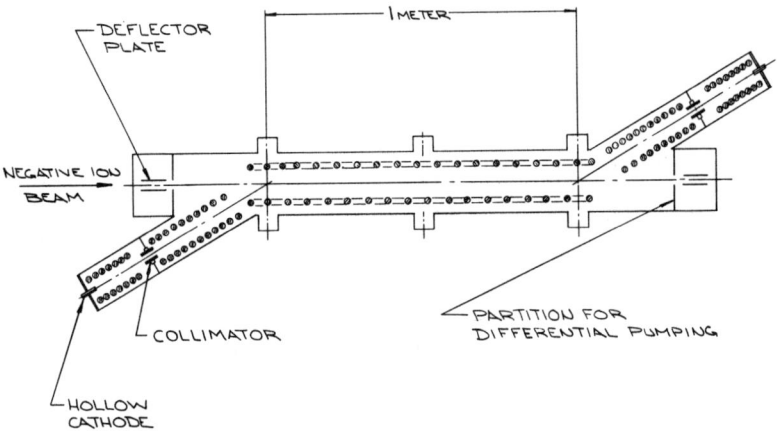

Fig. 2. Schematic of the plasma neutralizer.

RESULTS

In the previous study,[2] maximum neutral yields for gas neutralizers were determined to be 42, 27, 21, and 22% for Li^-, C^-, O^-, and Si^-, respectively, with little or no dependence on either beam energy (2-7 MeV) or target gas (N_2, CO_2, Ar). For the present study, the maximum neutral yields for Li^- at 3 MeV using an H_2 gas target were determined to be 50%. Under not yet optimal conditions, the hydrogen plasma target produced a higher neutral yield of 56%. The corresponding charge-state spectra are shown in Figures 3 and 4. Table I summarizes the preliminary results.

At this time, the plasma parameters have not been measured. Since the plasma density, i.e., target thickness, depends on the arc current, gas pressure, magnetic field strength, etc., the charge-state composition of the beam emerging from the neutralizer has been measured as a function of the arc current, while all other parameters were kept constant in each set of measurements. For plasma densities of interest to this subject matter, the plasma density in

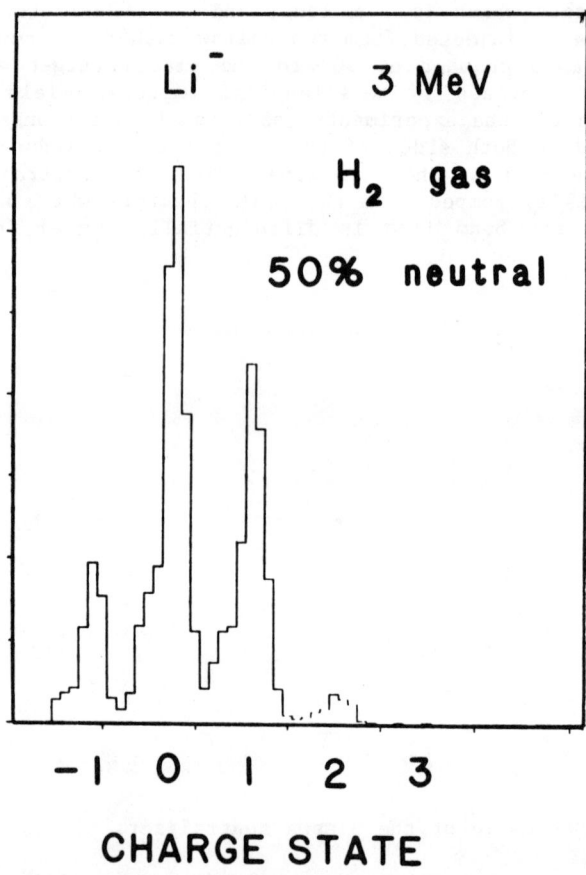

Fig. 3. Charge state spectra of Li^- at 3 MeV after passing through a hydrogen-gas neutralizer.

a hollow cathode discharge is proportional to the arc current. Furthermore, for plasma densities of 10^{12}–10^{13} cm^{-3} the dependence is linear.[3]

Our goal is to systematically study the neutralization efficiency of this type of plasma target to determine its maximum capabilities. This is not presently possible. We plan to modify the apparatus to allow for higher solenoidal magnetic fields, a shorter flight path between the foredeflector and the entrance to the neutralizer, and diagnostic probes to determine the plasma profile

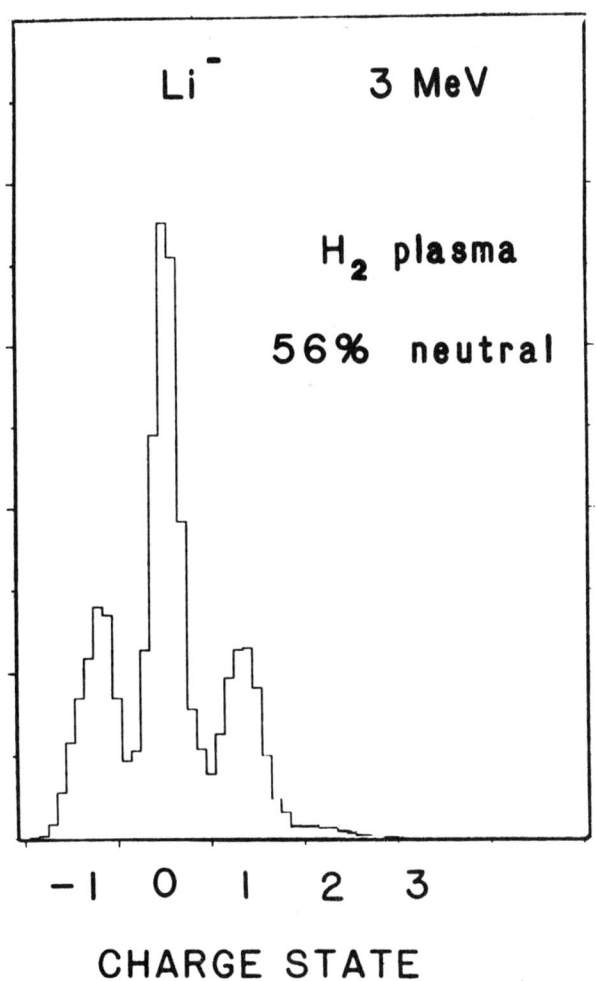

Fig. 4. Charge state spectra of Li$^-$ at 3 MeV after passing through a hydrogen plasma neutralizer.

along the length of the target. These improvements will enable us to reach higher plasma densities, minimize charge changing before the target, and ensure maximum overlap between the ion beam and densest part of the plasma. It will then be possible to make systematic measurements of neutralization versus plasma density and to ultimately optimize the performance of this type of plasma neutralizer.

Table I. Preliminary results using 3-MeV negative ions

Projectile	Neutralizer	Maximum Neutral Fraction
Si^-	Ar plasma*	27%
C^-	Ar plasma*	35%
C^-	H_2 plasma	43%
Li^-	H_2 gas	50%
Li^-	H_2 plasma	56%

*with no differential pumping between target cell and beam line.

ACKNOWLEDGEMENTS

We would like to thank Walt Hensel, Whitey Tramm, and the rest of the BNL Advanced Ion Source and Development Group and P. Thieberger, R. L. Lindgren, C. Carlson, M. Manni and the rest of the BNL Tandem Van de Graaff Operations Group for their fine technical assistance, support, and dedication. Work is supported by the US Department of Energy, Office of Fusion Energy, Division of Applied Plasma Physics, and Office of Basic Energy Sciences, Division of Chemical Sciences, Fundamental Interactions Branch, under Contract No. DE-AC02-76CH00016.

REFERENCES

1. L. Grisham, D. Post, H. Eubank, W. Stwalley, Bull. Am. Phys. Soc. 25, 873 (1980).
2. L. Grisham, D. Post, B. Johnson, K. Jones, J. Barrette, T. Kruse, I. Tserruya, Wang Da-Hai, Rev. Sci. Instrum. 53, 281 (1982).
3. A. Hershcovitch, V. Kovarik, Rev. Sci. Instrum. 54, 328 (1983).

DISCUSSION

DAGENHART: What value of the magnetic field did you have in the neutralizer coil?
HERSHCOVITCH: 220 G.
VERBEEK: How do you get negative beams out of the tandem?
HERSHCOVITCH: Negative ion source is in the terminal.
STIRLING: You mentioned you had a filament to help start or initiate the hollow cathode discharge. Could you elaborate on that?
HERSHCOVITCH: There are a number of ways to start a hollow cathode discharge. One can actually try to heat the cathode itself the hard way or another way, is to generate a small discharge with the cathode biased negatively so it starts to attract ions. When the ion bombardment is intense enough the cathode ignites. Then the filament is turned off and the cathode runs by itself.
GRAHAM: What ionization fraction do you anticipate in the hydrogen plasma?
HERSHCOVITCH: When we ran the cathode with differential pumping we had a background pressure of 3×10^{-6}. This corresponds to a neutral density of about 10^{11} cm^{-3}. I don't believe we can run the cathode with plasma densities below 10^{12} cm^{-3} since the discharge could not be sustained. From this one can estimate the ionization fraction to be about 90%. When we ran the cathode with a background pressure in the 10^{-4} range, chances are that the fraction was 50%.
COOPER: What was the energy of heavy negative ions?
HERSHCOVITCH: 3 MeV. We have the capability of varying the energy between 2 and 8 MeV. We would like to scan the energy.

TECHNOLOGY OF A LASER RESONATOR FOR THE PHOTODETACHMENT NEUTRALIZER*

V. Vanek, T. Hursman, D. Copeland and D. M. Goebel
TRW Inc., One Space Park, Redondo Beach, CA 90278

ABSTRACT

Technology of a laser resonator for the photodetachment neutralizer based on the COIL laser and folded resonator concept is presented. Current state of the art of component technologies (mirrors, windows, COIL laser) is assessed. A photodetachment resonator/COIL medium model is also presented to evaluate the neutralizer performance as a function of relevant resonator and medium parameters.

INTRODUCTION

Technology of a laser resonator for the photodetachment neutralizer (PDN) concept is presented. We consider a concept utilizing a Chemical Oxygen Iodine Laser (COIL) operating in a close cavity mode which provides a well defined interaction volume filled with laser photons. The negative ion beam (D^-) interacts with photons and consequently ions are neutralized by the photodetachment process. Laser power required for the photodetachment is small (1W per 1A of neutralized beam), however the amount of photons which must interact with the beam is large due to a small photodetachment cross section. This leads to a system design where laser produced photons circulate in a high Q closed optical resonator which yields a large optical thickness in the ion beam path. A closed folded laser resonator concept allows to build up the optical thickness to a sufficient value to obtain a high stripping coefficient, typically 0.9-0.95. The laser gain medium must supply all resonator losses (mirror, window, diffraction and scatter losses). In principle, these losses must be minimized to achieve a high stripping coefficient with the least laser power.

A schematic of the PDN resonator is shown in Figure 1. The COIL laser gain medium is isolated from the resonator optical train by a purge system. None of the COIL subsystems are shown (such as $O_2(^1\Delta)$ generator, pumps, etc.), instead subsystems relevant to the PDN concept are shown. The negative ion beam-laser photons interaction region is separated from the gain medium by a double plate solid window.

The following sections of this paper describe the mirror and window technologies schematically shown in Figure 1. The PDN resonator/COIL medium model is also presented in order to evaluate the neutralizer performance as a function of relevant resonator and medium parameters.

---

*Work supported by US DOE contract #W-7405-ENG-48, DOE through UC LLNL, P. O. 1340105.

0094-243X/84/1110568-17 $3.00 Copyright 1984 American Institute of Physics

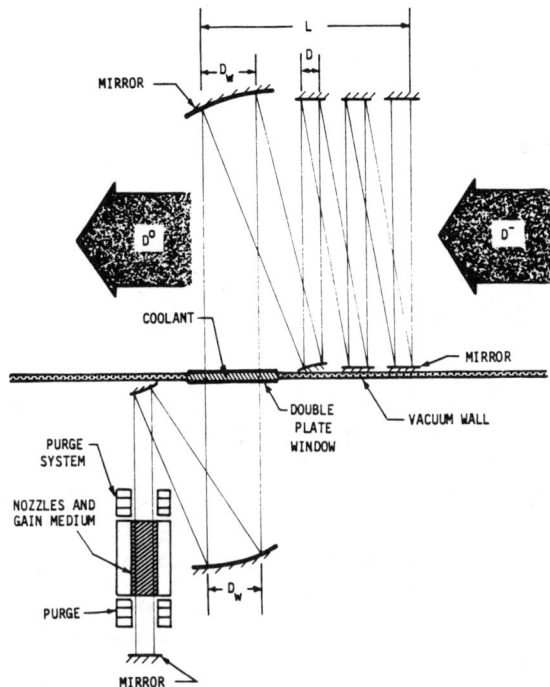

Fig. 1. Schematic of a COIL driven photo-detachment neutralizer.

MIRROR TECHNOLOGY ASSESSMENT

Laser Mirror Heat Exchanger Requirements

Mirrors for the PDN application must operate in a vacuum, be chemically inert, stabilized rapidly (1 second) upon initial exposure to high flux loading, maintain low-scatter surfaces, be figured with a surface figure error of less than $\lambda/10$ PV, be polishable to a surface finish of less than 10 Å RMS, and operate under absorbed laser flux intensities of greater than 300 watts/cm^2. Laser and PDN mirror cooling requirements depend upon the incident flux, angle of incidence, and the optical coating.

TRW has conducted a material figure-of-merit value engineering study, evaluated potential heat exchanger cooling concepts, determined heat exchanger manufacturing capability, performed 3-D interactive parametric thermal and structural mirror heat exchanger analysis and conducted high reflectivity and antireflecting optical coating survey.

Detailed summary of this engineering study on different substrate and faceplate materials of laser heat exchangers is beyond the scope of this paper, only the concepts and results relevant to the PDN application are described.

Mirror Heat Exchanger Distortions

Thermal deformations and distortions of the mirror surface must be minimized in order to maintain a high Q of the PDN resonator. An example of a heat exchanger with three-pass (three-level) cooling designed for laser applications is shown in Figure 2. The design features a thin faceplate with cooling channels immediately below the surface. This design allows the heat to be removed in close proximity to the point at which it is applied and isolates the substrates and mount from the heat load.

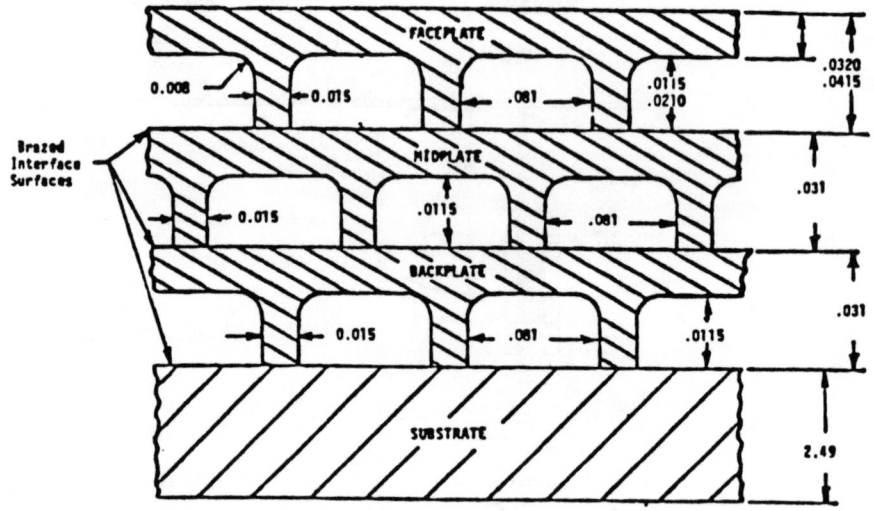

Fig. 2. Cross section of a typical three-pass laser mirror heat exchanger.

A typical non-uniform laser mirror irradiance profile is shown in Figure 3 with the predicted thermal surface distortion response.

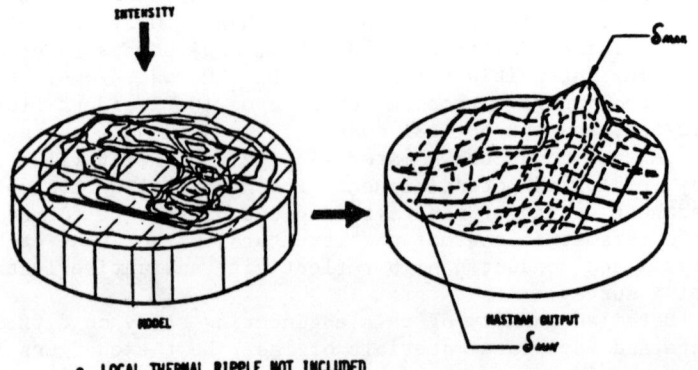

Fig. 3. Example of nonuniform irradiance causing surface deflections.

Flowing coolant introduces additional distortion effects into the mirror system: coolant jitter, pressure ripple, pressure tilt and astigmatism. Figure 4 illustrates the effect of pressure ripple above the coolant cavity channel that is filled with high pressure fluid. An overall surface tilt between the inlet and outlet sides results from the difference in the pressure ripple effect between the high pressure inlet side and low pressure outlet side. Thermal growth of the faceplate is greatest over the land area between the coolant channels and tends to cancel the effects of the pressure ripple.

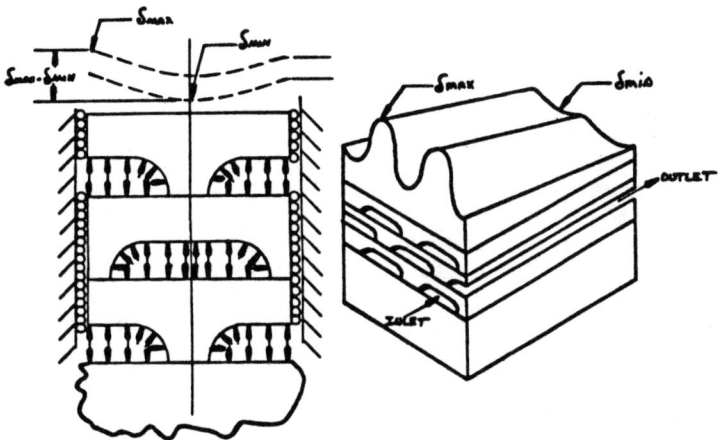

Fig. 4. Example of pressure tilt and ripple.

Thermal and pressure distortions of laser mirrors have been studied extensively at TRW because they generally have the greatest impact on mirror distortions and resulting degradation to the beam. TRW had developed a number of analysis codes that can be incorporated as units of the thermal and structural deformation analysis models for evaluating PDN mirror distortions.

The analysis of distortions is divided into two parts: (1) local distortions of the heat exchanger, and (2) overall distortions of the mirror and substrate. The local and overall mirror distortion analyses are combined to calculate the total mirror face distortion. TRW has developed a 3-D semiautomatic interactive first-order computer analysis program for assessing mirror distortions. This computer program is used to bound the problem and develop mirror heat exchanger concepts. For detailed analysis of mirror heat exchangers, finite element modeling is performed.

Selection of Mirror Heat Exchanger Materials

The selection of heat exchanger substrate and faceplate materials is a very significant issue if not the single most important issue in the development of mirror heat exchangers. The materials evaluation was performed which resulted in the:

a) Presenting the thermal/mechanical properties of candidate heat exchanger materials.
b) Evaluating the relative goodness between corresponding properties of the different materials.
c) Applying figure of merit equations to the material properties to determine the relative performance between materials.
d) Determining overall performance index for each material with a bias on mechanical performance.

In order to establish an overall performance index of each with a bias on mechanical performance, a weighting factor was applied to each normalized material property and figure of merit values then summed for each material. These values were then normalized to the largest value. The results of this exercise were to establish a rating from 0 to 1 for each candidate material. To summarize the results of this exercise:

a) Silicon had the best overall performance but was rejected due to recent failures in the development of Single Crystal Silicon Heat Exchangers.
b) Silicon Carbide (SiC) had the next highest rating but represented the best possible performance at this point in heat exchanger development.
c) Tungsten and Tungsten Carbide followed SiC and offer a potential due to processes now currently developed.
d) Moly was fourth and represented current state of heat exchanger manufacturing practices.
e) Copper had the worst performance of all the materials considered.

Based on the value engineering/figure of merit study we conclude that Silicon Carbide is used as a mirror heat exchanger material for the PDN application.

Projected Mirror Performance

Energy losses at the mirror are due to absorption in the coatings, scattering from the surface irregularities and mirror distortions. Causes of the latter effect were described in previous sections. The total integrated scatter loss (T.I.S.) depends on the surface roughness d, laser wavelength λ and angle of incidence

$$T.I.S. \sim (4\pi d \cos \theta)^2/\lambda^2.$$

Coating absorptances of less than 5×10^{-4} were achieved by several manufactures on Moly (OCLI, Coherent, Litton). OCLI designs with SiO_2/TiO_2 and ZnS/ThF_4 (lowest absorbers) have reflectivities R > 90% for all wavelengths. For silicon substrates, OCLI and Litton were best although Coherent also produced coatings with less than 5×10^{-4} absorptance.

Silicon Carbide material can be polished to 7A roughness which means that the scatter losses at 1.315 micron are less than 5×10^{-5}. Therefore, we conclude that for the PDN application, Silicon Carbide mirrors with reflectivity R = 0.9995 can be fabricated.

MATERIAL WINDOW TECHNOLOGY

The COIL laser gain medium (pressure 2 to 10 Torr) must be

separated from the laser flux-ion beam interaction region by a window since the ion beam requires a low pressure region (10^{-6} Torr) for its propagation.

Two candidate window concepts are suggested by TRW for the PDN: double plate liquid cooled material window and an aerodynamic window. Only the material window concept has been analyzed in detail. The window engineering analysis and major conclusions are described in the following sections.

Summary of Engineering Analysis

Three different material window concepts are analyzed:
Single plate, edge cooled
Double plate, gas cooled
Double plate, liquid cooled.

For the material and thermal analysis, the following requirements were specified:
Incident, one way flux 40 kW/cm^2
Beam footprint 3 to 6 x 10 cm
Wavelength 1.315 micron.

We have concluded through the material comparison analysis that the best material candidate is a clear sapphire (Al_2O_3). The thermal analysis has shown that only the double plate, liquid cooled concept is feasible provided that the incident beam footprint is expanded to 10 x 10 cm. The laser beam expansion schematically shown in Figure 1 is done by a combination of expanding and focusing mirrors on both sides of the window. This introduces at least four additional mirrors into the optical resonator. The beam shaping telescope reduces the one way flux to a level which can be thermally handled (12 kW/cm^2). The coolant is a Flourinert Electronic Fluid (FC-104).

Material Window for PDN

Material	Al_2O_3
Concept	Double plate liquid cooled
Coolant	FC-104 Flourinert Fluid, Index Matched to Sapphire
Irradiance	12 kW/cm^2
Footprint	10 x 10 cm
Expanding Telescope	Required

Double plate window transmissivity is still an open question because the absorption of FC-104 coolant is not known accurately enough. FC-104 manufacturer's data suggest transmissivity better than 99% for 1 cm thickness in infrared region (2 microns). However, a four digit accuracy of the absorption coefficient is required to fully evaluate this concept. The transmissivity of FC-104 will have to be measured at 1.315 microns in following phases of the PDN development. The window material is transparent and antireflection coatings are required to reduce the losses to about 2×10^{-4} per pass per coating. Index matching concept for FC-104 will have to be verified so the internal reflections are minimized.

PDN RESONATOR/COIL MEDIUM MODEL

A simple gain model for the COIL has been developed and applied to analyze the PDN concept. The objective of this work was to express the optical flux within the PDN resonator in terms of the chemical parameters describing the COIL medium and to examine the dependence of the optical thickness on the laser operating parameters.

Detailed description of this model is presented in References (1) and (2). In this paper we present its application to the PDN resonator.

COIL Medium Model

The model assumes that the injected iodine molecules have been completely dissociated and that iodine recombination can be ignored. This assumption is valid throughout the extent of the optical mode which is placed downstream from the nozzle exit plane by a few centimeters. Also since complete dissociation of the molecular iodine is assumed, the role of the O_2 singlet-sigma state is ignored and only the ground state of O_2 and the first excited electronic state of O_2 are considered. These assumptions imply that both the total number of iodine atoms and oxygen molecules are conserved or

$$[I] + [I^*] = [I^{tot}], \tag{1a}$$

and

$$[\delta] + [\sigma] = [O_2^{tot}], \tag{1b}$$

where $[I^*]$, $[I]$ and $[I^{tot}]$ denote the number densities of the $I(^2P_{1/2})$ and $I(^2P_{3/2})$ levels and the total number of I atoms, respectively, and $[\delta]$ and $[\sigma]$ denote the number densities of the singlet-delta and triplet sigma levels of the oxygen molecule, respectively.

To obtain an analytically integrable model both the temperature and the velocity of the flowing medium are assumed constant. The assumption that the velocity is constant corresponds to expansion of the gas in the cavity against the constant pressure, i.e., the so-called freejet condition. Finally, the effects of mixing I_2 and O_2 streams are neglected in this model.

Only stimulated emission and the pumping of the upper laser level are accounted for in this model. First consider the effects of stimulated emission. It is assumed that the flux circulating throughout the cavity is sufficiently large so that the gain is fully saturated. This ensures that the collisional deactivation rate of the upper level is negligible in comparison to the stimulated emission rate. The saturated gain is given by

$$g = \sigma([I^*] - \beta[I]), \tag{2a}$$

where σ denotes the effective stimulated emission cross section which is given by

$$\sigma = \frac{7}{12} \frac{\lambda^2}{8\pi} A \Phi, \tag{2b}$$

and here λ denotes the wavelength of the laser radiation, A denotes the appropriate Einstein A-coefficient and Φ denotes the lineshape function at line center. In Equation (2a) β denotes the ratio of the upper to lower electronic level degeneracies which is 1/2 for iodine. It is assumed that the saturated gain which satisfies $0 \le g \le g_o$, where g_o denotes the small signal gain, is constant along the direction of the flow. Physically this corresponds to a situation in which the flowing gas is enclosed in a Fabry-Perot resonator of reflectivity R with the saturated gain clamped at the threshold value, i.e., $g \equiv g_{thr} = -\ln(R)/2L_g$. Here L_g denotes the gain length of the medium.

The pump reaction producing the electronically excited iodine laser atoms which form the upper laser level is

$$O_2(\delta) + I \rightarrow O_2(\sigma) + I^* . \quad (3)$$

Because of the reaction small energy defect both the forward and reverse processes must be accounted for in an iodine medium model. With these assumptions the rate equation describing the steady-state population of the oxygen singlet-delta state is

$$u \frac{d}{dx} [\delta] = k_F [\delta][I] - k_R [\sigma][I^*] , \quad (4)$$

where u denotes the flow velocity, and k_F and k_R denotes the forward and reverse pump reaction rates, respectively. In the present work k_F has been taken to be 7.6×10^{-11} cm^3/molecules-sec and $k_R \equiv k_F/k_{eg}$ where the equilibrium constant of the reaction, k_{eg}, has been taken to be 0.75 exp (400/T) with the temperature T expressed in degrees Kelvin. Since the velocity, temperature, and iodine concentrations are independent of x under loaded conditions, Equations (1b) and (4) may be solved for the populations of $O_2(\delta)$ as a function of downstream distance.

Denoting the initial concentration of $O_2(\delta)$ at the nozzle exit plane by $[\delta_0]$ the solution of (4) may be expressed as

$$[\delta(x)] = [\delta_{ss}] + \{[\delta_0] - [\delta_{ss}]\}\exp(-\alpha x), \quad (5a)$$

where

$$[\delta_{ss}] = \frac{[O_2^{tot}]\{\beta[I^{tot}] + [N]\}}{(k_{eq} + \beta)\{[I^{tot}] - (k_{eq} - 1)[N]\}}, \quad (5b)$$

and

$$\alpha = \frac{k_R}{u(1+\beta)}(k_{eq} + \beta)\{[I^{tot}] - (k_{eq} - 1)[N]\}. \quad (5c)$$

It is worthwhile to examine this solution to determine the source of energy which maintains the inversion and provides the laser energy. In order for the $O_2(\delta)$ to remain constant along the flow direction the net forward reaction rate of the pump reaction must equal the net reverse reaction rate. Under these conditions reaction (3) is in equilibrium and the small signal gain as derived from Equations (1a) and (4) and the definition (2a) is

$$g_0 = \frac{\sigma[I^{tot}](k_F + \beta k_R)\{[\delta_0] - [\delta_{thr}]\}}{\frac{(1+\beta)}{\beta} k_R[\delta_{thr}] + (k_F - k_R)\{[\delta_0] - [\delta_{thr}]\}}, \qquad (6a)$$

where here

$$[\delta_{thr}] = \frac{\beta(k_R[O_2^{tot}])}{k_F + \beta k_R}. \qquad (6b)$$

The quantity $[\delta_{thr}]$ represents the density of $O_2(\delta)$ required to just maintain zero small signal gain, i.e., $g_0([\delta_{thr}]) \equiv 0$, and is just the amount of $O_2(\delta)$ required to maintain the pump reaction in chemical equilibrium. Finally, the model assumes that for every singlet-delta oxygen molecule lost a lasing photon is produced. Since deactivation processes have been assumed to be negligible in comparison to stimulated emission processes, energy conservation can be invoked to determine the laser photon density. The aforementioned assumptions imply that the rate of production of photons equals the rate of loss of the oxygen singlet-delta molecules or that

$$D_c[h\mu] = D_u[\delta], \qquad (7)$$

where $[h\mu]$ denotes the photon number density and D_v denotes the one-dimensional convective derivative of medium moving at speed v. In general, the photon energy per unit volume of the lasing medium is $E \equiv h\mu[h\mu]$ and the power per unit volume is the total rate of change of the E or

$$\frac{dP}{dV} = \frac{dE}{dt} \equiv D_c E. \qquad (8)$$

Thus, under steady-state conditions, we find upon substitution of equations (4), (5) and (7) into (8) that the power per unit volume of the gain medium is

$$\frac{dP}{dV} = h\mu\, u\, \alpha([\delta_0] - [\delta_{ss}])\exp(-\alpha x). \qquad (9)$$

Since the power per unit volume developed by a stimulated emission process is the product of the gain of the transition and the flux stimulating the transition (Reference 5), i.e., $dP/dV = gI$, the intracavity flux as a function of the downstream distance is

$$I(x) = \frac{1}{g} h\mu\, u\, \alpha\{[\delta_0] - [\delta_{ss}]\}\exp(-\alpha x), \qquad (10)$$

where it is to be remembered that $[\delta_{ss}]$ and α depend upon the loaded gain g. Equation (10) describes the saturation relationship between the loaded gain of the medium and the intracavity flux stimulating the medium. Observe that it is not the usual simple homogeneous saturation law often used in laser analysis (Reference 5).

Optical Analysis of a Folded Resonator for Application to PDN

In this section the COIL laser model is used to analyze a folded resonator concept for PDN. The analysis is generalized to account for the nonuniformity of the intensity along the flow direction (x) although it is still assumed that the optical beam is spatially uniform along the optical axis. Because the saturation relationship (10) is not the usual simple homogeneous law upon which Rigrod's analysis (Reference 6) is based, a Rigrod-like analysis accounting for this spatial dependence requires extensive numerical analysis. Such an analysis, although feasible, would not provide the qualitative insight obtained by the more analytical approach taken in the present work and is beyond the scope of paper. Neither the present analysis nor the Rigrod-like analysis would estimate the diffractive losses associated with the optical resonator. Two or three dimensional optics codes will be utilized to determine the diffractive losses (i.e., OISTAR code) in PDN development phases.

A schematic of the resonator configuration for the COIL photo-dissociation neutralizer is shown in Figure 1. The resonator is comprised of N mirrors arranged so that the optical beam is swept across the linear path of the negative ion beam N-1 times. The gain medium of length L_g is enclosed between a back mirror of reflectivity R_0 and a front window of transmissivity T. The remaining N-1 mirrors which optically fold the beam are assumed to be identical and to have a reflectivity R. For simplicity of this analysis the beam expander is not considered.

The geometric height of the optical mode, denoted here by H, is taken to be equal to the extent of the gain medium in the direction (y) perpendicular to both the flow and the optical axis. Assuming that any space between adjacent mirrors is small compared to the beam mode width, denoted here by D, the length of a single PDN system is

$$L_s = ND/2 , \tag{11a}$$

while the length of a PDN comprised of N_s systems, L_t, is

$$L_t = N_s L_s = N_s ND/2 . \tag{11b}$$

We will see further when the results are presented that a single PDN system (i.e., $N_s = 1$) can deliver a sufficient optical thickness, utilizing a current or near term COIL and mirror technologies. The optical thickness, denoted here by S, through which the negative ions pass is defined as

$$S = N_s \int_0^{L_s} \theta(x)\, dx = L_t \theta_{ave} , \tag{12a}$$

where

$$\theta_{ave} = \frac{1}{L_s} \int_0^{L_s} \theta(x)\, dx , \tag{12b}$$

and $\theta(x)$ denotes the optical intensity along the path of the ion-beam and θ_{ave} denotes the intensity averaged along this path for a single PDN system. If the effects of diffraction are neglected then as the laser mode propagates through the resonator, its profile, along the flow direction (x) in the gain medium will be repeated along each of the N-1 optical folds. Thus, denoting the average two-way intensity of the beam in the ith leg of the resonator by $\theta_i(x)$, we find

$$\theta_{ave} = \frac{1}{N/2} \sum_{i=1}^{N-1} \frac{1}{D} \int_0^D \theta_i(x)\, dx\,. \tag{13}$$

If the one-way flux in leg 1 of the resonator traveling from the window to mirror 1 is denoted by $\phi(x)$, then it is straightforward to show that the two-way flux in each of the $i = 1\ldots N-1$ legs is

$$\theta_i(x) = (R^{i-1} + R^{2N-3}/R^{i-1})\phi(x)\,, \tag{14}$$

and hence, substituting Equation (14) into (13) and performing the sum yields

$$\theta_{ave} = \phi_{ave} \frac{1}{N/2} \frac{(1 - R^{2N-2})}{(1 - R)} \tag{15a}$$

where

$$\phi_{ave} = \frac{1}{D} \int_0^D \phi(x)\, dx\,, \tag{15b}$$

As expected if the one-way flux in leg 1, $\phi(x)$, is assumed to be uniform along the flow, then Equations (15) reduce to the expressions derived by Fink (Reference 3). Since the one-way flux $\phi(x)$ is related to the two-way intercavity flux $I(x)$ introduced in the last section by $\phi(x) = 1/2\, T\, I(x)$, the average one-way flux in leg 1 is

$$\phi_{ave} = \frac{1}{2} T\, I_{ave}\,, \tag{16a}$$

where

$$I_{ave} = \frac{h\mu\, u\, \alpha\{[\delta_0] - [\delta_{ss}]\}}{g} \frac{(1 - \exp(-\alpha D))}{\alpha D}\,. \tag{16b}$$

It should be remembered that both $[\delta_{ss}]$ and α depend on the loaded gain of the medium which in turn depends on the resonator parameters.

Having derived the dependence of the optical thickness upon the medium parameters, it remains to determine the dependence of the loaded gain upon the resonator parameters. Within the approximations made in the present study, the folded PDN resonator is equivalent to the Fabry-Perot resonator shown schematically in Figure 5. This equivalent resonator consists of a back mirror of reflectivity R_0, a front window of transmissivity T, and a mirror of reflectivity $R_E \equiv R^{2N-3}$ which represents the N-1 external mirrors of the PDN resonator.

A simple analysis shows that the threshold gain to which the gain medium saturates when loaded is

$$g_{thr} = -\ln(R_0\, T^2\, R^{2N-3})/L_g\,. \tag{17}$$

Combining Equations (12), (15), and (16a), the optical thickness is

$$S = L_t \frac{T}{N} \frac{(1 - R^{2N-2})}{(1 - R)} I_{ave}\,, \tag{18a}$$

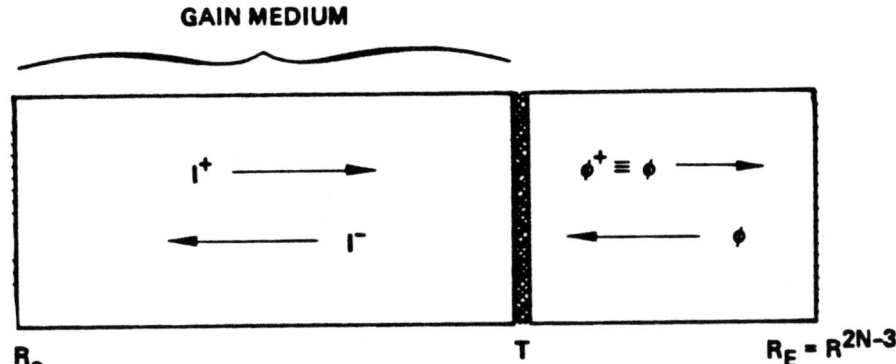

Fig. 5. Fabry-Perot equivalent resonator to the folded PDN resonator.

or upon substituting Equation (16b) for I_{ave} we find that

$$S = L_t \frac{T}{N} \frac{(1-R^{2N-2})}{(1-R)} \frac{h\mu u \alpha \{[\delta_0] - [\delta_{ss}]\}}{g_{thr}} \frac{(1-\exp(-\alpha D))}{\alpha D} . \qquad (18b)$$

Power Loss in the PDN Resonator

The primary loss of optical power is due to absorption and scattering at the window and mirror while diffractive loss occurs as the radiation propagates around the resonator. It is expected that the diffractive loss is negligible in comparison to these absorptive losses.

Scattering and absorption of the optical radiation also occurs in the gain medium but this is negligible because the COIL operating pressures are only a few torr. Finally, loss of optical energy to the negative ion beam is negligible. Using the notation of Figure 5, the power lost to the back mirror, P_{BM}, is the difference between the power leaving and entering the mirror or for N_s PDN systems

$$P_{BM} = N_s H \int_0^D (I^-(x) - I^+(x))dx = \frac{L_t H}{N}(1-R_0) I_{ave} . \qquad (19)$$

The power lost due to absorption in the window is the sum of the flux absorbed on each surface of the window or

$$P_W = N_s H (1-T) \int_0^D (I^+(x) + \phi^-(x))dx = \frac{L_t H}{N}(1-T)(1-R_E T)I_{ave}. (20)$$

Finally, the power lost in the N-1 external mirrors is the difference between the power entering and exiting the equivalent mirror R_E or

$$P_{EM} = N_s H \int_0^D (\phi^+(x) - \phi^-(x)) dx = \frac{L_t H}{N} (1 - R_E) T I_{ave} . \quad (21)$$

The total power loss in the PDN resonator is the sum of these contributions, Equations (19), (20), and (21), or

$$P_L = \frac{L_t H}{N} (1 - R_0 + 1 - T^2 R^{2N-3}) I_{ave} , \quad (22)$$

where the power loss is expressed in terms of the average intracavity flux. Next, we consider the power stored or "circulating" in the resonator which, for N_s PDN systems, is defined as

$$P_C \equiv N_s H \int_0^D I(x) dx = \frac{L_t H}{N/2} I_{ave} . \quad (23)$$

This definition will prove useful in deriving the cavity quality factor.

Finally, consider the cavity quality factor Q, and G factor introduced by Fink (References 3 and 4). The cavity Q is most generally defined as (Reference 5)

$$Q = \frac{\text{energy stored in the resonator}}{\text{energy dissipated in one optical cycle}} . \quad (24)$$

If the overall length of the resonator is denoted by L, then the energy stored during a single transit is $P_C(L/c)$, while the power loss per cycle is $P_L/(2\pi c/\mu)$ so that the Q of the resonator is

$$Q = \frac{2\pi L}{\mu} \frac{P_C}{P_L} = \frac{2\pi L}{\mu} \frac{1}{\frac{1}{2}(1 - R_0 + 1 - T^2 R^{2N-3})} , \quad (25)$$

This is a measure of the rate of decay of the power stored in the resonator. The larger the Q, the slower the decay of energy from the device. Notice that the optical cavity Q depends on the overall length of the resonator L which depends on the number of mirrors and the length of each leg. Now consider the G factor introduced by Fink. Using his definition (Reference 4) and Equations (18) and (22), we find that

$$G \equiv \frac{SH}{P_L} = \frac{T(1 - R^{2N-2})}{(1 - R)} \frac{Q}{2\pi L/\mu} . \quad (26)$$

The G factor is similar to the usual cavity Q factor but instead of measuring the ability of the entire resonator to store energy, it refers only to that portion of the resonator through which the ion beam passes. In other words, it is a measure of the ability of the PDN resonator to store optical energy within the path of the negative ion beam.

As an example, consider a PDN resonator with N = 10, T = 0.992,

$R_0 = R = 0.9995$. Combining Equations (25) and (26), the G factor is ~1400. For economic reasons G should be larger than 500 (see J. Fink, this conference paper). This value is still attainable for $T = 0.97$, thus leaving a sufficient room for uncertainty in T as discussed in a previous section (Window Technology).

RESULTS

The model of the COIL driven PDN device developed in the previous sections is used to examine the dependence of the optical thickness upon parameters of the medium and the resonator. To limit the study, only conditions typical of supersonic COIL device which are currently considered to be technically achievable have been considered. Although advanced $O_2(\delta)$ generator designs may yield as much as 80% $O_2(\delta)$, the initial fraction of O_2 which was assumed to be $O_2(\delta)$ was taken to be 60% which represents current generator technology. The relative mole fraction of the various species for the case considered here was $O_2/He/I_2 = 0.8/0.184/0.016$ so that the I_2 flow rate was 2% of the total O_2 flow rate. This is not necessarily an optimal gas mix. More detailed studies of the kinetics would be required to determine such an optimum mix for this application. The flow velocity was assumed to be 650 m/sec while the temperature of the medium was chosen as 200 K. Table I shows the small signal gain (SSG) as a function of pressure with relative mole fractions defined above.

TABLE I

Small Signal Gain (SSG) as a Function of Pressure

Pressure (Torr)	SSG (%/cm)
1	1.15
2	2.22
3	3.21
4	4.14

In all cases, the geometric mode size of the laser beam was 10 cm along the flow direction and 5 cm high perpendicular to the flow and optical axis. The back mirror and all of the external mirrors were assumed to be identical and have a reflectivity of 0.9995.

First consider the dependence of the optical thickness on the total gas pressure and gain length of the COIL medium.

For the conditions of interest $D \gg 1/\alpha$ which ensures that the finiteness of the mode will not penalize the resonator efficiency. Under these conditions, the optical thickness, Equation (18b), reduces to

$$S = \frac{N_s}{2} T \frac{(1 - R^{2N-2})}{(1 - R)} \frac{h\mu \, u \, \alpha \, \{[\delta_0] - [\delta_{ss}]\}}{g_{thr}}, \quad (27)$$

which indicates that the optical thickness depends linearly on both the pressure and the gain length. These conclusions are supported by

Figures 6 and 7 which show the optical thickness as a function of pressure and gain length for several values of the number of mirrors (N) and the window transmissivity (T). The two horizontal scales in Figure 6 indicate the dependence of the optical thickness, as

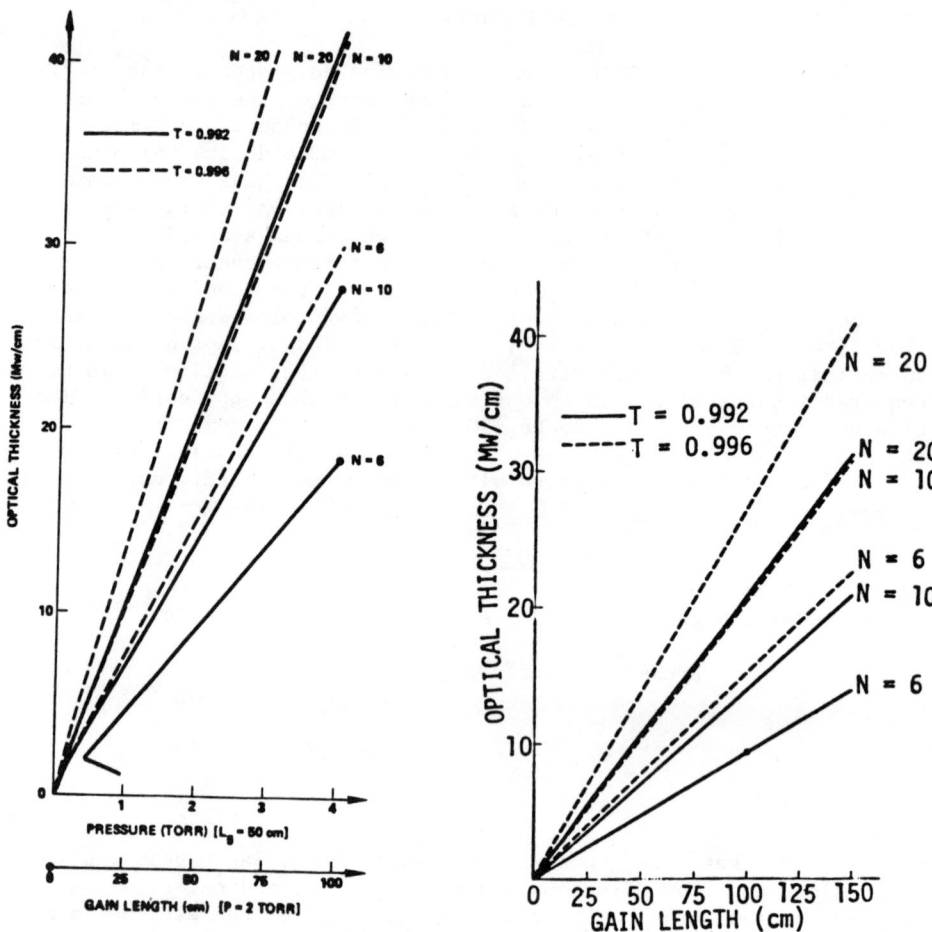

Fig. 6. Optical thickness as a function of gas at the gain length of 50 cm or equivalently as a function of gain length at a pressure of 2 Torr.

Fig. 7. Optical thickness as a function of gain length for a gas pressure of 1 Torr.

computed from Equation (18b), on the total medium pressure for a constant gain length of 50 cm or alternatively on the gain length for a total pressure of 2 Torr. The figures show that for a given number of mirrors the optical thickness decreases as T decreases. Also indicated is the decrease in optical thickness with decreasing N which is expected since the total length of the path of the ion beam through the resonator is proportional to N. Table I shows the small signal gain [Equations (6)] corresponding to each pressure although it should be remembered that mixing effects would reduce the available gain at each pressure. However, as indicated clearly by figures 6 and 7, the optical thickness required for a practical PDN device is certainly within the grasp of current COIL technology. Now consider the dependence of the optical thickness upon the number of mirrors (N) shown in Figure 8 for several different gain lengths (at 2 Torr total pressure) and window transmissivities. For a given length and window transmission there exists an optimum number of mirrors beyond which the optical thickness does not increase significantly. These results show that to achieve a given optical thickness, a larger number of mirrors is needed for shorter gain length device. Since the average flux incident on mirror 1 decreases as N increases, as shown in Figure 9, a large number of mirrors (N $\geq$ 10) is desirable to reduce the mirror fluxes. However, these advantages may be offset by increased mirror alignment complexity and cost. Data in Figure 9 define requirements on mirror heat exchangers for a given number of PDN mirrors, N.

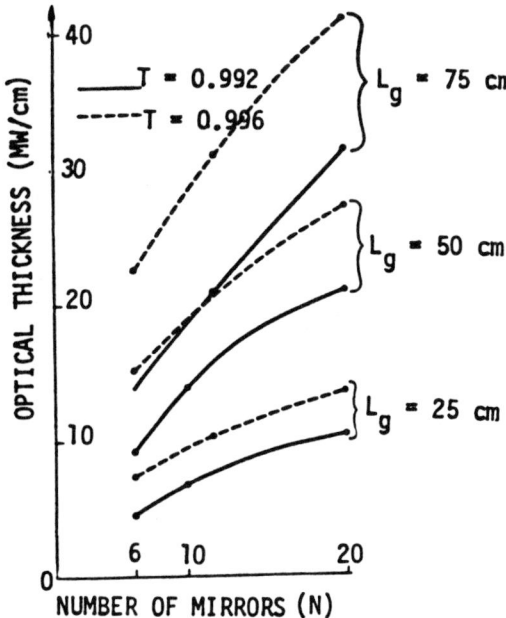
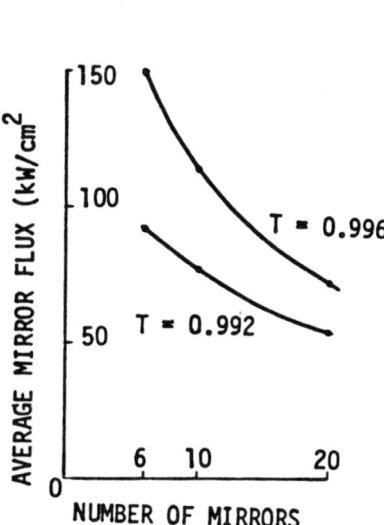

Fig. 8. Optical thickness as a function of the total number of mirrors in the resonator.

Fig. 9. Average flux incident on mirror 1 as a function of the total number of mirrors in the resonator.

CONCLUSIONS

We have shown that the present technology of the laser mirror heat exchangers is mature enough to support the claim that a high Q closed resonator for the PDN system can be built in the near future.

The window separating the interaction region from the laser gain medium will require additional development and tests to verify if its transmissivity satisfies the requirement $T \sim 0.99$. Finally the simple model and its application presented above show that the supersonic COIL technology can generate optical thickness in the interaction region large enough that only one PDN system (i.e., $N_5 = 1$) is required to achieve neutralization fraction 0.9 to 0.95 (i.e., $S \sim 10\text{-}20$ MW/cm).

REFERENCES

1. D. Copeland and V. Vanek, to be published.
2. Photodetachment Neutralizer Development--Laser Resonator Study, Final Report by V. Vanek et al., prepared for UCLLNL by TRW Inc., September 30, 1983, Contract Number W-7405-ENG-48, DOE.
3. J. H. Fink, LLNL Report, UCID - 19788 (1983).
4. J. H. Fink, LLNL Report, UCRL - 87301 (1982).
5. A. Yariv, "Quantum Electronics," 2nd Edition (Wiley, New York, 1975).
6. W. W. Rigrod, J. Appl. Phys, $\underline{36}$, 2487-2490 (1965).

BEAM SYSTEMS AND APPLICATIONS

Magnetic Fusion Energy
Heating Development Plan

H. Stanley Staten
Office of Fusion Energy
U.S. Department of Energy
Washington, DC

ABSTRACT

The future for neutral beam plasma heating is somewhat more cloudy than at the time of the last symposium. While considerable technical progress is evident, competition from Radio Frequency (RF) heating poses a challenge. Present plans to meet this challenge are presented. The US program is concentrating its efforts on a proof-of-principle demonstration effort with subsequent application to upgrades of Mirror Fusion Test Facility-B (MFTF-B) and possibly to Tokamak Fusion Core Experiment (TFCX).

I. INTRODUCTION

Neutral Beam Heating is the most used heating method in large magnetic fusion experiments. All large experiments depend upon neutral beam systems for some or all of the auxiliary heating needed to conduct experiments. Neutral beam systems are needed for specific functions in tandem mirror systems such as charge exchange pumping to maintain the desired plasma energy distribution function. RF heating is, however, the method of choice for future toroidal devices. Large RF heating experiments are planned for the Joint European Torus (JET) and JT-60. The reference heating method for the proposed TFCX is Ion Cyclotron Resonant Frequency (ICRF) heating with negative ion neutral beam heating as a backup.

It is important that the negative ion neutral beam community produce a useful and desirable product for the magnetic fusion program to justify the continued expenditure of significant resources over an extended period of time.

II. Neutral Beam System Requirements

The existing positive ion neutral beam systems being used in the largest, and most important, plasma confinement experiments have been the source of much learning by the fusion community. As a result, positive ion neutral beam systems for a fusion reactor are perceived to be too expensive and too difficult to operate and maintain. This later perception is particularly important in a fusion reactor environment. The neutral beam access to the plasma has been through a straight tube. This access allows the neutrons direct access to the complex internal structure of the neutral beam system. The resultant induced radioactivity of the internal components would make maintenance prohibitively difficult and expensive.

The cost of RF power generation equipment is considerably less than the cost of a neutral beam system. To first order the RF equipment is very similar to the existing neutral beam power supply equipment except that it operates at a lower voltage. The perceived advantages of RF heating are not only in cost, but in operation and maintenance. The RF wave launchers, while located at the plasma boundary, can conceivably be simple, cheap and easily removable for maintenance or replacement.

While the perceived advantages or disadvantages of RF plasma heating can be discussed in their own right, that is not the purpose of this paper. Rather, the discussion of RF heating is included to indicate that significant challenges lie ahead for the negative ion systems developer. Not only must the heating systems work, but they must have features that make them desirable. The competition for resources with the RF heating development activities will be keen.

III. Potential Needs

The appeal of neutral beams to the potential user is that the interaction of the beam with the plasma is essentially classical. Good experimental and theoretical agreement has allowed the systems designer to predict the power deposition profiles and to design systems that produce the desired heating within the limits of available beam energy. There is a question of Tokamak scaling when auxiliary heating is supplied, but this may well apply to all forms of auxiliary heating.

The major questions with respect to Tokamak applications are with respect to beam penetration and the capability to drive plasma currents. The first response to the beam penetration question has been to raise the beam energy, but this entails severe penalties in injection efficiency. As accelerated positive deuterium ion energy is increased beyond about 200 KeV, the efficiency of neutralizing the fast positively charged ions decreases to below 20 percent. (See Figure 1.) This low efficiency would necessitate a very large capital investment as well as increased operating costs. Above ~75 KeV/nucleon, negative ion injector systems, which have much higher neutralization efficiency, appear attractive. The energy of choice for Tokamak heating is in the 400-500 KeV range.

NEUTRALIZATION EFFICIENCY VERSUS ENERGY

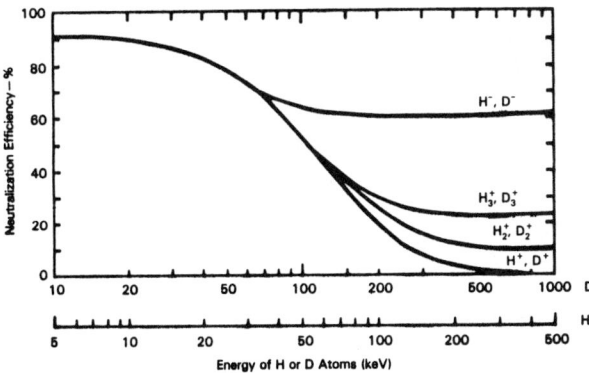

Figure 1. Neutralization Efficiency Versus Energy

Current in Tokamaks is presently provided by transformer action and by Lower Hybrid RF wave drive. Transformer drive is limited to pulse operation by the need to recharge the transformer flux. RF wave initiated and driven currents have been demonstrated, but the efficiency has been lower than desired for a reactor. A plasma density limit has also been observed above which RF current drive is not effective. While it has yet to be demonstrated, neutral beam current drive is predicted to be a viable option. The beam energies expected to be needed for a reactor are on the order of 1 MeV.

Tandem mirror fusion systems depend upon use of neutral beams to perform several functions. Not only are they used to heat the plasmas, but for specialized functions such as maintaining plasma potentials and non-isotropic distribution functions. While RF systems have been proposed and/or tested for some of these functions, neutral beams are still to be required for the others.

IV. Development Plans

The US negative ion development activity has undergone considerable change since the last symposium. The competition for source concept selection has been completed with the selection of the Lawrence Berkeley Laboratory (LBL) surface conversion source. The volume sources envisioned at the last symposium have been operated and are looked upon as the logical backup, if not replacement, of surface conversion sources. The Transverse Field Focusing (TFF) negative ion accelerator has been invented. Construction of a proof-of-principle system has been undertaken with initial 80 KeV tests beginning this winter and plans for operation at up to 160 KeV in about a year.

The program is conscious of the perceived disadvantages of neutral beams in a reactor environment and has initiated development of techniques that will minimize these concerns. As we have seen in other papers at this symposium, the TFF accelerator is capable of bending the negative ion beam around a corner and transporting it through a maze so that the source and accelerator can be shielded from the plasma; thus, overcoming one of the largest perceived disadvantages of a neutral beam system in a reactor.

LBL has been chosen to take the lead in development of negative ion neutral beam systems. LBL has just completed a comprehensive planning activity for the development of negative ion neutral beams. The plan includes a proof-of-principle demonstration at the 1-2 ampere, 160 KeV level where it can be done economically with existing positive ion test facilities. Based on a successful proof-of-principle demonstration, a dedicated 500 KeV facility utilizing the Neutral Beam Engineering Test Facility (NBETF) at LBL is planned. The overall schedules are shown in Figure 2. Note that this will require increasing the resource commitment in Fiscal Year 1985 beginning with the initiation of the NBETF negative ion conversion.

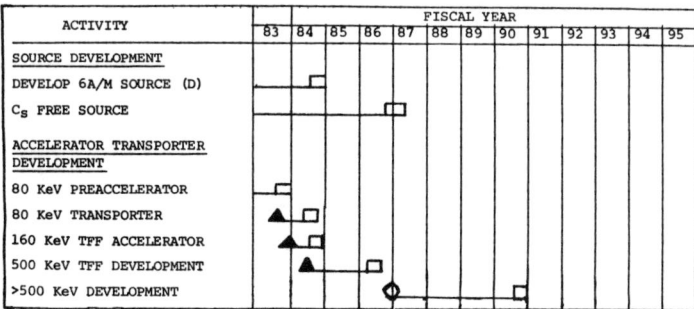

Figure 2. Overall US Negative Ion Neutral Beam Development Schedules.

The 1-2 ampere test module is likened to the 1/5 scale tests of positive ion neutral beam sources that were successful in developing the technology while minimizing the expenses of full size test facilities and sources. The successful completion of the 160 KeV proof-of principle demonstration should provide the confidence to plan for the commitment to the NBETF conversion and to applications.

The first application to a confinement experiment may occur as early as about 1990 with the application to an upgrade to the MFTF-B facility.

V. Summary

In summary, we have seen that the negative ion development activity has made considerable progress since the last symposium. A concept has been selected as the reference source, and a desirable alternate has been identified. A new acceleration concept has been invented and proof-of-principle systems demonstration initiated. Plans to convert a major facility, the NBETF, to test large systems, have been initiated. The next few years should prove to be exciting ones for those in negative ion development.

DISCUSSION

MORGAN: Could you just say something about who is working on RF, what is the effort there?
STATEN: The large RF technology programs are at Princeton University and Oak Ridge National Laboratory.
HERSHCOVITCH: I know there were, at least as of a few years ago, industrial development programs which were DOE funded, for example, Varian has been developing gyrotrons for EBT.
STATEN: The gyrotron program is continuing. Right now it is partly funded by the defense programs. There is an isotope separation process that depends on gyrotrons or plasma formation and it is partly funded by them and partly funded by our office. I did not mention that program. I would expect to see that continue roughly at the same level effort.

REQUIREMENTS FOR NEGATIVE ION BASED SYSTEMS
FROM USERS' POINT OF VIEW

L. D. Stewart
Plasma Physics Laboratory, Princeton University
Princeton, New Jersey 08544

ABSTRACT

This paper reviews the projected requirements of tokamak and mirror confinement devices for negative ion based neutral beam systems. The physics basis for the requirements are reviewed; the stated requirements are summarized for various mirror and tokamak devices, including MFTF-α+T, MARS, FED-A and FED-R; and changes in those stated requirements, based on improved calculations of beam-stopping cross sections, are estimated.

A. INTRODUCTION

The high beam energies required if neutral beams are used to heat, drive current in, or fuel reactor plasmas dictate negative ion based systems. Next generation confinement devices, now being designed, intended to lead us to reactors, in most cases are already suitable for high energy negative ion based systems. Thus, these devices could be used as intermediate applications for the beam technology, as well as intermediate steps in reactor confinement technology, if projected beam requirements match with projected available beam technology.

This paper summarizes present projections of future mirror and tokamak requirements for neutral beams in the energy range suitable for negative ions. The physics basis for the requirements are reviewed, the stated requirements are summarized for various mirror and tokamak devices, and changes in those requirements, based on recent, improved calculations of beam-stopping cross sections are estimated.

B. MIRROR USES[1,2]

1. Center-Cell Heating and Fueling

In many tandem mirror design studies, neutral beams are used to maintain the center-cell temperatures and density against energy and particle losses. In recent design studies however, there has been a tendency to use other technologies for this function. For example, as the recent Mirror Advanced Reactor Study (MARS) effort progressed, the 200 keV heating and fueling beams were replaced by ICRF for heating and by cold passing ions for fueling, principally because sufficient duct space could not be maintained in the design.[3,4]

2. Transition Pumping[3]

Center-cell ions passing through the transition region scatter in pitch angle and become trapped. Unless some of these ions are removed, the density will increase, leading to an undesirable increase of the potential in the transition region. Many of the trapped ions have turning points in the anchor, and these will increase the density and potential at the thermal barrier. The net effect is a decreased thermal insulation between the center-cell and the warm electrons, a cooling of these electrons, and a lowering of the confining potential of the center-cell ions.

A pump beam can be used to counter this effect. The beam energy and direction are chosen so that ions, resulting from charge-exchange of the beam atoms, pass into the central cell. The neutrals formed from the trapped ions pass radially out of the plasma.

Although the energy required of pump beams is generally low enough such that positive ion beams could be considered, very high energy purity is needed (because lower energy ions will fill the velocity space which is being pumped), and negative ion based systems satisfy this requirement.

As the MARS study evolved, the 94 keV pump beams were replaced by Drift Pumping.[3,4] However, it should be noted that the physics basis for this drift pumping concept is still relatively speculative.

3. Anchor Sloshing Ion Beams[5]

A related function of neutral beams which is essential in the present tandem mirror concept is to help form the thermal barriers by maintaining the sloshing ions. High energy ions are generated using neutral beams aimed slightly off the midplane of the end plugs. These ions are trapped in the magnetic well formed by the end plug magnets. The ions slosh back and forth, thus forming a potential minimum at the center of the magnetic well. The potential minimum separates the cold electrons in the central cell from the hot electrons in the end plugs. At the same time, low energy ions collect and fill out the ion distribution function, damping oscillations that cause microinstabilities. The potential minimum must be large enough to hold back the hot electrons while sustaining an adequate density of low energy ions.

The present design parameters for the Mirror Fusion Test Facility upgrade, MFTF-α+T,[6] and for MARS[4] include sloshing ion beams at energies in the negative ion range.

C. TOKAMAK USES[1,2]

1. Bulk Heating and Fueling

Neutral beams can be used to heat tokamak plasmas from temperatures characteristic of ohmically heated plasmas (~ 1 keV) to those required for ignition and/or sustained burn (~ 10 keV). This application is analagous to center-cell heating and fueling in tandem mirrors. The physics is well understood both experimentally and theoretically.

Although energetic neutral beams can provide local fueling, it is not possible to use high energy beams to totally fuel a reactor while still maintaining net power production: too much power is required for the beams. Therefore, reactors must be primarily fueled by a non-energy intensive method, such as gas puffs or high velocity frozen pellets. Beams might still be used to provide local fueling in the plasma center if it is found to be advantageous.

The beam energy required for tokamak reactor heating is generally agreed to be in the range of 150 to 500 keV. Most of the recent design studies for future large tokamaks, including the near-term Tokamak Fusion Core Experiment (TFCX), use ICRF heating and pellet fueling in place of neutral beams. The perceived advantages of RF heating techniques are less complexity, smaller size (adjacent to the tokamak), lower cost, less of a neutron-streaming problem, and nearer-term technology.

2. Current Drive

Neutral beams might be used to drive the toroidal plasma current that provides the confining poloidal field in a tokamak reactor, thus enabling a reduced dependence on the ohmic heating transformer, or possibly enabling steady-state operation. In tokamak design studies, a reduction in the performance requirements for the ohmic heating system leads to simplification, enhanced performance, and/or cost reduction. A tokamak reactor which can operate continuously may have reliability and cost advantages.

A disadvantage of neutral beam current drive is that it results in net toroidal momentum for the plasma. Experimentally, however, no deleterious effects have yet been observed, even with strongly unbalanced toroidal beam power input.[7] It should also be noted that the calculated neutral beam power requirements for current drive at full operating densities and temperatures lead to projections of high circulating power in reactors. However, theoretical and experimental current drive efficiencies for other candidate drivers, at these temperatures and densities, are no more promising.

Beam energies required for tokamak reactor current drive are in the range of 500 keV to 1 MeV. Most recent design studies of future large tokamaks, including the near-term TFCX, use lower hybrid current drive systems rather than neutral beams.

One advantage that neutral beams may have in tokamak applications is that it appears that a single system, 500 keV-1MeV neutral beams, could be used to both heat and drive the current in a tokamak reactor. In contrast, the presently preferred auxiliary heating technology, ICRF, probably will not efficiently drive current. The presently preferred current drive technology, lower hybrid frequency waves, is found experimentally and theoretically to be ineffective in heating the higher density and temperature tokamak plasmas characteristic of a reactor. Having one system do both the heating and the current drive may lead to simplification, and therefore improved reliability and reduced cost, in a tokamak reactor.

D. ENHANCEMENT OF THE BEAM-STOPPING CROSS SECTION

Recent work by Boley, et. al.[8,9] indicates that the beam-stopping cross sections, applied to calculations of energetic neutral beams of hydrogen isotopes incident on a plasma, should be increased to account for multistep processes involving excited atomic states.[10] Calculations of beam penetration have generally assumed this effect was small, and use only cross sections for ionization of ground-state neutrals.

For typical plasma and beam parameters of present large tokamak experiments, this enhancement is expected to be from 25% to 50%. The enhancement is predicted to increase with plasma density and with energy, and can be much larger for tokamak and mirror reactors.

In many tokamak designs, the beam energy is determined by the penetration requirements. Thus, in these cases, the enhancement of the stopping cross section will lead to specification of higher beam energies and, therefore, to more of a demand on beam technology.

For mirrors, the beam energy is generally fixed by requirements other than penetration, and, in fact, shine-through is usually a problem. Thus, in the mirror case, the enhancement of the stopping cross section may lead to lowered demands on the beam dump and the beam injector power capability, including source current capability, beam mechanical hardware, power supplies, and pumping, and may mean an improved mirror system power balance.

Specific examples of the changes in perceived user requirements, as a result of application of this recent work, are given below.

E. USER REQUIREMENTS

Machine designs or machine design studies which are current and which include requirements for negative ion based neutral beam systems are MFTF-α+T (tandem mirror), MARS (tandem mirror), FED-A (tokamak), and FED-R (tokamak). The parameters for these beam systems are presented and discussed below. The most serious next-device study now ongoing, the TFCX tokamak, presently considers negative ion based neutral beams as a backup to ICRH for plasma heating and has not established beam requirements.

1. MFTF-α+T Requirements[5,6]

The presently favored upgrade of MFTF-B, called MFTF-α+T, includes 200 keV beams to generate sloshing ions in the end plugs. A current of 8.4 A of 200 keV deuterium atoms is specified to be delivered to the plasma: one-half to each end plug. Additional requirements are the capability for running continuously for periods of 100 hours, compatibility with a radiation environment, and optics suitable for a relatively narrow, 14.7 cm, plasma target. MFTF-α+T is proposed for startup in 1992.

Applying the enhanced beam-stopping cross sections described above, to an MFTF-α+T end plug plasma with $\bar{n} = 5 \times 10^{12}$ cm^{-3} and $Z_{eff} = 1$, yields an estimate of an increase in the trapping cross section of about 30%. When the previous trapping fraction (0.15) is combined with the enhanced trapping, the incident current required is reduced to 6.6 A and the beam dump power is reduced by 0.36 MW. The required source current may be reduced more than linearly, because the source width could be reduced and the beam system is expected to be optics limited.

2. MARS Requirements[11]

The recently completed MARS tandem mirror reactor study invokes 475 keV neutral beams to provide sloshing ions for each of two anchors. A total of 5.7 MW is to be absorbed by the plasma, and 8.9 MW is calculated as the required incident power. Additional requirements are steady state operation, compatibility with a radiation environment, and very high availability: >97% for the total heating system, including beams, ECRH, ICRH, and all of the associated power supplies, pumps, controls, etc. In the design study, such high availability was considered reasonable because MARS is specified as a tenth-of-a-kind reactor.

Enhanced beam-stopping cross sections for the MARS anchors, with $\bar{n} = 5 \times 10^{13}$ cm^{-3} and $Z_{eff} = 1$, yield an estimated 40% increase in the trapping cross section. If it is assumed that deposition profiles are unaffected, and thus 5.7 MW is still the required absorbed power, only 7.5 MW of incident power is needed, and the beam dump requirement drops from 3.2 MW to 1.8 MW.

3. FED-A Requirements[12-14]

The FED-A tokamak study is a recent extension of the Fusion Engineering Device (FED) studies which is aimed at possible improvements in the original concept allowed by more agressive physics assumptions. A key physics assumption for FED-A is non-inductive current drive. The baseline current drive technology for the study is lower hybrid waves, while two alternative technologies considered are based on negative ion beam systems: 400 keV beams to provide a cyclic, "internal transformer" mode of current drive, and 800 keV beams to provide steady state current drive.

For the 400 keV, internal transformer case, the required power is 44 MW, and the beams are pulsed for 4 sec every 33 sec (plasma cycle time). For the 800 keV case, 50 MW of steady state beam power is specified to drive the 4 MA of plasma current. High availability and radiation compatibility are requirements in both cases, although in the cyclic case, the possibility exists for closing off the beam duct during most of the plasma burn phase.

The effect of the enhanced beam-stopping cross is quite dramatic for the 400 keV cyclic case. The Z_{eff} is purposely raised to 6 during the drive phase, and for the mean density $\bar{n} = 1 \times 10^{13}$ cm^{-3}, the stopping cross section would be enhanced by a factor of about 2.25. However, it is beam shine-through which determines the lower limit on the plasma density during the drive phase, and with an enhanced stopping cross section, this density can be lowered and the current drive efficiency will be raised.[15] The result is that the required beam power drops from 44 MW to 20 MW.

In the 800 keV steady-state case, the beam energy is determined by current drive considerations rather than penetration, so even though the stopping cross section is estimated to increase by about 75%, it is expected that penetration will still be sufficient and the beam requirements will be unchanged.[16]

4. FED-R Requirements[17]

The FED-R is a recently studied variation of FED utilizng resistive magnets, rather than superconducting, and aimed at providing a substantial fluence of fusion neutrons to irradiate a large test area in order to do blanket testing. A Stage II of this device, to be operable in the late 1990's, specifies 100 MW of incident neutral beams to both heat the plasma and drive the plasma current (4 MA). Half of the power is required to be 250 keV deuterium and half is to be 250 keV tritium. The blanket testing objective necessarily means that the beam systems must be radiation compatible and steady-state with high availability.

The beam energies were determined from penetration considerations, so enhanced beam-stopping cross sections will mean higher energy beams are required for proper penetration. For a mean density $\bar{n} = 8 \times 10^{13}$ cm^{-3} and $Z_{eff} = 1.5$, equivalent penetration will be provided with approximately 400 keV beams. It should be noted, however, that the FED-R design philosophy was to use near-term technology, so that some of this increment in beam energy might be absorbed in overall machine parameter and goal changes.

SUMMARY

User requirements are summarized in the Table. Listed are the approximate device startup date, the isotope, and the beam energy, current and power incident on the plasma, as dictated by the user. The modification to the energy, current and power, estimated to arise from recent re-calculations of beam-stopping cross sections, are also listed. Additional requirements for all of the listed systems are high availability and radiation compatibility. The FED-R 400 keV system requires 4 sec pulses every 33 sec, while the remaining systems are steady-state.

A dominant recent trend is a shift by all users to various RF techniques to replace beams. Tandem mirrors are more firmly committed to beams, with the principal application being sloshing ion beams, requiring 200 keV for the near-term and ~ 500 keV for reactors. The trend in tokamaks is towards specifying beams only as a technology back-up to RF.

In general, the effect of the enhanced stopping cross section appears to be less ambitious beam systems for tandem mirrors while for tokamaks it depends on the particular case: more attractive systems in some cases and somewhat less attractive systems in other cases.

ACKNOWLEDGMENT

This work was supported by the U. S. Department of Energy Contract No. DE-AC02-76-CHO-3073.

TABLE
USER REQUIREMENTS

	PUBLISHED REQUIREMENTS					REQUIREMENTS AS MODIFIED BY ENHANCED STOPPING		
	YEAR	ATOM	keV	A*	MW*	keV	A*	MW*
MFTF-α+T	1992	D	200	8.4	1.7	200	6.6	1.3
MARS	2020	D	475	19	8.9	475	16	7.5
FED-A[†]	~1995	D	400	110	44	400	50	20
FED-A[†]	~1995	D	800	63	50	800	63	50
FED-R	1998	D T	250 250	200 200	50 50	400 400	125 125	50 50

*Incident on the plasma
†Alternative approaches for FED-A

REFERENCES

1. "Plasma Heating Technology Program Plan," Oak Ridge National Laboratory Report, September 1981.
2. "The National Negative-Ion-Based Neutral Beam Development Plan," Lawrence Berkeley National Laboratory Report No. PUB-464, August 1983.
3. "Mirror Advanced Reactor Study Interim Design Report," Lawrence Livermore National Laboratory Report No. UCRL-3333, April 1983.
4. "Mirror Advanced Reactor Study Final Design Report," to be published.
5. "Neutral Beams for Mirrors," J. H. Fink, Lawrence Livermore National Laboratory Report No. UCRL-89166, Aug. 1983.
6. "Options to Upgrade the Mirror Fusion Test Facility," edited by K. I. Thomassen and J. N. Doggett, Lawrence Livermore National Laboratory Report No. UCID-19743, April, 1983.
7. "Neutral Beam Heating of Tokamaks: Past Performance and Future Applications," H. P. Eubank, presented at the American Vacuum Society Fusion Technology Division Meeting, Boston, November 1983.
8. "Enhancement of the Neutral Beam Stopping Cross Section in Fusion Plasmas due to Multistep Collision Processes," C. D. Boley, R. K. Janev, and D. E. Post, Princeton Plasma Physics Laboratory Report No. PPPL-2047, Oct. 1983.
9. "Enhancement of Beam Stopping Cross Sections Due to Multistep Processes," D. E. Post, R. K. Janev, and C. D. Boley, this conference.
10. "The Role of Excitation and Ionization Processes by Charged Particles in the Total Ionization of an Excited Neutral Beam," K. H. Berkner and J. R. Hiskes, presented at the Conference on Electronic and Atomic Collisions, Beograd, Yugoslavia, 1973.
11. "Neutral Beam Injector for 475 keV Mars Sloshing Ions," D. M. Goebel and J. H. Fink, this conference.
12. "FED-A: An Advanced Performance FED Based on Low Safety Factor and Current Drive," Fusion Engineering Design Center Report No. ORNL/FEDC-83/1, August, 1983.
13. "Continuous Tokamak Operation with an Internal Transformer," C. E. Singer and D. R. Mikkelsen, J. of Fusion Energy $\underline{3}$, 13 (1983).
14. "Optimization of Steady-State Beam-Driven Tokamak Reactors," D. R. Mikkelsen and C. E. Singer, Nuc. Tech./Fusion $\underline{4}$, 237 (1983).
15. C. E. Singer, private communication.
16. D. R. Mikkelsen, private communication.
17. "FED-R: A Fusion Engineering Device Utilizing Resistive Magnets," Fusion Engineering Design Center Report No. ORNL/FEDC-82/1, April, 1983.

DISCUSSION

JACQUOT: I would like to make a comment. With the neutron flux the problem of the insulators is a very important one in the RF scheme. It is also a disadvantage of the RF system, this problem of the insulators. There are ceramic windows for the wave guides, feed throughs, insulators for the antenna.

COOPER: Are you sure the MFTF - α + T pulse length is 100 hours and not 10 hours? I think we better check that, 10 hours is bad enough.

FINK: It is 100 hours.

COOPER: This is certainly a dramatic example of how one should not only understand the atomic physics but do it correctly. There is some tremendous money swinging back and forth here on the atomic physics of the trapping.

BARNETT: In view of the uncertainties of the cross sections used in the enhanced stopping, I don't think you should take too seriously the advantages you are going to gain by this in a 10 keV plasma.

STEWART: How about relative advantages?

BARNETT: Relative to what?

STEWART: When one says that the power goes from 8.9 MW to 7.4 MW or something, do we obtain that relative advantage?

BARNETT: I don't think you can believe it because you don't know some of these cross sections within that percentage. It will help some, but we don't know how much.

STATEN: Let me just put this a little bit in perspective. The idea of 100 MW of heating power boggles the mind. We did a study of power supplies a few years ago and we found that the whole electrical load at Cambridge, Mass. was only 32 MW. I took the numbers for the FED, we had 50 MW in the D beam, 50 MW in the T beam, at 400 keV, and that comes out to 250 A of negative ions we need to convert to neutrals and inject. We're talking about one ampere sources today. We're talking about going to multiples of a few, like 4 or 5 times that. There's still a long road ahead of us.

DAGENHART: Where did this 97% figure come from for the heating reliability? My understanding is that the availability of steam plants is only 75% and I would project that the nuclear cores is the thing that's got the most problems. Are we going to allow the heat exchanger people to take up the rest of the unreliability? That sounds unreasonable.

STEWART: It's part of the MARS design study and what they did was to assign reliabilities or availabilities. There is a lot of systems in fusion reactors, as you well know, and if one stacks up lot of 98's and 97's and so on, one gets down to the acceptable plant efficiency. They justify that very high number by saying, that this is the tenth of a kind, well in the next century, and we're going to know how to do that at that time.

EHLERS: I sometimes have to object to worrying at this stage about things like reliability, neutron sensitivity, and all these things until we've really proved that a fusion device can certainly do more than break even. It seems to me it's like having to build a 747 to prove the principle of flight which could just as well be done with the little plane that the Wright Brothers used.

McFARLAND: At the Los Angeles meeting, Doug Post showed some TV pictures of RF heating and the effect that they had on limiters and parts of the tokamak. I kind of agree with your last comments about how reliable they're going to be in the sense that there were large spots of impurities that were flaking off of the metal surfaces showing up as very bright spots in the plasma. I think that they just really got started in the problem areas that have to do with RF heating.

CONSIDERATIONS INVOLVED IN THE DESIGN OF NEGATIVE-ION-BASED NEUTRAL BEAM SYSTEMS

William S. Cooper
Lawrence Berkeley Laboratory
University of California
Berkeley, CA 94720

ABSTRACT

We consider the requirements and constraints for negative-ion-based neutral beam injection systems, and show how these are reflected in design considerations. We will attempt to develop a set of guidelines for users and developers to use to see how well (in a qualitative sense, at least) a particular neutral beam system fits a particular proposed need.

INTRODUCTION

A neutral beam injection system, whether based on negative ions or on positive ions, is a complex entity. It is composed of interacting and interlocking components, which cannot be designed separately in isolation and simply fitted together. Similarly, a neutral beam injection system is not simply an "add-on" to a reactor; each must take into account the constraints and limitations imposed by the other. I will try in the following discussion to outline, at least in a qualitative sense, how these constraints affect the design of a negative-ion-based neutral beam system.

APPLICATIONS

Neutral beams can perform several necessary functions for fusion reactors. The most obvious application is to heat the confined plasma. Heating of confined plasmas by beams of neutral atoms has been adequately demonstrated, both for mirror and for tokamak experiments; the highest temperatures reached to date in both cases have been achieved by neutral beam heating.

Another proposed application is to utilize the momentum transferred from the injection of neutral beams tangentially into a tokamak to drive a circulating current to aid in plasma confinement. This application for "current drive" has not yet been convincingly demonstrated.

Tandem mirror reactors will require the injection of neutral beams into the end plugs of the reactor to create a "sloshing ion" population to aid in the electrostatic confinement of ions in the axial direction. It is likely that this application will be the first test of negative-ion-based neutral beam systems on a reactor. We will consider the implications of this choice in subsequent discussions.

In the following discussion, we will examine the various system requirements and constraints, and consider the implications on the design of the neutral beam system. The information is also summarized in short form in Table I.

SYSTEM REQUIREMENTS, CONSTRAINTS, AND DESIGN IMPLICATIONS

A. FUNCTIONAL CONSTRAINTS

<u>Beam Power</u>: Next generation tokamaks, if they use neutral beams for heating or current drive, will require the injection of 20-50 MW of neutral beam power.[1] A more likely first application would be the injection of about 2 MW (1 MW into each end of the machine) into an upgrade of MFTF-B, the so-called MFTF-α+T version,[2] to create a sloshing ion population in the end plugs. This mirror need could be satisfied by neutral injector modules handling 1 or 2 MW per module; the tokamak requirements would require 5 to 10 MW per module. In any case, the neutral beam injection system will have to handle quite a few MW of power per injector, which means that the systems will have to be carefully protected from self-destruction.

<u>Beam energy</u>: Beam energy requirements range from 200 keV for the mirror application just discussed to 400 to 800 keV for heating or current drive of next-generation tokamaks. The implication of this requirement is that we must devise accelerators capable of accelerating deuterium ions to these energies without excessively frequent sparking. Since current-carrying capability is directly related to the electric field in the accelerator, a conservative design that leads to infrequent sparking will also result in a design with a low current-carrying capability compared to a less conservative design. A clear understanding of the requirements is necessary, as well as a compromise in one area or the other.

<u>Angular Divergence</u>: Access for neutral beams to a reactor is usually limited, in the case of tokamaks, by the size of apertures between magnetic field coils, and in the case of mirror machines, by this plus the small size of the plasma. In all cases, from a neutronics point of view, it is desirable to minimize the area of penetrations through the shielding of the reactor. Typical angular acceptances for tokamaks are $\pm 0.5°$ by $\pm 1°$. The plasma target in MFTF-α+T presents a smaller target: the target is 10 cm wide, at a distance of 9 meters. This target subtends a half-angle of only 0.32°. This small acceptance angle places stringent restrictions on the maximum transverse energy that the negative ions can have, and on the quality of the accelerating and transport systems. For the 200 keV MFTF- -T application, the maximum transverse energy acceptable, assuming perfect acceleration and transport, is 200 $\tan^2$ (0.32) keV, or 6.2 eV. This assumes no compression of the beam during transport and acceleration. With slot accelerators, the beam is typically compressed a factor of 3, which reduces the maximum tolerable ion

TABLE I
SYSTEM REQUIREMENTS, CONSTRAINTS, AND DESIGN IMPLICATIONS

CLASSIFICATION OF REQUIREMENT	REQUIREMENT	CONSTRAINTS AND DESIGN IMPLICATIONS
Functional	Inject 2-50 MW	Protect system against self-destruction
	Energy 200-800 keV	Balance current density against sparking
	Small target size	Ion transverse energies <1 to 5 eV, or Use high current density ion source
	Pulse length ⩾ 100 hours	Actively cool beamline components, Vacuum pumps capable of on-line regeneration
	>99% reliability	Conservative and simple design, Extensive testing of prototypes
	Beam purity	Develop ways to remove impurities from reactor, or from beam
Environmental	Radiation	Transport beam through shielding, Design for remote handling, Source/accelerator bakable, Tokamak and mirror injector designs likely to diverge
	Magnetic fields	Active or passive shielding must be provided
Economic	Space	Neutral beam injectors should occupy no more space than is required to assemble or dismantle reactor
	Efficiency	DC acceleraton rather than RF, Pump gas at high pressure, Sheet-like beams, Probable choice of laser photo-neutralizer, < 5% electrons in beam
	Cost	Should be competitive with RF systems

transverse energy by a factor of 3^2, to 0.7 eV. These numbers may be used as an initial check to see if a particular concept for a negative ion source might be suitable for this application. Surface-conversion sources, with transverse energies of about 5 eV, barely pass this test and then only if the source and accelerator can be oriented in the "good" direction. Volume-production sources, with transverse energies of about 0.5 eV,[4] will certainly be acceptable from this standpoint after they have been developed to the point of producing interesting quantities of negative ions. Sources producing negative ions with larger transverse energies are not necessarily excluded from consideration; if the currents produced are large enough, the fraction of ions with excessive transverse energy can always be thrown away, leaving only those ions with acceptable transverse energies.

Pulse lengths: Pulse lengths range from a proposed 100 hours for MFTF-α+T to weeks or months for a bona-fide reactor. Anything over a few seconds should be considered steady-state, which means that all portions of the beamline that beam particles could possibly hit should be water-cooled. Another implication of these long pulse lengths is that pumps used in the vacuum system must be capable of being recycled on-line. Reactors will use only a relatively few number of beamlines, sometimes only one per critical application (as in the case of the sloshing ion beams for MFTF-α+T), and so the possibility of simply shutting down the beamline for routine recycling of cryopumps does not exist. The time until recycling is required, in the case of cryopumps or other pumps that store deuterium, may be set either by the inventory of deuterium necessary to make an explosive mixture in the event of an up-to-air accident, or by the allowable tritium inventory in the beamline.

Reliability: This is a very critical area, to which, unfortunately, very little development effort has been devoted. As discussed above, since so few sources may be involved, each one must operate very reliably. Even the positive sloshing ion beams for MFTF-B require 99% reliability; the requirement will be even higher for a reactor. The way to achieve reliability is first by conservative design, which unfortunately runs counter to the simultaneous requirement of high performance, and second, by building as many units as is practical (or affordable), and running them to obtain reliable statistical data on failure modes. This is also not a palatable approach for first-of-a-kind reactor designers, and so I believe that this is going to remain a serious problem. We should continuously strive for simpler systems, for passive rather than active controls, and for systems that do not have important variables that are difficult to control. Cesium control in surface-conversion sources is one example of the latter problem, as is the problem of the simultaneous control of the primary beam, a metal vapor jet, and a second, high energy, accelerator in the case of systems utilizing double electron capture. One of the strongest arguments against neutral beam systems is their complexity, with the attendant problems of reliability.

Beam Purity: Negative ion beams typically contain a percent or so of atoms or molecules other than the desired D$^-$ ion. Typical impurities are O$^-$, O$_2^-$, OH$^-$, and CH$^-$. The operation of mirror machines is very sensitive to these impurities; the operation of tokamaks is not. In the worst case, mirror machines can only stand 10^{-6} of O in the D beam for 30-sec pulses (MFTF-B). Worse yet, the ions accumulate, which means the beam purity becomes even more critical for longer pulse lengths. Clearly some means must be found to remove the offending ions from the machine. Forms of resonant pumping have been proposed, but not yet tried. Although a suitable first step is to improve the beam purity by momentum selection or by some other means (such as choice of a wavelength in a photoneutralizer to selectively discriminate against conversion of impurity negative ions to neutrals), in the long run some other solution must be found. It is likely, therefore, that this constraint will be relaxed for neutral beam systems on functioning reactors.

B. ENVIRONMENTAL CONSTRAINTS

The requirements on performance of negative-ion-based neutral beam systems that we have discussed so far fall into the category of functional, first-order requirements on the injection systems. That is, the experiment or reactor will not function if these requirements are not met. Other classes of requirements and constraints fall into different categories, such as environmental (we are speaking of the problems associated with the environment of the beamline) or economical; we turn to each of these now in turn.

Radiation Environment: Fusion reactors produce neutrons. These neutrons will come out through the shielding penetrations that permit the neutral beams to enter the reactor; consideration must therefore be given to problems associated with activation of parts of the neutral beam injection system. One solution would be to make the neutral beam system remotely maintainable. To anyone who has seen a large negative ion source, with its multitude of electrical cables and water feed lines, or an installed and functioning beamline, this seems a difficult route to pursue. A much more attractive solution is to design the beamline, by suitable choice of materials, and by utilizing recent improvements in beam transport systems that can permit transporting the negative ion beam through a maze in the neutron shielding, so that the high-technology components are outside the shielding,[5] and so that the fraction of the beamline that has to be maintained remotely is minimized. Recent neutronics calculations indicate that hands-on maintenance of sources and accelerators so protected is probably feasible within two days of reactor shut-down.[6]

The neutral beam injector must be designed to be compatible with tritium operation. In extreme cases, the injector may have to produce tritium beams. Even if this is not the case, tritium from the reactor will leak into the neutral beam system and contaminate it. It is likely that the entire beamline will have to be bakable to 150 degrees C to purge the tritium from the beamline before opening it to the atmosphere; it is also possible that the entire beamline will have to be constructed with remote maintenance capability, in case the entire reactor hall is contaminated by a tritium spill.

Neutral beam systems have been criticised severely in comparison with competing RF systems on the grounds of difficulty of remote maintenance of complex components; this real or perceived disadvantage is important enough that designers of neutral beam systems should try very hard to find ways to alleviate the problem areas.

An important distinction is that the radiation environments for neutral beam systems are very different for tokamaks and mirrors. Since the neutral beam system (for sloshing ions, at least) on a mirror machine "sees" a region of the plasma that is not the primary neutron producer, the neutron flux into the beamline is lower than in the case of a tokamak, by a factor of a thousand or more. This means that it is easier to reduce the flux at the source to an acceptable level in the case of mirrors than in the case of tokamaks, and also that the development lines for the two systems are likely to diverge.

<u>Magnetic fields</u>: Neutral beam systems must operate in the fringe fields of the magnets used to confine the plasma. These fields can be up to 0.1 Tesla at the front end of the beamline. Sources typically can tolerate fields of only 10^{-3} of this field (1 Gauss), and ion trajectories also are influenced by stray fields, so magnetic shielding must be provided for most of the beamline. Eddy current effects can be used for short pulse systems, but not for 30 sec or longer pulse lengths. Shielding may be active (opposing fields generated by coils) or passive (magnetic materials or cryogenic superconducting materials), but it must be provided.

C. ECONOMIC CONSTRAINTS

We now turn to a number of economic constraints. These are constraints which if not satisfied, mean that the neutral beam system is too expensive to construct or operate, either because the entire reactor cannot compete against alternative power sources, or because the neutral beam system cannot compete against alternatives, such as RF systems.

<u>Space</u>: Free space around a reactor is at a premium. In some cases, the neutral beam system must fit into a vault already constructed. In others, building costs may increase because of excessive room needed by neutral beam systems surrounding the reactor. The reactor must be able to be taken apart, however, and so some space must be left around it to set down large components during assembly and disassembly. This requirement

varies from design to design, but as a rule of thumb, the beam designer should aim to occupy no more space with his injectors than is required for disassembly of the reactor. The argument here is that the neutral beam injectors would first be removed from the reactor hall to provide space for reactor components. This is actually an argument for high current-density negative ion sources. Although it appears that negative-ion neutral beam systems based on present and near-future technologies are at least commensurate with the size of projected large tokamaks,[5] the systems would certainly be improved if one could produce more current per injector.

Efficiency: A number of constraints fall into this category, although their inclusion may appear surprising at first. RF accelerators suffer in comparison with DC accelerators because of their poor power efficiency. The reason is that the RF systems do not actually spend a large fraction of the available time accelerating ions. Overall accelerator efficiencies may be only 1/2 to 1/3 of DC systems, and that is a very severe penalty to pay. The goal, therefore, is to push DC accelerator technology to its farthest limits.

Once the beam has been accelerated, it is important that it not be lost by gas collisions. The cross section for stripping of D^- in D_2 varies from 3×10^{-16} cm^2 at 200 keV to 2×10^{-16} cm^2 at 400 keV. Unfortunately, this is a very large cross-section, and the implication is that to reduce losses to a few percent over path lengths of several meters, background pressures should not be over a few times 10^{-6} Torr. There is an additional complication, since in cryopumped systems, the temperature of the thermally shielded walls and also the temperature of the background gas sink to about 100K,[7] which increases the gas density and the stripping losses. With a source gas efficiency of 10%, about the best so far achieved, to pump the gas coming just from the ion source at such a low pressure would require hundreds of square meters of cryopanels. The corollaries thus are that source gas efficiency should be improved, and that the gas should be pumped at relatively high pressure (10^{-4} Torr, for example). A scheme to transport the pre-accelerated ion beam through a differential pumping section becomes attractive; the beam then can be accelerated to higher energy in a lower pressure section of the beamline.

An additional, geometric, constraint results from this pumping requirement: since negative ion sources produce relatively low current densities of negative ions, to produce the total currents required, a large area beam is necessary. To remove the gas efficiently, the configuration of the beam must be such that the beam is thin in one dimension and long in the other; in addition, differential pumping is probably required.[8] Such a geometry also minimizes the gas flow from a gas neutralizer, and is a good match to a laser photoneutralizer, which will probably be the neutralizer of choice because of the potentially higher over-all system efficiency.[9]

In purely electrostatic accelerators, electrons will be accelerated along with the negative ions. In an efficient system, one would not want to invest more than a few percent of the system power in doing this. Therefore, control and removal of electrons before they reach high energies is important. This is a very important constraint; schemes that fail to remove practically all the electrons from the beam before final acceleration, or that invest too much power in acceleration of electrons, will not be serious contenders for use on reactors.

Cost: To the best of my knowledge, there has been no accurate study of the costs of the negative-ion-based neutral beam systems suitable for reactor application, so this is an open question. An example of the uncertainty is reflected in the choice of neutralizer. Gas targets certainly work, and can be accurately costed, but since the time until the first application of negative-ion-based neutral beam systems is so long, approximately 10 years, gas targets probably will not be the neutralizer of choice. It is more likely that a photodetachment neutralizer will be chosen, perhaps one employing a chemically excited laser. Development of such lasers is proceeding rapidly, and estimates of the cost of this component of the neutral beam system, if they existed, would be changing rapidly. Nevertheless, when such cost estimates are made, neutral beam systems will have a ready-made standard against which they will be judged, namely the cost of competing RF systems. It is therefore important for the designers of neutral beam systems to become and remain cognizant of developments and cost studies for RF systems.

AN EXAMPLE

As an example of the types of considerations necessary in designing a negative-ion-based neutral beam system, I would like to summarize briefly the thoughts that have so far gone into a typical conceptual design, the one utilizing TFF transport and accelerating systems and an LBL surface-conversion source illustrated in Figure 1, also discussed from the point of view of ion optics in another paper at this conference.[10] This design uses a Pierce type pre-accelerator operating at 80 keV; the 80 keV beam is transported through a matching and pumping section, and finally, is accelerated to 400 keV by a TFF accelerator. The choice of the pre-accelerator energy, 80 keV, resulted from a compromise between the conflicting desires of minimizing energy, so that beam loss due to stripping would take place at as low an energy as possible, and of avoiding high perveance pre-accelerator designs, which introduce undesirable aberrations. Once the accelerator was selected, the converter size was fixed -- the converter has to be large enough to illuminate the entrance to the 80 keV pre-accelerator uniformly, without vignetting. Otherwise, intolerable aberrations are introduced into the beam. This sets the size, then, of the ion source.

Design of the matching and pumping section is an iterative process, still going on. The transport electrodes must be as transparent as possible to gas, to facilitate pumping, yet still transport the beam without introducing aberrations. It appears that these requirements can be met in this design; losses up to the final accelerator are estimated at about 7%.

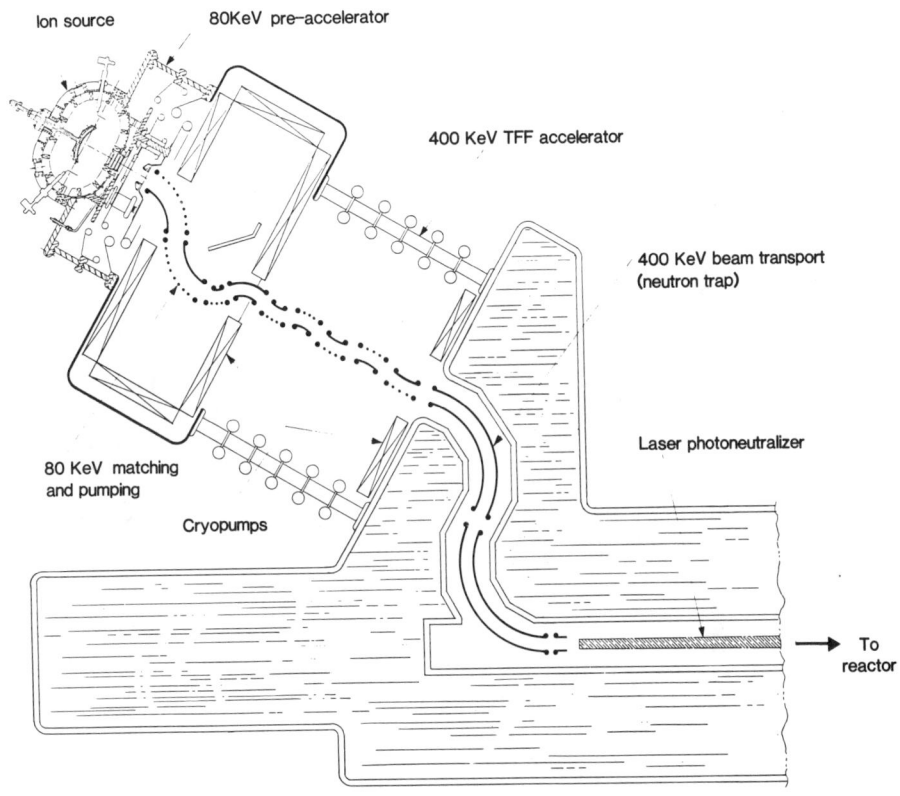

Fig. 1. Plan view of 400 keV TFF-based beam line.

Design of the transport and matching section cannot be fixed without examining ion trajectories all the way through the system. Ion trajectories have in fact been calculated self-consistently from the converter all the way through the final 400 keV transport section.[10] Uncertainties still remain: neutronics studies have not been carried out for this particular design, so we do not know for sure that the two bends shown in Figure 1 will be adequate to reduce the neutron flux at the source to the desired level. Another uncertainty has to do with the particulars of the laser photoneutralizer. To minimize losses, it is desirable to minimize the area of mirrors in the laser cavity resonator. If we reduce the final negative ion beam width to permit this, by changes in the design of the final transport section, the beam divergence will increase (Liouville's Theorem). Some balance must therefore be achieved between conflicting requirements of minimizing the laser power and satisfying the beam divergence requirement. Our study has not progressed to this state.

Nor have we completed an analysis of the pumping requirements. Pumping may have to be included in the final transport section or in the laser photoneutralizer section, or both. These pumps take up room -- their space requirement, plus the properties of materials used in their construction, will affect neutron transport and activation.

Finally, magnetic shielding has been ignored completely. We will have to return to this problem at a later date.

SUMMARY

It is not possible to produce a design of a negative-ion-based neutral beam system that will satisfy all requirements. Economic pressures will force each design to be tailored to the requirements of its particular reactor. The neutral beam designer must be aware of all the constraints, and I hope this discussion will provide a first step in this direction, so that the resulting neutral beam system will be at least approximately optimized for the particular application, and so that the designer will be able to avoid designing himself into traps along the way.

ACKNOWLEDGEMENT

The author would like to thank his colleagues at the Lawrence Berkeley Laboratory for numerous helpful discussions and contributions to the general problems of neutral beamline design, and especially to O. A. Anderson for permission to use Figure 1.

This work was supported by the Director, Office of Energy Research, Office of Fusion Energy, Development and Technology Division of the U.S. Department of Energy under Contract No. DE-AC03-76SF00098.

REFERENCES

1. W. S. Cooper and R. V. Pyle, Scientific Editors, "The National Negative-Ion-Based Neutral Beam Development Plan," Lawrence Berkeley Laboratory Report PUB-464 (August, 1983).
2. K. I. Thomassen and J. N. Doggett, Editors, "Options to Upgrade the Mirror Fusion Test Facility", Lawrence Livermore National Laboratory report UCID-19743 (April, 1983).
3. K. N. Leung, Lawrence Berkeley Laboratory, private communication.
4. R. L. York, R. R. Stevens Jr., K. N. Leung, and K. W. Ehlers, "Extraction of H^- Beams from a Magnetically Filtered Multicusp Source," Los Alamos National Laboratory report LA-9931 (October, 1983); submitted to Rev. Sci. Inst.
5. Y-K. M. Peng, P. H. Rutherford et al, "FED-A, An Advanced Performance FED Based on Low Safety Factor and Current Drive," Fusion Engineering Design Center report ORNL/FEDC-83/1 (August, 1983), pp. 4-82 through 4-86; also, O. A. Anderson et al, Lawrence Berkeley Laboratory Preprint LBL-14880 (August, 1982).
6. C. D. Henning, et al, "Mirror Advanced Reactor Study--Final Report," Lawrence Livermore National Laboratory Report UCRL-53333-83 (to be published December, 1983).
7. K. H. Berkner et al, "Performance Characteristics of NBSTF, The Prototype Neutral-Beamline for TFTR," Proceedings of the 9th Symposium on Engineering Problems of Fusion Research, Chicago, IL, October 26-29, 1981, p. 763.
8. C. F. Burrell and D.A. Goldberg, "Calculations of Gas Density in a Closely-Packed Multi-Channel Electrostatic Quadrupole (MESQ) Array," paper presented at the American Vacuum Society 29th National Symposium, Baltimore, Maryland, November 16-19, 1982; also, Lawrence Berkeley Laboratory report LBL-14631 (September, 1982).
9. J. H. Fink, "Laser-Neutralized Negative Ions as a Source of Neutral Beams for Magnetic Fusion Reactors," Lawrence Livermore National Laboratory Reportr UCRL-87301 (1982); submitted to Nucl. Tech. Fusion.
10. O. A. Anderson, "Transverse-Field Focusing Accelerator," this conference.

DISCUSSION

BARNETT: I guess that I have failed to appreciate why you have to go through all these nooks and crannies to get in. Why not make one bend and then put all your shielding in between?

COOPER: You might be able to. You want to avoid, as I understand it from Oscar Anderson, a series of bends that are around 127° because then you introduce resonances, so the thought was to stay below that. If you bend around too far then you've got your source in an awkward corner here.

BARNETT: Are these 90° bends?

COOPER: These are 90° bends. You get some attenuation of neutrons, perhaps by a factor of 100 per bend, and you just have to stack up enough to get the flux down to a tolerable level. You want this to be a narrow channel and in fact you'd like these (electrodes) to be I think not as long but 4 or 5 centimeters thick so the neutrons will scatter off of them out into the water.

JACQUOT: My first question is, what is the maximum value of the preacceleration voltage before the TFF focusing, and the second, what is the maximum temperature that has to be maintained in the source (for baking out tritium)?

O.A. ANDERSON: It's determined by the perveance that we can reach, so we chose 80 kV as giving a manageable perveance although it's pretty high. We think we can still have reasonable aberrations.

COOPER: I think that as far as the TFF accelerator or transport system goes, you can go to quite a bit higher energy, 200 kV if you wanted to, and the transport system would work just fine.

JACQUOT: What is the minimum energy for a good TFF transport?

COOPER: It's probably determined by geometry, or how small you can make the radius of curvature of electrodes. Would you like to comment on that too, Oscar?

O.A. ANDERSON: Well, if we go to higher than 80 kV then we waste a lot more power in electrons that we are dumping. If we go to much lower than 80 kV, we can't preaccelerate the current that we want to carry. That is why you wind up with that figure.

COOPER: About your second question, the only information I have on that came in fact from Garching. We were involved in a beam line study for their Zephyr program and they pointed out as a requirement of their beam line that it be bakeable to 150° C for several days to get the tritium out.

NEUTRAL BEAM INJECTOR FOR 475 keV MARS SLOSHING IONS

D. M. Goebel
TRW, One Space Park, Redondo Beach, CA 90278

G. W. Hamilton
Lawrence Livermore National Laboratory, Livermore, CA 94550

ABSTRACT

A neutral beam injector system which produces 5 MW of 475 keV $D°$ neutrals continuously on target has been designed. The beamline is intended to produce the sloshing ion distribution required in the end plug region of the conceptual MARS tandem mirror commercial reactor The injector design utilizes the LBL self-extraction negative ion source and Transverse Field Focusing (TFF) accelerator to generate a long, ribbon ion beam. A laser photodetachment neutralizer strips over 90% of the negative ions. Magnetic and neutron shield designs are included to exclude the fringe fields of the end plug and provide low activation by the neutron flux from the target plasma. The use of a TFF accelerator and photodetachment neutralizer produces a total system electrical efficiency of about 63% for this design.

INTRODUCTION

The two sloshing ion beamlines must continuously deliver a total of 8.86 MW of $D°$ at 475 keV into the 5.56 T point of both plug end cells of MARS at an angle of 70-80°. Of this incident power, 5.68 MW is trapped by the plasma for the calculated trapping fraction of 0.64. The remaining 3.18 MW of neutral beam power passes through the plasma and strikes two water cooled beam dumps some distance from machine axis. The sloshing ion neutrals are produced by the standard arrangement of ion source, accelerator, and neutralizer. The high beam energy dictates that negative ions be used to achieve reasonable neutralization efficiency. The un-neutralized ions in the beamline are deflected out of the neutral beam path and collected by internal beam dumps. The materials used in the dumps and other beamline components have been chosen for low activation wherever possible.

The fringing magnetic field from the plug yin-yang must be shielded from the ion source and ion beam regions prior to and during neutralization. The residual magnetic field in the beamline should not exceed 2-5 G for proper operation of the ion source and neglible deflection of the ion beam in the accelerator and neutralizer regions. The shield must exclude transverse magnetic fields of 0.5-0.6 T and axial fields of less than 0.1 T. For simplified maintenance of the beamline, it is desirable to attenuate the neutron flux from the plug plasma to levels where hands-on maintenance is possible within a reasonable time of a reactor shutdown. Depending on the materials selected for the beamline, the neutron shielding will therefore be required to limit the neutron flux to about 10^6 n/cm^2-sec at the ion source region. This will require an attenuation of about a factor of 10^5-10^7 by the shielding in and around the beamline. Increasing the distance of the sources from the target plasma will decrease the amount of magnetic and neutron sheilding required, but eventually lead to additional losses of the beam due to the finite angular divergence of the accelerator. The beamline length and position should be picked to minimize shielding and remote maintenance requirements without producing excessive neutral beam losses from scrape off on the duct walls and collimators due to optical divergence.

BEAMLINE ION SOURCE

The MARS sloshing ion beamline is illustrated in Figures 1 and 2. The components of the beamlines will be described in the following sections. The LBL self-extraction negative ion source shown in Figure 3 has been selected in the beamline design. The ion sources are two meters long (out of the figure) and utilize a cesiated molybdenum surface or volume production processes[3] to convert positive ions formed by an rf plasma generator in a magnetic multipole bucket containment geometry into negative ions. The bias voltage for accelerating the ions to the converter also accelerates the subsequently formed negative ions away from the surface. Geometric shaping of the converter focuses these D$^-$ ions into the extraction slit. Electrons in the plasma and from the converter surface are reflected from the accelerator entrance by a small transverse magnetic field produced by SmCo$_5$ magnets. The ions are pre-accelerated to 80 kV prior to injection into the TFF accelerator by a standard electrostatic single slot accelerator. The ion beam prior to acceleration is 1.5 cm wide x 200 cm long and has an average current density of 47 mA/cm^2. The ion source is therefore required to produce 7 A/m of source length. Compared to the published value of 5 A/m previously obtained, this seems like a very reasonable projection for MARS.

The gas efficiency of the ion source is assumed to be 30%, a slight improvement over the 20% published value. The neutral gas flowing from the source produces an average pressure in the pre-accelerator of about 2×10^{-4} torr. This pressure causes 15% of the ion beam to be lost due to stripping by the neutral gas. A TFF transport region at 80 keV with differential gas pumping will be used immediately after the preacceleration region to decrease the gas pressure to 10^{-6} torr before the high voltage accelerator. This introduces an additional 9% loss of the ion beam by charge exchange, but at the low preacceleration energy so that the average power loss is low. Charge exchange losses in the accelerator are below 1% at the 10^{-6} torr background pressure. The Low Energy Beam Transport (LEBT) system will be described in the next section.

One of the two sources seen in Figure 2 is for redundancy. At any time a single source will be used to produce the sloshing ions. A

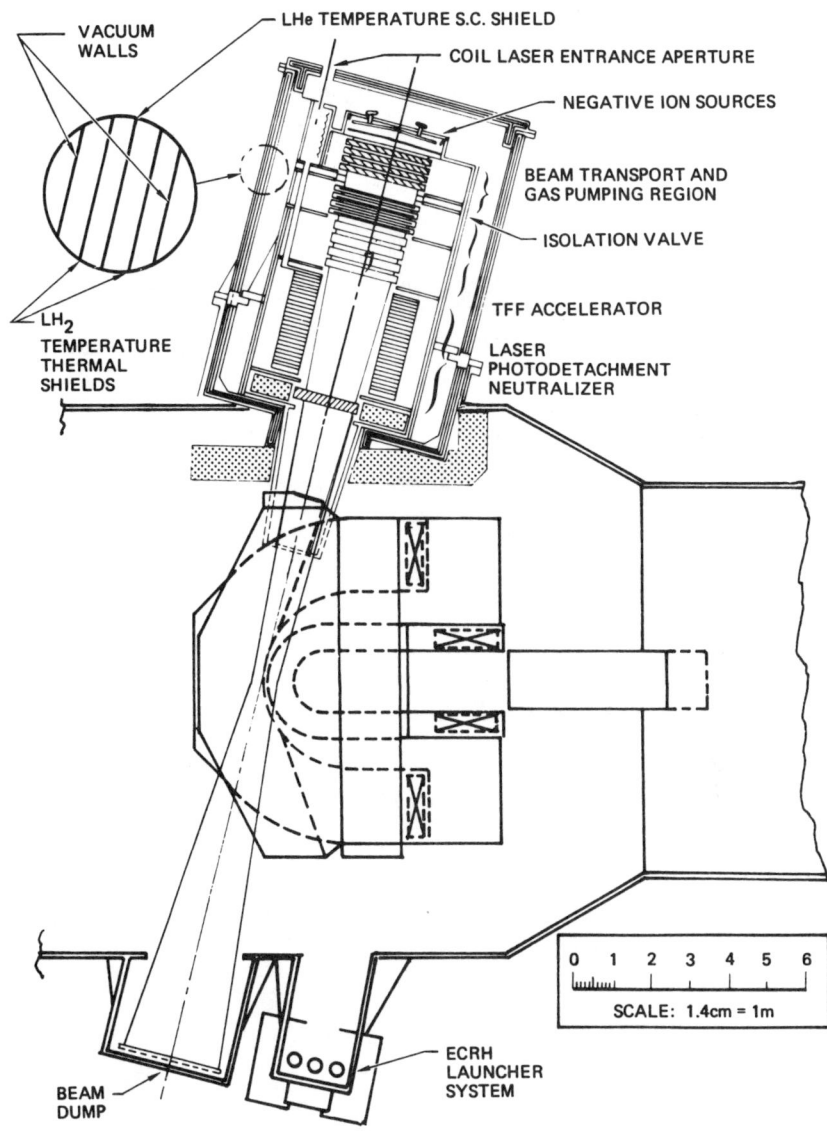

Figure 1. MARS Sloshing Ion Beam Line and Dump with Magnetic and Neutron Shielding

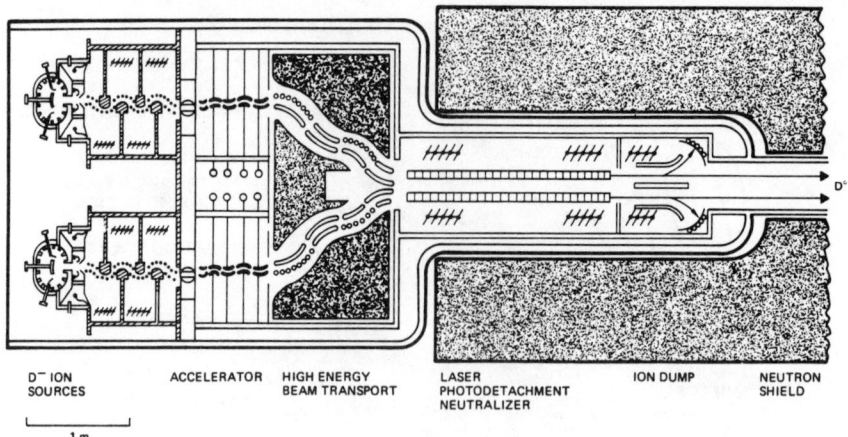

Figure 2. MARS Sloshing Ion Beam Line

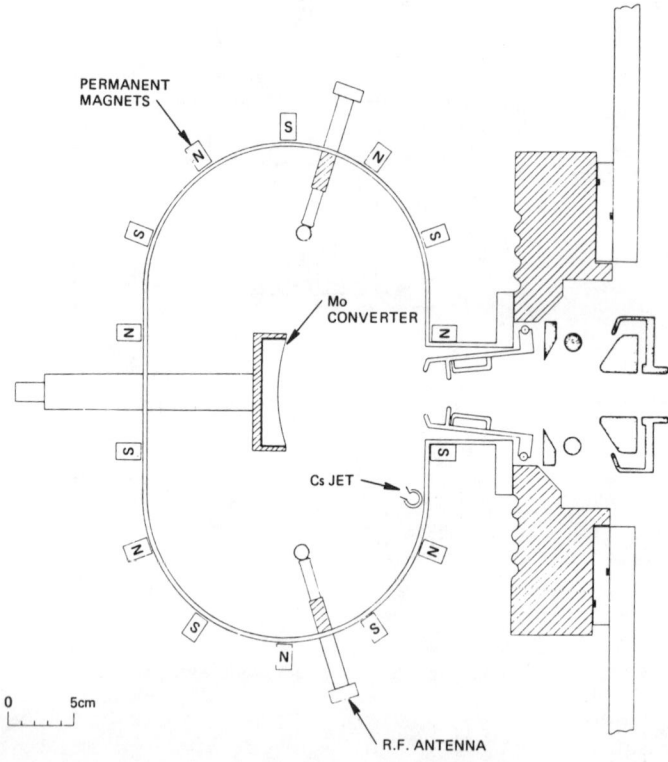

Figure 3. Schematic Diagram of the LBL Self-Extraction Negative Ion Source

built-in 50% redundancy is required because failure of one source effectively shuts down the reactor, and the time required to turn on a built-in conditioned source is considerably shorter than to open the facility to air for repairs of a single source. Isolation valves are provided for repair of one source without affecting the other. Remote maintenance can possibly permit repairs of one even while the other is operating.

ACCELERATOR

There are several accelerator options which can be utilized for the sloshing ion beam. Systems which use rf acceleration and quadrupole focusing, such as the LASNL-RFQ and the BNL-MEQALAC[4], provide high energy acceleration without very high voltages. These systems are conceptually safer and easier to protect and maintain than DC systems. However, the time averaged current density is low in the rf system, and the electrical efficiency is lower than in a DC system due to joule heating and power supply modulator inefficiencies. The standard DC multiple grid accelerators utilized today in positive ion systems could possibly be reconfigured to produce the 475 kV negative ion beam. These very high voltages, however, can contribute to stray electron and radiation induced breakdowns in the accelerator column which cause unreliable operation. The stored energy in each grid must also be minimized to eliminate melt down problems associated with the arc breakdowns. A DC ESQ (Electro-Static Quadrupole) system[5], in which the beam is focused by electrostatic quadrupoles between accelerator stages, reduces the probability of breakdown occuring over the entire accelerator column. The accelerator length is also adjustable by the quadrupole parameters so that reasonable voltage gradients can be achieved.

The quadrupole based accelerators are limited to sources which produce an array of multiple circular beamlets in order to provide space for the quadrupole electrodes. A high voltage accelerator which is compatible with the long slit beams of the self-extraction negative ion source is under development at LBL. This accelerator, called the TFF[6] (Transverse Field Focusing), utilizes alternating curved electrodes producing a transverse electric field to focus and transport the ion beam. A D.C. voltage applied between successive electrode pairs will also accelerate the ions. The advantages and disadvantages of the TFF system compared to the DC ESQ and MEQALAC are listed in Table 1. The TFF accelerator is a relatively new application of the well known electrostatic transverse electric field energy analyzers in use for many years. While development is required for use in the MARS beamline, the principles of the TFF system are well known and justify consideration for MARS.

The Transverse Field Focusing (TFF) system selected for MARS is a type of one-dimensional electrostatic strong-focusing system used for transport and acceleration of the charged beam from the ion source to

Table 1. Advantages and Disadvantages of Transverse Field Focusing (TFF)

Advantages over alternatives such as MESQ and MEQALAC

- Higher beam-carrying capacity (40-60 mA/cm^2, ~ 8 A/m)
- Compatibility with LBL sources of ribbon-shaped D$^-$ beams
- Compatibility with the laser photodetachment neutralizer
- Now under active study and development at LBL
- Moderate electric field requirements (20-50 kV/cm)
- Beam transport for several meters and around corners
- Compatibility with neutron sheilding.

Disadvantages

- Loss of 15-20% of the beam by gas collisions (mostly at low energy)
- Power losses by secondary electrons
- Versions of the Culham effect require adequate gas conductance
- Substantial requirements for space and for magnetic shielding
- New development is required.

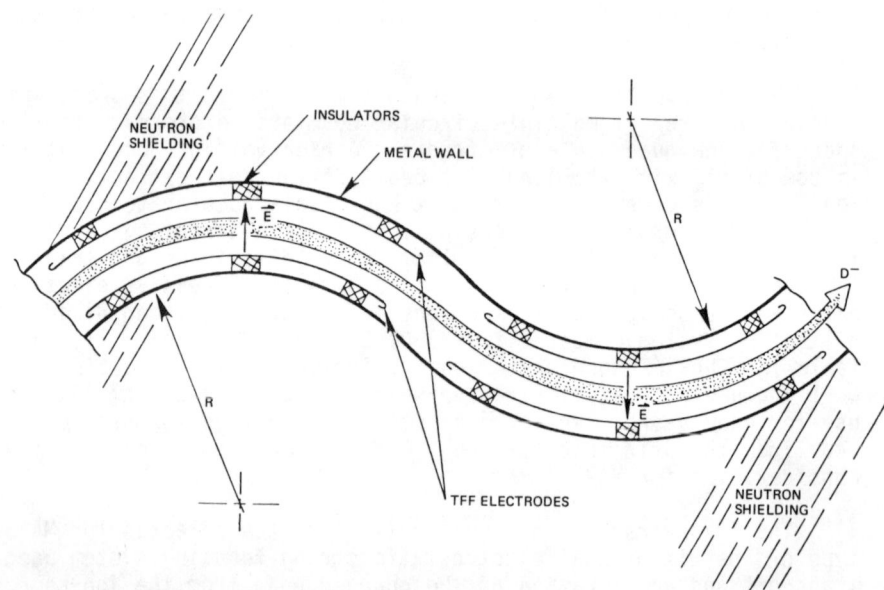

Figure 4. Transverse Field Focusing (TFF) Acelerator and Beam Transport System

the neutralizer. It is suitable for one-dimensional focusing of a ribbon-shaped ion beam by an electric field applied transversely. This causes the beam particles to follow circular arcs, as shown by Figure 4. In the absence of space charge the required transverse electric field is

$$E = mv^2/rq = 2W/r \qquad (1)$$

where W is the kinetic energy in eV. For spatial uniformity the electric field must be applied between two concentric cylindrical electrodes spaced a distance S apart, biased with voltage $\pm V_t$ relative to the potential of the equilibrium trajectory

$$E = 2V_t/S. \qquad (2)$$

A beam particle not on the equilibrium surface will be strongly focussed toward the equilibrium surface by the combination of electrostatic and centrifugal forces. This focusing theory has been published by Anderson[6]. The theory is supported by trajectory computtations, but has not yet been confirmed experimentally.

If the beam space charge is not negligible the space charge forces will tend to expand the beam in the transverse direction while the focusing forces resist the blow-up. The strength of the transverse electric field should be adjusted accordingly, with appropriate choices of the voltage V_t and dimensions r and S. This condition limits the beam-carrying ability of the TFF to about 40 to 60 mA/cm^2. This capacity is larger than the beam-carrying capacity of alternative focusing systems such as the various versions of dc or rf multipoles.

For most applications there is a maximum practical limit to the deflection angle caused by the transverse electric field. Therefore, our designs consist of a series of TFF stages as in Figure 2, which bend the beam alternately in one direction and then the other direction. The transverse focusing must be strong enough to compensate for the space charge blow-up in the gaps between focusing stages. If it is desired merely to transport the beam without acceleration the electrode potentials of adjacent stages are identical except for the reversal of polarities $\pm V_t$. To accelerate the beam, a longitudinal electric field E_z is applied between stages, superimposed upon the transverse electric field E. The acceleration increases the kinetic energy W and therefore requires increased values of r and V_t with each accel stage, in accordance with equations (1) and (2). The electric field intensities required to satisfy MARS requirements are not extreme, and we do not expect difficulties in holding high voltage. However, the ribbon-shaped beam will also have smaller space-charge forces tending to blow up the beam in the long direction. We believe these can be compensated for by slightly curving the electrodes in this direction.

Losses of power and of beam particles by collisions with background gas are potentially severe in a TFF system. For D⁻ ions the most important type of gas collision is collisional electron detachment. Most of the resulting D atoms will go tangentially to the outer (negative) electrode, where they will deposit energy and produce secondary electrons with a multiplication coefficient in the range 1 to 5. These secondary electrons will be attracted to an adjacent positive electrode, where they will arrive with energy 2 V_t. The beam transport and the gas pumping system must be designed such that the gas pressure is low enough that these losses are tolerable.

For the reasons above, a multi-stage differential pumping system is required to reduce the background gas pressure to about 10^{-6} torr before high voltage acceleration. The D⁻ ion source operates at a pressure of 10^{-3} torr, and the gas pressure must be reduced as quickly as possible to avoid loss of a large fraction of the D⁻ beam. Solutions to these problems have been achieved by the neutral beam group at LBL, Berkeley, for the FED-A Tokamak[9]. These solutions are adequate for MARS requirements. D⁻ ions are extracted from the source by a 80 kV pre-acceleration stage and then transported through a differential pumping system shown in Figure 2 by a Low Energy TFF Beam Transport system (LEBT). The differential pumping system consists of four stages, each with continuously-operated cryopanels. Monte Carlo computations of the gas flow indicate that the required pressure reduction will be attained, but that about 15% of the 80 keV D⁻ beam will be lost by collisional electron detachment, most likely in the pre-acceleration stage where the gas pressure is over 10^{-4} torr. This beam loss reduces the effective beam density, but does not constitute a serious power loss at this relatively low energy.

As shown by Figure 1 the D⁻ beam enters the low-pressure TFF accelerator to be accelerated from 80 keV to 475 keV. The accelerator consists of four stages, each of which is designed with the dimensions and the electrode potentials appropriate for the beam energy and density. The focus voltages V_t are progressively increased from about ±25 kV at low energy to about ±50 kV at 475 keV. The length of the four-stage accelerator is about 1 meter.

After the beam is accelerated to 475 keV, it is desirable to transport it through a neutron shield about 1 meter thick before the beam is neutralized by photodetachment. The neutron shield has the purpose of protecting the ion source and accelerator from neutron activation and neutron damage. Monte Carlo neutron flux computations and the neutron activation computations show the feasibility of hands-on maintenance, as opposed to remote maintenance. The High Energy Beam Transport (HEBT) consists of two large-radius TFF stages, each with a 90-degree bend, to attenuate the neutron flux. The overall attenuation must be at least about of factor of 10^6 to reduce the neutron activation sufficiently for hands-on maintenance. We define hands-on maintenance as a residual activity not greater than 2.5 mR/hr two days after reactor shut-down.

The negative ion beam energy, from the ion source through the different accelerator stages just described, is shown in Figure 5. The last 50 kV of acceleration between the HEBT and the neutralizer is provided to inhibit electron flow from any plasma in the neutralizer to the TFF focusing electrodes in the HEBT. The final accelerator performance parameters are shown in Table 2. While 26% of the initial ion beam current is lost in the accelerator due to charge exchange with the background gas, the total power lost is less than 10% of the final beam power. Other losses due to the ion source electrical efficiency, stripping fraction, and duct losses dominate the power expenditure required to make the sloshing ion neutral beam.

Focusing in the narrow dimension of the beam is rather critical because the plasma target in the end plug is only 28 cm thick in this direction. The beam must travel about 10 meters without further focusing after it leaves the HEBT. This imposes a maximum allowance beam divergence angle in this direction of ± 0.8 degrees.

The D⁻ beam produced by the ion source will have a maximum transverse kinetic energy of 6 eV because of the geometry of the self-extraction electrode. As the beam is accelerated it will probably be compressed in the thin direction, perhaps by a factor of two. Conservation of phase space requires that the transverse kinetic energy increase as a consequence of the beam compression. We estimate that the 6 eV transverse energy will thereby be increased to 24 eV. The maximum divergence angle in this direction will be

$$\theta = \tan^{-1}\left[(24/475.000)^{1/2}\right] = 0.41 \text{ degrees}, \tag{3}$$

which is adequate for the purpose.

The beam is transported and focused in the short direction by TFF, and is also made to converge in the long direction by shaping of the electrodes of the ion source and the TFF as seen in Figure 1. The width of the beam is about 2 meters at the ion source but only a few tens of centimeters at the plasma target. This beam convergence has the effect of intensifying the injection of sloshing ions and causing them to be trapped where their turning points are correct for generating the sloshing ion distrbution.

NEUTRALIZER

A system is required to neutralize the ions from the ion source and accelerator prior to injection into the MARS plasma. The fast neutral atoms from this neutralizer can be produced by stripping and change exchange in the gas or plasma cell, or stripping in a laser

Table 2. Accelerator Parameters

	Voltge (kV)	Pressure (torr)	Current Lost (%)	Power Lost (MW)
Source	475	10^{-3}	--	0.54
Pre-accelerator	395	2×10^{-4}	15	0.33
LEBT	395	$10^{-4} - 10^{-6}$	9	0.17
Accelerator	395 - 50	10^{-6}	1	0.06
HEBT	50	10^{-6}	1	0.09
Neutralizer	0	10^{-6}	5	0.49
Duct/Collimator	0	10^{-6}	5	0.47
TOTAL	475	-	36%	2.15 MW

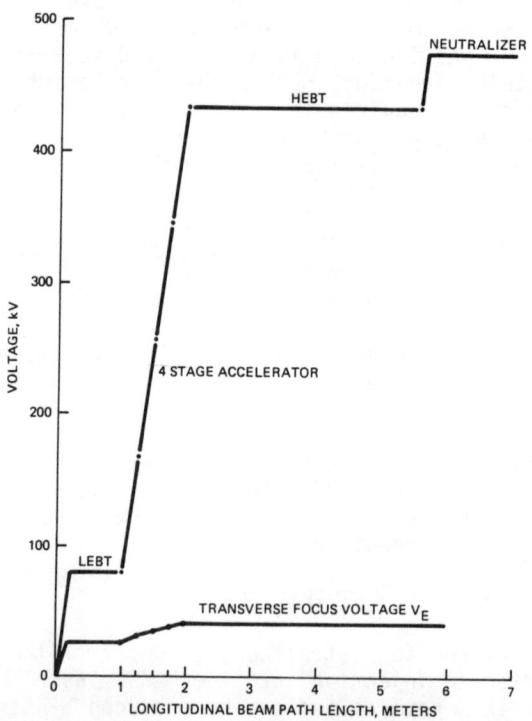

Figure 5. Beamline Voltage Profile for TFF Accelerator and Transport Stages

photodetachment neutralizer. These options are summarized in Table 3. The gas cell utilizes neutral gas from the ion source with additional injected gas if necessary to provide a target thickness adequate for neutralization. While 50-60% of the negative ions can be stripped without any power input, the gas cell produces a significant gas load for pumping and multiple ion species in the beam. The plasma cell neutralizes up to about 80% of the ions, but is undeveloped at this time due to problems in generating large, highly ionized plasmas of significant extent to achieve high stripping. Partially ionized plasmas are not useful due to excessive ionization of the fast neutral beam by the gas molecules in the beam path decreasing the stripping fraction to the gas cell level. The gas load and power requirements are unknown at this time for the plasma neutralizer.

A laser neutralizer[7,8] strips the negative ion by photodetachment in a laser resonator. Over 95% stripping of the negative ions is possible at the expense of about 1 MW of power required by the laser. Presently, this system produces no gas load for pumping in the beamline that is not isolated to the laser. A very transparent window provides gas and impurity isolation between the COIL (Chemical Oxygen-Iodine Laser) system and the beamline in present designs. The laser photodetachment neutralizer is under development for use with the LBL self-extraction sources.

The laser photodetachment neutralizer selected for MARS can achieve over 95% stripping of the ion beam emerging from the accelerator. A beam of D$^-$ ions traveling a lengh L at a speed v through a region illuminated by a total laser flux P will be stripped with a fraction f given by

$$f = 1 - \exp(-\sigma LP/h\nu v) \tag{4}$$

where σ is the photodetachment cross section and $h\nu$ is the photon energy. For a COIL laser, $\sigma/h\nu$ is 1.2×10^{-5} cm^2/erg. The ion beam can be passed directly through the laser resonator to maximize the laser flux. The interaction length can also be increased by folding the resonator with multiple mirrors as shown schematically in Figure 6. The stripping fraction for the MARS beam line as a function of the number of folds is shown in Figure 7. The 95% stripping fraction is achieved for a resonator of 40 folds and a power flux on the mirrors of 75 kW/cm^2. Cooled mirrors of SiC will be used with a reflectivity of greater than 0.999 to handle this high power flux.

The parameters of the folded laser resonator have been calculated by a model based on the Rigrod analysis.[10] These are summarized in Table 4. The laser gain is for the TRW supersonic COIL laser and the gain length is selected to produce the 75 kW/cm^2 for 40 folds of the resonator. A 70 kW COIL laser will be required for the beamline on each end cell. Electrical efficiencies of up to 16% have been previously estimated for this laser.[11] The effect of this laser

Table 3. Summary of Neutralizer Options

	Gas/Vapor Cell	Plasma Cell	Laser
Optimum Stripping Fraction	50-60%	~ 80%	> 95%
Exiting Species	$D°$, D^+, D^-	$D°$, D^+, D^-	$D°$, D^-
Gas load	> 10^{17} mol/cm² of neutralizer	< 10^{16} ions/cm²	None
Power Requirements	None	Unknown	0.5-1 MW
Level of Development	Developed	Not Developed	Under Development

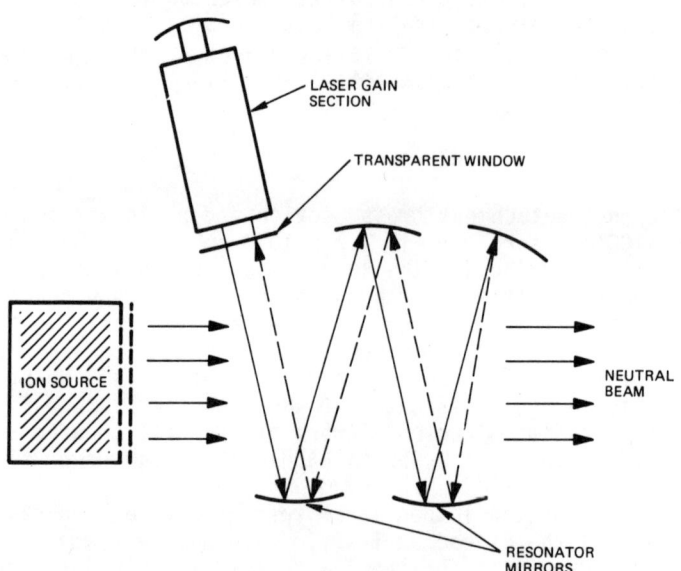

Figure 6. Folded Laser Photodetachment Resonator with Multiple Mirrors

Table 4. Laser Photodetachment Neutralizer Parameters

- Folded Chemical Oxygen Iodine Laser (COIL) Resonator

Mirror size	3 cm x 4 cm
Reflectivity	0.999
Window transmission	0.999
Gain	0.01 cm^{-1}
Gain length	400 cm
Scattering losses	0.2%
Power on mirrors	75 kW/cm^2
Neutralizer length	2m
Number of folds	40
Laser efficiency	10%
Laser power	70 kW
Wall power	0.7 MW (each)
Stripping fraction	95%

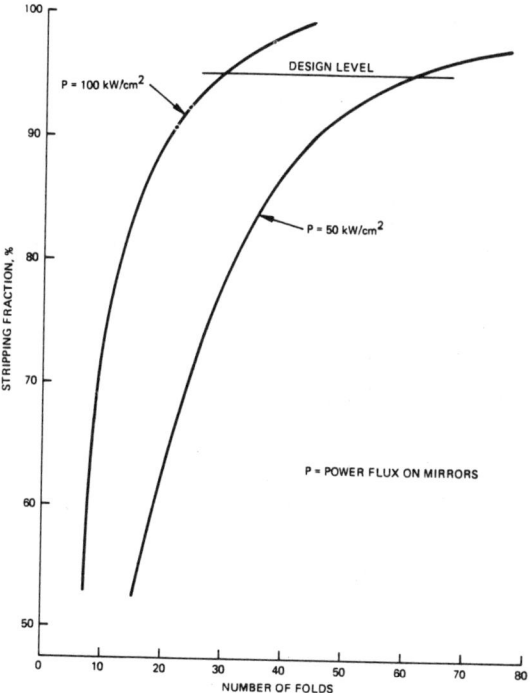

Figure 7. Laser Photodetachment Neutralizer Stripping Fraction vs. Number of Folds

electrical efficiency on the total electrical efficiency utilized in the system design is shown in Figure 8, and produces beamline efficiencies of greater than 70% excluding power supply losses.

The laser arrangement shown in Figure 1 consists of an isolated laser coupled to the rear of the beamline in a low magnetic field region. The laser light is sent to a series of mirrors at the ends of long tubes which act to protect the mirrors from any sputtered materials in the beamline. These tubes are fabricated from cooled aluminum for low activation. A separate set of mirrors are provided for use with the redundant ion source, but the laser beam from a single gain section can be moved to these mirrors during operation so that only one laser is required per end cell.

MAGNETIC SHIELD

The fringing field of the MARS magnets must be shielded away from the injection system to avoid peturbation of the ion beam trajectories and to avoid interference with the operation of the ion source discharge and self-extraction system. The fringing field within the ion source must be less than a few Gauss, and the fringing field in the neutralizer must be less than 5 Gauss. Within the TFF, the tolerance is somewhat larger because the electrostatic strong-focusing corrects for weak magnetic perturbations.

We propose to enclose the entire injection system in a cylindrical super-conducting magnetic shield to exclude the fringe magnetic fields plotted in Figure 9. Such a shield performed well in the Baseball II experiment[12] in a magnetic field of 0.15 Tesla. The cylindrical shield will be about 6 meters in diameter and 8 meters long. The shield will enclose both ends of the cylinder, except for apertures required for the beam, the structure, power, cooling, lasers, pumping, etc. The main shield will be a single sheet of Niobium-Titanium several millimeters thick, operating at 4 degrees Kelvin and engineered to exclude a maximum field of 0.6 Tesla with a superconducting current density of 100 kA/cm^2. The magnetic shield will be fabricated of sheet metal, joined by heliarc welding.

The superconducting shield will operate in a separate vacuum system, as shown in Figure 1. A steel structure is required inside the shield to resist the magnetic forces. Thermal shields are required at liquid nitrogen temperature to reduce the radiative heat load on the liquid-helium temperature superconductor. The superconducting shield is protected by neutron shielding.

To be conservative, a variable-pitch bucking coil will be installed outside the superconducting shield to buck out the fringing field component parallel to the beam axis. Because of the required penetrations through the main magnetic shield, supplementary shielding is required around the sensitive parts of the injection system. The

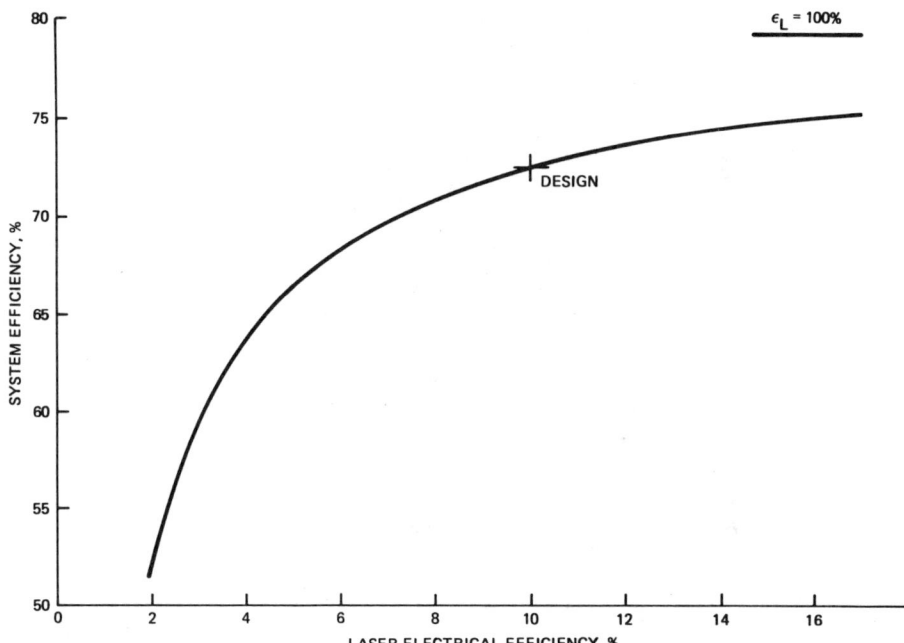

Figure 8. Sloshing Ion Beam Line Efficiency vs. Laser Efficiency

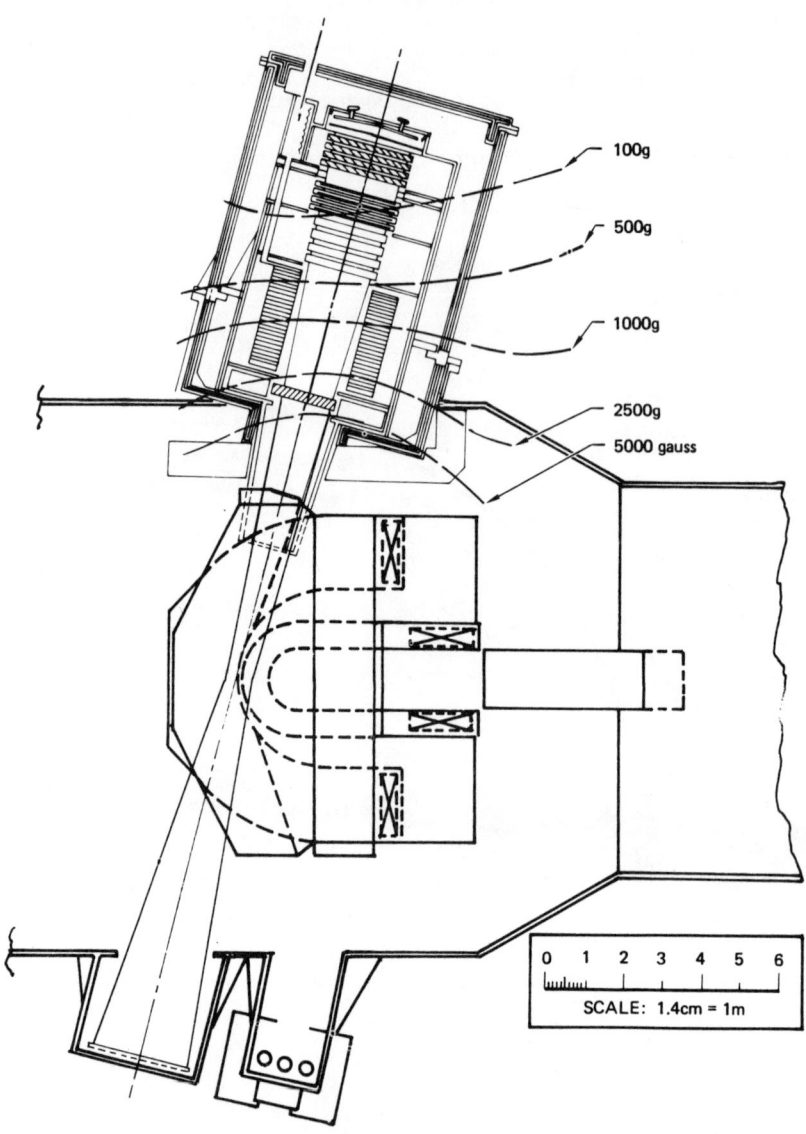

Figure 9. MARS Magnetic Field Profiles in Beam Line Location

supplementary shielding will consist of a box of mu-metal or magnetic material surrounding the ion source and a similar box surrounding the neutralizer. In each case the supplementary shields operate at the voltage of the components they are adjacent to.

TRAPPING FRACTION

In principle, the sloshing ions could be injected either at the point of maximum potential in the end cell or at the opposite turning point. Injection at the maximum potential has the liability of locally depositing a cold electron with each ionized neutral. The cold electron will be electrostatically trapped by the high potential and will require heating to several hundred keV. For this reason it is preferable to inject sloshing ions at the opposite turning point, although this requires a higher injection energy because of the potential difference. At this point the plasma cross-section is wide and thin, with semi-diameters 1.57 and 0.14 meters, respectively. To get a reasonable trapping fractions, the 475 keV beam will be injected in the long direction, which is parallel to the x-axis at the +z end of MARS.

In a three-dimensional magnetic field, the target point is actually a three-dimensional surface defined by the condition B (vacuum) = 5.56 T. The injection path is choosen to coincide as closely as possible with this surface. This is the reason Figure 1 shows the injection path at an angle of 75 degrees relative to the z axis. The path length for trapping of the sloshing ions is about 1.6 meters.

The beam of fast atoms will be trapped by ionizing collisions with four species of trapped particles: (charge exchange is negligible at these high relative speeds)

- 475 keV sloshing D^+ ions trapped in the thermal barrier

- 28 keV thermal D^+ and T^+ ions either trapped in the barrier or passing through

- 28 keV thermal He^{++}

- 24 keV thermal electrons

Table 5 shows the computation of the reaction rate $<n> <\sigma v>$ for ionization of the 475 keV neutrals by collisions with each of these species. Here $<n>$ is the average density of each of these species along the injection path and $<\sigma v>$ is the rate coefficient[14] for ionization, averaged over the distribution of relative speeds. If the density profile is quartic, the relationship between the average density $<n>$ and the z-axis density n is the following:

$$\langle n \rangle = 0.8 \, n \qquad (5)$$

Since the beam speed is much greater than the thermal speeds of the 28 keV ions, the relative speed is simply equal to the beam speed. In the case of the 475 keV D^+, the average relative speed is estimated to be larger by a factor of 1.4. The electron thermal speed is much greater than the beam speed and σ is available in the literature. From Table 5, the total reaction rate for trapping sloshing ions is 4.33×10^6 ions/s. The penetration depth for exponential beam attentuation is

$$\chi = v/(\text{reaction rate}) = \frac{6.77 \times 10^8}{4.33 \times 10^6} \qquad (6)$$
$$= 156 \text{ cm}$$

The portion of the beam path useful for trapping sloshing ions is about L = 160 cm. Therefore the trapping fraction is

$$f = 1 - \exp(-L/\chi) = 64.1\%. \qquad (7)$$

BEAM DUMPS

The beamline has an internal ion dump and an external neutral dump. The unneutralized ions in the neutral beam are removed by a parallel plate electrostatic deflector system toward the ion dump. Aluminum dumps are desired for low activation, and can be used if the power flux is not too great. Table 6 lists the parameters calculated for copper dumps at normal incidence to the beam. The power flux is low on both dumps, and the copper erosion acceptable. Data for aluminum sputtering at 475 kV is not available, but trends at lower energies indicate that aluminum sputters even less than copper.[13] The results for copper then represent a reasonable expected erosion for the MARS dumps. The low power fluxes permit low technology dumps such as back cooled plates to be used. Angling the dumps can further reduce this flux if necessary to produce reliable, long life dumps. The beam dumps are not a problem in this system due to the high stripping of the laser neutralizer and the long distance (approximately 20 m) from the accelerator to the neutral dump. If a gas neutralizer had been chosen, the internal ion dump would have had to continuously dissipate over 2 kW/cm^2 for at least one full power year.

RADIATION DAMAGE AND ACTIVATION

As discussed in Reference 9, one of the objectives of the Transverse Field Focusing System is to achieve radiation-hardening by means of the High Energy Beam Transport (HEBT), which transports the high-energy D^- beam through a radiation shield using a convoluted path

Table 5. Computation of Reaction Rate

Specie	D^+	D^+ & T^+	He^{++}	Electrons	Total
Temperaure, keV	475.	28.	28.	24.	
<n>, 10^{13} cm^{-3}	2.97	4.35	0.261	7.84	
v (relative), 10^{-8} cm/s	9.57	6.77	6.77	>> 6.77	
Sigma, ionization cross-section 10^{-16} cm^2	0.4	0.73	2.9		
<σ v>, 10^{-8} cm^3/s	3.83	4.98	19.6	0.65	
<n> <σ v>, reaction rate, 10^6 ionization/s	1.14	2.17	0.51	0.51	4.33

Table 6. MARS High Heat Flux Components - Neutral Beamlines

Beam Dump	Average Power Density (W/cm^2)	Total Power (MW)	Erosion Rate For Copper (cm/month)	Dump Life (Years)
Interior (ions)	190	0.3	1 x 10^{-4}	78
Exterior (neutrals)	480	2.26	2.4 x 10^{-4}	30

- Assumed 15% acceptable loss from 6 mm thick dump
- Blistering is not a problem
- Power peaking due to Gaussian Beam Profile is not expected to exceed 1.5 times the average
- Power levels and lifetime estimates are acceptable for MARS

to attenuate the neutrons streaming from the reactor plasma to the injector. If the neutron attentuation is adequate there will be no serious problems with neutron damage or neutron activation of the ion source and accelerator. The neutron shielding for MARS beamlines is shown in Figure 2. It is desired to provide the capability of hands-on maintenance of the ion source and accelerator. In other reactor studies hands-on maintenance has been defined as the capability of maintaining an apparatus such as an ion source with no special radiaton shield after a cooling-off period of 48 hours. However, it seems unlikely that economics would permit shutting down the reactor for 48 hours if a maintenance problems occurs. Therefore, if we use this definition of on-hands maintenance we must postulate a remote system for removal and replacement of an ion source. After the cooling-off period the ion source could be serviced with no further remote handling.

Preliminary computations indicate that if low-activation materials are chosen the allowable neutron flux at the injector is about 10^6 n/cm^2-s. The neutron flux where the injector duct enters the first wall will be about 10^{13} n/cm^2-s. Therefore the neutron attentuation in the neutron shield and in the HEBT must be approximately 10^7. In principal this attenuation is possible if the beam channel is sufficiently narrow and convoluted.

To verify the feasiblity of this neutron attenuation, three-dimensional Monte Carlo computations are required using a code such as MCNP[15]. After computation of the residual neutron flux, the resulting activation of the injector materials must be computed. These computations have been carried out in a preliminary version.[16]. The results indicate that on-hands maintenance will be marginal under MARS conditions.

A more exact computation of the neutron flux and activation is now in progress. Preliminary results indicate that the neutron attenuation is greatly improved if realistic three-dimensional boundary conditions are used and if materials are selected for their neutronics properties. Materials for the injector components have been selected for low activation, short decay times, and effective neutron shielding. The major components are as follows:

Neutron shielding	Tungsten
Structural	HT-9 stainless steel
High voltage insulation	Alumina
Electrical and thermal conductors	Aluminum

BEAMLINE PERFORMANCE

Once the beam has been formed, accelerated, and neutralized, the last loss mechanism is due to reionization or scrape off in the long duct leading toward the target. The reionization loss has been calculated by a simple gas evolution and build up code for the MARS geometry assuming a Gaussian beam profile. Figure 10 illustrates this duct loss for the angular divergence of the beam in the short dimension dimension at two gas pressures. The divergence previously calculated of less than 0.5 degrees means that pressures in the duct should be limited to below 5×10^{-6} torr for less than 5% loss. This is the baseline design allowance for this beam loss mechanism.

The final sloshing ion beam line design parameters are listed in Table 7. The 475 keV neutral beam line has a total electrical efficiency of about 63% including estimated power supply losses. Of the 8.86 MW injected into the end cells, 64% or 5.68 MW is absorbed by the plasma. The power transfer efficiency from the wall plug to the plasma is, therefore, $5.68/14.1 = 40\%$, in spite of a 71.5% efficiency in forming, neutralizing, and transporting the neutral beam. A higher trapping fraction will improve the overall power transfer efficiency to the plasma.

CONCLUSION

The MARS sloshing ion neutral beamline has been based on components under development or planned to be developed in the next few years. The following components must be developed to produce this efficient 475 keV beamline:

1. Continuously operating deuterium negative ion sources capable of producing 5-7 A/m of a sheet beam. This includes an rf plasma generator for long lifetime.

2. A focusing ion beam transport and acceleration system adaptable to the above ion source. The TFF system appears to be the best candidate at this time.

3. Laser photodetachment neutralizer for over 90% stripping of the ion beam.

4. Superconducting magnetic shield capable of excluding transverse fields of up to 0.6 T.

5. Continuously operating cryopumps.

6. Reliable beamline power supplies with a high availability.

The beam dump technology is presently available. If a gas neutralizer is ultimately used, then some type of energy recovery will

Table 7. MARS Sloshing Ion Beamline Parameters

Energy (keV)	475
Injected Power (MW)	8.86
Trapping Fraction	0.64
Trapped Power (MW)	5.68
Source Current Per End (A)	13.6
Source Size Per End	1.5 cm x 200 cm
Current Density at Source (mA/cm^2)	47
Current Density at Neutralizer (mA/cm^2)	34
Current at Target Per End (A)	9.3
Source Efficiency (kW/A)	20
Source Gas Efficiency (%)	30
Total Laser Power (MW) (2 systems required)	1.4
Beamline Power (MW)	12.4
Neutral Beam Efficiency (%)	71.5
Power Supply Efficiency (%)	86
Total Wall Plug Power (MW)	14.1
Total System Electrical Efficiency (%)	62.8
Overall Power Transfer Efficiency (%)	40.3

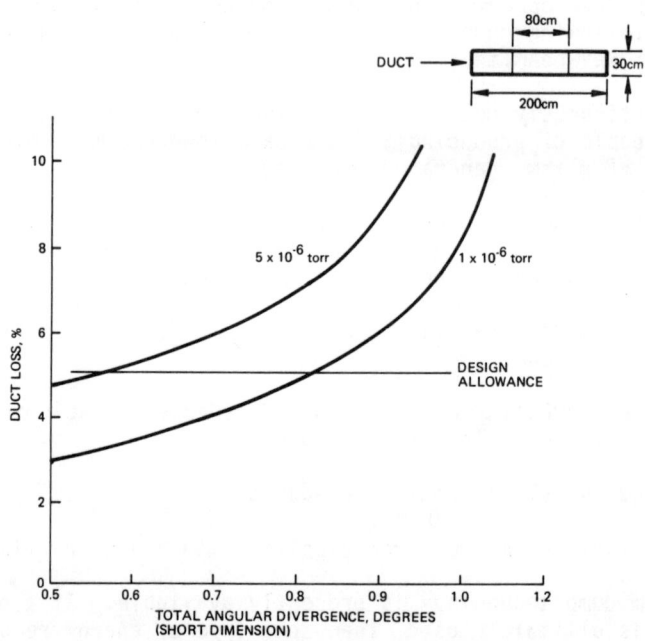

Figure 10. MARS Sloshing Ion Beam Line Total Duct Losses

have to be developed to protect the interior ion dump. In addition, research on the cleaning and conditioning of neutral beam components leading to faster turn-on times is necessary. The availability of future reactors with neutral beams will increase if conditioning of the surfaces is understood.

REFERENCES

1. K. N. Leung and K. W. Ehlers, Rev. Sci. Instrum., 53, 803 (1982).

2. K. W. Ehlers and K. N. Leung, Rev. Sci. Instrum., 53, 1423 (1982).

3. K. N. Leung, LBL, Private Communication.

4. G. M. Gammel, A. W. Maschke, R. M. Mobley, Rev. Sci. Instrum., 52, 971 (1981).

5. TRW Final Report on a Design Analysis of Supplemental Heating Systems, #DE-52058, Sept. 1981.

6. O. A. Anderson, D. A. Goldberg, W. S. Cooper, L. Soroka, Proc. of Advanced Accelerator Meeting, Santa Fe, N.M. (1983).

7. J. H. Fink, W. L. Barr, G. W. Hamilton, IEEE Trans. Plasma Sci. PS-7, 21 (1979).

8. M. W. McGeoch, Proc. of 2nd International Symp. on Production and Neutralization of Neg. Hydrogen Ions and Beams, Brookhaven, (1980), p304.

9. O. A. Anderson, W. S. Cooper, J. H. Fink, D. A. Goldberg, Proc. of 5th Topical Meeting on the Tech. of Fusion Energy, Knoxville, TN (1983).

10. W. W. Rigrod, J. Appl. Phys., 36, 2487 (1965).

11. J. H. Fink, Report, UCRL-87301, Sept. 1982.

12. C. D. Henning, R. L. Nelson, A. K. Chargin, B. S. Denhoy, and F. Harshbarger, "Engineering and Baseball Magnet System", in Proceedings of the 4th Symposiuim of Engineering Problems of Fusion Research, NRL, Washington, D.C., 1971 (IEEE Transactions on Nucl. Sci, Vol NS-18, 290 (1971)).

13. C. F. Barnett, et al., ORNL #5206, Feb, 1977.

14. R. L. Freeman and E. M. Jones, "Atomic Collision Processes in Plasma Physics Experiments", Culham Laboratory Report CLM-R 137.

15. "MCNP-A Genral Purpose Monte-Carlo Code for Neutron and Photon Transport", LA-7396-M (Rev) Version 2B, Los Alamos Monte-Carlo Group, Los Alamos National Laboratory (1981).

16. X. de Seynes, "Shielding Conditions for Neutral Beam Injection Systems", Thesis for degree of Master of Engineering, University of California, Berkeley (1983).

THE ROLE OF MULTISTEP COLLISION PROCESSES IN INCREASING
THE BEAM STOPPING CROSS SECTION FOR HIGH ENERGY NEUTRAL BEAMS

D. E. Post, R. K. Janev[†], and C. D. Boley[*]
Plasma Physics Laboratory, Princeton University
Princeton, New Jersey 08544

ABSTRACT

We have found that multistep processes involving excited atomic states can substantially increase the stopping cross section for a neutral hydrogen beam injected into a plasma. For the very large energies envisioned for negative ion based systems, the enhancement can be as large as a factor of two.

INTRODUCTION

The fusion interest in negative ion based neutral beam systems is largely due to the possibility that such systems offer the promise of high beam energy (> 100 keV/AMU) with reasonable efficiency. High beam energies are desirable since, at high energies, $\lambda \propto E/n_e$, where λ is the neutral beam mean free path, E is the beam energy, and n_e is the electron density, with the result that high energy beams can penetrate to the center of the plasma, thus supplying energy to the plasma center where the confinement is best. In addition, high beam energies are necessary for high potentials in mirror end cells.

The standard calculation of the beam deposition assumes that the neutral atoms in the beam are in the ground state. The fast atoms are ionized by electron and proton impact ionization and charge exchange, all from the ground state.

It turns out, however, that the cross sections for excitation and for ionization due to collisions with protons and impurity ions are of comparable magnitude. This can lead to the production of excited atoms in the neutral beam as it traverses the plasma. The ionization and charge-exchange cross sections scale as n^2, so multistep processes, in which an atom is first excited, then ionized or charge exchanged can become important, as recently pointed out by Wiesemann.[1] We have carried out a detailed calculation of this effect in which we find that excitation effects can produce significant enhancements to the stopping cross section for neutral beams injected into a fusion plasma.

[†]Permanent address: Institute of Physics, Belgrade, Yugoslavia.
[*]Permanent address: Argonne National Laboratory, Argonne, IL 60439.

MODEL

In our calculations, we represent the neutral beam as a sum of beams consisting of excited atoms with electrons in an excited state with principal quantum number n. Thus

$$I(x) = \sum_{n=1}^{N} I_n(x),$$

where x is the distance along the beam line, $I(x)$ is the total beam intensity, $I_n(x)$ is the intensity of the beam with principal quantum number n, and the sum is taken up to the Lorentz limit.

We solve an equation of the form

$$v_o \frac{dI_n(x)}{dx} = \sum_{n'=1}^{N} Q_{nn'} I_{n'}(x),$$

where v_o is the reaction rate matrix. The reaction rate $Q_{nn'}$ includes ionization and charge exchange out of each state n', deexcitation and excitation due to electron and ion collisions, and radiative decay.[2] Collisions with impurity ions are also included. The above equation is integrated numerically, with the initial condition that at x=0, only the ground state is populated.

The maximum principal quantum number N is given by the Lorentz ionization limit (N ~ 7 for typical conditions). Excitations to higher levels (n > N) are counted as ionizations.

RESULTS

We define an effective cross section for ionization of the beam as $\sigma_{eff} = 1/(\lambda/n_e)$, where n_e is the electron density and λ is the mean free path for penetration due to all processes. σ_{eff} is a decreasing function of energy (Fig. 1). The total cross section (without excitations) is slightly below the usual cross section due to a polynomial fit by Freeman and Jones,[3] used by most fusion researchers (Fig. 1,2). The proton ionization cross section has been remeasured and found to be slightly lower than the original value by Gilbody and co-workers.[4] We have characterized our results by an enhancement factor such that $\sigma = (1 + \delta)\sigma_{eff}$, where σ_{eff} is given in Fig. 1, and $\sigma = (1 + \delta)\sigma_{eff}$ includes only ionization and charge exchange from the ground state. δ can be cast as

$$\delta = \frac{\lambda_o - \lambda}{\lambda},$$

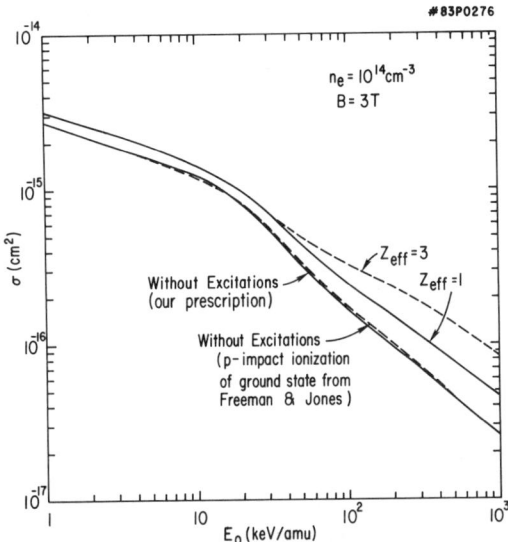

FIG. 1. Total beam stopping cross, defined as $\sigma_{eff} = 1/(\lambda n_e)$, without excitations. The electron temperature is E/10 for E < 100 keV, and 10 keV for E > 100 keV. The value of σ from the polynomial fits from Freeman and Jones[3] is shown for comparison. Also shown are values of σ_{eff} including excitations for Z_{eff} = 1 and Z_{eff} = 3.

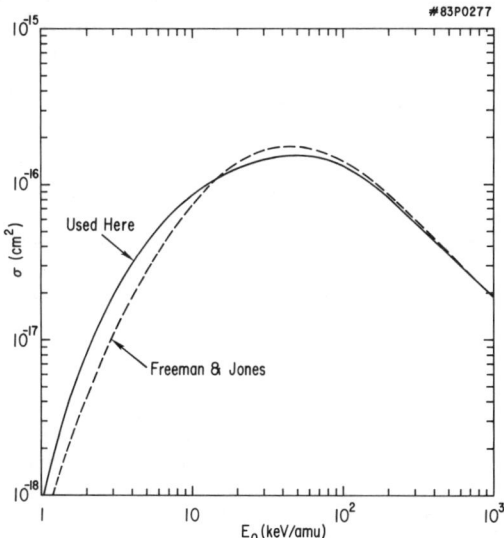

FIG. 2. A comparison of the proton impact ionization cross section used in our calculation based on Ref. 4 compared to the polynomial fit from Freeman and Jones.[3]

where $\lambda_o = 1/(n_e \sigma_{eff})$ is the mean free path in the absence of excitations and λ is the mean free path including excitations.

δ is a strong function of both density and energy (Fig. 3). δ is small at low energies, but increases substantially for energies above 50 keV/AMU. δ also increases with density, as would be expected for a multistep process which would depend on two successive collisions. δ also increases substantially for impure plasmas, increasing in some cases more than a factor of 3 as Z_{eff} goes from 1 to 5. In fact, δ is proportional to Z_{eff} (Fig. 4). It can also be noted that the excitation effects are much larger than the impurity effects due to ground state, charge exchange and ionization collisions with impurities.

INTERPRETATION AND CONCLUSIONS

Since many beam systems are planned to operate near the penetration limit, and since the cross section appears in the argument of a decaying exponential (i.e., $I = I_o \exp(-sn_e\sigma_{eff})$, where s is the path length along the beam), increases in σ of even 30% are important. For the beam systems used on current experiments where $E \lesssim 50$ keV/AMU, the enhancements are modest (~30%) for a clean plasma, and not increased by the addition of impurities. For TFTR energies (60 keV/AMU), the enhancements are larger (35-60%) depending on the impurity level and density. For the 80 keV/AMU beams for D-III and JET, the enhancement varies from 40 to 75%. For the energies being considered for negative ion based systems (200-400 keV/AMU), the enhancements are quite substantial (60-170%) depending again on the impurity level and density (the above numbers were quoted for $n_e = 10^{14} cm^{-3}$).

Since $\lambda \sim E/n_e$, the "loss" in beam penetration due to excitation processes has to be recovered by the experiment designer by either increasing E (probably more than linearly since δ increases with E), or decreasing n_e. While both of these tradeoffs may be possible, they will nonetheless require serious consideration. In the case of systems (such as mirrors) with $\lambda >$ a (the plasma size) where the upper limit on the beam energy is set by the requirement that a given fraction of the beam stop in the plasma, an enhancement in the stopping cross section will allow an increase in the beam energy or a decrease in the primary beam power.

SUMMARY

We have calculated the enhancement due to multistep processes of the stopping cross section for neutral hydrogen beams injected into plasmas. For reasonable parameters, enhancements of 50 to 75% or more are possible for high energy positive ion based systems. The effects are largest for the high energies characteristic of negative ion based systems, and range from 50% to 150% or more, depending on the plasma density, beam energy, and

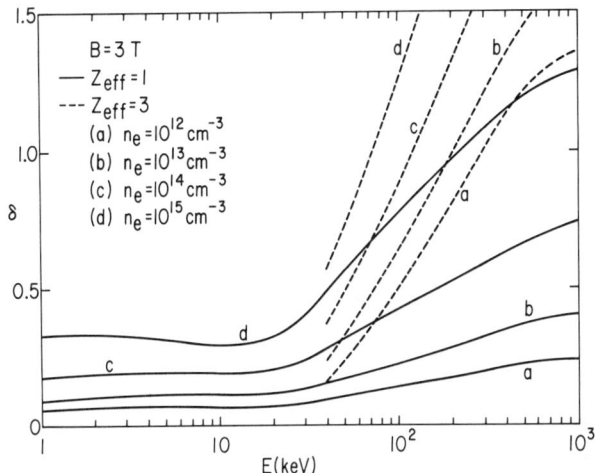

FIG. 3. Enhancement of the beam stopping cross section due to excitations, calculated for four electron densities as a function of beam energy. The electron temperature is taken as E/10 for E < 100 keV and as 10 keV for E > 100 keV. The solid curves are for Z_{eff} = 1, and the dashed curves are for Z_{eff} = 3 produced by carbon, oxygen, and iron in the ion density ratio 10:10:1.

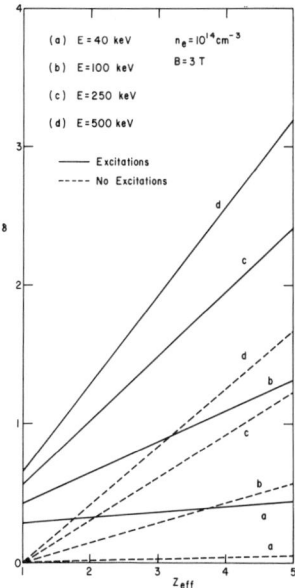

FIG. 4. The beam stopping cross section enhancement as a function of Z_{eff} for carbon, oxygen, and iron impurities in the ion density ratio 10:10:1, at four beam energies. The solid lines include the effects of excitations, while the dashed lines include only the effects of electron loss from the ground state.

impurity level. Since $\lambda \sim E/[(1+\delta)n_e]$, the penetration requirement of $\lambda \sim a$, the plasma minor radius, implies that E must be increased and n_e decreased if adequate neutral beam penetration is to be ensured. Thus, to achieve a given λ, the energy of a 500 keV D° beam system would have to be increased to 800 keV to 1.2 MeV, depending on the impurity level.

ACKNOWLEDGMENTS

The authors are grateful for discussions with D. Mikkelsen, K. Evans, and J. Hiskes. This work was supported by the U.S. Department of Energy Contract No. DE-AC02-76-CHO-3073.

REFERENCES

1. K. Wiesemann, Report 8-05-068, Institut fur Experimentalphysik II, Ruhr-Universitat, Bochum (1980).
2. C. Boley, R. Janev, D. Post, Princeton Plasma Physics Laboratory Report No. PPPl-2047, 1983.
3. R. Freeman and E. Jones, Culham Laboratory Report No. CLM-R-137, 1973, Culham Laboratory, Abingdon, England.
4. M. B. Shah and H. B. Gilbody, J. Phys. B <u>14</u>, 2361 (1981).

NEGATIVE ION BEAM REQUIREMENTS FOR COMPACT TORI

George H. Miley
Fusion Studies Laboratory
University of Illinois
103 S. Goodwin Avenue
Urbana, Illinois 61801

ABSTRACT

One potential application of negative ion beam sources involves injection into field-reversed fusion devices for both start-up and sustainment of plasma currents. High energies provide the prerequisite large orbit ions to drive these currents. A hybrid fluid-particle computational model is described that has been developed to study injection into a field-reversed mirror. Comparisons with the earlier experiments (2X-IIB) at the Lawrence Livermore National Laboratory are encouraging and support the concept of using high-energy injection in future devices.

INTRODUCTION

There are a variety of possible compact tori configurations, e.g., the field-reversed mirror (FRM), field-reversed theta pinch (FRTP), and spheromak (see Fig. 1). All have the common feature that internal plasma currents make a major contribution to the confining magnetic field, causing field reversal which, in turn, leads to enhanced stability and improved confinement characteristics. Thus, injection of neutral beams created from high-energy negative

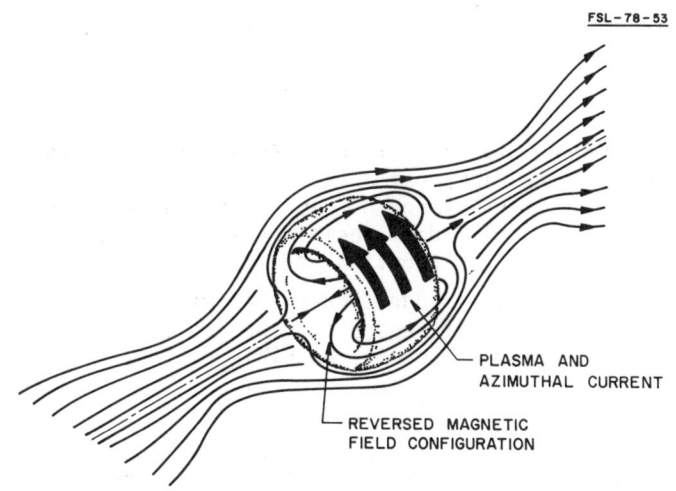

Figure 1. Typical Field-Reversed Mirror Plasma Geometry.

ion beams can potentially play an important role. First, intense beams can be used for start-up. For example, in addition to plasma heating, the injection of such beams into a conventional mirror provides a way to create field reversal, i.e., form a FRM.[1] Second, injected beams, if properly designed, can be used to maintain plasma currents, hence reversal, allowing steady-state operation.[2]

The plasma currents required are typically created by large orbit ions, e.g., orbits that completely encircle the plasma axis in the FRM. This requires high-energy ions, and beams. Negative ion systems appear to be the only way to meet these requirements and still offer a high efficiency.

The use of a negative ion beam system to drive the desired plasma current is more demanding than simply using a beam for plasma heating. Restrictions must be placed on the energy, beam profile, and angle of injection to achieve the necessary current profile. The present paper is intended to illustrate these considerations in some detail by describing a calculational technique that has been developed to study FRM start-up.

FRM START-UP CALCULATIONS

A one-dimensional radial code has been developed to model injection into an FRM. This code, FROST (Field Reversed One-dimensional STartup), allows for the inclusion of radial profiles for densities, currents, temperatures and fields.[3] Also included is a radial electric field. A hybrid model is employed: low temperature background ions and electrons are modelled using two-fluid equations and the higher energy ions are treated by a kinetic model. We will briefly consider each model in the following sections.

FLUID SECTION OF FROST

The background plasma section of FROST is needed for three reasons. First, the diamagnetic plasma current must be included. Second, a knowledge of background temperatures and densities as a function of radial position allows an accurate determination of the birth distribution and slowing down of the high energy ions. Finally, as suggested by Baldwin, et al.[4,5], electron currents in the field null region must be included since they can cancel out the local ion current, possibly preventing reversal.

FIELD CONFIGURATION

Figure 2 compares the field configuration implicit in FROST to the field shape in an "actual" FRM. The field lines in the code are all in the $\pm Z$ direction. Two cuts have been made from the

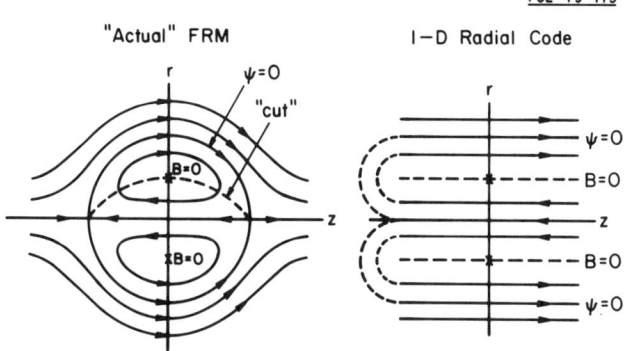

Figure 2. Magnetic Field Configuration.

field null to the separatrix in the "actual" FRM field lines, and the closed field lines have been opened up to form the 1-D configuration in FROST. Hence the closed-field flux surfaces, which are tori, are each mapped into two concentric cylinders. As a result, any closed-field flux surface has two radial positions. Thus, the background densities and temperatures are the same at each of these pairs of radii. As described later, to account for this coupling implicitly, a Lagrangian coordinate system with the flux coordinate, ψ, is used.

RADIAL FLUID EQUATIONS

The fluid equations used to model the background plasma, except in the region near the field null, are shown in Eulerian form in Fig. 3 for the closed region (open field equations are similar). These simplified equations are derived from the full set of fluid equations by first setting $\frac{\partial}{\partial z}$ and $\frac{\partial}{\partial \psi}$ to zero. Then inertial terms are dropped to eliminate unwanted MHD oscillations. Finally, the high-energy ion density is assumed to be much smaller than the background plasma density, as is typical for FRMs. This allows one to set the radial electric field equal to zero in the closed field region. (The radial electric field in the open field region is zero because of "line tying.")

The equation set involves "flow" equations along with pressure balance conditions. The magnetic field equations imply flux conservation - i.e., a background plasma particle always moves so as to remain on the same flux surface. The pressure balance equation relates the magnetic field, the plasma pressure and the ion ring current. The ring current (J_{ring}) is supplied by the high energy ion section of the code. The fluid approximation does not apply near the field null, where the low magnetic field permits a large

e⁻ continuity	$\frac{dn_e}{dt} + n_e \frac{1}{r}\frac{\partial}{\partial r}(rU_r) = S_e$
ion continuity	$\frac{dn_i}{dt} + n_i \frac{1}{r}\frac{\partial}{\partial r}(rU_r) = S_i$
e⁻ energy	$\frac{dp_e}{dt} + \frac{5}{3} P_e \frac{1}{r}\frac{\partial}{\partial r}(rU_r) = \frac{2}{3}\dot{\varepsilon}_e$
ion energy	$\frac{dp_i}{dt} + \frac{5}{3} P_i \frac{1}{r}\frac{\partial}{\partial r}(rU_r) = \frac{2}{3}\dot{\varepsilon}_i$
flux conservation	$\frac{dB_z}{dt} + B_z \frac{1}{r}\frac{\partial}{\partial r}(rU_r) = 0$
pressure balance	$\frac{\partial}{\partial r}\left(\frac{B_z^2}{2\mu_0} + P_e + P_i\right) = -j_{ring} B_z$

Figure 3. Radial Fluid Equations - Eulerian Form Closed Field Region. (approximations: $m_e/m_i \to 0$, $n_e \approx n_i \gg n_{ring}$, $E_r \to 0$)

electron and ion gyroradius. In this region a separate model for the plasma current density and temperature is incorporated using a single equation of motion that describes bulk electron dynamics. This treatment applies to the plasma within one thermal gyroexcursion from the field null. The null region is treated as unmagnetized; electrons are accelerated by the inductive electric field and collisions with ions; damping is provided by electron viscosity.

METHOD OF SOLUTION - LAGRANGIAN MESH

The fluid equations are solved using a Lagrangian mesh with the flux coordinate ψ as the independent variable. Plasma densities and temperatures for electrons and ions as well as the values of the magnetic field are tabulated over a grid of flux surfaces. The radii of all flux surfaces are also stored. During each time step the changes in the background plasma values are found by solving the pressure balance equation for the changes in the flux surface position.

EXAMPLE CALCULATIONS WITH 1-D FLUID CODE

The functional dependence of the ring current on r and t was pre-described for test runs of the fluid code. The ring current shape was assumed to be parabolic in r, lying between zero and 50 cm. The amplitude of the current rose linearly with time to a maximum of -8.7×10^2 A/cm^2 at 1 msec. Refueling sources were assumed to be directly proportional to the ion density (set equal to the ion density divided by 1 msec.) As ion refueling energy of 10.0 keV was assumed while the associated electrons were cold. To model the field null-region, the magnetic field there was assumed to be linear in r. The pressure balance equation $[p + B^2/2\mu_0] = -j_{ring}B$ was then integrated to give a condition on the total plasma pressure over the field null region.

Results are shown in Figures 4 and 5 where the magnetic field and the background densities and temperatures are plotted as a function of r for times 0, 2 and 5 msec. Fired mesh points in these figures are indicated by numbered crosses. This test started with an initially reversed background plasma ($B(r=0)/B_0 = -0.75$).

The initial ion and electron temperatures were equal in the closed field region, varying from 5 keV at the separatrix (92 cm) to 10 keV near the field null. The ion temperature in the open-field region is uniform at 2.5 keV except for a sharp drop to zero near the vacuum region (from r=97 cm to the wall). The open field

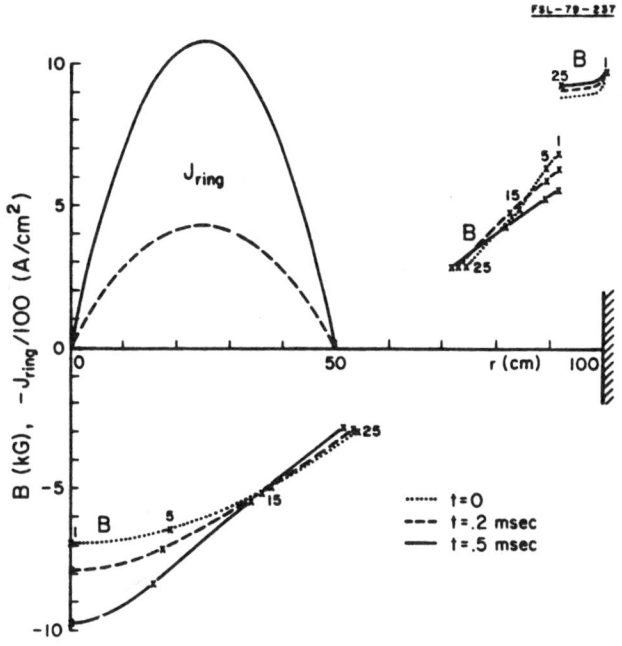

Figure 4. Profiles for the Magnetic Field and the Assumed Ring Current at 0.0, 0.2, and 0.5 msec.

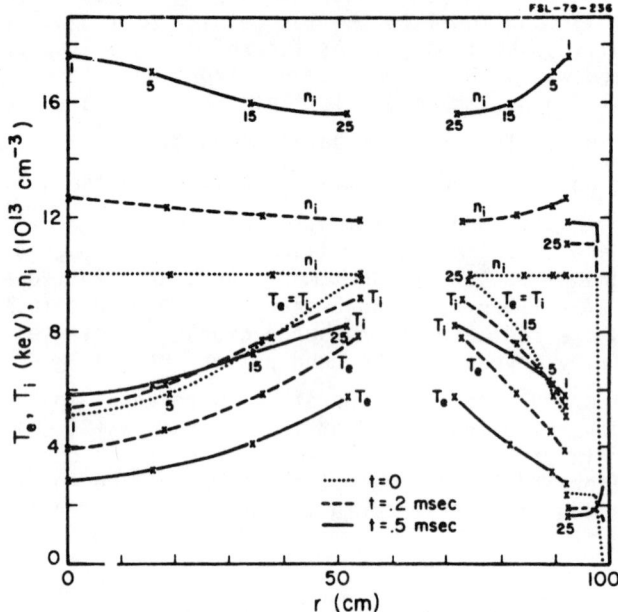

Figure 5. Profiles for the Background Electron Temperature, Ion Temperature, and Ion Density at 0.0, 0.2, and 0.5 msec.

electron temperature is fixed (by stream stabilization) at 100 eV. Ion and electron densities are initially uniform at 10^{14} cm^{-3}, except for the sharp drop at the edge of the plasma. The densities, temperatures and magnetic fields are all discontinuous at the separatrix but the plasma pressure plus the magnetic pressure is continuous there.

As the ring current increases, the centerline magnetic field is further reversed. However, the plasma diamagnetic current partially opposes the increasing ring current. If the change in the magnetic field were determined by the ring current alone, $B(r=0)$ would be -4.55 kG instead of only -2.8 kG. Since the greatest fractional increase in B occurs near the centerline, the plasma there and just inside the separatrix on the same flux surface are compressed more than elsewhere. This accounts for the positive gradient in density as one moves outward from the field null region to the centerline or separatrix.

Because there are no diffusive losses in the closed-field region, refueling causes a density buildup. The refueling energy (10 keV) for ions is close to the ion temperature (5 to 10 keV) so there is little change in ion temperature. However, since the refueling electrons are cold, the electron temperature falls in this example.

KINETIC MODEL OF ION RING

Tangential injection is used so that injected ions have large orbits that encircle the central axis, forming a "ring" of current. As the ring grows, its field cancels out the mirror field on axis, giving reversal. Since the ratio of the ion gyroradius to the plasma size is large, a kinetic model is desirable.

The one-dimensional kinetic ion model treats the plasma as azimuthally symmetric with no axial dependence, the radial dependence of the plasma parameters being computed in time. The time evolution of the distribution for ions from the beam is followed by a numerical multigroup treatment in the "constants" of motion: energy, E, and canonical angular momentum, $P_\phi = mv_\phi r + q\psi$.

With the assumptions of azimuthal symmetry and no axial velocity, the trajectory of the ions in each group is uniquely determined:

$$v_\phi = \frac{P_\phi - q\psi}{mr} \quad , \quad v_r = \pm \left(\frac{2E}{m} - v^2\right)^{1/2} \qquad (1)$$

Since the ion gyroperiod ($\sim 10^{-7}$ s) is so much shorter than the time scale for startup ($\sim 10^{-3}$ s), the ion orbits are averaged over a radial bounce to form density and current profiles for each group:

$$n(r) = \frac{1}{\tau}\left(\frac{1}{\pi r^2 v_r}\right) \quad , \quad j_\phi(r) = \frac{1}{\tau}\left(\frac{qV_\phi}{\pi r^2 v_r}\right) \; ; \; \tau \equiv \int dr/v_r \qquad (2)$$

The total ion current and density are found by a summation over these profiles, weighted by the number of particles per centimeter in each group.

The inductive electric field, E_ϕ, and collisions with electrons are treated as perturbations that gradually move particles to adjacent groups in E and P_ϕ. For example, energy losses from the inductive electric field and from collisions with electrons are averaged in radius using the group profiles as weights. These integrated energy loss rates then cause particles to move downward in energy groups. The injected beam attenuation, i.e., the ion source rate, are computed on a rectangular grid across the beam footprint. Ionization and charge exchange reactions are produced by collisions with both the fluid target plasma and the large-orbit ion groups. All charge exchange neutrals are assumed to leave the plasma. Knowledge of the beam particle energy and the magnetic flux function, $\psi(r)$, then permits the source profile in E-P_ϕ space to be computed.

To begin each time step, the number of particles per centimeter deposited by the beam in each group is found. Next, the total ion density and current are computed from the new ion distribution. Then, the self-consistent magnetic and inductive electric fields are updated. To complete the cycle, particles are moved in phase space due to collisions with electrons and the inductive electric field.

In a typical example, we considered an 80-keV 100 A/cm neutral beam injected 10 cm off-axis through a 10-cm beamport into a 30-cm radius target plasma. The backgound density profile was taken to be cubic, going to zero at the 30-cm wall. The average plasma density of the target was 2.5×10^{13} cm^{-3}, and the temperature was 100 eV. The vacuum magnetic field was 6 kG. Field reversal was achieved after 0.12 ms. The total ion current at this time was -6.16 kA. The large-orbit ions made up ~30% of the plasma ions.

In summary, with the large-orbit ion distribution described by E and P_ϕ, the characteristic time scale for the simulation is the ion slowing down time, rather than the cyclotron period used in particle tracking codes. Unlike the previous application of this technique to describe fusion products, our model computes the ion current, which reverses the field, directly from the distribution function. The self-consistent fields are then found through Ampere's and Faraday's Laws.

MERGER TO FORM FROST

A flow diagram of the merged codes is shown in Fig. 6. Coupling between the fluid background and the high energy ring is provided by a quasi-static pressure balance which treats both fluid and ring contribution to the plasma current:

$$\frac{\partial}{\partial \psi}\left[\left(\frac{B^2}{2\mu_0}\right) + P_{fluid}\right] = -\frac{J_{ring}}{r} \qquad (3)$$

In FROST, time steps follow the sequence:
1) Ion ring currents and densities are calculated by the high energy ion model.
2) Based on this current, the solution to the pressure balance is found by solving the corresponding matrix equation over the Lagrangian mesh. Here, a Gaussian elimination algorithm is used.
3) The values of flux surface velocities (U_r) are computed at each grid point. The plasma density, pressure and magnetic field are then updated.
4) The time step is incremented and steps 1-3 repeated.

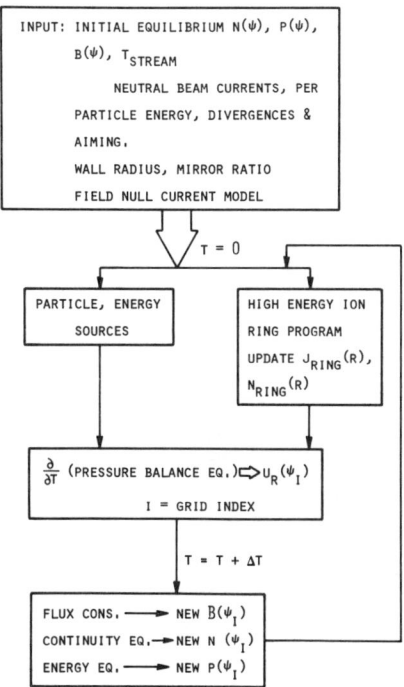

Figure 6. Flowchart for Combined Fluid Code and High Energy Ion Model.

APPLICATION OF FROST TO 2X-IIB

Start-up of a FRM via neutral beam injection was attempted on 2X-IIB at Livermore[2] (1975-1979), but the greatest field depression observed was $B_{applied}/B = 0.90$. Failure to reverse was blamed on low electron temperature ($T_e \sim 100$ eV) and insufficient beam current (up to 500 A @ 20 kV was used). The experiment was then reconfigured into β-II, which used a Marshall gun plasma to form the target plasma. Still, the beam and target requirements for reversal in 2X-IIB remain uncertain. A central issue is the role of electron return currents just before the field lines close.

In order to model a 2X-IIB reversal experiment, the null current model described earlier has been used in FROST. Figure 7 depicts results where a 15 A/cm neutral beam was aimed 4.5 cm off-axis into a plasma stream of density $10^{13}/cm^3$, temperature 0.1 keV and radius 12 cm. The vacuum field was 4.65 kG and the wall radius was 12 cm. Without the electron current model reversal occurs after 0.95 msec of injection. With this current included, incipient reversal still occurs, but is delayed by ~ 0.065 msec. More importantly, the rate at which the magnetic field is decreased on axis is only ~ 20% of the case with no electron null model.

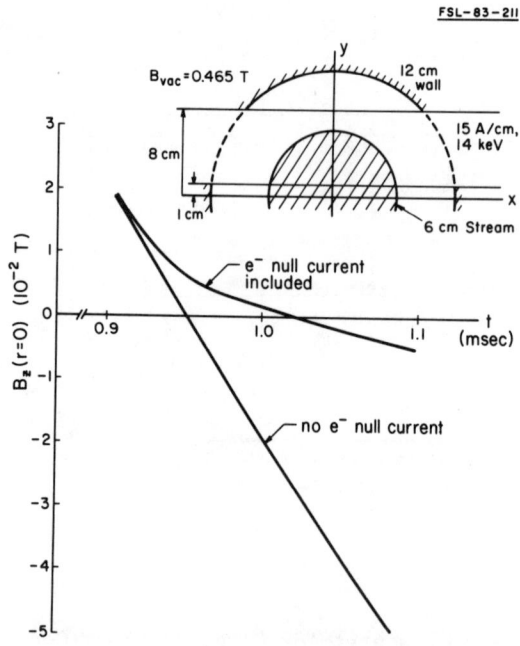

Figure 7. Simulation of a 2X-IIB reversal experiment using the hybrid code FROST. The central magnetic field is shown as a function of time, both with and without the null electron current model. The rate of reversal with electron dynamics is only .175 of the rate predicted for ion dynamics only.

SUMMARY

Electron return currents at a field null pose difficulty in FRM start-up and steady-state operation. Consequently, we have developed a hybrid model that not only describes their dynamics but also treats high energy injected ions separately from the background "fluid" plasmas.

A preliminary application of the model has simulated 2X-IIB reversal experiment. We find that electron viscosity prevents complete cancellation of the ion current, allowing initial reversal to occur, although at a much reduced rate. However, the electron viscousity is highly sensitive to temperature; for electron temperature above ~ 1 keV it is ineffective for 2X-IIB-like parameters. If electrons in the newly created closed-field plasma heat up too rapidly, reversal is thwarted.

These results are very encouraging for beam driven reversed configurations. The 2X-IIB device used relatively low (~20 keV) injected beams; however, much higher energies (>100 keV) would be required for both start-up and sustainment of currents in reactor-scale devices. Consequently this appears to be an important potential future application of negative ion beams.

ACKNOWLEDGEMENTS

This work was supported by the Office of Magnetic Fusion Energy under contract DOE DEAC02 76ET52040. Much of the work described here was carried by Thomas Haill and Robert Stark, graduate assistants in the Fusion Studies Laboratory at the University of Illinois.

REFERENCES

1. J. H. Hammer and H. L. Berk, Nuclear Fusion, 22, 1 89-104 (1982).
2. W. C. Turner, et al., Nuclear Fusion, 19, 1011 (1979).
3. R. A. Stark and G. H. Miley, Trans. Am. Nucl. Soc. 43, 390 (1982).
4. D. E. Baldwin and N. E. Rensick, Comments Plasma Phys. Cont. Fusion, 4, 55 (1978).
5. D. E. Baldwin and T. K. Fowler, "A Reassessment of the Requirements to Obtain Field Reversal in Mirror Machines," UCID-17691 (Dec. 1977).
6. T. A. Haill, R. A. Stark, and G. H. Miley, Trans. Am. Nucl. Soc. 39, 491-492 (1981).
7. H. J. Willenberg, J. Comp. Phys. 34, 330 (1980).
8. H. J. Willenberg, Nucl. Fusion, 19, 313 (1979).

DISCUSSION

CORNELIUS: I have seen at least one proposal where a molecular ion beam would be used rather than a neutral beam. Could you perhaps comment on the desirability or undesirability of that? Then instead of using a neutral beam stripped to an ion beam you use a molecular ion beam where you strip to ions and perhaps dissociate into two ions.

MILEY: That reminds me of the old DCX days at Oak Ridge where that type of experiment was tried and the molecular beam was dissociated in an arc. I feel that from the efficiency point of view this method is less desirable. Also, from studies that were done earlier, many problems were found with that approach. I think that the negative ion beam approach ultimately has a better efficiency. If one can tailor the negative ion beam profile based on prior knowledge of the density profile of the target, one can probably place the beam into the device and obtain the desired more easily with negative ions than you could with the technique of dissociation. That is my personal opinion and I would have to admit more study is needed for a definitive answer to your question.

APPLICATIONS AND DEVELOPMENT REQUIREMENTS FOR MULTI-MeV LIGHT ATOM BEAMS

L. R. Grisham, D. E. Post, and D. R. Mikkelsen
Princeton University, Plasma Physics Laboratory
P.O. Box 451, Princeton, N. J. 08544

ABSTRACT

Energetic neutral beams of various light atoms have been proposed for a number of fusion applications including diagnostics, heating and current drive in tokamaks, and to maintain the electrostatic potential in tandem mirror end cells. These beams would be formed from negative ions which would be accelerated to energies in the range of 1 MeV/amu and then neutralized. We review these applications, the problems in their implementation, and what developments in the way of neutralizers, sources, and accelerators would enhance their feasibility.

INTRODUCTION

In this paper we will review briefly the fusion applications that have been studied for beams of light atoms at energies in the MeV range. Producing usable systems to carry out any of these applications would involve significant development. However, the degree of innovation required varies considerably with the application, being determined in part by the power needed, but even more fundamentally by the requisite power efficiency. In the following sections we will discuss the potential applications in the order of their increasing demands upon system efficiency.

A DIAGNOSTIC FOR FAST CONFINED ALPHA PARTICLES

In terms of power and efficiency, the least demanding role in which a multi-MeV light atom beam could find application is as a means to measure the velocity distribution of the fast alpha particles slowing down in the plasma after being born in D-T fusion reactions. It is important to ascertain whether these alpha particles do indeed slow down classically through coulomb collisions with constituents of the bulk plasma, since it is this slowing down process which is expected to provide the primary heating in fusion reactors after start up. If a significant portion of the alpha particles were lost through collective effects such as drift waves prior to thermalization, the scaling of plasma and auxiliary heating parameters would be altered, and, in addition, provision would have to be made to protect the first wall from augmented erosion. In a device such as TFTR, which is designed to only reach scientific breakeven ($Q = 1$), the alpha particles will account for only 20% of the heating even if they relinquish all of their energy to the plasma. This will be insufficient to make the quality of alpha confinement apparent from the plasma energy balance alone.

Accordingly one needs another way to measure the velocity distribution of the alphas as they slow down.

Such a measurement could be made by neutralizing the alpha particles so that they would escape from the plasma,[1,2] after which they could be stripped and energy analyzed. The cross sections for double charge exchange decline rapidly at velocities greater than 3×10^8 cm sec^{-1}, whereas the alpha particles are born at a velocity of 1.3×10^9 cm/sec. Consequently, a neutral doping beam will mainly neutralize only alpha particles with velocities near its own, and in order to measure the higher portion of the distribution one needs a probe beam able to approach the alpha birth velocity (corresponding to an energy of 880 keV/amu).

Beam energies this high mandate the use of negative ions as the precursors of the neutrals in the doping beam. In order to minimize the required energy of the doping beam, it is desirable to use the lightest multi-electron atom that is practical. At first glance this would appear to be ^{3}He. However, this would not be easily usable due to the fact that the ^{3}He^0 formed from ^{3}He$^-$ will be predominantly in a long-lived meta-stable state.[3] The relevant atomic cross sections for this state are unknown, and they would be extremely difficult to measure with sufficient accuracy. It has been suggested[4] that one might still be able to use ^{3}He by simply allowing ^{3}He$^-$ to decay to ^{3}He^0. The estimate is that at 1.7 MeV, ~24% neutralization will be achieved after a flight path of 30 meters. The length required is, however, dependent upon the lifetimes of the shorter lived components of He$^-$, which are rather poorly known. In any event a drift region of several tens of meters would be highly inconvenient for use on existing facilities such as TFTR, and it would place severe constraints upon the beam divergence after the accelerator.

The next lightest usable atom, ^{6}Li, would result in a neutral beam after stripping which would be virtually entirely in the ground state by the time it entered the plasma. Consequently, ^{6}Li appears to be usable as a probe beam, but it requires a beam energy of 5.3 MeV to match the fusion alpha birth velocity. However, a maximum beam energy of 4.0 MeV would allow most of the high energy portion of the alpha distribution to be probed.

As a gauge of this diagnostic's sensitivity to major alterations in the alpha slowing down spectrum, we have calculated[2] the flux of neutralized alpha particles which would result from double charge exchange for two different alpha particle slowing down distributions.

The results are given in reference 2, where we show that the different distributions give very different count rates as well as different variations in the rate as a function of beam energy. The forward peaking arises because alpha particles traveling in about the same direction as the beam are more likely to have low velocities relative to the beam than are alpha particles less paralel trajectories. It appears that a Li^0 beam of 45 mA would provide sufficient signal. A current of this magnitude could be produced from a 100 mA Li$^-$ beam with a gas neutralizer.

BEAMS FOR TANDEM MIRROR END PLUGS

Several groups have performed designs[5,6] for tandem mirror fusion reactors which require 200-500 keV neutral hydrogen beams along with ECRH heating to maintain the electrostatic potential barrier in the end cell. These would require considerable currents of H^- (20-22 A per end cell if gas neutralizers were used and if there were no transmission losses). An alternative approach[7] which would reduce the required current would involve the use of beams of light atoms with $Z \geq 3$ at energies of 0.5-1 MeV per amu. These beams would be formed from negative ions that had been RF-accelerated and then neutralized in a thin gas cell, a plasma cell, or possibly, a photodetachment cell. Upon entering the plasma the atoms quickly pass through successive ionization stages until they are fully stripped. The end plug plasma is fueled and heated by the beam ions, and comes to consist primarily of the fully ionized injected isotope. When the $Z \geq 3$ ions scatter into the end plug loss cone they escape through the end opposite the central cell because the electric field pulling ions through the mirror is higher there. It appears that impurity accumulation in the central cell due to the light atom beams will be slight and will pose much less of a problem than will the helium ash from fusion reactions.[7] Calculations in ref. 7 indicate that the energy gains in mirror reactions using beams of Li, C, or O can be as good as those for schemes using H beams.

Reference 7 considered the use of beams of 5 MeV lithium, 5 MeV carbon, or 10 MeV oxygen in place of beams of hydrogen for the end cells of the WITAMIR-1 tandem mirror reactor design. In this design each end cell required just over 13 A of neutral hydrogen current at 500 keV. If, on the other hand, the high energy $Z \geq 3$ beams were used, each end cell would require an equivalent neutral current of only 0.3 A using lithium, 0.7 A using carbon, or 0.22 A using oxygen. If a thin cell of N_2 or CO_2 were used as the neutralizer, then the corresponding required negative ion currents from the accelerator (neglecting transmission losses) would be about 0.7 A of Li^-, 2.7 A of C^-, or 1.1 A of O^-. The requirements for Li^- and C^- probably exceed what could be supplied by a single source, so several sources and accelerator channels would be required. The C^- current requirement could probably be decreased if the energy were increased to 10 or 12 MeV. A single O^- source and accelerator channel could probably supply the oxygen current requirement. (See ref. 8 for a discussion of the technology for light atom beams).

HEATING AND CURRENT DRIVE IN TOKAMAKS

In the past Dawson and McKenzie[9] and Grand[10] proposed injection of positive ions (B^+, Ne^+, etc.) as an alternative to neutral hydrogen injection. The ions would be accelerated to about 1 MeV/amu, at which energy the nuclei, once fully stripped in the plasma, would be as well confined as the fusion product alpha particles. As the ions undergo successive ionizations they drift further into the plasma. The use of elements heavier than hydrogen allows one to increase the energy carried per particle without degradation in confinement for energies up to 1-2 MeV/amu. This then reduces the current required to achieve a given power. Positive ion beams are technologically attractive, since the source can be relatively simple and no neutralizer or ion dump is required. However, transporting the ions across the spatially and temporally varying magnetic fields surrounding the plasma is a difficult problem which has not been solved.

As an alternative, and as a natural extension of the beam required for the alpha diagnostic, we have considered the possibility of using light atom beams to heat tokamak plasmas (see ref. 8). The attraction is, of course, that as neutrals these beams experience fewer difficulties in transversing the region around the plasma. The accompanying difficulty is that it is more difficult and less efficient to make neutrals. Since one is limited by confinement constraints to energies in the region of 1-2 MeV/amu, there is a premium on going to higher masses. However, in going to very massive elements the ion's confinement in its early stages of ionization degrades, since its Larmor radius is then very large. Accordingly, in our past studies we primarily considered possible negative ions with $A \leq 40$.[8]

Figure 1 shows that these beams should be effective in heating an INTOR plasma. This shows the calculated heating deposition profile for 16 MeV oxygen injected as O^0, and for comparison 46 MeV sodium injected as Na^+, 2 MeV D^0, and 150 keV D^0. We also show the initial ionization profile of $O^0 \to O^+$. The O^0 itself does not penetrate to the plasma center, but the subsequent inward drifting results in strong central peaking by the time the oxygen becomes O^{+8}. Since almost all the energy transfer occurs after the oxygen is fully ionized, the heating is also concentrated in the core. We see that the 46 MeV Na^+ and the 2 MeV D^0 (the final deposition profiles are shown) both heat the plasma core about as well as the O^0. Thus, if one eventually chose to use one of these approaches, the choice would probably be determined more by technical considerations then by plasma interactions. In this example the Na^+ was assumed to be introduced at the edge of the plasma, without reference to how it arrived there.

Light atom beams could also be used to drive current in a tokamak. However, the current drive efficiency of such beams relative to high energy D^0 beams depends critically upon how substantial trapped electron effects are. If they are insignificant, then the heavier atoms are more efficient, but 2-3 MeV D^0 becomes more attractive the larger the trapped electron effects are.[8] At this time it is not possible to say how important trapped electron effects will be.

TECHNOLOGICAL CONSIDERATIONS

The technological base for light atom beams has been discussed in ref. 8. Figure 2 shows one possible configuration of a light atom beam system.

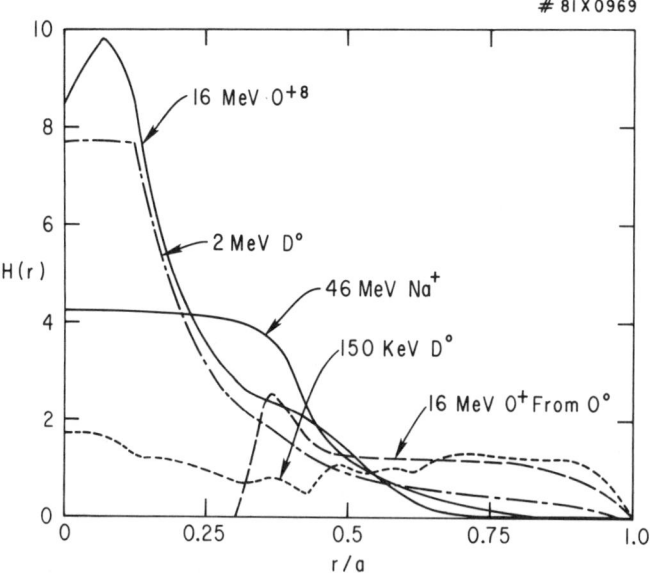

Figure 1. Heating deposition profiles in an INTOR-sized plasma for 2 MeV D^0 (made from D^-), 46 MeV Na^+, 150 keV D^0 (made from D^+ yielding neutral power fractions of 60% at full energy, 24% at one-half energy and 16% at one-third energy), 2 MeV D^0 made from D^-, the initial birth profile of the 16 MeV O^+ resulting from the ionization of coinjected O^0, and the final profile of the O^{+8} resulting from injection of the O^0. [Ref. 8].

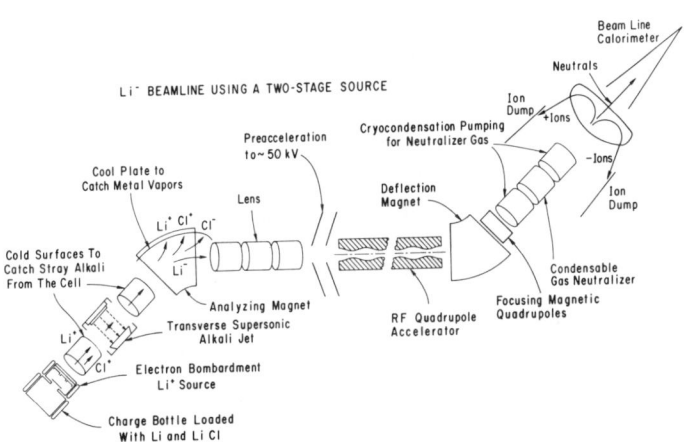

Figure 2. Conceptual design (not to scale) of a beamline to produce multi-MeV Li^0. [Ref. 8].

It appears probable that ion sources could be developed to produce ~100 mA of Li⁻, ampere level currents of O⁻, and hundreds of mA of C⁻ or Si⁻. A Li⁻ current of 100 mA would be quite sufficient for the alpha diagnostic, but taken in conjunction with its low mass, it would probably result in too little power per beam for the other applications. An ampere of O⁻, on the other hand, that was accelerated to 2 MeV/amu would carry 32 MW. This would be satisfactory for heating purposes, but only if most of it could be neutralized.

The combinations of voltage and current requirements seem to preclude the use of electrostatic accelerators. However, as discussed in ref. 8, electric-quadrupole-focussed RF accelerators should be suitable for all of these applications. A larger-throated version of the RF quadrupole accelerators (RFQ) being developed at LASL[11] appears to be appropriate for the alpha diagnostic. For the other applications, either a RFQ or a MEQALAC[9] (originally developed at BNL) would be applicable, depending upon the type of source to which it was mated.

The simplest type of neutralizer would be a thin cell of some easily condensed gas. Recent measurements show that at 6-7 MeV ~46% of Li⁻ can be neutralized using nitrogen,[12,13] and that as much as 54% can be neutralized if one is willing to use hydrogen (which has a higher vapor pressure). However, the maximum neutral fractions are much lower for heavier ions[12]: ~27% for C⁻, ~20% for O⁻, and ~22% for Si⁻ when the neutralizer gas is something easily condensable such as argon, nitrogen, or carbon dioxide. This efficiency might be improved somewhat if hydrogen were substituted as the neutralizer, although no measurements have been done with this. Clearly a more efficient neutralization method would be desirable. Lorentz neutralization by passage through an electric or a magnetic field appears infeasible at these low velocities (1.3-1.8 x 10^9 cm sec⁻¹); as the ions became neutralized after transversing different path lengths through the field the divergence would be greatly increased. A plasma neutralizer should give greater neutral fractions of all these elements, and experiments are presently underway at BNL to obtain such measurements.[14] In principle the most desirable type of neutralizer might be one which used photodetachment and, indeed, one should be feasible [2,8] for Li⁻ with extensions of existing laser technology. Li⁻ is less bound than H⁻ (0.62 eV versus 0.75 eV) and the photodetachment cross section for Li⁻ has been calculated to be quite similar in overall shape to that of H⁻, but larger[15]. Accordingly, any photodetachment neutralizer which works for D⁻ should neutralize an even larger fraction of Li⁻. Unfortunately the other negative ions we have mentioned - C⁻, O⁻, and Si⁻ - all have larger electron affinities (which is why they are easier to produce) and would require a laser in the 3000-6500 Å region, where no efficient high power CW laser systems presently operate. This situation may change, however, when free electron lasers are perfected.

CONCLUSION

The single biggest problem of multi-MeV light atom beams is their power efficiency, which with existing technologies would be fairly low. The efficiency of a 1-2 A RFQ has been projected to be 35-38% (wall plug to beam)[16]. With a gas neutralizer a Li^0 beam would then have an overall efficiency of ~17%, and for heavier ions this would decline to ~9%. The 17% efficiency for lithium would be acceptable for the alpha diagnostic, and the system efficiency could approach that of the accelerators with a chemical laser photodetachment neutralizer of the type being studied for D^- beams.[17] For the mirror plugging application these beams might be acceptable with their presently projected poor power efficiencies simply because the plug beams account for a very small fraction of the power flow in a tandem mirror reactor, so the power consumption and increased investment in power supplies would not be overwhelming if they were deemed to be easier to develop than higher current 0.5 MeV D^0 systems.

With extrapolations of present technology it appears that Li^- might have an acceptable efficiency (~35%) for tokamak heating if used with a chemical laser photodetachment cell similar to one for D^-. However, for Li^0 beams to offer any attractiveness in simplicity over D^0 beams it would be necessary to increase the Li^- current well beyond the presently anticipated 100 mA so that the power per beam system could be increased beyond the 0.6-1.2 MW which this would imply. The heavier and more tightly bound negative ions (C^-, O^-, Si^-) would be attractive for tokamak heating if efficient lasers of suitable wavelengths become available, or if plasma neutralizers should prove to be efficient. Current drive in a tokamak would require even greater efficiency, and requires more information about trapped particle effects.

The attractiveness of multi-MeV light atom beams for the more power-intensive applications could be greatly improved if developments were to occur in a number of areas. Any development of more efficient plasma or photodetachment neutralizers would be significant, as would alternative accelerators with greater efficiency. In addition, development of high current sources of weakly bound negative ions which could be photodetached by existing chemical lasers would be highly welcome. A possible high current Li^- source based on molecular photodissociation has been proposed,[18] but not pursued. For heating or current drive an even more desirable development would be a Na^- source capable of at least a few hundred mA. Na^- is bound by only 0.46 eV, has a calculated photodetachment cross section similar to that of Li^-, and could thus be easily neutralized with any photodetachment cell suitable for D^-. With a mass of 23, sodium could be accelerated to 30-50 MeV and still be confined, so a beam of a few hundred mA could carry a large amount of power.

ACKNOWLEDGMENTS

We appreciate useful discussions with T. Boyd, J. M. Dawson, K. W. Ehlers, G. A. Emmert, H. P. Eubank, K. N. Leung, A. W. Maschke, J. F. Santarius, R. H. Stokes, D. A. Swensen, and T. P. Wangler. Work Supported by U.S.D.O.E. Contract No. DE-AC02-76-CH03073.

REFERENCES

1. D. E. Post, D. R. Mikkelsen, R. A. Hulse, L. D. Stewart, and J. C. Weisheit, J. Fusion Energy 1, 129 (1981).
2. L. R. Grisham, D. E. Post, D. R. Mikkelsen, Nuclear Technology/Fusion 3, 121 (1983).
3. A. S. Schlacter, BAPS 25, 698 (1980).
4. K. N. Sato, M. Sasao, Annual Review April 1981 - March 1981, Institute of Plasma Physics, Nagoya, Japan, 52 (1982).
5. G. Carlson, B. Arfin, W. L. Barr, B. M. Boghosian, et.al., Tandem Mirror Reactor with Thermal Barriers, UCRL-52836 (1979).
6. B. Badger, K. Audenaerde, J. Beyer, et.al., WITAMIR-I, A Tandem Mirror Reactor Study, VWFDM-400, Univ. of Wis. (1980).
7. D. E. Post, L. R. Grisham, J. F. Santarius, and G. A. Emmert Nucl. Fusion 23, 3 (1983).
8. L. R. Grisham, D. E. Post, D. R. Mikkelsen, and H. P. Eubank, Nuclear Technology/Fusion 2, 199 (1982).
9. J. M. Dawson and K. R. McKenzie, "Heating of Tokamak Plasmas by Means of Energetic Ion Beams with $Z > 1$," PPG-470, Plasma Physics Dept., U. of Cal., L.A. (1980).
10. Pierre Grand, private communication (1979).
11. T. P. Wangler and R. H. Stokes, IEEE Trans. On Nuc. Sci. NS-28, 1494 (1981).
12. L. R. Grisham, D. E. Post, B. M. Johnson, K. W. Jones, et.al., Rev. Sci. Instru. 53, 281 (1982).
13. J. P. Aldridge and J. D. King, "Charge-Exchange Cross-Sections for Li^- Ions at 6 MeV," LA-8682-MS, Los Alamos National Laboratory (1981).
14. A Herschcovitz, B. M. Johnson, et.al., this conference.
15. D. W. Norcross and D. L. Morres, in Atomic Physics 3:Proc. Third Int. Conf. Atomic Phy., Boulder, Colo., August 1972, S. J. Smith and G. K. Walter, Eds., Plenum Press, New York (1973).
16. R. H. Stokes and T. P. Wangler, private communication (1981).
17. J. L. Fink, this conference.
18. W. C. Stwalley, private communication (1980).

DISCUSSION

<u>JACQUOT</u>: When you consider O^{8+}, is oxygen injected as neutral and then ionized step by step or is O^{8+} injected directly?

<u>GRISHAM</u>: It is actually injected as neutral oxygen and the initial distribution for O^0 going to O^+ is not at all peaked on the axis; in fact, it's peaked rather far out. All of the heavier atom injection schemes rely upon the fact that the particle drifts inward as it undergoes orbit compression. Each time it becomes successively more ionized, its gyro radius decreases and there is a lot of forward interior peaking that occurs after that. The initial deposition curve for O^0 going to O^+ is shown on the first figure. In fact, more than 99% of the power transferred to the plasma occurs after ions are fully ionized. The curve for Na^+ is the final curve, once the ion is completely ionized.

PANEL SESSION: WHERE DO WE GO FROM HERE?

W.S. Cooper (Moderator), J.W. Fink, C. Jacquot,
H.S. Staten, L.D. Stewart

COOPER: First, the title of this panel is "Where Do We Go From Here?" and the answer was suggested to me by one of you not long ago: we go to Washington for money. But the problem is, what do we do when we get there? Maybe we can find out. I want to make just a couple of comments to set the tone of this. We really have a tough job because the very first negative ion application on any fusion experiment is already much more difficult than any positive ion application has been or is likely to be. The energy is higher, reliability has to be very high, the pulse length is long, it has to operate in a radiation environment. Such a beam line might be needed in sometime like 10 years, which from the developer's point of view seems a terribly short time to accomplish that goal, yet it is far enough in the future that the machine builders and designers can think, well, they can do anything in less than 10 years, we'll just shove that goal ahead; that tends to be what has happened in this program. As you've heard a number of times, we don't have to worry about positive versus negative ions anymore. That problem has been resolved and we have to consider only RF versus negative ions. So I thought we might start here by asking Joel Fink to tell us a little more detail about, the apparently, at the moment, most likely first application of negative ion based neutral beams, namely an upgrade of MFTF.

FINK: Well, I guess like everybody else I'm tired. But the device that was of particular interest was MFTF alpha + T and this is a strikingly interesting concept. Just to give you a background of what this machine is supposed to do, it is for the 1990's, 1992. It uses deuterium and tritium, 80 keV injection to make a driven mirror in the center of the cell. One of the prime objectives of the machine is to be the first neutron generating machine, admittedly at low level, but still to get some experience with what a reactor would be like. The sloshing ion beams we're talking about, the 80 keV beams, are to be positive ions and are not of too much interest except that one of them is supposed to be tritium and we have not made any tritium beams. We really don't know too much about the tritium performance with high voltage. What really would occur, we'll be finding out. It is also supposed to operate continuously for 100 hours. Having finished that we can look at the negative ion beams. Having the liberty of standing in Livermore and not having to build the darn thing I could use my imagination and I have used it. I was pointing out at the time that we had a bad problem in this 15 centimeter plasma width. The question was, how far back did the source have to be and, if you take the reasonable angle, could you actually hit that target. Then it got even worse than that because the designers decided that the shielding that went around the coils was going to give us a 15 centimeter slot in which to get the 15 centimeter beam through and Lord help us if the coils vibrate! You'll burn!

I didn't really want to lecture, but to get some kind of a feeling of what we're looking at, I might say we were concerned with the separation of positive and negative ions and we're very much aware that this coil gives us a magnetic field out of the board, so that looked like duck soup. The field of the coils themselves will permit us to do that. Then the question was what would this look like and here we took advantage of Oscar Anderson's TFF accelerator and the concept here was that maybe you could pick that up with a crane. We're looking at a neutron trap here; I've heard it called a glory hole and the idea was that the intense neutrons coming in through this hole would rattle around inside here and therefore reduce the flux of neutrons that were rattling around here (pointing to viewgraph). That seems to be a reasonable assumption. I know the statements that I have heard were very favorable on this approach. Larry Ruby in Berkeley had a student do a study. There's one other feature here of significance--

JACQUOT: That means that the source itself is at high voltage, minus 200 kV, insulated from the box where the cryopumps are.

FINK: Yes, as far as it's concerned, the question of the spacing in here, I might show you the next generation diagrams to give you a little better feeling. This is to guide and more or less what we're looking for and where we're going. There's another very significant point in this that isn't very clear. Because the pressure in this region has to be so low to prevent loss, the neutralizer, which was envisioned to be a gas neutralizer, pours gas into this region which has to be pumped along with the gas coming out of the source. But in addition to that we had a question of gas coming out into the reactor. In the tandem mirror, where the present concept was fueling by pellets, the calculations of the amount of gas that would evaporate from the pellets over this long path have shown that this would make a tremendous pumping load. I don't know if you remember the mirror diagrams from years ago where the energy recovery electrodes were so large. Well, they have been conceptually reduced to a smaller size and in the process of so doing we became aware of the fact that around the plasma there is a halo which is quite dense. The numbers that I was given and that I started out with showed that you could live with something like 10^{-5} torr outside the plasma and that none of that low temperature gas would penetrate the plasma because it would be ionized in this halo. Since they were drawing the halo out the ends, to be pumped, they would in essence be drawing and pumping at the same time. Now some experiments seem to indicate that they have seen that type of phenomena. And the question was, could they take all the pellet gas, which turned out to be 10 to 20 times more than anything I could dream of putting in with a neutral beam injector. So I'm generous, I let them have it. In essence the gas coming out of the neutralizer is assumed to flow into the halo to be drawn out. The point is that the gas is ionized in the halo and thermally wraps around the magnetic field lines and is drawn to the end, to be pumped at an equivalent pressure like 10^{-2} torr, and they believe they can pick this up with some type of rotary pump, which is a considerable difference from what they had before. At present, of course, all of this is being analyzed in an attempt to be confirmed. But it is one of the most, I think, positive aspects relative to beam technology we have on the horizon, a new one at any rate. It is the first time I ever heard of the fusion reactor being kind to anybody and giving us a break. But if indeed this is true I think this is another big step forward in terms of neutral beam technology. In that respect I didn't have courage but I wondered if there was a hole in the wall where I could draw that gas out too, but some day . . . . In the Oak Ridge designs which were built on these kinds of concepts, the sources, to answer your question, are supposed to be surrounded by SF_6 chambers so that indeed you could do something like that. They were not as comfortable with the glory hole but I think it will be introduced, that they are getting convinced. The associated top view shows that there was one beam on top and one underneath, and they were bent into the common plane to go into the neutralizer. There was another concept using four beams. The point was that the 14.7 or 15 cm width is a tremendous restraint on the source as originally envisioned and the problem was to shove everything up. Shoving everything up into the reactor region made the magnetic shielding that much more difficult. There is no question that what is going to happen to sources is, we're going to want to jam them as close to reactors as we can. The floor space is valuable. The TFF with its bends gives a tremendous advantage from the structure point of view as compared with the head on concept. So I don't know whether that answers the question but you see one concept of a direction trying to take advantage of the technology we see.

STEWART: The enhanced stopping cross section business in that particular problem not only helps with the total amount of power, because you get to cut down on the source requirement and maybe shrink the source and ease off the optics problems so there is more than a proportional effect on the source requirements.

FINK: The optics problem was based upon the divergence angle and we've looked at the source from the Berkeley point of view, using the exit aperture as beam skimmer, similar to Oak Ridge where they were talking about a cooler beam because they were skimming the higher transverse energy particles. I was trying to see whether I could come up with an advantage in losing the ions at low energy rather than at high, but I couldn't beat the idea of bringing the source in closer. No question that that is the way to go and this is in essence a battle to do that.

COOPER: This question "Where do we go from here" has both practical answers and political answers, and they are to a large extent interrelated. I don't know which is the most interesting. I would like to ask Stan Staten if he could give us his comments on the probability that something like this will in fact be the next goal or the ultimate goal of the negative ion beam.

STATEN: That's a hard question to answer because it involves so many variables. How well is the tandem mirror concept going to work? There have been some recent problems in the tandem mirror trying to observe some things on TMX and they appear to be associated with vacuum leaks because up to this point every time they looked for one of them and have gone into the next stage in the experiments, one of the new features that they were trying to test, it's been there, and it's worked more or less as predicted. So at this point we feel pretty comfortable with the tandem mirror concept. But it's a rather complex concept and how well will MFTF-B work, I really don't know. I think it tends to come down also to the question of RF and how well or not well will it work. But I think if you look at the things under our control it comes back under the point I was trying to make in the paper I gave. I'm not very good at looking very far ahead. I mean, these reactor studies we do are invaluable to us. They tend to point us in the right direction, make us worry about the right things, make sure we don't develop things that just simply are not applicable, but I end up worrying about fighting for the FY '85 budget, which is what I will be doing over the next couple of months back in the office. The pressure is to do various things and it seems to me that the probability will be higher if we are successful in the next say three years, two years. Nothing succeeds like success, and I'm very impressed with how far we've come. Six years ago, this symposium, I guess, was really pretty much dominated by double charge exchange systems or, at least, half of the symposium was devoted to them. We didn't see much of that this time. We saw some, but we saw a great interest in a concept that six years ago was really unique in only one place. Marthe Bacal walked in and said, Hey people, I have negative ions in my discharge and I have a lot of them. A lot of people said, gee, how can that be? I've watched this community at large, and I mean really the international community, work on that idea, people from various places talk to her, visit her laboratory, participate in joint experiments with her. The first couple of days of this symposium have been, I don't want to say completely dominated, but better than half of it has been on that concept. It offers great potential and so where is that going to be in another three years, at the next symposium, I really don't know. If we can make a source that works and works relatively reliably at reasonable current densities, in quasicontinuous operation, even better if we could do it without cesium, which the volume source offers, and if we can demonstrate some reasonable system, not isolated components here and there, but things that machine designers can look at and have some measure of confidence that they can commit to a big experiment with some expectation that it will work, then I think the likelihood that we'll do something like this is rather high. You've got to realize that people committed to the TFTR back when we had 20 keV, 2X-II sources running. We had some small 10 x 10's on ATC and what did we have on ORMAC, a 15 centimeter diameter or so, a 10 centimeter source. People looked at 120 keV and they were rather frightened by it. But there was enough success, if you will, proof of principle under their belt that they felt, first of all we have to do it. That's one thing that I'd like to compliment the fusion program as a whole on: as they see something that has to be done, they tend to go out and do it. As I watch other organizations, they tend to study things to death and I'll give a specific example, the gyrotron. DOE beat out DOD because DOE said we have to do it, we'll go do it. And I watched people at various defense laboratories put committees together and study it. As a result, the U.S. and in particular the U.S. Fusion Program holds the world record for power, pulse length, and combined joules, from gyrotrons because we decided we had to do it and we did it. I think that's a characteristic of the fusion program. We pick a challenge and we say we need to do that and we don't spend too much time worrying about why we can't. We worry about how we do it and how do the creative ideas support it. The gyrotrons we're running today are not the gyrotrons that the people first tried to run. The negative ions we're running now are not the negative ions we were running six years ago either. Again, I don't know the answer to your question, Bill, but I think if we're successful over the next two to three years and can generate the confidence on the part of a broader user community that there is some real credibility in this, there's a high probability that something like these will get used on a major device.

COOPER: One problem we obviously have is that our potential users' program has to succeed in addition to our own. Let's say MFTF succeeds and TFTR,Q=1, succeeds, then I can see an unresolved competition between mirrors and tokamaks. Do you think, Stan, I know you can't see that far in the future probably, do you really think it would be likely that there would be major upgrades of both, in both areas, or would some decision be made one way or the other in 1988 or so?

STATEN: The national fusion plan as it was published a while back really showed a strong competition between tokamaks and mirror systems. People are relying less on that. It's not time to do a competition between the two yet because MFTF has not run, has not operated, and yet it is time to consider very seriously, and get started with the serious design of the next tokamak, and so if I sense what's going on correctly, we are tending to depend less on, let's wait until we can decide, but more on how can we take advantage of either one of them to do the sort of things that need to be done in the next few years. Again I think negative ions will be used where they have an advantage, where they are needed. I don't know whether that answers your question or not.

STEWART: It seems to me that you're talking about a fairly long term thing and you're also talking about the FY 85 budget and it seems that the negative ion program maybe is in a fairly precarious position with respect to next year's budget and budgets coming up. There are certainly forces on the overall fusion budget. I am talking about this country now, to hold them at least constant. I don't know what your plan is for '85, maybe even decrease it and at the same time there's this thing called TFCX that is being taken quite seriously by the fusion program, and that may present two problems. First of all, the direct competition for funds. I think the level is at something like $7 million this year, but projected to go up to something like $20 million next year, and if we're in a flat or decreasing budget, the money has to come from some place. Second of all, on our R&D priorities, as you said, negative ions are for TFCX a backup for ICRF heating. Cooper points out that negative ions can't be a backup if the development program isn't funded because if it's not funded, when the time comes it won't be there, so it won't be a backup. So in principle TFCX supports negative ions but if you look at it on a priority basis, certainly coming before that is going to be lower hybrid or ICRH even within the heating, and there are a lot of other R&D items that have to come along, the walls and the limiters, and diverters and some superconducting magnets and tritium systems and remote handling. So in fact, might there not be a problem in the near term with funding of negative ions?

STATEN: Well, there is a problem if the budget remains level. Again, I'm not party to all of the high level discussions that go on but there was a lot of dissatisfaction with the program plan for fusion and a lot of argument about this and a lot of argument about that. A lot of argument about whether it's engineering orientation or physics orientation, or a shoot-out between tokamaks or mirrors. Finally, this last summer and fall, during the dialogue in Congress, it finally came down to the dissatisfaction with simply one and only one thing, the budget level. It is money. It is not how you phrase something in a plan or anything else, it is simply the budget level. There are pressures to want it to go up and I think there is a feeling that as we perceive the TFCX, we are going to need some sort of a budget increase. Not as large as the fusion engineering act of a few years ago talked about and not in '85, but still a budget increase. Let me tell you what strategy I've been using in house to defend negative ions. It's sort of as follows: that we assume that RF is the heating method of choice for TFCX and that in the meantime TFCX won't be built for a few years. Let's give negative ions a real solid try. Let's see if we can't pull off this proof of principle sort of thing so that it can be a credible backup. But then if you want to be a devil's advocate, you assume that in three or four years maybe RF has found some fundamental problem, or people would want to exercise the neutral beam backup more seriously than they might otherwise do just by wanting the concept. They know they could use it if they had to, and further let's suppose we've had negative ions problems. One advantage in maintaining a reasonable negative ion program is we can hopefully maintain a reasonable cadre of people that were expert in the positive ion work and so if you assume the pessimistic RF and pessimistic negative ions, you can still use that to defend the negative ion program because you have to keep those people around so that they then can go back and pick up where we stopped on positive ion work. People are willing to accept that as an argument. So I would project negative ion budgets, maybe not so much in dollars, but the level of effort, the number of people working in the negative ion development to not change much over the next year or two. I would expect to see the dollar resources to go up as we commit to (negative ions), if you will take the NBETF out of the positive ion column in the budget and put it in the negative ion column. That brings the people from that facility, the operators, the technicians, the engineers, and all those associated with it and puts them under the negative ion budget so in that sense I would expect to see the budget go up, some in '85 and rather significantly in '86. But that doesn't mean that you would see twice the effort in a base program. That's my own guess, reading what the budgets are and what I think I can defend, and trying to read the politics of the office, read the tea leaves, or whatever. I could be wrong.

COOPER: Claude, could you give us a European viewpoint on negative ions?

JACQUOT: To commence, one is political, the other is technical. The first one is the position. It is my feeling that the position about negative ions in Europe is very bad because we have no mirror device, which means that it is not necessary for us to search for ions for a mirror device. Second, if I understand correctly, negative ions will not be needed for the next generation of tokamaks. The second difficulty for the moment is that there are many laboratories in Europe which work on negative ions at a very low level. One is in Amsterdam, one in Culham, one in Karlsruhe, one in Grenoble, and at the CNRS, Ecole Polytechnique. It seems that for European scientists working on negative ions it is important to collaborate together on a big program, for example, on a source of negative ions. The second point, very important to me, is to try to collaborate with the U.S. people on negative ions because it is such a lot of money to build a test stand of 500 kV, 10 A. It is my feeling that it is very difficult to develop a test stand like that in Europe without a device which uses negative ions. And now I have a technical comment on the source. Because my English is poor, I have a viewgraph and I would like to present in this way my impression of this conference. The different negative ion sources are the volume source, the converter source at LBL, SITEX, the hollow cathode discharge source, and the charge exchange source. Perhaps we have to add some numbers here because I do not have all the numbers in the talk but people working on SITEX and hollow cathodes can give these numbers. In the LBL source, which is of the converter type, a current I_H+ equal to 10 A falls on the converter and we extract 1 A from the converter with an efficiency H^-/H^+ of 10%; the arc power is 18 kW at 120 V, 150 A, and there is some power for the filaments. The pressure is 10^{-3} torr and the efficiency of this source is 18 kW/A. For the volume source, if I make a rough calculation on the illumination of a surface with an area of 20 cm by 20 cm and the transparency of 50%, I can put there about 1000 holes of 3 mm diameter. That is going to correspond to 2 or 3 A of negative ions with an arc power of 56 kW, 350 A, and 160 V, plus the filaments. The pressure is of the order of 4.5×10^{-3} torr, but the efficiency, if we compare the source with the converter and the volume source, is practically the same, of the order of 18 kW/A. The emittance is good. Perhaps you can put the same number for the SITEX and HCD but now let me speak about the charge exchange source. We produce on the monogrids 20 A of H^+. The transparency for low voltage is 0.2, in the best case perhaps 0.3. The conversion efficiency for cesium is 0.3. That means that with 20 A of H^+, we get 1 A of H^- and that the ratio of H^-/H^+ is 6%, but we have 8 kW of RF in the cavity to produce 20 A of H^+. The total efficiency of a klystron at 8 GHz is approximately 50%. That means that you have 16 kW in the tube but the main difference with the other approach where there are 18 kW in the source itself, is that although we have 16 kW in the tube, we have just 8 kW in the source with no filaments in this case. The pressure is 10^{-3} torr. If I take into account the real electric power for a charge exchange source, I get 16 kW/A. That is, in my opinion, my feeling now that all the sources operate with practically the same efficiency. Again, the volume source has the same efficiency as the converter type sources and as the charge exchange source, but probably the best quality of the volume source is a good emittance. I have no results. That's just my feeling. Let's ask the people from Oak Ridge or from BNL.

COOPER: I'd like to make one comment, Claude. I think the best possible thing you could do for cooperation with the U.S. would be to convince NET to use negative ion based neutral beams. I'd like to invite questions and encourage questions. I can't believe you people are enjoying this so much that you're holding your breath and won't interrupt. Why don't we throw the session open to general discussion?

VERBEEK: Has anybody any idea about a surface which could replace this cesium surface? It was mentioned so often during these days but nobody had any idea.

COOPER: Ken, you might comment on that. I think K. Leung has tried low work function materials which did not perform as well as the cesiated molybdenum although the work functions were similar. There is more involved than just the work functions.

EHLERS: That's true, Bill. We tried a large variety of things like lanthanum hexaboride, both hot and cold. The types of cathodes that you buy, that are used in tubes for example, are aluminum and strontium. None of these have performed up to the values that you get when you run molybdenum and cesium. I don't know that this is really due to the fact that the work function is less or whether we've done something wrong. The proof of the pudding always comes down to the eating and when we eat we find we get very little production from these things. I agree with you. Your comment is a very good one, and we certainly have tried to follow up along that very line because the gains would be great if you could mill a converter out of some magic material, stick it in there, and then do away with all the cesium and still get the current. But we haven't found it yet.

PRELEC: About three years ago we tried lanthanum hexaboride which was doped with cesium, not just covered, but doped with cesium. It was unfortunately left standing for maybe a year in a cabinet and then Ady Hershcovitch put it on the hollow cathode discharge and eventually he did get some H^- production although it was not as good as with cesium on molybdenum. We don't know why. Was it because the surface was contaminated or just in principle it doesn't work? We have also in mind a whole list of exotic materials, but except for the list we didn't do anything.

HERSHCOVITCH: I'd like to follow up on Krsto's comment. W. Kunnmann from our chemistry department, who prepared that sample, told me that the doping was not uniform due to the process used. The next step is to prepare samples that would be used on the small hollow cathode discharge system where the porous molybdenum was tested. Kunnmann prepared all kinds of samples, one was by using a different method of doping and now he could insure the uniformity. The first one was tried with cesium, then lanthanum hexaboride doped with sodium which did not perform well and then we were supposed to test the lanthanum hexaboride doped with cesium again. Unfortunately he doped the wrong side, the surface facing the other side, and we never tried to test it again. We had a nice controlled experiment which had the capability also to measure the work function with a laser.

FINK: I would like to point out that at long last the negative ion program has an enemy. I think it has not been proper before for the negative ion program to actually go on the offensive and say what's wrong with RF but our RF people have been taking that attitude for several years and if we look at our assets and think of the competitive situation relative to development of reliable RF, we're not bad off at all. The idea that you are going to efficiently transport high power RF energy from down in the basement up to the fifth floor is nonsense. They are going to have their power supplies in the same vicinity as we have. The idea of developing a high power, high vacuum tube with a high reliability right off the top of your head, I tell you you won't do it, I spent 10 years in the business. The most marvelous thing about the present sources is the tendency toward ruggedness. The amazing thing about Ehlers' source is that the apertures are heavy, they're solid chunks of metal. There is a filament in it, I grant you that, but as far as it is concerned there are ideas of beams supplementing it but this is fundamentally a heavy device, and fundamentally something that would last. If you take, as I say, my own experience of having seen a 75¢ radio tube developed at $100,000 in which we spent racks and racks of life-testing in order to find out what was wrong with our designs, we haven't done a damn thing yet as far as it is concerned when you imagine what it will take from an industrial sense to take the most advanced source we have, the positive ion source, and really turn it into a work horse. You haven't even begun. As far as I am concerned, we know the RF people haven't begun either and that they are envisioning developments that are 50 years old and say, that you just turn on this power supply and it works, that obviously is not the case.

COOPER: I'd like to comment on that with a question to Stan Staten. Stan, you commissioned an RF development plan similar to the negative ion development plan we've done. I agree with Joel Fink that RF people freely admit that they have lots of problems because I read their plan. There was a table in it that indicated problem areas and in a large number of them, there were some marked that meant serious problems and others marked that meant they couldn't even think of a solution. My question is why don't people believe that?

STATEN: People see RF working on tokamaks right now. They see 3 MW being used on PLT. They're going to try to push that to 6 MW next year. Can they? I don't know. The technology program developed the new feedthrough for them which had been their limiting component thus far. PLT may be an exception. For those of you not familiar, it is heated with ion cylotron resonant heating. The impurities experienced with RF heating have not been bad. They have been maybe a little worse, like they had been a little worse than if you have used pure co-injected neutral beams, they have not been as bad as if you used pure counter-injected neutral beams. To the first order, they're in the same ballpark as beams if you have a mix of co- and counter beams. But when I was at Princeton a month or so ago for a review of PLT, the day before there was a presentation by Adaum of TFR and I don't know if his experience is more typical of others but they have lots of impurities. Does that mean that PLT is charmed and when they try to double the power they're going to see lots of impurities? I really don't know. Are they going to be able to run those antennas at a megawatt apiece? Right up on the surface of that plasma, 2 or 3 cm from the edge of the plasma? I don't know. That's why I said in my talk that I don't know how to tell you how good or bad ICRH is going to be. We can do studies and we can say what we need to work on and what's good and what's bad, but the only answer that I know of is the one that when you go out and try to do it you see what happens and how tough the problems are and which ones you can solve and which ones you can't. All I can say is watch it for a while. Let's see what happens.

EHLERS: Joel Fink's comment is sort of the type of thinking that was going on in my head when I made the comment some time ago this afternoon that we would never had learned how to fly if the Wright Brothers had been forced to design a 747 just to prove the principle of flight. Sometimes we get involved in that type of logic, in other words, we're neutron hardening because the RF people say they won't have a neutron problem. So we take that as a personal affront and we say, gee, we've got to neutron harden. Well, sure we do, but we don't have to do it in the first generation device. We've got to be aware of the problem. We've got to know it's there and we've got to have ways to solve it but I don't think we should go out and spend the type of money it takes to solve it before we've ever proven that we can make negative ions work. For example, we built and went to great expense, and I don't think it was wasted effort, to put alumina insulators in our APIS accelerator. That cost us a lot of money and true we learned a lot. Nevertheless we spent a lot of money where we could have used epoxy which we know works perfectly well and do our neutron hardening after we've solved the basic problem, namely will it fly. I must confess that two years ago the last thing I wanted to hear about was neutralizers because they're far down the line. I know it is a problem, we have got to solve it but I am not ready to contemplate it yet. This year I'm ready to think about neutralizers. I think we haven't moved that far forward.

PRELEC: I would like to comment on the table C. Jacquot showed a few minutes ago. We get 10 A of positive ion current on the converter with about 5 to 6 kW in the hollow cathode discharge, compared to 18 kW in the other sources.

BACAL: I also have some comments on the figures there. I mean if I take 40 mA/cm^2 and 400 cm^2, I obtain 16 A and not as low as 3 A.

JACQUOT: I have assumed that holes have a diameter of 3 mm or an area of 6 mm^2, and that the transparency is 50%. At Los Alamos, they have obtained 2.5 mA of H$^-$ current per hole, which corresponds to a current density of 40 mA/cm^2. In a total area of 20 cm x 20 cm, about 1000 holes could be drilled, so that the total current is 2 to 3 A.

BACAL: I think we can do something better than 2 or 3 A from that area. Next I would like to comment on the figure of 4.5 millitorr. I discussed with R. York why they worked at 4.5 millitorr and he told me that at 5×10^{-4}, with 10 times less gas, you could obtain a good yield of negative ions. So I think we have only started studying those designs of the source and we shouldn't be blocked by a figure which is actually preliminary and the question I would like to ask is what will we do? What would we do to really design a good volume source in the U.S. or together?

COOPER: I think you're in a good position to answer that, Marthe.

BACAL: Well, let's talk about that. Thank you, Stan, for your pleasant comments. Thank you to all of those who came and worked with us and made many sacrifices, such as coming to Paris! Will you keep coming, or will it stop? I think we should continue and keep in touch as often as possible and do things together. I'd like to ask you people who have some money to distribute to help us continue in doing these things.

DAGENHART: I just want to fill in one or two of those numbers on your table, not all of them. We can do that later. It seemed that a couple of the things we've sort of come up on may be a little different and we're currently working on this. First of all, about the power efficiency of the arc for producing negative ions. When we started off several years ago, it was something like 180 kW/A. Now we're down to the number of like 5 kW/A and I can certainly see how that can be cut in half. We're not utilizing all of our converter so that seems to be a factor; I don't know how much the other ones can be changed. Certainly there was a progression there just by working on the system a little bit. We started out with burning things up because of the electron problem and managed to get from 300% of electrons down to 5% or thereabouts, with deuterium, similar to the Berkeley experience. I think our negative ion conversion efficiencies on the converter are not different. Our best efficiency we measured was about 14% and the pressure was a little bit higher, about 4 millitorr. But the thing that really seems to be different is the emittance. It looks like we have got a substantially lower emittance number. We have discussed that with some of the Berkeley people here and we are going to have to resolve this. While there seem to be four or five substantially different things, like differences in the discharge, there is a substantial difference in 0.7 eV transverse temperature vs. 7 or 5 eV or whatever the number was there. I don't know. That didn't fill in all of your table. We've gotten up to three-quarters of an ampere.

PRELEC: I think that at this panel meeting our objective is not so much to compare one source with the other but to summarize the overall progress in the negative ion source developement. We can say that in spite of very restrictive budgets, the progress achieved in the last three years in the design of negative hydrogen ion sources has been spectacular. From small pulsed devices we have designed three versions of possible candidates for a steady state device and we can challenge any positive ion source to show similar characteristics, let's say, steady state operation over hours or days. This is one comment. The second comment concerns the statement we heard earlier this afternoon about the pendulum for negative ions swinging back and forth. It is a fact that the period of oscillation is short if you have to handle say only milliamperes of current or watts of power. But once you begin to talk about megawatts and steady state operation, then you should realize that the period of oscillation becomes very long and we have to keep that in mind.

JACQUOT: I would just like to say that my feeling was not to prove that one source is better than the other but to prove that the volume source gives the same result as the source with cesium. I don't believe in 16 A of negative ions because if you have 16 A, with positive ions we have 160 A extracted, and if we have 160 A extracted of positive ions with a surface of 20 cm x 20 cm, it is clear that it is the best positive ion source in the world. But Marthe, I will say again that this source, the volume source, is as good as the source with cesium. That is an important point. However, I don't know if it is so good that we can stop immediately all the activities on all the other sources with cesium. I don't believe that. I think that this source is a proof that we can get a current with the same power in the arc. But my feeling is, from the paper by York, Stevens, Leung, and Ehlers where they say that they have to work at 4.5×10^{-3} torr to get the 38 mA, that means that we have to increase the gas efficiency of this source.

BACAL: It is not mentioned in the paper, but I have asked him why did he work at 4.5 millitorr and he told me, it was because he worked in a pulsed mode and it was a method of starting the discharge. Maybe we can't extrapolate that at all to a steady state source, but we should understand how this source operates at lower pressures in spite of his difficulties to start it. I can tell you, that it is not very difficult to operate it at lower pressures steady state.

COOPER: The nice thing about a steady state source is that you can do anything you want at the beginning, to start it and then you can change things.

EHLERS: I'd like to ask Stan Staten a question. Stan, where do you think the total fusion program sits in the American government right now? Are we losing favor, are we holding our own, or moving forward? I can't believe the latter.

STATEN: I don't know if I can judge that. The Secretary of Energy in a recent major speech of his, I think to the ANS meeting in San Francisco, made the comment that fusion cannot expect to keep as large a budget as it has without making major progress and the U.S. fusion budget is just under half a billion dollars a year. But also when we complain that we don't get the budget increases we had wanted, that the Magnetic Fusion Engineering Act said we should have, we have been told that we have been very fortunate not to be cut because if you look at other major energy programs in the Unites States, they have gone from several hundred million dollars a year to maybe a few tens of millions of dollars a year in the last few years. A number of programs have. If you look at the Clinch River breeder reactor, it was simply stopped within the last few weeks. So, what's going to happen to us in the future? I can't tell you. If we continue to enjoy a fairly comfortable energy climate, if we are able to get oil, if we are not scared by that, and to me that is a rather frightening situation in that area of the world, but if somehow that doesn't become any more precarious than it is now, then I would expect that all energy programs are going to have a rough time arguing for increased budgets. But, on the other hand, somebody might discover something that is much more serious, like suggested in recent articles on CO_2 problems in the atmosphere, or maybe the Middle East can be a problem in oil again, and you remember it was back in 1973, and about that time people were willing to spend money on any energy scheme that looked like it might solve the problem. I just don't know how to answer your questions. It has depended on just so many things. I would expect over the next few years to see a reasonably stable program and I think that we have a reasonably stable negative ion program over the next few years. But I am afraid I cannot predict much beyond the next few; as an optimist I would like to see it go up and I'd like to see the next big device made and I'd even like to see it heated by negative ions.

DAGENHART: The point I was trying to make is that with all these problems we have, quite likely the source that we finally wind up with may not look anything like the sources that we're looking at right now. By working on the problems many of them can be gotten rid of. I certainly would like to see a volume source instead of one with cesium and as you work on these problems you can get rid of a lot of them as you understand the physics of them. That's the point I was trying to make.

COOPER: I think that's a very good attitude. May I suggest that we continue this at dinner or later over a beer. I would just like to make a final comment: it is clear that we should think positively about negative ions. Stan Staten is absolutely correct that good ideas and good results will do more good than anything else. I would like to thank all of you for coming to this panel, and especially to thank the four panel members. As my last official act for the day, I would also like to thank Krsto and his staff for putting on such a fine conference.

POLARIZED H⁻ IONS AND SOURCES

PRINCIPLES OF PRODUCTION OF H⁻ POLARIZED IONS

W. Haeberli
University of Wisconsin-Madison, Wisconsin 53706

ABSTRACT

The basic principles which underly the design of sources for polarized H⁻ ions are reviewed, in order to provide background information for the research papers which follow.

1. INTRODUCTION

Over the last 20 years a number of methods have been developed to produce beams of polarized ions for injection into accelerators, in order to study the spin-dependence in nuclear reactions. Work on negative polarized ions was carried out primarily at nuclear physics laboratories which had available tandem accelerators. The best ion sources installed on such accelerators produce H⁻ and D⁻ intensities of a few microamperes with beam polarizations of 80 to 90 percent. Because negative ions are also preferred for acceleration in synchrotrons and some of the medium energy accelerators (meson factories at LAMPF and TRIUMF), there has over the last few years been increased interest in the development of sources producing higher beam intensities. Another possible application, still very speculative and requiring much higher beam intensities than presently available, is the loading of controlled thermonuclear reactors with polarized fuel. Depending on the reactions and polarization states involved, polarized fuel would enhance the rate of desired reactions, suppress the rate of undesired reactions, or reduce the neutron flux in a particular direction.

The present session consists of a number of talks which detail recent advances and new ideas in this field. It seemed reasonable to assume that most participants of this symposium are not specialists in the physics of spin-polarized beams, so that a brief introduction to the basic methods might be useful.

2. HYDROGEN ATOMS IN A MAGNETIC FIELD

The production of polarized beams in most cases involves at some state the interaction of hydrogen atoms with a magnetic field. Because the proton magnetic moment is 660 times smaller than the moment μ_e of the electron, one would expect that the energy W of the hydrogen atom in a magnetic field would be essentially $W = \pm \mu_e B$, depending on whether the electron spin projection is $m_j = \pm 1/2$. This is indeed the case for the states labeled 1 and 3 (fig. 1), for which the proton spin is "parallel" to the electron spin, i.e. the states of total angular momentum along B of $m_F = +1$ (state 1) and $m_F = -1$ (state 3). The proton spin projection m_I for these two states is $m_I = \pm 1/2$, and correspondingly the proton polarization $P = \pm 1$, independent of B (fig. 2). The wave functions of these

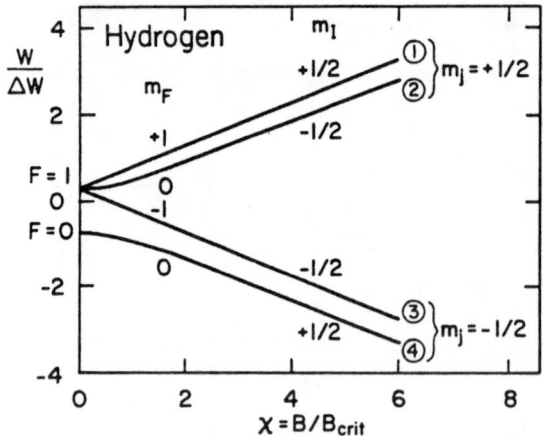

Fig. 1. Energy level diagram of the hydrogen atom in a magnetic field. The energy W is measured in terms of the zero-field splitting $\Delta W = h \times 1420.4$ MHz. The magnetic field is measured in terms of $B_c = 50.7$ mT = 507 Gauss. The figure is from ref. 1.

"pure" states can be represented symbolically as $|m_j, m_I>$:

$$\left. \begin{array}{l} \text{state 1} = |+\tfrac{1}{2}, +\tfrac{1}{2}> \\ \text{state 2} = |-\tfrac{1}{2}, -\tfrac{1}{2}> \end{array} \right\} \text{pure states} \quad (1)$$

For states 2 and 4, the proton spin is "opposite" to the electron spin, adding up to total angular projection $m_F = 0$. To understand the behavior of these two states it is useful to remember that the electron feels a field of a few hundred gauss from the proton magnetic moment. The crucial parameter is $\chi = B/B_c$, where B is the external field and $B_c = 50.7$ mT = 507 Gauss is the "critical field". For $B \ll B_c$, the electron is exposed only to the magnetic field of the proton. Electron and proton precess about one another keeping their spins opposite, so that the proton and electron polarization each oscillates with the Larmor frequency, the time average of the proton polarization being zero (see fig. 2 for $\chi \ll 1$). For the same reason, the effective magnetic moment $\mu_{eff} = -dW/dB$ of the atoms in states 2 and 4 approach zero for $\chi \ll 1$, as seen in fig. 1. These states are represented as a linear combination of wave functions with electron spin m_j opposite to the proton spin m_I, i.e. $|\tfrac{1}{2}, -\tfrac{1}{2}>$ and $|-\tfrac{1}{2}, +\tfrac{1}{2}>$:

$$\left. \begin{array}{l} \text{state 2} = \sin\theta \, |-\tfrac{1}{2}, +\tfrac{1}{2}> + \cos\theta \, |+\tfrac{1}{2}, -\tfrac{1}{2}> \\ \text{state 4} = \cos\theta \, |-\tfrac{1}{2}, +\tfrac{1}{2}> - \cos\theta \, |+\tfrac{1}{2}, -\tfrac{1}{2}> \end{array} \right\} \begin{array}{l} \text{mixed} \\ \text{states} \end{array} \quad (2)$$

where $\tan 2\theta = 1/\chi$. In zero external field, $\theta = 45°$, while for a "strong" external field ($\chi \gg 1$) $\theta \to 0$. Thus in a strong field, the polarization of states 2 and 4 approaches -1 and +1, respectively.

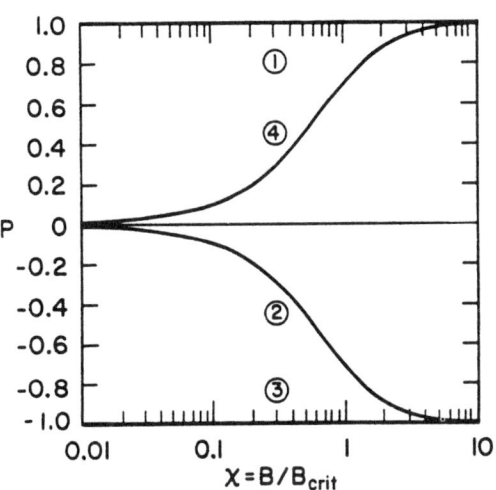

Fig. 2. Polarization of the proton in the hydrogen atom as a function $\chi = B/B_c$, where $B_c = 50.7$ mT = 507 Gauss. The numbering of states corresponds to fig. 1. The figure is from ref. 1.

One might expect that in a strong external field, the state of highest energy is the state for which all magnetic moments are lined up opposite to the external field, i.e. the state $|+ 1/2, - 1/2 >$. Fig. 1 shows that this is <u>not</u> the case. The <u>reason</u> is that for the external fields we are considering here the proton still primarily feels the very strong field caused by the electron magnetic moment (17.4 T) and not the external field. This very strong internal field, of course, also is the reason for the relatively large energy change (fig. 1) associated with flipping the proton spin, in spite of the smallness of the proton magnetic moment.

In most cases, the production of polarized beams starts out by separating atoms according to their electron spin, since one wants to take advantage of the large magnetic moment of the electron. One may prepare, for instance, atoms with $m_j = + 1/2$ (states 1 and 2) by the methods described in the following section. The achievement of large nuclear polarization requires separate measures which will be discussed in sect. 4.

3. PRODUCTION OF POLARIZED ATOMS

<u>Magnetic Separation</u>: If hydrogen atoms are exposed to a magnetic field gradient, they will want to minimize their potential energy W. This means (see fig. 1) that $m_j = + 1/2$ atoms are driven out of the B-field, $m_j = 1/2$ atoms are pulled into the B-field. If $\chi \gg 1$, the energy difference is $\Delta W = \pm \mu_e B$. The relevant dimensionless quantity is this energy difference compared to the thermal energy of the atom

$$\frac{\mu_e B}{kT} = 0.67 \frac{B \text{ (Tesla)}}{T \text{ (K)}} \qquad (3)$$

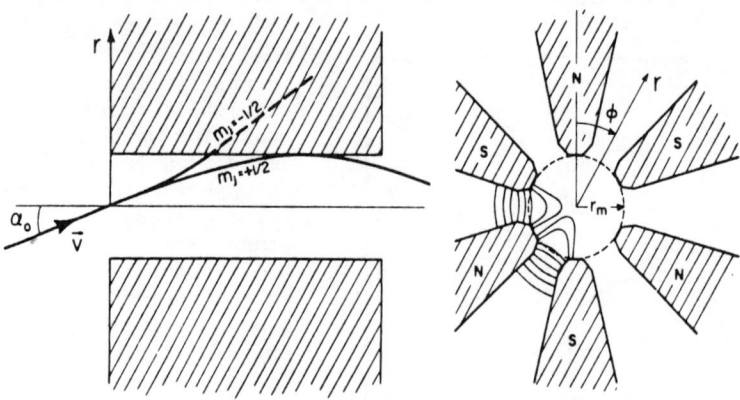

Fig. 3. Schematic representation of spin selection in a six-pole magnet. Atoms with electron spin $m_j = +1/2$ experience a force toward the axis and pass through the magnet as long as their radial velocity at entry is not too large. Atoms with $m_j = -1/2$ are removed from the beam (dashed line).

Most operating ion sources use the <u>atomic beam</u> method for spin separation. Hydrogen gas is dissociated in an RF discharge and the beam of atoms (temperature T) emerging from a nozzle (e.g. 2 mm diameter) is directed along the axis of a four- or six-pole magnet (fig. 3). Since $|B|$ increases monotonically from zero on the axis to B_m at the pole tip radius r_m, atoms with $m_j = +1/2$ are driven toward the axis, $m_j = -1/2$ atoms are pushed outward and get lost. The acceptance angle $\Delta\Omega$ is given by the condition that the kinetic energy associated with the radial component of velocity of the atom entering the magnet be $< \mu_e B_m$. Consequently, the pole tip field B_m determines the solid angle of acceptance $\Delta\Omega$:

$$\Delta\Omega \simeq \mu_e B_m / kT, \qquad (4)$$

where on the right-hand side of eq. (4) a multiplicative factor (2.09 for a six-pole magnet) is omitted (see ref. 1). Note that even for a cooled atomic beam (e.g., T = 70 K) $\mu_e B/kT$ is small (e.g. 10^{-2} for B_m = 1T). Lowering the atomic beam temperature not only increases the acceptance angle of the magnet but also improves the ionization efficiency because the lower velocity increases the dwell time in the ionizer. On the other hand, with increasing acceptance angle the divergence of the beam after leaving the magnet is also increased, unless measures are taken to reduce either the velocity spread of the beam or the chromatic aberrations of the magnet system. Some of these problems are addressed in the paper by Grüebler below.

At low temperatures (T < 1K), magnetic separation can be accomplished in principle by an arrangement such as proposed by Kleppner and Greytak[2] (fig. 4), where cold H atoms are fed e.g. into a 10 T solenoid. Since in this case $\mu_e B/kT \gg 1$, the $m_j = +1/2$ atoms will be rapidly pushed out of the strong field region, and $m_j = -1/2$ atoms will be confined to the inside of the solenoid. While high densities of polarized H atoms have been obtained in closed cells at low temperatures, the realization of storage cells accessible from the outside will require the solution of certain problems connected with heat load and maintenance of coatings (superfluid He) to prevent recombination. Several methods to circumvent these problems have been discussed by Kleppner (ref. 2 and paper in this session).

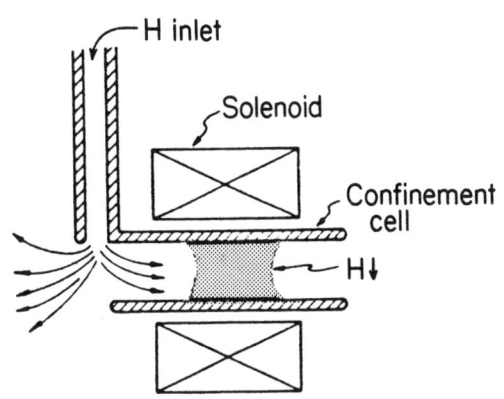

Fig. 4. Low temperature cell for confinement of polarized hydrogen atoms in a solenoid.

Pickup of Polarized Electrons: In 1957, Zavoiskii[3] proposed that fast (keV) polarized hydrogen atoms can be produced by passing a beam of protons through a magnetized ferromagnetic foil where they pick up polarized electrons (fig. 5). The principal advantage of a fast beam is that it can be ionized with high efficiency by a second charge exchange collision. To circumvent the serious technical problems (heating, radiation damage, scattering) associated with the use of a foil as the donor of polarized electrons, I subsequently proposed to use as donor a polarized paramagnetic gas (optically pumped

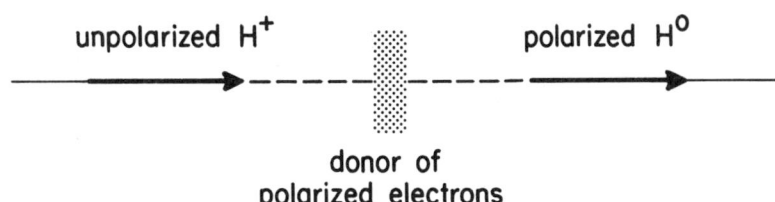

Fig. 5. Schematic diagram showing production of a beam of fast polarized H atoms by pickup of polarized electrons. The energy of the proton beam can be assumed to be of the order keV.

alkali vapor or polarized H atoms in a storage cell[1,4]). The method was first seriously analyzed by Anderson[5], who considered as donor of polarized electrons optically-pumped Na vapor. The current status of sources based on this method will be discussed by Anderson later today.

Lamb-Shift Method: When fast (keV) protons pass through Cs vapor, neutralization leads to a large extent not to the ground state of the H atom, but to the n = 2 excited state. Those atoms which are in the 2S metastable excited state can readily be polarized by destroying the 2S atoms which have $m_j = -1/2$, but not those which have $m_j = +1/2$. This is accomplished simply by passing the 2S atoms through a suitable magnetic field region (57.5 mT) where the $m_j = -1/2$ component of the 2S state is degenerate with the $m_j = +1/2$ component of the nearby 2P-state, so that decay to the ground state ("quenching") results from mixing of the 2S-state with the 2P state. The Lamb-Shift method will not be discussed in this session, even though it is widely used, because it is unlikely that it will fill the need for future high intensity sources.

4. NUCLEAR POLARIZATION

Thermal Atomic Beam: Hydrogen atoms separated according to electron spin m_j contain equal populations of atoms in a pure state (state 1 or 3) and in a mixed state (state 2 or 4). Figure 2 shows that in a strong field ($\chi \gg 1$) the net nuclear polarization is zero, since the nuclear polarization of the mixed state is opposite to that of the pure state. However, in a weak field ($\chi \ll 1$), the nuclear polarization of the mixed state is zero, so that the net nuclear polarization of the ensemble is 1/2. In making the transition from a strong-field region (e.g. the six-pole magnet in fig. 3), to a weak-field region (e.g. the region beyond the exit of the six-pole), the changes in field direction need to be slow compared to the Larmor precession angular velocity of the atom in the field (adiabatic transition from strong field to weak field). For a beam of thermal velocities even the rapid field variations of the exit of a six-pole magnet are plenty slow to meet the criterion. This method was used in the very first experiment[6] to produce polarized negative ions by Grüebler, Schwandt and myself some 20 years ago, but this method is no longer used. Instead, one now invariably equips atomic-beam devices with RF transition units, since they offer the advantage of large and reversible beam polarization. The effect of the so-called adiabatic 2 → 4 transition is to interchange the populations of the (occupied) state 2 and the (unoccupied) state 4. After passing through the RF transition the atomic beam now contains states 1 and 4, giving proton polarization P = +1 if taken adiabatically to a strong field. Similarly, a 1 → 3 transition will give P = -1.

Similar transition have been devised for deuterons to obtain, at will, vector or tensor polarization of ±1. These transitions are simple and highly effective (> 98% transition probability) for thermal beams, but not applicable to fast (keV) beams, because the fast atoms do not spend enough time in a cavity of reasonable size.

Fast Beams: For a beam of fast (keV) neutral hydrogen atoms containing states 1 and 2, large nuclear polarization is rather easily obtained by the sudden field reversal method[7] (Sona transitions). Assume that a H^o beam polarized in electron spin along z, is travelling in the z-direction and that at some point, z_o, the magnetic field B_z is reversed (fig. 6). The field configuration shown may be realized, e.g. by two opposing solenoids coaxial with z. Figure 6 shows that after the sudden reversal of B_z, the nuclear polarization in a strong magnetic field is P = +1. Note that on the right in fig. 6, the projections m_I are labeled with respect to the +z direction, no matter whether B_z is positive or negative. Elementary magnetostatics (div B = 0) tells us that there must also be present in the vicinity of z_o a radical field component B_r which increases with distance r from the z-axis: $B_r = (1/2)(\partial B_z/\partial z)_{z_o}$.

Thus the further from the axis the atom travels, the less sudden is the field reversal. The criterion again is the angular velocity of the change in field direction which the atom sees relative to the Larmor precession angular velocity of the atom in the field. Sudden field reversal is employed for sources which produce a fast H^o beam by pickup of polarized electrons. The method is simple and effective, but for deuterons does not permit separate choice of vector and tensor polarization, except for Lamb-Shift sources where one can employ repeated quenching. For Lamb-Shift sources, methods (spin-filters) have also been developed to select a single hyperfine state.

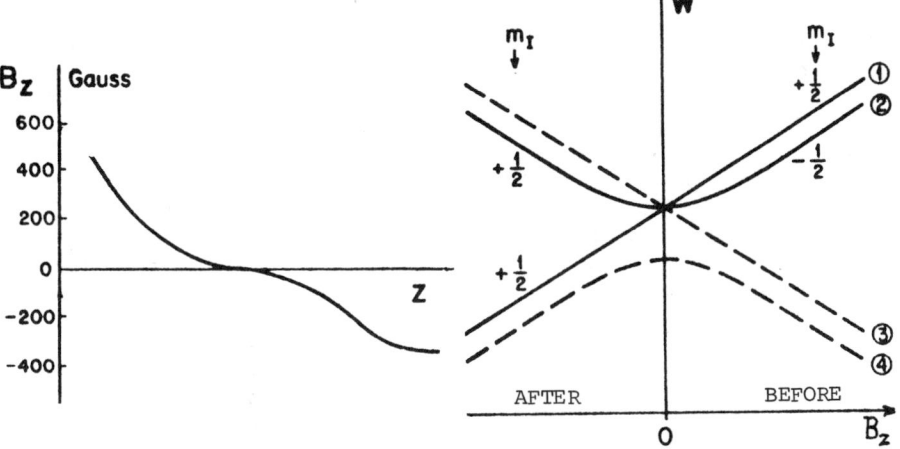

Fig. 6. Schematic representation of the sudden field reversal method (Sona transitions) to produce large nuclear polarization in hydrogen atoms initially polarized in electron spin only (states 1 and 2). On the left, the magnetic field configuration, on the right the occupied states before and after the sudden field reversal (solid lines). The dashed states are not occupied. The label m_I refers to the proton spin projection with respect to +z.

Confinement Cell: For atoms confined in a solenoid (fig. 4), net nuclear polarization is obtained if one of the two hyperfine states present is removed from the cell. Kleppner and Greytak[2] have discussed two mechanisms. At high densities, it is found that the mixed state is removed spontaneously through recombination. Recombination requires that two hydrogen atoms have opposite electron spin, since H_2 is a singlet state. Thus atoms in the pure state cannot recombine with one another, while atoms in the mixed state have a small admixture of a state with opposite electron spin even in a strong magnetic field, so that they will eventually recombine. The resulting H_2 will condense on the walls of the cell, since H_2 has a negligible vapor pressure at the temperature considered here (0.3 K). The second possibility, which does not require high density, is to induce transitions between hyperfine states by electron spin resonance. If the electron spin of the mixed state is reversed, the mixed state atoms will be expelled from the strong field region, leaving behind atoms completely polarized in proton spin.

5. IONIZATION

Double Charge Exchange: Polarized thermal atomic beams are usually ionized by electron bombardment to produce polarized protons. Negative ions are produced by double charge exchange of the protons, e.g. in an alkali vapor at a few keV energy (fig. 7). Ionization of the mixed states after RF transitions of course requires a sufficiently strong magnetic field in the electron bombardment ionizer to reach the desired proton polarization (fig. 2). Depolarization in electron impact ionization itself is negligible, because the transient magnetic fields at the nucleus are small and short in duration[1]. However, in charge exchange from H^+ to H^- (or vice versa), partial depolarization can occur if the atom remains in the intermediate H^o state for too long a time, since in H^o the proton is coupled to an unpolarized electron. Depolarization is prevented by applying a strong magnetic field ($\chi \gg 1$) in the charge exchange region. Once the second electron is picked up, the problem ceases because the electrons in H^- have no net angular momentum. For charge exchange in a foil, the transit time is so short compared to the Larmor time that no precautions are required. The paper by Grüebler contains more details and current performance figures for this ionization method.

Direct Charge Transfer (Colliding Beams): Another method for ionization of thermal polarized H^o to H^- is shown schematically in fig. 8. The polarized atoms are bombarded with a beam of unpolarized atoms or ions which have a favorable cross section for charge transfer. The reactions which I originally proposed[8] are:

$$\vec{H}^o \text{ (thermal)} + Cs^o \text{ (fast)} \rightarrow \vec{H}^- + Cs^+$$
$$\vec{H}^o \text{ (thermal)} + D^- \text{ (fast)} \rightarrow \vec{H}^- + D^o.$$
(5)

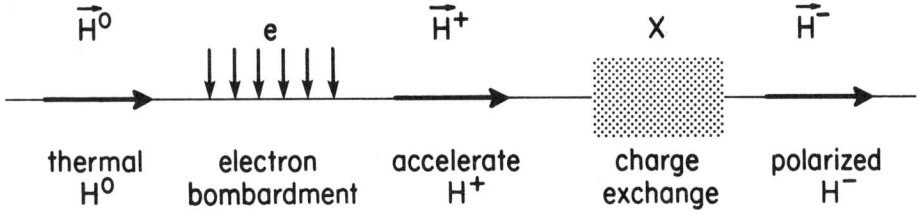

Fig. 7. Ionization of a beam of thermal H atoms by electron bombardment, followed by double charge exchange in a gas or vapor X to form negative polarized ions. Magnetic fields needed to prevent depolarization are not shown.

For the first reaction, the cross section reaches a maximum for ~ 60 keV Cs^0 energy, while for the second reaction a low D^- energy (< 1 keV) is favorable. Only the first reaction has been used so far, and as expected no depolarization has been observed in the charge exchange collision. Recent progress on this type of ion source will be presented later today by Sluyters et al., who will discuss the results obtained with their new source for the Brookhaven AGS. The second of the reactions above for sufficiently low D^- energies has a larger cross section than the first, but requires attention to space charge problems so that the newly formed slow H^- ions are not lost. Experiments to ionize H^0 with D^- bombardment are in preparation at Brookhaven.

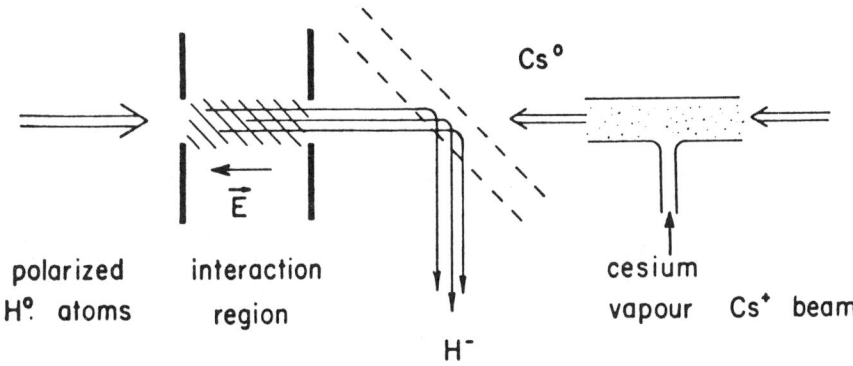

Fig. 8. Schematic diagram of arrangement for direct conversion of a polarized atomic beam to polarized negative ions (colliding beam method).

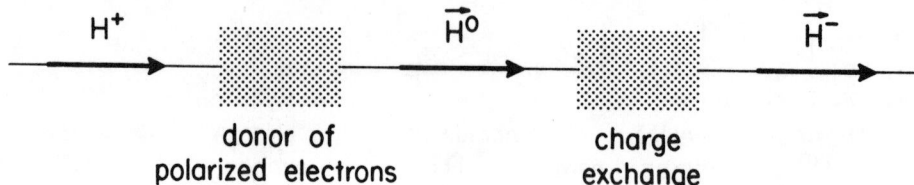

Fig. 9. Ionization of <u>fast</u> polarized hydrogen atoms by charge exchange to form polarized $\overline{H^-}$ ions. The fast polarized atoms are produced by pickup of polarized electrons. Magnetic fields needed to avoid depolarization and for sudden field reversal are not shown.

<u>Ionization of Fast Atoms</u>: As mentioned above, the advantage of producing <u>fast</u> polarized H^0 atoms is that they are easily ionized with relatively high efficiency by a charge exchange reaction (fig. 9). In practice, one deals with beam energies of ~ 5 keV and Na vapor as the charge exchange medium. Again a strong magnetic field is required over the charge exchange region. In fig. 9, the H^0 beam should be assumed to pass through a sudden field reversal region. The present status of this type of source is reviewed below by Anderson.

<u>Ionization of H(2S) Atoms</u>: For the operation of Lamb-Shift sources it is essential to ionize selectively the polarized H(2S) atoms and not the large background of atoms in the ground state. It was shown by Donnally and Sawyer[9] that the reaction

$$H^0(2S) + Ar \rightarrow H^- + Ar^+ \qquad (6)$$

is highly selective near 500 eV energy. The reason for the selectivity is well understood in terms of the binding energies of the electrons involved.

CONCLUSION

The brief review presented here only describes those principles which have gained importance in the construction of practical sources for polarized negative hydrogen ions. No actual design or performance figures are included here, because these are covered in the following progress reports on ion source development work. For a broader review of the subject, we refer to the proceedings of the Symposium on Polarization Phenomena in Nuclear Reactions[10-12], the proceedings of the Symposium on High Energy Physics with Polarized Beams and Polarized Targets[13-15], and the proceedings of two Ann Arbor workshops on the subject of polarized ion sources[16,17].

REFERENCES

1. W. Haeberli, Ann. Rev. Nucl. Sci. $\underline{17}$, 373 (1967).
2. D. Kleppner and T.J. Greytak, High Energy Spin Physics-1982 (AIP Conf. Proc. No. 95) p. 546.
3. E.K. Zavoiskii, Soviet Physics (JETP) $\underline{5}$, 603 (1957).
4. W. Haeberli, Second Int. Symp. on Polarization Phenomena of Nucleons (Karlsruhe 1965), Experimentia Supplement $\underline{12}$, 64 (1966).
5. L.W. Anderson, Nucl. Instr. Meth. $\underline{167}$, 363 (1979).
6. W. Grüebler, W. Haeberli and P. Schwandt, Phys. Rev. Lett. $\underline{12}$, 595 (1964).
7. P.G. Sona, Energ. Nucl. $\underline{14}$, 295 (1967).
8. W. Haeberli, Nucl. Instr. Meth. $\underline{62}$, 355 (1968).
9. B.L. Donnally and W. Sawyer, Phys. Rev. Lett. $\underline{15}$, 439 (1965).
10. Polarization Phenomena in Nuclear Reactions-1970, (H.H. Barschall and W. Haeberli eds.) University of Wisconsin Press (Madison, 1971).
11. Polarization Phenomena in Nuclear Reactions-1975, (W. Grüebler and V. König, eds.) Birkhäuser (Basel, 1976).
12. Polarization Phenomena in Nuclear Physics-1980 (G.G. Ohlsen, ed.) AIP Conf. Proc. No. 69 (New York, 1981).
13. High Energy Physics with Polarized Beams and Polarized Targets-1978 (G.H. Thomas, ed.), AIP Conf. Proc. No. 51 (New York, 1979).
14. High Energy Physics with Polarized Beams and Polarized Targets-1980 (C. Joseph and J. Soffer, eds.) Birkhäuser (Basel, 1981).
15. High Energy Spin Physics-1982 (G.M. Bunce, ed.) AIP Conf. Proc. No. 95 (New York, 1983).
16. Higher Energy Polarized Proton Beams (A.D. Krisch and A.J. Salthouse, eds.) AIP Conf. Proc. No. 42 (New York, 1978).
17. Polarized Proton Ion Sources (A.D. Krisch and A.T.M. Lin, eds.) AIP Conf. Proc. No. 80 (New York, 1982).

PRODUCTION OF POLARIZED H⁻ IONS
USING LASER OPTICAL PUMPING*

L. W. Anderson
Dept. of Physics, Univ. of Wisconsin, Madison, WI 53706

ABSTRACT

Laser optical pumping can be used to produce a polarized alkali vapor target in a high magnetic field. A H^+ beam incident on the polarized alkali target is partially neutralized by capturing a polarized electron. The electron spin polarization is transferred to the nuclear spin via a Sona diabadatic transition and the beam is then partially converted to negative ions in an unpolarized alkali target. Mori and coworkers have constructed a source of this type and have obtained a 35 AμH⁻ ion beam with a nuclear polarization of 0.60.

INTRODUCTION

In 1957 Zavoiskii[1] suggested that polarized H^+ or H^- ion beams can be produced as follows. A fast H^+ ion beam is incident on a magnetized ferromagnetic foil. After passing through the foil some of the H^+ ions are neutralized by the pickup of a polarized electron forming fast polarized H^0 atoms. Because of the hyperfine interaction the fast H^0 atoms will have a non-zero time average nuclear polarization. The fast H^0 atoms are then incident on a second foil where they are partially converted into either polarized H^+ or H^- ions. Haeberli[2] pointed out that some of the problems associated with the charge-transfer production of polarized ions would be solved if the ferromagnetic foil were replaced by an electron spin polarized gas or vapor target such as a stored polarized H^0 target or an optically pumped polarized alkali target. The use of an optically pumped polarized alkali target has been analyzed by Anderson.[3,4]

Mori et al.[5] have recently constructed and tested an optically pumped polarized H⁻ ion source. This ion source has produced an H⁻ ion current of 35 μA with a nuclear polarization of 0.6. The nuclear polarization was measured using the ^{6}Li(^{1}H, ^{4}He)^{3}He reaction. Because of the large H⁻ ion currents and high nuclear polarization obtained the optically pumped polarized H⁻ ion source appears very promising for future use with nuclear accelerators. Briefly the optically pumped ion source work as follows.

---

*This work supported by U.S. Dept. of Energy, Office of High Energy and Nuclear Physics, Division of Nuclear Physics, under Grant No. DE-AC02-81ER4001.

A beam of H$^+$ ions is extracted from an ECR ion source, which is in a large magnetic field. The H$^+$ ion beam is incident on a Na vapor target with an energy of 5 keV. This target, which is in the same large magnetic field as the ECR ion source and which is colinear with the ion beam axis, is electron spin polarized by optical pumping with a dye laser beam. In this target some of the H$^+$ ions are neutralized by the reaction H$^+$ + $\vec{Na}^0 \rightarrow \vec{H}^0$ + Na$^+$ where the arrow indicates that the polarized electron of the Na target is transferred to the H^0 atom that is produced in the reaction. The fast H^0 atom emerges from the first target and enters a second Na vapor target. The magnetic field at the second target is directed oppositely from the field in the first target. The fast electron spin polarized H^0 atoms pass diabatically through a region of near zero field between the two targets. This diabatic passage through zero field transfers the electron spin polarization to the nuclear spin.[6] Some of the fast nuclear spin polarized H^0 atoms are converted to polarized H$^-$ ions in the second target.

The reason an optically pumped H$^-$ ion source produces large polarized ion currents can be understood from the following crude estimate of the potential H$^-$ ion current.[3,4] A 1W dye laser operating at a wavelength of 589.6 nm produces 3 x 10^{18} photons/sec. In a high magnetic field (I decoupled from S in the 3s level) 1.5 photons are required on the average to polarize a Na atom. Thus 2 x 10^{18} Na atoms can be polarized per sec. At a target temperature of 600°K the average time between wall collisions is about 10^{-5} sec assuming the target is a long tube about 1 cm in diameter. Thus even if the Na atoms are completely depolarized at each wall collision a target with π = 10^{13} atoms/cm^2 is possible. The long target assures that little polarization is lost by effusion out the ends of the tube. The charge transfer cross section for the reaction H$^+$ + Na^0 $\rightarrow$ H^0 + Na$^+$ is 6 x 10^{-15} cm^2 at 5 keV[7-9] so that about 6% of the incident H$^+$ ion beam picks up a polarized electron. Since the equilibrium fraction of H$^-$ ions emerging from the second Na target is 7.3% at 5 keV[7-10] about 4 x 10^{-3} of the incident H$^+$ beam emerges from the second target as polarized H$^-$ ions. Thus a current of about 4 μA of polarized H$^-$ ions per mA of incident H$^+$ ions is expected. Both the technical problems involved in producing polarized H$^-$ ions by charge transfer in an optically pumped target and possible improvements in the optically pumped polarized H$^-$ ion source are discussed in this paper.

THE H$^-$ ION POLARIZATION

The reaction H$^+$ + Na^0 $\rightarrow$ H^0 + Na$^+$ is nearly resonant to the n = 2 level. Calculations indicate that more than 99% of the fast H^0 atoms are formed in the n = 2 level.[11] Both experiment and theory indicate that for 5 keV incident energy about 80% of the H^0 atoms are formed in the 2p level and about 20% are formed in the 2s level.[9,11] The electron capture leading to the formation of fast H^0

atoms in the n = 2 level has serious implications for the nuclear polarization of H^- ions produced by this type of ion source because in a low magnetic field the electron spin polarization of the H^0 atom is partially lost in the radiative decay that leads to the ground level. This loss of polarization would not be present if the H^0 atoms were formed in the ground level rather than in the n = 2 level. We call the electron spin polarization of the optically pumped Na vapor target P_e. If the H^0 atom is formed in the 2p level then the electron spin polarization, P, of the ground level H^0 atom following the 2p → 1s transition is $P = 0.41\ P_e$. Electron capture into the 2s level leads to a similar polarization loss since the 2s level decays primarily by electric field mixing with the 2p level. The loss of polarization can be avoided by the use of a magnetic field at the first target strong enough to decouple L and S in the n = 2 level of the H^0 atom. Fig. 1 shows as a function of the magnetic field, H, the electron spin

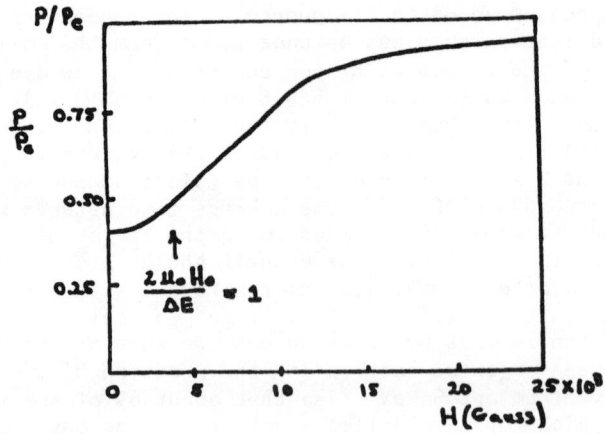

Figure 1. P/P_e vs H.

polarization of the ground level H^0 atom formed if an electron of polarization P_e is captured forming a H^0 atom in the 2p level followed by radiative decay to the ground level.[12,13,14] A field of 10 kG or more is needed to prevent the loss of a substantial fraction of the electron spin polarization in the radiative decay. As discussed later in this paper a field as large as 10 kG causes severe problems in the ion optics of the neutralization process.

In the calculation of P/P_e as a function of H (as shown in Fig.1) it was assumed that the collision producing the H^0 atom in the 2p level results in equal populations of the $m_L = \pm 1$ and 0 sublevels. Hinds et al.[12] have calculated P/P_e as a function of H including the possibility that the collisions producing the H^0 atom in the 2p level may produce a population of the $m_L = 0$ sublevel different than the populations of the $m_L = \pm 1$ sublevels. An interesting result of their calculation is that if $m_L = 0$ sublevel is populated differently than $m_L = \pm 1$ sublevels in the reaction $H^+ + Na^0 \to H^0 + Na^+$ then the final nuclear polarization of the H^- ion is not reversed by reversing the electron spin polarization of the Na target. Their paper also shows how to treat the depolarization that results when the H^0 atoms are formed in the 2s level which decays by mixing with the 2p level due to the motional electric field seen by the H^0 atom as it moves through the fringing magnetic field at the first Na target.

Mori et al.[5] have used a magnetic field of about 9×10^3 Gauss at the optically pumped target. One would expect from Fig. 1 that increasing the magnetic field to about 2×10^4 Gauss would increase the ion beam polarization by about 20%.

THE MAGNETIC FIELD PROBLEM

If a parallel beam of H^+ ions enters the Na target and neutralizes at the center then a particle in the neutral beam acquires a tangential velocity, v_t, given by

$$\frac{v_t}{v_z} = \frac{eHr}{2mv_z}$$

where r is the radius of the particle, H is the magnetic field, m is the mass of the proton and v_z is the incident velocity.[15] For $H = 10^4$ gauss and $r = 0.5$ cm we calculate $v_t/v_z = 0.25$ at 5 keV incident energy. Clearly $v_t/v_z = 0.25$ is unacceptable since most of the beam after neutralization will strike the Na target tube. Thus one must operate the ion source and Na target in a colinear geometry and in the same large magnetic field. An electron cyclotron resonance (ECR) ion source can be operated in a high magnetic field. Mori et al.[5] has used a 16.5 GHz ECR ion source operated colinear with and in the same magnetic field as a Na target to eliminate the blow up of the H^- ion beam produced.

THE OPTICAL PUMPING OF A Na TARGET

A Na vapor target can be polarized by the absorption of circularly polarized D_1 radiation (the $3^2S_{1/2} \to 3^2P_{1/2}$ transition). In high field the Na vapor can be polarized by the absorption of light by the $m_S = -1/2$ levels but not the $m_S = 1/2$ levels even if the light is not circularly polarized.

The hyperfine-Zeeman energy levels of Na are shown in Fig. 2. For ^{23}Na the nuclear spin is $I = 3/2$. In a large magnetic field the ground level of Na has 8 sublevels corresponding to the various values of m_S and m_I. The zero field hyperfine separation in ^{23}Na is $\Delta\nu_{HFS} = 1.77 \times 10^9$ Hz. The Doppler width for the $3^2S_{1/2} \to 3^2P_{1/2}$

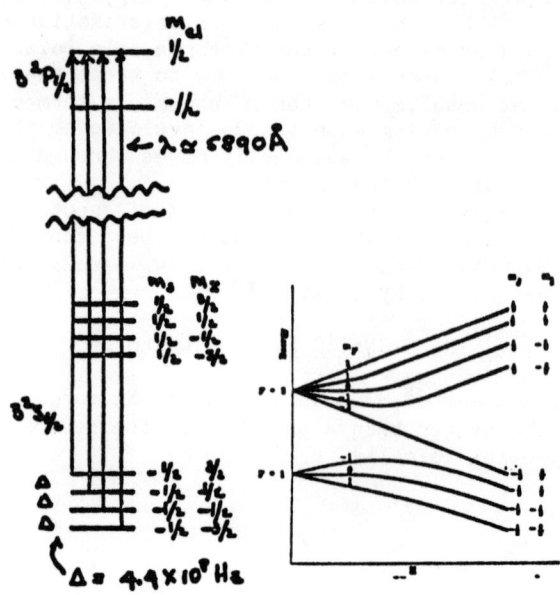

Figure 2. Relevant energy levels of Na.

absorption in Na is $\Delta\nu_D = 1.7 \times 10^9$ Hz at 600° K. Each of the 4 sublevels with $m_S = -1/2$, and $m_I = -3/2, -1/2, 1/2,$ or $3/2$ is separated from the next nearest sublevel by $\Delta\nu = \Delta\nu_{HFS}/4 = 443$ MHz. In order to optically pump Na in a high field it is necessary for the laser to interact with Na atoms in all 4 sublevels with $m_S = -1/2$, m_I and with a Doppler profile for each level. Thus the laser must interact with atoms having a distribution of absorption frequencies with a width of about
$\Delta\nu = (3/4)\Delta\nu_{HFS} + \Delta\nu_D \cong 3 \times 10^9$ Hz. There are several possible methods of obtaining frequency coverage so that all the atoms in a bandwidth of $\Delta\nu = 3 \times 10^9$ Hz are optically pumped.

THE USE OF HIGH INTENSITY SINGLE MODE LASERS

Mori et al.[5] have used two single frequency stabilized ring dye lasers to optically pump the Na target. A laser burns a hole in an inhomogeneous line. The width of the hole is
$\Delta\nu_{hole} = \Delta\nu_{homo} \sqrt{1 + I/I_S}$ where $\Delta\nu_{homo}$ is the width of the homogeneous packet, I is the laser intensity, and I_S is the saturation intensity. For Na $\Delta\nu_{homo} = \frac{1}{2\pi\tau} = 10^7$ Hz where τ is the radiative lifetime of the 3p level. For an optical pumping

situation Feld et al.[16] have shown that the saturation intensity is given by

$$I_s = \frac{\hbar\omega}{\sigma\tau} \frac{(1+\tau/T)}{(2+\Gamma T)}$$

where ω is the angular frequency of the transition, σ is the optical absorption cross section at line center for a homogeneous packet, T is the radiation interaction time, and Γ is the decay rate from the excited level into the non-absorbing level. For D_1 pumping of Na in a large magnetic field using circular polarized light $\Gamma = 1/(3\tau)$. For D_1 pumping using circularly polarized light $\sigma = \lambda^2/\pi$. We take T as the time for a Na atom to transverse a 1 cm diameter tube so that $T = 10^{-5}$s. For this case $I_s = 10^{-4}$ W/cm^2. If a 1W single mode laser with a beam diameter of 1 cm is used for the optical pumping then $\Delta\nu_{hole} = 1.1 \times 10^9$ Hz. By using 3-4 ring dye lasers operating at 1W power it is possible to optically pump with good frequency coverage a region with a bandwidth of $(3-4) \times 10^9$ Hz.

Imprisonment of resonance radiation can limit the target density. Radiation trapping becomes important for target densities high enough that the mean free path of a photon is comparable to the radius of the target. Imprisonment of the resonance radiation does not depolarize the Na target if H is large enough but does result in more photons per Na atom being required to polarize the target than are required for a less dense target where trapping is not important. The D_1 line absorption cross section at line center for a 600°K Na vapor is $\sigma \approx 10^{-12}$ cm^2. If the Na target has a radius r = 0.5 cm then we estimate that imprisonment of the resonance radiation becomes important for a Na density of $(2-4) \times 10^{12}$ atoms/cm^3. If the target is 20 cm long this corresponds to $\pi = (2-4) \times 10^{13}$ atoms/cm^2.

Cornelius et al.[17] have studied the use of a single frequency dye laser at $\lambda = 589.6$nm to optically pump a Na vapor target in a magnetic field of 3500G. They use the Faraday rotation of a second laser operating at $\lambda = 589.3$nm to measure both the polarization and the target thickness (in atoms/cm^2) of their Na target. Their results are encouraging. Using a single frequency laser with a beam diameter of 3mm and with an intensity less than 4W/cm^2 to pump a Na target they obtain a $P_e > 0.5$ for $\pi = 2 \times 10^{13}$ atoms/cm^2 and $P_e = 0.32$ for $\pi = 6 \times 10^{13}$ atoms/cm^2. They have made calculations of P_e as a function of the laser power and the target thickness for a magnetic field of 3500G and a laser beam diameter of 3mm. They also calculate that using 2 dye lasers each with a power of 2.5W and a beam diameter of 3mm that a target with $\pi = 5 \times 10^{13}$ atoms/cm^2 can be pumped so that $P_e = 0.92$. They point out that a ^{39}K target ($\pi = 5 \times 10^{13}$ atom/cm^2) optically pumped using a laser with $\lambda = 769.9$nm, with 1W power, and with 3mm beam diameter will achieve $P_e = 0.9$. This is possible because ^{39}K has a smaller hyperfine separation and a smaller Doppler width than ^{23}Na.

Mori et al.[5] have used two single mode ring dye lasers to optically pump their Na target. They find that they produce a larger polarization at low target thickness than at high target thickness. It is probable that if they use 3 or 4 lasers rather than 2 then their target polarization will increase thereby increasing the nuclear polarization of the negative ions.

The discussion of optical pumping has assumed that the spin depolarization time of the Na is equal to the time between successive wall collisions. If a wall coating can be found that prevents depolarization then high values of the target polarization and thickness should be possible using existing dye lasers. Such a coating should have a low vapor pressure and good structural strength at the operating temperature. We have measured the spin depolarization time for Na bouncing from stainless steel, or polyethelyne surfaces.[18] Our target has a 1.27 cm i.d. and a length of 15.24 cm with the Na entering the target at the center. We find that for the metal surface the spin depolarization time is $T_1 \simeq 8\mu s$ which indicates complete depolarization at each bounce. With the polyethelyene we find $T_1 \simeq 125 \mu s$ which indicates at least 15 bounces are required for depolarization. A depolarization time this long is consistent with some loss of polarization by effusion from the ends of the target. Unfortunately polyethelyene can be used only at lower temperatures than are needed for a Na target of useful thickness. It may be that a ^{39}K target can be used with polyethelyene walls. More research is needed to determine if there are materials that prevent depolarization and are useful with thick Na or ^{39}K targets.

OTHER METHODS OF OPTICAL PUMPING

Anderson has analyzed other methods of optical pumping for Na targets.[4] These include the use of velocity changing collisions to obtain frequency coverage,[19] the use of mode locked lasers to obtain frequency coverage, and the use of multimode lasers to obtain frequency coverage. These methods all show promise especially if surfaces that prevent depolarization are obtained. More research on these methods of optical pumping is needed.

CONCLUSIONS

The use of optical pumping to produce polarized H^- ions has been demonstrated experimentally by Mori et al.[5] Possible improvements in the optical pumping H^- ion source must be investigated before the ultimate H^- ion currents and polarizations are known.

REFERENCES

1. E. K. Zavoiskii, Soviet Physics JETP $\underline{5}$, 338 (1957).
2. W. Haeberli, Proceedings 2nd Int. Symp. Polarization Phenomena in Nuclear Reactions (P. Huber and H. Schopper, eds.) Birkhauser Basel 1966, p.64.
3. L. W. Anderson, Nucl. Instr. and Methods $\underline{167}$, 363 (1979).
4. L. W. Anderson, IEEE Trans. on Nucl. Sci. $\underline{NS30}$, 1051 (1983).
5. Y. Mori, K. Ikegami, Z. Igarashi, A. Takagi, and S. Fukumoto, Proc. of the Workshop on Intense Polarized Proton Ions Sources at Vancouver, to be published; also Y. Mori, private communication.
6. P. G. Sona, Energia Nucl. $\underline{14}$, 295 (1970).
7. C. J. Anderson, A. M. Howald, and L. W. Anderson, Nucl. Instr. and Methods $\underline{165}$, 583 (1979).
8. W. Grübler, P. A. Schmelzbach, V. König, and H. Marmier, Helv. Phys. Acta. $\underline{43}$, 254 (1970).
9. T. Nagata, J. Phys. Soc. Jap. $\underline{48}$, 2068 (1980).
10. A. S. Schlachter, K. R. Stalder, and J. W. Stearns, Phys. Rev. A $\underline{22}$ 2494 (1980).
11. M. Kimura, R. E. Olson, and J. Pascale, Phys. Rev. A$\underline{26}$, 1138 (1982).
12. E. A. Hinds, W. D. Cornelius, and R. L. York, Nucl. Instr. and Methods $\underline{189}$, 599 (1981).
13. L. W. Anderson, Proceedings of the Joint Mexican U. S. Conference on Negative Ions, (edited by I. Alvarez and C. Cisneros), Mexico City, Mexico (1981), p. 281.
14. Y. Mori, K. Ito, A. Takagi, and S. Fukumoto, Bull. Am. Phys. Soc. $\underline{26}$, 129 (1981).
15. G. G. Ohlsen, J. L. McKibben, R. R. Stevens, Jr., and G. P. Lawrence, Nucl. Instr. and Methods $\underline{73}$, 45 (1969).
16. M. S. Feld, M. M. Burns, T. V. Kuhl, P. G. Pappas, and D. E. Murnick, Optics Letters $\underline{5}$, 79 (1980); see also P. G. Pappas, M. M. Burns, D. D. Hinshelwood, M. S. Feld, and D. E. Murnick, Phys. Rev. A$\underline{21}$, 1955 (1980).
17. W. D. Cornelius, D. J. Taylor, and R. L. York, and E. A. Hinds, Phys. Rev. Letters $\underline{49}$, 870 (1982).
18. D. R. Swenson and L. W. Anderson, unpublished.
19. P. G. Pappas, R. L. Forbes, W. W. Quivers, Jr., R. R. Dasari, M. S. Feld, and D. E. Murnick, Phys. Rev. Lett. $\underline{47}$, 236 (1981).

DISCUSSION

McFARLAND: What kind of currents were you suggesting with the refrigerator and the hydrogen beam? You were talking about the 10^{16}, 10^{17} target thickness. What kind of polarized currents are possible under those conditions?

L.W. ANDERSON: I think that's very speculative. It seems to me that you could achieve currents of use for nuclear accelerators if you could produce targets like this in a fairly straightforward fashion. If you were to try to use something that would be useful for the fusion where you wanted amperes of current, you'd have to have a current of amperes of particles incident on a thick target. There may be severe heat loads and all sorts of other problems but it seems to me that it's not completely out of the question to hope for such a thing.

KLEPPNER: In the method you just described you're doing the polarization transfer at high magnetic fields. What about the problems of getting these ions in and out of that magnetic field?

L.W. ANDERSON: That was the reason that I showed you that the neutrals are incident. If you can have neutrals incident and neutrals emerging that is not a problem and presumably the only emittance growth would occur due to scattering inside the target. If you had to get them in and out of the magnetic field, then I think the ion source has to be in the same large magnetic field as the target and colinear with it and you have to extract neutrals.

EHLERS: In that last process, in the case of neutrals, what sort of neutral energies could you tolerate?

L.W. ANDERSON: The curves that I just showed were for 5 keV and if you want to produce negative ions, I don't think you can go much above 5 keV. The negative equilibrium fraction from the second target is too small at energies much above 5 keV. If you want to produce neutrals or positive ions I think you can go up to 50 or perhaps 100 keV.

EHLERS: I have a second question, namely in the last process does the same logic hold for deuterium where the polarization is a bit more complicated?

L.W. ANDERSON: Deuterium is more tricky. You can polarize the electron on a deuterium atom but if you can go through a Sona transition from a B field this does not lead to the prospect of getting whatever polarization you want. Deuterium has both a vector and tensor polarization. It's nice to be able to change between these and to get the highest possible vector and tensor polarization but simple Sona transitions do not lead to that. With fast beams RF transitions are difficult. You do get some polarization with deuterium using a Sona transition. I don't want to say you don't get any nuclear polarization but you don't have the simple ease of varying the polarization.

COOPER: Could you expand further on that? What sort of currents do you think would be possible for deuterium?

L.W. ANDERSON: I've already tried to skirt the question. The reason I showed this was to indicate that if you can make these targets there are logical things to try to get high currents but I don't know what it would lead to. I would be very worried that these particles may scatter, put excessive heat loads on cryogenic apparatus or lead to other problems that might be very severe. But if you could make such a target at least there's a way to try something.

PRODUCTION OF INTENSE BEAMS OF POLARIZED NEGATIVE HYDROGEN IONS BY DOUBLE CHARGE EXCHANGE IN ALKALI VAPOURS

W. Grüebler and P.A. Schmelzbach
Institut für Mittelenergiephysik, Eidg. Techn. Hochschule,
CH-8093 Zürich, Switzerland

ABSTRACT

In the last few years substantial progress on the performance of the electron impact ionizer and the double charge exchanger has been made in our laboratory. A proton or deuteron beam intensity of over 100 µA has been observed in the positive stage of the source. A polarized negative ion intensity of 6 µA was obtained after the charge exchange. After acceleration in the ETH EN-tandem accelerator 2 to 3 µA of polarized deuterons with 90% polarization could be focussed through a 3 mm diameter collimator on a target. This is the highest value ever reported for an accelerated beam.

At present we investigate improvements of the atomic beam stage. With presently available techniques some 100 µA of polarized H⁻ and D⁻ ions can be obtained with this type of source. New developments suggest that it is possible to reach in the future a beam intensity in the mA range.

1. INTRODUCTION

Intense beams of polarized negative ions have become very important for use in high energy cyclotrons or synchrotrons because of the ease with which these beams can be inserted or extracted by stripping the electrons in thin foil[1,2]. Today also polarized neutrons are produced mainly by nuclear reactions induced by polarized proton or deuteron beams.

On the other hand, the recent suggestion that great benefits can be expected by the injection of polarized hydrogen ions in fusion reactors[3] shows the general large interest in very intense sources of polarized hydrogen ions.

All production methods for polarized ions have in common that at first a polarized neutral atomic beam is prepared, which subsequently will be ionized selectively to positive or negative ions. The distinction between the different types of sources is based on the method by which the neutral beam is produced and polarized, and the process used to ionize the neutral atoms. The latter includes electron bombardment ionization, charge transfer and charge exchange. A review of these different methods is given in ref.[4].

A schmeatic diagram of the ground state atomic beam source is shown in fig. 1. Hydrogen atoms are generated by dissociating molecules at low pressure in a rf discharge. An atomic beam of thermal velocity is formed by a nozzle. This beam is polarized in passing through an inhomogeneous magnetic field (usually either a quadrupole or sextupole). Rf transitions are induced between the hyperfine states of the atoms in order to select different nuclear polarization states and to provide rapid spin reversal. Positive ions are produced by electron bombardment. An adder canal can be used to convert the positive to negative ions by double charge exchange.

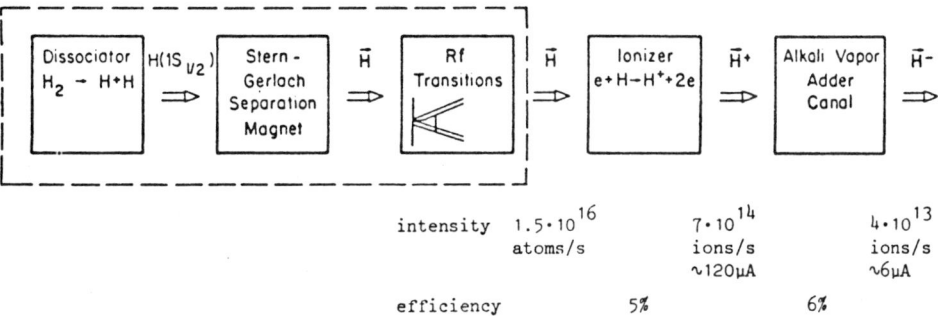

Fig. 1. *Schematic diagram of the atomic beam source for polarized hydrogen ions.*

Fig. 1 shows also the number of particles produced at the different points of the present operational ETH polarized ion source and the efficiencies we obtain in the conversion of the polarized atoms to positive and negative ions. While these values are already outstanding, the source shows excellent promises for further considerable improvements. In this paper we present the status of the different components of the ETH source and we shall discuss the possibilities to gain one or two order of magnitude in beam intensity.

2. ATOMIC BEAM APPARATUS

A review of the ground state atomic beam stage of polarized hydrogen ion sources and the basic problems are discussed extensively in ref.[5], and we refer for the notation to this reference. The ETH atomic beam apparatus has been described in details in ref.[6]. It is equipped with 2→4 and weak field rf transitions for the H^0 atoms, and 3→5, 2→6 and weak field rf transitions for the D^0 atoms. For many years, Stern-Gerlach separation and focussing were provided by a 50 cm long sextupole magnet. The optimum atomic beam intensity was measured to be $1.5 \cdot 10^{16}$ atoms/sec in a cross sectional area of about 1 cm². Recently, a short (10 cm) sextupole compressor magnet was added to the apparatus[7]. While the separation sextupole magnet should be redesigned in order to take full advantage of this configuration, the improvement of the atomic beam quality was such that an increase of the ion beam intensity by a factor 1.5 was obtained in this way. It should be pointed out here that this apparatus was developped more than a decade ago. With the incorporation of new knowledges on beam production and transport, an important increase of its performance is expected . These improvements will be discussed in the last section.

3. ELECTRON BOMBARDMENT IONIZER

In the last several years large improvements have been achieved in the electron bombardment ionizer technique. This fact is reflected by the increase of the positive ion beam intensity from about 10 µA to more than 100 µA . The 10 times higher efficiency is obtained by a longer ionization region, a better location of the electron gun and the ion extraction system, a more elaborate control of the potential of the ionization column and the shaping of the magnetic field.

A schematic representation of the new ETH ionizer[8], which has incorporated these technical features, is shown in fig. 2. Since the aim of this source is the production of polarized negative hydrogen ions for the injection in a tandem accelerator the ionization region and the succeeding electrodes are held at -50 kV in order to provide the necessary injection energy. For technical convenience, the vacuum housing and the coils of the solenoid are at ground potential. The technical details are explained in the figure caption. After the extraction of the ions from the ionizing region, located between electron gun and electron reflector, they are accelerated to 15 keV, focussed and slowed down to the energy of 5 keV necessary for the charge exchange in sodium vapour. The ionizer is used for the production of polarized protons and deuterons with the atomic beam apparatus described in the preceding section. Positive ions with an intensity of more than 100 µA could be extracted and transported to the charge exchanger, i.e. the efficiency of the ionizer reaches a value of about 5%.

While the older conventional ionizers already showed several sets of tuning parameters with similar output currents, this multiplicity has further increased in the new type of ionizer. As a tuning parameter like the electron emission current, the magnetic field shape, an electrode potential either from the electron gun system or the ion extraction region is varied, stepwise changes or sharp maxima together with hysteresis effects can be observed. Since it is not possible to pass from one mode to another in a continuous way, it is obviously more difficult to tune the ionizer into the stable operation mode with the maximum beam current output. However, since the operating conditions are highly reproducible, setup for routine operation is even easier than with older, less sophisticated devices. The starting operation is generally reduced to energizing the main switches and setting the filament current. The other parameters require only minimum adjustment.

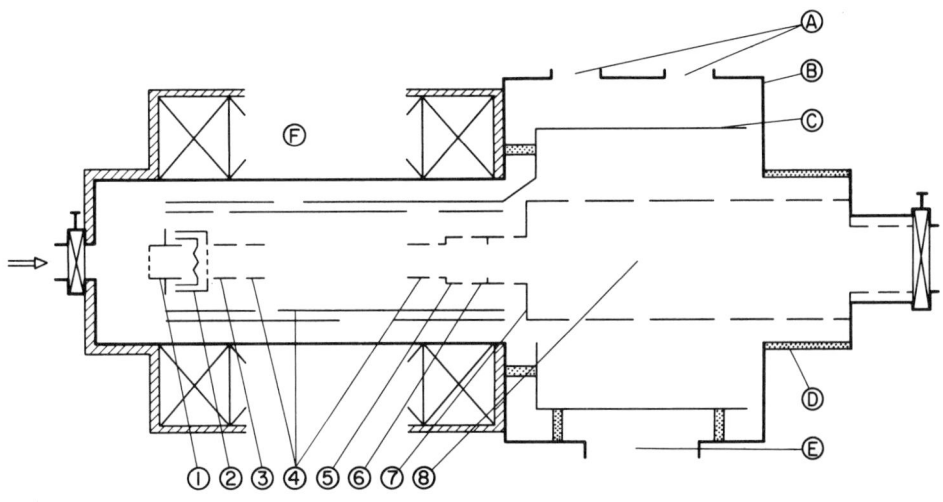

Fig. 2. Schematic representation of the ETH ionizer (not to scale).

General
A: pumping ports (two 900 ℓ/sec orbitron pumps)
B: vacuum housing
C: electrode supporting system
D: insulator
E: pumping port (5000 ℓ/sec LN_2-baffled diffusion pump
F: solenoid coil

Electrode system
1: electron repelling
2: filament and grid
3: electron acceleration and ion repelling
4: ionization column potential
5: electron reflexion and ion extraction
6: preacceleration
7-8: acceleration, beam forming and transport to charge exchanger

4. CHARGE EXCHANGER

Several design consideration has to be taken into account for the construction of a charge exchange device. Double charge exchange efficiency of up to 30% for Cs as donor has been obtained[9,10], however this process is strongly depending on the velocity of the positive ions entering the exchange canal. Since high efficiency electron impact ionizers require a high extraction voltage (~15kV) the positive ions have to be decelerated to quite low energy in order to obtain the full benefit of the charge exchange. This is in the case of Cs 0.5 to 1 keV (cf. fig. 4). It is obvious that in this process the beam is blown up and space charge problems become serious. Therefore in the past other alkali metals have been used since at higher energy more favourable results are obtained. Since more than a decade we have used at our laboratory sodium at an operational energy of

5 keV. We had restricted in this device[11] the diameter of the exchange canal to 1.0 cm in order to limit the loss of sodium and avoid the contamination of the ionizer and the following acceleration tube by the sodium vapour. Due to the geometrical dimension of the exchange canal only 1/3 of the available positive ion beam could effectively be used. Another disadvantage was the obstruction which occured after a few days of operation by sodium deposits on the cold diaphragms at the entrance and exit of the charge exchanger.

In a new design the conditions in charge exchange yield and minimum loss of sodium have been optimized for a 5 keV beam. Further a larger sodium storage container and the use of a recirculation system was required in order to obtain long operational periods without maintenance. It is well known that the charge exchange has to be carried out in a strong magnetic field, thus preventing depolarization of the hydrogen atoms in the intermediate neutral state. These considerations led to the design shown in fig. 3. The sodium container held at 280°C is placed in the center. The 1.4 cm diameter

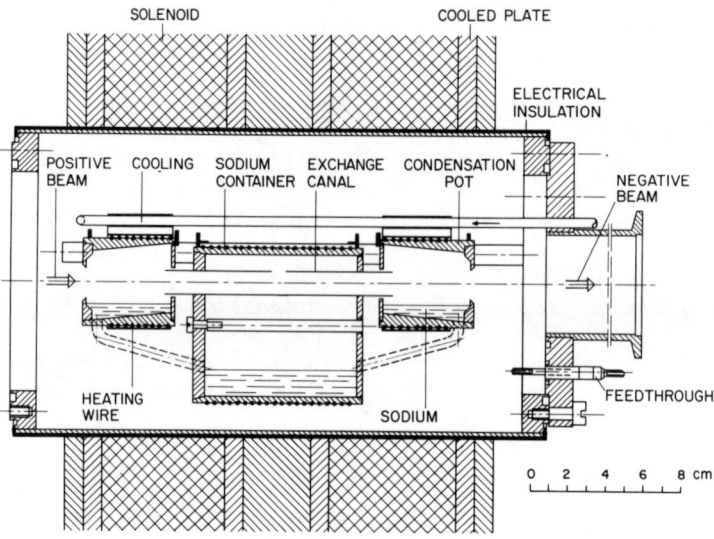

Fig. 3. Cross section of the charge exchanger.

charge exchange tube is located excentrically in the oven and is fed with sodium vapour from a central opening. The conductance of this tube prevents an exaggerated loss of sodium. Sodium vapour leaving the canal is catched at both ends in condensation pots, which are held at 150°C by a combined heating and cooling system. Provisions have been made that the liquid sodium collected in these pots flows back in the central container. At the moment this has not yet technically accomplished in the construction shown in fig. 3 but has worked satisfactorily in an earlier version[4]. With this new charge exchanger, maintenance will be required only after several hundred hours of operation. The transmission of the device has been doubled, leading to the production of up to 6 µA of negative ions.

The effective efficiency of 6% of the double charge exchange indicated in fig. 1 is determined, besides the transmission problem, mainly by the chosen restriction of the entrance energy of the positive ions of 5 keV, which is an easy attainable energy by decelerating the ions from the higher ionizer extraction energy. At this energy, the charge exchange in sodium vapour yields the largest intensity of negative ions (cf. fig. 4). However, recently new investigation of charge exchange in alkaline earth has shown, that in these vapours,

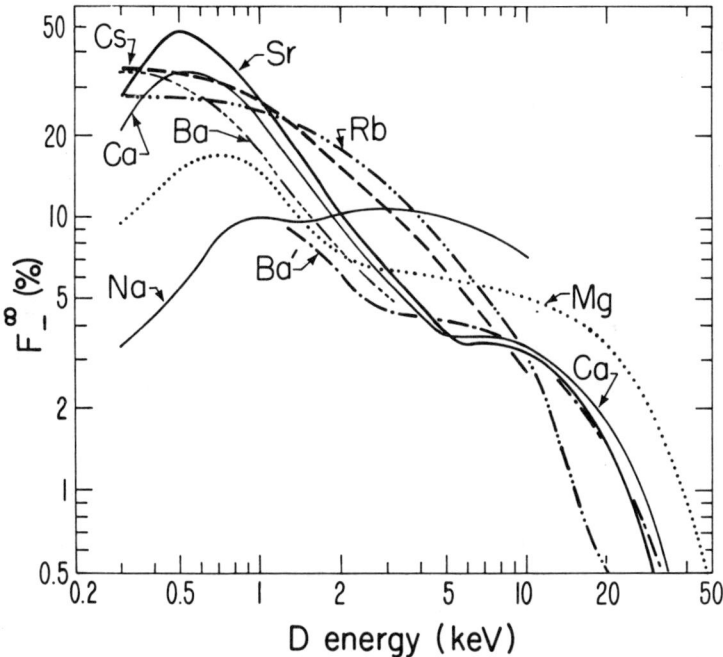

Fig. 4. A collection of equilibrium double charge exchange yields F_-^∞ for different alkali and alkaline earth vapours as a function of deuteron energy (from ref.[12]).

yields of negative ions of up to 50% can be reached[12]. These new results are shown in fig. 4 together with the corresponding alkali metal results. The design of an appropriate deceleration system, which brings the ions to the required low energy between a few hundred eV and 1 keV in a high transmission oven, could improve the double charge exchange yield by approximately a factor 4.

An interesting approach in this direction is the design of a new type of neutral injector line based on negative deuterons, for a Tokamak fusion reactor[13]. Positive deuterons from a ECR source are converted to negative ions by passing through a supersonic cesium jet at an energy of 1 keV. The whole production line is immersed in a more or less uniform magnetic field which improves the beam transport between the source and the double charge exchanger. A schematic diagram of this Grenoble design is shown in fig. 5. A charge exchange efficiency of 20 to 25% is measured in this device. Details are

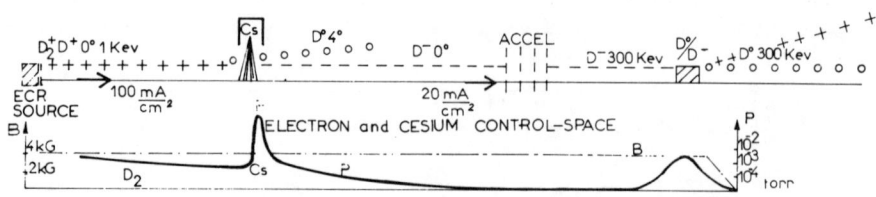

Fig. 5. The magnetized cesium D^- line with supersonic cesium cell and ECR ion source (from ref. 13).

given in ref. 13. For situations where the use of alkali vapours is too dangerous, xenon can be used with reasonable conversion yield. Between an energy of 2 and 4 keV a H^- yield from 4 to 6% is experimentally observed[13].

5. FURTHER IMPROVEMENTS

5.1. Dissociator of the atomic beam apparatus

While the electron bombardment ionizer and the charge exchanger are presently at a state-of-the art level, the weakest part of our source is the atomic beam stage. For this reason we have build a test bench for an atomic beam apparatus where the connection between rf discharge tube and accommodator is made by a very short teflon tube. The hydrogen molecules are dissociated at room temperature in a Pyrex tube by a 27 MHz rf discharge and the produced atoms are transferred through a short Teflon tubing to a copper accommodator which for some experiments is coated with a thin Teflon film. The accommodator is cooled by a closed cycle ^{4}He-refrigerator. A short nozzle, a skimmer and a collimator at the entrance in the magnet chamber form the particle beams. Two sextupole magnets allow us to study the effect to the desired atomic states. A third sextupole magnet is at present replaced by a velocity selector in order to study the velocity distribution of the produced atomic or molecular beams. These investigations are carried out for judging the efficiency of the cooling, the properties

of the beam forming elements and the degree of dissociation obtained
in the atomic beam. The measurement of the density of the beams is
accomplished by a quadrupole mass spectrometer with a cross beam ion
source. The intensity of the beams is measured in a compression tube
containing an ionization gauge. The velocity measurements have been
tested with molecular hydrogen and deuterium beams as well as with helium beams. The results of these investigations show that the cooling
of the particles in the accommodator behaves as expected. A typical
example for a cooled atomic beam is shown in fig. 6. The experimentally determied width (FWHM) is about half of the width of the Maxwell distribution. The increase of the most probable velocity compared
to the accommodator temperature and the decrease of the width arises
from the gasdynamical character of the beam forming and expansion of
the atomic gas from the accommodator into the vacuum.

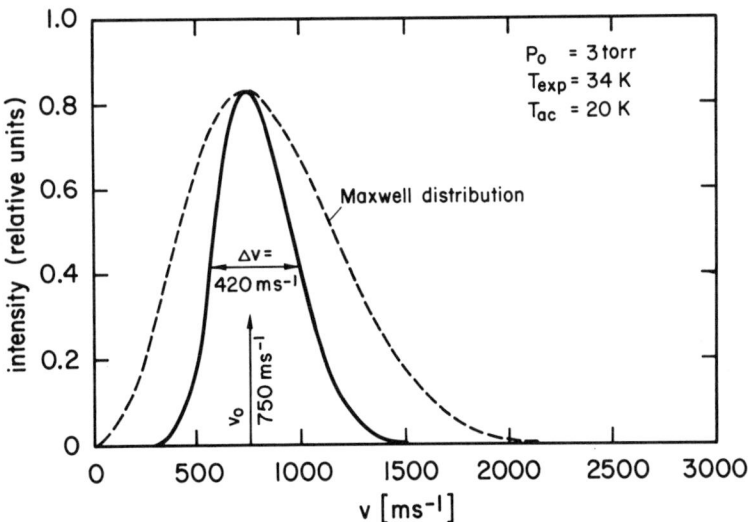

Fig. 6. *The velocity distribution of an atomic hydrogen beam measured at the ETH atomic beam apparatus. Temperature of the accommodator 20K. Pressure in the dissociator $p_0 = 3$ torr.*

In our experimental arrangement we measure the gas flow and the
pressure in the dissociator. This gas at room temperature can approximately be considered as stagnant. For our geometry the conditions in
the dissociator are almost independent of the temperature in the accommodator. In this case a constant flow of atoms is supplied. In the
accommodator region, the density and pressure are given by the flow
conditions at the entrance and exit apertures. Therefore, provided one
has a constant gas flow through the system, the pressure in the

accommodator varies essentially as $\sqrt{T}$ and the density as $1/\sqrt{T}$. Hence, the intensity of the atomic beam stays constant as a function of the temperature. The result of the measurement with a Teflon coated accommodator and for a pressure $p_o = 3$ torr in the dissociator is seen in fig. 7. In the temperature region between 100 K and 300 K the atomic

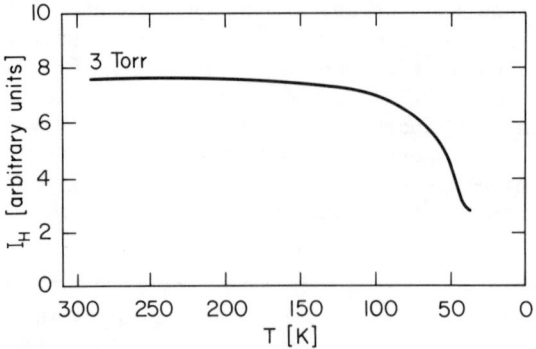

Fig. 7. Intensity of the atomic beam as a function of the accommodator temperature for $p_o = 3$ torr in the dissociator.

beam intensity is nearly constant. At lower temperature a fast intensity drop is observed reaching at 50 K about half of the original value. A similar behaviour is observed for the density n of the atomic beam. This is shown in fig. 8. After an increase of about a factor 2.5 due to the lower velocity of the beam a sharp decrease is measured in

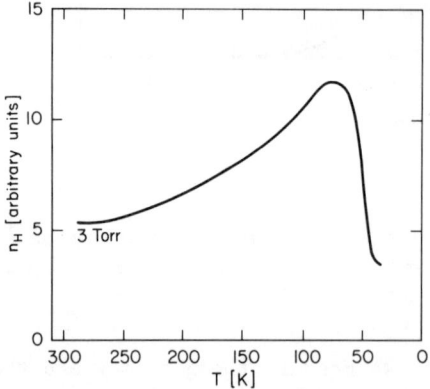

Fig. 8. Density of the atomic beam as a function of the accommodator temperature for $p_O = 3$ torr in the dissociator.

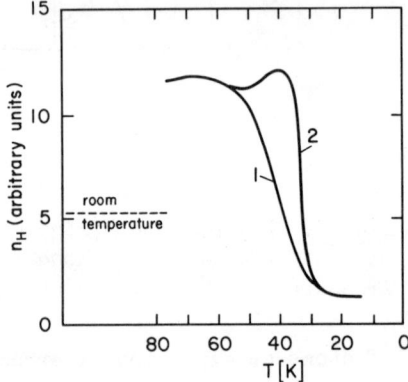

Fig. 9. Behaviour of the density of the atomic beam at low accommodator temperature. Curve 1 without treatment. Curve 2 with H_2 treatment.

the temperature region of 50 K. Intensive investigations with atomic
and molecular beams show that the intensity and density drop is mainly
caused by recombination on the contaminated accommodator surface.
Covering the surface of the accommodator by a material with low recombination coefficient at this temperature is able to prevent the strong
recombination in this temperature region. Different methods to improve
the surface conditions have been tried successfully. As an example the
result of a molecular hydrogen treatment at the lowest reachable temperature of 16 K is shown in fig. 9. The curve 1 shows the behaviour
without treatment as seen in fig. 8. After flushing for several
minutes with H_2 and warming up, the beam density reaches a new stable
maximum of the same value but at a substantially lower temperature.
This behaviour is indicated by curve 2 in fig. 9. Investigations with
different gases, show that surfaces with better properties than the
one naturally obtained under the operating vacuum conditions can be
produced and maintained and that a cooled atomic beam of about 30 K
can be produced with a density substantially larger than at room temperature. The narrow velocity width of this beam will facilitate
strongly the design of the separation magnets in order to obtain a
large solid angle of acceptance.

5.2. The separation magnets

The separation magnets - mostly of the sextupole type - produce
electron spin polarized hydrogen beams. In order to obtain nuclear polarization rf transitions have to be induced between a populated and
an empty substate. A sextupole magnet acts on hydrogen atoms not only
as a separator of electronic substates but also as a thick focussing
lens. Since the atomic beam entering the separation magnet has not a
single velocity, but a velocity distribution, achromatic focussing has
to be applied in order to get the atoms accepted by the separation
magnet into the ionizer. This is achieved by a system of sextupole
lenses. Complication arises from the different starting conditions
related to an extended geometry of the source of atoms.

At ETH we have developed two types of sextupole magnets 10 and
15 cm long, which are designed in a modular technique such that the
pole pieces can be changed in a simple way. Computer programs allow
the calculation of acceptance diagrams and of the trajectories of
electron polarized atoms for different starting conditions (radius,
angle and velocity) for lens systems of up to four sextupole magnets.
Fig. 10 shows a typical example of the behaviour of the trajectories
of atoms starting on the axis with an angle of 5° and velocities between 525 ms^{-1} and 975 ms^{-1} (which corresponds about to Δv in fig. 6).
Most of the atoms are 40 cm after leaving the magnet system within
a reasonable diameter, however, a part of atoms from the velocity
spectrum is lost due to overfocussing of these particles. Misleading
results can be obtained if not enough trajectories are investigated.
So, if only the three velocities 525, 750 and 925 ms^{-1} would be considered one would conclude erroneously that a perfectly convergent

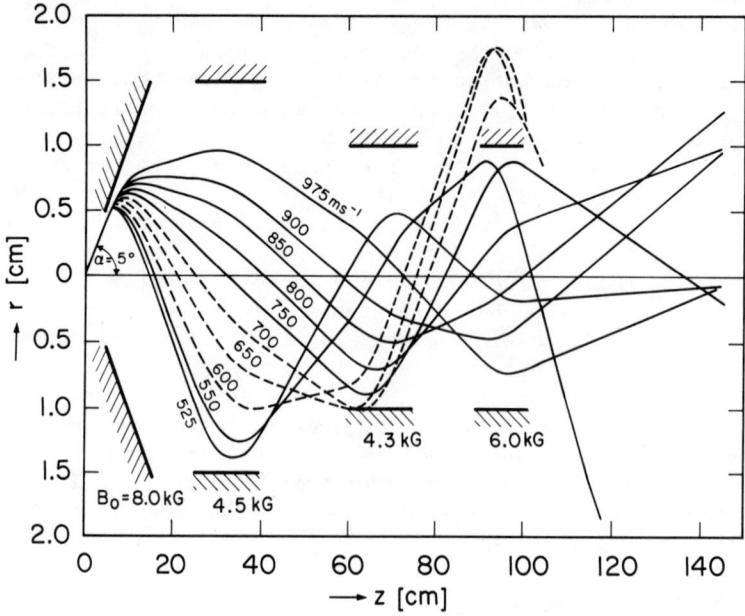

Fig. 10. Calculated trajectories of hydrogen atoms starting at velocities between 525 and 975 ms^{-1} and an angle of 5° on the axis. The plot gives the radial distance of the atoms from the symmetry axis of the magnets.

beam would be obtained in the ionization region. Intensive computer simulations for optimizing the geometry of the sextupole lens system, using the velocity distribution found from cooled beams and for realistic starting conditions, are performed in order to obtain a maximum acceptance and transmission of the separation magnets for conditions allowing a good matching of the beam to the subsequent elements, i.e. rf transitions and ionizer. In this case the enlarged solid angle which is attainable with cooled beams will lead to 20 or 30 times increased polarized ion beam intensity.

5.3. ECR ionizer

A most efficient method to ionize light and heavy ion is the use of electron cyclotron resonance (ECR) ionizer[14]. In this type of ionizer a plasma of cold ions with hot electrons is produced in the ionization region. The applied electron cyclotron resonance conditions produce a high electron density, which can be 10^3 times higher than in a normal electron impact ionizer. Electron energy of several keV can be obtained. Gas efficiency $\eta_g = I^+/(I^+ + I_g)$ of approximately 70% has been observed for H_2 molecules.

The application of electron cyclotron resonance conditions (e.g. B_z = 3.75 kG and ν = 10 GHz) to the ionization region, however, seems very dangerous, because similar conditons are necessary for transitions from the m_j = +1/2 states to the m_j = -1/2 states in the hydrogen atoms. Such transitions certainly would produce strong depolarization if the residence time would be relatively long. A remedy for depolarization could be a strong mismatch of the adiabatic transition conditions[15].

The high efficiency of the ECR ionizer is very interesting for the application in an atomic beam source. The low extraction energy also meets directly the requirement for a subsequent double charge exchange in order to produce polarized H^- ions. Experimental investigations should be carried out in order to study the severity of the depolarization suggested above.

6. CONCLUSIONS

The ground state atomic beam souces with double charge exchanger currently produce polarized DC beams of 6 µA H^- and D^- ions with high polarization and good beam emittance. In a pulsed mode an increase of beam current of a factor 2 to 3 has been observed.

Cooling the atomic beam to low temperature and the use of a properly designed multistage separation magnet system will increase the beam current by a factor 20 or 30, which is by far the largest improvement for the near future. This development requires systematic investigations of the cooling procedure and the formation of the atomic beam. Since not only dissociator and magnet system are interrelated but the ionizer and possible charge exchange device as well, the matching of all these components has to be taken into account for the design of a high intensity polarized negative hydrogen ion source.

Improvement of the double charge exchange method for higher yield can be expected by the use of other alkali or alkaline earth vapours using low energy H^+ beams. The ECR ionizer should be of particular interest, as the positive beam is extracted at low energy, which fits well with the requirement of a double charge exchanger. An estimated increase of the ionization efficiency of a factor 5 should be possible for an ECR ionizer and a further gain factor 4 can be obtained in future charge exchange devices. The presently known techniques suggest future polarized negative ion beam intensity in the mA region delivered by polarized negative ion sources for accelerators.

REFERENCES

1. C. Hojvat et al., IEEE Trans. Nucl. Sci. **26**, 3149 (1979)
2. C.W. Potts, IEEE Trans. Nucl. Sci. **24**, 1385 (1977)
3. R.M. Kulsrud, H.P. Furth, E.J. Valeo and M. Goldhaber, Phys. Rev. Lett. **49**, 1248 (1982)
4. W. Grüebler and P.A. Schmelzbach, Fifth Int. Symp. on Polarized Phenomena in Nuclear Physics, Santa Fé, G.G. Ohlsen et al., Eds., AIP Proc. No 69 (AIP, New York 1981) p. 848
5. W. Grüebler, in Polarized Proton Ion sources, A.D. Krisch and A.T.M. Lin, Eds., AIP Proc. No 80 (AIP, New York 1982) p. 53
6. R. Risler, W. Grüebler, V. König and P.A. Schmelzbach, Nucl. Inst. Meth. **121**, 425 (1974)
7. H.F. Glavish, Proc. Fourth Int. Symp. on Polarization Phenomena in Nuclear Reactions, W. Grüebler and V. König, Eds., (Birkhäuser Verlag Basel and Stuttgart 1976) p.844
8. P.A. Schmelzbach, W. Grüebler, V. König and B. Jenny, Nucl. Inst. Meth. **186**, 655 (1981)
9. W. Grüebler, P.A. Schmelzbach, V. König and P. Marmier, Helv. Phys. Acta **43**, 254 (1970)
   W. Grüebler, P.A. Schmelzbach, V. König and P. Marmier, Phys. Lett. **29A**, 440 (1969)
10. A.S. Schlachter, K.R. Stadler and J.W. Stearns, Phys. Rev. **A22**, 2494 (1980)
11. W. Grüebler, V. König and P.A. Schmelzbach, Nucl. Inst. Meth. **86**, 127 (1970)
12. R.H. McFarland, A.S. Schlachter, J.W. Stearns, B. Liu and R.E. Olson, Phys. Rev. **A26**, 775 (1982)
13. R. Geller, B. Jacquot, C. Jacquot and P. Sermet, Nucl. Inst. Meth. **175**, 261 (1980)
14. R. Geller and B. Jacquot, Nucl. Inst. Meth. **184**, 293 (1981)
15. A. Abragam and J.M. Winter, Phys. Rev. Lett. **1**, 374 (1958)

DISCUSSION

L.W. ANDERSON: I did not understand what your conclusion was on the effect of resonances in the ECR source. Would it or would it not depolarize?

GRUEBLER: There are two possibilities. You have to think first in hydrogen. In hydrogen I have the feeling one would have depolarization because one gets the resonance of the 2 to 4 transition. The question is, naturally, if you can you obtain in principle 100% of this RF transition, from 2 to 4. Another possibility would be to make an additional Stern-Gerlach separation by first making the transition from 2 to 4 and use only a single state, namely the state 1, the transition 1 to 3 is forbidden, and you could work with this but you would lose a factor of two. With the deuterons you have other transition conditions and then probably it would work. I mean that one would have to find out what other frequencies one would have in the RF cavity, in the ECR ionizer. Maybe C. Jacquot can comment.

JACQUOT: I would just like to make a comment. I am very pragmatic about the utilization of the ECR source to ionize polarized atom beams. But for example, for fusion applications, if we put polarized ions in a tokamak we have there a lot of electric and magnetic fields and whether the degree of polarization will stay or not is an important question. In some papers it has been envisaged to enhance the cross section for fusion reactions by using polarized ion beams, but if we inject polarized ions into the reactor, these also have a high plasma density, RF fields, and magnetic fields. I don't know if the ECR ion source can depolarize or not, we have to try.

HAEBERLI: I have a question related to the ionization in an ECR source. What is the situation as far as the effort is concerned that has been made for a number of years at Saclay to achieve a high ionization efficiency with an ECR device?

JACQUOT: It was Dr. Deschamps, but I don't know exactly what he did with the ECR source. Finally, I believe, he stopped the experiment.

HAEBERLI: It was abandoned and the ionizer was replaced with the type that Dr. Gruebler talked about, but I was wondering what the fundamental difficulties were in that application.

JACQUOT: I don't know.

OPTICS FOR A SPIN-POLARIZED HYDROGEN ATOMIC BEAM

Daniel Kleppner*

Laboratoire de Spectroscopie Herzienne
École Normale Supérieur
24 rue Lhomond, Paris 75005, France

ABSTRACT

Cold atomic hydrogen from a spin-polarized hydrogen atomic beam source can be focused by simple solenoids. In contrast to conventional magnetic deflection systems, the focusing of cold hydrogen can be so efficient that a large fraction of the atoms are brought into a highly collimated beam. The beam can be employed in the production of polarized H^-, or for use in a jet.

INTRODUCTION

Spin-polarized hydrogen is a cold relatively dense gas of atomic hydrogen. Molecular recombination is prevented by achieving essentially 100% electron spin polarization. This is accomplished by bringing the atoms into thermal equilibrium in a magnetic field of 5-10T, at a temperature of 0.4K or lower [1,2]. The atoms are confined along the axis of a cell by the magnetic field gradients, and confined radially by the helium covered walls of the cell. At modest densities, in the range of 10^{15}-10^{16} cm^{-3}, the gas can be stable for many hours. At the highest density so far achieved, 4×10^{18} cm^{-3}, the lifetime is limited by a three-body recombination process [3,4] to approximately one second.

Spin-polarized hydrogen has potential applications to the creation of polarized sources and targets of hydrogen and deuterium [5,6,7]. One approach is to use the cell of spin-polarized hydrogen as the source for a proton polarized atomic beam. As pointed out by Niinikoski [5], by irradiating the hydrogen with microwave radiation at the frequency of an electron spin-flip transition, the atoms are transferred into a single hyperfine state which is repelled by the magnetic field. If the storage cell is open-ended, the atoms are expelled from the apparatus along the axis of the solenoid, and can be formed into an atomic beam. The beam achieve should be capable of higher intensity than possible by conventional methods.

---

* Permanent address: Physics Dept., MIT, Cambridge, MA 02139

0094-243X/84/1110720-16 $3.00 Copyright 1984 American Institute of Physics

Considerations which govern the operation of such a source include the supply of an adequate flux of spin-polarized atoms, recombination processes, design of a "film burner" to control the superfluid helium film which lines the cell, disposal of atoms in the unwanted hyperfine state, the systematics of the microwave resonance process and the kinetics of ejection. These matters will be addressed elsewhere. In this paper we discuss how the expelled atoms can be formed into an atomic beam by using the focusing power of simple magnetic fields.

The technique of deflecting atoms magnetically is well known in atomic and molecular physics [8]. Some of the most refined beam focusing systems have been designed for applications to polarized proton sources, and are referred to elsewhere in these proceedings and the proceedings of earlier conferences on polarized sources [9] and high energy spin physics [10]. However, at the very low temperatures of spin polarized hydrogen, the beam optics differ radically from conventional systems, and the subject must be rethought. As one might expect, it turns out to be much easier to focus a cold hydrogen beam than a conventional atomic beam. This paper discusses some of the general considerations governing the optics of very cold atomic beams.

A good indicator of the difference between cold atomic beams and conventional atomic beams is the expression for the effective acceptance solid angle Ω_{eff} of the most commonly used focusing element, a hexapole magnet

$$\Omega_{eff} = C\frac{\mu B}{kT} \qquad (1)$$

where μ is the Bohr magnetron, B is the field at the pole tips (approximately 1 tesla), and T is the temperature of the source. In conventional atomic hydrogen sources, T is typically 70K-300K. C is a numerical constant which is approximately two. For a well designed system, $\Omega_{eff} \sim 3 \times 10^{-3}$; often it is much less.

For cold hydrogen, $\mu B/kT \gg 1$ and Eq. (1) longer applies. The assumptions underlying Eq. (1) are invalid, and one must return to first principles.

GENERAL CONSIDERATIONS

The major elements of a cold hydrogen atomic beam source are shown in Fig. 1. Hydrogen in hyperfine states <u>a</u> and <u>b</u> (see insert, Fig. 1) are confined in a helium-lined cell at temperature T_c in a magnetic field with peak intensity B. Microwave signals induce the transition <u>a</u>→<u>d</u> somewhere near the center of the cell; the <u>d</u> state atoms diffuse into a field gradient and are expelled. Their motion from the magnet is restricted by a thermalizer at temperature T_t, which contains an aperture on the axis. Atoms flowing out of the aperture constitute the source of the atomic beam. The atoms are accelerated by the magnetic field. The acceleration tends to focus the beam; the radial forces of the diverging field enhance this effect, and the solenoid behaves like a fairly powerful lens. The emerging beam is highly directed compared to a beam from a conventional effusive source. Furthermore, the final kinetic energy, approximately μB, has only a small spread due to the thermal motion because $\mu B \gg kT_t$. Hence, the beam is highly monoenergetic compared to a conventional thermal beam. This greatly enhances the efficiency of subsequent focusing elements.

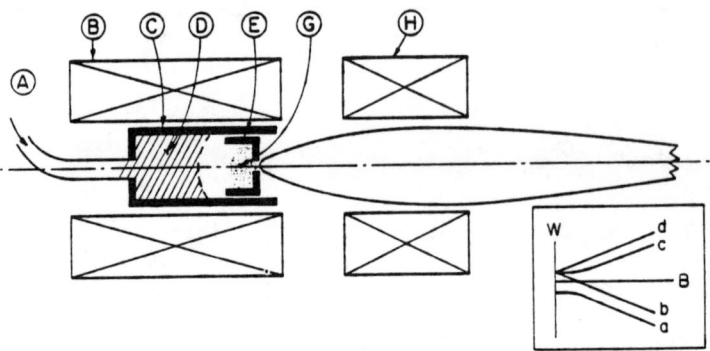

Fig. 1. Schematic diagram of spin-polarized hydrogen atomic beam source. A. H inlet; B. 7-10T solenoid; C. Storage cell, coated with ^{4}He at temperature $T_c \sim 0.4 K$; D. Spin polarized hydrogen in states <u>a</u> and <u>b</u> (see insert); E. Thermalizer and aperture at temperature T_t; G. Atoms in state <u>d</u> being ejected form the cell; H. focusing solenoid. Not shown: microwave source to drive the <u>a</u>→<u>d</u> transition. Atoms in state <u>b</u> could also be ejected by driving the <u>b</u>→<u>a</u> transition; the resulting beam would have the opposite nuclear polarization.

The atomic beam can be collimated or refocused by a conventional hexapole magnetic. A better focusing element, however, is simply a short solenoid.

Although details of the cell and the hydrogen production are not discussed here, brief comment may be in order on the distinction between the cell temperature T_c and thermalizer temperature T_t. In order to suppress recombination in the cell, it is desirable to keep the cell temperature relatively high, perhaps 0.4-0.5K. (The destructive processes generally occur on the walls, and their rates vary as $\exp[2E_{ads}/kT]$. E_{ads} is the adsorption energy of H on ^{4}He, approximately 1K.) On the other hand, the beam quality is enhanced by having the ratio $\mu B/KT_t$ large, so that it is desirable to have T_t as low as possible. The atoms spend relatively short time in the thermalizer region, allowing a low temperature in that region. For a pulsed source, in which the atoms are accumulated in the cell and then rapidly released, there can be significant advantages to having $T_t < T_c$. For a continuous source, a single temperature may be desirable.

FOCUSING OF A SOURCE IN A SEMI-INFINITE SOLENOID

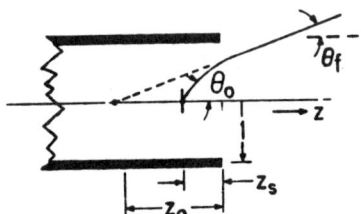

Fig. 2. Geometry of the source. The heavy lines represent the end of a semi-infinite solenoid: the atoms effuse through an aperture at Z_s.

The source geometry is defined in Fig. 2. The source is located at distance Z_s from the end of an ideal semi-infinite solenoid, whose radius is taken to be 1 unit. The magnetic field along the axis is

$$B(z) = \frac{B_o}{2}\left[1 - \frac{z}{(1+z^2)^{1/2}}\right] . \qquad (2)$$

The atoms are assumed to emerge from the aperture as in a conventional effusive source. The total flux from the source is [8]

$$I_{tot} = n\bar{v}A_s/4 \text{ atoms/sec} \tag{3}$$

where n is the density in the source, $\bar{v} = \sqrt{\frac{8}{\pi}\frac{kT}{m}}$ is the mean speed, and A is the aperture area. The angular flux density is

$$I(\Omega) = I_{tot}\frac{\cos\theta}{\pi} \text{ atoms/sec/ster} \tag{4}$$

The fraction of atoms within a cone of half angle θ is

$$F_s(\theta) = \sin^2\theta. \tag{5}$$

Two effects tend to collimate the emerging atomic beam. The dominant acceleration is longitudinal; increasing the longitudinal velocity while leaving the transverse velocity constant causes the trajectory to shift closer to the axis. The shift is enhanced by the radial forces, which decrease the transverse velocity. The first effect is most important. By neglecting the radial focusing, one obtains a useful expression for the flux density in the beam. Using Eq. (4) it is found that the mean kinetic energy of transverse motion in a beam is kT. Assuming, for the moment, that the atoms move randomly with energy exactly equal to kT, then the maximum angle θ_f' subtended by the ejected beam is

$$\theta_f' = \arctan\left[\frac{kT}{\mu B}\right]^{1/2}. \tag{6}$$

The solid angle subtended by the beam is

$$\Omega_f = \pi\theta_f'^2 \approx \pi\frac{kT}{\mu B}. \tag{7}$$

An effusive source emits atoms into $\Omega_o = 2\pi$ ster. The "focusing power" of the solenoid source is

$$P = \frac{\Omega_o}{\Omega_f} = \frac{2\pi}{\pi\theta_f^2} = 2\frac{\mu B}{kT} \tag{8}$$

where the last expression is valid for $kT \ll \mu B$. This result quantifies the evident fact that the colder the source, the more highly collimated the beam. It can also be shown, using the same approximation, that the angular flux density in the emerging beam is

$$I(\theta_f) = I_o \cos\theta_f \, (1+\frac{\mu B}{kT}) \qquad \theta_f < \theta'_f$$
$$= 0 \qquad \theta_f > \theta'_f \qquad (9)$$

and that the fraction of atoms within angle θ is

$$F_f = \sin^2\theta_f / \sin^2\theta'_f \qquad (10)$$
$$= 0 \qquad \theta_f > \theta'_f \, .$$

We turn now to a more careful analysis. The speed distribution from a thermal source is given by

$$f_o(v) = 2\alpha^{-4} v^3 \exp[-v^2/\alpha^2] \qquad (11)$$

where $\alpha = \sqrt{2kT/M}$ is the most probable speed in the source. $T = T_t$ is the temperature of the source thermalizer: the subscript can be dropped without ambiguity.

The distribution of energies in the emerging beam is given by

$$g_o(E) = E(kT)^{-2} \exp[-E/kT] \qquad (12)$$

The mean energy is

$$<E> = \int_0^\infty E g(E) dE = 2kT \, . \qquad (13)$$

If the atoms are expelled into free space from a point where the magnetic field is B, then they gain energy $E_o = \mu B$ and the energy distribution function becomes

$$g_f(E) = [\frac{E-E_o}{kT}]\exp[\frac{E-E_o}{kT}], \qquad E > E_o .$$
$$= 0 \qquad\qquad\qquad\qquad E < E_o \qquad (14)$$

The mean energy is $\mu B + 2kT$ and the fractional spread in energy is $\frac{2kT}{\mu B + 2kT}$. As $\mu B/kT$ is increased, the beam becomes increasingly monoenergetic.

Atoms emerging from the source at angle θ_o are deflected forward and emerge at a final angle $\theta_f \leq \theta_o$, as shown in Fig. 2. The final angle depends on the energy of the atom, the position of the source and the exact field profile. θ_f cannot be calculated using a simple impact approximation since the kinetic energy varies rapidly as the atoms emerge, and far off-axis trajectories can be important. Consequently, we have chosen to calculate the orbits numerically, using methods described in the appendix. After inspecting many orbits it was found that the most favorable location for the source aperture is approximately at $Z_s = -1/2$, as shown in Fig. 2. To simplify the presentation, the value $z_s = -1/2$ was held as a fixed parameter. Orbits were plotted as a function of the reduced kinetic energy ε, the ratio of the initial kinetic energy to the magnetic field energy at the aperture.

$$\varepsilon = Mv_o^2/2\mu B, \qquad (15)$$

(Note that the maximum field, far inside the solenoid, is 1.38B). Results are shown in Fig. 3. Curves are given for $\theta_o = 45°$, which from Eq. (3) subtends 50% of the atoms, and $\theta_o = 90°$, which contains all the atoms. (The 75% curve, $\theta_o = 60°$, lies close to the 90° curve.) For small energies θ_f varies as $E^{1/2}$, as expected from Eq. 7. The actual focusing power turns out to be slightly stronger than the approximate value given by Eq. (8). If we write

$$P = C \times \frac{\pi^2}{4} \mu \frac{B}{kT}, \qquad (16)$$

then the constant C is approximately 1.1 at small energy, increasing to 1.2 at $\varepsilon = 0.25$.

Thermal sources have a relatively broad spread of energy. Introducing the temperature-field energy ratio

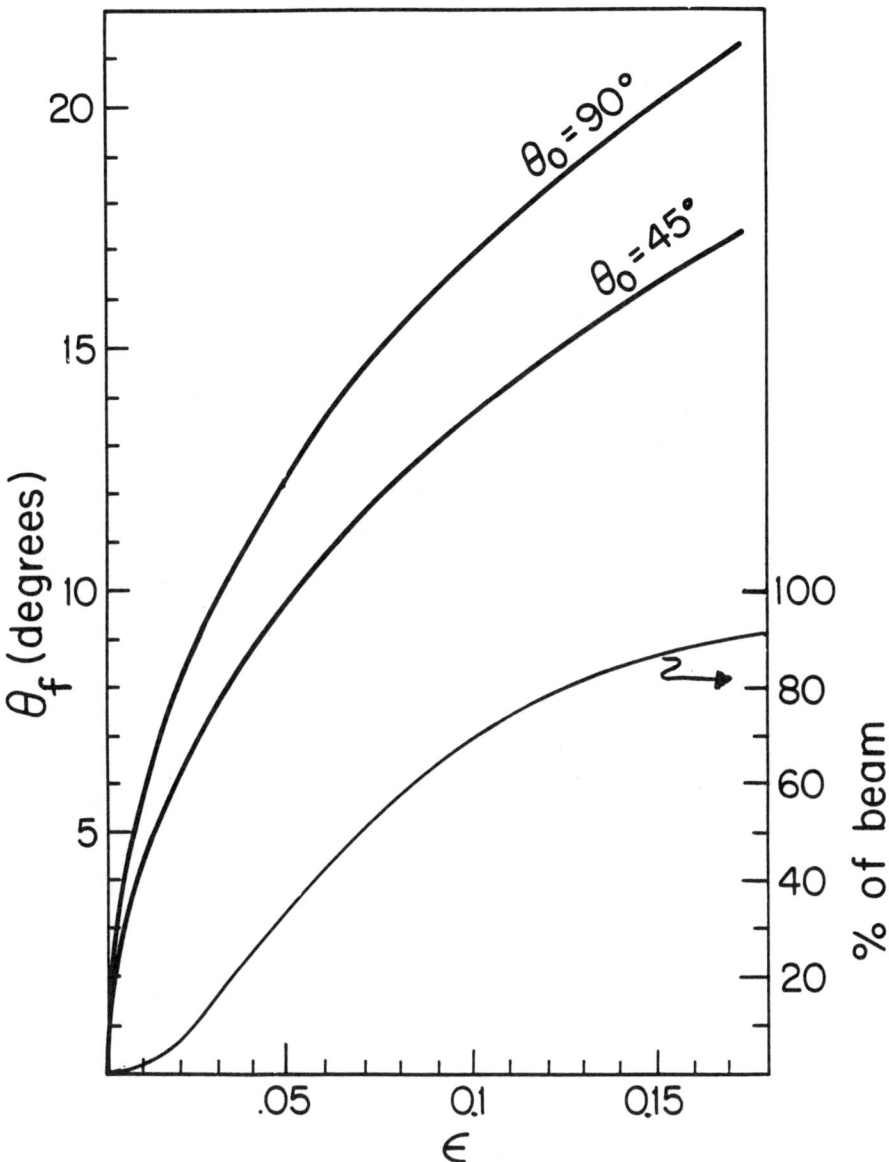

Fig. 3, heavy lines: emergence angle θ_f as a function of $\varepsilon = Mv_o^2/2\mu B$. Curve for $\theta_o = 45°$ subtends half the atom from the source, $\theta_o = 90°$ subtends all the atoms. Light lines: fraction of atoms having kinetic energy less than ε for T = 0.2K, B = 10T.

$$r = kT/\mu B \tag{17}$$

then from Eq. 15 it can be shown that the fraction of atoms with energy less than ε is

$$g(\varepsilon) = 1 - (1+\varepsilon/r)\exp(-\varepsilon/r) \tag{18}$$

Equation (18) is plotted in Fig. 3 for B = 7T and T = 0.2K (r=0.042). For example, the graphs indicate that all of the atoms with ε<0.1 lie within a cone given by θ_f = 16.5°, and half lie within a cone given by 13.5°. For the field-temperature ratio illustrated, 70% of the atoms have ε<0.1.

The simple solenoid behaves like a lens in that the trajectories in the far field appear to be straight lines eminating from an efective source psotion z_s located behind the actual source at z_s (see Fig. 2). However, the system has some "aberrations" since the location of the effective source position Z_o is a function of θ_o. This is shown in Fig. 4 for various values of the temperature field ratio r. For θ_o less than 60°, which includes 75% of the atoms, the aberrations is not severe: the spread in Z_o is about 10%.

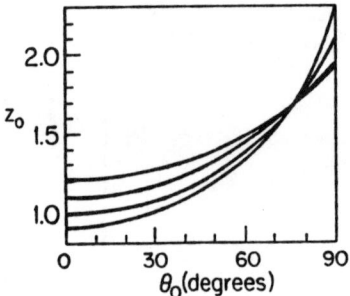

Fig. 4. Effective source location Z_o (see Fig. 2) as a function of θ_o, for ratios of ε = $Mv_o^2/2\mu B$. Values for ε are, in order of descending curves at left, epsilon = .04, .07, 1.4 and 2.6.

The focusing properties of a real solenoid source should be reasonably described by these results using a suitable choice for the effective radius. The possibility exists for shaping the field so as to enhance the collimation. The major gain probably lies in reducing the effective radius, R, that is, by changing the scale size of the optics. Denoting the maximum gradient of the solenoid by B'_{max}, then from Eq. (2)

$$R = B_o/2B'_{max} \tag{19}$$

For a fixed field intensity, the effective radius decreases inversely as the gradient.

FOCUSING BY A SHORT SOLENOID

Cold hydrogen atoms emerging from a solenoid in one of the upper hyperfine states are focused by the forward axial acceleration and the inward radial acceleration. If these atoms then enter a short solenoid they are retarded until they pass through the midplane, and then accelerated. The axial force defocuses the atoms as they enter and then refocuses them as they leave, with little net effect. The radial force, however, always focus the atoms. (This is because the force depends on the product of the gradient and the direction of the field; for the longitudinal field the gradient reverses while the direction stays fixed, but for the radial field both of these change sign.) As a result, a short solenoid can behave like a powerful lens.

The solenoid lens only operates in the region $r = Mv_o^2/2\mu B > 1$; otherwise the atoms are reflected. For large r the solenoid becomes an ideal lens of long focal length. However, in practice it is the short focal length region which is most useful.

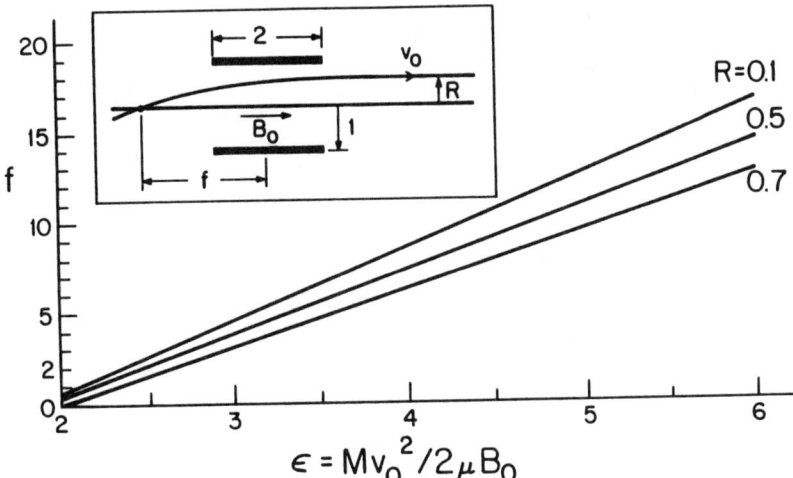

Fig. 5. Focal length (in units of solenoid radius) for a short solenoid, as a function of ϵ.

Figure 5 gives results for the focal length of a solenoid with length as a function of ε, for various value of the radius. If the atoms originate from an aperture in a solenoid at field B_s, then $\varepsilon \simeq B_s/B_o$, where B_o is the maximum field in the focusing solenoid.

For large values of ε one expects a linear relation between the focal length and ε by simple arguments based on the impulse approximation. It is unexpected, however, that a linear relation will hold even for small values of ε, but the plots in Fig. 5 indicate that such is the case, though the energy intercept is close to $\varepsilon = 2$, rather than to $\varepsilon = 0$ as would be predicted by the impulse approximation. Such behavior undoubtedly has a simple physical explanation; possibly the reason could be discovered by study of the orbit equation.

Figure 5 also reveals that the solenoid lens displays spatial aberration; the focal length decreases slightly with radius. Typically it varies less within 15% in the range $0<R<0.5$.

To form a highly collimated beam the solenoid lens would be matched to the source as shown in Fig. 1. If the largest radius to be used is R_m, then the incident angle to the solenoid is $\theta_s \simeq R_m/f$. Setting this equal to the largest angle from the source, θ_f, yields $f = R_m/\theta_f$. Taking, for instance, $R_m = 0.5$ and $\theta_f = 10°$ then $f = 2.9$. From Fig. 5 it follows that r is about 2.6. If the source magnet is 10T, the focuser field should be 3.3T. If f is taken to be about four times the source radius, then the source and focuser magnets would have about the same radius. (Note, however, to allow for orbits of radius R_m, the physical radius of the focuser must be at least half of the effective radius.) By increasing the strength of the focuser, or bringing it closer to the source, the beam could be refocused to a target.

In some circumstances it might be desirable to focus the atomic beam to a fan-shaped target, rather than to a circular shape. A number of magnetic configurations can accomplish this including a quadrupole formed by a pair of parallel conductors with currents in the same direction. The atomic beam would be perpendicular to the conductors, passing through the midpoint.

COMMENTS

This elementary treatment has dealt with the source solenoid and the focusing solenoid as independent elements. The orbit tracing technique outlined in the Appendix is quite suitable for dealing with a complete system of arbitrarily shaped cylindrically symmetric fields. In designing practical systems, one would want to undertake such a program, rather than to treat the individual magnets as isolated elements. Nevertheless, it is hoped that the results presented here provide some insight into the major design parameters, and the sort of performance one might expect.

The author wishes to thank Harald F. Hess for helpful conversations, Jean Michel Raimond for assistance with the computer and Thomas J. Greytak for many helpful comments on the manuscript. This work was carried out as part of a research program at M.I.T. supported by DOE, grant #De-AC02-76ER03069.

APPENDIX

Orbit Tracing in Cylindrically Symmetric Field

The Hamiltonian for hydrogen in a magnetic field is

$$H = ha(\bar{I}\cdot\bar{S}) - g_S \mu \bar{S}\cdot\bar{B} - g_I' \mu \bar{I}\cdot\bar{B} \tag{A1}$$

where μ is the Bohr magnetron and the remaining symbols have their conventional meaning. The second term dominates the energy in fields large compared to .05T. In this case the hyperfine coupling has no effect on the magnetic force and the Hamiltonian can be taken as

$$H = 2\mu \bar{S}\cdot\bar{B} \tag{A2}$$

For the $\underline{c}$ and $\underline{d}$ hyperfine states (see Fig. 1), $m_s = 1/2$, and the energy is

$$W = \mu |\vec{B}| \tag{A3}$$

The force is

$$F = -\mu \vec{\nabla} |\vec{B}| \tag{A4}$$

The field is assumed to have cylindrical symmetry.

$$\vec{B} = B_\rho \hat{\rho} + B_z \hat{z} . \tag{A5}$$

The components of the force are

$$F_\rho = -\mu \left[\frac{\partial B_\rho}{\partial \rho} \frac{B_\rho}{B} + \frac{\partial B_z}{\partial \rho} \frac{B_z}{B} \right] . \tag{A6}$$

$$F_z = -\mu \left[\frac{\partial B_\rho}{\partial z} \frac{B_\rho}{B} + \frac{\partial B_z}{\partial z} \frac{B_z}{B} \right] .$$

If the field is known everywhere on the z axis it can be calculated off axis as a power series in ρ. Designating the field on the axis at point z by B, and the gradients along the axis at that point by B', B", ..., then successive applications of $\nabla \cdot \bar{B} = 0$ and $\nabla \times \bar{B} = 0$ generate the following expansion through terms of order ρ^4.

$$\begin{aligned} B_z &= B - (\rho^2/4)B'' + (\rho^4/64)B^{iv} \\ B_\rho &= -(\rho/2)B' + (\rho^3/16)B''' \\ \frac{\partial B_z}{\partial z} &= B' - (\rho^2/4)B''' \\ \frac{\partial B_z}{\partial \rho} &= \frac{\partial B_\rho}{\partial z} = (\rho/2)B'' + (\rho^3/16)B^{iv} \\ \frac{\partial B_\rho}{\partial \rho} &= -(1/2)B' + (3/16)\rho^2 B''' \end{aligned} \tag{A7}$$

By introducing Eq. (A7) in (A6), the acceleration is obtained. These can be integrated to find the motion, or combined to obtain an orbit equation. The former method has the advantage of also yielding the kinetic energy, which gives a useful check on the accuracy of the integration, and of handling double-valued orbits which occur when an atom is reflected by potential barrier.

REFERENCES

1. I.F. Silvera, Proceedings, 16th International Conference on Low Temperature Physics, Physica 109 & 110B, 1499 (1982).

2. T.J. Greytak and D. Kleppner, École d'Été de Physique Théorique, Les Houches, France, 1982 (to be published).

3. R. Sprick, J.T.M. Walraven and I.F. Silvera, Phys. Rev. Lett. 51, 479 (1983); 51, 972(E) (1983).

4. H.F. Hess, D.A. Bell, G.P. Kochanski, R.W. Cline, T.J. Greytak and D. Kleppner, Phys. Rev. Lett. 51, 483 (1983).

5. T.O. Niinikoski, Proc. Int. Symp. on High-Energy Physics with Polarized Beams and Polarized Targets, Lausanne, 1980 (EXS 38, Birkhauser Verlag, Basle and Stuttgart, 1981), p. 191.

6. D. Kleppner, Polarized Proton Ion Sources, (Ann Arbor, 1981); ed., A.D. Krisch and A.T.M. Lin, American Institute of Physics Conference Proceedings No. 80, (1982), p. 111.

7. D. Kleppner and T.J. Greytak, High Energy Spin Physics, (Brookhaven 1982), ed., G.M. Bunce, American Institute of Physics Conference Proceedings, No. 95, (1983), p. 546.

8. N.F. Ramsey, Molecular Beams, Oxford, 1956.

9. Polarized Proton Ion Sources, (Ann Arbor, 1981); ed. A.D. Krisch and A.T.M. Lin, AIP Conference Proceedings No. 80, AIP N.Y. 1982.

10. High Energy Spin Physics, (Brookhaven 1982), ed., G.M. Bunce, AIP Conference Proceedings, No. 95, (1983), p. 546.

DISCUSSION

L.W. ANDERSON: Can you estimate how many atoms you could eject out of one of these targets and store in a room temperature target without atoms losing their polarization? What kind of fluxes of atoms could you get into a storage target that wouldn't have to be at these ultra low temperatures?

KLEPPNER: Well, we're not talking about just making a high throughput atomic beam, but I think throughputs of 10^{17} s^{-1} to 10^{18} s^{-1} should be feasible. But let me ask you, if you want to store atoms in a room temperature target what are you going to do with them?

L.W. ANDERSON: I'd like to try to use a low B field, to pass them through a low B field target if it's possible.

KLEPPNER: But it has to be open because you want to get things through it. I think that the density that you get in the target is not going to be larger than the density you get in the beam. You're not going to get a huge increase in density. If you have a target, it has to have tubes collimating the beam going in so that all atoms are going in one direction. They rattle around and the tubes help inhibit the flux coming out in the ratio of the pumping speed going in to the pumping speed going out. You will get an increase in density but it's very hard to make that ratio enormous, although it might help.

L.W. ANDERSON: Haeberli and the group at Wisconsin did store atoms with a substantially increased density.

HAEBERLI: An order of magnitude. But, as you say, it depends on gas conductances.

KLEPPNER: For instance, if you're sending in cold hydrogen, as soon as it warms up, the speed goes up and that works against you. What kind of target thickness are you talking about here? Are you talking about 10^{17}?

L.W. ANDERSON: In some cases, but in other cases as low as 10^{15} to 10^{16}.

KLEPPNER: The beams I'm talking about have densities of about 10^{14} and they are quite well collimated. If you want to use 30 cm or maybe even more, you really have quite a high target thickness there.

L.W. ANDERSON: You're saying that the nl is bigger in the beam than you might get in a target.

KLEPPNER: I think it can be big. The only difficulty, and it is also the reason why I didn't show drawings of one of Haeberli's ionizers, is that if you are going to send very fast ions down the beam they are going to end up back in our refrigerator. But in fact, with these focusers I'm sure you can put a catcher in the center so you really don't need to worry about that problem. I think that getting these densities in an external beam is possible and it seems to me that it makes life much much simpler, simply because the walls aren't there.

CORNELIUS: Aside from the problem of getting some power at 350 GHz, which corresponds to a 100 kG field, aren't you also near the ECR resonance again, in which case you'll turn all your hydrogen into a plasma when you put some power in there?

KLEPPNER: Well, first of all, the ECR resonance we're talking about is the cyclotron resonance in this field. The hyperfine splitting is fairly sizeable for these atoms, so that will take the spin frequency and the cyclotron frequency far from each other. I haven't worried about the problem, frankly. You've got to get some electrons in there to start off, and we don't want electrons in here. We might get them if one is doing Anderson's multiple scattering in there, which may lead to the production of some ions and electrons and all hell might break lose. I haven't worried about those problems yet.

STATUS AND FUTURE PLANS FOR THE BNL POLARIZED H⁻ SOURCE*

Th. Sluyters, J. Alessi and A. Kponou
Accelerator Department, Brookhaven National Laboratory
Associated Universities, Inc., Upton, NY 11973

In 1982, when Haeberli described[1] the design and performance of his 3 µA polarized negative hydrogen source, he predicted that the colinear colliding beam source had the potential to produce H⁻ beam currents well in excess of 10 µA. The recently constructed AGS source, which is similar to Haeberli's system, has reached peak beam currents in excess of 25 µA, while operating in the pulsed mode. Standard operation of the AGS machine is 10 µA in beam pulses of 0.5 ms each two seconds. These "intense" beams have been achieved by cooling the atomic beam from room temperature to 110°K and by increasing the cesium ion current from 2-3 mA to the 10-15 mA level. Higher polarized beam currents are expected with relatively simple modifications in the design.

The source design is shown in Figures 1 and 2. Polarized hydrogen atoms are produced in a ground state atomic beam source (Figure 1) consisting of an rf dissociator, sextupoles and rf transition cavities. Subsequently the atoms pick up an electron

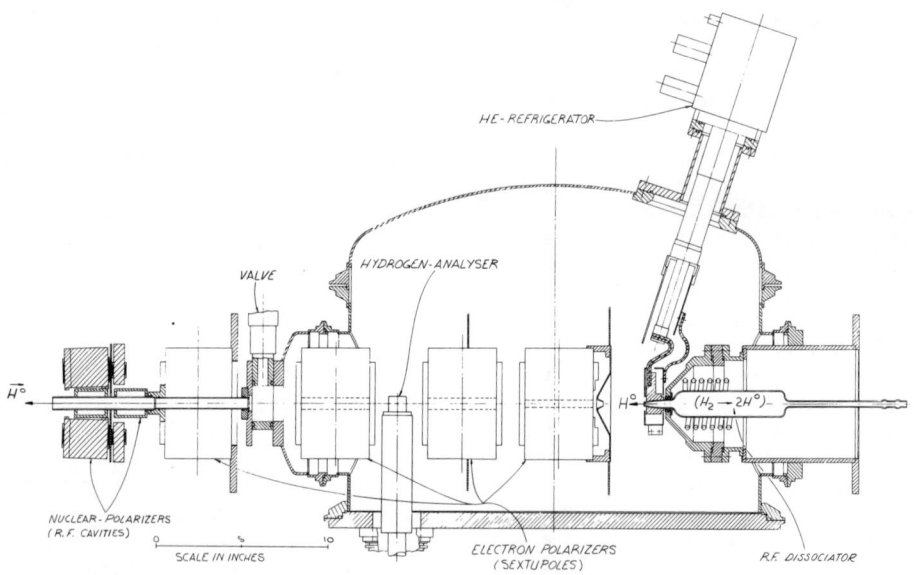

Figure 1 - The polarized atomic beam line.

---

*Work performed under the auspices of the U.S. Department of Energy.

in the interaction region by charge exchange with a 45 keV Cs°
beam moving in the opposite direction (Figure 2). The emerging
polarized H⁻ ions are accelerated to 20 keV and removed from the
source by a 90° electrostatic deflector.

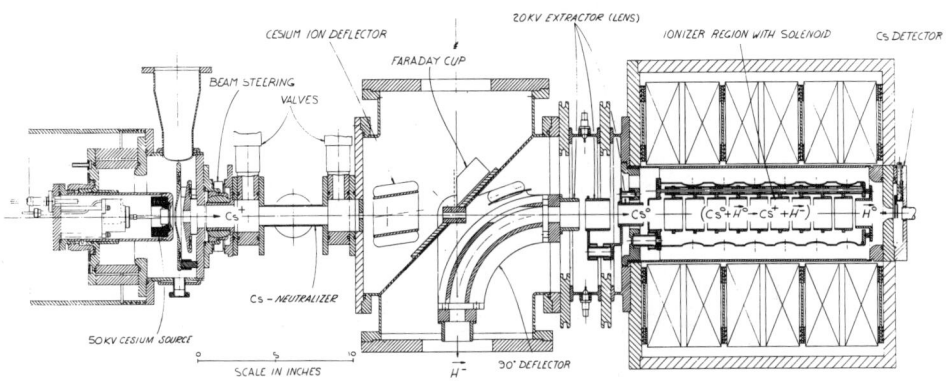

Figure 2 - The neutral cesium beam line with interaction
region and polarized H⁻ extraction system.

The first main difference between the BNL design and the
Wisconsin unit is that the BNL source is pulsed and has extra
cooling at the front end (nozzle) of the rf dissociator. The
cooling is by means of a closed cycle helium refrigerator. Velocity measurements revealed cooling of the atomic beam to only
110°K, limited by the design of the nozzle and the capacity of the
refrigerator. The density of the cold atomic beam is a factor of
two higher than a thermalized beam at room temperature. This
increase in density can be explained qualitatively by the ratio of
the atomic velocities $[v(300°)/v(110°) = 1.65]$ and the improved
acceptance angle of the first sextupole magnet.

The other improvement responsible for the larger polarized
beam currents is a new cesium source, whose design is based on
surface ionization using curved porous tungsten.[3] A unique feature of this source is the operation of the porous tungsten in the
pulsed mode rather than in the d.c. mode.[4] In between pulses
cesium accumulates on the tungsten emission surface, so that only
a low cesium flow rate (0.1 mg/hr) is required. This feature of
short pulsed sources minimizes problems associated with Cs coating of electrodes and insulators, loading on power supplies, etc.
It is possible to maintain a pulse shape with a relatively flat
portion of at least the required 0.5 millisecond each two seconds
(see Figure 3). The pulse shape is a sensitive function of the
ionizer temperature (about 1000°C) and the Cs coverage of the

emitting surface at the start of the pulse. This coverage, a
fraction of a monolayer, is adjusted by the temperature of the Cs
boiler, which is usually less than 100°C.

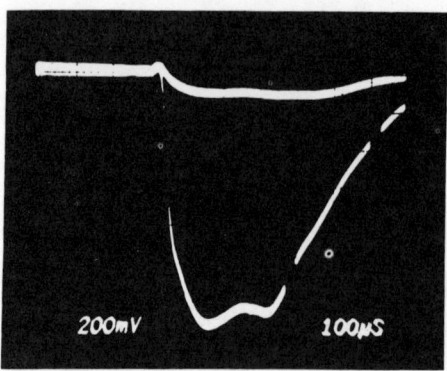

Figure 3 - The bottom trace is the polarized H⁻ beam.
The top trace represents the unpolarized
H⁻ beam, which has been recently reduced
to much less than 7.5%.

Figure 4 shows the layout of the three-electrode cesium
source. The 0.2 cm thick tungsten ionizer has a radius of 1.9 cm.
The extraction gap can be adjusted with the source in operation.
A mechanical swivel can steer the cesium beam into the collision
chamber, which is located about 1 meter from the source. Figure 5
shows the details of the electrode geometry as well as its voltage
pulsing scheme. This arrangement suppresses cesium ion emission
in between pulses from the cesium ion emitter and backstreaming
electrons from the beam line.

The emittance of the H⁻ beam was estimated in the vertical
plane with a simple, two-slit, emittance detector located 0.5 m
from the spherical inflector. The emittance for a 9 µA and 20 keV
beam is 8.9 mm.mrad $\sqrt{MeV}$, a value very close to the emittance
values measured more accurately by Haeberli for a 3 µA beam.[1]

The source is installed on the linac of the Alternating
Gradient Synchrotron. Routine operation will start soon.

Further increase in beam intensity can be expected a) with
increased cesium beam neutralization efficiency, which is presently only 40-50%, b) by more efficient cooling of the nozzle of the
dissociator, c) by tapered design of the first sextupole magnet,
and d) by improving the vacuum around the rf transition cavities.

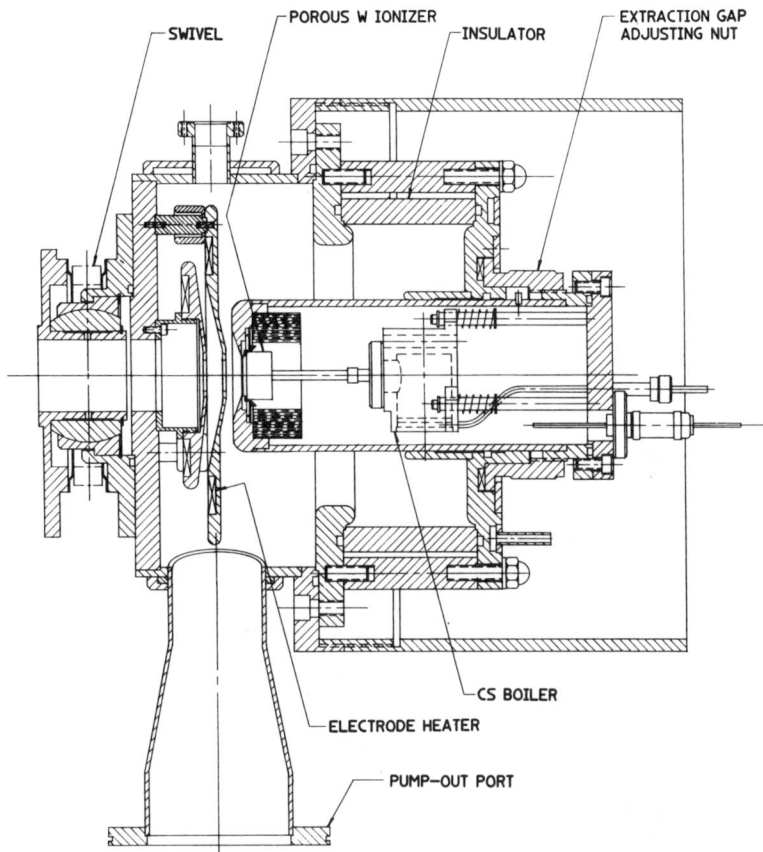

Figure 4 - The 10-15 mA cesium ion source.

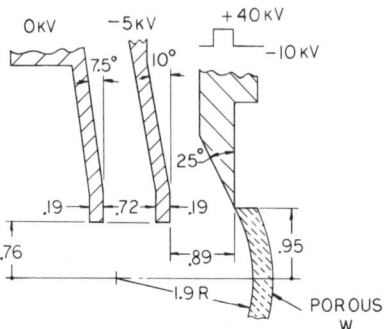

Figure 5 - The electrode geometry of the cesium ion source and its voltage pulsing scheme.

REFERENCES

1. W. Haeberli, et al., Nucl. Instr. and Meth. 196 (1982), 319.
2. P.F. Schultz, E.F. Parker and J.J. Madsen, in <u>Polarization Phenomena in Nuclear Physics</u>, 1980; eds. G.G. Ohlsen, et al., AIP Conf. Proc. No. <u>69</u>, Part 2 (Am. Inst. of Phys., NY, (1981), 909.
3. T. Wise and W. Haeberli, Ann Arbor, 1981, AIP Conf. Proceed. No. 80.
4. J. Alessi, Proc. Conf. on Low Energy Ions and Beams-3, Loughborough, U.K., 1983, to be published in Vacuum.

DISCUSSION

CORNELIUS: Dissociators are not 100% efficient in dissociating the hydrogen molecule. I would think that eventually you would form a freeze plug in your nozzle. Are you planning on doing anything to take care of that?

SLUYTERS: The point is that we are pulsing our system. If we calculate the amount of hydrogen we need, we can operate for a very long period before you start to have trouble with clogging the nozzle. In a steady state operation the lifetime will be indeed relatively short.

HAEBERLI: If you will permit me, I would like to ask a somewhat technical question about some of the things you said. When you describe the very nice improvement you got in intensity over the design we have been using at Wisconsin, you commented on the cesium ionizer being different in that it uses a porous tungsten ionizer and an adjustable gap. Now both of these features were present in our source as well, so I did not really quite understand in what way other than the pulsing, which is a fundamental difference, you may have made improvements in the cesium part.

SLUYTERS: There is not much difference between the Wisconsin and Brookhaven's cesium source hardware. We have installed this new pulsed cesium gun on our polarized source and obtained the improved performance. It is also a fact that the density of the cold atomic hydrogen beam of the BNL polarized ion source is at least two times higher than the density of the room temperature beam.

IONIZATION OF POLARIZED HYDROGEN ATOMS*

James G. Alessi
Accelerator Department, Brookhaven National Laboratory,
Associated Universities, Inc., Upton, NY 11973

ABSTRACT

Methods are discussed for the production of polarized H^- ions from polarized atoms produced in ground state atomic beam sources. Present day sources use ionizers of two basic types--electron ionizers for $\vec{H}^+$ production followed by double charge exchange in a vapor, or direct $\vec{H}^-$ production by charge exchange of H^0 with Cs^0. Both methods have ionization efficiencies of less than 0.5%. Ionization efficiencies in excess of 10% may be obtained in the future by the use of a plasma ionizer plus charge exchange in Cs or Sr vapor, or ionization by resonant charge exchange with a self-extracted D^- beam from a ring magnetron or HCD source.

I. INTRODUCTION

Presently, all polarized ion sources produce at some stage in the process polarized H^0 atoms, which must then be ionized to produce polarized H^+ or H^-. In some sources these polarized atoms are at high energies (550 eV in Lamb shift, 5 keV in optical pumping). Other "ground state" sources produce the polarized atoms at thermal energies (< .03 eV). In the ionization of the fast atoms in Lamb shift sources one is very restricted. One wants a "selective" ionization process, where metastable atoms are ionized with a much larger probability than the ionization of ground state hydrogen atoms. This is best accomplished for polarized H^- production by passing the beam through an argon gas cell, with an efficiency of $\vec{H}^0 \rightarrow \vec{H}^-$ of ~7%. In optical pumping, where one also has fast $\vec{H}^0$ atoms, the ionization is done by passing the beam through a vapor target (sodium). The ionization of thermal polarized atoms can be done in a variety of ways, and the ionization techniques continue to be improved. Only the application of various ionization techniques to these ground state atomic beam sources will be discussed in this paper. Other recent reviews of ground state ionizers have been given by Haeberli[1] and Grüebler[2].

The following section will give some general requirements of ionizers for ground state polarized ion sources, covering some things you can and cannot do when working with a polarized atomic beam. Then, in Section III, the first part of a two step process for $\vec{H}^-$ production, that being the formation of $\vec{H}^+$, will be discussed. This will include electron ionizers and "plasma" ionizers. Section IV will then cover the conversion of $\vec{H}^+$ to $\vec{H}^-$ by charge exchange in vapors and surface conversion. The direct conversion of $\vec{H}^0$ to $\vec{H}^-$ by charge exchange with a Cs^0 beam or D^-, will be discussed in Section V. Finally, Section VI will briefly summarize the existing methods and the possible future improvements.

---

*Work performed under the auspices of the U.S. Department of Energy.

II. GENERAL CONSIDERATIONS

Before discussing particular ionization techniques presently used on polarized ion sources, some basic information will be presented which might be of interest when considering whether an ionization scheme might or might not work. What are some considerations in determining the feasibility of an approach?

When one looks at the flux of polarized atoms produced by present ground state sources (a few x 10^{16} atoms/sec) and considering both future improvements due to cooling of the atomic beam and $\vec{H}^-$ intensities desired in the future (25 mA for the AGS, for example) one sees the need for as high an ionization efficiency as possible (at least until one has an ultracold atom source[3]). In addition, because the thermal atomic beam is achromatic and one cannot form a parallel beam, the ionizer should be as close as possible to the atomic beam section. Polarized atoms can be transported some distance by confining them with surfaces having a low recombination rate for hydrogen. For example, in one experiment, polarized atomic hydrogen from an atomic beam source traveled through a 10 cm long, 1 cm diameter teflon coated tube, into a teflon coated target vessel, where each atom experienced an estimated 900 wall collisions, and no depolarization was measured.[4]

The direction of polarization of the atoms will follow changing magnetic fields as long as the field changes are adiabatic, i.e., changes in the field have to be slow compared to the Larmor precession time of the magnetic moment in that magnetic field. Because the beam has thermal velocities, this condition is nearly always automatically satisfied. It is only when the field gets very close to zero, and the Larmor precession time in this field becomes long, that one may run into problems. One frequently encounters depolarization when the atomic beam travels in a region where a magnetic field passes through zero.

Depending on how the polarized atoms are prepared, ionization can occur in either a strong magnetic field (> 1.5 kG) or a weak "guide field" (a few gauss) which merely serves to keep the direction of polarization well defined. In strong field ionizers, the maximum final polarization depends on the strength of this field (P = 90, 95, and 97% at 1.0, 1.5, and 2.0 kG, respectively). To get P > 50% in a weak field ionizer, one must have a state selection magnet after an rf transition unit[5].

Once the ionization occurs, the polarized ions maintain the polarization direction they had at the point of ionization (i.e., the direction of B at the point of ionization). Therefore, the ionization must be done in a region of uniform field. After ionization, electric fields, and magnetic field parallel to the polarization direction will not change the direction of polarization of the atom, even though the direction of motion of the atom may change. Magnetic fields transverse to the direction of polarization will cause the polarization direction to precess by a well defined amount, and causes no basic problem unless different parts of the beam see different transverse B, causing variations in the polarization direction across the beam. This can occur, for example, in the fringing magnetic field at the exit of a strong field ionizer.

Loss of nuclear polarization during ionization or charge exchange is not a problem unless the process takes you through a stage where you have $\vec{H}^0$ without a strong magnetic field for a time comparable to the Larmor precession time. In experiments originally done to see if one could accelerate a $\vec{D}^-$ beam in a Tandem Van deGraaff, it was demonstrated that one could go from $\vec{D}^+$ to $\vec{D}^-$ and also from $\vec{D}^-$ to $\vec{D}^+$ without loss of polarization by passing the respective beams through thin carbon foils[6,7]. In both cases, there was no external magnetic field, but the transit times through the foil were short compared to the Larmor precession frequency. However, in the same experiments, there was a loss of polarization in going from $\vec{D}^-$ to $\vec{D}^+$ in a gas stripper, with no applied magnetic fields, because during the intermediate step the D^0 was depolarized. A gas stripper in a strong magnetic field (B > 1.5 kG) causes negligible depolarization.

Another consideration when discussing ionizers is a reduction in polarization of the extracted beam due to ionization and extraction of unpolarized hydrogen coming from background gas (molecular hydrogen coming from the atomic beam stage, water vapor, hydrocarbons). Typical atomic beam densities are only a few x 10^{11} cm^{-3} in the ionizer, so a background H_2 pressure in the 10^{-6} Torr range is not negligible. In addition, the ionization of background gas is aggravated if the volume over which one is ionizing is much larger than the extent of the atomic beam. Even in cases where the method of ionization does not readily produce H^+ from H_2, one has to worry about dissociation of H_2 on hot surfaces (filaments) and subsequent ionization of H. Problems will also occur in ionizers where there is recycling of gas between a discharge and metal walls. Therefore, one tries to make the ionizer as "open" a structure as possible.

A final point concerning "strong field" ionizers is an increase in the beam emittance due to the ionization in and subsequent extraction from the magnetic field. This growth in normalized emittance is[8]:

$$\Delta \varepsilon_N = \frac{e}{2mc} BR^2 \pi \text{ m-rad} \qquad (1)$$

where m is the mass of the particle, B the field in the ionizer, and R the radius of the beam, all in mks units. This is often a major contribution to the final beam emittance.

III. PRODUCTION OF POSITIVE IONS FROM POLARIZED ATOMS ($\vec{H}^0 \rightarrow \vec{H}^+$)

The efficiency of an electron ionizer, η, can be expressed as

$$\eta = \frac{I(\vec{H}^+)}{I(\vec{H}^0)} = 1 - e^{-(J_{e^-}\sigma l/ev_{\vec{H}^0})} \qquad (2)$$

where J_{e^-} is the electron current density, l is the ionizer length,

744

σ is the ionization cross section, and $v_{\vec{H}0}$ is the velocity of the polarized atoms. The cross section for ionization by electron impact has a maximum of 7×10^{-17} at ~ 60 v, and falls off approximately as $1/V$ at higher energies. However, since the space charge limited current of the electron beam increases as $V^{3/2}$, the ionization efficiency typically improves as the electron energy is increased.

Early ionizers used an electron beam in a weak magnetic field, and had poor efficiencies (~ 10^{-4}). Subsequently, by doing the ionization in a magnetic field of 1-2 kG, a higher ionization efficiency was obtained (a few x 10^{-3}). The field confines electrons, increases their effective path length in the ionizer, and along with suitably biased electrodes, allows multiple passes of the electrons through the ionizer. Further improvements in the design of these ionizers (contouring the magnetic field, improving the electrode configuration) has now raised the efficiency to about 5%. Figure 1 shows a schematic of an ANAC, Inc. "super ionizer"[9], and a

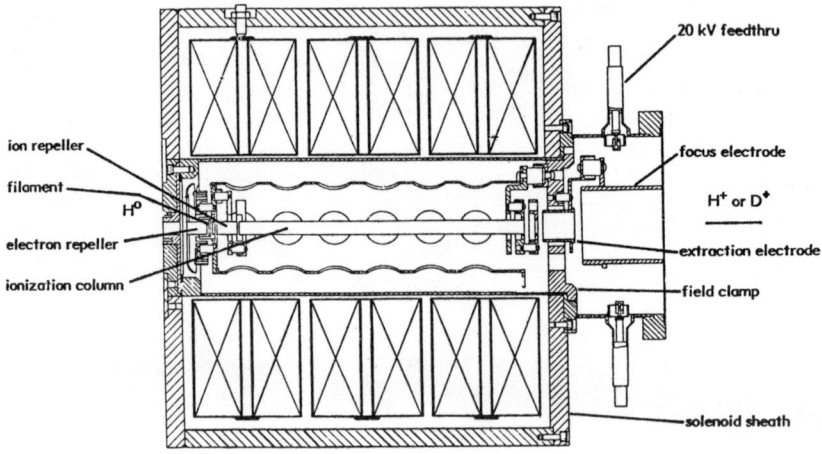

Fig. 1. The ANAC, Inc. "super ionizer."

similar ionizer in use at ETH, Zurich is described in another paper in these proceedings[10,11]. The electron current density in the ETH ionizer is several A/cm^2, the ionization length is 35 cm, and 120 μA of $\vec{H}^+$ has been extracted.

All the above mentioned ionizers use a hot cathode to supply electrons. At Bonn University, a Penning ionizer is used[12], in which the ionization is done in a superconducting solenoid with magnetic fields of up to 75 kG. A self sustaining discharge occurs at pressures of 10^{-7} to 10^{-6} torr. While an ionization efficiency of approximately 6% is obtained, the strong magnetic field causes the beam emittance to be very large.

It may be possible in some cases to improve the ionization by adding a "buffer" gas. While it is sometimes observed that as the background pressure in an electron ionizer increases the extracted current drops, this trend may reverse as one continues to increase the pressure. Recently, experiments were described in which the effect of buffer gases was studied.[13] Extracted hydrogen current was measured for a fixed background hydrogen pressure (2.2×10^{-4} Torr). Adding argon increased the extracted hydrogen by a factor of 40 due to ionization by plasma electrons, and gave an ionization efficiency of 8.3%. It is not clear, however, what one would actually gain over existing ionizers, since the ratio of extracted current to background gas density is similar to the ratio of extracted $\vec{H}^+$ to $\vec{H}^0$ density presently obtained in the ETH ionizer.

It has recently been suggested that $\vec{H}^0$ ionization using a hollow cathode discharge (HCD) plasma may be very effective[14]. Attractive features of the HCD are that it can operate in magnetic fields from a few hundred gauss to several kilogauss, it can produce a plasma of 1-2 cm^2 cross section and 10 cm to > 1 meter length, and with differential pumping it can operate at pressures down to ~ 10^{-5} Torr. One can easily get an electron current density of 10 A/cm^2, and the electron energy is under 100 volts. In addition, plasma densities from 10^{12}-10^{14} cm^{-3} can be obtained, depending on the background pressure. One could imagine feeding an HCD-produced plasma (argon, for example), into a configuration similar to an electron ionizer (l = 35 cm), and with a similar extraction geometry. Due to the increased ionization cross section at the operating energy, and the increased electron current density, the ionization efficiency calculated from equation 2 is greater than 50%. If the HCD were operated with deuterium, symmetric charge exchange between the polarized hydrogen atoms and the plasma D^+ ions would lead to a significantly higher ionization efficiency, while the extracted D^+ current would be no greater than the $\vec{H}^+$ current, and could be magnetically separated. Configurations using an HCD ionizer are presently being studied at BNL, and it is likely that a very high ionization efficiency can be obtained in a plasma of only 10 to 20 cm length. A possible difficulty is efficient ionization of H_2 coming from the atomic beam, resulting in a large unpolarized component in the beam.

The use of an rf discharge ion source to ionize polarized atoms has been tried in the past. In one case[15], polarized deuterium was injected into a pyrex system (coated with silane to reduce recombination). D^+ ions extracted from the rf discharge (with a helium buffer gas) were not polarized, and it appeared that most of these ions came from ionization of gas coming from the walls of the vessel. In another attempt[16], $\vec{D}^0$ was injected into a pyrex vessel, again with a helium buffer gas. The extracted beam was partially polarized when operated in a magnetic field of 350 gauss. More recently, it has been suggested that an electron cyclotron resonance (ECR) ion source[17] may be an efficient ionizer for polarized atoms. This source has a high electron density (10^{12} cm^{-3}), and electron energy of several keV. The ionization efficiency of this source is high, and as with the HCD, by operating with deuterium one could benefit from symmetric charge exchange. A concern with an ionizer of this type is that the similarity of the electron cyclotron resonance conditions to the conditions for inducing transitions between hyperfine levels in the hydrogen atom, could lead to strong depolarization[2]. It has been pointed out, however, that the conditions are close to those needed for 2 → 4 transitions in the Rabi diagram, and hence depolarization might be avoided by injecting a polarized atomic beam in a pure 1 state[18]. If depolarization in the ECR discharge is not a problem, a configuration as is used to produce H^-/D^- for fusion[19]--i.e., the H^+ extraction and charge exchange to H^- in Cs vapor, all immersed in a magnetic field--is essentially what would be required for a polarized H^- source.

Finally, ionization using a fast proton beam by the reaction $H^+ + \vec{H}^0 \rightarrow H^0 + \vec{H}^+$ has been tried[20], but the efficiency was only 0.1%. This was done using a 10 keV proton beam, decelerated to 4.7 keV in the interaction region, and with a 75 kG superconducting solenoid magnet to cause a spiraling of the proton beam and also to act as a mirror to reflect the beam.

IV. PRODUCTION OF NEGATIVE IONS FROM POLARIZED PROTONS ($\vec{H}^+ \rightarrow \vec{H}^-$)

In order to obtain the efficiency of producing $\vec{H}^-$ from $\vec{H}^0$, one must multiply the ionization efficiencies discussed in the previous section by the efficiency for subsequent conversion of $\vec{H}^+$ to $\vec{H}^-$. This double charge exchange is typically done by passing the beam through a vapor target. As mentioned in section II, this charge exchange must occur in a strong magnetic field to prevent depolarization.

Equilibrium negative ion yields have been measured for H or D in a variety of vapor targets[21,22]. One sees from these measurements, for example, that one could get ~ 7% H^- conversion efficiency at 5 keV in Na, ~ 20% at 1 keV in Rb, ~ 30% at 500 eV in Cs or Sr, and almost 50% ionization in Sr at 250 eV. The difficulty as one goes down in beam energy is that the $\vec{H}^+$ current entering the vapor target decreases due to beam optics considerations. Na is often used as the charge exchange vapor, at a beam energy of ~ 5 keV. The Na charge exchange cell used at ETH is shown in figure 2[23]. They obtain a conversion efficiency of $\vec{H}^+$ to $\vec{H}^-$ of 6%, and are limited by their $\vec{H}^+$ beam optics. Of course, one would like to gain a factor of ~ 5 by going to lower energies and using Sr or Cs vapor. A beam optical system such as that developed at Grenoble for neutral beam heating[19] might be applicable. By having a strong magnetic field all the way from the source to beyond the neutralizer, they improve the confinement and space charge neutralization of the positive ion beam. In this way, they have been able to get a conversion efficiency of 22% of 1 keV D^+ to D^-. On the other hand, if one can substantially increase the $\vec{H}^+$ current by improvements in both the atomic beam and in the $\vec{H}^0 \rightarrow \vec{H}^+$ ionizer, it will become even more difficult to decrease the beam energy to gain in $\vec{H}^+ \rightarrow \vec{H}^-$ conversion. At higher currents, where the $\vec{H}^+$ current through the vapor cell would be proportional to $V^{3/2}$, the product of the conversion efficiency (values given above) times $\vec{H}^+$ current into the cell ($\propto V^{3/2}$) would increase with increasing beam energy.

It was mentioned in section II that due to the short transit time during charge exchange, a thin foil could be used with no external magnetic field. The conversion efficiency of a 40 keV $\vec{H}^+$ beam to $\vec{H}^-$ in a 5 µg/cm^2 carbon foil was ~ 1%[24].

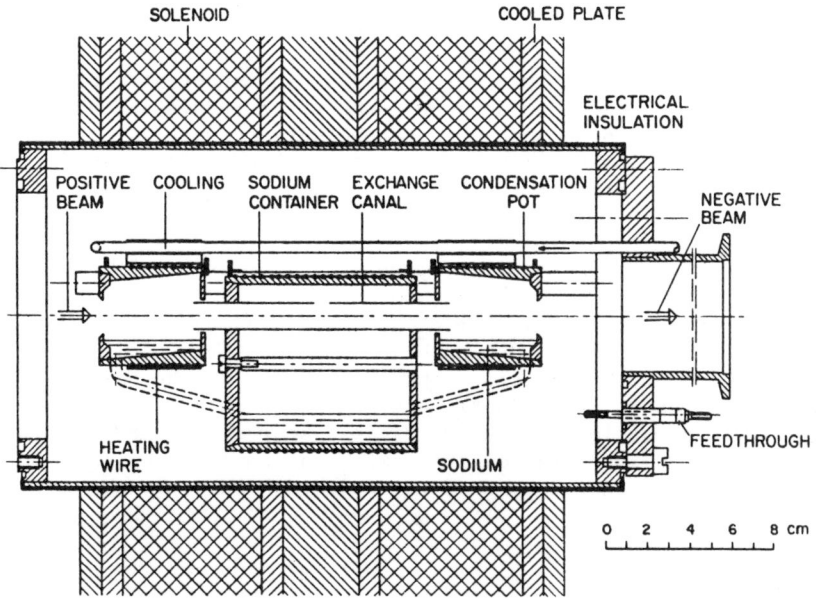

Fig. 2. The ETH Na charge exchange cell.

Another possibility may be surface conversion of $\vec{H}^+$ to $\vec{H}^-$. This process is often used in high current, unpolarized negative ion sources. At FOM, measurements of the conversion of H^+ to H^- in grazing incidence ($\theta > 80°$ with respect to normal) on cesiated monocrystalline W(110) surfaces have shown total conversion efficiencies of up to 37% at 200 eV[25]. At 1 keV, the conversion efficiency was 21%. The conversion efficiency in grazing incidence on polycrystalline W was 12% at 400 eV and 10% at 1 keV[26]. One essentially has specular reflection of the particles from the surface. The interaction time here is very short, and it seems likely that the polarization of an incident $\vec{H}^+$ beam would be preserved in this type of interaction. Some experiments which may be relevant in this regard have to do with the production of polarized Li^+ (and Na^+) by surface ionization of polarized atoms on a hot W-O surface. In one case the surface ionization occurred in a "strong" magnetic field, and only a slight depolarization was observed even though the hold-up times of the particles on the surface were many orders of magnitude larger than the Larmor period[27]. Polarization was lost when the magnetic field was turned off. In another experiment the polarization after surface ionization of polarized Li and Na was measured, again with an external magnetic field[28]. Here, however, depolarization was observed as the surface residence time increased (lower surface temperatures). While $\vec{H}^-$ surface production does not offer any higher efficiency at a given energy than what can be achieved with a vapor target, there may be some source configurations where surface conversion would be more suitable.

One might consider the combination of a plasma ionizer and surface conversion for producing polarized H⁻ ions in some "traditional" surface plasma source. However, it seems less likely that polarization would be preserved in normal incidence conversion occurring in surface-plasma sources, where desorption is an important process. In addition, one would need an open geometry to reduce recycling of gas from walls. Even if some such scheme were to work, it is unlikely that one would end up with a higher overall $\vec{H}^0 \to \vec{H}^-$ ionization efficiency than with some of the other configurations discussed in this paper.

V. PRODUCTION OF NEGATIVE IONS FROM POLARIZED ATOMS ($\vec{H}^0 \to \vec{H}^-$)

The reaction $X^0 + \vec{H}^0 \to X^+ + \vec{H}^-$ was suggested by Haeberli[29] as a way of directly producing $\vec{H}^-$ from $\vec{H}^0$. Subsequently, a polarized source based on this ionization scheme was built at the University of Wisconsin[30], and more recently a source with this same type ionizer has become operational at BNL[31]. These sources use Cs⁰ as the ionizing beam. A schematic of the configuration is shown in figure 3. Cs⁺ ions are produced by surface ionization on a porous tungsten

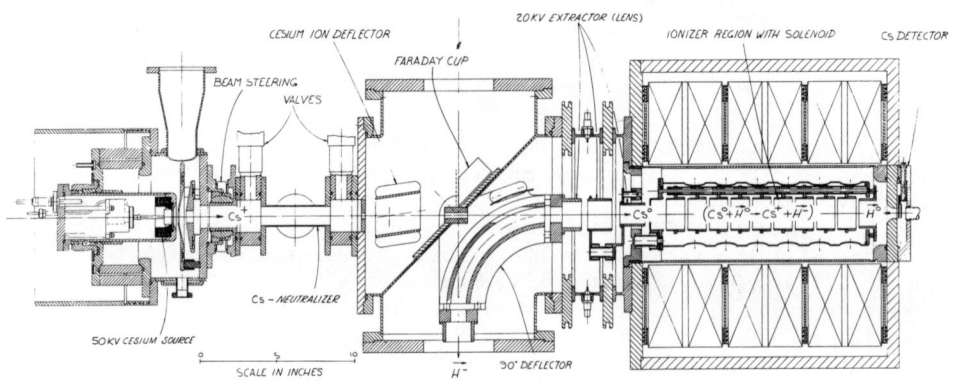

Fig. 3. The BNL Cs beam ionizer arrangement.

ionizer (1.9 cm diameter), and accelerated to 40-55 kV. The Cs beam is then neutralized in a Cs vapor neutralizer and travels into the interaction region, where it is collinear with the $\vec{H}^0$ beam. The interaction region is ~ 35 cm long, and has an axial magnetic field of 1.5-2 kG. $\vec{H}^-$ ions formed by charge transfer are accelerated by a weak axial electric field (< 1.5 v/cm) toward the extraction end, where the ions are accelerated to 20 keV, focused, and electrostatically deflected by 90°. A nice feature of this technique is that one has two neutral beams, and therefore no problems with space charge forces. The efficiency of this type ionizer is given by

$$\eta \approx J_{Cs} \sigma l / e v_{\vec{H}^0} \qquad (3)$$

where J_{Cs} is the Cs^0 current density in the ionization region, l is the ionizer length, and $v_{\vec{H}^0}$ is the atomic beam velocity.

The $\vec{H}^-$ current is proportional to the intensity of the Cs^0 beam overlapping the atomic beam in the interaction region. The atomic beam has a diameter of ~ 1 cm in this ionization region. With the neutralizer, electrostatic mirror, and $\vec{H}^-$ extraction electrodes in between, the distance from the exit of the Cs source to the far end of the interaction region turns out to be ~ 90 cm in the Wisconsin source and ~ 120 cm in the BNL source. One is left with fairly stringent optical requirements for the Cs beam (~ 1/2° convergence). At Wisconsin, where the source operates dc, they obtain 2.5 mA of neutral Cs within 1 cm diameter at the far end of the interaction region, and with an improved source, 10 mA Cs^0 in 1 cm diameter at an equivalent distance on a test stand. At BNL, on a test stand we typically measured 8-10 mA of Cs^+ on a 1.6 cm diameter Faraday Cup located 1.2 meters from the source. Since only a pulsed H^- beam is required for the AGS, the Cs beam from the porous emitter is pulsed. It has been found that when operating in this mode, the Cs intensity is limited by pulse width requirements[33].

From present day performance, $J_{Cs} \approx$ 5-10 mA/cm^2. The maximum cross section for the charge exchange reaction with Cs^0 is ~ 8 x 10^{-16} cm^2 [32]. Therefore, from eq. 3 one could expect an ionization efficiency of 0.5-1.0%. At Wisconsin, with their less intense Cs source and an uncooled atomic beam, the efficiency is about .1%, and at BNL, with the Cs neutralizer performance still unsatisfactory, the efficiency is roughly estimated to be ~ .3% (the atomic beam flux is uncertain). Unpolarized H^- coming from ionization of background H_2 hydrocarbons, water vapor, etc. is small, due to the small cross sections involved. At Wisconsin, the extracted H^- with the dissociator turned off is only 0.3% of the polarized current. At BNL, a somewhat larger H^- background is observed which seems to come from Cs beam hitting metal surfaces around the ionization region.

Improved ionization efficiency due to an increased Cs beam intensity should be possible. The Cs beam energy could be increased to improve the extractor optics (the charge transfer cross section drops fairly slowly). Focusing elements could be added between the Cs source and neutralizer (plasma lens or electrostatic quadrupoles). Also, one could increase the optical acceptance of the interaction region by decreasing the distance between the source and interaction region.

Use of the resonant charge exchange reaction $D^- + \vec{H}^0 \rightarrow D^0 + \vec{H}^-$ was originally suggested by Haeberli[29] as an attractive ionization technique due to its large cross section, and the availability of intense D^- sources. D^- rather than H^- is chosen as the ionizing beam because it allows one to later mass analyze and separate the $\vec{H}^-$ from the unpolarized D^- ions. The difficulty encountered in this approach is that space charge forces from the intense ionizing beam could cause the blowup of the $\vec{H}^-$ ions, which have thermal velocities. One solution which has been suggested is to confine the ions with a very strong magnetic field, but in this case the $\vec{H}^-$ emittance would become very large. Other possibilities for compensating the space charge are to inject (heavy) positive ions into the ionization region along with the D⁻, or merely to use properly biased electrodes at the ends of the ionization region to trap positive ions formed in the ionizer[34]. While no ionizer using the above reaction as yet exists, an experiment is presently beginning at BNL to test an ionizer based on a pulsed D⁻ magnetron source. The basic idea[35] is shown in figure 4.

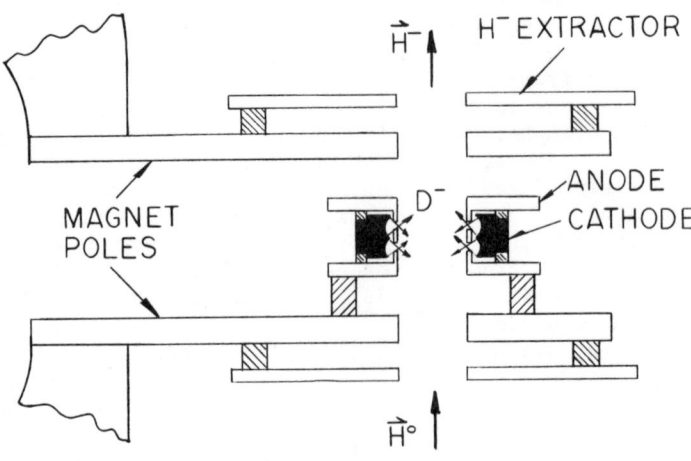

Fig. 4. The ring-magnetron ionizer under development at BNL.

One has a ring shaped magnetron source, with the anode on the inside, operating in a magnetic field of ~ 1.5 kG. D⁻ ions produced on the cathode emerge through slits in the anode, directed toward the center of the ring. They have an energy of ~ 200 eV (where the cross section for the charge exchange is 6×10^{-15} cm²). The polarized atoms pass through the center of this source, where they are ionized by charge exchange with D⁻. After extraction, the $\vec{H}^-$ and D⁻ ions will be magnetically separated. Because there is no extraction field inside the ring of the magnetron, plasma D⁺ ions can diffuse from the magnetron discharge and contribute to the space charge compensation of the D⁻ ions and any $\vec{H}^-$ ions formed. The pulsed magnetron source is designed to deliver 0.5 A of D⁻ ions to the center of the ring. In estimating the efficiency of this ionizer[35], destruction of $\vec{H}^-$ by collisions with D⁺, D⁰, charge exchange with fast D⁰, etc. were included. The overall ionization efficiency is estimated to be 8%, although scattering of $\vec{H}^-$ and $\vec{H}^0$ in the high pressure outside the magnetron could reduce the effective efficiency to ~ 1%. Based on present estimates, the conversion efficiency in an optimized magnetron design could be as high as 16% (neglecting gas scattering). An advantage of this ionizer is the short ionization length (1-2 cm). Because the polarized atomic beam is very achromatic, one should be able to get more atoms through this ionizer than what one can get through the much longer Cs beam or electron ionizers.

An alternative ionizer, which could operate steady state and at much lower pressures, would use an HCD D⁻ source³⁶ operating inside a solenoidal magnetic field. The $\vec{H}^0$ beam would then travel along the source slit and again charge exchange with 100-200 eV self-extracted D⁻ ions. If necessary, a gas such as krypton could be added to aid the space charge neutralization of the beam. By allowing the D⁻ ions to travel a short distance from the source before interacting with the $\vec{H}^0$ beam, and relying on Kr⁺ rather than D⁺ to provide space charge neutralization, the ionization efficiency would be improved because an important destruction process (mutual neutralization of $\vec{H}^-$ and D⁺) would be reduced.

VI. SUMMARY

The two methods presently producing the highest intensity H⁻ beams from ground state sources, an electron ionizer with Na double charge exchange, or charge exchange with a Cs beam, both have ionization efficiencies less than 0.5%. Neglecting improvements coming from reducing the velocity of the atomic beam, the efficiencies of both techniques can still be improved by careful development. However, somewhat radical variations of these two basic methods seem to offer more dramatic improvements. For example, an HCD plasma ionizer could improve the $\vec{H}^0 \rightarrow \vec{H}^+$ ionization efficiency by an order of magnitude. The use of Cs or Sr for $\vec{H}^+$ to $\vec{H}^-$ conversion in a system similar to that developed at Grenoble could give an additional improvement of at least a factor of 4. The combined effect would be an $\vec{H}^0 \rightarrow \vec{H}^-$ conversion efficiency of ~ 20%. Resonant charge exchange with a D⁻ beam in a ring magnetron type configuration could give efficiencies of 1-10%. With a D⁻ beam produced from an HCD based source, the efficiency could be greater than 20%, and in principle, offers the highest conversion efficiency of the techniques considered here. Thus, one can be optimistic about the prospects for improving ionizers for polarized H⁻ sources. However, much development and attention to details has gone into existing ionizers, and one must not underestimate the work involved in making a practical ionizer based on a new approach. In addition, improvements must be made in the atomic beam intensity in order to get > 10 mA of polarized H⁻.

REFERENCES

1. W. Haeberli, Proc. Polarized Proton Ion Sources, Ann Arbor, 1981, A. D. Krisch and A. T. M. Lin eds., AIP Conf. Proc. No. 80, p. 8.
2. W. Grüebler, Proc. Workshop on High Intensity Polarized Proton Sources, TRIUMF, Vancouver, 1983 (to be published in AIP Conf. Proc.).
3. D. Kleppner and T. J. Greytak, Proc. Symp. High Energy Spin Physics, Brookhaven, 1982, G. M. Bunce ed., AIP Conf. Proc. No. 95, p. 546.
4. M. D. Barker, G. Caskey, C. A. Gossett, W. Haeberli, D. G. Mavis, P. A. Quin, S. Riedhauser, J. Sowinski, and J. Ulbricht,

Proc. Fifth Int. Symp. on Polar. Phenom. in Nucl. Phys., Santa Fe, 1980, G. G. Ohlsen, R. E. Brown, N. Jarmie, W. W. McNaughton, and G. M. Hale eds., AIP Conf. Proc. No. 69, p. 931.
5. H. F. Glavish, Proc. 4th Int. Symp. on Polar. Phenom. in Nucl. Reactions, W. Grüebler and V. König eds., (Birkhäuser, Basel, 1976), p. 844.
6. W. Grüebler, W. Haeberli, and P. Schwandt, Phys. Rev. Lett. 12, 595 (1964).
7. W. Haeberli, W. Grüebler, P. Extermann, and P. Schwandt, Phys. Rev. Lett. 15, 267 (1965).
8. G. G. Ohlsen, J. L. McKibben, R. R. Stevens, Jr., and G. P. Lawrence, Nucl. Instr. Meth. 73, 45 (1969).
9. H. F. Glavish, IEEE Trans. Nucl. Sci., NS-26, 1517 (1979).
10. W. Grüebler and P. A. Schmelzbach (published in these proceedings).
11. P. A. Schmelzbach, W. Grüebler, V. König, and B. Jenny, Nucl. Instr. Meth. 186, 655 (1981).
12. A. Kruger, H.-G. Mathews, S. Penselin, and A. Weinig, Nucl. Instr. Meth. 138, 201 (1976).
13. J. Ishikawa, A. Motamed Ektessabi, and T. Takagi, Nucl. Instr. Meth. 207, 487 (1983).
14. K. Prelec, AGS H$^-$ Tech Note No. 75, Brookhaven Nat. Lab. (1983).
15. W. Grüebler, P. Schwandt, T. J. Yule, and W. Haeberli, Nucl. Instr. Meth. 41, 245 (1965).
16. M. Heyman, P. Delpierre, and R. Sene, Proc. 2nd Int. Symp. on Polar. Phenom. of Nucleons, Karlsruhe, 1965, P. Huber and H. Schopper eds. (Birkhäuser Verlag Basel, 1966) p. 97.
17. R. Geller, C. Jacquot, and P. Sermet, Proc. Symp. on the Production and Neutralization of Negative Hydrogen Ions and Beams, K. Prelec ed., Brookhaven, 1977, BNL 50727, p. 173.
18. W. Grüebler and P. A. Schmelzbach, Proc. Workshop on High Intensity Polarized Proton Sources, TRIUMF, Vancouver, 1983 (to be published in AIP Conf. Proc.).
19. M. Delaunay, J. L. Foucher, R. Geller, C. Jacquot, P. Ludwig, F. Mazhari, E. Ricard, J. C. Rocco, P. Sermet, F. Zadworny, Proc. 2nd Int. Symp. on Prod. and Neut. of Neg. Hydrogen Ions and Beams, Brookhaven, 1980, Th. Sluyters ed., BNL 51304, p. 255.
20. W. Hammon and A. Weinig, Nucl. Instr. Meth. 130, 23 (1975).
21. A. S. Schlachter, K. R. Stadler, and J. W. Stearns, Phys. Rev. A22, 2494 (1980).
22. R. H. McFarland, A. S. Schlachter, J. W. Stearns, B. Liu, and R. E. Olson, Phys. Rev. A26, 775 (1982).
23. W. Grüebler and P. A. Schmelzbach, Nucl. Instr. Meth. 212, 1 (1983).
24. W. Haeberli, Ann. Rev. Nucl. Sci., 17, 373 (1967).
25. J. N. M. van Wunnik, J. J. C. Geerlings, E. H. A. Granneman, and J. Los, Surface Sci., to be published.
26. P. J. M. van Bommel, J. J. C. Geerlings, J. N. M. van Wunnik, P. Massmann, E. H. A. Granneman, and J. Los, submitted for publication in J. Appl. Phys.
27. E. Steffens, W. Dreves, H. Ebinghaus, M. Köhne, F. Fielder, P. Egelhof, G. Engelhardt, D. Kassen, R. Schäfer, W. Weiss, and D. Fick, Nucl. Instr. Meth., 143, 409 (1977).

28. R. Böttger, B. Bauer, P. Egelhof, K.-H. Möbius, Z. Moroz, E. Steffens, G. Tungate, W. Dreves, I. Koenig, and D. Fick, Proc. Fifth Int. Symp. on Polar. Phenom. in Nucl. Phys., Santa Fe, 1980, G. G. Ohlsen, R. E. Brown, N. Jarmie, W. W. McNaughton, and G. M. Hale eds., AIP Conf. Proc. No. 69, p. 979.
29. W. Haeberli, Nucl. Instr. Meth. $\underline{62}$, 355 (1968).
30. W. Haeberli, M. D. Barker, C. A. Gossett, D. G. Mavis, P. A. Quin, J. Sowinski, T. Wise, Nucl. Instr. Meth. $\underline{196}$, 319 (1982).
31. Th. Sluyters, J. G. Alessi, and A. Kponou (published in these proceedings).
32. T. Nagata, J. Phys. Soc. Jpn. $\underline{48}$, 2068 (1980).
33. J. G. Alessi, Proc. Conf. Low Energy Ion Beams-3, Loughborough, U.K., 1983 (proceedings to be published in Vacuum).
34. H. F. Glavish, Proc. Workshop on Higher Energy Polarized Proton Beams, Ann Arbor, 1977, AIP Conf. Proc. No. 42, p. 47.
35. J. G. Alessi, Th. Sluyters, and A. Hershcovitch, Proc. Workshop on High Intensity Polarized Proton Sources, TRIUMF, Vancouver, 1983 (to be published in AIP Conf. Proc.).
36. A. Hershcovitch and K. Prelec, Rev. Sci. Instr. $\underline{52}$, 1459 (1981).

DISCUSSION

VERBEEK: You extract your polarized negative ions by a 90° deflection. What happens to the spin while you bend the beam? Does it stay in space or . . . ?
ALESSI: It stays in space.
VERBEEK: The other question is what do you want to have? Do you want to have the polarization transverse to the trajectory or . . .
ALESSI: You can make it whatever you want. As long as you have a well defined direction of polarization then, by putting a bend in, for instance, you can get the polarization direction proper. You can also put a magnetic field perpendicular to the spin to precess the polarization. So in different applications they may want the polarization direction different. That's no problem. If you have a polarized H⁻ beam you can change the direction of polarization.
FINK: In the fusion community, they're talking about maintaining a polarized plasma as a concept and shooting in polarized beams. If you could combine the polarized positive ion discharge in some way could you get negative ions by the volume process?
ALESSI: I think for the volume process to work you need to start with the molecules, and for the techniques I'm describing the polarization has to be done with the atoms.

THE PRODUCTION OF POLARIZED NEGATIVE ION BEAMS BY "COLLISIONAL PUMPING"*

L.W. Anderson
University of Wisconsin, Madison, Wisconsin 53706

S.N. Kaplan, R.V. Pyle, L. Ruby, A.S. Schlachter, and J.W. Stearns
Lawrence Berkeley Laboratory, University of California,
Berkeley, California 94720

ABSTRACT

The production of polarized negative ion beams by "collisional pumping" is described. Collisional pumping utilizes repeated charge changing collisions in a thick electron-spin-polarized gas or vapor target to form a polarized fast atom beam. The polarized fast atom beam is then partially converted into a polarized negative ion beam in a vapor target. Analysis is presented for a hydrogen beam passing through either a thick polarized H atom target or a thick polarized alkali target. Large polarizations and large currents may be possible.

INTRODUCTION

This paper describes the production of polarized hydrogen negative ion beams using repeated charge-changing collisions in a thick electron-spin polarized gas or vapor target to produce a polarized fast atom beam. We call this process "collisional pumping".[1,2] The fast polarized atoms are subsequently converted to polarized negative ions in a second target.

Polarized negative ions are useful for nuclear scattering experiments. Polarized negative ions may also be useful in the production by electron detachment of high-energy polarized atoms for injection into fusion reactors. Kulsrud et al.[3] have shown that polarized reacting particles in a fusion reactor can be used advantageously to modify either reaction rates or angular distributions of reaction products. One method of fueling and heating a reactor would be by the injection of multiampere (equivalent) neutral beams of nuclear polarized fast atoms.[4] We show that collisional pumping is a possible method of producing beams of intense nuclear-polarized negative ions.

---

*This work supported by U.S. Dept. of Energy, Office of High Energy and Nuclear Physics, Division of Nuclear Physics, under Grant No. DE-AC02-81ER4001 (U.W.) and by The director of Energy Research, Office of Fusion Energy, Applied Plasma Physics Division of the U. S. Dept. of Energy under Contract No. DE-AC03-76F0098 (LBL).

The production of polarized negative ions by collisional pumping requires a thick polarized gas or vapor target. There has recently been much progress on producing such targets. Cline et.al.[5] have demonstrated that it is possible to produce a cold dense polarized atomic hydrogen target in a large magnetic field using cryogenic methods. There are good prospects for producing a dense polarized atomic-hydrogen target in a low field by ejecting atoms from a low temperature high field target using rf transitions. Happer[6] and coworkers have proposed using spin-exchange optical pumping with high intensity lasers to produce a high flux of polarized atomic hydrogen. It may also be possible to produce dense polarized alkali-vapor targets using laser optical pumping.[7-10]

This paper shows that collisional pumping permits production of intense nuclear polarized fast negative-ion beams provided a dense polarized gas or vapor target is available.

COLLISIONAL PUMPING USING A HYDROGEN TARGET WITH $B<<B_c$.

The production of polarized negative ions using collisional pumping is accomplished in two steps. In the first step fast unpolarized H^+ ions or H^0 atoms are converted into fast polarized H^0 atoms using collision pumping; in the second step the polarized H^0 atoms are converted into polarized H^- ions. We first describe collisional pumping using a dense polarized target in a low magnetic field ($B<<B_c$). In order to make the discussion concrete consider beam of 5 keV H^+ ions incident on a polarized atomic-hydrogen target with a target electron spin polarization of 1. Following the capture of a polarized electron by H^+, the hyperfine interaction transfers some of the electron spin polarization into nuclear spin polarization. A subsequent electron-loss collision does not depolarize the nucleus. A sucession of electron capture and loss collisions "pumps" the nuclear polarization up to nearly the target electron-spin polarization, provided the collision frequency is less than the hyperfine frequency. We describe the composition of the fast-hydrogen beam as follows:

$$dH^+_{\frac{1}{2}}/d\pi = -\sigma_{+o}H^+_{\frac{1}{2}} + \sigma_{o+}(H^0_{11} + \tfrac{1}{2}H^0_{10} + \tfrac{1}{2}H^0_{00})$$

$$dH^+_{-\frac{1}{2}}/d\pi = -\sigma_{+o}H^+_{-\frac{1}{2}} + \sigma_{o+}(H^0_{1-1} + \tfrac{1}{2}H^0_{10} + \tfrac{1}{2}H^0_{00})$$

$$dH^0_{11}/d\pi = \sigma_{o+}H^0_{11} + \sigma_{+o}H^+_{\frac{1}{2}}$$

$$dH^0_{10}/d\pi = -\sigma_{o+}H^0_{10} + \sigma_{+o}H^+_{-\frac{1}{2}} \qquad (1)$$

$$dH^0_{1-1}/d\pi = -\sigma_{o+}H^0_{1-1}$$

$$dH^0_{00}/d\pi = -\sigma_{o+}H^0_{00} + \sigma_{+o}H^+_{-\frac{1}{2}}$$

where $\sigma_{+o} = 1.1 \times 10^{-15}$ cm^2 [11] and $\sigma_{o+} = 0.04 \times 10^{-15}$ cm^2 [12] are electron-capture and electron-loss cross sections at 5 keV; $H^+_{\frac{1}{2}}$ and $H^+_{-\frac{1}{2}}$ are fractional populations of spin-up and spin-down protons; $H^0_{Fm_F}$ are fractional populations of the atoms in the low-field atomic eigenstates of total angular momentum F and its projection in the magnetic field direction m_F; and π is the integral of the target atomic density over the path length. The H$^-$ fraction is negligible at 5 keV. For an unpolarized proton beam incident on the target the neutral fraction of the beam leaving the target is

$$f = H^0_{Fm_F} = f_\infty (1 - e^{-\tau}) \tag{2}$$

where $f_\infty = \sigma_{+o}/(\sigma_{+o} + \sigma_{o+})$ and $\tau = (\sigma_{+o} + \sigma_{o+})\pi$. The nuclear polarization of the neutral beam is

$$P = (H^0_{11} - H^0_{1-1})/f = 1 - f_\infty (e^{-\gamma_-\tau} - e^{-\gamma_+\tau})/[(\gamma_+ - \gamma_-)f] \tag{3}$$

where $\gamma_\pm = 1/2 \pm \sqrt{1-2f_\infty(1-f_\infty)}/2$. A calculation that considers a target that is not 100% electron-spin polarized and which includes some depolarization in the electron-capture process gives

$$P = P_t \frac{1-\epsilon}{1+\epsilon}[1 - f_\infty \frac{1-\epsilon}{2}(e^{-\gamma_-\tau} - e^{-\gamma_+\tau})/[(\gamma_+ - \gamma_-)f] \tag{4}$$

where ϵ is the fraction of electron-capture collisions that lead to nuclear depolarization, and P_t is the target polarization. For small ϵ the thick-target polarization $P \simeq P_t(1-2\epsilon)$.

Electron, and therefore nuclear, depolarization, occurs via radiative decay from n = 2 or higher atomic levels produced in the electron capture. We estimate that no more than 4% of the fast neutral atoms are produced in the n = 2 levels[12] for 5 keV protons incident on our atomic hydrogen target. Most of the capture is into the 2p level. Calculations show[13] that the electron spin polarization of atoms in the 2p level is reduced from 1.0 to 0.41 in decaying to the 1s level, if the atoms decay in a low magnetic field. Capture into higher n levels decreases as n^{-3} and can be neglected. Based on these considerations, we estimate that $\epsilon \leq 0.03$.

Polarized beams of deuterium or tritium atoms can also be produced using collisional pumping with D$^+$ or T$^+$ incident on a thick polarized atomic hydrogen target.

Figure 1 shows both the calculated neutral fraction and the nuclear polarization of the fast hydrogen atoms as a function of target thickness, for 5 keV unpolarized protons incident on a target with polarization P_t = 1.00 and depolarization factor ϵ = 0.03. Each fast particle makes 10-100 collisions in the polarized target; the production rate of polarized target atoms must be comparable.

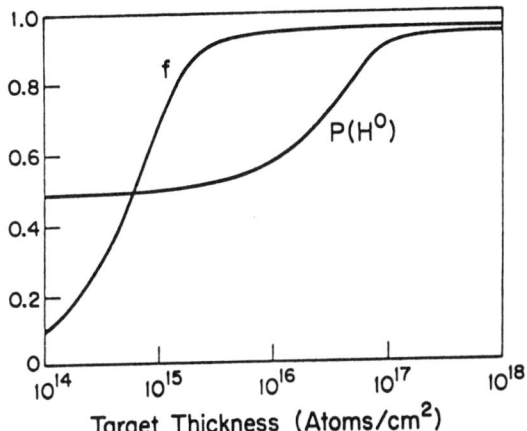

Fig. 1. Neutral fraction, f, and nuclear polarization of the neutral fraction, $P(H^0)$, as a function of target thickness for 5 keV H^+ incident on an atomic hydrogen target with electron spin polarization $P_t=1$. The polarized target is in a low magnetic field ($B<<B_c$).

COLLISIONAL PUMPING USING A HYDROGEN TARGET WITH $B>>B_c$

(A) Fast H^0 Incident

We consider an unpolarized beam of fast hydrogen atoms incident on a thick electron-spin polarized hydrogen target in a high magnetic field ($B \simeq 100kG$). The electron and nuclear angular momenta are decoupled in both the polarized target H atoms and in the fast atoms because of the large magnetic field, hence the nuclear spin plays no role. The fast H atoms will repeatedly lose and capture electrons during their passage through the target. The H atoms emerging from the target will have a high electron-spin polarization parallel to the target polarization. The equations governing this process are

$$dH^0_\alpha/d\pi = -\sigma_{o+}H^0_\alpha + P_t(1-\varepsilon)\sigma_{+o}H^+ + \frac{1}{2}\sigma_{+o}H^+ + \frac{1}{2}Q_t(1-\varepsilon)\sigma_{+o}H^+$$

$$dH^0_\beta/d\pi = -\sigma_{o+}H^0_\beta + \frac{1}{2}\sigma_{+o}H^+ - \frac{1}{2}Q_t(1-\varepsilon)\sigma_{+o}H^+ \quad (5)$$

$$dH^+/d\pi = -\sigma_{+o}H^+ + \sigma_{o+}(H^0_\alpha + H^0_\beta)$$

where H^0_α and H^0_β are the fractional populations of fast H^0 with electron spins parallel and antiparallel to the magnetic field; H^+ is the fraction of protons in the beam; P_t is the target electron-spin polarization; and $Q_t = 1-P_t$. Negative ions can be neglected. For the initial condition $H^0_{0\alpha} = H^0_{0\beta} = 1/2$, the neutral fraction is

$$f = H^0_\alpha + H^0_\beta = f^\infty + (1-f^\infty)e^{-\tau} \qquad (6)$$

The electron-spin polarization of the fast neutral atoms in the beam is

$$P_e = (H^0_\alpha - H^0_\beta)f = P_t(1-\varepsilon)[f^\infty + (1-f^\infty)e^{-\tau} - e^{(1-f^\infty)\tau}]/f. \qquad (7)$$

The depolarization parameter, ε, is expected to be much smaller than in the low-field case, because, in a 100 kG magnetic field, not only are I and J decoupled, but also L and S, in all n levels for atomic hydrogen. There is little depolarization as H atoms radiatively decay from n = 2 and higher levels to the n = 1 level, so $\varepsilon \simeq 0$. The electron spin polarization of the fast atoms can be converted into nuclear polarization by a Sona diabatic transition.[14] Figure 2 shows both the neutral fraction and the nuclear polarization after a Sona transition, as a function of target thickness, for 5 keV H^0 atoms incident on a target with polarization $P_t = 1.0$. The electron polarization of the fast beam prior to the Sona transition is the same as $P(H^0)$.

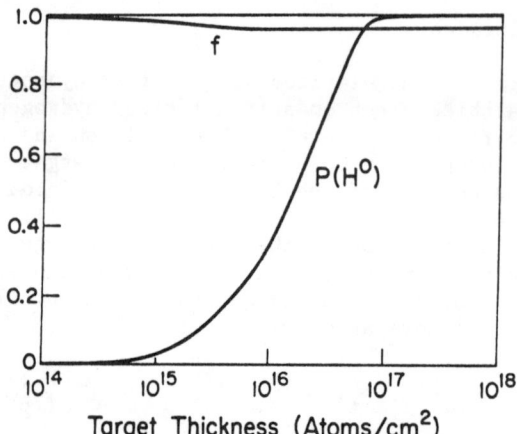

Fig. 2. Neutral fraction, f, and nuclear polarization after a Sona transition of the neutral fraction, $P(H^0)$, as a function of target thickness for 5 keV H^0 incident on an atomic hydrogen target with $P_t = 1$. The polarized target is in a high magnetic field ($B \simeq 100kG$).

The emittance of the fast beam is expected to increase only slightly as the beam passes through the high field target, because the particles entering and exiting the high magnetic field of the target are neutral and because the particle energy is high.

(B) Fast H^+ Incident

When 5 keV H^+ ions are incident on a thick polarized target the beam fractions are obtained using Eq. (5) with the initial condition that $H^+ = 1.0$ at $\pi = 0$. The neutral fraction is given by

$$f = f^\infty(1-e^{-\tau})$$

and the electron spin polarization of the fast neutral atoms in the beam is

$$P_e = P_t(1-\varepsilon).$$

Again ε is expected to be very small. The electron spin polarization can be converted into nuclear polarization by a Sona transition. Figure 3 shows the neutral beam fraction and the nuclear polarization after a Sona transition as a function of target thickness.

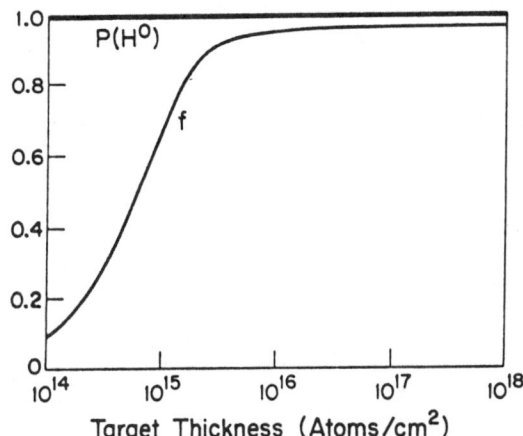

Fig. 3. Neutral fraction, f, and the nuclear polarization after a Sona transition of the neutral fraction, $P(H^o)$, as a function of the target thickness for 5 keV H^+ incident on an atomic hydrogen target with $P_t = 1$. The polarized target is in a high magnetic field (B $\simeq$ 100kG).

For this case with fast H^+ ions incident on a polarized target in high magnetic field, the ion source must be colinear with the target and in the same magnetic field to prevent unacceptable increase in the beam emittance due to charge exchange in a large magnetic field.

COLLISIONAL PUMPING USING AN ALKALI TARGET WITH $B<<B_c$

If a thick polarized alkali target with electron-spin polarization $P_t = 1$ can be produced for a low-energy H^+ beam entering the target, collisional pumping occurs, producing at equilibrium a beam that is primarily H^0 atoms in the $F = 1$, $m_F = 1$ state. This result comes about because, for low-energy hydrogen ions in an alkali vapor target the neutral atoms can only form negative ions, and, if a neutral H atom is in the 1,1 state, it cannot capture another electron from an electron-polarized target atom to form H^-. (The H^- ion exists only in a $1s^2$ state, where the electrons have oppositely directed spins.) In the limit of a thick target with an electron polarization of 1, the entire beam emerging from the target is in the 1,1 state.[2]

CONVERSION OF FAST POLARIZED ATOMS INTO NEGATIVE IONS

A nuclear-polarized fast neutral beam can be partially converted into fast polarized negative ions by charge-changing collisions in an alkali or alkaline earth vapor in a large magnetic field ($B>B_c$). If the target is very thick the negative-ion fraction is equal to the equilibrium negative fraction.[15] However this may not be the best method of converting the fast atom beam into negative ions. Following the production of fast polarized atoms with either a dense polarized atomic-hydrogen target in a low field, or a dense polarized alkali target in a low field, a higher negative-ion yield than the equilibrium yield can be obtained by using a polarized alkali target with an electron-spin polarization opposite to the electron spin-polarization of the fast polarized atomic hydrogen beam and with a target thickness chosen to maximize the negative ion yield.[2] This occurs because the H^- ion is a singlet state; the electron captured by the polarized atom must have spin opposite to the spin of the atom. Figure 4 shows the negative ion fraction produced when fast H^0 with electron-spin polarization 1.0 is incident at 5 keV on a Na target with electron-spin polarization -1.0. The use of a polarized target for conversion of the fast polarized H atom beam into negative ions is not useful after a Sona transition because the fast H^0 atoms in the beam have only nuclear polarization, not electron-spin polarization.

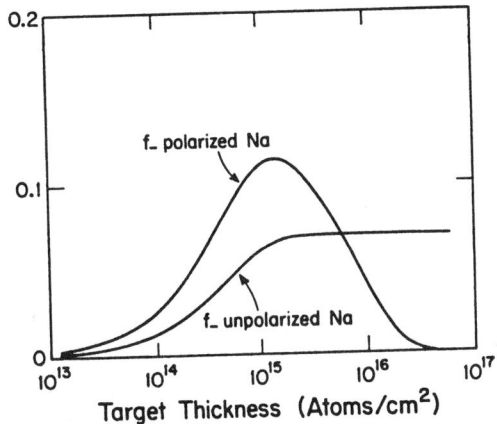

Fig. 4. Negative fraction, f_-, for fast H^0 with electron-spin polarization 1.0 incident on a Na target with electron spin polarization -1.0.

CONCLUSION

The production of an intense highly polarized fast negative ion beam should be possible using collisional pumping in thick polarized targets. Polarized hydrogen-atom gas with density greater than 10^{17} atoms/cm^3 electron-spin polarization greater than 99% has been produced.[5] It is also possible at the present time to produce intense H^+ or H^0 beams at 5 keV and higher energies. If thick electron-polarized H^0 or alkali targets can be produced at a rate of 10^{20} -10^{21} atoms per second, we anticipate that intense fast polarized negative ion beams can be made.

REFERENCES

1. L. W. Anderson, S. N. Kaplan, R. V. Pyle, L. Ruby, A. S. Schlachter, and J. W. Stearns, submitted for publication.
2. L. W. Anderson, S. N. Kaplan, R. V. Pyle, L. Ruby, A. S. Schlachter and J. W. Stearns, submitted for publication.
3. R.M. Kulsrud, H.P. Furth, E.J. Valeo, and M. Goldhaber, Phys. Rev. Lett. 49, 1248 (1982).
4. D. Post and R. Pyle, "Neutral Particle Beam Production and Injection," in Atomic and Molecular Processes in Controlled Thermonuclear Fusion, C. J. Joachain and D. E. Post, Editors, Plenum Press (1983).
5. R. W. Cline, T. J. Greytak and D. Kleppner, Phys. Rev. Lett. 47, 1195 (1981); R. W. Cline, D. A. Smith, T. J. Greytak, and D. Kleppner, Phys. Rev. Lett. 45, 2117 (1980); and D. Kleppner and T. J. Greytak, in Proceedings of

the 5th Intern. Conf. on High Energy Spin Physics, ed. by
G. M. Bunce (AIP Conf. Proc. No. 95, New York, 1983), p. 546.
6. W. Happer, in Proceedings of the Workshop on Polarized Proton
Ion Sources, Vancouver, 1983 (AIP Conf. Proc., to be
published).
7. L. W. Anderson, IEEE Trans. Nucl. Sci. NS-30, 1051 (1983);
L. W. Anderson, Nucl. Instr. Methods 167, 363 (1979);
L. W. Anderson and G. A. Nimmo, Phys. Rev. Lett. 42, 1520
(1979).
8. P. G. Pappas, R. A. Forber, W. W. Quivers, Jr., R. R. Desari,
M. S. Feld, and D. E. Murnick, Phys. Rev. Lett. 47, 236 (1981).
9. W. D. Cornelius, D. J. Taylor, R. L. York, and E. A. Hinds,
Phys. Rev. Lett. 49, 870 (1982).
10. Y. Mori, in Proceedings of the Workshop on Polarized Proton
Sources, Vancouver, 1983 (AIP Conf. Proc., to be published).
11. G. W. McClure, Phys. Rev. 148, 47 (1966); W. L. Fite,
R. F. Stebbings, D. G. Hummer, and R. T. Brackman,
Phys. Rev. 119, 663 (1960). 12. M. R. Flannery,
Phys. Rev. 183, 241 (1969); T. J. Morgan, J. Geddes, and
H. B. Gilbody, J. Phys. B 6, 2118 (1973); J. E. Bayfield,
Phys. Rev. 185, 105 (1969).
13. E. A. Hinds, W. D. Cornelius, and R. L. York,
Nucl. Instr. Methods 189, 599 (1981).
14. P. G. Sona, Energ. Nucl. (Milan) 14, 295 (1967).
15. A. S. Schlachter, Proc. of the 2nd International Symp. of the
Production and Neutralization of Neg. Hydrogen Ions and Beams,
Brookhaven Natl. Lab. Upton, New York, ed. by Th. Sluyters
(1980) pg. 42.

GENERATION OF INTENSE POLARIZED BEAMS BY
SELECTIVE NEUTRALIZATION OF NEGATIVE IONS*

A.I. Hershcovitch
Brookhaven National Laboratory, Upton, NY 11973

E.A. Hinds
Yale University, Dept. of Physics, New Haven, CT 06520

ABSTRACT

A novel scheme is proposed. This method is based on selective neutralization by laser of negative hydrogen ions in a magnetic field. This selectivity is based on the fact that the final state of the neutralized atom depends on nuclear polarization in the magnetic field. A two-scenario approach is to be followed: one in which the resulting neutral atom is in the ground state, and in the other the neutral atom is in the n = 2 level. Limiting factors are discussed. The main advantages of this scheme are the availability of multi-ampere negative ion sources and the possibility to neutralize negative ions with very high efficiency.

INTRODUCTION

There has been demand[1] for spin polarized protons for high energy research, and there may be need[2] for intense "high" energy neutral deuterium beams with polarized nuclei for fusion reactors.
All methods[3] presently being pursued start by either polarizing the electron of the hydrogen atom or by producing nuclear and electron spin polarized atomic gas. The next step is either to polarize the nuclear spin in one case or to eject a particular spin state from a gas in the other case. Finally, this nuclear polarized atomic beam has to be converted to either a positive or negative ion beam before it enters an accelerator. The major drawbacks of all present methods is the difficulty in producing a proper polarized atomic beam, which has both high density and high velocity (to avoid space charge effects and collisional destruction). In addition, the efficiency of converting polarized H_o to polarized H^- (or even H^+) is rather poor.

This novel approach to the production of H^- ions with polarized protons utilizes selective neutralization, by laser, of only one proton spin state of an H^- beam in a magnetic field, resulting in an H^- beam and an atomic hydrogen beam with polarized protons. The two beams can then be easily separated in a section with curved magnetic field. Figure 1 shows possible schematic arrangements which can be used. The advantages of this one-step

---

*Work performed under the auspices of the U.S. Department of Energy.

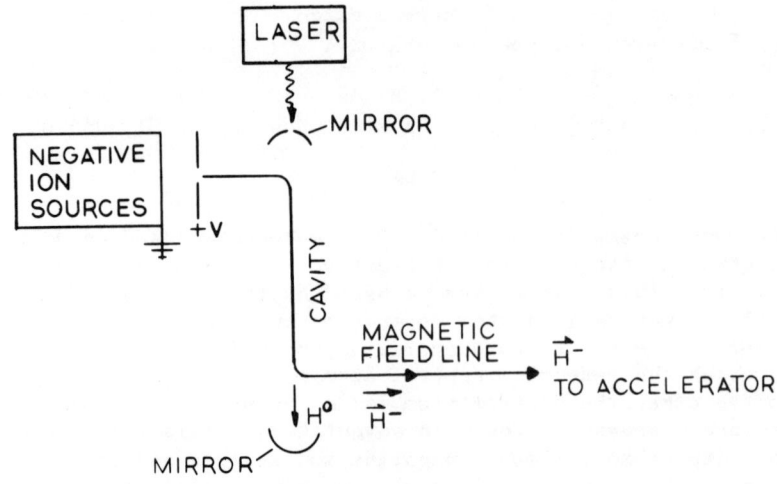

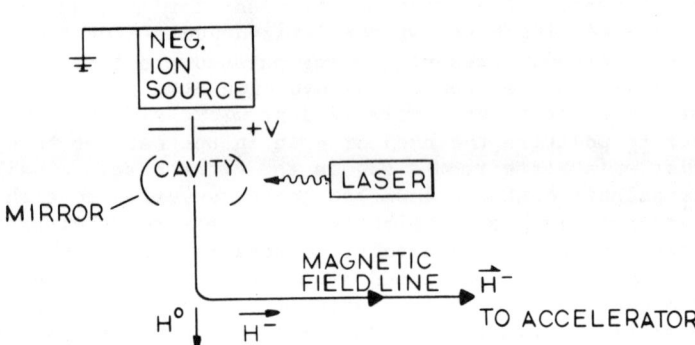

Figure 1. Schematic arrangements of selective neutralization systems. To minimize laser power optimization of interaction length (a), and/or cavity (b) can de done.

method are obvious: multiamperes of H^- beams have been produced both pulsed[4] and steady state.[5] Also, laser neutralization of H^- beams can be done with 100% efficiency.[6]

SELECTIVE NEUTRALIZATION

The concept is based on the following principle: when H^- ions are subjected to a uniform solenoidal magnetic field, one of the electrons in each ion will have its spin aligned in the direction of the magnetic field, while the other electron will have its spin in the opposite direction. The proton spin of each of the H^- ions will be aligned either with or opposite to the magnetic field. Since for magnetic fields below 10^8 Gauss, the H^- ion has only one bound state,[7,8] there will be two populations of H^- ions between which the sole difference is the orientation of the proton spin. The idea is to selectively neutralize only one population of H^- ions resulting in a population of H^- ions who have spin down and a population of H atoms with proton spin up. The negative ions can then be seperated from the neutrals with curvature in the magnetic field.

Selective neutralization can work if there are proton spin dependent states which are preferentially formed, i.e. selective neutralization of H^- ions will be the result of selective formation of H_o atoms. There are two possibilities which we would like to explore: formation of H^o atoms in the ground state and formation of excited hydrogen atoms (in the $n = 2$ level). The advantage of operating near the threshold (H^- atoms in the ground state) is the availability of lasers; however, the photodetachment cross section is low.[9] While the cross section for the formation of excited hydrogen atoms is eleven orders of magnitude higher,[10] lasers at that wavelength of 1135Å have short pulses.

Since more data exists for photodetachment of H^- ions which result in ground state H^o atoms, including experimental verification,[11] this case will be used in this paper to demonstrate how the scheme works. Although extensive quantitative data about photodetachment of H^- ions exists,[9,11] the effects of an external magnetic field were never incorporated like the work which was done for the photodetachment of S^- in the presence of a magnetic field.[12] The main difference between photodetachment in presence and absence of a magnetic field is the quantization of the energy of the free electron in a magnetic field (into Landau levels). But, due to the electron dipole magnetic moment μ and the external magnetic field B, there will be an energy shift from each Landau level of $\mu \cdot B$. Therefore, in each shifted level, there could be either an electron spin up from one Landau level, or an electron spin down from the Landau level above it, i.e. $\mu \cdot B$ is added to or subtracted from each Landau level. Since the lowest energy level in the ground state hydrogen atom is electron spin down and proton spin up, the lowest energy state of a detached H^- ion is an H_o atom with electron spin down, proton spin up, and a free electron spin up. Figure 2 shows combined energy levels of a stripped H^- ion in a magnetic field of a few hundreds Gauss, i.e., the added energy levels of the free electron and the H_o atom in the ground state. The Landau levels refer to the energy levels of a free electron in a magnetic field. These levels are

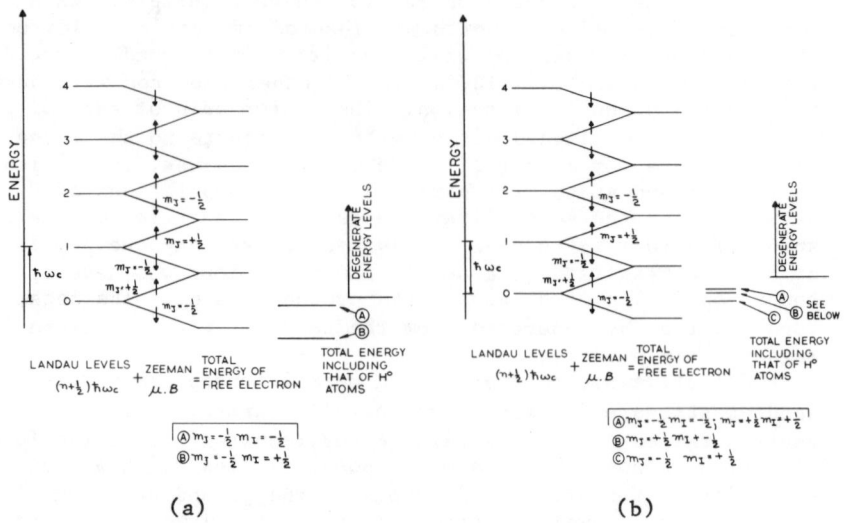

Figure 2. Energy levels of the stripped electron and H° atom for magnetic fields below (a) or above (b) the critical field.

quantized[12] perpendicular to the field. Please note that by degenerate levels it is meant that the levels are either truly degenerate or very closely spaced.

RESOLUTION

Qualitatively, in view of the above discussion, the possibility of selectively detaching the spin up electron from H⁻ ions with proton spin up exists. In frequency units, the energy difference between hydrogen atoms with proton spin down and proton spin up (with the electron spin down) is over 400 MHz for magnetic fields of a few hundred Gauss to about 1 KG. Therefore, a resolution of about 100 MHz should be sufficient. Present day commercial lasers have resolutions much better than 100 MHz.

Stark broadening would not be important for this scheme due to low density and thermal spread of the H⁻ beam. Since the laser line can be made arbitrarily narrow (for cw or "long pulses"), the effective Doppler broadening of the laser line as seen by H⁻ ions due to their thermal spread becomes the limiting factor.

In the absence of any acceleration, the Doppler width of a beam from a conventional H⁻ source is much larger than any hyperfine separation or splitting obtainable, e.g. at 1 eV the Doppler width is about 8.5×10^9 Hz. However, if the beam is accelerated, a phenomenon known as kinematic compression occurs. In simpler words, there is a reduction in Doppler width due to

Doppler shift, which can be easily compensated for by adjusting the laser frequency. The factor by which Doppler width is reduced[13] $R = 1/2(kT/eU)^{1/2}$, where T is the beam temperature and U is the accelerating potential. Thus, the reduced Doppler width becomes $\Delta\nu = \Delta\nu(0) R$. In the case of a typical BNL H⁻ source with an H⁻beam, whose thermal spread is about 4 eV, extracted at 20 kV, $\Delta\nu \sim$ 120 MHz. This Doppler broadening is already acceptable and it can easily be further reduced. Spectroscopists have been employing this effect to resolve hyperfine structure with an accuracy which exceeds theoretical calculations.[14] We plan to use a Penning source whose thermal spread is about 1 eV. The selective neutralization can then be done at beam energy of about 1 KeV.

LASER POWER NEEDED

For this scheme to be of interest, one proton spin orientation of H⁻ ions must be fully neutralized in order to achieve a very high degree of polarization. Therefore, the probability of photodetachment $p \gtrsim 1$ is required. The photon flux Γ(photons/cm²-sec) needed can be estimated from

$$p \simeq \sigma \Gamma \tau \qquad (1)$$

Where, σ is the photodetachment cross section and τ is the interaction time. At 1 KeV for an interaction region of about 4m, $\tau \sim 10^{-5}$ sec. At threshold, the photodetachment cross section when extrapolated from experimental[11] and theoretical results is only 10^{-24} cm². Solution of equation (1) yield $\Gamma \sim 10^{29}$ photons/cm²-sec. Assume a beam cross section $A \simeq 0.1$ cm², the required power P for 0.75 eV photons is

$$P = A\Gamma h\nu \simeq 10^{11} \text{ watt.} \qquad (2)$$

There is no hope in achieving this power level in the foreseen future. The problem in this case of a single photon absorption stems from low cross section near threshold while it peaks at about 1.5 eV photons. By contrast, the double photon absorption cross section, which requires the use of 3.2 micron laser, peaks at threshold. Since the H⁻ ion has a very high degree of dynamic polarizability, the cross section for double photon absorption must be high at threshold. It is presently being estimated.

An alternative option is to use 10.93 eV photons (i.e. 1135 Å laser), whose photodetachment products are a free electron and an H atom excited in the n = 2 level. At this photon energy, the cross section has a very sharp resonance whose magnitude is 1.4×10^{-15} cm². Using this value of cross section in equation 1 and a $h\nu$ of 10.93 eV in equation 2 the needed power becomes

$$P = 12.5 \text{ watt} \qquad (3)$$

Lyman Alpha lasers have achieved power levels of 100 watt,[15] for 10's of nsec. Utilization of such a laser in the configuration of Fig. 1b should work.

CONCLUSION

The scheme described in this paper is in principle feasible. Multiampere of negative ions have been produced and kinematic compression can be used to resolve hyperfine structure with an accuracy which exceeds theoretical calculations. For future work, a two scenario approach is to be followed: use a commercial 32000 Å laser if double photon absorption has the large cross section expected, i.e. leave the H^- in the $n = 1$ level. Or, use an available 1135 Å laser (with the resulting $H^°$ in the $n = 2$ level). The latter would limit future uses of this scheme to short pulses, unless free electron lasers become a reality in which case the sky is the limit.

REFERENCES

1. C. Bournely, E. Leader and J. Soffer. Physics Reports $\underline{59}$, 96 (1980).
2. R.M. Kulsrud, H.P. Furth, E.J. Valeo and M. Goldhaber. Phys. Rev. Lett. $\underline{49}$, 1248 (1982).
3. W. Haeberli. Polarized Ion Sources, AIP Conference Proceedings, No. 51, Particle and Fields Subseries No. 17, G.H. Thomas, Editor, p. 269 (1978). Also see paper by W. Haeberli in these proceedings.
4. Yu.I. Bel'chenko, V.G. Dudnikov. Proceedings XVth International Conference on Phenomena in Ionized Gases (ICPIG), Minss, USSR, July 1981.
5. K.N. Leung and K.W. Ehlers. Rev. Sci. Inst. $\underline{53}$, 803 (1982).
6. H.H. Fink and A.M. Frank. Lawrence Livermore Laboratory Report UCRL-16844 (1975).
7. H. Bethe and E. Salpeter. <u>Quantum Mechanics of One- and Two-Electron Atoms</u>, Academic Press, N.Y. 1975, p. 154.
8. R. Henry, R. O'Connell, E. Smith, G. Chammugam, and K. Rajagopal, Phys. Rev. $\underline{D9}$, 329 (1974).
9. M. Nascimento and W. Goddard III. Phys. Rev. $\underline{A16}$, 1559 (1977).
10. J. Broad and W. Reinhardt. Phys. Rev. $\underline{A1}$, 2159 (1976).
11. D. Feldmann, Z. Naturforsch $\underline{25A}$, 621 (1970).
12. W. Blumberg, W. Iano, and D. Larson. Physical Review $\underline{D19}$, 139 (1979).
13. S.L. Kaufman. Optics Communications $\underline{17}$, 309 (1976).
14. W. Wing, G. Ruff. W. Lamb, and S. Spezeski. Phys. Rev. Lett. $\underline{36}$, 1488 (1976).
15. R. Mahon, T.J. McDrath and D.W. Koopman. Appl. Phys. Lett. $\underline{33}$, 305 (1978).

OPTICALLY PUMPED POLARIZED H⁻ ION SOURCE

Y. Mori, K. Ikegami, A. Takagi, Z. Igarashi, S. Fukumoto
W. Cornelius* and R. York*
National Laboratory for High Energy Physics
* Los Alamos National Laboratory

ABSTRACT

Optically pumped polarized H⁻ ion source has been developed for the 12 GeV proton synchrotron at KEK. Sodium atoms are electron-spin polarized by two single frequency dye lasers. Protons, which are yielded by an ECR ion source, receive the polarized electrons from the sodium atoms and are polarized by conventional zero-crossing method. The hydrogen atoms become H⁻ ions by charge-exchange reaction with sodium vapor. Beam intensities and polarizations so far achieved are 15 µA with 60 % and 25 µA with 40 % at 350 keV. The intensity increases recently to 30 µA and the beam is accelerated to 20 MeV by the linac and injected in the 500 MeV booster synchrotron. The polarization is 45 ∿ 56 % at 20 MeV.

INTRODUCTION

An optically pumped polarized H⁻ ion source has been developed for the acceleration of a polarized proton beam in the 12 GeV synchrotron at KEK. The polarized ion source utilizes charge-exchange reactions between fast H⁺ ions and electron-spin polarized sodium atoms. The idea of this type polarized ion source was proposed by W. Haeberli[1] and L.W. Anderson[2] examined the possibility of using a

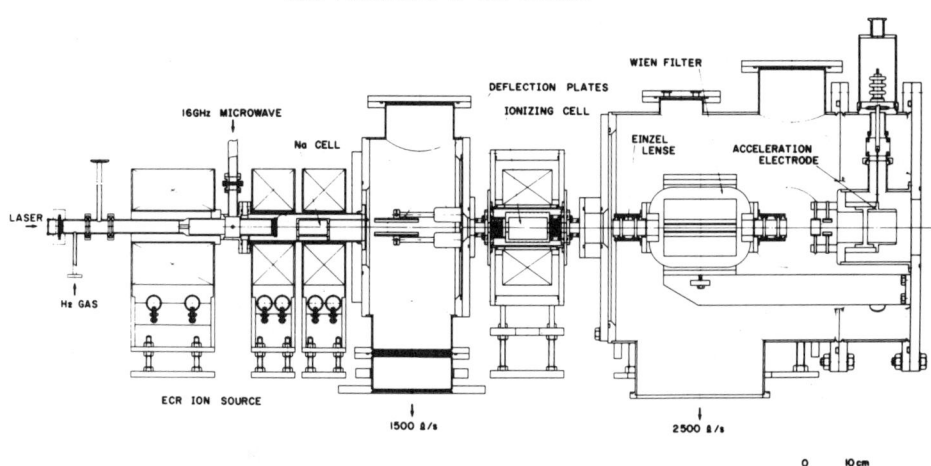

Fig. 1 Schematic-layout of optically pumped polarized H⁻ ion source.

dye laser. There are two difficulties to succeed this type polarized ion source: depolarization during the charge-exchange reaction and efficient optical pumping of sodium atoms without buffer gas. In order to overcome these difficulties, we use an ECR ion source and single frequency dye lasers.

Recently, the first beam acceleration test was performed in the 20 MeV linac and 500 MeV booster synchrotron. The polarized H^- ion beam was successfully accelerated to 20 MeV and injected in the booster synchrotron.

ION SOURCE AND BEAM ACCELERATION

Fig. 1 shows the schematic layout of the optically pumped polarized H^- ion source. The ion source consists of a 16.5 GHz ECR ion source, charge-exchange cell of optically pumped sodium atoms, ionizing cell and single frequency laser system. The laser system is shown in Fig. 2. Two single frequency dye lasers are used and the beam from each laser is combined and led to the ion source in the 750 kV high voltage dome. A new 16.5 GHz ECR ion source has been developed to produce an intense low energy (5 keV) H^+ beam at a high magnetic field of 1 T. The extracted H^+ ions capture polarized electrons from optically pumped sodium atoms and form electron polarized hydrogen atoms in the same magnetic field. The high magnetic field reduces the depolarization due to spin-orbit coupling of 2 P state of hydrogen atom. Calculations show that about 70 % of electron polarization of sodium atoms transfers to hydrogen atoms even if all hydrogen atoms are in 2 P state after the charge-exchange reactions. Electron-spin polarization of hydrogen atoms is transfered to nuclear polarization by a diabatic transition in a zero-crossing magnetic field. We have made some calculations about an optimum field gradient at the zero-crossing point and found that it should be less than 2 G/cm for reducing a noticeable depolarization.

Optical pumping was performed by using two sets of single frequency dye laser. Fig. 3 shows the typical experimental results of electron polarization of sodium atoms for various target thickness. At small target thickness of 1×10^{13} atoms/cm^2, the polarization achieved more than 90 % by two lasers, however, it decreased gradually by increasing the target thickness. Theoretical calculations agree with the experimental values at the small target thickness, however, they show large discrepancies at large target thickness. These discrepancies

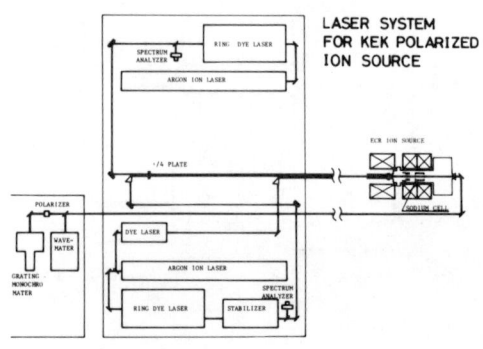

Fig. 2 Laser system for the polarized ion source.

might be explained by a radiation trapping process of optical pumping.

Measurements of nuclear polarization was initially performed by using p(^{6}Li, ^{3}He)α reaction at 350 keV and the H$^-$ beam of 15 μA of 60 % polarization and 25 μA of 40 % polarization were obtained[3].

By recent improvement for the ECR ion source, we could get 30 μA polarized H$^-$ ion current at the exit of the ion source.

In this October, we accelerated polarized H$^-$ ion beam to 750 keV and injected in the linac. Although magnets and others were not fully optimized, 750 keV, 3 μA beam was transported by a beam line of 40 m long and injected into the linac. One-third of the injected protons were accelerated up to 20 MeV.

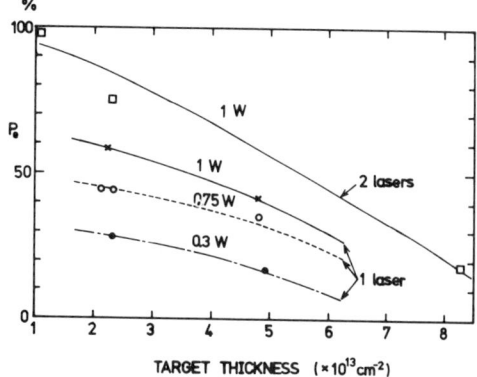

Fig. 3 Electron polarization of optically pumped sodium atoms.

Beam polarization at 20 MeV was measured by p-carbon elastic scattering and the polarization of 45 ∿ 56 ± 8 % was obtained.

CONCLUSION

The beam from the optically pumped polarized ion source was successfully accelerated up to 20 MeV by the linac. Observed polarization at 20 MeV was 45 ∿ 56 ± 8 %. It was proven that the optically pumped polarized ion source fundamentally works, and the first beam test was successfully finished. Obviously, further development work on the ion source is still needed. Especially, improvement for the ECR ion source is very important to increase the beam intensity and polarization.

We appreciate very much Profs. Nishikawa and Kamei for their encouragements and members of Accelerator Division for their operation and tuning of the synchrotron.

REFERENCES

(1) W. Haeberli: Proc. 2nd Int. Symp. on Polarization Phenomena in Nuclear Reactions, 64 (1966).
(2) L.W. Anderson: Nucl. Inst. Meth., 167, 363 (1979).
(3) Y. Mori et al.: to be published in Proc. 2nd Workshop Intense polarized Ion Source, TRIUMF, 1983.

CONCLUDING REMARKS

K. Ehlers

I look at the clock and I see that we are ahead of schedule and that allows me to talk perhaps a little bit longer. By the same token, I will make certain that I don't talk too long because I have a plane to catch along with the rest of you.

Before about 1951 there was essentially no interest at all on the part of the physics community in negative hydrogen ions. In 1951 Louis Alvarez proposed the swindletron system, this is what we called it then: you start with a negative ion at ground potential and you accelerate it, strip it, and then reaccelerate it back to ground and you gain a factor of 2 in energy or twice the power supply potential. In a sense that sort of turned things around. For example, shortly thereafter the isochronous cyclotron came along, a cyclotron where B goes up with the radius instead of down and focusing is done by other means. Shortly thereafter they found that the orbits at the outer edge of the cyclotron were so close together that it was almost impossible to slip a septum in between to extract the beam. But they realized that by accelerating negative ions you could put in a thin foil, strip the ions from negative to positive, and they would come out all by themselves: a remarkable new development and it's used today in the medical isotope industry. The ring accelerators came along. They were finding it rather difficult to inject positive ions into these machines and a contribution by our friend Dimov and his boss Budker, namely the charge exchange injection method, was proposed: you start with negative ions, bring them in with a couple of bump magnets, strip them, and bring them into the closed equilibrium orbit. This has taken over and I think most all the new ring accelerators from here on out are going to be using negative ions and not positive ions. Neutral beam users, of course, have a great interest in negative ions and that's why I'm here as well a goodly number of us. Star Wars has come along with a possible interest in negative ions. Polarized proton beams are becoming of interest again and it's rather remarkable that for the first time I think in our history of this symposium, we have representatives from all of the four groups that I just mentioned, gathered under one roof.

I remember our first meeting six years ago, in 1977, when this series of symposia first started. I came as an observer because I really wasn't in the negative ion game. The rules of that particular symposium were that if you didn't participate or didn't talk you didn't get to attend. So because I was there, I got the job of delivering the summary remarks and I have not been able to escape that chore since. Our second meeting, three years ago, in 1980, was a very good meeting as was this meeting. I can't say that we've had any meeting that surpassed all the data and information that has come through this particular meeting. We now have to think about the next meeting in 1986. The chairman of Brookhaven's Accelerator Department did say that he would like to see it here at BNL. He didn't really use those words but he did say, "the next meeting here at Brookhaven in 1986." We've got to be sure that happens. This series of symposia should not stop at this point and I think it's the duty of all of us to be certain that in 1986, be it the fall or some other time, that the next meeting be held. It's possible we might want to change the time to a point when there aren't so many meetings. Many of us have been on the road for three weeks now. But we've got to make sure that this series of symposia continues. Krsto tells me that even today, Brookhaven is receiving requests for the proceedings of the first symposium. Those proceedings are pretty valuable and if you look through the literature they are very well quoted.

Well, going backward then through this symposium with a summary, today as you know, we heard a very excellent presentation of activity in the field of polarized protons. We've seen some new ideas and I really think most of us here have been listening with interest because fusion has an interest in polarized protons. Of course we have to change that current from 10 microamperes to 1 ampere at least, to be of great interest. That's a big extrapolation but we'll be watching this field very closely. I must say that I am not a polarized proton man though I was associated with the field 20 years ago or so when the 80" cyclotron first wanted polarized proton beams. I worked in the field long enough to feel that I almost understood it. That is, I almost understood it until somebody showed me a deuterium chart and I completely gave up at that point. It's a very difficult field and I commend those of you who work in it.

We had a good session on neutralizers, accelerators, and systems yesterday. We now find that people are actually underway building various types of accelerators and this is important. We're going to need a lot of work in this field because before too long, we feel we may have the source problems solved, at least to the point where we can go ahead with the next step. I keep reminding myself that three years ago I wasn't really interested in hearing too much about accelerators. I didn't want to hear anything at all about neutralizers because I was too engrossed in trying to figure out how to make a source work. Why worry about these steps that are so far down the line that if you don't make the source work they're not important. But this year I am actually of a mind where I was interested in listening to the neutralizer papers.

We had papers on TFF or the transverse field focusing, the RFQ, the MEQALAC for both DC and RF type of operation, and each of these accelerators offers a good bit for a specific type of a source geometry and I'm very glad to see developmental activity on all. We had some poster sessions. Now, that is a new item for this series of symposia. We had about 27 posters. I had a paper myself so I didn't get to see them all but the few I did get to see were of extremely high quality and I think it's a very nice feature to add to this symposium. The fact that I didn't get to see the remaining posters doesn't really worry me too much because each of those posters will be written up in the proceedings and Krsto tells me, those proceedings very likely will be out in something like four months.

We had our usual great games with fundamental processes. Jim Peterson reported to us the finding of a new molecular helium negative ion, highly improbable and it's made from an equally highly improbable molecular helium plus ion. He tells me he's going to publish this and I really worry about that because the minute he does, some reactor designer is going to decide that if he had 100 amperes of these, it would be exactly what he needs to make his machine break even.

Tom Morgan and Fred Schlachter told us that they now have all the alkaline earth physics wrapped up. I think that's real nice, but I hasten to add that we threw that process out about three years ago, so it's real nice physics but it's damn poor engineering. However, I must admit I remember talking with Bob McFarland back in Berkeley when he was working with strontium and I remember telling him he wasn't really doing us any favors finding 50% in the negative ion fraction at 200 volts and that if he really wanted to make his mark in the world he should get us 50% at 30 kilovolts and then we'll be interested in listening. But it was nice physics.

Dr. Wadehra came in with a paper showing that if we vibrationally excite hydrogen molecules up to a sizeable level, maybe 6 or so, that when we make the H^- ions by dissociative attachment, the ions have a very low energy, like 0.4 eV maximum. Now, that's the sort of number we've been finding experimentally and I think that's real great. At least he used the right formulas.

I was very impressed with Arnold Karo's pac-man presentation of vibrational excitation and it's probably the first time I've had a real good feel as to what vibrational excitation is all about. I just wish he had a few LED's running around there indicating where the electron was at the time. But our theory session was very good and I think the work for the theory people is going to be well staked out for the next three years.

So we come down to sources. Our remaining efforts in double charge exchange throughout the world are the work at Grenoble--we had a report on that--and the work in Russia by Semashko. I know very little about what's going on in the latter since Afganistan. My last information was that they were having difficulty sorting the electrons out of the accelerated beam. Needless to say, in the negative ion game, electrons are not our friend and they're all over the place. I hasten to add that I hope the double charge exchange work throughout the world does keep up because we should have some effort in all three of the methods by which we generate negative ions, otherwise we may miss some little effect that may pull us through.

Surface production has certainly come of age. Needless to say, it's the only method thus far demonstrated where we can make DC beams of H^- ions above 1 A. It's got its problems as do all the methods but it looks like a viable candidate and a good bit of work remains to be done.

I saved volume to the last because I think volume turned out to be the highlight of this meeting. You might ask just what's happened in volume, why has it become of great interest again? After all, when I look back, I generated 45 or 50 mA/cm^2 of volume generated H^- ions over 20 years ago. That particular source is still in use today and our friends from Triumph still use it, as I recall, in their Triumph variable energy cyclotron. Well, twenty years ago and even today I found absolutely no way to extrapolate that source from the 4 or 5 mA region up to the ampere region. It had several basic things wrong. First, it required an extremely high pressure, well in excess of 100 millitorr, and that's a long way from the pressure you could allow to run a large area system. Secondly, the ion optics were just not amenable to large area type systems. What's really happened today is that the magnetic filter has turned the field around. It has allowed us to extract negative ions from large area ion sources of the type we use for positive ion neutral beam systems. There are some important gains coming along with it, namely we find that the ions are cold. The optics should be better rather than worse than our present proton systems. Needless to say, we get rid of cesium and that is a big gain. As a matter of fact, when I was at APS last week in Los Angeles giving a poster, a number of people came by of the user types and I heard them utter, "well you know, maybe we better consider negative ions again," and I think what volume has done for us is to make our program look more credible. In other words, people are not quite so scared of systems that look not too dissimilar from the systems they have already seen, rather than systems that incorporate new science and technology with cesium and surface effects which very few really understand. So what's to be done then on volume. Well, I think the whole world is going to start working on volume. I hope we have the chance at Berkeley to continue. There's a good bit to do. We've got to wipe out the remaining electrons. Another order of improvement in electron to H^- ratio would be a big step forward. Two orders would be delightful, but I'll settle for one order of magnitude, to be very frank with you. We've got to optimize the filter. All the work done so far has been with filter geometries that were optimized for the production of high atomic species. It could well be that the filter fields could double and we'd get better gains. It will cut the electron numbers down, but to optimize the filter it would be awful nice to know how the negative ions are being made. I think the conclusion of this meeting is that they are being made by dissociative attachment of vibrationally excited molecules. I'm not sure I thoroughly agree. Twenty years ago I said the process required molecular ions so I don't disagree there. But I said it required hot electrons and that's a little bit in conflict with the cold electrons that are needed for dissociative attachment. I thoroughly think that if dissociative attachment is the key, and certainly some of the negative ions are produced that way, some of the negative ions are going to be produced by all the known processes. If that is the key, we've got to make the exit region or the extraction region of the source longer. If it turns out that polar dissociation is really the key, then we're going to find out that we need hot electrons all the way, hot electrons to vibrationally excite the molecules, and to polar dissociate them. I really think, my seat of the pants feeling is that polar dissociation is the key and we need hot electrons all the way. So both the theorists and ourselves should work on this as it is going to effect the design of the source. If we have time we certainly will be putting in variable energy electrons in the various parts of this ion source to see just which energy is the most important. When we find that, we'll certainly let you know.

In the halls, we find that we've got even budgets at best, ahead possibly reduced budgets and this means that we found a lot of talk about cooperation. Certainly we've talked about it with people from France and elsewhere and I think it's a very good topic of conversation. We've got to cooperate among ourselves. That's really the beauty of a symposium like this in that we do get together for a week, we know each other, and we're all friends and we can cooperate. We at Berkeley are the so-called lead laboratory for negative ion neutral beam development. We've had the opportunity to promote cooperation this past year and it's worked wonderfully well. For example, the entire volume results can be largely credited to the work that people from Los Alamos Laboratory, Rob York and Ralph Stevens provided for us. We'd never been able to do that without them. We had a great cooperative program this year with F.O.M. We've had some cooperation with Karlsruhe, a little bit of cooperation with Japan, and I can see that cooperation is one word that we've going to have to adopt, particularly if the budget is reduced. You know money doesn't solve the problems. It helps, but it still takes good ideas and hard work.

Well, we come down to the time for credits. I don't really know how I can even start to thank Krsto. If you've never been in charge of a meeting like this, you don't realize how much work it really is. He's had this meeting to organize more than once so he's getting to be an expert at it. But this meeting was really a delightful meeting and I looked forward to it. I start looking forward to the next meeting as I leave to go back to Kennedy, and to the airport. We had a delightful banquet. All the speakers for reasons that I don't understand were able to stay on time. The organization seemed exceptionally good and at this point I think we should all give Krsto a great hand of applause. In addition I want to thank the girls at the desk and the fellows here with the microphones. Great diligence has been displayed all week and in return I'd like to thank the administration here at Brookhaven for being so nice to us.

Now just a few words from me. I'm not going to be at your next meeting, so Krsto, you've got to find a new boy for this job. We're all getting older and a number of us will be retiring before then. It's time some of us go sit on the sidelines and watch the game go on, but as I look around here at the team that remains, the varsity if you will, for this total negative ion program, it looks to me that the talent is here and there is not too much doubt that we'll make the playoffs. When we do get to the playoffs, we're going to have to play with the RF team and I think we have a good leg up. I don't think we're going to have too many worries but it is going to take hard work. If I don't miss my guess, RF's going to fumble down near the goal line and it's going to give us a great chance to score. RF isn't that great panacea that those who don't know it think it is. It's plagued with impurities and problems with getting power into the plasma and I'm beginning to think that probably their biggest problem could be the fact that RF's going to juggle plasma potentials and I don't think the whole field of fusion really has come to recognize how important plasma potentials are in the whole game of fusion. You know when you retire it's easy to say these things. At any rate, I want to wish all of you a good trip back to the airport and also remind you that I think the worst thing about retiring is that I will miss seeing people like you from this point on. When you have so many friends from all over the country and all over the world that you may never see again that's the hardest thing about retirement.

Now to conclude this symposium I'd like to introduce our man of the hour who is in charge of this Third International Symposium on the Production and Neutralization of Negative Ions and Beams, Dr. Krsto Prelec.

APPENDICES

LIST OF PARTICIPANTS

Dr. James G. Alessi
Brookhaven National Laboratory
Accelerator Department
Bldg. 930
Upton, NY 11973

Dr. Paul Allison
Los Alamos National Laboratory
AT-2, Mail Stop H818
P.O. Box 1663
Los Alamos, NM 87545

Dr. L.W. Anderson
Physics Department
University of Wisconsin
Madison, WI 53706

Dr. Oscar A. Anderson
Lawrence Berkeley Laboratory
Bldg. 4
Berkeley, CA 94720

Dr. Martha Bacal
Laboratorie de Physique
 des Milieux Ionises
Ecole Polytechnique
F-91228 Palaiseau Cedex,
FRANCE

Dr. Clarence Barnett
Oak Ridge National Laboratory
Bldg. 6003
P.O. Box X
Oak Ridge, TN 37830

Dr. George Basbas
Physical Review Letters
P.O. Box 1000
Ridge, NY 11961

Dr. Carmen Cisneros
Instituto de Fisica UNAM
Apdo Postal 20-364
01000 Mexico, D.F.

Dr. Michael J. Coggiola
Stanford Research Institute
 International
333 Ravenswood Avenue
Menlo Park, CA 94025

Dr. William S. Cooper
Lawrence Berkeley Laboratory
University of California
Berkeley, CA 94720

Dr. Wayne Cornelius
Los Alamos National Laboratory
AT-4, MS H821
Los Alamos, NM 87545

Dr W. Kelly Dagenhart
Oak Ridge National Laboratory
Bldg. 9201-2
Box Y
Oak Ridge, TN 37830

Mr. Basil DeVito
Brookhaven National Laboratory
Accelerator Department
Bldg. 911A
Upton, NY 11973

Dr. Kenneth W. Ehlers
Lawrence Berkeley Laboratory
Bldg. 4.
University of California
Berkeley, CA 94720

Dr. Joel W. Fink
Negion, Inc.
4023 East Avenue
Hayward, CA 94542

Dr. Dan M. Goebel
TRW
One Space Park
R1/2128
Redondo Beach, CA 90278

Dr. William G. Graham
Physics Department
New Univesity of Ulster
Coleraine BT 52 1SA
Northern Ireland

Dr. Pierre Grand
Brookhaven National Laboratory
Bldg. 129
Upton, NY 11973

Dr. Ernst H. A. Granneman
FOM Institute for Atomic &
  Molecular Physics
Kruislaan 407
1098 SJ Amsterdam
The Netherlands

Dr. Larry R. Grisham
Plasma Physics Laboratory
Princeton University
P.O. Box 451
Princeton, NJ 08544

Dr. Willy Grüebler
Laboratorium fur Kernphysik
ETH Honngerberg
CH-8083 Zurich
Switzerland

Dr. Willy Haeberli
Department of Physics
The University of Wisconsin
Madison, WI 53706

Dr. Goran Hellblom
DRFC/SIG 85X
F-38041 Grenoble CEDEX
FRANCE

Dr. P.R.W. Henkes
Kernforschungszentrum Karlsruhe
Institut fur Kernverfahrenstechnik
Postfach 3640
D-7500 Karlsruhe 1
West Germany

Dr. Ady Hershcovitch
Brookhaven National Laboratory
Accelerator Department
Bldg. 911B
Upton, NY 11973

Dr. Charles Hill
P.S. Division/Linac
CERN
1211 Geneva 23
SWITZERLAND

Dr. John R. Hiskes
Lawrence Livermore Laboratory
Mail Stop L-630
P.O. Box 5511
Livermore, CA 94550

Dr. Hsiao-Chaun Hseuh
Brookhaven National Laboratory
Accelerator Department
Bldg. 911A
Upton, NY 11973

Dr. Claude Jacquot
Centre D'Etudes Nucleaires
Dept. de Physique du Plasma
  et de la Fusion Controlee
85 X 38041 Grenoble Cedex
FRANCE

Dr. Kaneko
KEK
Oho-Cho, Tsukuba-Gun
Ibaraki-Ken 305
JAPAN

Dr. Arnold M. Karo
Dept. of Chemistry & Materials Science
Lawrence Livermore Laboratory
P.O. Box 808
L-329
Livermore, CA 94550

Dr. Hans-Michael Katsch
Universitat Essen - Gesamthochschule
Fachbereich 7
Physik
Universitatsstrabe 5
D 4300 Essen 1
West Germany

Dr. Vincent Kovarik
Brookhaven National Laboratory
Accelerator Department
Bldg. 911B
Upton, NY 11973

Dr. Ahovi Kponou
Brookhaven National Laboratory
Accelerator Department
Bldg. 911B
Upton, NY 11973

Dr. Robert Larson
Brookhaven National Laboratory
Accelerator Department
Bldg. 911B
Upton, NY 11973

Dr. Ka-Ngo Leung
Lawrence Berkeley Laboratory
Bldg. 4
University of California
Berkeley, CA 94720

Dr. A.F. Lietzke
Lawrence Berkeley Laboratory
University of California
Berkeley, CA 94720

Robert H. McFarland
U.S. Dept. of Energy
ER-542, MS G-226, GTN
Washington, D.C. 20545

Dr. Malcolm W. McGeoch
AVCO Everett Research
  Laboratory, Inc. (AERL)
2385 Revere Beach Parkway
Everett, MA 02149

Mr. R.B. McKenzie-Wilson
Brookhaven National Laboratory
Accelerator Department
Bldg. 911A
Upton, NY 11973

Dr. H. Harvey Michels
United Technologies Research Center
East Hartford, CT 06108

Dr. George H. Miley, Chairman
Nuclear Engineering Program
University of Illinois at
  Urbana-Champaign
214 Nuclear Engineering Laboratory
103 S. Goodwin
Urbana, IL 61801

Dr. Tom J. Morgan
Physics Department
Wesleyan University
Middletown, CT 06457

Mr. Ken Moses
Jaycor
2811 Wilshire Blvd.
Suite 690
Santa Monica, CA 90403

Mr. Michael Norris
AFWL
Kirkland Air Force Base
New Mexico 87117

Dr. Ordean Oen
Oak Ridge National Laboratory
P.O. Box X
Oak Ridge, TN 37830

Dr. Robert Palmer
IRT Corporation
P.O. Box 80817
San Diego, CA 92138

Dr. James R. Peterson
Stanford Research
 Institute/International
333 Ravenswood Avenue
Menlo Park, CA 94025

Dr. D.E. Post
Princeton Plasma Physics Lab.
Princeton University
Princeton, NJ 08544

Dr. C. Powell
Stevens Inst. of Technology
Castle Point
Hoboken, NJ 07030

Dr. Krsto Prelec
Brookhaven National Laboratory
Accelerator Department
Bldg. 911B
Upton, NY 11973

Dr. Henry L. Pugh, Jr.,
 Capt. USAF
AFOSR
Bolling Air Force Base
Washington, D.C. 20332

Dr. J. Claude Rocco
Centre D'Etudes Nucleaires
Dept. de Physique du Plasma
et de la Fusion Controlee
85 X 38041 Grenoble Cedex
FRANCE

Professor E. Salzborn
Institut fur Kernphysik
Universitat Giessen
Leihgesterner WEG 217
D-6300 Giessen
WEST GERMANY

Dr. Gerd Schilling
Princeton University Plasma
 Physics Laboratory
P.O. Box 451
Princeton, NJ 08544

Dr. Alfred S. Schlachter
Lawrence Berkeley Laboratory
Bldg. 5
University of California
Berkeley, CA 94720

Dr. Charles Schmidt
Fermilab
MS-307
Batavia, IL 60510

Mr. H Stanley Staten
U.S. Department of Energy
Office of Fusion Energy
MS G-234, ER-531
Component Development Branch
Washington, D.C. 20545

Dr. Paul Schmor
Triumf
4004 Wesbrook Mall
University of British Columbia
Vancouver, BC V6T 2A3
Canada

J. Warren Stearns
University of California
Lawrence Berkeley Laboratory
Bldg. 5
Berkeley, CA 74720

Dr. Milos Seidl
Dept. of Physics/Engineering Physics
Stevens Institute of Technology
Castle Point Station
Hoboken, NJ 07030

Dr. Ralph R. Stevens, Jr.
Los Alamos National Laboratory
MP-DO, MS-H823
Los Alamos, NM 87545

Dr. Joseph Sherman
University of California
Los Alamos Scientific Laboratory
P.O. Box 1663
MS H-818
Los Alamos, NM 87545

Dr. Larry Stewart
Princeton Plasma Physics Laboratory
Exxon Nuclear Company
Princeton University
P.O. Box 451
Princeton, NJ 08544

Dr. Theo Sluyters
Brookhaven National Laboratory
Accelerator Department
Bldg. 911B
Upton, NY 11973

Dr. Will L. Stirling
Oak Ridge National Laboratory
Bldg. 9201-2, Y-12 Area
P.O. Box X
Oak Ridge, TN 37830

Dr. H. Vernon Smith
Los Alamos National Laboratory
AT-2, Mail Stop H818
P.O. Box 1663
Los Alamos, NM 87545

Dr. R.J. Turnbull
University of Illinois at
 Urbana-Champaign
Department of Electrical Engineering
1406 West Green Street
Urbana, IL 61801

Dr. Santosh K. Srivastava
Jet Propulsion Laboratory
California Institute of Technology
4800 Oak Grove Drive
Pasadena, CA 91109

Dr. Vladislav Vanek
TRW, Inc.
Plasma Physics Department
Defense and Space Systems Group
Bldg. R1, Room 1078
One Space Park, Mail Stop R1-2144
Redondo Beach, CA 90278

Dr. H. Verbeek
Max Planck Institut fur
 Plasma Physik
D-8046 Garching
WEST GERMANY

Dr. Gen Wada
Hitachi Works
Hitachishi
Ibaraki-Ken
JAPAN

Dr. Jogindra M. Wadehra
Department for Physics-Astronomy
Wayne State University
Detroit, MI 48202

Mr. Nikita Wells
Rand Corporation
2100 M Street N.W.
Washington, D.C. 20037

Dr. J. H. Whealton
Oak Ridge National Laboratory
Bldg. 9201-2
P.O. Box X
Oak Ridge, TN 37830

Dr. Richard L. Witkover
Brookhaven National Laboratory
Accelerator Department
Bldg. 911A
Upton, NY 11973

Dr. Rob York
Los Alamos National Laboratory
MP-DO, MS-H823
Los Alamos, NM 87545

AUTHOR INDEX

Alessi, J.....736,741
Allison, P.....458,511
Alvarez, I.....140
Anderson, L.W.....696,754
Anderson, O.A..... 473

Bacal, M... 31,105,313,418
Bae, Y.K.....90
Banic, G.M.....353
Barber, G.C..... 353
Bel'chenko, Y.I.....363
Bergstrom, J.B..... 438
Boley, C.D..... 641
Bruneteau, A.M..... 31

Chubb, S.R..... 184
Cisneros, C..... 140
Coggiola, M.J..... 90,239
Cooper, W.S..... 605,669
Copeland, D..... 568
Cornelius, W..... 769

Dagenhart, W.K.....353,450,524
Dammertz, G..... 55
de Urquijo, J..... 140
DeBoni, T.M..... 197
Delaunay, M..... 438
Derevyankin, G.E..... 376
Dimov, G.I..... 363
Donaghy, J.J..... 450
Dudnikov, V.G..... 376

Ebel, F..... 162
Eckstein, W..... 273
Ehlers, K.W.. 67,227,258,265,344,773

Fink, J.H..... 547,669
Freeman, A.J..... 184
Fukumoto, S..... 520,769

Geerlings, J.J.C.....206
Geller, R..... 438

Giordano, S.T..... 505
Goebel, D.M..... 568,617
Graham, W.G..... 113
Granneman, E.H.A.. 206,313,492
Green, T.S..... 55,429
Greer, J.A..... 220
Grisham, L.R..... 561,659
Grüebler, W..... 706
Guttman, J.L..... 132

Haeberli, W..... 685
Hagena, O.F..... 96
Hamilton, G.W..... 617
Hardy, R.J..... 197
Hellblom, G..... 438
Henkes, P.R.W..... 96
Hershcovitch, A.I.. 561,763
Hillion, F..... 418
Hinds, E.A.....763
Hiskes, J.R.. 3,125,184,197,313
Hobbs, R.H..... 118
Holmes, A.J.T..... 55,429
Hopman, H.J..... 206,492
Hursman, T..... 568

Igarashi, Z..... 520,769
Ikegami, K..... 520,769

Jacquot, C..... 438,669
Janev, R.K..... 641
Johnson, B.M..... 561
Jones, K.W..... 561

Kaplan, S.N..... 754
Karo, A.M... 3,125,184,197,313
Katsch, H.-M.....254
Kistemaker, J.....492
Klein, H.....492
Kleppner, D..... 720
Klingelhöfer, R.....96
Kovarik, V.J.....561
Kponou, A..... 736
Krevet, B..... 96
Kubota, C..... 520

Leung, K.N.. 67,105,227,258,265,344
Lietzke, A.F..... 344
Los, J..... 206
Ludwig, P..... 438

Maschke, A.W..... 489
McCollough, D.H..... 524,533
McGaffey, R.W..... 524,533
McGeoch, M.W..... 291
Meron, M..... 561
Meszaros, P.S..... 533
Michels, H.H..... 118
Mikkelsen, D.R..... 659
Miley, G.H..... 647
Morales, A..... 140
Morgan, T.J..... 82,149
Mori, Y..... 520,769
Moser, H.O..... 96

Nachman, M..... 418
Nardi, V..... 463

Oen, O.S..... 171
Olwell, K.D..... 197
Orient, O.J..... 56

Palmer, R.L..... 281
Pauli, R..... 438
Peterson, J.R..... 90,239
Ponte, N.S..... 353
Post, D.E..... 641,659
Powell, C..... 463
Prelec, K..... 333,561
Pyle, R.V..... 247,754

Raridon, R.J..... 524,533
Robinson, M.T..... 171
Rocco, J.C..... 438
Ruby, L..... 754

Salzborn, E..... 162
Schempp, A..... 492

Schlachter, A.S...149,300,754
Schlier, R.E..... 291
Schmelzbach, P.A..... 706
Schneider, P.J..... 273
Seidl, M..... 220,313
Sermet, R..... 438
Sherman, J.D..... 458,511
Siebenlist, F..... 492
Sluyters, Th..... 736
Smith, Jr., H.V..... 458
Srivastava, S.K..... 56
Staten, H.S..... 587,669
Stearns, J.W..... 247,754
Steckelmacher, W..... 418
Stevens, Jr., R.R..... 410
Stewart, L.D..... 594,669
Stirling, W.L...353,450,524

Takagi, A..... 520,769
Thomae, R.W..... 492
Turnbull, R.J..... 132

van Amersfoort, P.W..... 492
van Bommel, P.J.M..... 206,258
van Wunnik, J.N.M..... 206
Vanek, V..... 568
Verbeek, H..... 273

Wada, M..... 247,313
Wadehra, J.M..... 46,313
Walther, S.R..... 132
Weis, T..... 492
Whealton, J.H.. 353,450,524,533
Wiesemann, K..... 254
Wilhelmsson, H..... 438
Wimmer, E..... 184
Witkover, R.L..... 398

York, R.L..... 410,769

Zadworny, F..... 438

THIRD INTERNATIONAL SYMPOSIUM ON THE PRODUCTION AND NEUTRALIZATION OF NEGATIVE IONS AND BEAMS

Brookhaven National Laboratory
Associated Universities, Inc.
Upton, New York 11973

Berkner Hall
November 14-18, 1983

PROGRAM

Monday, November 14, Room B

8:30   WELCOME

Session I: Fundamental Processes – Volume

Chairperson: M. Bacal

9:00   Hiskes/Karo   Volume Generation of Negative Ions in High Density Hydrogen Discharges (30')

        Leung/Ehlers   Volume H^- Ion Production Experiments at LBL (30')

        COFFEE

10:30   Wadehra   Negative Ion Production Via Dissociative Attachment to H_2 (30')

        Bacal/Bruneteau H^- Production and Destruction Mechanisms in Hydrogen Low Pressure Discharges (30')

        Srivastava/Orient   Polar Dissociation as a Source of Negative Ions (20')

12:00   LUNCH

(Monday continued)

Session II: Fundamental Processes - Volume

Chairman: J.R. Hiskes

2:00	Holmes/Dammertz/ Green	H^- and Electron Production in a Magnetic Multipole Source (25')
	Morgan	Negative Ion Neutralization Accompanied by Excitation (20')
	Schlachter/ Morgan	Formation of H^- by Charge Transfer in Alkaline-Earth Vapors (30')
	COFFEE	
3:30	Bae/Coggiola/ Peterson	A Search for H_2^-, H_3^-, and Other Metastable Negative Ions (20')
	Schlachter	Formation of Negative Ions by Charge Transfer: He^- to Cl^-
	Hagena/Henkes/ Klingelhöfer/ Krevet/Moser	Research at the IKVT/KfK Related to the Development of an H^- Beam Line (20')
5:00	Wine and Cheese (Berkner Hall, Lobby)	

Tuesday, November 15, Room B

Session I: Fundamental Processes - Surface

Chairman: K.W. Ehlers

9:00	Oen/Robinson	Ion Backscattering From Layered Targets (25')
	Freeman/Wimmer/ Chubb/Hiskes/ Karo	Cs/Transition Metal Composite Surfaces: First Principles Calculations of High Z, Low Work Function Systems (25')
	Karo/Hiskes/ Olwell/DeBoni/ Hardy	De-excitation and Equipartition in H_2-Surface Collisions (15')
	COFFEE	
10:30	Granneman/ Geerlings/van Wunnik/van Bommel/ Hopman/Los	H^- and Li^- Formation by Scattering H^+, H_2^+, and Li^+ From Cesiated Tungsten Surfaces (30')
	Greer/Seidl	Sputtering Yields of Negative Hydrogen Ions (30')
	Ehlers/Leung	Effects of Cesium in the Plasma of the Surface Conversion H^- Source (20')
12:00	LUNCH	

(Tuesday, continued)

2:00 - 4:00   Session IV: Poster Papers - Room A

Graham	Vacuum Ultraviolet Emission and H^- Production in a Low Pressure Hydrogen Plasma
Michels/Hobbs	Dissociative-Recombination of $e + H_3^+$. An Analysis of Reaction Product Channels
Hiskes/Karo	Generation of Vibrationally Excited H_2 Molecules by H_2^+ Wall Collisions
Turnbull/Walther/ Guttman	Generation of Vibrationally Excited Hydrogen for Use in a Negative Ion Source
McGeoch/Schlier	Experimental Investigation of Volume Li^- Production
Ebel/Salzborn	Charge Exchange of Protons and Hydrogen Atoms in Na, K, Rb-Vapor Targets
Alvarez/Morales/ de Urquijo/Cisneros	Interference Effects in Negative Ion Formation
Coggiola/Peterson	Systematic Investigation of Negative Ion Production From Low Work Function Surfaces
Wada/Pyle/Stearns	Work Function Dependence of Surface Produced H^- in the Presence of a Plasma
Katsch/Wiesemann	Plasma-Surface Interaction Involved in H^- Generation
van Bommel/Leung/ Ehlers	Production of H^- Ions From Polycrystalline and Single Crystal (110) Tungsten and Molybdenum Surfaces
Leung/Ehlers	H^- Production From Different Metallic Converter Surfaces
Verbeek/Eckstein/ Schneider	The Negative Fraction of Deuterium and Helium Scattered from a Sodium Surface
Palmer	Observation of H^- by Surface Chemi-Ionization on W(110)
Nardi/Powell	The Plasma Focus as a Source of Collimated Beams of Negative Ion Clusters and of Neutral Deuterium Atoms

(Tuesday, continued)

2:00 - 4:00   Session IV: Poster Papers - Room D

Bacal/Leung	H^- Density in a Tandem Multicusp Discharge
Bacal/Hillion/Nachman/ Steckelmacher	Progress in Developing a 'Volume' Hydrogen Negative Ion Source
Stirling/Dagenhart/ Whealton/Donaghy	Normalized Emittance of SITEX Negative Ion Source
Bel'chenko/Dimov (presented by J. Alessi, BNL)	Pulsed Multiampere Source of Negative Hydrogen Ions
Derevyankin/ Dudnikov (presented by J. Alessi, BNL)	Production of High Brightness H^- Beams in Surface Plasma Sources
Smith/Allison/ Sherman	A Scaled, Circular-Emitter Penning SPS for Intense H^- Beams
Hershcovitch/Hinds	Generation of Intense Polarized Beams by Selective Neutralization of Negative Ions
Mori/Ikegami/ Takagi/Igarashi/ Fukumoto/Cornelius/ York	Optically Pumped Polarized H^- Ion Source
Takagi/Mori/Igarashi/ Ikegami/Kubota/ Fukumoto (presented by W. Cornelius, LANL)	Accelerated Beam From Cusp H^- Ion Source
Vanek/Hursman/ Copeland/Goebel	Technology of a Laser Resonator for Photodetachment Neutralizer
Goebel/Hamilton	Neutral Beam Injector for 475 keV MARS Sloshing Ions
Post/Janev/Boley	The Role of Multistep Collision Processes in Increasing the Beam Stopping Cross Section for High Energy Neutral Beams

4:00 - 6:00 PANEL SESSION ON FUNDAMENTAL PROCESSES, Room B

        Moderator: J.R. Hiskes

Wednesday, November 16, Room B

Session V: H⁻ Ion Sources

Chairman: K. Prelec

9:00	Prelec	Report on the BNL H⁻ Ion Source Development (30')
	Witkover	Operational Experience with the BNL Magnetron H⁻ Source (15')
	Lietzke/Ehlers/ Leung	The Status of $\gtrsim$ 1 Ampere H⁻ Ion Source Development at the Lawrence Berkeley Laboratory
	COFFEE	
10:30	Dagenhart/Stirling/ Banic/Barber/ Ponte/Whealton	Short Pulse Operation With the SITEX Negative Ion Source (30')
	Holmes/Green	Extraction and Acceleration of H⁻ Ions From a Magnetic Multipole Source (20')
	York/Stevens, Jr.	Development of a Multicusp H⁻ Ion Source for Accelerator Applications (15')
	Delaunay/Geller/ Jacquot/Ludwig/ Sermet/Rocco Zadworny/Bergstrom/ Hellblom/Pauli/ Wilmelmsson	Large Negative Ion Source for Energetic Neutral Beams (30')
	Miley	Negative Ion Beam Requirements for Compact Tori (30')
12:45	LUNCH	
	FREE AFTERNOON (sports, tours, etc.)	
6:30	COCKTAILS (Harbor Hills Country Club, Port Jefferson)	
7:30	SYMPOSIUM BANQUET (Harbor Hills Country Club, Port Jefferson)	

Thursday, November 17, Room B

Session VI: Acceleration

Chairman: Th. Sluyters

9:00	O.A. Anderson	Transverse-field Focusing Accelerator (30')
	Giordano	The BNL RFQ (30')
	COFFEE	
10:15	Granneman/Thomae/ Siebenlist/van Amersfoort/Hopman/ Kistemaker/Klein/ Schempp/Weis	The Amsterdam "MEQALAC" RF Acceleration System (20')
	Allison/Sherman	Operating Experiences with a 100 keV, 100 mA H⁻ Injector (20')
	Whealton/Raridon/ McGaffey/ McCollough/ Stirling/Dagenhart	2D Accelerator Design for SITEX Negative Ion Source (15')
	McGaffey/ Meszaros/ Whealton/Raridon/ McCollough	End Effects in Slot Extraction for a SITEX Negative Ion Source (15')
12:00	LUNCH	

(Thursday, continued)

Session VII: Neutralizers; Neutral Beam Systems

Chairman: W.S. Cooper

1:30	Fink	Photodetachment Technology (30')
	Hershcovitch/ Johnson/Kovarik/ Meron/Jones/ Prelec/Grisham	Neutralization Efficiency of Plasma Targets for High Energy Negative Ions (30')
	COFFEE	
3:00	Staten	Magnetic Fusion Energy Heating Development Plan (30')
	Stewart	Requirements for Negative Ion Based Systems from Users' Point of View (30')
	Cooper	Considerations Involved in the Design of Negative-Ion-Based Neutral Beam Systems (30')
	Grisham/Post/ Mikkelsen	Applications and Development Requirements for Multi-MeV Light Atom Beams (20')
5:00 - 7:00		PANEL: Where Do We Go From Here? Moderator: W.S. Cooper

Friday, November 18, Room B

Production of Polarized H^- Ions

Chairman: W. Haeberli

8:30	Haeberli	Principles of Production of H^- Polarized Ions (30')
	L.W. Anderson	Production of Polarized H^- Ions Using Laser Optical Pumping (30')
	Grüebler/ Schmelzbach	Production of Intense Beams of Polarized Negative Hydrogen Ions by Double Charge Exchange in Alkali Vapours (30')
	COFFEE	
10:15	Kleppner	Optics for a Spin-Polarized Hydrogen Atomic Beam (30')
	Sluyters/ Alessi/Kponou	Status and Future Plans for the BNL Polarized H^- Source (20')
	Alessi	Ionization of Polarized Hydrogen Atoms (20')
12:00	Ehlers	CONCLUDING REMARKS

AIP Conference Proceedings

		L.C. Number	ISBN
No.1	Feedback and Dynamic Control of Plasmas	70-141596	0-88318-100-2
No.2	Particles and Fields - 1971 (Rochester)	71-184662	0-88318-101-0
No.3	Thermal Expansion - 1971 (Corning)	72-76970	0-88318-102-9
No.4	Superconductivity in d-and f-Band Metals (Rochester, 1971)	74-18879	0-88318-103-7
No.5	Magnetism and Magnetic Materials - 1971 (2 parts) (Chicago)	59-2468	0-88318-104-5
No.6	Particle Physics (Irvine, 1971)	72-81239	0-88318-105-3
No.7	Exploring the History of Nuclear Physics	72-81883	0-88318-106-1
No.8	Experimental Meson Spectroscopy - 1972	72-88226	0-88318-107-X
No.9	Cyclotrons - 1972 (Vancouver)	72-92798	0-88318-108-8
No.10	Magnetism and Magnetic Materials - 1972	72-623469	0-88318-109-6
No.11	Transport Phenomena - 1973 (Brown University Conference)	73-80682	0-88318-110-X
No.12	Experiments on High Energy Particle Collisions - 1973 (Vanderbilt Conference)	73-81705	0-88318-111-8
No.13	π-π Scattering - 1973 (Tallahassee Conference)	73-81704	0-88318-112-6
No.14	Particles and Fields - 1973 (APS/DPF Berkeley)	73-91923	0-88318-113-4
No.15	High Energy Collisions - 1973 (Stony Brook)	73-92324	0-88318-114-2
No.16	Causality and Physical Theories (Wayne State University, 1973)	73-93420	0-88318-115-0
No.17	Thermal Expansion - 1973 (lake of the Ozarks)	73-94415	0-88318-116-9
No.18	Magnetism and Magnetic Materials - 1973 (2 parts) (Boston)	59-2468	0-88318-117-7
No.19	Physics and the Energy Problem - 1974 (APS Chicago)	73-94416	0-88318-118-5
No.20	Tetrahedrally Bonded Amorphous Semiconductors (Yorktown Heights, 1974)	74-80145	0-88318-119-3
No.21	Experimental Meson Spectroscopy - 1974 (Boston)	74-82628	0-88318-120-7
No.22	Neutrinos - 1974 (Philadelphia)	74-82413	0-88318-121-5
No.23	Particles and Fields - 1974 (APS/DPF Williamsburg)	74-27575	0-88318-122-3
No.24	Magnetism and Magnetic Materials - 1974 (20th Annual Conference, San Francisco)	75-2647	0-88318-123-1
No.25	Efficient Use of Energy (The APS Studies on the Technical Aspects of the More Efficient Use of Energy)	75-18227	0-88318-124-X

No.26	High-Energy Physics and Nuclear Structure - 1975 (Santa Fe and Los Alamos)	75-26411	0-88318-125-8
No.27	Topics in Statistical Mechanics and Biophysics: A Memorial to Julius L. Jackson (Wayne State University, 1975)	75-36309	0-88318-126-6
No.28	Physics and Our World: A Symposium in Honor of Victor F. Weisskopf (M.I.T., 1974)	76-7207	0-88318-127-4
No.29	Magnetism and Magnetic Materials - 1975 (21st Annual Conference, Philadelphia)	76-10931	0-88318-128-2
No.30	Particle Searches and Discoveries - 1976 (Vanderbilt Conference)	76-19949	0-88318-129-0
No.31	Structure and Excitations of Amorphous Solids (Williamsburg, VA., 1976)	76-22279	0-88318-130-4
No.32	Materials Technology - 1976 (APS New York Meeting)	76-27967	0-88318-131-2
No.33	Meson-Nuclear Physics - 1976 (Carnegie-Mellon Conference)	76-26811	0-88318-132-0
No.34	Magnetism and Magnetic Materials - 1976 (Joint MMM-Intermag Conference, Pittsburgh)	76-47106	0-88318-133-9
No.35	High Energy Physics with Polarized Beams and Targets (Argonne, 1976)	76-50181	0-88318-134-7
No.36	Momentum Wave Functions - 1976 (Indiana University)	77-82145	0-88318-135-5
No.37	Weak Interaction Physics - 1977 (Indiana University)	77-83344	0-88318-136-3
No.38	Workshop on New Directions in Mossbauer Spectroscopy (Argonne, 1977)	77-90635	0-88318-137-1
No.39	Physics Careers, Employment and Education (Penn State, 1977)	77-94053	0-88318-138-X
No.40	Electrical Transport and Optical Properties of Inhomogeneous Media (Ohio State University, 1977)	78-54319	0-88318-139-8
No.41	Nucleon-Nucleon Interactions - 1977 (Vancouver)	78-54249	0-88318-140-1
No.42	Higher Energy Polarized Proton Beams (Ann Arbor, 1977)	78-55682	0-88318-141-X
No.43	Particles and Fields - 1977 (APS/DPF, Argonne)	78-55683	0-88318-142-8
No.44	Future Trends in Superconductive Electronics (Charlottesville, 1978)	77-9240	0-88318-143-6
No.45	New Results in High Energy Physics - 1978 (Vanderbilt Conference)	78-67196	0-88318-144-4
No.46	Topics in Nonlinear Dynamics (La Jolla Institute)	78-057870	0-88318-145-2
No.47	Clustering Aspects of Nuclear Structure and Nuclear Reactions (Winnepeg, 1978)	78-64942	0-88318-146-0
No.48	Current Trends in the Theory of Fields (Tallahassee, 1978)	78-72948	0-88318-147-9
No.49	Cosmic Rays and Particle Physics - 1978 (Bartol Conference)	79-50489	0-88318-148-7

No. 50	Laser-Solid Interactions and Laser Processing - 1978 (Boston)	79-51564	0-88318-149-5
No. 51	High Energy Physics with Polarized Beams and Polarized Targets (Argonne, 1978)	79-64565	0-88318-150-9
No. 52	Long-Distance Neutrino Detection - 1978 (C.L. Cowan Memorial Symposium)	79-52078	0-88318-151-7
No. 53	Modulated Structures - 1979 (Kailua Kona, Hawaii)	79-53846	0-88318-152-5
No. 54	Meson-Nuclear Physics - 1979 (Houston)	79-53978	0-88318-153-3
No. 55	Quantum Chromodynamics (La Jolla, 1978)	79-54969	0-88318-154-1
No. 56	Particle Acceleration Mechanisms in Astrophysics (La Jolla, 1979)	79-55844	0-88318-155-X
No. 57	Nonlinear Dynamics and the Beam-Beam Interaction (Brookhaven, 1979)	79-57341	0-88318-156-8
No. 58	Inhomogeneous Superconductors - 1979 (Berkeley Springs, W.V.)	79-57620	0-88318-157-6
No. 59	Particles and Fields - 1979 (APS/DPF Montreal)	80-66631	0-88318-158-4
No. 60	History of the ZGS (Argonne, 1979)	80-67694	0-88318-159-2
No. 61	Aspects of the Kinetics and Dynamics of Surface Reactions (La Jolla Institute, 1979)	80-68004	0-88318-160-6
No. 62	High Energy e^+e^- Interactions (Vanderbilt, 1980)	80-53377	0-88318-161-4
No. 63	Supernovae Spectra (La Jolla, 1980)	80-70019	0-88318-162-2
No. 64	Laboratory EXAFS Facilities - 1980 (Univ. of Washington)	80-70579	0-88318-163-0
No. 65	Optics in Four Dimensions - 1980 (ICO, Ensenada)	80-70771	0-88318-164-9
No. 66	Physics in the Automotive Industry - 1980 (APS/AAPT Topical Conference)	80-70987	0-88318-165-7
No. 67	Experimental Meson Spectroscopy - 1980 (Sixth International Conference, Brookhaven)	80-71123	0-88318-166-5
No. 68	High Energy Physics - 1980 (XX International Conference, Madison)	81-65032	0-88318-167-3
No. 69	Polarization Phenomena in Nuclear Physics - 1980 (Fifth International Symposium, Santa Fe)	81-65107	0-88318-168-1
No. 70	Chemistry and Physics of Coal Utilization - 1980 (APS, Morgantown)	81-65106	0-88318-169-X
No. 71	Group Theory and its Applications in Physics - 1980 (Latin American School of Physics, Mexico City)	81-66132	0-88318-170-3
No. 72	Weak Interactions as a Probe of Unification (Virginia Polytechnic Institute - 1980)	81-67184	0-88318-171-1
No. 73	Tetrahedrally Bonded Amorphous Semiconductors (Carefree, Arizona, 1981)	81-67419	0-88318-172-X
No. 74	Perturbative Quantum Chromodynamics (Tallahassee, 1981)	81-70372	0-88318-173-8

No. 75	Low Energy X-ray Diagnostics-1981 (Monterey)	81-69841	0-88318-174-6
No. 76	Nonlinear Properties of Internal Waves (La Jolla Institute, 1981)	81-71062	0-88318-175-4
No. 77	Gamma Ray Transients and Related Astrophysical Phenomena (La Jolla Institute, 1981)	81-71543	0-88318-176-2
No. 78	Shock Waves in Condensed Matter - 1981 (Menlo Park)	82-70014	0-88318-177-0
No. 79	Pion Production and Absorption in Nuclei - 1981 (Indiana University Cyclotron Facility)	82-70678	0-88318-178-9
No. 80	Polarized Proton Ion Sources (Ann Arbor, 1981)	82-71025	0-88318-179-7
No. 81	Particles and Fields - 1981: Testing the Standard Model (APS/DPF, Santa Cruz)	82-71156	0-88318-180-0
No. 82	Interpretation of Climate and Photochemical Models, Ozone and Temperature Measurements (La Jolla Institute, 1981)	82-071345	0-88318-181-9
No. 83	The Galactic Center (Cal. Inst. of Tech., 1982)	82-071635	0-88318-182-7
No. 84	Physics in the Steel Industry (APS.AISI, Lehigh University, 1981)	82-072033	0-88318-183-5
No. 85	Proton-Antiproton Collider Physics - 1981 (Madison, Wisconsin)	82-072141	0-88318-184-3
No. 86	Momentum Wave Functions - 1982 (Adelaide, Australia)	82-072375	0-88318-185-1
No. 87	Physics of High Energy Particle Accelerators (Fermilab Summer School, 1981)	82-072421	0-88318-186-X
No. 88	Mathematical Methods in Hydrodynamics and Integrability in Dynamical Systems (La Jolla Institute, 1981)	82-072462	0-88318-187-8
No. 89	Neutron Scattering - 1981 (Argonne National Laboratory)	82-073094	0-88318-188-6
No. 90	Laser Techniques for Extreme Ultraviolt Spectroscopy (Boulder, 1982)	82-073205	0-88318-189-4
No. 91	Laser Acceleration of Particles (Los Alamos, 1982)	82-073361	0-88318-190-8
No. 92	The State of Particle Accelerators and High Energy Physics(Fermilab, 1981)	82-073861	0-88318-191-6
No. 93	Novel Results in Particle Physics (Vanderbilt, 1982)	82-73954	0-88318-192-4
No. 94	X-Ray and Atomic Inner-Shell Physics-1982 (International Conference, U. of Oregon)	82-74075	0-88318-193-2
No. 95	High Energy Spin Physics - 1982 (Brookhaven National Laboratory)	83-70154	0-88318-194-0
No. 96	Science Underground (Los Alamos, 1982)	83-70377	0-88318-195-9

No. 97	The Interaction Between Medium Energy Nucleons in Nuclei-1982 (Indiana University)	83-70649	0-88318-196-7
No. 98	Particles and Fields - 1982 (APS/DPF University of Maryland)	83-70807	0-88318-197-5
No. 99	Neutrino Mass and Gauge Structure of Weak Interactions (Telemark, 1982)	83-71072	0-88318-198-3
No. 100	Excimer Lasers - 1983 (OSA, Lake Tahoe, Nevada)	83-71437	0-88318-199-1
No. 101	Positron-Electron Pairs in Astrophysics (Goddard Space Flight Center, 1983)	83-71926	0-88318-200-9
No. 102	Intense Medium Energy Sources of Strangeness (UC-Santa Cruz, 1983)	83-72261	0-88318-201-7
No. 103	Quantum Fluids and Solids - 1983 (Sanibel Island, Florida)	83-72440	0-88318-202-5
No. 104	Physics,Technology and the Nuclear Arms Race (APS Baltimore-1983)	83-72533	0-88318-203-3
No. 105	Physics of High Energy Particle Accelerators (SLAC Summer School, 1982)	83-72986	0-88318-304-8
No. 106	Predictability of Fluid Motions (La Jolla Institute, 1983)	83-73641	0-88318-305-6
No. 107	Physics and Chemistry of Porous Media (Schlumberger-Doll Research, 1983)	83-73640	0-88318- 306-4
No. 108	The Time Projection Chamber (TRIUMF, Vancouver, 1983)	83-83445	0-88318-307-2
No. 109	Random Walks and Their Applications in the Physical and Biological Sciences (NBS/La Jolla Institute, 1982)	84-70208	0-88318-308-0
No. 110	Hadron Substructure in Nuclear Physics (Indiana University, 1983)	84-70165	0-88318-309-9
No. 111	Production and Neutralization of Negative Ions and Beams (3rd Int'l Symposium, Brookhaven, 1983)	84-70379	0-88318-310-2
No. 112	Particles and Fields-1983 (APS/DPF, Blacksburg, VA)	84-70378	0-88318-311-0

"Anthrozoology – It's such a broad interdisciplinary field! No wonder this new edited book by Aubrey Fine and his associates has over 50 chapters: something for everyone. I love seeing chapters by pioneers like Erika Friedmann and Susan Cohen, but also new names writing on animal topics related to developmental disabilities, crises, popular culture, leisure, cinema, ecotourism, and, of course, wildlife. This book will give anthrozoology new prominence in disciplines where it has a significant role but is still unknown by many people."

Lynette A. Hart, MA, PhD, *Professor of Anthrozoology and Animal Behavior, University of California, Davis, School of Veterinary Medicine*

"If you are looking for a current and comprehensive resource on anthrozoology and human-animal interactions, this handbook is it! Chapters reflect the diverse and interdisciplinary field of anthrozoology, going beyond animal-assisted interventions to also address ethics, animal cognition and sentience, animal abuse, human-wildlife interactions, working and emotional support animals, and more. Authored by accomplished leaders in their content areas, this handbook will be a valuable resource for those interested in the field and more experienced researchers and practitioners."

Sandra B. Barker, PhD, NCC, LPC, *Professor Emeritus of Psychiatry, EMDR Certified Therapist and Consultant, Founding Director and Senior Advisor, School of Medicine Center for Human-Animal Interaction, Virginia Commonwealth University*

"*The Routledge International Handbook of Human-Animal Interactions and Anthrozoology* is a book many of us have been waiting for! This extraordinary resource, assembled under the guidance of the field's most committed luminaries, covers numerous issues at the forefront of the transdisciplinary field of Human Animal Interaction, critical animal studies and Anthrozoology.

With a focused emphasis on fairness and justice and forming reciprocal relationships with other animals, this timely and incredibly important compendium to the field of Human-Animal Interaction includes over 50 chapters encompassing the work of more than 100 leading thinkers, scholars, clinicians, researchers, and human-animal bond academic centers. The field of Human-Animal Interaction and the Anthrozoology literature, more than ever, has needed a go-to source offering contemporary understanding of the many intersectional and interconnected considerations of our relationships with other non-human members of our families and communities. This book is by far, the most comprehensive array of diverse and thought-provoking literature on this topic, making this a landmark publishing and academic achievement for promoting the relationship between people and other animals."

Philip Tedeschi, LCSW, *Clinical Professor, Institute for Human-Animal Connection, Affiliated Faculty and Director Emeritus, Institute for Animal Sentience and Protection*

"What a pleasant experience: reading this Handbook on Human-Animal Interactions and Anthrozoology. In a natural and careful way, the editors collected relevant and actual topics to be filled in by experienced scholars and practitioners to teach students and newbies in this wonderful field. It provides a wealth of information and addresses the multidisciplinary character of the field in a transparent way.

The Routledge International Handbook of Human-Animal Interactions and Anthrozoology is not only interesting for students and newbies, however is also great and accessible information for all of us working or interested in the field of human-animal interactions. It is updating us on the latest scientific insights, challenges, debates and best practices in this area.

I trust this handbook will find its way in many colleges and universities and on bookshelves of many libraries and individuals. Learning about so many perspectives of the human-animal interactions will enrich us and will help us to understand each other and will make the world a little bit better for humans and animals."

Prof. em.dr. Marie-Jose Enders-Slegers, *Open University, the Netherlands, President I.A.H.A.I.O. (International Association of Human Animal Interaction Organizations)*